A First Course
in the
Finite Element Method
Using Algor™

Second Edition

Daryl L. Logan
University of Wisconsin–Platteville

BROOKS/COLE

THOMSON LEARNING™

Australia • Canada • Mexico • Singapore • Spain • United Kingdom • United States

BROOKS/COLE

THOMSON LEARNING

Sponsoring Editor: *Bill Stenquist*
Marketing Team: *Tom Ziolkowski, Laura Hubrich, Ericka Thompson, Chris Kelly*
Editorial Coordinator: *Shelley Gesicki*
Production Editor: *Tom Novack*
Production Service: *Matrix Productions*
Manuscript Editor: *Carol Dean*
Permissions Editor: *Sue Ewing*
Interior Design: *John Edeen*
Cover Design: *Cheryl Carrington*

Cover Illustration: *Courtesy of Algor, Inc.*
Interior Illustration: *Asco Typesetters*
Photo Researcher: *Sarah Evertson*
Print Buyer: *Kristine Waller*
Typesetting: *Asco Typesetters*
Cover Printing: *R. R. Donnelley & Sons, Crawfordsville*
Printing and Binding: *R. R. Donnelley & Sons, Crawfordsville*

For more information about this or any other Brooks/Cole products, contact:
BROOKS/COLE
511 Forest Lodge Road
Pacific Grove, CA 93950 USA
www.brookscole.com
1-800-423-0563 (Thomson Learning Academic Resource Center)

Printed in United States of America

10 9 8 7 6 5 4 3 2 1

Library of Congress Cataloging-in-Publication Data
Logan, Daryl L.
    A first course in the finite element method using algor / Daryl L. Logan.
        p.   cm.
    Includes bibliographical references and index.
    ISBN 0-534-38068-9
    1. Finite element method—Data processing. 2. Algor. I. Title.
TA347.F5 L65 2000
620′.001′51535—dc21   00-031281

Contents

6 Frame and Grid Equations   276

7 Development of the Plane Stress and Plane Strain Stiffness Equations   386

14 Fluid Flow 701

15 Thermal Stress 732

16 Structural Dynamics and Time-Dependent Heat Transfer 778

17 Plate Bending Element   862

Appendix A   Matrix Algebra   883

Appendix B   Methods for Solution of Simultaneous Linear Equations   897

Appendix C   Equations from Elasticity Theory   919

Appendix D   Equivalent Nodal Forces   927

Appendix E   Principle of Virtual Work   930

Appendix F   Basics of Algor   934

Answers to Selected Problems   943

Index   965

Preface

The purpose of this second edition of the book is again to provide a simple, basic approach to the finite element method that can be understood by both undergraduate and graduate students without the usual prerequisites (such as structural analysis) required by most available texts in this area. The book is written primarily as a basic learning tool for the undergraduate student in civil and mechanical engineering whose main interest is in stress analysis and heat transfer. However, the concepts are presented in sufficiently simple form so that the book serves as a valuable learning aid for students of other backgrounds, as well as for practicing engineers. The text is geared toward those who want to apply the finite element method to solve practical physical problems.

General principles are presented for each topic, followed by traditional applications of these principles, which are in turn followed by computer applications where relevant. This approach is taken to illustrate concepts used for computer analysis of large-scale problems.

The book proceeds from basic to advanced topics and can be suitably used in a two-course sequence. Topics include basic treatments of (1) simple springs and bars, leading to two- and three-dimensional truss analysis; (2) beam bending, leading to plane frame and grid analysis and space frame analysis; (3) elementary plane stress/strain elements, leading to more advanced plane stress/strain elements; (4) axisymmetric stress; (5) isoparametric formulation of the finite element method; (6) three-dimensional stress; (7) heat transfer and fluid mass transport; (8) basic fluid mechanics; (9) thermal stress; (10) time-dependent stress and heat transfer, and (11) plate bending.

Additional features include how to handle inclined or skewed supports, beam element with nodal hinge, beam element arbitrarily located in space, and the concept of substructure analysis.

This edition includes a thorough treatment of one of the most frequently used commercially available computer systems, the Algor finite element analysis system. A description of the Algor system is included in Appendix F. Detailed step-by-step procedures for using the most current release (Release 12) of Algor software to solve numerous problems, from trusses to three-dimensional stress to transient heat transfer, are included throughout the text.

The direct approach, the principle of minimum potential energy, and Galerkin's residual method are introduced at various stages, as required, to develop the equations needed for analysis.

Appendices include (1) basic matrix algebra used throughout the text, (2) solution methods for simultaneous equations, (3) basic theory of elasticity, (4) the principle of virtual work, and (5) basics of the Algor system.

More than 110 examples appear throughout the text. Seventy examples are solved "longhand" to illustrate the concepts, and 45 are solved with the Algor system. The Algor solutions are incorporated directly into relevant places in the text to create a natural extension of basic principles to longhand examples and then to computer program examples. More than 300 end-of-chapter problems are provided to reinforce concepts. Answers to many problems are included in the back of the book. Those end-of-chapter problems to be solved by using the computer programs are marked with a computer symbol.

New features of this edition include new, updated Release 12 Algor computer program examples replacing the outdated earlier release examples, more in-depth explanations of the Algor program and concepts behind each specific application, a new chapter on plate bending, additional information on modeling, interpreting results, and comparing finite element solutions to analytical solutions, and additional design-type problems added to Chapters 4, 6, 8, 12, and 17.

Following is an outline of suggested topics for a first course (approximately 43 lectures, 50 minutes each) in which this textbook is used.

Topic	Number of Lectures
Appendix A	1
Appendix B	1
Chapter 1	2
Chapter 2	3
Chapter 3, Sections 3.1–3.10	5
Exam 1	1
Appendix F	2
Chapter 4	2
Chapter 5, Sections 5.1–5.6	4
Chapter 6, Sections 6.1–6.3, 6.7	4
Exam 2	1
Chapter 7	4
Chapter 8	4
Chapter 12	3
Chapter 13, Sections 13.1–13.8, 13.10–13.11	5
Exam 3	1

This outline can be used in a one-semester course for undergraduate and graduate students in civil and mechanical engineering. (If a total stress analysis emphasis is desired, Chapter 13 can be replaced, for instance, with material from Chapters 9 and 10, Chapter 11, or some of Chapters 16 and 17. The rest of the text can be finished in a second semester course.

I express my deepest appreciation to the staff at Brooks/Cole Publishing Company, especially Bill Stenquist and Shelley Gesicki, as well as Merrill Peterson, for their assistance in producing this new edition.

I am grateful to Ted Belytschko for his excellent teaching of the finite element method, which aided me in writing this text. I also appreciate the encouragement given to me by W. Charles Paulsen, the former vice president of Algor Inc., to "Algorize" the text by incorporating numerous examples solved with Algor. These examples complement the subject matter quite clearly.

I thank the many students who used the notes that developed into this text. I am especially grateful to Ron Cenfetelli, Barry Davignon, Konstantinos Kariotis, Howard Koswara, Hidajat Harintho, Hairi Salemganesan, Joe Kesari, Yanping Lu, and Khailan Zhang for checking and solving problems in the first two editions of the text. Thanks also to Kevin Bischel, Troy Kollmann, and Jeffery Makovec for solving numerous examples using the Algor program. Finally, a special thanks to my students for solving many of the Algor example problems using Release 12 of the Algor computer program. To further reduce errors in the step-by-step solutions to the Algor example problems, I have personally solved many of the Algor examples included in the text.

Finally, very special thanks to my family, Diane, Kathy, Daryl Jr., and Paul, for their many sacrifices during the development of this second edition.

Daryl L. Logan

Notation

English Symbols

a_i	generalized coordinates (coefficients used to express displacement in general form)
A	cross-sectional area
$\underline{B}$	matrix relating strains to nodal displacements or relating temperature gradient to nodal temperatures
c	specific heat of a material
$\underline{C}'$	matrix relating stresses to nodal displacements
C	direction cosine in two dimensions
$C_x,\ C_y,\ C_z$	direction cosines in three dimensions
$\underline{d}$	element and structure nodal displacement matrix, both in global coordinates
$\underline{\hat{d}}$	local-coordinate element nodal displacement matrix
D	bending rigidity of a plate
$\underline{D}$	matrix relating stresses to strains
e	exponential function
E	modulus of elasticity
$\underline{f}$	global-coordinate nodal force matrix
$\underline{\hat{f}}$	local-coordinate element nodal force matrix
$\underline{f}_b$	body force matrix
$\underline{f}_h$	heat transfer force matrix
$\underline{f}_q$	heat flux force matrix
$\underline{f}_Q$	heat source force matrix
$\underline{f}_s$	surface force matrix
$\underline{F}$	global-coordinate structure force matrix
$\underline{F}_c$	condensed force matrix
$\underline{F}_i$	global nodal forces
$\underline{F}_0$	equivalent force matrix

$\underline{g}$	temperature gradient matrix or hydraulic gradient matrix
G	shear modulus
h	heat-transfer (or convection) coefficient
i, j, m	nodes of a triangular element
I	principal moment of inertia
$\underline{J}$	Jacobian matrix
k	spring stiffness
$\underline{k}$	global-coordinate element stiffness or conduction matrix
$\underline{k}_c$	condensed stiffness matrix, and conduction part of the stiffness matrix in heat-transfer problems
$\hat{\underline{k}}$	local-coordinate element stiffness matrix
$\underline{k}_h$	convective part of the stiffness matrix in heat-transfer problems
$\underline{K}$	global-coordinate structure stiffness matrix
K_{xx}, K_{yy}	thermal conductivities (or permeabilities, for fluid mechanics) in the x and y directions, respectively
L	length of a bar or beam element
m	maximum difference in node numbers in an element
$m(x)$	general moment expression
m_x, m_y, m_{xy}	moments in a plate
$\hat{\underline{m}}$	local mass matrix
$\hat{m}_i$	local nodal moments
$\underline{M}$	global mass matrix
$\underline{M}^*$	matrix used to relate displacements to generalized coordinates for a linear-strain triangle formulation
$\underline{M}'$	matrix used to relate strains to generalized coordinates for a linear-strain triangle formulation
n_b	bandwidth of a structure
n_d	number of degrees of freedom per node
$\underline{N}$	shape (interpolation or basis) function matrix
N_i	shape functions
p	surface pressure (or nodal heads in fluid mechanics)
p_r, p_z	radial and axial (longitudinal) pressures, respectively
P	concentrated load
$\hat{\underline{P}}$	concentrated local force matrix
q	heat flow (flux) per unit area or distributed loading on a plate
$\bar{q}$	rate of heat flow
q^*	heat flow per unit area on a boundary surface
Q	heat source generated per unit volume or internal fluid source
Q^*	line or point heat source
Q_x, Q_y	transverse shear line loads on a plate

r, θ, z	radial, circumferential, and axial coordinates, respectively
R	residual in Galerkin's integral
R_b	body force in the radial direction
R_{ix}, R_{iy}	nodal reactions in x and y directions, respectively
s, t, z'	natural coordinates attached to isoparametric element
S	surface area
t	thickness of a plane element or a plate element
t_i, t_j, t_m	nodal temperatures of a triangular element
T	temperature function
T_∞	free-stream temperature
$\underline{T}$	displacement, force, and stiffness transformation matrix
$\underline{T}_i$	surface traction matrix in the i direction
u, v, w	displacement functions in the x, y, and z directions, respectively
U	strain energy
ΔU	change in stored energy
v	velocity of fluid flow
$\hat{V}$	shear force in a beam
w	distributed loading on a beam or along an edge of a plane element
W	work
x_i, y_i, z_i	nodal coordinates in the x, y, and z directions, respectively
$\hat{x}, \hat{y}, \hat{z}$	local element coordinate axes
x, y, z	structure global or reference coordinate axes
$\underline{X}$	body force matrix
X_b, Y_b	body forces in the x and y directions, respectively
Z_b	body force in longitudinal direction (axisymmetric case) or in the z direction (three-dimensional case)

Greek Symbols

α	coefficient of thermal expansion
$\alpha_i, \beta_i, \gamma_i, \delta_i$	used to express the shape functions defined by Eq. (7.2.10) and Eqs. (12.2.5)–(12.2.8)
δ	spring or bar deformation
ε	normal strain
$\underline{\varepsilon}_T$	thermal strain matrix
$\kappa_x, \kappa_y, \kappa_{xy}$	curvatures in plate bending
ν	Poisson's ratio
ϕ_i	nodal angle of rotation or slope in a beam element
π_p	total potential energy

π_h	functional for heat-transfer problem
ρ	mass density of a material
ρ_w	weight density of a material
ω	angular velocity
Ω	potential energy of forces
ϕ	fluid head or potential
σ	normal stress
$\underline{\sigma}_T$	thermal stress matrix
τ	shear stress
θ	angle between the x axis and the local $\hat{x}$ axis for two-dimensional problems
θ_p	principal angle
$\theta_x, \theta_y, \theta_z$	angles between the global x, y, and z axes and the local $\hat{x}$ axis, respectively, or rotations about the x and y axes in a plate
$\underline{\Psi}$	general displacement function matrix

Other Symbols

$\dfrac{d(\ )}{dx}$	derivative of a variable with respect to x
dt	time differential
$(\dot{\ })$	the dot over a variable denotes that the variable is being differential with respect to time
$[\ ]$	denotes a rectangular or a square matrix
$\{\ \}$	denotes a column matrix
$(\_)$	the underline of a variable denotes a matrix
$(\hat{\ })$	the hat over a variable denotes that the variable is being described in a local coordinate system
$[\ ]^{-1}$	denotes the inverse of a matrix
$[\ ]^{T}$	denotes the transpose of a matrix
$\dfrac{\partial(\ )}{\partial x}$	partial derivative with respect to x
$\dfrac{\partial(\ )}{\partial\{d\}}$	partial derivative with respect to each variable in $\{d\}$
■	denotes the end of the solution of an example problem

1

Introduction

Prologue

The finite element method is a numerical method for solving problems of engineering and mathematical physics. Typical problem areas of interest in engineering and mathematical physics that are solvable by use of the finite element method include structural analysis, heat transfer, fluid flow, mass transport, and electromagnetic potential.

For problems involving complicated geometries, loadings, and material properties, it is generally not possible to obtain analytical mathematical solutions. Analytical solutions are those given by a mathematical expression that yields the values of the desired unknown quantities at any location in a body (here total structure or physical system of interest) and are thus valid for an infinite number of locations in the body. These analytical solutions generally require the solution of ordinary or partial differential equations, which, because of the complicated geometries, loadings, and material properties, are not usually obtainable. Hence we need to rely on numerical methods, such as the finite element method, for acceptable solutions. The finite element formulation of the problem results in a system of simultaneous algebraic equations for solution, rather than requiring the solution of differential equations. These numerical methods yield approximate values of the unknowns at discrete numbers of points in the continuum. Hence this process of modeling a body by dividing it into an equivalent system of smaller bodies or units (finite elements) interconnected at points common to two or more elements (nodal points or nodes) and/or boundary lines and/or surfaces is called *discretization*. In the finite element method, instead of solving the problem for the entire body in one operation, we formulate the equations for each finite element and combine them to obtain the solution of the whole body.

Briefly, the solution for structural problems typically refers to determining the displacements at each node and the stresses within each element making up the structure that is subjected to applied loads. In nonstructural problems, the nodal unknowns may, for instance, be temperatures or fluid pressures due to thermal or fluid fluxes.

This chapter first presents a brief history of the development of the finite element method. You will see from this historical account that the method has become a practical one for solving engineering problems only in the past 40 years (paralleling the developments associated with the modern high-speed electronic digital computer).

This historical account is followed by an introduction to matrix notation; then we describe the need for matrix methods (as made practical by the development of the modern digital computer) in formulating the equations for solution. This section discusses both the role of the digital computer in solving the large systems of simultaneous algebraic equations associated with complex problems and the development of numerous computer programs based on the finite element method. Next, a general description of the steps involved in obtaining a solution to a problem is provided. This description includes discussion of the types of elements available for a finite element method solution. Various representative applications are then presented to illustrate the capacity of the method to solve problems, such as those involving complicated geometries, several different materials, and irregular loadings. Chapter 1 also lists some of the advantages of the finite element method in solving problems of engineering and mathematical physics. Finally, we present numerous features of computer programs based on the finite element method.

▲ 1.1 Brief History   ▲

This section presents a brief history of the finite element method as applied to both structural and nonstructural areas of engineering and to mathematical physics. References cited here are intended to augment this short introduction to the historical background.

    The modern development of the finite element method began in the 1940s in the field of structural engineering with the work by Hrennikoff [1] in 1941 and McHenry [2] in 1943, who used a lattice of line (one-dimensional) elements (bars and beams) for the solution of stresses in continuous solids. In a paper published in 1943 but not widely recognized for many years, Courant [3] proposed setting up the solution of stresses in a variational form. Then he introduced piecewise interpolation (or shape) functions over triangular subregions making up the whole region as a method to obtain approximate numerical solutions. In 1947 Levy [4] developed the flexibility or force method, and in 1953 his work [5] suggested that another method (the stiffness or displacement method) could be a promising alternative for use in analyzing statically redundant aircraft structures. However, his equations were cumbersome to solve by hand, and thus the method became popular only with the advent of the high-speed digital computer.

    In 1954 Argyris and Kelsey [6, 7] developed matrix structural analysis methods using energy principles. This development illustrated the important role that energy principles would play in the finite element method.

    The first treatment of two-dimensional elements was by Turner et al. [8] in 1956. They derived stiffness matrices for truss elements, beam elements, and two-dimensional triangular and rectangular elements in plane stress and outlined the procedure commonly known as the *direct stiffness method* for obtaining the total structure stiffness matrix. Along with the development of the high-speed digital computer in the early 1950s, the work of Turner et al. [8] prompted further development of finite element stiffness equations expressed in matrix notation. The phrase *finite element* was introduced by Clough [9] in 1960 when both triangular and rectangular elements were used for plane stress analysis.

A flat, rectangular-plate bending-element stiffness matrix was developed by Melosh [10] in 1961. This was followed by development of the curved-shell bending-element stiffness matrix for axisymmetric shells and pressure vessels by Grafton and Strome [11] in 1963.

Extension of the finite element method to three-dimensional problems with the development of a tetrahedral stiffness matrix was done by Martin [12] in 1961, by Gallagher et al. [13] in 1962, and by Melosh [14] in 1963. Additional three-dimensional elements were studied by Argyris [15] in 1964. The special case of axisymmetric solids was considered by Clough and Rashid [16] and Wilson [17] in 1965.

Most of the finite element work up to the early 1960s dealt with small strains and small displacements, elastic material behavior, and static loadings. However, large deflection and thermal analysis were considered by Turner et al. [18] in 1960 and material nonlinearities by Gallagher et al. [13] in 1962, whereas buckling problems were initially treated by Gallagher and Padlog [19] in 1963. Zienkiewicz et al. [20] extended the method to visco-elasticity problems in 1968.

In 1965 Archer [21] considered dynamic analysis in the development of the consistent-mass matrix, which is applicable to analysis of distributed-mass systems such as bars and beams in structural analysis.

With Melosh's [14] realization in 1963 that the finite element method could be set up in terms of a variational formulation, it began to be used to solve nonstructural applications. Field problems, such as determination of the torsion of a shaft, fluid flow, and heat conduction, were solved by Zienkiewicz and Cheung [22] in 1965, Martin [23] in 1968, and Wilson and Nickel [24] in 1966.

Further extension of the method was made possible by the adaptation of weighted residual methods, first by Szabo and Lee [25] in 1969 to derive the previously known elasticity equations used in structural analysis and then by Zienkiewicz and Parekh [26] in 1970 for transient field problems. It was then recognized that when direct formulations and variational formulations are difficult or not possible to use, the method of weighted residuals may at times be appropriate. For example, in 1977 Lyness et al. [27] applied the method of weighted residuals to the determination of magnetic field.

In 1976 Belytschko [28, 29] considered problems associated with large-displacement nonlinear dynamic behavior, and improved numerical techniques for solving the resulting systems of equations.

A relatively new field of application of the finite element method is that of bio-engineering [30, 31]. This field is still troubled by such difficulties as nonlinear materials, geometric nonlinearities, and other complexities still being discovered.

From the early 1950s to the present, enormous advances have been made in the application of the finite element method to solve complicated engineering problems. Engineers, applied mathematicians, and other scientists will undoubtedly continue to develop new applications. For an extensive bibliography on the finite element method, consult the work of Kardestuncer [32], Clough [33], or Noor [54].

▲ 1.2 Introduction to Matrix Notation ▲

Matrix methods are a necessary tool used in the finite element method for purposes of simplifying the formulation of the element stiffness equations, for purposes of long-hand solutions of various problems, and, most important, for use in programming the methods for high-speed electronic digital computers. Hence matrix notation represents a simple and easy-to-use notation for writing and solving sets of simultaneous algebraic equations.

Appendix A discusses the significant matrix concepts used throughout the text. We will present here only a brief summary of the notation used in this text.

A **matrix** *is a rectangular array of quantities arranged in rows and columns that is often used as an aid in expressing and solving a system of algebraic equations.* As examples of matrices that will be described in subsequent chapters, the force components $(F_{1x}, F_{1y}, F_{1z}, F_{2x}, F_{2y}, F_{2z}, \ldots, F_{nx}, F_{ny}, F_{nz})$ acting at the various nodes or points $(1, 2, \ldots, n)$ on a structure and the corresponding set of nodal displacements $(d_{1x}, d_{1y}, d_{1z}, d_{2x}, d_{2y}, d_{2z}, \ldots, d_{nx}, d_{ny}, d_{nz})$ can both be expressed as matrices:

$$\{F\} = \underline{F} = \begin{Bmatrix} F_{1x} \\ F_{1y} \\ F_{1z} \\ F_{2x} \\ F_{2y} \\ F_{2z} \\ \vdots \\ F_{nx} \\ F_{ny} \\ F_{nz} \end{Bmatrix} \qquad \{d\} = \underline{d} = \begin{Bmatrix} d_{1x} \\ d_{1y} \\ d_{1z} \\ d_{2x} \\ d_{2y} \\ d_{2z} \\ \vdots \\ d_{nx} \\ d_{ny} \\ d_{nz} \end{Bmatrix} \tag{1.2.1}$$

The subscripts to the right of F and d identify the node and the direction of force or displacement, respectively. For instance, F_{1x} denotes the force at node 1 applied in the x direction. The matrices in Eqs. (1.2.1) are called *column matrices* and have a size of $n \times 1$. The brace notation $\{ \ \}$ will be used throughout the text to denote a column matrix. The whole set of force or displacement values in the column matrix is simply represented by $\{F\}$ or $\{d\}$. A more compact notation used throughout this text to represent any rectangular array is the underlining of the variable; that is, $\underline{F}$ and $\underline{d}$ denote general matrices (possibly column matrices or rectangular matrices— the type will become clear in the context of the discussion associated with the variable).

The more general case of a known rectangular matrix will be indicated by use of the bracket notation [ ]. For instance, the element and global structure stiffness matrices $[k]$ and $[K]$, respectively, developed throughout the text for various element types (such as those in Figure 1–1 on page 9), are represented by square matrices given as

$$[k] = \underline{k} = \begin{bmatrix} k_{11} & k_{12} & \cdots & k_{1n} \\ k_{21} & k_{22} & \cdots & k_{2n} \\ \vdots & \vdots & & \vdots \\ k_{n1} & k_{n2} & \cdots & k_{nn} \end{bmatrix} \tag{1.2.2}$$

and
$$[K] = \underline{K} = \begin{bmatrix} K_{11} & K_{12} & \cdots & K_{1n} \\ K_{21} & K_{22} & \cdots & K_{2n} \\ \vdots & \vdots & & \vdots \\ K_{n1} & K_{n2} & \cdots & K_{nn} \end{bmatrix} \qquad (1.2.3)$$

where, in structural theory, the elements k_{ij} and K_{ij} are often referred to as *stiffness influence coefficients*.

You will learn that the global nodal forces $\underline{F}$ and the global nodal displacements $\underline{d}$ are related through use of the global stiffness matrix $\underline{K}$ by

$$\underline{F} = \underline{K}\underline{d} \qquad (1.2.4)$$

Equation (1.2.4) is called the *global stiffness equation* and represents a set of simultaneous equations. It is the basic equation formulated in the stiffness or displacement method of analysis. Using the compact notation of underlining the variables, as in Eq. (1.2.4), should not cause you any difficulties in determining which matrices are column or rectangular matrices.

To obtain a clearer understanding of elements K_{ij} in Eq. (1.2.3), we use Eq. (1.2.1) and write out the expanded form of Eq. (1.2.4) as

$$\begin{Bmatrix} F_{1x} \\ F_{1y} \\ \vdots \\ F_{nz} \end{Bmatrix} = \begin{bmatrix} K_{11} & K_{12} & \cdots & K_{1n} \\ K_{21} & K_{22} & \cdots & K_{2n} \\ \vdots & & & \\ K_{n1} & K_{n2} & \cdots & K_{nn} \end{bmatrix} \begin{Bmatrix} d_{1x} \\ d_{1y} \\ \vdots \\ d_{nz} \end{Bmatrix} \qquad (1.2.5)$$

Now assume a structure to be forced into a displaced configuration defined by $d_{1x} = 1, d_{1y} = d_{1z} = \cdots d_{nz} = 0$. Then from Eq. (1.2.5), we have

$$F_{1x} = K_{11} \qquad F_{1y} = K_{21}, \ldots, F_{nz} = K_{n1} \qquad (1.2.6)$$

Equations (1.2.6) contain all elements in the first column of $\underline{K}$. In addition, they show that these elements, $K_{11}, K_{21}, \ldots, K_{n1}$, are the values of the full set of nodal forces required to maintain the imposed displacement state. In a similar manner, the second column in $\underline{K}$ represents the values of forces required to maintain the displaced state $d_{1y} = 1$ and all other nodal displacement components equal to zero. We should now have a better understanding of the meaning of stiffness influence coefficients.

Subsequent chapters will discuss the element stiffness matrices $\underline{k}$ for various element types, such as bars, beams, and plane stress. They will also cover the procedure for obtaining the global stiffness matrices $\underline{K}$ for various structures and for solving Eq. (1.2.4) for the unknown displacements in matrix $\underline{d}$.

Using matrix concepts and operations will become routine with practice; they will be valuable tools for solving small problems longhand. And matrix methods are crucial to the use of the digital computers necessary for solving complicated problems with their associated large number of simultaneous equations.

▲ 1.3 Role of the Computer ▲

As we have said, until the early 1950s, matrix methods and the associated finite element method were not readily adaptable for solving complicated problems. Even though the finite element method was being used to describe complicated structures, the resulting large number of algebraic equations associated with the finite element method of structural analysis made the method extremely difficult and impractical to use. However, with the advent of the computer, the solution of thousands of equations in a matter of minutes became possible.

The development of the computer resulted in computational program development. Numerous special-purpose and general-purpose programs have been written to handle various complicated structural (and nonstructural) problems. Programs, such as the Algor program [46], illustrate the elegance of the finite element method and reinforce understanding of it.

To use the computer, the analyst, having defined the finite element model, inputs the information into the computer. This information may include the position of the element nodal coordinates, the manner in which elements are connected, the material properties of the elements, the applied loads, boundary conditions, or constraints, and the kind of analysis to be performed. The computer then uses this information to generate and solve the equations necessary to carry out the analysis.

▲ 1.4 General Steps of the Finite Element Method ▲

This section presents the general steps included in a finite element method formulation and solution to an engineering problem. We will use these steps as our guide in developing solutions for structural and nonstructural problems in subsequent chapters.

For simplicity's sake, for the presentation of the steps to follow, we will consider only the structural problem. The nonstructural heat-transfer and fluid mechanics problems and their analogies to the structural problem are considered in Chapters 13 and 14.

Typically, for the structural stress-analysis problem, the engineer seeks to determine displacements and stresses throughout the structure, which is in equilibrium and is subjected to applied loads. For many structures, it is difficult to determine the distribution of deformation using conventional methods, and thus the finite element method is necessarily used.

There are two general approaches associated with the finite element method. One approach, called the *force*, or *flexibility*, *method*, uses internal forces as the unknowns of the problem. To obtain the governing equations, first the equilibrium equations are used. Then necessary additional equations are found by introducing compatibility equations. The result is a set of algebraic equations for determining the redundant or unknown forces.

The second approach, called the *displacement*, or *stiffness*, *method*, assumes the displacements of the nodes as the unknowns of the problem. For instance, compatibility conditions requiring that elements connected at a common node, along a common edge, or on a common surface before loading remain connected at that node, edge, or surface after deformation takes place are initially satisfied. Then the governing equations are expressed in terms of nodal displacements using the equations of equilibrium and an applicable law relating forces to displacements.

These two approaches result in different unknowns (forces or displacements) in the analysis and different matrices associated with their formulations (flexibilities or stiffnesses). It has been shown [34] that, for computational purposes, the displacement (or stiffness) method is more desirable because its formulation is simpler for most structural analysis problems. Furthermore, a vast majority of general-purpose finite element programs have incorporated the displacement formulation for solving structural problems. Consequently, only the displacement method will be used throughout this text.

The finite element method involves modeling the structure using small interconnected elements called *finite elements*. A displacement function is associated with each finite element. Every interconnected element is linked, directly or indirectly, to every other element through common (or shared) interfaces, including nodes and/or boundary lines and/or surfaces. By using known stress/strain properties for the material making up the structure, one can determine the behavior of a given node in terms of the properties of every other element in the structure. The total set of equations describing the behavior of each node results in a series of algebraic equations best expressed in matrix notation.

We now present the steps, along with explanations necessary at this time, used in the finite element method formulation and solution of a structural problem. The purpose of setting forth these general steps now is to expose you to the procedure generally followed in a finite element formulation of a problem. You will easily understand these steps when we illustrate them specifically for springs, bars, trusses, beams, plane frames, plane stress, axisymmetric stress, three-dimensional stress, heat transfer, and fluid flow in subsequent chapters. We suggest that you review this section periodically as we develop the specific element equations.

Keep in mind that the analyst must make decisions regarding dividing the structure or continuum into finite elements and selecting the element type or types to be used in the analysis (step 1), the kinds of loads to be applied, and the types of boundary conditions or supports to be applied. The other steps, 2–7, are carried out automatically by a computer program.

Step 1 Discretize and Select the Element Types

Step 1 involves dividing the body into an equivalent system of finite elements with associated nodes and choosing the most appropriate element type to model most closely the actual physical behavior. The total number of elements used and their variation in size and type within a given body are primarily matters of engineering judgment. The elements must be made small enough to give usable results and yet large enough to reduce computational effort. Small elements (and possibly higher-order elements) are generally desirable where the results are changing rapidly, such as where changes in geometry occur; large elements can be used where results are relatively constant. We will have more to say about discretization guidelines in later chapters, particularly in Chapter 8, where the concept becomes quite significant. The discretized body or mesh is often created with mesh-generation programs or pre-processor programs available to the user.

The choice of elements used in a finite element analysis depends on the physical makeup of the body under actual loading conditions and on how close to the actual behavior the analyst wants the results to be. Judgment concerning the appropriateness of one-, two-, or three-dimensional idealizations is necessary. Moreover, the choice of the most appropriate element for a particular problem is one of the major tasks that must be carried out by the designer/analyst. Elements that are commonly employed in practice—most of which are considered in this text—are shown in Figure 1–1.

The primary line elements [Figure 1–1(a)] consist of bar (or truss) and beam elements. They have a cross-sectional area but are usually represented by line segments. In general, the cross-sectional area within the element can vary, but throughout this text it will be considered to be constant. These elements are often used to model trusses and frame structures (see Figure 1–2 on page 15, for instance). The simplest line element (called a *linear element*) has two nodes, one at each end, although higher-order elements having three nodes or more (called *quadratic, cubic*, etc. *elements*) also exist. The line elements are the simplest of elements to consider and will be discussed in Chapters 2, 3, 5, and 6 to illustrate many of the basic concepts of the finite element method.

The basic two-dimensional (or plane) elements [Figure 1–1(b)] are loaded by forces in their own plane (plane stress or plane strain conditions). They are triangular or quadrilateral elements. The simplest two-dimensional elements have corner nodes only (linear elements) with straight sides or boundaries (Chapter 7), although there are also higher-order elements, typically with midside nodes (called *quadratic elements*) and curved sides (Chapters 9 and 11). The elements can have variable thicknesses throughout or be constant. They are often used to model a wide range of engineering problems (see Figures 1–3 and 1–4 on pages 16 and 17).

The most common three-dimensional elements [Figure 1–1(c)] are tetrahedral and hexahedral (or brick) elements; they are used when it becomes necessary to perform a three-dimensional stress analysis. The basic three-dimensional elements (Chapter 12) have corner nodes only and straight sides, whereas higher-order elements with midedge nodes (and possible midface nodes) have curved surfaces for their sides.

The axisymmetric element [Figure 1–1(d)] is developed by rotating a triangle or quadrilateral about a fixed axis located in the plane of the element through 360°. This element (described in Chapter 10) can be used when the geometry and loading of the problem are axisymmetric.

Step 2   Select a Displacement Function

Step 2 involves choosing a displacement function within each element. The function is defined within the element using the nodal values of the element. Linear, quadratic, and cubic polynomials are frequently used functions because they are simple to work with in finite element formulation. However, trigonometric series can also be used. For a two-dimensional element, the displacement function is a function of the coordinates in its plane (say, the x-y plane). The functions are expressed in terms of the nodal unknowns (in the two-dimensional problem, in terms of an x and a y component).

(a) Simple line element typically used to represent a bar or beam element

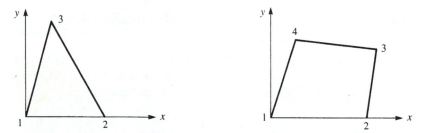

(b) Simple two-dimensional elements typically used to represent plane stress/strain

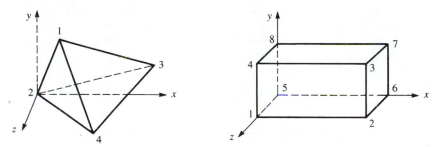

(c) Simple three-dimensional elements typically used to represent three-dimensional stress

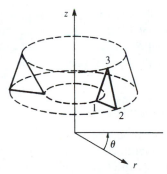

(d) Simple axisymmetric element used for axisymmetric problems

Figure 1–1   Various types of finite elements

The same general displacement function can be used repeatedly for each element. Hence the finite element method is one in which a continuous quantity, such as the displacement throughout the body, is approximated by a discrete model composed of a set of piecewise-continuous functions defined within each finite domain or finite element.

Step 3 Define the Strain/Displacement and Stress/Strain Relationships

Strain/displacement and stress/strain relationships are necessary for deriving the equations for each finite element. In the case of one-dimensional deformation, say, in the x direction, we have strain ε_x related to displacement u by

$$\varepsilon_x = \frac{du}{dx} \tag{1.4.1}$$

for small strains. In addition, the stresses must be related to the strains through the stress/strain law—generally called the *constitutive law*. The ability to define the material behavior accurately is most important in obtaining acceptable results. The simplest of stress/strain laws, Hooke's law, which is often used in stress analysis, is given by

$$\sigma_x = E\varepsilon_x \tag{1.4.2}$$

where σ_x = stress in the x direction and E = modulus of elasticity.

Step 4 Derive the Element Stiffness Matrix and Equations

Initially, the development of element stiffness matrices and element equations was based on the concept of stiffness influence coefficients, which presupposes a background in structural analysis. We now present alternative methods used in this text that do not require this special background.

Direct Equilibrium Method

According to this method, the stiffness matrix and element equations relating nodal forces to nodal displacements are obtained using force equilibrium conditions for a basic element, along with force/deformation relationships. Because this method is most easily adaptable to line or one-dimensional elements, Chapters 2, 3, and 5 illustrate this method for spring, bar, and beam elements, respectively.

Work or Energy Methods

To develop the stiffness matrix and equations for two- and three-dimensional elements, it is much easier to apply a work or energy method [35]. The principle of virtual work (using virtual displacements), the principle of minimum potential energy, and Castigliano's theorem are methods frequently used for the purpose of derivation of element equations.

The principle of virtual work outlined in Appendix E is applicable for any material behavior, whereas the principle of minimum potential energy and Castigliano's theorem are applicable only to elastic materials. Furthermore, the principle of virtual work can be used even when a potential function does not exist. However, all three principles yield identical element equations for linear-elastic materials; thus which method to use for this kind of material in structural analysis is largely a matter of convenience and personal preference. We will present the principle of minimum potential energy—probably the best known of the three energy methods mentioned here—in detail in Chapters 2 and 3, where it will be used to derive the spring and bar element equations. We will further generalize the principle and apply it to the beam element in Chapter 5 and to the plane stress/strain element in Chapter 7. Thereafter, the principle is routinely referred to as the basis for deriving all other stress-analysis stiffness matrices and element equations given in Chapters 9, 10, and 12.

For the purpose of extending the finite element method outside the structural stress analysis field, a **functional**[1] (a function of another function) analogous to the one to be used with the principle of minimum potential energy is quite useful in deriving the element stiffness matrix and equations (see Chapters 13 and 14 on heat transfer and fluid flow, respectively). For instance, letting π denote the functional and $f(x, y)$ denote a function f of two variables x and y, we then have $\pi = \pi(f(x, y))$, where π is a function of the function f.

Methods of Weighted Residuals

The methods of weighted residuals are useful for developing the element equations; particularly popular is Galerkin's method. These methods yield the same results as the energy methods wherever the energy methods are applicable. They are especially useful when a functional such as potential energy is not readily available. The weighted residual methods allow the finite element method to be applied directly to any differential equation.

Galerkin's method is introduced in Chapter 3. Then it is used to derive the bar element equations in Chapter 3 and the beam element equations in Chapter 5 and to solve the combined heat-conduction/convection/mass transport problem in Chapter 13. For more information on the use of the methods of weighted residuals, see Reference [36]; for additional applications to the finite element method, consult References [37] and [38].

Using any of the methods just outlined will produce the equations to describe the behavior of an element. These equations are written conveniently in matrix

---

[1] A functional is an integral expression that implicitly contains differential equations that describe the problem. A typical functional is of the form $I(u) = \int F(x, u, u')\, dx$ where $u(x), x$, and F are real so that $I(u)$ is also a real number.

form as

$$
\begin{Bmatrix} f_1 \\ f_2 \\ f_3 \\ \vdots \\ f_n \end{Bmatrix} = \begin{bmatrix} k_{11} & k_{12} & k_{13} & \dots & k_{1n} \\ k_{21} & k_{22} & k_{23} & \dots & k_{2n} \\ k_{31} & k_{32} & k_{33} & \dots & k_{3n} \\ \vdots & & & & \vdots \\ k_{n1} & & & \dots & k_{nn} \end{bmatrix} \begin{Bmatrix} d_1 \\ d_2 \\ d_3 \\ \vdots \\ d_n \end{Bmatrix} \tag{1.4.3}
$$

or in compact matrix form as

$$
\{f\} = [k]\{d\} \tag{1.4.4}
$$

where $\{f\}$ is the vector of element nodal forces, $[k]$ is the element stiffness matrix, and $\{d\}$ is the vector of unknown element nodal degrees of freedom or generalized displacements, n. Here generalized displacements may include such quantities as actual displacements, slopes, or even curvatures. The matrices in Eq. (1.4.4) will be developed and described in detail in subsequent chapters for specific element types, such as those in Figure 1–1.

Step 5   Assemble the Element Equations to Obtain the Global or Total Equations and Introduce Boundary Conditions

The individual element equations generated in step 4 can now be added together using a method of superposition (called the *direct stiffness method*)—whose basis is nodal force equilibrium—to obtain the global equations for the whole structure. Implicit in the direct stiffness method is the concept of continuity, or compatibility, which requires that the structure remain together and that no tears occur anywhere in the structure.

The final assembled or global equation written in matrix form is

$$
\{F\} = [K]\{d\} \tag{1.4.5}
$$

where $\{F\}$ is the vector of global nodal forces, $[K]$ is the structure global or total stiffness matrix, and $\{d\}$ is now the vector of known and unknown structure nodal degrees of freedom or generalized displacements. It can be shown that at this stage, the global stiffness matrix $[K]$ is a singular matrix because its determinant is equal to zero. To remove this singularity problem, we must invoke certain boundary conditions (or constraints or supports) so that the structure remains in place instead of moving as a rigid body. Further details and methods of invoking boundary conditions are given in subsequent chapters. At this time it is sufficient to note that invoking boundary or support conditions results in a modification of the global Eq. (1.4.5). We also emphasize that the applied known loads have been accounted for in the global force matrix $\{F\}$.

Step 6 Solve for the Unknown Degrees of Freedom
(or Generalized Displacements)

Equation (1.4.5), modified to account for the boundary conditions, is a set of simultaneous algebraic equations that can be written in expanded matrix form as

$$\begin{Bmatrix} F_1 \\ F_2 \\ \vdots \\ F_n \end{Bmatrix} = \begin{bmatrix} K_{11} & K_{12} & \ldots & K_{1n} \\ K_{21} & K_{22} & \ldots & K_{2n} \\ \vdots & & & \vdots \\ K_{n1} & K_{n2} & \ldots & K_{nn} \end{bmatrix} \begin{Bmatrix} d_1 \\ d_2 \\ \vdots \\ d_n \end{Bmatrix} \tag{1.4.6}$$

where now n is the structure total number of unknown nodal degrees of freedom. These equations can be solved for the ds by using an elimination method (such as Gauss's method) or an iterative method (such as the Gauss–Seidel method). These two methods are discussed in Appendix B. The ds are called the *primary unknowns*, because they are the first quantities determined using the stiffness (or displacement) finite element method.

Step 7 Solve for the Element Strains and Stresses

For the structural stress-analysis problem, important secondary quantities of strain and stress (or moment and shear force) can be obtained because they can be directly expressed in terms of the displacements determined in step 6. Typical relationships between strain and displacement and between stress and strain—such as Eqs. (1.4.1) and (1.4.2) for one-dimensional stress given in step 3—can be used.

Step 8 Interpret the Results

The final goal is to interpret and analyze the results for use in the design/analysis process. Determination of locations in the structure where large deformations and large stresses occur is generally important in making design/analysis decisions. Postprocessor computer programs help the user to interpret the results by displaying them in graphical form.

▲ 1.5 Applications of the Finite Element Method ▲

The finite element method can be used to analyze both structural and nonstructural problems. Typical structural areas include

1. Stress analysis, including truss and frame analysis, and stress concentration problems typically associated with holes, fillets, or other changes in geometry in a body
2. Buckling
3. Vibration analysis

Nonstructural problems include

1. Heat transfer
2. Fluid flow, including seepage through porous media
3. Distribution of electric or magnetic potential

Finally, some biomechanical engineering problems (which may include stress analysis) typically include analyses of human spine, skull, hip joints, jaw/gum tooth implants, heart, and eye.

We now present some typical applications of the finite element method. These applications will illustrate the variety, size, and complexity of problems that can be solved using the method and the typical discretization process and kinds of elements used.

Figure 1–2 illustrates a control tower for a railroad. The tower is a three-dimensional frame comprising a series of beam-type elements. The 48 elements are labeled by the circled numbers, whereas the 28 nodes are indicated by the uncircled numbers. Each node has three rotation and three displacement components associated with it. The rotations (θs) and displacements (ds) are called the *degrees of freedom*. Because of the loading conditions to which the tower structure is subjected, we have used a three-dimensional model.

The finite element method used for this frame enables the designer/analyst quickly to obtain displacements and stresses in the tower for typical load cases, as required by design codes. Before the development of the finite element method and the computer, even this relatively simple problem took many hours to solve.

The next illustration of the application of the finite element method to problem solving is the determination of displacements and stresses in an underground box culvert subjected to ground shock loading from a bomb explosion. Figure 1–3 shows the discretized model, which included a total of 369 nodes, 40 one-dimensional bar or truss elements used to model the steel reinforcement in the box culvert, and 333 plane strain two-dimensional triangular and rectangular elements used to model the surrounding soil and concrete box culvert. With an assumption of symmetry, only half of the box culvert need be analyzed. This problem requires the solution of nearly 700 unknown nodal displacements. It illustrates that different kinds of elements (here bar and plane strain) can often be used in one finite element model.

Another problem, that of the hydraulic cylinder rod end shown in Figure 1–4, was modeled by 120 nodes and 297 plane strain triangular elements. Symmetry was also applied to the whole rod end so that only half of the rod end had to be analyzed, as shown. The purpose of this analysis was to locate areas of high stress concentration in the rod end.

Figure 1–5 shows a chimney stack section that is four form heights high (or a total of 32 ft high). In this illustration, 584 beam elements were used to model the vertical and horizontal stiffeners making up the formwork, and 252 flat-plate elements were used to model the inner wooden form and the concrete shell. Because of the irregular loading pattern on the structure, a three-dimensional model was necessary. Displacements and stresses in the concrete were of prime concern in this problem.

Figure 1–6 shows the finite element discretized model of a proposed steel die used in a plastic film-making process. The irregular geometry and associated po-

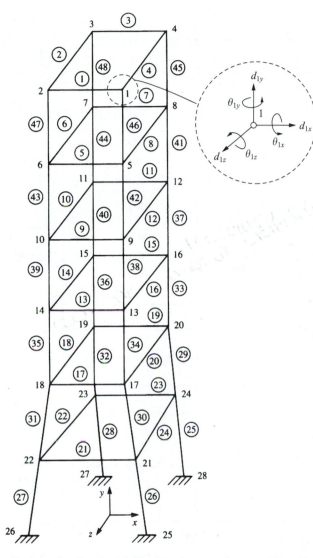

Figure 1–2 Discretized railroad control tower (28 nodes, 48 beam elements) with typical degrees of freedom shown at node 1, for example

tential stress concentrations necessitated use of the finite element method to obtain a reasonable solution. Here 240 axisymmetric elements were used to model the three-dimensional die.

Figure 1–7 illustrates the use of a three-dimensional solid element to model a swing casting for a backhoe frame. The three-dimensional hexahedral elements are necessary to model the irregularly shaped three-dimensional casting. Two-dimensional models certainly would not yield accurate engineering solutions to this problem.

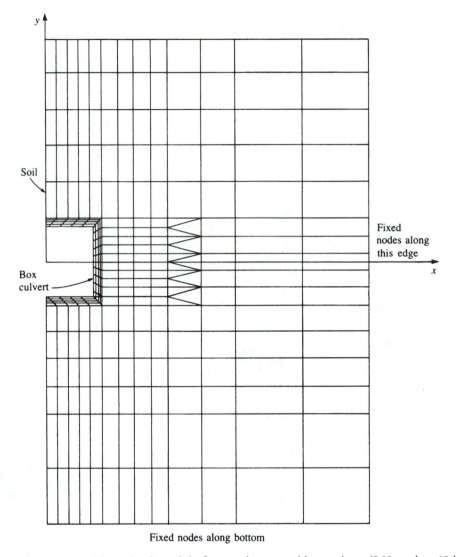

Figure 1–3 Discretized model of an underground box culvert (369 nodes, 40 bar elements, and 333 plane strain elements) [39]

Figure 1–8 illustrates a two-dimensional heat-transfer model used to determine the temperature distribution in earth subjected to a heat source—a buried pipeline transporting a hot gas.

Finally, Figure 1–9 shows a three-dimensional finite element model of a pelvis bone with an implant, used to study stresses in the bone and the cement layer between bone and implant.

The preceding illustrations suggest the kinds of problems that can be solved by the finite element method. Additional guidelines concerning modeling techniques will be provided in Chapter 8.

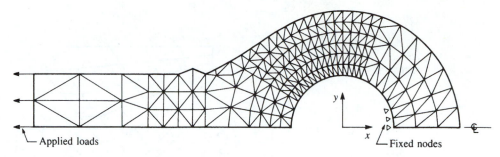

Figure 1–4 Two-dimensional analysis of a hydraulic cylinder rod end (120 nodes, 297 plane strain triangular elements)

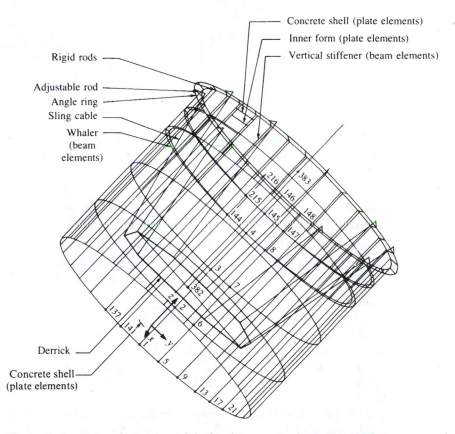

Figure 1–5 Finite element model of a chimney stack section (end view rotated 45°) (584 beam and 252 flat-plate elements)

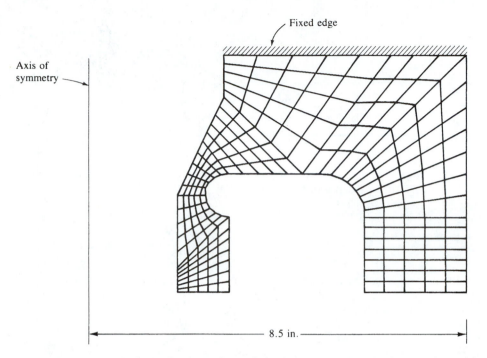

Figure 1–6 Model of a high-strength steel die (240 axisymmetric elements) used in the plastic film industry [40]

▲ 1.6 Advantages of the Finite Element Method ▲

As previously indicated, the finite element method has been applied to numerous problems, both structural and nonstructural. This method has a number of advantages that have made it very popular. They include the ability to

1. Model irregularly shaped bodies quite easily
2. Handle general load conditions without difficulty
3. Model bodies composed of several different materials because the element equations are evaluated individually
4. Handle unlimited numbers and kinds of boundary conditions
5. Vary the size of the elements to make it possible to use small elements where necessary
6. Alter the finite element model relatively easily and cheaply
7. Include dynamic effects
8. Handle nonlinear behavior existing with large deformations and nonlinear materials

The finite element method of structural analysis enables the designer to detect stress, vibration, and thermal problems during the design process and to evaluate design changes *before* the construction of a possible prototype. Thus confidence in the ac-

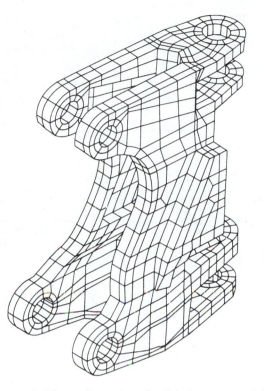

Figure 1–7   Three-dimensional solid element model of a swing casting for a backhoe frame

ceptability of the prototype is enhanced. Moreover, if used properly, the method can reduce the number of prototypes that need to be built.

Even though the finite element method was initially used for structural analysis, it has since been adapted to many other disciplines in engineering and mathematical physics, such as fluid flow, heat transfer, electromagnetic potentials, soil mechanics, and acoustics [22–24, 27, 42–44].

▲ 1.7 Computer Programs for the Finite Element Method ▲

There are two general computer methods of approach to the solution of problems by the finite element method. One is to use large commercial programs, many of which have been configured to run on personal computers (PCs); these general-purpose programs are designed to solve many types of problems. The other is to develop many small, special-purpose programs to solve specific problems. In this section, we will discuss the advantages and disadvantages of both methods. We will then list some of the available general-purpose programs and discuss some of their standard capabilities.

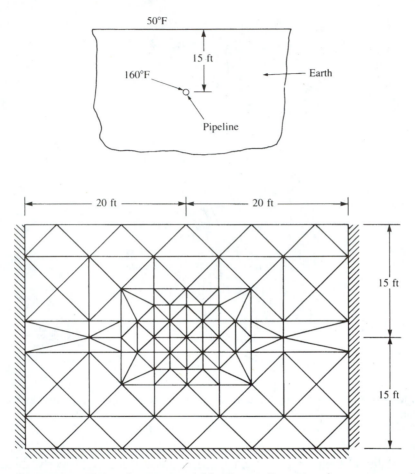

Figure 1–8 Finite element model for a two-dimensional temperature distribution in the earth

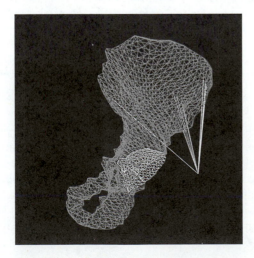

Figure 1–9 Finite element model of a pelvis bone with an implant (over 5000 solid elements were used in the model) (© Thomas Hansen/Courtesy of Harrington Arthritis Research Center, Phoenix, Arizona) [41]

Some advantages of general-purpose programs:

1. The input is well organized and is developed with user ease in mind. Users do not need special knowledge of computer software or hardware. Preprocessors are readily available to help create the finite element model.
2. The programs are large systems that often can solve many types of problems of large or small size with the same input format.
3. Many of the programs can be expanded by adding new modules for new kinds of problems or new technology. Thus they may be kept current with a minimum of effort.
4. With the increased storage capacity and computational efficiency of PCs, many general-purpose programs can now be run on PCs.
5. Many of the commercially available programs have become very attractive in price and can solve a wide range of problems [45, 46].

Some disadvantages of general-purpose programs:

1. The initial cost of developing general-purpose programs is high.
2. General-purpose programs are less efficient than special-purpose programs because the computer must make many checks for each problem, some of which would not be necessary if a special-purpose program were used.
3. Many of the programs are proprietary. Hence the user has little access to the logic of the program. If a revision must be made, it often has to be done by the developers.

Some advantages of special-purpose programs:

1. The programs are usually relatively short, with low development costs.
2. Small computers are able to run the programs.
3. Additions can be made to the program quickly and at a low cost.
4. The programs are efficient in solving the problems they were designed to solve.

The major disadvantage of special-purpose programs is their inability to solve different classes of problems. Thus one must have as many programs as there are different classes of problems to be solved.

There are numerous vendors supporting finite element programs, and the interested user should carefully consult the vendor before purchasing any software. However, to give you an idea about the various commercial personal computer programs now available for solving problems by the finite element method, we present a partial list of existing programs.

1. Algor [46]
2. ANSYS [47]
3. COSMOS/M [48]
4. STARDYNE [49]
5. IMAGES-3D [50]

6. MSC/NASTRAN [51]
7. SAP90 [52]
8. GT-STRUDL [53]

Standard capabilities of many of the listed programs are provided in the preceding references and in Reference [45]. These capabilities include information on

1. Element types available, such as beam, plane stress, and three-dimensional solid
2. Type of analysis available, such as static and dynamic
3. Material behavior, such as linear-elastic and nonlinear
4. Load types, such as concentrated, distributed, thermal, and displacement (settlement)
5. Data generation, such as automatic generation of nodes, elements, and restraints (most programs have preprocessors to generate the mesh for the model)
6. Plotting, such as original and deformed geometry and stress and temperature contours (most programs have postprocessors to aid in interpreting results in graphical form)
7. Displacement behavior, such as small and large displacement and buckling
8. Selective output, such as at selected nodes, elements, and maximum or minimum values

All programs include at least the bar, beam, plane stress, plate-bending, and three-dimensional solid elements, and most now include heat-transfer analysis capabilities.

Complete capabilities of the programs are best obtained through program reference manuals, such as References [46–53]. The intention of this new edition is to blend finite element concepts with the many capabilities of the Algor program [46], one of the most widely used programs in both industry and academia. Therefore, the Algor system is described and illustrated throughout the text wherever computer examples are discussed, and a general description of the Algor system is included in Appendix F. The combination of theory, longhand solutions, and Algor solutions to many example problems provides a thorough pedagogical approach to understanding of the finite element method for solving a variety of complex engineering problems.

▲ **References**

[1] Hrennikoff, A., "Solution of Problems in Elasticity by the Frame Work Method," *Journal of Applied Mechanics*, Vol. 8, No. 4, pp. 169–175, Dec. 1941.

[2] McHenry, D., "A Lattice Analogy for the Solution of Plane Stress Problems," *Journal of Institution of Civil Engineers*, Vol. 21, pp. 59–82, Dec. 1943.

[3] Courant, R., "Variational Methods for the Solution of Problems of Equilibrium and Vibrations," *Bulletin of the American Mathematical Society*, Vol. 49, pp. 1–23, 1943.

[4] Levy, S., "Computation of Influence Coefficients for Aircraft Structures with Discontinuities and Sweepback," *Journal of Aeronautical Sciences*, Vol. 14, No. 10, pp. 547–560, Oct. 1947.

[5] Levy, S., "Structural Analysis and Influence Coefficients for Delta Wings," *Journal of Aeronautical Sciences*, Vol. 20, No. 7, pp. 449–454, July 1953.

[6] Argyris, J. H., "Energy Theorems and Structural Analysis," *Aircraft Engineering*, Oct., Nov., Dec. 1954 and Feb., Mar., Apr., May 1955.

[7] Argyris, J. H., and Kelsey, S., *Energy Theorems and Structural Analysis*, Butterworths, London, 1960 (collection of papers published in *Aircraft Engineering* in 1954 and 1955).

[8] Turner, M. J., Clough, R. W., Martin, H. C., and Topp, L. J., "Stiffness and Deflection Analysis of Complex Structures," *Journal of Aeronautical Sciences*, Vol. 23, No. 9, pp. 805–824, Sept. 1956.

[9] Clough, R. W., "The Finite Element Method in Plane Stress Analysis," *Proceedings*, American Society of Civil Engineers, 2nd Conference on Electronic Computation, Pittsburgh, PA, pp. 345–378, Sept. 1960.

[10] Melosh, R. J., "A Stiffness Matrix for the Analysis of Thin Plates in Bending," *Journal of the Aerospace Sciences*, Vol. 28, No. 1, pp. 34–42, Jan. 1961.

[11] Grafton, P. E., and Strome, D. R., "Analysis of Axisymmetric Shells by the Direct Stiffness Method," *Journal of the American Institute of Aeronautics and Astronautics*, Vol. 1, No. 10, pp. 2342–2347, 1963.

[12] Martin, H. C., "Plane Elasticity Problems and the Direct Stiffness Method," *The Trend in Engineering*, Vol. 13, pp. 5–19, Jan. 1961.

[13] Gallagher, R. H., Padlog, J., and Bijlaard, P. P., "Stress Analysis of Heated Complex Shapes," *Journal of the American Rocket Society*, Vol. 32, pp. 700–707, May 1962.

[14] Melosh, R. J., "Structural Analysis of Solids," *Journal of the Structural Division*, Proceedings of the American Society of Civil Engineers, pp. 205–223, Aug. 1963.

[15] Argyris, J. H., "Recent Advances in Matrix Methods of Structural Analysis," *Progress in Aeronautical Science*, Vol. 4, Pergamon Press, New York, 1964.

[16] Clough, R. W., and Rashid, Y., "Finite Element Analysis of Axisymmetric Solids," *Journal of the Engineering Mechanics Division*, Proceedings of the American Society of Civil Engineers, Vol. 91, pp. 71–85, Feb. 1965.

[17] Wilson, E. L., "Structural Analysis of Axisymmetric Solids," *Journal of the American Institute of Aeronautics and Astronautics*, Vol. 3, No. 12, pp. 2269–2274, Dec. 1965.

[18] Turner, M. J., Dill, E. H., Martin, H. C., and Melosh, R. J., "Large Deflections of Structures Subjected to Heating and External Loads," *Journal of Aeronautical Sciences*, Vol. 27, No. 2, pp. 97–107, Feb. 1960.

[19] Gallagher, R. H., and Padlog, J., "Discrete Element Approach to Structural Stability Analysis," *Journal of the American Institute of Aeronautics and Astronautics*, Vol. 1, No. 6, pp. 1437–1439, 1963.

[20] Zienkiewicz, O. C., Watson, M., and King, I. P., "A Numerical Method of Visco-Elastic Stress Analysis," *International Journal of Mechanical Sciences*, Vol. 10, pp. 807–827, 1968.

[21] Archer, J. S., "Consistent Matrix Formulations for Structural Analysis Using Finite-Element Techniques," *Journal of the American Institute of Aeronautics and Astronautics*, Vol. 3, No. 10, pp. 1910–1918, 1965.

[22] Zienkiewicz, O. C., and Cheung, Y. K., "Finite Elements in the Solution of Field Problems," *The Engineer*, pp. 507–510, Sept. 24, 1965.

[23] Martin, H. C., "Finite Element Analysis of Fluid Flows," *Proceedings of the Second Conference on Matrix Methods in Structural Mechanics*, Wright-Patterson Air Force Base, Ohio, pp. 517–535, Oct. 1968. (AFFDL-TR-68-150, Dec. 1969; AD-703-685, N.T.I.S.)

[24] Wilson, E. L., and Nickel, R. E., "Application of the Finite Element Method to Heat Conduction Analysis," *Nuclear Engineering and Design*, Vol. 4, pp. 276–286, 1966.

[25] Szabo, B. A., and Lee, G. C., "Derivation of Stiffness Matrices for Problems in Plane

Elasticity by Galerkin's Method," *International Journal of Numerical Methods in Engineering*, Vol. 1, pp. 301–310, 1969.

[26] Zienkiewicz, O. C., and Parekh, C. J., "Transient Field Problems: Two-Dimensional and Three-Dimensional Analysis by Isoparametric Finite Elements," *International Journal of Numerical Methods in Engineering*, Vol. 2, No. 1, pp. 61–71, 1970.

[27] Lyness, J. F., Owen, D. R. J., and Zienkiewicz, O. C., "Three-Dimensional Magnetic Field Determination Using a Scalar Potential. A Finite Element Solution," *Transactions on Magnetics*, Institute of Electrical and Electronics Engineers, pp. 1649–1656, 1977.

[28] Belytschko, T., "A Survey of Numerical Methods and Computer Programs for Dynamic Structural Analysis," *Nuclear Engineering and Design*, Vol. 37, No. 1, pp. 23–34, 1976.

[29] Belytschko, T., "Efficient Large-Scale Nonlinear Transient Analysis by Finite Elements," *International Journal of Numerical Methods in Engineering*, Vol. 10, No. 3, pp. 579–596, 1976.

[30] Huiskies, R., and Chao, E. Y. S., "A Survey of Finite Element Analysis in Orthopedic Biomechanics: The First Decade," *Journal of Biomechanics*, Vol. 16, No. 6, pp. 385–409, 1983.

[31] *Journal of Biomechanical Engineering*, Transactions of the American Society of Mechanical Engineers, (published quarterly) (1st issue published 1977).

[32] Kardestuncer, H., ed., *Finite Element Handbook*, McGraw-Hill, New York, 1987.

[33] Clough, R. W., "The Finite Element Method After Twenty-Five Years: A Personal View," *Computers and Structures*, Vol. 12, No. 4, pp. 361–370, 1980.

[34] Kardestuncer, H., *Elementary Matrix Analysis of Structures*, McGraw-Hill, New York, 1974.

[35] Oden, J. T., and Ripperger, E. A., *Mechanics of Elastic Structures*, 2nd ed., McGraw-Hill, New York, 1981.

[36] Finlayson, B. A., *The Method of Weighted Residuals and Variational Principles*, Academic Press, New York, 1972.

[37] Zienkiewicz, O. C., *The Finite Element Method*, 3rd ed., McGraw-Hill, London, 1977.

[38] Cook, R. D., Malkus, D. S., and Plesha, M. E., *Concepts and Applications of Finite Element Analysis*, 3rd ed., Wiley, New York, 1989.

[39] Koswara, H., *A Finite Element Analysis of Underground Shelter Subjected to Ground Shock Load*, M.S. Thesis, Rose-Hulman Institute of Technology, 1983.

[40] Greer, R. D., "The Analysis of a Film Tower Die Utilizing the ANSYS Finite Element Package," M.S. Thesis, Rose-Hulman Institute of Technology, Terre Haute, Indiana, May 1989.

[41] Koeneman, J. B., Hansen, T. M., and Beres, K., "The Effect of Hip Stem Elastic Modulus and Cement/Stem Bond on Cement Stresses," 36th Annual Meeting, Orthopaedic Research Society, Feb. 5–8, 1990, New Orleans, Louisiana.

[42] Girijavallabham, C. V., and Reese, L. C., "Finite-Element Method for Problems in Soil Mechanics," *Journal of the Structural Division*, American Society of Civil Engineers, No. Sm2, pp. 473–497, Mar. 1968.

[43] Young, C., and Crocker, M., "Transmission Loss by Finite-Element Method," *Journal of the Acoustical Society of America*, Vol. 57, No. 1, pp. 144–148, Jan. 1975.

[44] Silvester, P. P., and Ferrari, R. L., *Finite Elements for Electrical Engineers*, Cambridge University Press, Cambridge, England, 1983.

[45] Falk, H., and Beardsley, C. W., "Finite Element Analysis Packages for Personal Computers," *Mechanical Engineering*, pp. 54–71, Jan. 1985.

[46] Algor Interactive Systems, 260 Alpha Drive, Pittsburgh, PA 15238.

[47] Swanson, J. A., ANSYS-Engineering Analysis Systems User's Manual, Swanson Analysis Systems, Inc., Johnson Rd., P.O. Box 65, Houston, PA 15342.

[48] COSMOS/M, Structural Research & Analysis Corp., 12121 Wilshire Blvd., Los Angeles, CA 90025.

[49] STARDYNE, Research Engineers, Inc., 22700 Savi Ranch Pkwy, Yorba Linda, CA 92687.

[50] Celestial Software, 125 University Ave., Berkeley, CA 94710.

[51] MSC/NASTRAN, MacNeal-Schwendler Corp., 600 Suffolk St., Lowell, MA 01854.

[52] Computers & Structures, Inc., 1995 University Ave., Berkeley, CA 94704.

[53] Web site http://ce6000.ce.gatech.edu/

[54] Noor, A. K., "Bibliography of Books and Monographs on Finite Element Technology," *Applied Mechanics Reviews*, Vol. 44, No. 6, pp. 307–317, June 1991.

▲ Problems

1.1 Define the term *finite element*.

1.2 What does *discretization* mean in the finite element method?

1.3 In what year did the modern development of the finite element method begin?

1.4 In what year was the direct stiffness method introduced?

1.5 Define the term *matrix*.

1.6 What role did the computer play in the use of the finite element method?

1.7 List and briefly describe the general steps of the finite element method.

1.8 What is the displacement method?

1.9 List four common types of finite elements.

1.10 Name three commonly used methods for deriving the element stiffness matrix and element equations. Briefly describe each method.

1.11 To what does the term *degrees of freedom* refer?

1.12 List five typical areas of engineering where the finite element method is applied.

1.13 List five advantages of the finite element method.

2

Introduction to the Stiffness (Displacement) Method

Introduction

This chapter introduces some of the basic concepts on which the direct stiffness method is founded. The linear spring is introduced first because it provides a simple yet generally instructive tool to illustrate the basic concepts. We begin with a general definition of the stiffness matrix and then consider the derivation of the stiffness matrix for a linear-elastic spring element. We next illustrate how to assemble the total stiffness matrix for a structure comprising an assemblage of spring elements by using elementary concepts of equilibrium and compatibility. We then show how the total stiffness matrix for an assemblage can be obtained by superimposing the stiffness matrices of the individual elements in a direct manner. The term *direct stiffness method* evolved in reference to this technique.

After establishing the total structure stiffness matrix, we illustrate how to impose boundary conditions—both homogeneous and nonhomogeneous. A complete solution including the nodal displacements and reactions is thus obtained. (The determination of internal forces is discussed in Chapter 3 in connection with the bar element.)

We then introduce the principle of minimum potential energy, apply it to derive the spring element equations, and use it to solve a spring assemblage problem. We will illustrate this principle for the simplest of elements (those with small numbers of degrees of freedom) so that it will be a more readily understood concept when applied, of necessity, to elements with large numbers of degrees of freedom in subsequent chapters.

▲ 2.1 Definition of the Stiffness Matrix ▲

Familiarity with the stiffness matrix is essential to understanding the stiffness method. We define the stiffness matrix as follows: *For an element, a* **stiffness matrix** $\hat{\underline{k}}$ *is a matrix such that* $\hat{\underline{f}} = \hat{\underline{k}}\hat{\underline{d}}$, *where* $\hat{\underline{k}}$ *relates local-coordinate* $(\hat{x}, \hat{y}, \hat{z})$ *nodal displacements* $\underline{\hat{d}}$ *to local forces* $\hat{\underline{f}}$ *of a single element.* (Throughout this text, the underline notation denotes a

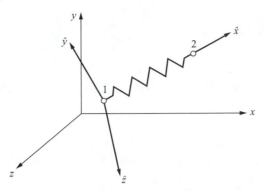

Figure 2–1 Local $(\hat{x}, \hat{y}, \hat{z})$ and global (x, y, z) coordinate systems

matrix, and the ˆ symbol denotes quantities referred to a local-coordinate system set up to be convenient for the element as shown in Figure 2–1.)

For a continuous medium or structure comprising a series of elements, a stiffness matrix $\underline{K}$ relates global-coordinate (x, y, z) nodal displacements $\underline{d}$ to global forces $\underline{F}$ of the whole medium or structure. (Lowercase letters such as x, y, and z without the ˆ symbol denote global-coordinate variables.)

▲ 2.2 Derivation of the Stiffness Matrix for a Spring Element ▲

Using the direct equilibrium approach, we will now derive the stiffness matrix for a one-dimensional linear spring—that is, a spring that obeys Hooke's law and resists forces only in the direction of the spring. Consider the linear spring element shown in Figure 2–2. Reference points 1 and 2 are located at the ends of the element. These reference points are called the *nodes* of the spring element. The local nodal forces are $\hat{f}_{1x}$ and $\hat{f}_{2x}$ for the spring element associated with the local axis $\hat{x}$. The local axis acts in the direction of the spring so that we can directly measure displacements and forces along the spring. The local nodal displacements are $\hat{d}_{1x}$ and $\hat{d}_{2x}$ for the spring element. These nodal displacements are called the *degrees of freedom* at each node. Positive directions for the forces and displacements at each node are taken in the positive $\hat{x}$ direction as shown from node 1 to node 2 in the figure. The symbol k is called the *spring constant* or *stiffness* of the spring.

Analogies to actual spring constants arise in numerous engineering problems. In Chapter 3, we see that a prismatic uniaxial bar has a spring constant $k = AE/L$, where A represents the cross-sectional area of the bar, E is the modulus of elasticity, and L is the bar length. Similarly, in Chapter 6 we show that a prismatic circular-cross-section bar in torsion has a spring constant $k = JG/L$, where J is the polar moment of inertia and G is the shear modulus of the material. For one-dimensional heat conduction (Chapter 13), $k = AK_{xx}/L$, where K_{xx} is the thermal conductivity of

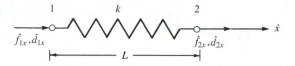

Figure 2–2 Linear spring element with positive nodal displacement and force conventions

the material, and for one-dimensional fluid flow through a porous medium (Chapter 14), $k = AK_{xx}/L$, where K_{xx} is the permeability coefficient of the material.

We now want to develop a relationship between nodal forces and nodal displacements for a spring element. This relationship will be the stiffness matrix. Therefore, we want to relate the nodal force matrix to the nodal displacement matrix as follows:

$$\left\{ \begin{array}{c} \hat{f}_{1x} \\ \hat{f}_{2x} \end{array} \right\} = \left[\begin{array}{cc} k_{11} & k_{12} \\ k_{21} & k_{22} \end{array} \right] \left\{ \begin{array}{c} \hat{d}_{1x} \\ \hat{d}_{2x} \end{array} \right\} \tag{2.2.1}$$

where the element stiffness coefficients k_{ij} of the $\hat{k}$ matrix in Eq. (2.2.1) are to be determined. Recall from Eqs. (1.2.5) and (1.2.6) that k_{ij} represent the force F_i in the ith degree of freedom due to a unit displacement d_j in the jth degree of freedom while all other displacements are zero. That is, when we let $d_j = 1$ and $d_k = 0$ for $k \neq j$, force $F_i = k_{ij}$.

We now use the general steps outlined in Section 1.4 to derive the stiffness matrix for the spring element in this section (while keeping in mind that these same steps will be applicable later in the derivation of stiffness matrices of more general elements) and then to illustrate a complete solution of a spring assemblage in Section 2.3. Because our approach throughout this text is to derive various element stiffness matrices and then to illustrate how to solve engineering problems with the elements, step 1 now involves only selecting the element type.

Step 1   Select the Element Type

Consider the linear spring element (which can be an element in a system of springs) subjected to resulting nodal tensile forces T (which may result from the action of adjacent springs) directed along the spring axial direction $\hat{x}$ as shown in Figure 2–3, so

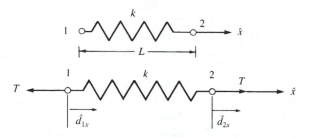

Figure 2–3 Linear spring subjected to tensile forces

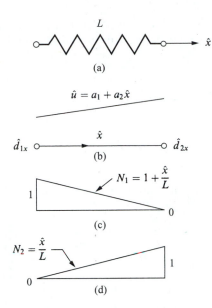

Figure 2–4 (a) Spring element showing plots of (b) displacement function $\hat{u}$ and shape functions (c) N_1 and (d) N_2 over domain of element

as to be in equilibrium. The local $\hat{x}$ axis is directed from node 1 to node 2. We represent the spring by labeling nodes at each end and by labeling the element number. The original distance between nodes before deformation is denoted by L. The material property (spring constant) of the element is k.

Step 2   Select a Displacement Function

We must choose in advance the mathematical function to represent the deformed shape of the spring element under loading. Because it is difficult, if not impossible at times, to obtain a closed form or exact solution, we assume a solution shape or distribution of displacement within the element by using an appropriate mathematical function. The most common functions used are polynomials.

Because the spring element resists axial loading only with the local degrees of freedom for the element being displacements $\hat{d}_{1x}$ and $\hat{d}_{2x}$ along the $\hat{x}$ direction, we choose a displacement function $\hat{u}$ to represent the axial displacement throughout the element. Here a linear displacement variation along the $\hat{x}$ axis of the spring is assumed [Figure 2–4(b)], because a linear function with specified endpoints has a unique path. Therefore,

$$\hat{u} = a_1 + a_2\hat{x} \tag{2.2.2}$$

In general, the total number of coefficients a is equal to the total number of degrees of freedom associated with the element. Here the total number of degrees of freedom is two—an axial displacement at each of the two nodes of the element (we present further discussion regarding the choice of displacement functions in Section 3.2). In matrix form, Eq. (2.2.2) becomes

$$\hat{u} = \begin{bmatrix} 1 & \hat{x} \end{bmatrix} \begin{Bmatrix} a_1 \\ a_2 \end{Bmatrix} \tag{2.2.3}$$

We now want to express $\hat{u}$ as a function of the nodal displacement $\hat{d}_{1x}$ and $\hat{d}_{2x}$. We achieve this by evaluating $\hat{u}$ at each node and solving for a_1 and a_2 from Eq. (2.2.2) as follows:

$$\hat{u}(0) = \hat{d}_{1x} = a_1 \tag{2.2.4}$$

$$\hat{u}(L) = \hat{d}_{2x} = a_2 L + \hat{d}_{1x} \tag{2.2.5}$$

or, solving Eq. (2.2.5) for a_2,

$$a_2 = \frac{\hat{d}_{2x} - \hat{d}_{1x}}{L} \tag{2.2.6}$$

Upon substituting Eqs. (2.2.4) and (2.2.6) into Eq. (2.2.2), we have

$$\hat{u} = \left(\frac{\hat{d}_{2x} - \hat{d}_{1x}}{L}\right)\hat{x} + \hat{d}_{1x} \tag{2.2.7}$$

In matrix form, we express Eq. (2.2.7) as

$$\hat{u} = \begin{bmatrix} 1 - \dfrac{\hat{x}}{L} & \dfrac{\hat{x}}{L} \end{bmatrix} \begin{Bmatrix} \hat{d}_{1x} \\ \hat{d}_{2x} \end{Bmatrix} \tag{2.2.8}$$

or

$$\hat{u} = \begin{bmatrix} N_1 & N_2 \end{bmatrix} \begin{Bmatrix} \hat{d}_{1x} \\ \hat{d}_{2x} \end{Bmatrix} \tag{2.2.9}$$

Here

$$N_1 = 1 - \frac{\hat{x}}{L} \qquad \text{and} \qquad N_2 = \frac{\hat{x}}{L} \tag{2.2.10}$$

are called the *shape functions* because the N_i's express the shape of the assumed displacement function over the domain ($\hat{x}$ coordinate) of the element when the ith element degree of freedom has unit value and all other degrees of freedom are zero. In this case, N_1 and N_2 are linear functions that have the properties that $N_1 = 1$ at node 1 and $N_1 = 0$ at node 2, whereas $N_2 = 1$ at node 2 and $N_2 = 0$ at node 1. See Figure 2–4(c) and (d) for plots of these shape functions over the domain of the spring element. Also, $N_1 + N_2 = 1$ for any axial coordinate along the bar. (Section 3.2 further explores this important relationship.) In addition, the N_i's are often called *interpolation functions* because we are interpolating to find the value of a function between given nodal values. The interpolation function may be different from the actual function except at the endpoints or nodes, where the interpolation function and actual function must be equal to specified nodal values.

Step 3 Define the Strain/Displacement and Stress/Strain Relationships

The tensile forces T produce a total elongation (deformation) δ of the spring. The typical total elongation of the spring is shown in Figure 2–5. Here $\hat{d}_{1x}$ is a negative value because the direction of displacement is opposite the positive $\hat{x}$ direction, whereas $\hat{d}_{2x}$ is a positive value.

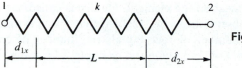

Figure 2–5   Deformed spring

The deformation of the spring is then represented by

$$\delta = \hat{d}_{2x} - \hat{d}_{1x} \tag{2.2.11}$$

From Eq. (2.2.11), we observe that the total deformation is the difference of the nodal displacements in the $\hat{x}$ direction.

For a spring element, we can relate the force in the spring directly to the deformation. Therefore, the strain/displacement relationship is not necessary here.

The stress/strain relationship can be expressed in terms of the force/deformation relationship instead as

$$T = k\delta \tag{2.2.12}$$

Now, using Eq. (2.2.11) in Eq. (2.2.12), we obtain

$$T = k(\hat{d}_{2x} - \hat{d}_{1x}) \tag{2.2.13}$$

Step 4   Derive the Element Stiffness Matrix and Equations

We now derive the spring element stiffness matrix. By the sign convention for nodal forces and equilibrium, we have

$$\hat{f}_{1x} = -T \quad \hat{f}_{2x} = T \tag{2.2.14}$$

Using Eqs. (2.2.13) and (2.2.14), we have

$$T = -\hat{f}_{1x} = k(\hat{d}_{2x} - \hat{d}_{1x})$$
$$T = \hat{f}_{2x} = k(\hat{d}_{2x} - \hat{d}_{1x}) \tag{2.2.15}$$

Rewriting Eqs. (2.2.15), we obtain

$$\hat{f}_{1x} = k(\hat{d}_{1x} - \hat{d}_{2x})$$
$$\hat{f}_{2x} = k(\hat{d}_{2x} - \hat{d}_{1x}) \tag{2.2.16}$$

Now expressing Eqs. (2.2.16) in a single matrix equation yields

$$\left\{ \begin{array}{c} \hat{f}_{1x} \\ \hat{f}_{2x} \end{array} \right\} = \left[\begin{array}{cc} k & -k \\ -k & k \end{array} \right] \left\{ \begin{array}{c} \hat{d}_{1x} \\ \hat{d}_{2x} \end{array} \right\} \tag{2.2.17}$$

This relationship holds for the spring along the $\hat{x}$ axis. From our basic definition of a stiffness matrix and application of Eq. (2.2.1) to Eq. (2.2.17), we obtain

$$\hat{k} = \left[\begin{array}{cc} k & -k \\ -k & k \end{array} \right] \tag{2.2.18}$$

as the stiffness matrix for a linear spring element. Here $\underline{\hat{k}}$ is called the *local stiffness matrix* for the element. We observe from Eq. (2.2.18) that $\underline{\hat{k}}$ is a symmetric (that is, $k_{ij} = k_{ji}$) square matrix (the number of rows equals the number of columns in $\underline{\hat{k}}$). Appendix A gives more description and numerical examples of symmetric and square matrices.

Step 5   Assemble the Element Equations to Obtain the Global Equations and Introduce Boundary Conditions

The global stiffness matrix and global force matrix are assembled using nodal force equilibrium equations, force/deformation and compatibility equations from Section 2.3, and the direct stiffness method described in Section 2.4. This step applies for structures composed of more than one element such that

$$\underline{K} = [K] = \sum_{e=1}^{N} \underline{k}^{(e)} \quad \text{and} \quad \underline{F} = \{F\} = \sum_{e=1}^{N} \underline{f}^{(e)} \tag{2.2.19}$$

where $\underline{k}$ and $\underline{f}$ are now element stiffness and force matrices expressed in a global reference frame. (Throughout this text, the $\sum$ sign used in this context does not imply a simple summation of element matrices but rather denotes that these element matrices must be assembled properly according to the direct stiffness method described in Section 2.4.)

Step 6   Solve for the Nodal Displacements

The displacements are then determined by imposing boundary conditions, such as support conditions, and solving a system of equations, $\underline{F} = \underline{K}\underline{d}$, simultaneously.

Step 7   Solve for the Element Forces

Finally, the element forces are determined by back-substitution, applied to each element, into equations similar to Eqs. (2.2.16).

▲ **2.3 Example of a Spring Assemblage**   ▲

Structures such as trusses, building frames, and bridges comprise basic structural components connected together to form the overall structures. To analyze these structures, we must determine the total structure stiffness matrix for an interconnected system of elements. Before considering the truss and frame, we will determine the total structure stiffness matrix for a spring assemblage by using the force/displacement matrix relationships derived in Section 2.2 for the spring element, along with fundamental concepts of nodal equilibrium and compatibility. Step 5 above will then have been illustrated.

We will consider the specific example of the two-spring assemblage shown in Figure 2–6*. This example is general enough to illustrate the direct equilibrium

---

* Throughout this text, element numbers in figures are shown with circles around them.

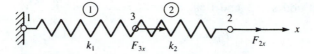

Figure 2–6 Two-spring assemblage

approach for obtaining the total stiffness matrix of the spring assemblage. Here we fix node 1 and apply axial forces for F_{3x} at node 3 and F_{2x} at node 2. The stiffnesses of spring elements 1 and 2 are k_1 and k_2, respectively. The nodes of the assemblage have been numbered 1, 3, and 2 for further generalization because sequential numbering between elements generally does not occur in large problems.

The x axis is the global axis of the assemblage. The local $\hat{x}$ axis of each element coincides with the global axis of the assemblage.

For element 1, using Eq. (2.2.17), we have

$$\left\{ \begin{array}{c} f_{1x} \\ f_{3x} \end{array} \right\} = \left[\begin{array}{cc} k_1 & -k_1 \\ -k_1 & k_1 \end{array} \right] \left\{ \begin{array}{c} d_{1x}^{(1)} \\ d_{3x}^{(1)} \end{array} \right\} \tag{2.3.1}$$

and for element 2, we have

$$\left\{ \begin{array}{c} f_{3x} \\ f_{2x} \end{array} \right\} = \left[\begin{array}{cc} k_2 & -k_2 \\ -k_2 & k_2 \end{array} \right] \left\{ \begin{array}{c} d_{3x}^{(2)} \\ d_{2x}^{(2)} \end{array} \right\} \tag{2.3.2}$$

Furthermore, elements 1 and 2 must remain connected at common node 3 throughout the displacement. This is called the *continuity* or *compatibility requirement*. The compatibility requirement yields

$$d_{3x}^{(1)} = d_{3x}^{(2)} = d_{3x} \tag{2.3.3}$$

where, throughout this text, the superscript in parentheses above d refers to the element number to which they are related. Recall that the subscripts to the right identify the node and the direction of displacement, respectively, and that d_{3x} is the node 3 displacement of the total or global spring assemblage.

Free-body diagrams of each element and node (using the established sign conventions for element nodal forces in Figure 2–2) are shown in Figure 2–7.

Based on the free-body diagrams of each node shown in Figure 2–7 and the fact that external forces must equal internal forces at each node, we can write nodal equi-

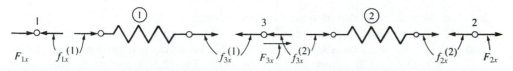

Figure 2–7 Nodal forces consistent with element force sign convention

librium equations at nodes 3, 2, and 1 as

$$F_{3x} = f_{3x}^{(1)} + f_{3x}^{(2)} \tag{2.3.4}$$

$$F_{2x} = f_{2x}^{(2)} \tag{2.3.5}$$

$$F_{1x} = f_{1x}^{(1)} \tag{2.3.6}$$

where F_{1x} results from the external applied reaction at the fixed support.

Here Newton's third law, of equal but opposite forces, is applied in moving from a node to an element associated with the node. Using Eqs. (2.3.1)–(2.3.3) in Eqs. (2.3.4)–(2.3.6), we obtain

$$F_{3x} = (-k_1 d_{1x} + k_1 d_{3x}) + (k_2 d_{3x} - k_2 d_{2x})$$

$$F_{2x} = -k_2 d_{3x} + k_2 d_{2x} \tag{2.3.7}$$

$$F_{1x} = k_1 d_{1x} - k_1 d_{3x}$$

In matrix form, Eqs. (2.3.7) are expressed by

$$\begin{Bmatrix} F_{3x} \\ F_{2x} \\ F_{1x} \end{Bmatrix} = \begin{bmatrix} k_1 + k_2 & -k_2 & -k_1 \\ -k_2 & k_2 & 0 \\ -k_1 & 0 & k_1 \end{bmatrix} \begin{Bmatrix} d_{3x} \\ d_{2x} \\ d_{1x} \end{Bmatrix} \tag{2.3.8}$$

Rearranging Eq. (2.3.8) in numerically increasing order of the nodal degrees of freedom, we have

$$\begin{Bmatrix} F_{1x} \\ F_{2x} \\ F_{3x} \end{Bmatrix} = \begin{bmatrix} k_1 & 0 & -k_1 \\ 0 & k_2 & -k_2 \\ -k_1 & -k_2 & k_1 + k_2 \end{bmatrix} \begin{Bmatrix} d_{1x} \\ d_{2x} \\ d_{3x} \end{Bmatrix} \tag{2.3.9}$$

Equation (2.3.9) is now written as the single matrix equation

$$\underline{F} = \underline{K}\underline{d} \tag{2.3.10}$$

where $\underline{F} = \begin{Bmatrix} F_{1x} \\ F_{2x} \\ F_{3x} \end{Bmatrix}$ is called the *global nodal force matrix*, $\underline{d} = \begin{Bmatrix} d_{1x} \\ d_{2x} \\ d_{3x} \end{Bmatrix}$ is called the *global nodal displacement matrix*, and

$$\underline{K} = \begin{bmatrix} k_1 & 0 & -k_1 \\ 0 & k_2 & -k_2 \\ -k_1 & -k_2 & k_1 + k_2 \end{bmatrix} \tag{2.3.11}$$

is called the *total* or *global* or *system stiffness matrix*.

In summary, to establish the stiffness equations and stiffness matrix, Eqs. (2.3.9) and (2.3.11), for a spring assemblage, we have used force/deformation relationships Eqs. (2.3.1) and (2.3.2), compatibility relationship Eq. (2.3.3), and nodal force equilibrium Eqs. (2.3.4)–(2.3.6). We will consider the complete solution to this example problem after considering a more practical method of assembling the total stiffness matrix in Section 2.4 and discussing the support boundary conditions in Section 2.5.

▲ 2.4 Assembling the Total Stiffness Matrix by Superposition (Direct Stiffness Method) ▲

We will now consider a more convenient method for constructing the total stiffness matrix. This method is based on proper superposition of the individual element stiffness matrices making up a structure (also see References [1] and [2]).

Referring to the two-spring assemblage of Section 2.3, the element stiffness matrices are given in Eqs. (2.3.1) and (2.3.2) as

$$
\underline{k}^{(1)} =
\begin{array}{c}
 \\
\begin{array}{cc} d_{1x} & d_{3x} \end{array} \\
\left[\begin{array}{cc} k_1 & -k_1 \\ -k_1 & k_1 \end{array} \right]
\begin{array}{c} d_{1x} \\ d_{3x} \end{array}
\end{array}
\qquad
\underline{k}^{(2)} =
\begin{array}{c}
\begin{array}{cc} d_{3x} & d_{2x} \end{array} \\
\left[\begin{array}{cc} k_2 & -k_2 \\ -k_2 & k_2 \end{array} \right]
\begin{array}{c} d_{3x} \\ d_{2x} \end{array}
\end{array}
\qquad (2.4.1)
$$

Here the d_{ix}'s written above the columns and next to the rows in the $\underline{k}$'s indicate the degrees of freedom associated with each element row and column.

The two element stiffness matrices, Eqs. (2.4.1), are not associated with the same degrees of freedom; that is, element 1 is associated with axial displacements at nodes 1 and 3, whereas element 2 is associated with axial displacements at nodes 2 and 3. Therefore, the element stiffness matrices cannot be added together (superimposed) in their present form. To superimpose the element matrices, we must expand them to the order (size) of the total structure (spring assemblage) stiffness matrix so that each element stiffness matrix is associated with all the degrees of freedom of the structure. To expand each element stiffness matrix to the order of the total stiffness matrix, we simply add rows and columns of zeros for those displacements not associated with that particular element.

For element 1, we rewrite the stiffness matrix in expanded form so that Eq. (2.3.1) becomes

$$
k_1
\begin{array}{c}
\begin{array}{ccc} d_{1x} & d_{2x} & d_{3x} \end{array} \\
\left[\begin{array}{ccc} 1 & 0 & -1 \\ 0 & 0 & 0 \\ -1 & 0 & 1 \end{array} \right]
\end{array}
\left\{ \begin{array}{c} d_{1x}^{(1)} \\ d_{2x}^{(1)} \\ d_{3x}^{(1)} \end{array} \right\}
=
\left\{ \begin{array}{c} f_{1x}^{(1)} \\ f_{2x}^{(1)} \\ f_{3x}^{(1)} \end{array} \right\}
\qquad (2.4.2)
$$

where, from Eq. (2.4.2), we see that $d_{2x}^{(1)}$ and $f_{2x}^{(1)}$ are not associated with $\underline{k}^{(1)}$. Similarly, for element 2, we have

$$
k_2
\begin{array}{c}
\begin{array}{ccc} d_{1x} & d_{2x} & d_{3x} \end{array} \\
\left[\begin{array}{ccc} 0 & 0 & 0 \\ 0 & 1 & -1 \\ 0 & -1 & 1 \end{array} \right]
\end{array}
\left\{ \begin{array}{c} d_{1x}^{(2)} \\ d_{2x}^{(2)} \\ d_{3x}^{(2)} \end{array} \right\}
=
\left\{ \begin{array}{c} f_{1x}^{(2)} \\ f_{2x}^{(2)} \\ f_{3x}^{(2)} \end{array} \right\}
\qquad (2.4.3)
$$

Now, considering force equilibrium at each node results in

$$
\left\{ \begin{array}{c} f_{1x}^{(1)} \\ 0 \\ f_{3x}^{(1)} \end{array} \right\} + \left\{ \begin{array}{c} 0 \\ f_{2x}^{(2)} \\ f_{3x}^{(2)} \end{array} \right\} = \left\{ \begin{array}{c} F_{1x} \\ F_{2x} \\ F_{3x} \end{array} \right\} \tag{2.4.4}
$$

where Eq. (2.4.4) is really Eqs. (2.3.4)–(2.3.6) expressed in matrix form. Using Eqs. (2.4.2) and (2.4.3) in Eq. (2.4.4), we obtain

$$
k_1 \begin{bmatrix} 1 & 0 & -1 \\ 0 & 0 & 0 \\ -1 & 0 & 1 \end{bmatrix} \left\{ \begin{array}{c} d_{1x}^{(1)} \\ d_{2x}^{(1)} \\ d_{3x}^{(1)} \end{array} \right\} + k_2 \begin{bmatrix} 0 & 0 & 0 \\ 0 & 1 & -1 \\ 0 & -1 & 1 \end{bmatrix} \left\{ \begin{array}{c} d_{1x}^{(2)} \\ d_{2x}^{(2)} \\ d_{3x}^{(2)} \end{array} \right\} = \left\{ \begin{array}{c} F_{1x} \\ F_{2x} \\ F_{3x} \end{array} \right\} \tag{2.4.5}
$$

where, again, the superscripts on the d's indicate the element numbers. Simplifying Eq. (2.4.5) results in

$$
\begin{array}{ccc} d_1 & d_2 & d_3 \end{array}
$$
$$
\begin{bmatrix} k_1 & 0 & -k_1 \\ 0 & k_2 & -k_2 \\ -k_1 & -k_2 & k_1 + k_2 \end{bmatrix} \left\{ \begin{array}{c} d_{1x} \\ d_{2x} \\ d_{3x} \end{array} \right\} = \left\{ \begin{array}{c} F_{1x} \\ F_{2x} \\ F_{3x} \end{array} \right\} \tag{2.4.6}
$$

Here the superscripts indicating the element numbers associated with the nodal displacements have been dropped because $d_{1x}^{(1)}$ is really d_{1x}, $d_{2x}^{(2)}$ is really d_{2x}, and, by Eq. (2.3.3), $d_{3x}^{(1)} = d_{3x}^{(2)} = d_{3x}$, the node 3 displacement of the total assemblage. Equation (2.4.6), obtained through superposition, is identical to Eq. (2.3.9).

The expanded element stiffness matrices in Eqs. (2.4.2) and (2.4.3) could have been added directly to obtain the total stiffness matrix of the structure, given in Eq. (2.4.6). This reliable method of directly assembling individual element stiffness matrices to form the total structure stiffness matrix and the total set of stiffness equations is called the *direct stiffness method*. It is the most important step in the finite element method.

For this simple example, it is easy to expand the element stiffness matrices and then superimpose them to arrive at the total stiffness matrix. However, for problems involving a large number of degrees of freedom, it will become tedious to expand each element stiffness matrix to the order of the total stiffness matrix. To avoid this expansion of each element stiffness matrix, we suggest a direct, or short-cut, form of the direct stiffness method to obtain the total stiffness matrix. For the spring assemblage example, the rows and columns of each element stiffness matrix are labeled according to the degrees of freedom associated with them as follows:

$$
\begin{array}{cc} d_{1x} & d_{3x} \end{array} \qquad\qquad\qquad \begin{array}{cc} d_{3x} & d_{2x} \end{array}
$$
$$
\underline{k}^{(1)} = \begin{bmatrix} k_1 & -k_1 \\ -k_1 & k_1 \end{bmatrix} \begin{array}{c} d_{1x} \\ d_{3x} \end{array} \qquad \underline{k}^{(2)} = \begin{bmatrix} k_2 & -k_2 \\ -k_2 & k_2 \end{bmatrix} \begin{array}{c} d_{3x} \\ d_{2x} \end{array} \tag{2.4.7}
$$

$\underline{K}$ is then constructed simply by directly adding terms associated with degrees of freedom in $\underline{k}^{(1)}$ and $\underline{k}^{(2)}$ into their corresponding identical degree-of-freedom locations in $\underline{K}$ as follows. The d_{1x} row, d_{1x} column term of $\underline{K}$ is contributed only by element 1, as only element 1 has degree of freedom d_{1x} [Eq. (2.4.7)], that is, $k_{11} = k_1$. The d_{3x} row,

d_{3x} column of $\underline{K}$ has contributions from both elements 1 and 2, as the d_{3x} degree of freedom is associated with both elements. Therefore, $k_{33} = k_1 + k_2$. Similar reasoning results in $\underline{K}$ as

$$
\underline{K} = \begin{array}{c} \begin{array}{ccc} d_{1x} & d_{2x} & d_{3x} \end{array} \\ \begin{bmatrix} k_1 & 0 & -k_1 \\ 0 & k_2 & -k_2 \\ -k_1 & -k_2 & k_1 + k_2 \end{bmatrix} \begin{array}{c} d_{1x} \\ d_{2x} \\ d_{3x} \end{array} \end{array}
\tag{2.4.8}
$$

Here elements in $\underline{K}$ are located on the basis that degrees of freedom are ordered in increasing node numerical order for the total structure. Section 2.5 addresses the complete solution to the two-spring assemblage in conjunction with discussion of the support boundary conditions.

▲ 2.5 Boundary Conditions ▲

We must specify boundary (or support) conditions for structure models such as the spring assemblage of Figure 2–6, or $\underline{K}$ will be singular; that is, the determinant of $\underline{K}$ will be zero, and its inverse will not exist. This means the structural system is unstable. Without our specifying adequate kinematic constraints or support conditions, the structure will be free to move as a rigid body and not resist any applied loads.

Boundary conditions are of two general types. Homogeneous boundary conditions—the more common—occur at locations that are completely prevented from movement; nonhomogeneous boundary conditions occur where finite nonzero values of displacement are specified, such as the settlement of a support.

To illustrate the two general types of boundary conditions, let us consider Eq. (2.4.6), derived for the spring assemblage of Figure 2–6. We first consider the case of homogeneous boundary conditions. Hence all boundary conditions are such that the displacements are zero at certain nodes. Here we have $d_{1x} = 0$ because node 1 is fixed. Therefore, Eq. (2.4.6) can be written as

$$
\begin{bmatrix} k_1 & 0 & -k_1 \\ 0 & k_2 & -k_2 \\ -k_1 & -k_2 & k_1 + k_2 \end{bmatrix} \begin{Bmatrix} 0 \\ d_{2x} \\ d_{3x} \end{Bmatrix} = \begin{Bmatrix} F_{1x} \\ F_{2x} \\ F_{3x} \end{Bmatrix}
\tag{2.5.1}
$$

Equation (2.5.1), written in expanded form, becomes

$$
k_1(0) + (0)d_{2x} - k_1 d_{3x} = F_{1x}
$$
$$
0(0) + k_2 d_{2x} - k_2 d_{3x} = F_{2x}
\tag{2.5.2}
$$
$$
-k_1(0) - k_2 d_{2x} + (k_1 + k_2)d_{3x} = F_{3x}
$$

where F_{1x} is the unknown reaction and F_{2x} and F_{3x} are known applied loads.

Writing the second and third of Eqs. (2.5.2) in matrix form, we have

$$
\begin{bmatrix} k_2 & -k_2 \\ -k_2 & k_1 + k_2 \end{bmatrix} \begin{Bmatrix} d_{2x} \\ d_{3x} \end{Bmatrix} = \begin{Bmatrix} F_{2x} \\ F_{3x} \end{Bmatrix}
\tag{2.5.3}
$$

We have now effectively partitioned off the first column and row of $\underline{K}$ and the first row of $\underline{d}$ and $\underline{F}$ to arrive at Eq. (2.5.3).

For homogeneous boundary conditions, Eq. (2.5.3) could have been obtained directly by deleting the row and column of Eq. (2.5.1) corresponding to the zero-displacement degrees of freedom. Here row 1 and column 1 are deleted because one is really multiplying column 1 of $\underline{K}$ by $d_{1x} = 0$. However, F_{1x} is not necessarily zero and can be determined once d_{2x} and d_{3x} are solved for.

After solving Eq. (2.5.3) for d_{2x} and d_{3x}, we have

$$\begin{Bmatrix} d_{2x} \\ d_{3x} \end{Bmatrix} = \begin{bmatrix} k_2 & -k_2 \\ -k_2 & k_1 + k_2 \end{bmatrix}^{-1} \begin{Bmatrix} F_{2x} \\ F_{3x} \end{Bmatrix} = \begin{bmatrix} \dfrac{1}{k_2} + \dfrac{1}{k_1} & \dfrac{1}{k_1} \\ \dfrac{1}{k_1} & \dfrac{1}{k_1} \end{bmatrix} \begin{Bmatrix} F_{2x} \\ F_{3x} \end{Bmatrix} \tag{2.5.4}$$

Now that d_{2x} and d_{3x} are known from Eq. (2.5.4), we substitute them in the first of Eqs. (2.5.2) to obtain the reaction F_{1x} as

$$F_{1x} = -k_1 d_{3x} \tag{2.5.5}$$

We can express the unknown nodal force at node 1 (also called the *reaction*) in terms of the applied nodal forces F_{2x} and F_{3x} by using Eq. (2.5.4) for d_{3x} substituted into Eq. (2.5.5). The result is

$$F_{1x} = -F_{2x} - F_{3x} \tag{2.5.6}$$

Therefore, for all homogeneous boundary conditions, we can delete the rows and columns corresponding to the zero-displacement degrees of freedom from the original set of equations and then solve for the unknown displacements. This procedure is useful for hand calculations. (However, Appendix B.4 presents a more practical, computer-assisted scheme for solving the system of simultaneous equations.)

We now consider the case of nonhomogeneous boundary conditions. Hence some of the specified displacements are nonzero. For simplicity's sake, let $d_{1x} = \delta$, where δ is a known displacement (Figure 2–8), in Eq. (2.4.6). We now have

$$\begin{bmatrix} k_1 & 0 & -k_1 \\ 0 & k_2 & -k_2 \\ -k_1 & -k_2 & k_1 + k_2 \end{bmatrix} \begin{Bmatrix} \delta \\ d_{2x} \\ d_{3x} \end{Bmatrix} = \begin{Bmatrix} F_{1x} \\ F_{2x} \\ F_{3x} \end{Bmatrix} \tag{2.5.7}$$

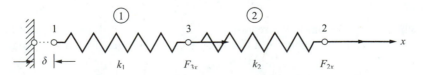

Figure 2–8 Two-spring assemblage with known displacement δ at node 1

Equation (2.5.7) written in expanded form becomes

$$k_1\delta + 0d_{2x} - k_1 d_{3x} = F_{1x}$$
$$0\delta + k_2 d_{2x} - k_2 d_{3x} = F_{2x} \qquad (2.5.8)$$
$$-k_1\delta - k_2 d_{2x} + (k_1 + k_2)d_{3x} = F_{3x}$$

where F_{1x} is now a reaction from the support that has moved an amount δ. Considering the second and third of Eqs. (2.5.8) because they have known right-side nodal forces F_{2x} and F_{3x}, we obtain

$$0\delta + k_2 d_{2x} - k_2 d_{3x} = F_{2x}$$
$$-k_1\delta - k_2 d_{2x} + (k_1 + k_2)d_{3x} = F_{3x} \qquad (2.5.9)$$

Transforming the known δ terms to the right side of Eqs. (2.5.9) yields

$$k_2 d_{2x} - k_2 d_{3x} = F_{2x}$$
$$-k_2 d_{2x} + (k_1 + k_2)d_{3x} = +k_1\delta + F_{3x} \qquad (2.5.10)$$

Rewriting Eqs. (2.5.10) in matrix form, we have

$$\begin{bmatrix} k_2 & -k_2 \\ -k_2 & k_1 + k_2 \end{bmatrix} \begin{Bmatrix} d_{2x} \\ d_{3x} \end{Bmatrix} = \begin{Bmatrix} F_{2x} \\ k_1\delta + F_{3x} \end{Bmatrix} \qquad (2.5.11)$$

Therefore, when dealing with nonhomogeneous boundary conditions, we cannot initially delete row 1 and column 1 of Eq. (2.5.7), corresponding to the nonhomogeneous boundary condition, as indicated by the resulting Eq. (2.5.11) because we are multiplying each element by a nonzero number. Had we done so, the $k_1\delta$ term in Eq. (2.5.11) would have been neglected, resulting in an error in the solution for the displacements. For nonhomogeneous boundary conditions, we must, in general, transform the terms associated with the known displacements to the right-side force matrix before solving for the unknown nodal displacements. This was illustrated by transforming the $k_1\delta$ term of the second of Eqs. (2.5.9) to the right side of the second of Eqs. (2.5.10).

We could now solve for the displacements in Eq. (2.5.11) in a manner similar to that used to solve Eq. (2.5.3). However, we will not further pursue the solution of Eq. (2.5.11) because no new information is to be gained.

However, on substituting the displacement back into Eq. (2.5.7), the reaction now becomes

$$F_{1x} = k_1\delta - k_1 d_{3x} \qquad (2.5.12)$$

which is different than Eq. (2.5.5) for F_{1x}.

At this point, we summarize some properties of the stiffness matrix in Eq. (2.5.7) that are also applicable to the generalization of the finite element method.

1. $\underline{K}$ is symmetric, as is each of the element stiffness matrices. If you are familiar with structural mechanics, you will not find this symmetry property surprising. It can be proved by using the reciprocal laws described in such References as [3] and [4].

2. $\underline{K}$ is singular, and thus no inverse exists until sufficient boundary conditions are imposed to remove the singularity and prevent rigid body motion.
3. The main diagonal terms of $\underline{K}$ are always positive. Otherwise, a positive nodal force F_i could produce a negative displacement d_i—a behavior contrary to the physical behavior of any actual structure.

In general, specified support conditions are treated mathematically by partitioning the global equilibrium equations as follows:

$$\left[\begin{array}{c|c} \underline{K}_{11} & \underline{K}_{12} \\ \hline \underline{K}_{21} & \underline{K}_{22} \end{array}\right] \left\{ \begin{array}{c} \underline{d}_1 \\ \hline \underline{d}_2 \end{array} \right\} = \left\{ \begin{array}{c} \underline{F}_1 \\ \hline \underline{F}_2 \end{array} \right\} \tag{2.5.13}$$

where we let $\underline{d}_1$ be the unconstrained or free displacements and $\underline{d}_2$ be the specified displacements. From Eq. (2.5.13), we have

$$\underline{K}_{11}\underline{d}_1 = \underline{F}_1 - \underline{K}_{12}\underline{d}_2 \tag{2.5.14}$$

and $$\underline{F}_2 = \underline{K}_{21}\underline{d}_1 + \underline{K}_{22}\underline{d}_2 \tag{2.5.15}$$

where $\underline{F}_1$ are the known nodal forces and $\underline{F}_2$ are the unknown nodal forces at the specified displacement nodes. $\underline{F}_2$ is found from Eq. (2.5.15) after $\underline{d}_1$ is determined from Eq. (2.5.14). In Eq. (2.5.14), we assume that $\underline{K}_{11}$ is no longer singular, thus allowing for the determination of $\underline{d}_1$.

To illustrate the stiffness method for the solution of spring assemblages we now present the following examples.

Example 2.1

For the spring assemblage with arbitrarily numbered nodes shown in Figure 2–9, obtain (a) the global stiffness matrix, (b) the displacements of nodes 3 and 4, (c) the reaction forces at nodes 1 and 2, and (d) the forces in each spring. A force of 5000 lb is applied at node 4 in the x direction. The spring constants are given in the figure. Nodes 1 and 2 are fixed.

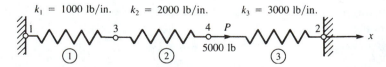

Figure 2–9 Spring assemblage for solution

(a) We begin by making use of Eq. (2.2.18) to express each element stiffness matrix as follows:

$$\underline{k}^{(1)} = \begin{array}{cc} 1 & 3 \\ \left[\begin{array}{cc} 1000 & -1000 \\ -1000 & 1000 \end{array}\right] & \begin{array}{c} 1 \\ 3 \end{array} \end{array} \qquad \underline{k}^{(2)} = \begin{array}{cc} 3 & 4 \\ \left[\begin{array}{cc} 2000 & -2000 \\ -2000 & 2000 \end{array}\right] & \begin{array}{c} 3 \\ 4 \end{array} \end{array}$$

(2.5.16)

$$\underline{k}^{(3)} = \begin{array}{cc} 4 & 2 \\ \left[\begin{array}{cc} 3000 & -3000 \\ -3000 & 3000 \end{array}\right] & \begin{array}{c} 4 \\ 2 \end{array} \end{array}$$

where the numbers above the columns and next to each row indicate the nodal degrees of freedom associated with each element. For instance, element 1 is associated with degrees of freedom d_{1x} and d_{3x}. Also, the local $\hat{x}$ axis coincides with the global x axis for each element.

Using the concept of superposition (the direct stiffness method), we obtain the global stiffness matrix as

$$\underline{K} = \underline{k}^{(1)} + \underline{k}^{(2)} + \underline{k}^{(3)}$$

or

$$\underline{K} = \begin{array}{cccc} 1 & 2 & 3 & 4 \\ \left[\begin{array}{cccc} 1000 & 0 & -1000 & 0 \\ 0 & 3000 & 0 & -3000 \\ -1000 & 0 & 1000+2000 & -2000 \\ 0 & -3000 & -2000 & 2000+3000 \end{array}\right] & \begin{array}{c} 1 \\ 2 \\ 3 \\ 4 \end{array} \end{array}$$

(2.5.17)

(b) The global stiffness matrix, Eq. (2.5.17), relates global forces to global displacements as follows:

$$\begin{Bmatrix} F_{1x} \\ F_{2x} \\ F_{3x} \\ F_{4x} \end{Bmatrix} = \begin{bmatrix} 1000 & 0 & -1000 & 0 \\ 0 & 3000 & 0 & -3000 \\ -1000 & 0 & 3000 & -2000 \\ 0 & -3000 & -2000 & 5000 \end{bmatrix} \begin{Bmatrix} d_{1x} \\ d_{2x} \\ d_{3x} \\ d_{4x} \end{Bmatrix}$$

(2.5.18)

Applying the homogeneous boundary conditions $d_{1x} = 0$ and $d_{2x} = 0$ to Eq. (2.5.18), substituting applied nodal forces, and partitioning the first two equations of Eq. (2.5.18) (or deleting the first two rows of $\{F\}$ and $\{d\}$ and the first two rows and columns of $\underline{K}$ corresponding to the zero-displacement boundary conditions), we obtain

$$\begin{Bmatrix} 0 \\ 5000 \end{Bmatrix} = \begin{bmatrix} 3000 & -2000 \\ -2000 & 5000 \end{bmatrix} \begin{Bmatrix} d_{3x} \\ d_{4x} \end{Bmatrix}$$

(2.5.19)

Solving Eq. (2.5.19), we obtain the global nodal displacements

$$d_{3x} = \frac{10}{11} \text{ in.} \qquad d_{4x} = \frac{15}{11} \text{ in.}$$

(2.5.20)

(c) To obtain the global nodal forces (which include the reactions at nodes 1 and 2), we back-substitute Eqs. (2.5.20) and the boundary conditions $d_{1x} = 0$ and

$d_{2x} = 0$ into Eq. (2.5.18). This substitution yields

$$
\begin{Bmatrix} F_{1x} \\ F_{2x} \\ F_{3x} \\ F_{4x} \end{Bmatrix} = \begin{bmatrix} 1000 & 0 & -1000 & 0 \\ 0 & 3000 & 0 & -3000 \\ -1000 & 0 & 3000 & -2000 \\ 0 & -3000 & -2000 & 5000 \end{bmatrix} \begin{Bmatrix} 0 \\ 0 \\ \frac{10}{11} \\ \frac{15}{11} \end{Bmatrix}
$$

(2.5.21)

Multiplying matrices in Eq. (2.5.21) and simplifying, we obtain the forces at each node

$$
F_{1x} = \frac{-10,000}{11} \text{ lb} \qquad F_{2x} = \frac{-45,000}{11} \text{ lb} \qquad F_{3x} = 0
$$

$$
F_{4x} = \frac{55,000}{11} \text{ lb}
$$

(2.5.22)

From these results, we observe that the sum of the reactions F_{1x} and F_{2x} is equal in magnitude but opposite in direction to the applied force F_{4x}. This result verifies equilibrium of the whole spring assemblage.

(d) Next we use local element Eq. (2.2.17) to obtain the forces in each element.

Element 1

$$
\begin{Bmatrix} \hat{f}_{1x} \\ \hat{f}_{3x} \end{Bmatrix} = \begin{bmatrix} 1000 & -1000 \\ -1000 & 1000 \end{bmatrix} \begin{Bmatrix} 0 \\ \frac{10}{11} \end{Bmatrix}
$$

(2.5.23)

Simplifying Eq. (2.5.23), we obtain

$$
\hat{f}_{1x} = \frac{-10,000}{11} \text{ lb} \qquad \hat{f}_{3x} = \frac{10,000}{11} \text{ lb}
$$

(2.5.24)

A free-body diagram of spring element 1 is shown in Figure 2–10(a). The spring is subjected to tensile forces given by Eqs. (2.5.24). Also, $\hat{f}_{1x}$ is equal to the reaction force F_{1x} given in Eq. (2.5.22). A free-body diagram of node 1 [Figure 2–10(b)] shows this result.

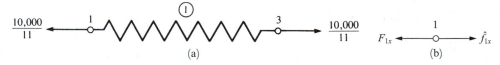

(a)   (b)

Figure 2–10 (a) Free-body diagram of element 1 and (b) free-body diagram of node 1.

Element 2

$$
\begin{Bmatrix} \hat{f}_{3x} \\ \hat{f}_{4x} \end{Bmatrix} = \begin{bmatrix} 2000 & -2000 \\ -2000 & 2000 \end{bmatrix} \begin{Bmatrix} \frac{10}{11} \\ \frac{15}{11} \end{Bmatrix}
$$

(2.5.25)

Simplifying Eq. (2.5.24), we obtain

$$\hat{f}_{3x} = \frac{-10,000}{11} \text{ lb} \qquad \hat{f}_{4x} = \frac{10,000}{11} \text{ lb} \qquad (2.5.26)$$

A free-body diagram of spring element 2 is shown in Figure 2–11. The spring is subjected to tensile forces given by Eqs. (2.5.26).

Figure 2–11 Free-body diagram of element 2

Element 3

$$\left\{ \begin{matrix} \hat{f}_{4x} \\ \hat{f}_{2x} \end{matrix} \right\} = \left[\begin{matrix} 3000 & -3000 \\ -3000 & 3000 \end{matrix} \right] \left\{ \begin{matrix} \frac{15}{11} \\ 0 \end{matrix} \right\} \qquad (2.5.27)$$

Simplifying Eq. (2.5.27) yields

$$\hat{f}_{4x} = \frac{45,000}{11} \text{ lb} \qquad \hat{f}_{2x} = \frac{-45,000}{11} \text{ lb} \qquad (2.5.28)$$

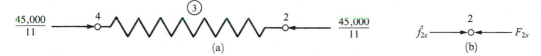

Figure 2–12 (a) Free-body diagram of element 3 and (b) free-body diagram of node 2

A free-body diagram of spring element 3 is shown in Figure 2–12(a). The spring is subjected to compressive forces given by Eqs. (2.5.28). Also, $\hat{f}_{2x}$ is equal to the reaction force F_{2x} given in Eq. (2.5.22). A free-body diagram of node 2 (Figure 2–12b) shows this result. ■

Example 2.2

For the spring assemblage shown in Figure 2–13, obtain (a) the global stiffness matrix, (b) the displacements of nodes 2–4, (c) the global nodal forces, and (d) the local element forces. Node 1 is fixed while node 5 is given a fixed, known displacement $\delta = 20.0$ mm. The spring constants are all equal to $k = 200$ kN/m.

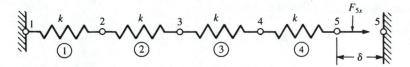

Figure 2–13 Spring assemblage for solution

(a) We use Eq. (2.2.18) to express each element stiffness matrix as

$$\underline{k}^{(1)} = \underline{k}^{(2)} = \underline{k}^{(3)} = \underline{k}^{(4)} = \begin{bmatrix} 200 & -200 \\ -200 & 200 \end{bmatrix} \tag{2.5.29}$$

Again using superposition, we obtain the global stiffness matrix as

$$\underline{K} = \begin{bmatrix} 200 & -200 & 0 & 0 & 0 \\ -200 & 400 & -200 & 0 & 0 \\ 0 & -200 & 400 & -200 & 0 \\ 0 & 0 & -200 & 400 & -200 \\ 0 & 0 & 0 & -200 & 200 \end{bmatrix} \frac{\text{kN}}{\text{m}} \tag{2.5.30}$$

(b) The global stiffness matrix, Eq. (2.5.30), relates the global forces to the global displacements as follows:

$$\begin{Bmatrix} F_{1x} \\ F_{2x} \\ F_{3x} \\ F_{4x} \\ F_{5x} \end{Bmatrix} = \begin{bmatrix} 200 & -200 & 0 & 0 & 0 \\ -200 & 400 & -200 & 0 & 0 \\ 0 & -200 & 400 & -200 & 0 \\ 0 & 0 & -200 & 400 & -200 \\ 0 & 0 & 0 & -200 & 200 \end{bmatrix} \begin{Bmatrix} d_{1x} \\ d_{2x} \\ d_{3x} \\ d_{4x} \\ d_{5x} \end{Bmatrix} \tag{2.5.31}$$

Applying the boundary conditions $d_{1x} = 0$ and $d_{5x} = 20$ mm ($= 0.02$ m), substituting known global forces $F_{2x} = 0$, $F_{3x} = 0$, and $F_{4x} = 0$, and partitioning the first and fifth equations of Eq. (2.5.31) corresponding to these boundary conditions, we obtain

$$\begin{Bmatrix} 0 \\ 0 \\ 0 \end{Bmatrix} = \begin{bmatrix} -200 & 400 & -200 & 0 & 0 \\ 0 & -200 & 400 & -200 & 0 \\ 0 & 0 & -200 & 400 & -200 \end{bmatrix} \begin{Bmatrix} 0 \\ d_{2x} \\ d_{3x} \\ d_{4x} \\ 0.02 \text{ m} \end{Bmatrix} \tag{2.5.32}$$

We now rewrite Eq. (2.5.32), transposing the product of the appropriate stiffness coefficient multiplied by the known displacement to the left side.

$$\begin{Bmatrix} 0 \\ 0 \\ 4 \text{ kN} \end{Bmatrix} = \begin{bmatrix} 400 & -200 & 0 \\ -200 & 400 & -200 \\ 0 & -200 & 400 \end{bmatrix} \begin{Bmatrix} d_{2x} \\ d_{3x} \\ d_{4x} \end{Bmatrix} \tag{2.5.33}$$

Solving Eq. (2.5.33), we obtain

$$d_{2x} = 0.005 \text{ m} \qquad d_{3x} = 0.01 \text{ m} \qquad d_{4x} = 0.015 \text{ m} \qquad (2.5.34)$$

(c) The global nodal forces are obtained by back-substituting the boundary condition displacements and Eqs. (2.5.34) into Eq. (2.5.31). This substitution yields

$$F_{1x} = (-200)(0.005) = -1.0 \text{ kN}$$
$$F_{2x} = (400)(0.005) - (200)(0.01) = 0$$
$$F_{3x} = (-200)(0.005) + (400)(0.01) - (200)(0.015) = 0 \qquad (2.5.35)$$
$$F_{4x} = (-200)(0.01) + (400)(0.015) - (200)(0.02) = 0$$
$$F_{5x} = (-200)(0.015) + (200)(0.02) = 1.0 \text{ kN}$$

The results of Eqs. (2.5.35) yield the reaction F_{1x} opposite that of the nodal force F_{5x} required to displace node 5 by $\delta = 20.0$ mm. This result verifies equilibrium of the whole spring assemblage.

(d) Next, we make use of local element Eq. (2.2.17) to obtain the forces in each element.

Element 1

$$\begin{Bmatrix} \hat{f}_{1x} \\ \hat{f}_{2x} \end{Bmatrix} = \begin{bmatrix} 200 & -200 \\ -200 & 200 \end{bmatrix} \begin{Bmatrix} 0 \\ 0.005 \end{Bmatrix} \qquad (2.5.36)$$

Simplifying Eq. (2.5.36) yields

$$\hat{f}_{1x} = -1.0 \text{ kN} \qquad \hat{f}_{2x} = 1.0 \text{ kN} \qquad (2.5.37)$$

Element 2

$$\begin{Bmatrix} \hat{f}_{2x} \\ \hat{f}_{3x} \end{Bmatrix} = \begin{bmatrix} 200 & -200 \\ -200 & 200 \end{bmatrix} \begin{Bmatrix} 0.005 \\ 0.01 \end{Bmatrix} \qquad (2.5.38)$$

Simplifying Eq. (2.5.38) yields

$$\hat{f}_{2x} = -1 \text{ kN} \qquad \hat{f}_{3x} = 1 \text{ kN} \qquad (2.5.39)$$

Element 3

$$\begin{Bmatrix} \hat{f}_{3x} \\ \hat{f}_{4x} \end{Bmatrix} = \begin{bmatrix} 200 & -200 \\ -200 & 200 \end{bmatrix} \begin{Bmatrix} 0.01 \\ 0.015 \end{Bmatrix} \qquad (2.5.40)$$

Simplifying Eq. (2.5.40), we have

$$\hat{f}_{3x} = -1 \text{ kN} \qquad \hat{f}_{4x} = 1 \text{ kN} \qquad (2.5.41)$$

Element 4

$$\begin{Bmatrix} \hat{f}_{4x} \\ \hat{f}_{5x} \end{Bmatrix} = \begin{bmatrix} 200 & -200 \\ -200 & 200 \end{bmatrix} \begin{Bmatrix} 0.015 \\ 0.02 \end{Bmatrix} \qquad (2.5.42)$$

Simplifying Eq. (2.5.42), we obtain

$$\hat{f}_{4x} = -1 \text{ kN} \qquad \hat{f}_{5x} = 1 \text{ kN} \qquad (2.5.43)$$

You should draw free-body diagrams of each node and element and use the results of Eqs. (2.5.35)–(2.5.43) to verify both node and element equilibria. ∎

Finally, to review the major concepts presented in this chapter, we solve the following example problem.

Example 2.3

(a) Using the ideas presented in Section 2.3 for the system of linear elastic springs shown in Figure 2–14, express the boundary conditions, the compatibility or continuity condition similar to Eq. (2.3.3), and the nodal equilibrium conditions similar to Eqs. (2.3.4)–(2.3.6). Then formulate the global stiffness matrix and equations for solution of the unknown global displacement and forces. The spring constants for the elements are $k_1, k_2,$ and k_3; P is an applied force at node 2.

(b) Using the direct stiffness method, formulate the same global stiffness matrix and equation as in part (a).

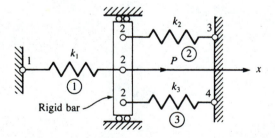

Figure 2–14 Spring assemblage for solution

(a) The boundary conditions are

$$d_{1x} = 0 \qquad d_{3x} = 0 \qquad d_{4x} = 0 \qquad (2.5.44)$$

The compatibility condition at node 2 is

$$d_{2x}^{(1)} = d_{2x}^{(2)} = d_{2x}^{(3)} = d_{2x} \qquad (2.5.45)$$

The nodal equilibrium conditions are

$$F_{1x} = f_{1x}^{(1)}$$

$$P = f_{2x}^{(1)} + f_{2x}^{(2)} + f_{2x}^{(3)}$$

$$F_{3x} = f_{3x}^{(2)}$$

$$F_{4x} = f_{4x}^{(3)}$$

$$(2.5.46)$$

where the sign convention for positive element nodal forces given by Figure 2–2 was used in writing Eqs. (2.5.46). Figure 2–15 shows the element and nodal force free-body diagrams.

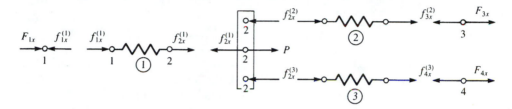

Figure 2–15 Free-body diagrams of elements and nodes of spring assemblage of Figure 2–14

Using the local stiffness matrix Eq. (2.2.17) applied to each element, and compatibility condition Eq. (2.5.45), we obtain the total or global equilibrium equations as

$$F_{1x} = k_1 d_{1x} - k_1 d_{2x}$$

$$P = -k_1 d_{1x} + k_1 d_{2x} + k_2 d_{2x} - k_2 d_{3x} + k_3 d_{2x} - k_3 d_{4x}$$

$$F_{3x} = -k_2 d_{2x} + k_2 d_{3x}$$

$$F_{4x} = -k_3 d_{2x} + k_3 d_{4x}$$

$$(2.5.47)$$

In matrix form, we express Eqs. (2.5.47) as

$$
\begin{Bmatrix} F_{1x} \\ P \\ F_{3x} \\ F_{4x} \end{Bmatrix} =
\begin{bmatrix}
k_1 & -k_1 & 0 & 0 \\
-k_1 & k_1 + k_2 + k_3 & -k_2 & -k_3 \\
0 & -k_2 & k_2 & 0 \\
0 & -k_3 & 0 & k_3
\end{bmatrix}
\begin{Bmatrix} d_{1x} \\ d_{2x} \\ d_{3x} \\ d_{4x} \end{Bmatrix}
\qquad (2.5.48)
$$

Therefore, the global stiffness matrix is the square, symmetric matrix on the right side of Eq. (2.5.48). Making use of the boundary conditions, Eqs. (2.5.44), and then considering the second equation of Eqs. (2.5.47) or (2.5.48), we solve for d_{2x} as

$$d_{2x} = \frac{P}{k_1 + k_2 + k_3} \qquad (2.5.49)$$

We could have obtained this same result by deleting rows 1, 3, and 4 in the $\underline{F}$ and $\underline{d}$ matrices and rows and columns 1, 3, and 4 in $\underline{K}$, corresponding to zero displacement, as previously described in Section 2.4, and then solving for d_{2x}.

Using Eqs. (2.5.47), we now solve for the global forces as

$$F_{1x} = -k_1 d_{2x} \qquad F_{3x} = -k_2 d_{2x} \qquad F_{4x} = -k_3 d_{2x} \qquad (2.5.50)$$

The forces given by Eqs. (2.5.50) can be interpreted as the global reactions in this example. The negative signs in front of these forces indicate that they are directed to the left (opposite the x axis).

(b) Using the direct stiffness method, we formulate the global stiffness matrix. First, using Eq. (2.2.18), we express each element stiffness matrix as

$$
\underline{k}^{(1)} = \begin{matrix} d_{1x} & d_{2x} \\ \begin{bmatrix} k_1 & -k_1 \\ -k_1 & k_1 \end{bmatrix} \end{matrix} \quad
\underline{k}^{(2)} = \begin{matrix} d_{2x} & d_{3x} \\ \begin{bmatrix} k_2 & -k_2 \\ -k_2 & k_2 \end{bmatrix} \end{matrix} \quad
\underline{k}^{(3)} = \begin{matrix} d_{2x} & d_{4x} \\ \begin{bmatrix} k_3 & -k_3 \\ -k_3 & k_3 \end{bmatrix} \end{matrix} \qquad (2.5.51)
$$

where the particular degrees of freedom associated with each element are listed in the columns above each matrix. Using the direct stiffness method as outlined in Section 2.4, we add terms from each element stiffness matrix into the appropriate corresponding row and column in the global stiffness matrix to obtain

$$
\underline{K} = \begin{matrix} \quad d_{1x} & \quad d_{2x} & d_{3x} & d_{4x} \\ \begin{bmatrix} k_1 & -k_1 & 0 & 0 \\ -k_1 & k_1 + k_2 + k_3 & -k_2 & -k_3 \\ 0 & -k_2 & k_2 & 0 \\ 0 & -k_3 & 0 & k_3 \end{bmatrix} \end{matrix} \qquad (2.5.52)
$$

We observe that each element stiffness matrix $\underline{k}$ has been added into the location in the global $\underline{K}$ corresponding to the identical degree of freedom associated with the element $\underline{k}$. For instance, element 3 is associated with degrees of freedom d_{2x} and d_{4x}; hence its contributions to $\underline{K}$ are in the 2–2, 2–4, 4–2, and 4–4 locations of $\underline{K}$, as indicated in Eq. (2.5.52) by the k_3 terms.

Having assembled the global $\underline{K}$ by the direct stiffness method, we then formulate the global equations in the usual manner by making use of the general Eq. (2.3.10), $\underline{F} = \underline{K}\underline{d}$. These equations have been previously obtained by Eq. (2.5.48) and therefore are not repeated. ∎

Another method for handling imposed boundary conditions that allows for either homogeneous (zero) or nonhomogeneous (nonzero) prescribed degrees of freedom is called the *penalty method*. This method is easy to implement in a computer program.

Consider the simple spring assemblage in Figure 2–16 subjected to applied forces F_{1x} and F_{2x} as shown. Assume the vertical displacement at node 1 to be forced to be $d_{1x} = \delta$.

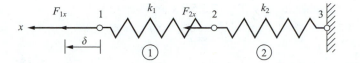

Figure 2–16 Spring assemblage used to illustrate the penalty method

We add another spring (often called a boundary element) with a large stiffness k_b to the assemblage in the direction of the nodal displacement $d_{1x} = \delta$ as shown in Figure 2–17. This spring stiffness should have a magnitude about 10^6 times that of the largest k_{ii} term.

Figure 2–17 Spring assemblage with a boundary spring element added at node 1

Now we add the force $k_b\delta$ in the direction of d_{1x} and solve the problem in the usual manner as follows.

The element stiffness matrices are

$$\underline{k}^{(1)} = \begin{bmatrix} k_1 & -k_1 \\ -k_1 & k_1 \end{bmatrix} \qquad \underline{k}^{(2)} = \begin{bmatrix} k_2 & -k_2 \\ -k_2 & k_2 \end{bmatrix} \tag{2.5.53}$$

Assembling the element stiffness matrices using the direct stiffness method, we obtain the global stiffness matrix as

$$\underline{K} = \begin{bmatrix} k_1 + k_b & -k_1 & 0 \\ -k_1 & k_1 + k_2 & -k_2 \\ 0 & -k_2 & k_2 \end{bmatrix} \tag{2.5.54}$$

Assembling the global $\underline{F} = \underline{K}\underline{d}$ equations and invoking the boundary condition $d_{3x} = 0$, we obtain

$$\begin{Bmatrix} F_{1x} + k_b\delta \\ F_{2x} \\ F_{3x} \end{Bmatrix} = \begin{bmatrix} k_1 + k_b & -k_1 & 0 \\ -k_1 & k_1 + k_2 & -k_2 \\ 0 & -k_2 & k_2 \end{bmatrix} \begin{Bmatrix} d_{1x} \\ d_{2x} \\ d_{3x} = 0 \end{Bmatrix} \tag{2.5.55}$$

Solving the first and second of Eqs. (2.5.55), we obtain

$$d_{1x} = \frac{F_{2x} - (k_1 + k_2)d_{2x}}{-k_1} \tag{2.5.56}$$

and

$$d_{2x} = \frac{(k_1 + k_b)F_{2x} + F_{1x}k_1 + k_b\delta k_1}{k_b k_1 + k_b k_2 + k_1 k_2} \tag{2.5.57}$$

Now as k_b approaches infinity, Eq. (2.5.57) simplifies to

$$d_{2x} = \frac{F_{2x} + \delta k_1}{k_1 + k_2} \tag{2.5.58}$$

and Eq. (2.5.56) simplifies to

$$d_{1x} = \delta \tag{2.5.59}$$

These results match those obtained by setting $d_{1x} = \delta$ initially.

In using the penalty method, a very large element stiffness should be parallel to a degree of freedom as is the case in the preceding example. If k_b were inclined, or were placed within a structure, it would contribute to both diagonal and off-diagonal coefficients in the global stiffness matrix $\underline{K}$. This condition can lead to numerical difficulties in solving the equations $\underline{F} = \underline{K}\underline{d}$. To avoid this condition, we transform the displacements at the inclined support to local ones as described in Section 3.9.

▲ **2.6 Potential Energy Approach to Derive Spring Element Equations** ▲

One of the alternative methods often used to derive the element equations and the stiffness matrix for an element is based on the principle of *minimum potential energy*. (The use of this principle in structural mechanics is fully described in Reference [4].) This method has the advantage of being more general than the method given in Section 2.2, which involves nodal and element equilibrium equations along with the stress/strain law for the element. Thus the principle of minimum potential energy is more adaptable to the determination of element equations for complicated elements (those with large numbers of degrees of freedom) such as the plane stress/strain element, the axisymmetric stress element, the plate bending element, and the three-dimensional solid stress element.

Again, we state that the principle of virtual work (Appendix E) is applicable for any material behavior, whereas the principle of minimum potential energy is applicable only for elastic materials. However, both principles yield the same element equations for linear-elastic materials, which are the only kind considered in this text. Moreover, the principle of minimum potential energy, being included in the general category of *variational methods* (as is the principle of virtual work), leads to other variational functions (or functionals) similar to potential energy that can be formulated for other classes of problems, primarily of the nonstructural type. These other problems are generally classified as *field problems* and include, among others, torsion of a bar, heat transfer (Chapter 13), fluid flow (Chapter 14), and electric potential.

Still other classes of problems, for which a variational formulation is not clearly definable, can be formulated by *weighted residual methods*, of which Galerkin's method is the most often used. We will describe Galerkin's method and its application to a bar element in Section 3.12. (For more information on weighted residual methods, also consult References [5–7].)

Here we present the principle of minimum potential energy as used to derive the bar element equations. We will illustrate this concept by applying it to the simplest of elements in hopes that the reader will then be more comfortable when applying it to handle more complicated element types in subsequent chapters.

The total potential energy π_p of a structure is expressed in terms of displacements. In the finite element formulation, these will generally be nodal displacements such that $\pi_p = \pi_p(d_1, d_2, \ldots, d_n)$. When π_p is minimized with respect to these displacements, equilibrium equations result. For the spring element, we will show that the same nodal equilibrium equations $\hat{\underline{k}}\hat{\underline{d}} = \hat{\underline{f}}$ result as previously derived in Section 2.2.

We first state the principle of minimum potential energy as follows:

> Of all the geometrically possible shapes that a body can assume, the true one, corresponding to the satisfaction of stable equilibrium of the body, is identified by a minimum value of the total potential energy.

To explain this principle, we must first explain the concepts of potential energy and of a stationary value of a function. We will now discuss these two concepts.

Total potential energy is defined as *the sum of the internal strain energy U and the potential energy of the external forces Ω; that is,*

$$\pi_p = U + \Omega \tag{2.6.1}$$

Strain energy is the capacity of internal forces (or stresses) to do work through deformations (strains) in the structure; Ω is the capacity of forces such as body forces, surface traction forces, and applied nodal forces to do work through deformation of the structure.

Recall that a linear spring has force related to deformation by $F = kx$, where k is the spring constant and x is the deformation of the spring (Figure 2–18).

The differential internal work (or strain energy) dU in the spring for a small change in length of the spring is the internal force multiplied by the change in displacement through which the force moves, given by

$$dU = F\,dx \tag{2.6.2}$$

Now we express F as

$$F = kx \tag{2.6.3}$$

Using Eq. (2.6.3) in Eq. (2.6.2), we find that the differential strain energy becomes

$$dU = kx\,dx \tag{2.6.4}$$

The total strain energy is then given by

$$U = \int_0^x kx\,dx \tag{2.6.5}$$

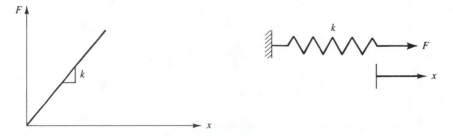

Figure 2–18 Force/deformation curve for linear spring

Upon explicit integration of Eq. (2.6.5), we obtain

$$U = \tfrac{1}{2}kx^2 \qquad (2.6.6)$$

Using Eq. (2.6.3) in Eq. (2.6.6), we have

$$U = \tfrac{1}{2}(kx)x = \tfrac{1}{2}Fx \qquad (2.6.7)$$

Equation (2.6.7) indicates that the strain energy is the area under the force/deformation curve.

The potential energy of the external force, being opposite in sign from the external work expression because the potential energy of the external force is lost when the work is done by the external force, is given by

$$\Omega = -Fx \qquad (2.6.8)$$

Therefore, substituting Eqs. (2.6.6) and (2.6.8) into (2.6.1), yields the total potential energy as

$$\pi_p = \tfrac{1}{2}kx^2 - Fx \qquad (2.6.9)$$

The concept of a *stationary value* of a function G (used in the definition of the principle of minimum potential energy) is shown in Figure 2–19. Here G is expressed as a function of the variable x. The stationary value can be a maximum, a minimum, or a neutral point of $G(x)$. To find a value of x yielding a stationary value of $G(x)$, we use differential calculus to differentiate G with respect to x and set the expression equal to zero, as follows:

$$\frac{dG}{dx} = 0 \qquad (2.6.10)$$

An analogous process will subsequently be used to replace G with π_p and x with discrete values (nodal displacements) d_i. With an understanding of variational calculus (see Reference [8]), we could use the first variation of π_p (denoted by $\delta\pi_p$, where δ denotes arbitrary change or variation) to minimize π_p. However, we will avoid the details of variational calculus and show that we can really use the familiar differential calculus to perform the minimization of π_p. To apply the principle of minimum potential energy—that is, to minimize π_p—we take the *variation* of π_p, which is a function of nodal displacements d_i defined in general as

$$\delta\pi_p = \frac{\partial\pi_p}{\partial d_1}\delta d_1 + \frac{\partial\pi_p}{\partial d_2}\delta d_2 + \cdots + \frac{\partial\pi_p}{\partial d_n}\delta d_n \qquad (2.6.11)$$

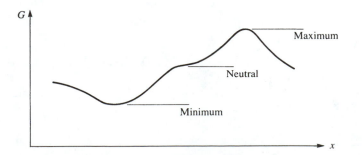

Figure 2–19 Stationary values of a function

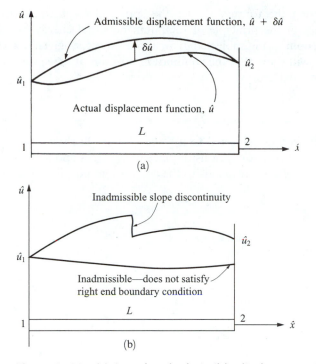

Figure 2–20 (a) Actual and admissible displacement functions and (b) inadmissible displacement functions

The principle states that equilibrium exists when the d_i define a structure state such that $\delta\pi_p = 0$ (change in potential energy $= 0$) for arbitrary admissible variations in displacement δd_i from the equilibrium state. An *admissible variation* is one in which the displacement field still satisfies the boundary conditions and interelement continuity. Figure 2–20(a) shows the hypothetical actual axial displacement and an admissible one for a bar with specified boundary displacements $\hat{u}_1$ and $\hat{u}_2$. Figure 2–20(b) shows inadmissible functions due to slope discontinuity between endpoints 1 and 2 and due to failure to satisfy the right end boundary condition of $\hat{u}(L) = \hat{u}_2$. Here $\delta\hat{u}$ represents the variation in $\hat{u}$. In the general finite element formulation, $\delta\hat{u}$ would be replaced by δd_i. This implies that any of the δd_i might be nonzero. Hence, to satisfy $\delta\pi_p = 0$, all coefficients associated with the δd_i must be zero independently. Thus,

$$\frac{\partial\pi_p}{\partial d_i} = 0 \quad (i = 1, 2, 3, \ldots, n) \quad \text{or} \quad \frac{\partial\pi_p}{\partial\{d\}} = 0 \quad (2.6.12)$$

where n equations must be solved for the n values of d_i that define the static equilibrium state of the structure. Equation (2.6.12) shows that for our purposes throughout this text, we can interpret *the variation of* π_p as a compact notation equivalent to differentiation of π_p with respect to the unknown nodal displacements for which π_p is expressed. For linear-elastic materials in equilibrium, the fact that π_p is a minimum is shown, for instance, in Reference [4].

Before discussing the formulation of the spring element equations, we now illustrate the concept of the principle of minimum potential energy by analyzing a single-degree-of-freedom spring subjected to an applied force, as given in Example 2.4. In this example, we will show that the equilibrium position of the spring corresponds to the minimum potential energy.

Example 2.4

For the linear-elastic spring subjected to a force of 1000 lb shown in Figure 2–21, evaluate the potential energy for various displacement values and show that the minimum potential energy also corresponds to the equilibrium position of the spring.

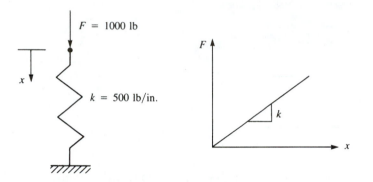

Figure 2–21 Spring subjected to force; load/displacement curve

We evaluate the total potential energy as

$$\pi_p = U + \Omega$$

where $$U = \tfrac{1}{2}(kx)x \quad \text{and} \quad \Omega = -Fx$$

We now illustrate the minimization of π_p through standard mathematics. Taking the variation of π_p with respect to x, or, equivalently, taking the derivative of π_p with respect to x (as π_p is a function of only one displacement x), as in Eqs. (2.6.11) and (2.6.12), we have

$$\delta\pi_p = \frac{\partial\pi_p}{\partial x}\delta x = 0$$

or, because δx is arbitrary and might not be zero,

$$\frac{\partial\pi_p}{\partial x} = 0$$

Using our previous expression for π_p, we obtain

$$\frac{\partial\pi_p}{\partial x} = 500x - 1000 = 0$$

or $$x = 2.00 \text{ in.}$$

This value for x is then back-substituted into π_p to yield

$$\pi_p = 250(2)^2 - 1000(2) = -1000 \text{ lb-in.}$$

which corresponds to the minimum potential energy obtained in Table 2–1 by the following searching technique. Here $U = \frac{1}{2}(kx)x$ is the strain energy or the area under the load/displacement curve shown in Figure 2–21, and $\Omega = -Fx$ is the potential energy of load F. For the given values of F and k, we then have

$$\pi_p = \frac{1}{2}(500)x^2 - 1000x = 250x^2 - 1000x$$

We now search for the minimum value of π_p for various values of spring deformation x. The results are shown in Table 2–1. A plot of π_p versus x is shown in Figure 2–22,

Table 2–1 Total potential energy for various spring deformations

Deformation x, in.	Total Potential Energy π_p, lb-in.
−4.00	8000
−3.00	5250
−2.00	3000
−1.00	1250
0.00	0
1.00	−750
2.00	−1000
3.00	−750
4.00	0
5.00	1250

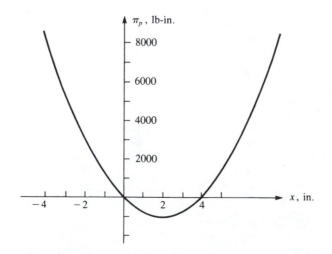

Figure 2–22 Variation of potential energy with spring deformation

where we observe that π_p has a minimum value at $x = 2.00$ in. This deformed position also corresponds to the equilibrium position because $(\partial \pi_p / \partial x) = 500(2) - 1000 = 0$. ∎

We now derive the spring element equations and stiffness matrix using the principle of minimum potential energy. Consider the linear spring subjected to nodal forces shown in Figure 2–23. Using Eq. (2.6.9) reveals that the total potential energy is

$$\pi_p = \tfrac{1}{2}k(\hat{d}_{2x} - \hat{d}_{1x})^2 - \hat{f}_{1x}\hat{d}_{1x} - \hat{f}_{2x}\hat{d}_{2x} \tag{2.6.13}$$

where $\hat{d}_{2x} - \hat{d}_{1x}$ is the deformation of the spring in Eq. (2.6.9). The first term on the right in Eq. (2.6.13) is the strain energy in the spring. Simplifying Eq. (2.6.13), we obtain

$$\pi_p = \tfrac{1}{2}k(\hat{d}_{2x}^2 - 2\hat{d}_{2x}\hat{d}_{1x} + \hat{d}_{1x}^2) - \hat{f}_{1x}\hat{d}_{1x} - \hat{f}_{2x}\hat{d}_{2x} \tag{2.6.14}$$

The minimization of π_p with respect to each nodal displacement requires taking partial derivatives of π_p with respect to each nodal displacement such that

$$\frac{\partial \pi_p}{\partial \hat{d}_{1x}} = \frac{1}{2}k(-2\hat{d}_{2x} + 2\hat{d}_{1x}) - \hat{f}_{1x} = 0$$

$$\frac{\partial \pi_p}{\partial \hat{d}_{2x}} = \frac{1}{2}k(2\hat{d}_{2x} - 2\hat{d}_{1x}) - \hat{f}_{2x} = 0 \tag{2.6.15}$$

Simplifying Eqs. (2.6.15), we have

$$k(-\hat{d}_{2x} + \hat{d}_{1x}) = \hat{f}_{1x}$$

$$k(\hat{d}_{2x} - \hat{d}_{1x}) = \hat{f}_{2x} \tag{2.6.16}$$

In matrix form, we express Eq. (2.6.16) as

$$\begin{bmatrix} k & -k \\ -k & k \end{bmatrix} \begin{Bmatrix} \hat{d}_{1x} \\ \hat{d}_{2x} \end{Bmatrix} = \begin{Bmatrix} \hat{f}_{1x} \\ \hat{f}_{2x} \end{Bmatrix} \tag{2.6.17}$$

Because $\{\hat{f}\} = [\hat{k}]\{\hat{d}\}$, we have the stiffness matrix for the spring element obtained from Eq. (2.6.17):

$$[\hat{k}] = \begin{bmatrix} k & -k \\ -k & k \end{bmatrix} \tag{2.6.18}$$

Figure 2–23 Linear spring subjected to nodal forces

As expected, Eq. (2.6.18) is identical to the stiffness matrix obtained in Section 2.2, Eq. (2.2.18).

We considered the equilibrium of a single spring element by minimizing the total potential energy with respect to the nodal displacements (see Example 2.4). We also developed the finite element spring element equations by minimizing the total potential energy with respect to the nodal displacements. We now show that the total potential energy of an entire structure (here an assemblage of spring elements) can be minimized with respect to each nodal degree of freedom and that this minimization results in the same finite element equations used for the solution as those obtained by the direct stiffness method.

Example 2.5

Obtain the total potential energy of the spring assemblage (Figure 2–24) for Example 2.1 and find its minimum value. The procedure of assembling element equations can then be seen to be obtained from the minimization of the total potential energy.

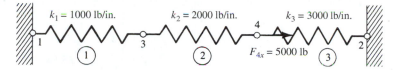

Figure 2–24 Spring assemblage solved by energy principle

Using Eq. (2.6.10) for each element of the spring assemblage, we find that the total potential energy is given by

$$\pi_p = \sum_{e=1}^{3} \pi_p^{(e)} = \frac{1}{2} k_1 (d_{3x} - d_{1x})^2 - f_{1x}^{(1)} d_{1x} - f_{3x}^{(1)} d_{3x}$$
$$+ \frac{1}{2} k_2 (d_{4x} - d_{3x})^2 - f_{3x}^{(2)} d_{3x} - f_{4x}^{(2)} d_{4x} \qquad (2.6.19)$$
$$+ \frac{1}{2} k_3 (d_{2x} - d_{4x})^2 - f_{4x}^{(3)} d_{4x} - f_{2x}^{(3)} d_{2x}$$

Upon minimizing π_p with respect to each nodal displacement, we obtain

$$\frac{\partial \pi_p}{\partial d_{1x}} = -k_1 d_{3x} + k_1 d_{1x} - f_{1x}^{(1)} = 0$$

$$\frac{\partial \pi_p}{\partial d_{2x}} = k_3 d_{2x} - k_3 d_{4x} - f_{2x}^{(3)} = 0$$

$$\frac{\partial \pi_p}{\partial d_{3x}} = k_1 d_{3x} - k_1 d_{1x} - k_2 d_{4x} + k_2 d_{3x} - f_{3x}^{(1)} - f_{3x}^{(2)} = 0 \qquad (2.6.20)$$

$$\frac{\partial \pi_p}{\partial d_{4x}} = k_2 d_{4x} - k_2 d_{3x} - k_3 d_{2x} + k_3 d_{4x} - f_{4x}^{(2)} - f_{4x}^{(3)} = 0$$

In matrix form, Eqs. (2.6.20) become

$$
\begin{bmatrix}
k_1 & 0 & -k_1 & 0 \\
0 & k_3 & 0 & -k_3 \\
-k_1 & 0 & k_1 + k_2 & -k_2 \\
0 & -k_3 & -k_2 & k_2 + k_3
\end{bmatrix}
\begin{Bmatrix}
d_{1x} \\
d_{2x} \\
d_{3x} \\
d_{4x}
\end{Bmatrix}
=
\begin{Bmatrix}
f_{1x}^{(1)} \\
f_{2x}^{(3)} \\
f_{3x}^{(1)} + f_{3x}^{(2)} \\
f_{4x}^{(2)} + f_{4x}^{(3)}
\end{Bmatrix}
\tag{2.6.21}
$$

Using nodal force equilibrium similar to Eqs. (2.3.4)–(2.3.6), we have

$$
f_{1x}^{(1)} = F_{1x}
$$

$$
f_{2x}^{(3)} = F_{2x}
$$

$$
f_{3x}^{(1)} + f_{3x}^{(2)} = F_{3x}
$$

$$
f_{4x}^{(2)} + f_{4x}^{(3)} = F_{4x}
\tag{2.6.22}
$$

Using Eqs. (2.6.22) in (2.6.21) and substituting numerical values for k_1, k_2, and k_3, we obtain

$$
\begin{bmatrix}
1000 & 0 & -1000 & 0 \\
0 & 3000 & 0 & -3000 \\
-1000 & 0 & 3000 & -2000 \\
0 & -3000 & -2000 & 5000
\end{bmatrix}
\begin{Bmatrix}
d_{1x} \\
d_{2x} \\
d_{3x} \\
d_{4x}
\end{Bmatrix}
=
\begin{Bmatrix}
F_{1x} \\
F_{2x} \\
F_{3x} \\
F_{4x}
\end{Bmatrix}
\tag{2.6.23}
$$

Equation (2.6.23) is identical to Eq. (2.5.18), which was obtained through the direct stiffness method. The assembled Eqs. (2.6.23) are then seen to be obtained from the minimization of the total potential energy. When we apply the boundary conditions and substitute $F_{3x} = 0$ and $F_{4x} = 5000$ lb into Eq. (2.6.23), the solution is identical to that of Example 2.1. ■

▲ References

[1] Turner, M. J., Clough, R. W., Martin, H. C., and Topp, L. J., "Stiffness and Deflection Analysis of Complex Structures," *Journal of the Aeronautical Sciences*, Vol. 23, No. 9, pp. 805–824, Sept. 1956.

[2] Martin, H. C., *Introduction to Matrix Methods of Structural Analysis*, McGraw-Hill, New York, 1966.

[3] Hsieh, Y. Y., *Elementary Theory of Structures*, 2nd ed., Prentice-Hall, Englewood Cliffs, NJ, 1982.

[4] Oden, J. T., and Ripperger, E. A., *Mechanics of Elastic Structures*, 2nd ed., McGraw-Hill, New York, 1981.

[5] Finlayson, B. A., *The Method of Weighted Residuals and Variational Principles*, Academic Press, New York, 1972.

[6] Zienkiewicz, O. C., *The Finite Element Method*, 3rd ed., McGraw-Hill, London, 1977.

[7] Cook, R. D., Malkus, D. S., and Plesha, M. E., *Concepts and Applications of Finite Element Analysis*, 3rd ed., Wiley, New York, 1989.

[8] Forray, M. J., *Variational Calculus in Science and Engineering*, McGraw-Hill, New York, 1968.

Problems

2.1 **a.** Obtain the global stiffness matrix K of the assemblage shown in Figure P2–1 by superimposing the stiffness matrices of the individual springs. Here k_1, k_2, and k_3 are the stiffnesses of the springs as shown.
b. If nodes 1 and 2 are fixed and a force P acts on node 4 in the positive x direction, find an expression for the displacements of nodes 3 and 4.
c. Determine the reaction forces at nodes 1 and 2.
(*Hint:* Do this problem by writing the nodal equilibrium equations and then making use of the force/displacement relationships for each element as done in the first part of Section 2.4. Then solve the problem by the direct stiffness method.)

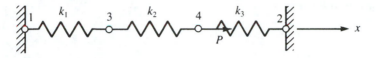

Figure P2–1

2.2 For the spring assemblage shown in Figure P2–2, determine the displacement at node 2 and the forces in each spring element. Also determine the force F_3. Given: Node 3 displaces an amount $\delta = 1$ in. in the positive x direction because of the force F_3 and $k_1 = k_2 = 1000$ lb/in.

Figure P2–2

2.3 **a.** For the spring assemblage shown in Figure P2–3, obtain the global stiffness matrix by direct superposition.
b. If nodes 1 and 5 are fixed and a force P is applied at node 3, determine the nodal displacements.
c. Determine the reactions at the fixed nodes 1 and 5.

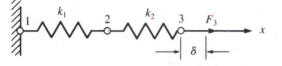

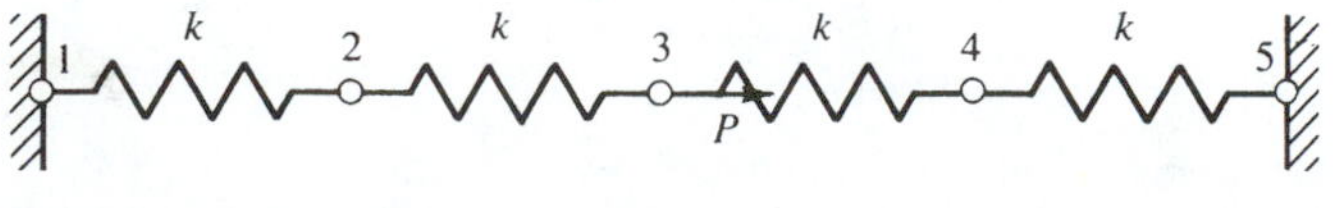

Figure P2–3

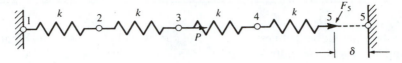

Figure P2–4

2.4 Solve Problem 2.3 with $P = 0$ (no force applied at node 3) and with node 5 given a fixed, known displacement of δ as shown in Figure P2–4.

2.5–2.12 For the spring assemblages shown in Figures P2–5–P2–12, determine the nodal displacements, the forces in each element, and the reactions. Use the direct stiffness method for all problems.

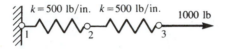

Figure P2–5

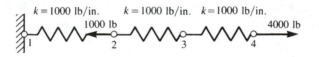

Figure P2–6

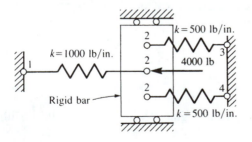

Figure P2–7

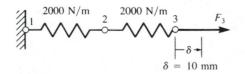

Figure P2–8

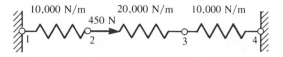

Figure P2–9

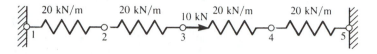

Figure P2–10

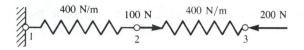

Figure P2–11

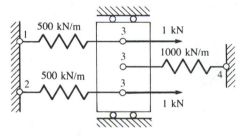

Figure P2–12

2.13 Use the principle of minimum potential energy developed in Section 2.6 to solve the spring problems shown in Figure P2–13. That is, plot the total potential energy for variations in the displacement of the free end of the spring to determine the minimum potential energy. Observe that the displacement that yields the minimum potential energy also yields the stable equilibrium position.

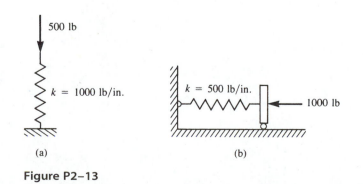

(a)   (b)

Figure P2–13

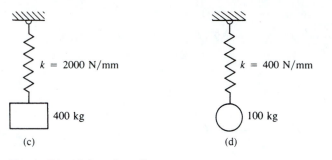

(c) (d)

Figure P2–13 (*continued*)

2.14 Reverse the direction of the load in Example 2.4 and recalculate the total potential energy. Then use this value to obtain the equilibrium value of displacement.

2.15 The nonlinear spring in Figure P2–15 has the force/deformation relationship $f = k\delta^2$. Express the total potential energy of the spring, and use this potential energy to obtain the equilibrium value of displacement.

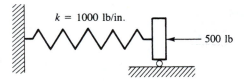

Figure P2–15

2.16–2.17 Solve Problems 2.7 and 2.12 by the potential energy approach (see Example 2.5).

3

Development of Truss Equations

Introduction

Having set forth the foundation on which the direct stiffness method is based, we will now derive the stiffness matrix for a linear-elastic bar (or truss) element using the general steps outlined in Chapter 1. We will include the introduction of both a local coordinate system, chosen with the element in mind, and a global or reference coordinate system, chosen to be convenient (for numerical purposes) with respect to the overall structure. We will also discuss the transformation of a vector from the local coordinate system to the global coordinate system, using the concept of transformation matrices to express the stiffness matrix of an arbitrarily oriented bar element in terms of the global system. We will solve two example plane truss problems to illustrate the procedure of establishing the total stiffness matrix and equations for solution of a structure.

Next we extend the stiffness method to include space trusses. We will develop the transformation matrix in three-dimensional space and analyze a space truss. Then we describe the concept of symmetry and its use to reduce the size of a problem and facilitate its solution. We will use an example truss problem to illustrate the concept and then describe how to handle inclined, or skewed, supports.

We will then use the principle of minimum potential energy and apply it to rederive the bar element equations. We then compare a finite element solution to an exact solution for a bar subjected to a linear varying distributed load. Finally, we will introduce Galerkin's residual method and then apply it to derive the bar element equations.

▲ 3.1 Derivation of the Stiffness Matrix for a Bar Element in Local Coordinates ▲

We will now consider the derivation of the stiffness matrix for the linear-elastic, constant cross-sectional area (prismatic) bar element shown in Figure 3–1. The derivation here will be directly applicable to the solution of pin-connected trusses. The bar is

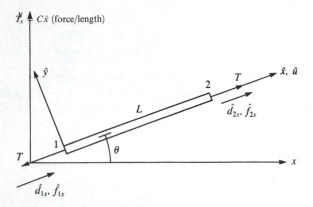

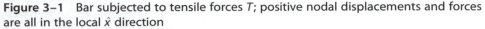

Figure 3–1 Bar subjected to tensile forces T; positive nodal displacements and forces are all in the local $\hat{x}$ direction

subjected to tensile forces T directed along the local axis of the bar and applied at nodes 1 and 2. Here we have introduced two coordinate systems: a local one $(\hat{x}, \hat{y})$ with $\hat{x}$ directed along the length of the bar and a global one (x, y) assumed here to be best suited with respect to the total structure. Proper selection of global coordinate systems is best demonstrated through solution of two- and three-dimensional truss problems as illustrated in Sections 3.6 and 3.7. Both systems will be used extensively throughout this text.

The bar element is assumed to have constant cross-sectional area A, modulus of elasticity E, and initial length L. The nodal degrees of freedom are local axial displacements (longitudinal displacements directed along the length of the bar) represented by $\hat{d}_{1x}$ and $\hat{d}_{2x}$ at the ends of the element as shown in Figure 3–1.

From Hooke's law [Eq. (a)] and the strain/displacement relationship [Eq. (b) or Eq. (1.4.1)], we write

$$\sigma_x = E\varepsilon_x \tag{a}$$

$$\varepsilon_x = \frac{d\hat{u}}{d\hat{x}} \tag{b}$$

From force equilibrium, we have

$$A\sigma_x = T = \text{constant} \tag{c}$$

for a bar with loads applied only at the ends. (We will consider distributed loading in Section 3.9.) Using Eq. (b) in Eq. (a) and then Eq. (a) in Eq. (c) and differentiating with respect to $\hat{x}$, we obtain the differential equation governing the linear-elastic bar behavior as

$$\frac{d}{d\hat{x}}\left(AE\frac{d\hat{u}}{d\hat{x}}\right) = 0 \tag{d}$$

where $\hat{u}$ is the axial displacement function along the element in the $\hat{x}$ direction and A and E are written as though they were functions of $\hat{x}$ in the general form of the differential equation, even though A and E will be assumed constant over the whole length of the bar in our derivations to follow.

The following assumptions are used in deriving the bar element stiffness matrix:

1. The bar cannot sustain shear force; that is, $\hat{f}_{1y} = 0$ and $\hat{f}_{2y} = 0$.
2. Any effect of transverse displacement is ignored.
3. Hooke's law applies; that is, axial stress σ_x is related to axial strain ε_x by $\sigma_x = E\varepsilon_x$.
4. No intermediate applied loads.

The steps previously outlined in Chapter 1 are now used to derive the stiffness matrix for the bar element and then to illustrate a complete solution for a bar assemblage.

Step 1   Select the Element Type

Represent the bar by labeling nodes at each end and in general by labeling the element number (Figure 3–1).

Step 2   Select a Displacement Function

Assume a linear displacement variation along the $\hat{x}$ axis of the bar because a linear function with specified endpoints has a unique path. These specified endpoints are the nodal values $\hat{d}_{1x}$ and $\hat{d}_{2x}$. (Further discussion regarding the choice of displacement functions is provided in Section 3.2 and References [1–3].) Then

$$\hat{u} = a_1 + a_2\hat{x} \tag{3.1.1}$$

with the total number of coefficients a_i always equal to the total number of degrees of freedom associated with the element. Here the total number of degrees of freedom is two—axial displacements at each of the two nodes of the element. Using the same procedure as in Section 2.2 for the spring element, we express Eq. (3.1.1) as

$$\hat{u} = \left(\frac{\hat{d}_{2x} - \hat{d}_{1x}}{L}\right)\hat{x} + \hat{d}_{1x} \tag{3.1.2}$$

In matrix form, Eq. (3.1.2) becomes

$$\hat{u} = [N_1 \quad N_2]\begin{Bmatrix} \hat{d}_{1x} \\ \hat{d}_{2x} \end{Bmatrix} \tag{3.1.3}$$

with shape functions given by

$$N_1 = 1 - \frac{\hat{x}}{L} \qquad N_2 = \frac{\hat{x}}{L} \tag{3.1.4}$$

The linear displacement function $\hat{u}$ [Eq. (3.1.2)], plotted over the length of the bar element, is shown in Figure 3–2. The bar is shown with the same orientation as in Figure 3–1.

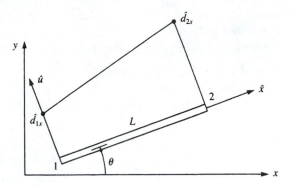

Figure 3–2 Displacement $\hat{u}$ plotted over the length of the element

Step 3 Define the Strain/Displacement and Stress/Strain Relationships

The strain/displacement relationship is

$$\varepsilon_x = \frac{d\hat{u}}{d\hat{x}} = \frac{\hat{d}_{2x} - \hat{d}_{1x}}{L} \tag{3.1.5}$$

where Eqs. (3.1.3) and (3.1.4) have been used to obtain Eq. (3.1.5), and the stress/strain relationship is

$$\sigma_x = E\varepsilon_x \tag{3.1.6}$$

Step 4 Derive the Element Stiffness Matrix and Equations

The element stiffness matrix is derived as follows. From elementary mechanics, we have

$$T = A\sigma_x \tag{3.1.7}$$

Now, using Eqs. (3.1.5) and (3.1.6) in Eq. (3.1.7), we obtain

$$T = AE\left(\frac{\hat{d}_{2x} - \hat{d}_{1x}}{L}\right) \tag{3.1.8}$$

Also, by the nodal force sign convention of Figure 3–1,

$$\hat{f}_{1x} = -T \tag{3.1.9}$$

When we substitute Eq. (3.1.8), Eq. (3.1.9) becomes

$$\hat{f}_{1x} = \frac{AE}{L}(\hat{d}_{1x} - \hat{d}_{2x}) \tag{3.1.10}$$

Similarly,
$$\hat{f}_{2x} = T \tag{3.1.11}$$

or, by Eq. (3.1.8), Eq. (3.1.11) becomes

$$\hat{f}_{2x} = \frac{AE}{L}(\hat{d}_{2x} - \hat{d}_{1x}) \tag{3.1.12}$$

Expressing Eqs. (3.1.10) and (3.1.12) together in matrix form, we have

$$\begin{Bmatrix} \hat{f}_{1x} \\ \hat{f}_{2x} \end{Bmatrix} = \frac{AE}{L} \begin{bmatrix} 1 & -1 \\ -1 & 1 \end{bmatrix} \begin{Bmatrix} \hat{d}_{1x} \\ \hat{d}_{2x} \end{Bmatrix} \tag{3.1.13}$$

Now, because $\hat{f} = \underline{\hat{k}}\hat{d}$, we have, from Eq. (3.1.13),

$$\underline{\hat{k}} = \frac{AE}{L} \begin{bmatrix} 1 & -1 \\ -1 & 1 \end{bmatrix} \tag{3.1.14}$$

Equation (3.1.14) represents the stiffness matrix for a bar element in local coordinates. In Eq. (3.1.14), AE/L for a bar element is analogous to the spring constant k for a spring element.

Step 5 Assemble Element Equations to Obtain Global or Total Equations

Assemble the global stiffness and force matrices and global equations using the direct stiffness method described in Chapter 2 (see Section 3.6 for an example truss). This step applies for structures composed of more than one element such that (again)

$$\underline{K} = [K] = \sum_{e=1}^{N} \underline{k}^{(e)} \quad \text{and} \quad \underline{F} = \{F\} = \sum_{e=1}^{N} \underline{f}^{(e)} \tag{3.1.15}$$

where now all local element stiffness matrices $\underline{\hat{k}}$ must be transformed to global element stiffness matrices $\underline{k}$ (unless the local axes coincide with the global axes) before the direct stiffness method is applied as indicated by Eq. (3.1.15). (This concept of coordinate and stiffness matrix transformations is described in Sections 3.3 and 3.4.)

Step 6 Solve for the Nodal Displacements

Determine the displacements by imposing boundary conditions and simultaneously solving a system of equations, $\underline{F} = \underline{K}\underline{d}$.

Step 7 Solve for the Element Forces

Finally, determine the strains and stresses in each element by back-substitution of the displacements into equations similar to Eqs. (3.1.5) and (3.1.6).

We will now illustrate a solution for a one-dimensional bar problem.

Example 3.1

For the three-bar assemblage shown in Figure 3–3, determine (a) the global stiffness matrix, (b) the displacements of nodes 2 and 3, and (c) the reactions at nodes 1 and 4. A force of 3000 lb is applied in the x direction at node 2. The length of each element is

30 in. Let $E = 30 \times 10^6$ psi and $A = 1$ in² for elements 1 and 2, and let $E = 15 \times 10^6$ psi and $A = 2$ in² for element 3. Nodes 1 and 4 are fixed.

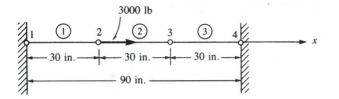

3000 lb

Figure 3–3 Three-bar assemblage

(a) Using Eq. (3.1.14), we find that the element stiffness matrices are

$$
\underline{k}^{(1)} = \underline{k}^{(2)} = \frac{(1)(30 \times 10^6)}{30}
\begin{bmatrix} 1 & -1 \\ -1 & 1 \end{bmatrix} = 10^6
\begin{matrix} 1 & 2^{(1)} \\ 2 & 3^{(2)} \end{matrix}
\begin{bmatrix} 1 & -1 \\ -1 & 1 \end{bmatrix} \frac{\text{lb}}{\text{in.}}
$$

$$
\underline{k}^{(3)} = \frac{(2)(15 \times 10^6)}{30}
\begin{bmatrix} 1 & -1 \\ -1 & 1 \end{bmatrix} = 10^6
\begin{matrix} 3 & 4 \\ \end{matrix}
\begin{bmatrix} 1 & -1 \\ -1 & 1 \end{bmatrix} \frac{\text{lb}}{\text{in.}}
$$

(3.1.16)

where, again, the numbers above the matrices in Eqs. (3.1.16) indicate the displacements associated with each matrix. Assembling the element stiffness matrices by the direct stiffness method, we obtain the global stiffness matrix as

$$
\underline{K} = 10^6
\begin{matrix} d_{1x} & d_{2x} & d_{3x} & d_{4x} \end{matrix}
\begin{bmatrix}
1 & -1 & 0 & 0 \\
-1 & 1+1 & -1 & 0 \\
0 & -1 & 1+1 & -1 \\
0 & 0 & -1 & 1
\end{bmatrix} \frac{\text{lb}}{\text{in.}}
$$

(3.1.17)

(b) Equation (3.1.17) relates global nodal forces to global nodal displacements as follows:

$$
\begin{Bmatrix} F_{1x} \\ F_{2x} \\ F_{3x} \\ F_{4x} \end{Bmatrix} = 10^6
\begin{bmatrix}
1 & -1 & 0 & 0 \\
-1 & 2 & -1 & 0 \\
0 & -1 & 2 & -1 \\
0 & 0 & -1 & 1
\end{bmatrix}
\begin{Bmatrix} d_{1x} \\ d_{2x} \\ d_{3x} \\ d_{4x} \end{Bmatrix}
$$

(3.1.18)

Invoking the boundary conditions, we have

$$
d_{1x} = 0 \qquad d_{4x} = 0
$$

(3.1.19)

Using the boundary conditions, substituting known applied global forces into Eq. (3.1.18), and partitioning equations 1 and 4 of Eq. (3.1.18), we solve equations 2 and 3 of Eq. (3.1.18) to obtain

$$\begin{Bmatrix} 3000 \\ 0 \end{Bmatrix} = 10^6 \begin{bmatrix} 2 & -1 \\ -1 & 2 \end{bmatrix} \begin{Bmatrix} d_{2x} \\ d_{3x} \end{Bmatrix} \tag{3.1.20}$$

Solving Eq. (3.1.20) simultaneously for the displacements yields

$$d_{2x} = 0.002 \text{ in.} \qquad d_{3x} = 0.001 \text{ in.} \tag{3.1.21}$$

(c) Back-substituting Eqs. (3.1.19) and (3.1.21) into Eq. (3.1.18), we obtain the global nodal forces, which include the reactions at nodes 1 and 4, as follows:

$$F_{1x} = 10^6(d_{1x} - d_{2x}) = 10^6(0 - 0.002) = -2000 \text{ lb}$$

$$F_{2x} = 10^6(-d_{1x} + 2d_{2x} - d_{3x}) = 10^6[0 + 2(0.002) - 0.001] = 3000 \text{ lb}$$

$$F_{3x} = 10^6(-d_{2x} + 2d_{3x} - d_{4x}) = 10^6[-0.002 + 2(0.001) - 0] = 0 \tag{3.1.22}$$

$$F_{4x} = 10^6(-d_{3x} + d_{4x}) = 10^6(-0.001 + 0) = -1000 \text{ lb}$$

The results of Eqs. (3.1.22) show that the sum of the reactions F_{1x} and F_{4x} is equal in magnitude but opposite in direction to the applied nodal force of 3000 lb at node 2. Equilibrium of the bar assemblage is thus verified. Furthermore, Eqs. (3.1.22) show that $F_{2x} = 3000$ lb and $F_{3x} = 0$ are merely the applied nodal forces at nodes 2 and 3, respectively, which further enhances the validity of our solution. ■

3.2 Selecting Approximation Functions for Displacements

Consider the following guidelines, as they relate to the one-dimensional bar element, when selecting a displacement function. (Further discussion regarding selection of displacement functions and other kinds of approximation functions (such as temperature functions) will be provided in Chapter 5 for the beam element, in Chapter 7 for the constant-strain triangular element, in Chapter 9 for the linear-strain triangular element, and in Chapter 13 for the heat-transfer problem. More information is also provided in References [1–3].)

1. Common approximation functions are usually polynomials such as the simplest one that gives the linear variation of displacement given by Eq. (3.1.1) or equivalently by Eq. (3.1.3), where the function is expressed in terms of the shape functions.
2. The approximation function should be continuous within the bar element. The simple linear function for $\hat{u}$ of Eq. (3.1.1) certainly is continuous within the element. Therefore, the linear function yields continuous values of $\hat{u}$ within the element and prevents openings, overlaps, and jumps because of the continuous and smooth variation in $\hat{u}$ (Figure 3–4).
3. The approximating function should provide interelement continuity

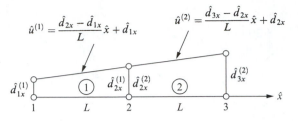

$$\hat{u}^{(1)} = \frac{\hat{d}_{2x} - \hat{d}_{1x}}{L}\hat{x} + \hat{d}_{1x}$$

$$\hat{u}^{(2)} = \frac{\hat{d}_{3x} - \hat{d}_{2x}}{L}\hat{x} + \hat{d}_{2x}$$

Figure 3–4 Interelement continuity of a two-bar structure

for all degrees of freedom at each node for discrete line elements and along common boundary lines and surfaces for two- and three-dimensional elements. For the bar element, we must ensure that nodes common to two or more elements remain common to these elements upon deformation and thus prevent overlaps or voids between elements. For example, consider the two-bar structure shown in Figure 3–4. For the two-bar structure, the linear function for $\hat{u}$ [Eq. (3.1.2)] within each element will ensure that elements 1 and 2 remain connected; the displacement at node 2 for element 1 will equal the displacement at the same node 2 for element 2; that is, $\hat{d}_{2x}^{(1)} = \hat{d}_{2x}^{(2)}$. This rule was also illustrated by Eq. (2.3.3). The linear function is then called a *conforming*, or *compatible, function* for the bar element because it ensures the satisfaction both of continuity between adjacent elements and of continuity within the element.

4. The approximation function should allow for rigid-body displacement and for a state of constant strain within the element. The one-dimensional displacement function [Eq. (3.1.1)] satisfies these criteria because the a_1 term allows for rigid-body motion (constant motion of the body without straining) and the $a_2\hat{x}$ term allows for constant strain because $\varepsilon_x = d\hat{u}/d\hat{x} = a_2$ is a constant. (This state of constant strain in the element can, in fact, occur if elements are chosen small enough.) The simple polynomial Eq. (3.1.1) satisfying this fourth guideline is then said to be *complete* for the bar element. Completeness of a function is a necessary condition for convergence to the exact answer, for instance, for displacements and stresses (Figure 3–5) (see

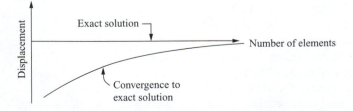

Figure 3–5 Convergence to the exact solution for displacement as the number of elements of a finite element solution is increased

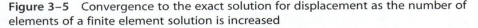

Reference [3]). Figure 3–5 illustrates monotonic convergence toward an exact solution for displacement as the number of elements in a finite element solution is increased. Monotonic convergence is then the process in which successive approximation solutions (finite element solutions) approach the exact solution consistently without changing sign or direction.

The idea that the interpolation (approximation) function must allow for a rigid-body displacement means that the function must be capable of yielding a constant value (say, a_1), because such a value can, in fact, occur. Therefore, we must consider the case

$$\hat{u} = a_1 \tag{3.2.1}$$

For $\hat{u} = a_1$ requires nodal displacements $\hat{d}_{1x} = \hat{d}_{2x}$ to obtain a rigid-body displacement. Therefore

$$a_1 = \hat{d}_{1x} = \hat{d}_{2x} \tag{3.2.2}$$

Using Eq. (3.2.2) in Eq. (3.1.3), we have

$$\hat{u} = N_1 \hat{d}_{1x} + N_2 \hat{d}_{2x} = (N_1 + N_2)a_1 \tag{3.2.3}$$

From Eqs. (3.2.1) and (3.2.3), we then have

$$\hat{u} = a_1 = (N_1 + N_2)a_1 \tag{3.2.4}$$

Therefore, by Eq. (3.2.4), we obtain

$$N_1 + N_2 = 1 \tag{3.2.5}$$

Thus Eq. (3.2.5) shows that the displacement interpolation functions must add to unity at every point within the element so that $\hat{u}$ will yield a constant value when a rigid-body displacement occurs.

▲ 3.3 Transformation of Vectors in Two Dimensions ▲

In many problems it is convenient to introduce both local and global (or reference) coordinates. Local coordinates are always chosen to represent the individual element conveniently. Global coordinates are chosen to be convenient for the whole structure.

Given the nodal displacement of an element, represented by the vector **d** in Figure 3–6, we want to relate the components of this vector in one coordinate system to components in another. For general purposes, we will assume in this section that **d** is not coincident with either the local or the global axis. In this case, we want to relate global displacement components to local ones. In so doing, we will develop a transformation matrix that will subsequently be used to develop the global stiffness matrix for a bar element. We define the angle θ to be positive when measured counterclockwise from x to $\hat{x}$. We can express vector displacement **d** in both global and local

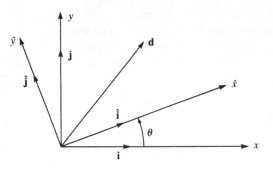

Figure 3–6 General displacement vector **d**

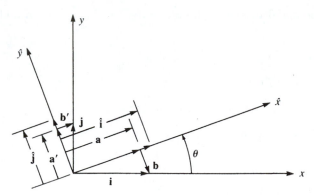

Figure 3–7 Relationship between local and global unit vectors

coordinates by

$$\mathbf{d} = d_x\mathbf{i} + d_y\mathbf{j} = \hat{d}_x\hat{\mathbf{i}} + \hat{d}_y\hat{\mathbf{j}} \tag{3.3.1}$$

where **i** and **j** are unit vectors in the x and y global directions and $\hat{\mathbf{i}}$ and $\hat{\mathbf{j}}$ are unit vectors in the $\hat{x}$ and $\hat{y}$ local directions. We will now relate **i** and **j** to $\hat{\mathbf{i}}$ and $\hat{\mathbf{j}}$ through use of Figure 3–7.

Using Figure 3–7 and vector addition, we obtain

$$\mathbf{a} + \mathbf{b} = \mathbf{i} \tag{3.3.2}$$

Also, from the law of cosines,

$$|\mathbf{a}| = |\mathbf{i}| \cos \theta \tag{3.3.3}$$

and because **i** is, by definition, a unit vector, its magnitude is given by

$$|\mathbf{i}| = 1 \tag{3.3.4}$$

Therefore, we obtain

$$|\mathbf{a}| = 1 \cos \theta \tag{3.3.5}$$

Similarly,

$$|\mathbf{b}| = 1 \sin \theta \tag{3.3.6}$$

Now $\mathbf{a}$ is in the $\hat{\mathbf{i}}$ direction and $\mathbf{b}$ is in the $-\hat{\mathbf{j}}$ direction. Therefore,

$$\mathbf{a} = |\mathbf{a}|\hat{\mathbf{i}} = (\cos\theta)\hat{\mathbf{i}} \tag{3.3.7}$$

and
$$\mathbf{b} = |\mathbf{b}|(-\hat{\mathbf{j}}) = (\sin\theta)(-\hat{\mathbf{j}}) \tag{3.3.8}$$

Using Eqs. (3.3.7) and (3.3.8) in Eq. (3.3.2) yields

$$\mathbf{i} = \cos\theta\hat{\mathbf{i}} - \sin\theta\hat{\mathbf{j}} \tag{3.3.9}$$

Similarly, from Figure 3–7, we obtain

$$\mathbf{a'} + \mathbf{b'} = \mathbf{j} \tag{3.3.10}$$

$$\mathbf{a'} = \cos\theta\hat{\mathbf{j}} \tag{3.3.11}$$

$$\mathbf{b'} = \sin\theta\hat{\mathbf{i}} \tag{3.3.12}$$

Using Eqs. (3.3.11) and (3.3.12) in Eq. (3.3.10), we have

$$\mathbf{j} = \sin\theta\hat{\mathbf{i}} + \cos\theta\hat{\mathbf{j}} \tag{3.3.13}$$

Now, using Eqs. (3.3.9) and (3.3.13) in Eq. (3.3.1), we have

$$d_x(\cos\theta\hat{\mathbf{i}} - \sin\theta\hat{\mathbf{j}}) + d_y(\sin\theta\hat{\mathbf{i}} + \cos\theta\hat{\mathbf{j}}) = \hat{d}_x\hat{\mathbf{i}} + \hat{d}_y\hat{\mathbf{j}} \tag{3.3.14}$$

Combining like coefficients of $\hat{\mathbf{i}}$ and $\hat{\mathbf{j}}$ in Eq. (3.3.14), we obtain

$$d_x\cos\theta + d_y\sin\theta = \hat{d}_x$$

and
$$-d_x\sin\theta + d_y\cos\theta = \hat{d}_y \tag{3.3.15}$$

In matrix form, Eqs. (3.3.15) are written as

$$\left\{\begin{matrix} \hat{d}_x \\ \hat{d}_y \end{matrix}\right\} = \begin{bmatrix} C & S \\ -S & C \end{bmatrix} \left\{\begin{matrix} d_x \\ d_y \end{matrix}\right\} \tag{3.3.16}$$

where $C = \cos\theta$ and $S = \sin\theta$.

Equation (3.3.16) relates the global displacement $\underline{d}$ to the local displacement $\underline{\hat{d}}$. The matrix

$$\begin{bmatrix} C & S \\ -S & C \end{bmatrix} \tag{3.3.17}$$

is called the *transformation* (or *rotation*) *matrix*. For an additional description of this matrix, see Appendix A. It will be used in Section 3.4 to develop the global stiffness matrix for an arbitrarily oriented bar element and to transform global nodal displacements and forces to local ones.

Now, for the case of $\hat{d}_y = 0$, we have, from Eq. (3.3.1),

$$d_x\mathbf{i} + d_y\mathbf{j} = \hat{d}_x\hat{\mathbf{i}} \tag{3.3.18}$$

Figure 3–8 shows $\hat{d}_x$ expressed in terms of global x and y components. Using trigonometry and Figure 3–8, we then obtain the magnitude of $\hat{d}_x$ as

$$\hat{d}_x = Cd_x + Sd_y \tag{3.3.19}$$

Equation (3.3.19) is equivalent to equation 1 of Eq. (3.3.16).

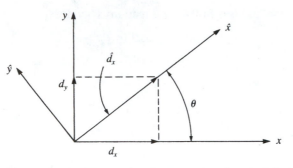

Figure 3-8 Relationship between local and global displacements

Example 3.2

The global nodal displacements at node 2 have been determined to be $d_{2x} = 0.1$ in. and $d_{2y} = 0.2$ in. for the bar element shown in Figure 3-9. Determine the local $\hat{x}$ displacement at node 2.

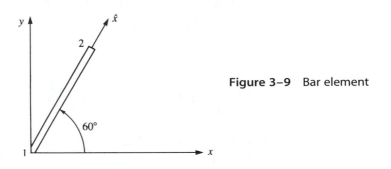

Figure 3-9 Bar element

Using Eq. (3.3.19), we obtain

$$\hat{d}_{2x} = (\cos 60°)(0.1) + (\sin 60°)(0.2) = 0.223 \text{ in.} \qquad ■$$

▲ 3.4 Global Stiffness Matrix ▲

We will now use the transformation relationship Eq. (3.3.16) to obtain the global stiffness matrix for a bar element. We need the global stiffness matrix of each element to assemble the total global stiffness matrix of the structure. We have shown in Eq. (3.1.13) that for a bar element in the local coordinate system,

$$\left\{ \begin{array}{c} \hat{f}_{1x} \\ \hat{f}_{2x} \end{array} \right\} = \frac{AE}{L} \begin{bmatrix} 1 & -1 \\ -1 & 1 \end{bmatrix} \left\{ \begin{array}{c} \hat{d}_{1x} \\ \hat{d}_{2x} \end{array} \right\} \qquad (3.4.1)$$

or

$$\underline{\hat{f}} = \underline{\hat{k}}\underline{\hat{d}} \qquad (3.4.2)$$

We now want to relate the global element nodal forces f to the global nodal displacements d for a bar element arbitrarily oriented with respect to the global axes as was shown in Figure 3–1. This relationship will yield the global stiffness matrix k of the element. That is, we want to find a matrix k such that

$$\begin{Bmatrix} f_{1x} \\ f_{1y} \\ f_{2x} \\ f_{2y} \end{Bmatrix} = \underline{k} \begin{Bmatrix} d_{1x} \\ d_{1y} \\ d_{2x} \\ d_{2y} \end{Bmatrix} \tag{3.4.3}$$

or, in simplified matrix form, Eq. (3.4.3) becomes

$$\underline{f} = \underline{k}\underline{d} \tag{3.4.4}$$

We observe from Eq. (3.4.3) that a total of four components of force and four of displacement arise when global coordinates are used. However, a total of two components of force and two of displacement appear for the local-coordinate representation of a spring or a bar, as shown by Eq. (3.4.1). By using relationships between local and global force components and between local and global displacement components, we will be able to obtain the global stiffness matrix. We know from transformation relationship Eq. (3.3.15) that

$$\hat{d}_{1x} = d_{1x} \cos \theta + d_{1y} \sin \theta \tag{3.4.5}$$

$$\hat{d}_{2x} = d_{2x} \cos \theta + d_{2y} \sin \theta$$

In matrix form, Eqs. (3.4.5) can be written as

$$\begin{Bmatrix} \hat{d}_{1x} \\ \hat{d}_{2x} \end{Bmatrix} = \begin{bmatrix} C & S & 0 & 0 \\ 0 & 0 & C & S \end{bmatrix} \begin{Bmatrix} d_{1x} \\ d_{1y} \\ d_{2x} \\ d_{2y} \end{Bmatrix} \tag{3.4.6}$$

or as

$$\hat{\underline{d}} = \underline{T}^*\underline{d} \tag{3.4.7}$$

where

$$\underline{T}^* = \begin{bmatrix} C & S & 0 & 0 \\ 0 & 0 & C & S \end{bmatrix} \tag{3.4.8}$$

Similarly, because forces transform in the same manner as displacements, we have

$$\begin{Bmatrix} \hat{f}_{1x} \\ \hat{f}_{2x} \end{Bmatrix} = \begin{bmatrix} C & S & 0 & 0 \\ 0 & 0 & C & S \end{bmatrix} \begin{Bmatrix} f_{1x} \\ f_{1y} \\ f_{2x} \\ f_{2y} \end{Bmatrix} \tag{3.4.9}$$

Using Eq. (3.4.8), we can write Eq. (3.4.9) as

$$\hat{\underline{f}} = \underline{T}^*\underline{f} \tag{3.4.10}$$

Now, substituting Eq. (3.4.7) into Eq. (3.4.2), we obtain

$$\underline{f} = \underline{\hat{k}}\underline{T}^*\underline{d} \tag{3.4.11}$$

and using Eq. (3.4.10) in Eq. (3.4.11) yields

$$\underline{T}^*\underline{f} = \underline{\hat{k}}\underline{T}^*\underline{d} \tag{3.4.12}$$

However, to write the final expression relating global nodal forces to global nodal displacements for an element, we must invert $\underline{T}^*$ in Eq. (3.4.12). This is not immediately possible because $\underline{T}^*$ is not a square matrix. Therefore, we must expand $\underline{\hat{d}}, \underline{\hat{f}}$, and $\underline{\hat{k}}$ to the order that is consistent with the use of global coordinates even though $\hat{f}_{1y}$ and $\hat{f}_{2y}$ are zero. Using Eq. (3.3.16) for each nodal displacement, we thus obtain

$$\begin{Bmatrix} \hat{d}_{1x} \\ \hat{d}_{1y} \\ \hat{d}_{2x} \\ \hat{d}_{2y} \end{Bmatrix} = \begin{bmatrix} C & S & 0 & 0 \\ -S & C & 0 & 0 \\ 0 & 0 & C & S \\ 0 & 0 & -S & C \end{bmatrix} \begin{Bmatrix} d_{1x} \\ d_{1y} \\ d_{2x} \\ d_{2y} \end{Bmatrix} \tag{3.4.13}$$

or

$$\underline{\hat{d}} = \underline{T}\underline{d} \tag{3.4.14}$$

where

$$\underline{T} = \begin{bmatrix} C & S & 0 & 0 \\ -S & C & 0 & 0 \\ 0 & 0 & C & S \\ 0 & 0 & -S & C \end{bmatrix} \tag{3.4.15}$$

Similarly, we can write

$$\underline{\hat{f}} = \underline{T}\underline{f} \tag{3.4.16}$$

because forces are like displacements—both are vectors. Also, $\underline{\hat{k}}$ must be expanded to a 4×4 matrix. Therefore, Eq. (3.4.1) in expanded form becomes

$$\begin{Bmatrix} \hat{f}_{1x} \\ \hat{f}_{1y} \\ \hat{f}_{2x} \\ \hat{f}_{2y} \end{Bmatrix} = \frac{AE}{L} \begin{bmatrix} 1 & 0 & -1 & 0 \\ 0 & 0 & 0 & 0 \\ -1 & 0 & 1 & 0 \\ 0 & 0 & 0 & 0 \end{bmatrix} \begin{Bmatrix} \hat{d}_{1x} \\ \hat{d}_{1y} \\ \hat{d}_{2x} \\ \hat{d}_{2y} \end{Bmatrix} \tag{3.4.17}$$

In Eq. (3.4.17), because $\hat{f}_{1y}$ and $\hat{f}_{2y}$ are zero, rows of zeros corresponding to the row numbers $\hat{f}_{1y}$ and $\hat{f}_{2y}$ appear in $\underline{\hat{k}}$. Now, using Eqs. (3.4.14) and (3.4.16) in Eq. (3.4.2), we obtain

$$\underline{T}\underline{f} = \underline{\hat{k}}\underline{T}\underline{d} \tag{3.4.18}$$

Equation (3.4.18) is Eq. (3.4.12) expanded. Premultiplying both sides of Eq. (3.4.18) by $\underline{T}^{-1}$, we have

$$\underline{f} = \underline{T}^{-1}\underline{\hat{k}}\underline{T}\underline{d} \tag{3.4.19}$$

where $\underline{T}^{-1}$ is the *inverse* of $\underline{T}$. However, it can be shown (see Problem 3.28) that

$$\underline{T}^{-1} = \underline{T}^{T} \qquad (3.4.20)$$

where $\underline{T}^{T}$ is the *transpose* of $\underline{T}$. The property of square matrices such as $\underline{T}$ given by Eq. (3.4.20) defines $\underline{T}$ to be an orthogonal matrix. For more about orthogonal matrices, see Appendix A. The transformation matrix $\underline{T}$ between rectangular coordinate frames is orthogonal. This property of $\underline{T}$ is used throughout this text. Substituting Eq. (3.4.20) into Eq. (3.4.19), we obtain

$$\underline{f} = \underline{T}^{T}\hat{\underline{k}}\underline{T}\underline{d} \qquad (3.4.21)$$

Equating Eqs. (3.4.4) and (3.4.21), we obtain the global stiffness matrix for an element as

$$\underline{k} = \underline{T}^{T}\hat{\underline{k}}\underline{T} \qquad (3.4.22)$$

Substituting Eq. (3.4.15) for $\underline{T}$ and the expanded form of $\hat{\underline{k}}$ given in Eq. (3.4.17) into Eq. (3.4.22), we obtain $\underline{k}$ given in explicit form by

$$\underline{k} = \frac{AE}{L} \begin{bmatrix} C^2 & CS & -C^2 & -CS \\ & S^2 & -CS & -S^2 \\ & & C^2 & CS \\ \text{Symmetry} & & & S^2 \end{bmatrix} \qquad (3.4.23)$$

Now, because the trial displacement function Eq. (3.1.1) was assumed piecewise-continuous element by element, the stiffness matrix for each element can be summed by using the direct stiffness method to obtain

$$\sum_{e=1}^{N} \underline{k}^{(e)} = \underline{K} \qquad (3.4.24)$$

where $\underline{K}$ is the total stiffness matrix and N is the total number of elements. Similarly, each element global nodal force matrix can be summed such that

$$\sum_{e=1}^{N} \underline{f}^{(e)} = \underline{F} \qquad (3.4.25)$$

$\underline{K}$ now relates the global nodal forces $\underline{F}$ to the global nodal displacements $\underline{d}$ for the whole structure by

$$\underline{F} = \underline{K}\underline{d} \qquad (3.4.26)$$

Example 3.3

For the bar element shown in Figure 3–10, evaluate the global stiffness matrix with respect to the *x-y* coordinate system. Let the bar's cross-sectional area equal 2 in.^{2}, length equal 60 in., and modulus of elasticity equal 30×10^6 psi. The angle the bar makes with the x axis is $30°$.

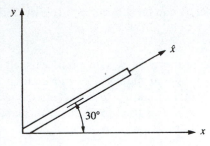

Figure 3–10 Bar element for stiffness matrix evaluation

To evaluate the global stiffness matrix $\underline{k}$ for a bar, we use Eq. (3.4.23) with angle θ defined to be positive when measured counterclockwise from x to $\hat{x}$. Therefore,

$$\theta = 30° \qquad C = \cos 30° = \frac{\sqrt{3}}{2} \qquad S = \sin 30° = \frac{1}{2}$$

$$\underline{k} = \frac{(2)(30 \times 10^6)}{60} \begin{bmatrix} \dfrac{3}{4} & \dfrac{\sqrt{3}}{4} & \dfrac{-3}{4} & \dfrac{-\sqrt{3}}{4} \\[2mm] & \dfrac{1}{4} & \dfrac{-\sqrt{3}}{4} & \dfrac{-1}{4} \\[2mm] & & \dfrac{3}{4} & \dfrac{\sqrt{3}}{4} \\[2mm] \text{Symmetry} & & & \dfrac{1}{4} \end{bmatrix} \frac{\text{lb}}{\text{in.}} \qquad (3.4.27)$$

Simplifying Eq. (3.4.27), we have

$$\underline{k} = 10^6 \begin{bmatrix} 0.75 & 0.433 & -0.75 & -0.433 \\ & 0.25 & -0.433 & -0.25 \\ & & 0.75 & 0.433 \\ \text{Symmetry} & & & 0.25 \end{bmatrix} \frac{\text{lb}}{\text{in.}} \qquad (3.4.28)$$ ■

▲ 3.5 Computation of Stress for a Bar in the x-y Plane   ▲

We will now consider the determination of the stress in a bar element. For a bar, the local forces are related to the local displacements by Eq. (3.1.13) or Eq. (3.4.17). This equation is repeated here for convenience.

$$\begin{Bmatrix} \hat{f}_{1x} \\ \hat{f}_{2x} \end{Bmatrix} = \frac{AE}{L} \begin{bmatrix} 1 & -1 \\ -1 & 1 \end{bmatrix} \begin{Bmatrix} \hat{d}_{1x} \\ \hat{d}_{2x} \end{Bmatrix} \qquad (3.5.1)$$

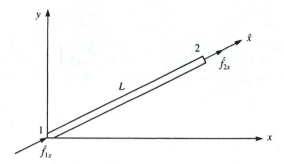

Figure 3–11 Basic bar element with positive nodal forces

The usual definition of axial tensile stress is axial force divided by cross-sectional area. Therefore, axial stress is

$$\sigma = \frac{\hat{f}_{2x}}{A} \tag{3.5.2}$$

where $\hat{f}_{2x}$ is used because it is the axial force that pulls on the bar as shown in Figure 3–11. By Eq. (3.5.1),

$$\hat{f}_{2x} = \frac{AE}{L}[-1 \quad 1]\left\{ \begin{array}{c} \hat{d}_{1x} \\ \hat{d}_{2x} \end{array} \right\} \tag{3.5.3}$$

Therefore, combining Eqs. (3.5.2) and (3.5.3) yields

$$\underline{\sigma} = \frac{E}{L}[-1 \quad 1]\hat{\underline{d}} \tag{3.5.4}$$

Now, using Eq. (3.4.7), we obtain

$$\underline{\sigma} = \frac{E}{L}[-1 \quad 1]\underline{T}^*\underline{d} \tag{3.5.5}$$

Equation (3.5.5) can be expressed in simpler form as

$$\underline{\sigma} = \underline{C}'\underline{d} \tag{3.5.6}$$

where, when we use Eq. (3.4.8),

$$\underline{C}' = \frac{E}{L}[-1 \quad 1]\begin{bmatrix} C & S & 0 & 0 \\ 0 & 0 & C & S \end{bmatrix} \tag{3.5.7}$$

After multiplying the matrices in Eq. (3.5.7), we have

$$\underline{C}' = \frac{E}{L}[-C \quad -S \quad C \quad S] \tag{3.5.8}$$

Example 3.4

For the bar shown in Figure 3–12, determine the axial stress. Let $A = 4 \times 10^{-4}$ m^2, $E = 210$ GPa, and $L = 2$ m, and let the angle between x and $\hat{x}$ be 60°. Assume the global displacements have been previously determined to be $d_{1x} = 0.25$ mm, $d_{1y} = 0.0$, $d_{2x} = 0.50$ mm, and $d_{2y} = 0.75$ mm.

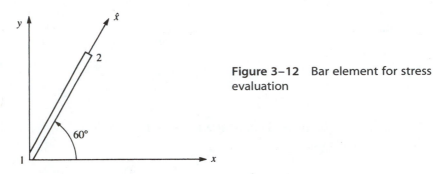

Figure 3–12 Bar element for stress evaluation

We can use Eq. (3.5.6) to evaluate the axial stress. Therefore, we first calculate $\underline{C}'$ from Eq. (3.5.8) as

$$\underline{C}' = \frac{210 \times 10^6 \text{ kN/m}^2}{2 \text{ m}} \begin{bmatrix} -\dfrac{1}{2} & -\dfrac{\sqrt{3}}{2} & \dfrac{1}{2} & \dfrac{\sqrt{3}}{2} \end{bmatrix} \qquad (3.5.9)$$

where we have used $C = \cos 60° = \frac{1}{2}$ and $S = \sin 60° = \sqrt{3}/2$ in Eq. (3.5.9). Now $\underline{d}$ is given by

$$\underline{d} = \begin{Bmatrix} d_{1x} \\ d_{1y} \\ d_{2x} \\ d_{2y} \end{Bmatrix} = \begin{Bmatrix} 0.25 \times 10^{-3} \text{ m} \\ 0.0 \\ 0.50 \times 10^{-3} \text{ m} \\ 0.75 \times 10^{-3} \text{ m} \end{Bmatrix} \qquad (3.5.10)$$

Using Eqs. (3.5.9) and (3.5.10) in Eq. (3.5.6), we obtain the bar axial stress as

$$\sigma_x = \frac{210 \times 10^6}{2} \begin{bmatrix} -\dfrac{1}{2} & -\dfrac{\sqrt{3}}{2} & \dfrac{1}{2} & \dfrac{\sqrt{3}}{2} \end{bmatrix} \begin{Bmatrix} 0.25 \\ 0.0 \\ 0.50 \\ 0.75 \end{Bmatrix} \times 10^{-3}$$

$$= 81.32 \times 10^3 \text{ kN/m}^2 = 81.32 \text{ MPa} \qquad \blacksquare$$

▲ 3.6 Solution of a Plane Truss ▲

We will now illustrate the use of equations developed in Sections 3.4 and 3.5, along with the direct stiffness method of assembling the total stiffness matrix and equations, to solve the following plane truss example problems. *A **plane truss** is a structure composed of bar elements that all lie in a common plane and are connected by frictionless pins. The plane truss also must have loads acting only in the common plane.*

Example 3.5

For the plane truss composed of the three elements shown in Figure 3–13 subjected to a downward force of 10,000 lb applied at node 1, determine the x and y displacements at node 1 and the stresses in each element. Let $E = 30 \times 10^6$ psi and $A = 2$ in.^{2} for all elements. The lengths of the elements are shown in the figure.

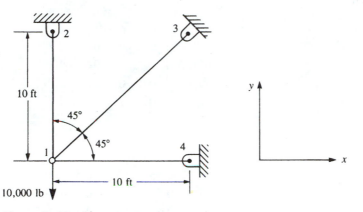

Figure 3–13 Plane truss

First, we determine the global stiffness matrices for each element by using Eq. (3.4.23). This requires determination of the angle θ between the global x axis and the local $\hat{x}$ axis for each element. In this example, the direction of the $\hat{x}$ axis for each element is taken in the direction *from* node 1 *to* the other node. The node numbering is arbitrary for each element. However, once the direction is chosen, the angle θ is then established as positive when measured counterclockwise from positive x to $\hat{x}$. For element 1, the local $\hat{x}$ axis is directed from node 1 to node 2; therefore, $\theta^{(1)} = 90°$. For element 2, the local $\hat{x}$ axis is directed from node 1 to node 3 and $\theta^{(2)} = 45°$. For element 3, the local $\hat{x}$ axis is directed from node 1 to node 4 and $\theta^{(3)} = 0°$. It is convenient to construct Table 3–1 to aid in determining each element stiffness matrix.

There are a total of eight nodal components of displacement, or degrees of freedom, for the truss before boundary constraints are imposed. Thus the order of the total stiffness matrix must be 8×8. We could then expand the $\underline{k}$ matrix for each element to the order 8×8 by adding rows and columns of zeros as explained in the first

Table 3–1 Data for the truss of Figure 3–12

Element	$\theta°$	C	S	C^2	S^2	CS
1	90°	0	1	0	1	0
2	45°	$\sqrt{2}/2$	$\sqrt{2}/2$	$\frac{1}{2}$	$\frac{1}{2}$	$\frac{1}{2}$
3	0°	1	0	1	0	0

part of Section 2.4. Alternatively, we could label the rows and columns of each element stiffness matrix according to the displacement components associated with it as explained in the latter part of Section 2.4. Using this latter approach, we construct the total stiffness matrix $\underline{K}$ simply by adding terms from the individual element stiffness matrices into their corresponding locations in $\underline{K}$. This approach will be used here and throughout this text.

For element 1, using Eq. (3.4.23), along with Table 3–1 for the direction cosines, we obtain

$$\underline{k}^{(1)} = \frac{(30 \times 10^6)(2)}{120} \begin{array}{c} \begin{array}{cccc} d_{1x} & d_{1y} & d_{2x} & d_{2y} \end{array} \\ \begin{bmatrix} 0 & 0 & 0 & 0 \\ 0 & 1 & 0 & -1 \\ 0 & 0 & 0 & 0 \\ 0 & -1 & 0 & 1 \end{bmatrix} \end{array} \qquad (3.6.1)$$

Similarly, for element 2, we have

$$\underline{k}^{(2)} = \frac{(30 \times 10^6)(2)}{120 \times \sqrt{2}} \begin{array}{c} \begin{array}{cccc} d_{1x} & d_{1y} & d_{3x} & d_{3y} \end{array} \\ \begin{bmatrix} 0.5 & 0.5 & -0.5 & -0.5 \\ 0.5 & 0.5 & -0.5 & -0.5 \\ -0.5 & -0.5 & 0.5 & 0.5 \\ -0.5 & -0.5 & 0.5 & 0.5 \end{bmatrix} \end{array} \qquad (3.6.2)$$

and for element 3, we have

$$\underline{k}^{(3)} = \frac{(30 \times 10^6)(2)}{120} \begin{array}{c} \begin{array}{cccc} d_{1x} & d_{1y} & d_{4x} & d_{4y} \end{array} \\ \begin{bmatrix} 1 & 0 & -1 & 0 \\ 0 & 0 & 0 & 0 \\ -1 & 0 & 1 & 0 \\ 0 & 0 & 0 & 0 \end{bmatrix} \end{array} \qquad (3.6.3)$$

The common factor of $30 \times 10^6 \times 2/120$ $(= 500{,}000)$ can be taken from each of Eqs. (3.6.1)–(3.6.3). After adding terms from the individual element stiffness matrices into their corresponding locations in $\underline{K}$, we obtain the total stiffness matrix as

$$\underline{K} = (500{,}000) \begin{array}{c} \begin{array}{cccccccc} d_{1x} & d_{1y} & d_{2x} & d_{2y} & d_{3x} & d_{3y} & d_{4x} & d_{4y} \end{array} \\ \begin{bmatrix} 1.354 & 0.354 & 0 & 0 & -0.354 & -0.354 & -1 & 0 \\ 0.354 & 1.354 & 0 & -1 & -0.354 & -0.354 & 0 & 0 \\ 0 & 0 & 0 & 0 & 0 & 0 & 0 & 0 \\ 0 & -1 & 0 & 1 & 0 & 0 & 0 & 0 \\ -0.354 & -0.354 & 0 & 0 & 0.354 & 0.354 & 0 & 0 \\ -0.354 & -0.354 & 0 & 0 & 0.354 & 0.354 & 0 & 0 \\ -1 & 0 & 0 & 0 & 0 & 0 & 1 & 0 \\ 0 & 0 & 0 & 0 & 0 & 0 & 0 & 0 \end{bmatrix} \end{array} \qquad (3.6.4)$$

The global $\underline{K}$ matrix, Eq. (3.6.4), relates the global forces to the global displacements. We thus write the total structure stiffness equations, accounting for the applied force at node 1 and the boundary constraints at nodes 2–4 as follows:

$$\left\{\begin{array}{c} 0 \\ -10{,}000 \\ F_{2x} \\ F_{2y} \\ F_{3x} \\ F_{3y} \\ F_{4x} \\ F_{4y} \end{array}\right\} = (500{,}000) \begin{bmatrix} 1.354 & 0.354 & 0 & 0 & -0.354 & -0.354 & -1 & 0 \\ 0.354 & 1.354 & 0 & -1 & -0.354 & -0.354 & 0 & 0 \\ 0 & 0 & 0 & 0 & 0 & 0 & 0 & 0 \\ 0 & -1 & 0 & 1 & 0 & 0 & 0 & 0 \\ -0.354 & -0.354 & 0 & 0 & 0.354 & 0.354 & 0 & 0 \\ -0.354 & -0.354 & 0 & 0 & 0.354 & 0.354 & 0 & 0 \\ -1 & 0 & 0 & 0 & 0 & 0 & 1 & 0 \\ 0 & 0 & 0 & 0 & 0 & 0 & 0 & 0 \end{bmatrix}$$

$$\times \left\{\begin{array}{c} d_{1x} \\ d_{1y} \\ d_{2x} = 0 \\ d_{2y} = 0 \\ d_{3x} = 0 \\ d_{3y} = 0 \\ d_{4x} = 0 \\ d_{4y} = 0 \end{array}\right\} \qquad (3.6.5)$$

We could now use the partitioning scheme described in the first part of Section 2.5 to obtain the equations used to determine unknown displacements d_{1x} and d_{1y}—that is, partition the first two equations from the third through the eighth in Eq. (3.6.5). Alternatively, we could eliminate rows and columns in the total stiffness matrix corresponding to zero displacements as previously described in the latter part of Section 2.5. Here we will use the latter approach; that is, we eliminate rows and column 3–8 in Eq. (3.6.5) because those rows and columns correspond to zero displacements. (Remember, this direct approach must be modified for nonhomogeneous boundary conditions as was indicated in Section 2.5.) We then obtain

$$\left\{\begin{array}{c} 0 \\ -10{,}000 \end{array}\right\} = (500{,}000) \begin{bmatrix} 1.354 & 0.354 \\ 0.354 & 1.354 \end{bmatrix} \left\{\begin{array}{c} d_{1x} \\ d_{1y} \end{array}\right\} \qquad (3.6.6)$$

Equation (3.6.6) can now be solved for the displacements by multiplying both sides of the matrix equation by the inverse of the 2×2 stiffness matrix or by solving the two equations simultaneously. Using either procedure for solution yields the displacements

$$d_{1x} = 0.414 \times 10^{-2} \text{ in.} \qquad d_{1y} = -1.59 \times 10^{-2} \text{ in.}$$

The minus sign in the d_{1y} result indicates that the displacement component in the y direction at node 1 is in the direction opposite that of the positive y direction based on the assumed global coordinates, that is, a downward displacement occurs at node 1.

Using Eq. (3.5.6) and Table 3–1, we determine the stresses in each element as follows:

$$\sigma^{(1)} = \frac{30 \times 10^6}{120} [0 \quad -1 \quad 0 \quad 1] \begin{Bmatrix} d_{1x} = 0.414 \times 10^{-2} \\ d_{1y} = -1.59 \times 10^{-2} \\ d_{2x} = 0 \\ d_{2y} = 0 \end{Bmatrix} = 3965 \text{ psi}$$

$$\sigma^{(2)} = \frac{30 \times 10^6}{120\sqrt{2}} \begin{bmatrix} \frac{-\sqrt{2}}{2} & \frac{-\sqrt{2}}{2} & \frac{\sqrt{2}}{2} & \frac{\sqrt{2}}{2} \end{bmatrix} \begin{Bmatrix} d_{1x} = 0.414 \times 10^{-2} \\ d_{1y} = -1.59 \times 10^{-2} \\ d_{3x} = 0 \\ d_{3y} = 0 \end{Bmatrix}$$

$$= 1471 \text{ psi}$$

$$\sigma^{(3)} = \frac{30 \times 10^6}{120} [-1 \quad 0 \quad 1 \quad 0] \begin{Bmatrix} d_{1x} = 0.414 \times 10^{-2} \\ d_{1y} = -1.59 \times 10^{-2} \\ d_{4x} = 0 \\ d_{4y} = 0 \end{Bmatrix} = -1035 \text{ psi}$$

We now verify our results by examining force equilibrium at node 1; that is, summing forces in the global x and y directions, we obtain

$$\sum F_x = 0 \quad (1471 \text{ psi})(2 \text{ in}^2)\frac{\sqrt{2}}{2} - (1035 \text{ psi})(2 \text{ in}^2) = 0$$

$$\sum F_y = 0 \quad (3965 \text{ psi})(2 \text{ in}^2) + (1471 \text{ psi})(2 \text{ in}^2)\frac{\sqrt{2}}{2} - 10,000 = 0 \quad \blacksquare$$

Example 3.6

For the two-bar truss shown in Figure 3–14, determine the displacement in the y direction of node 1 and the axial force in each element. A force of $P = 1000$ kN is applied at node 1 in the positive y direction while node 1 settles an amount $\delta = 50$ mm in the negative x direction. Let $E = 210$ GPa and $A = 6.00 \times 10^{-4}$ m^2 for each element. The lengths of the elements are shown in the figure.

We begin by using Eq. (3.4.23) to determine each element stiffness matrix.

Element 1

$$\cos \theta^{(1)} = \frac{3}{5} = 0.60 \qquad \sin \theta^{(1)} = \frac{4}{5} = 0.80$$

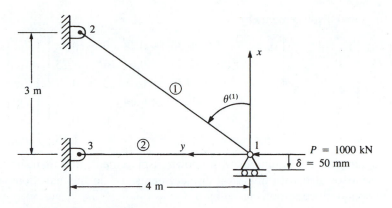

Figure 3–14   Two-bar truss

$$\underline{k}^{(1)} = \frac{(6.0 \times 10^{-4} \text{ m}^2)(210 \times 10^6 \text{ kN/m}^2)}{5 \text{ m}} \begin{bmatrix} 0.36 & 0.48 & -0.36 & -0.48 \\ & 0.64 & -0.48 & -0.64 \\ & & 0.36 & 0.48 \\ \text{Symmetry} & & & 0.64 \end{bmatrix} \quad (3.6.7)$$

Simplifying Eq. (3.6.7), we obtain

$$\underline{k}^{(1)} = (25,200) \begin{matrix} d_{1x} & d_{1y} & d_{2x} & d_{2y} \\ \begin{bmatrix} 0.36 & 0.48 & -0.36 & -0.48 \\ & 0.64 & -0.48 & -0.64 \\ & & 0.36 & 0.48 \\ \text{Symmetry} & & & 0.64 \end{bmatrix} \end{matrix} \quad (3.6.8)$$

Element 2

$$\cos \theta^{(2)} = 0.0 \qquad \sin \theta^{(2)} = 1.0$$

$$\underline{k}^{(2)} = \frac{(6.0 \times 10^{-4})(210 \times 10^6)}{4} \begin{bmatrix} 0 & 0 & 0 & 0 \\ & 1 & 0 & -1 \\ & & 0 & 0 \\ \text{Symmetry} & & & 1 \end{bmatrix} \quad (3.6.9)$$

$$\underline{k}^{(2)} = (25,200) \begin{matrix} d_{1x} & d_{1y} & d_{3x} & d_{3y} \\ \begin{bmatrix} 0 & 0 & 0 & 0 \\ & 1.25 & 0 & -1.25 \\ & & 0 & 0 \\ \text{Symmetry} & & & 1.25 \end{bmatrix} \end{matrix} \quad (3.6.10)$$

where, for computational simplicity, Eq. (3.6.10) is written with the same factor (25,200) in front of the matrix as Eq. (3.6.8). Superimposing the element stiffness matrices, Eqs. (3.6.8) and (3.6.10), we obtain the global $\underline{K}$ matrix and relate the global forces to global displacements by

$$
\begin{Bmatrix} F_{1x} \\ F_{1y} \\ F_{2x} \\ F_{2y} \\ F_{3x} \\ F_{3y} \end{Bmatrix} = (25{,}200) \begin{bmatrix} 0.36 & 0.48 & -0.36 & -0.48 & 0 & 0 \\ & 1.89 & -0.48 & -0.64 & 0 & -1.25 \\ & & 0.36 & 0.48 & 0 & 0 \\ & & & 0.64 & 0 & 0 \\ & & & & 0 & 0 \\ \text{Symmetry} & & & & & 1.25 \end{bmatrix} \begin{Bmatrix} d_{1x} \\ d_{1y} \\ d_{2x} \\ d_{2y} \\ d_{3x} \\ d_{3y} \end{Bmatrix} \tag{3.6.11}
$$

We can again partition equations with known displacements and then simultaneously solve those associated with unknown displacements. To do this partitioning, we consider the boundary conditions given by

$$
d_{1x} = \delta \qquad d_{2x} = 0 \qquad d_{2y} = 0 \qquad d_{3x} = 0 \qquad d_{3y} = 0 \tag{3.6.12}
$$

Therefore, using Eqs. (3.6.12), we partition equation 2 from equations 1, 3, 4, 5, and 6 of Eq. (3.6.11) and are left with

$$
P = 25{,}200(0.48\delta + 1.89 d_{1y}) \tag{3.6.13}
$$

where $F_{1y} = P$ and $d_{1x} = \delta$ were substituted into Eq. (3.6.13). Expressing Eq. (3.6.13) in terms of P and δ allows these two influences on d_{1y} to be clearly separated. Solving Eq. (3.6.13) for d_{1y}, we have

$$
d_{1y} = 0.000021P - 0.254\delta \tag{3.6.14}
$$

Now, substituting the numerical values $P = 1000$ kN and $\delta = -0.05$ m into Eq. (3.6.14), we obtain

$$
d_{1y} = 0.0337 \text{ m} \tag{3.6.15}
$$

where the positive value indicates horizontal displacement to the left.

The local element forces are obtained by using Eq. (3.4.11). We then have the following.

Element 1

$$
\begin{Bmatrix} \hat{f}_{1x} \\ \hat{f}_{2x} \end{Bmatrix} = (25{,}200) \begin{bmatrix} 1 & -1 \\ -1 & 1 \end{bmatrix} \begin{bmatrix} 0.60 & 0.80 & 0 & 0 \\ 0 & 0 & 0.60 & 0.80 \end{bmatrix} \begin{Bmatrix} d_{1x} = -0.05 \\ d_{1y} = 0.0337 \\ d_{2x} = 0 \\ d_{2y} = 0 \end{Bmatrix} \tag{3.6.16}
$$

Performing the matrix triple product in Eq. (3.6.16) yields

$$
\hat{f}_{1x} = -76.6 \text{ kN} \qquad \hat{f}_{2x} = 76.6 \text{ kN} \tag{3.6.17}
$$

Element 2

$$
\begin{Bmatrix} \hat{f}_{1x} \\ \hat{f}_{3x} \end{Bmatrix} = (31{,}500) \begin{bmatrix} 1 & -1 \\ -1 & 1 \end{bmatrix} \begin{bmatrix} 0 & 1 & 0 & 0 \\ 0 & 0 & 0 & 1 \end{bmatrix} \begin{Bmatrix} d_{1x} = -0.05 \\ d_{1y} = 0.0337 \\ d_{3x} = 0 \\ d_{3y} = 0 \end{Bmatrix} \tag{3.6.18}
$$

Performing the matrix triple product in Eq. (3.6.18), we obtain

$$\hat{f}_{1x} = 1061 \text{ kN} \qquad \hat{f}_{3x} = -1061 \text{ kN} \qquad (3.6.19)$$

Verification of the computations by checking that equilibrium is satisfied at node 1 is left to your discretion. ◼

3.7 Transformation Matrix and Stiffness Matrix for a Bar in Three-Dimensional Space

We will now derive the transformation matrix necessary to obtain the general stiffness matrix of a bar element arbitrarily oriented in three-dimensional space as shown in Figure 3–15. Let the coordinates of node 1 be taken as x_1, y_1, and z_1, and let those of node 2 be taken as x_2, y_2, and z_2. Also, let θ_x, θ_y, and θ_z be the angles measured from the global x, y, and z axes, respectively, to the local $\hat{x}$ axis. Here $\hat{x}$ is directed along the element from node 1 to node 2. We must now determine $\underline{T}^*$ such that $\underline{\hat{d}} = \underline{T}^* \underline{d}$. We begin the derivation of $\underline{T}^*$ by considering the vector $\hat{\mathbf{d}} = \mathbf{d}$ expressed in three dimensions as

$$\hat{d}_x \hat{\mathbf{i}} + \hat{d}_y \hat{\mathbf{j}} + \hat{d}_z \hat{\mathbf{k}} = d_x \mathbf{i} + d_y \mathbf{j} + d_z \mathbf{k} \qquad (3.7.1)$$

where $\hat{\mathbf{i}}$, $\hat{\mathbf{j}}$, and $\hat{\mathbf{k}}$ are unit vectors associated with the local $\hat{x}$, $\hat{y}$, and $\hat{z}$ axes, respectively, and $\mathbf{i}$, $\mathbf{j}$, and $\mathbf{k}$ are unit vectors associated with the global x, y, and z axes. Taking the dot product of Eq. (3.7.1) with $\hat{\mathbf{i}}$, we have

$$\hat{d}_x + 0 + 0 = d_x(\hat{\mathbf{i}} \cdot \mathbf{i}) + d_y(\hat{\mathbf{i}} \cdot \mathbf{j}) + d_z(\hat{\mathbf{i}} \cdot \mathbf{k}) \qquad (3.7.2)$$

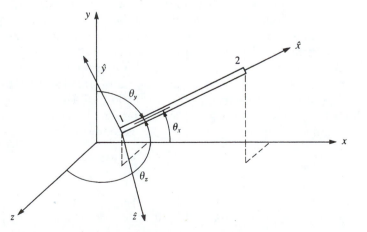

Figure 3–15 Bar in three-dimensional space

and, by definition of the dot product,

$$\hat{\mathbf{i}} \cdot \mathbf{i} = \frac{x_2 - x_1}{L} = C_x$$

$$\hat{\mathbf{i}} \cdot \mathbf{j} = \frac{y_2 - y_1}{L} = C_y \tag{3.7.3}$$

$$\hat{\mathbf{i}} \cdot \mathbf{k} = \frac{z_2 - z_1}{L} = C_z$$

where $\qquad L = [(x_2 - x_1)^2 + (y_2 - y_1)^2 + (z_2 - z_1)^2]^{1/2}$

and $\qquad C_x = \cos \theta_x \qquad C_y = \cos \theta_y \qquad C_z = \cos \theta_z \tag{3.7.4}$

Here $C_x, C_y,$ and C_z are the projections of $\hat{\mathbf{i}}$ on $\mathbf{i}, \mathbf{j},$ and $\mathbf{k}$, respectively. Therefore, using Eqs. (3.7.3) in Eq. (3.7.2), we have

$$\hat{d}_x = C_x d_x + C_y d_y + C_z d_z \tag{3.7.5}$$

For a vector in space directed along the $\hat{x}$ axis, Eq. (3.7.5) gives the components of that vector in the global $x, y,$ and z directions. Now, using Eq. (3.7.5), we can write $\underline{\hat{d}} = \underline{T}^* \underline{d}$ in explicit form as

$$\left\{ \begin{matrix} \hat{d}_{1x} \\ \hat{d}_{2x} \end{matrix} \right\} = \begin{bmatrix} C_x & C_y & C_z & 0 & 0 & 0 \\ 0 & 0 & 0 & C_x & C_y & C_z \end{bmatrix} \begin{Bmatrix} d_{1x} \\ d_{1y} \\ d_{1z} \\ d_{2x} \\ d_{2y} \\ d_{2z} \end{Bmatrix} \tag{3.7.6}$$

where $\qquad \underline{T}^* = \begin{bmatrix} C_x & C_y & C_z & 0 & 0 & 0 \\ 0 & 0 & 0 & C_x & C_y & C_z \end{bmatrix} \tag{3.7.7}$

is the transformation matrix, which enables the local displacement matrix $\underline{\hat{d}}$ to be expressed in terms of displacement components in the global coordinate system.

We showed in Section 3.4 that the global stiffness matrix (the stiffness matrix for a bar element referred to global axes) is given in general by $\underline{k} = \underline{T}^T \underline{\hat{k}} \underline{T}$. This equation will now be used to express the general form of the stiffness matrix of a bar arbitrarily oriented in space. In general, we must expand the transformation matrix in a manner analogous to that done in expanding $\underline{T}^*$ to $\underline{T}$ in Section 3.4. However, the same result will be obtained here by simply using $\underline{T}^*$, defined by Eq. (3.7.7), in place of $\underline{T}$. Then $\underline{k}$ is obtained by using the equation $\underline{k} = (\underline{T}^*)^T \underline{\hat{k}} \underline{T}^*$ as follows:

$$\underline{k} = \begin{bmatrix} C_x & 0 \\ C_y & 0 \\ C_z & 0 \\ 0 & C_x \\ 0 & C_y \\ 0 & C_z \end{bmatrix} \frac{AE}{L} \begin{bmatrix} 1 & -1 \\ -1 & 1 \end{bmatrix} \begin{bmatrix} C_x & C_y & C_z & 0 & 0 & 0 \\ 0 & 0 & 0 & C_x & C_y & C_z \end{bmatrix} \tag{3.7.8}$$

Simplifying Eq. (3.7.8), we obtain the explicit form of $\underline{k}$ as

$$
\underline{k} = \frac{AE}{L}
\begin{bmatrix}
C_x^2 & C_x C_y & C_x C_z & -C_x^2 & -C_x C_y & -C_x C_z \\
 & C_y^2 & C_y C_z & -C_x C_y & -C_y^2 & -C_y C_z \\
 & & C_z^2 & -C_x C_z & -C_y C_z & -C_z^2 \\
 & & & C_x^2 & C_x C_y & C_x C_z \\
 & & & & C_y^2 & C_y C_z \\
\text{Symmetry} & & & & & C_z^2
\end{bmatrix}
\tag{3.7.9}
$$

You should verify Eq. (3.7.9). First, expand $\underline{T}^*$ to a 6×6 square matrix in a manner similar to that done in Section 3.4 for the two-dimensional case. Then expand $\hat{\underline{k}}$ to a 6×6 matrix by adding appropriate rows and columns of zeros (for the $\hat{d}_z$ terms) to Eq. (3.4.17). Finally, perform the matrix triple product $\underline{k} = \underline{T}^T \hat{\underline{k}} \underline{T}$ (see Problem 3.44).

Equation (3.7.9) is the basic form of the stiffness matrix for a bar element arbitrarily oriented in three-dimensional space. We will now analyze a simple space truss to illustrate the concepts developed in this section. We will show that the direct stiffness method provides a simple procedure for solving space truss problems.

Example 3.7

Analyze the space truss shown in Figure 3–16. The truss is composed of four nodes, whose coordinates (in inches) are shown in the figure, and three elements, whose cross-

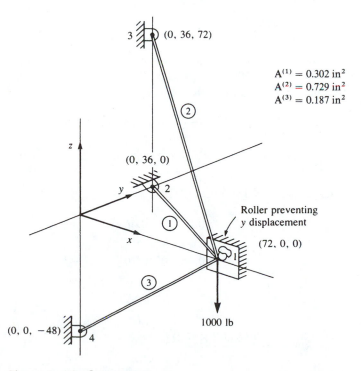

Figure 3–16 Space truss

sectional areas are given in the figure. The modulus of elasticity $E = 1.2 \times 10^6$ psi for all elements. A load of 1000 lb is applied at node 1 in the negative z direction. Nodes 2–4 are supported by ball-and-socket joints and thus constrained from movement in the x, y, and z directions. Node 1 is constrained from movement in the y direction by the roller shown in Figure 3–16.

Using Eq. (3.7.9), we will now determine the stiffness matrices of the three elements in Figure 3–16. To simplify the numerical calculations, we first express $\underline{k}$ for each element, given by Eq. (3.7.9), in the form

$$\underline{k} = \frac{AE}{L} \begin{bmatrix} \underline{\lambda} & -\underline{\lambda} \\ -\underline{\lambda} & \underline{\lambda} \end{bmatrix} \tag{3.7.10}$$

where $\underline{\lambda}$ is a 3×3 submatrix defined by

$$\underline{\lambda} = \begin{bmatrix} C_x^2 & C_x C_y & C_x C_z \\ C_y C_x & C_y^2 & C_y C_z \\ C_z C_x & C_z C_y & C_z^2 \end{bmatrix} \tag{3.7.11}$$

Therefore, determining $\underline{\lambda}$ will sufficiently describe $\underline{k}$.

Element 3

The direction cosines of element 3 are given, in general, by

$$C_x = \frac{x_4 - x_1}{L^{(3)}} \qquad C_y = \frac{y_4 - y_1}{L^{(3)}} \qquad C_z = \frac{z_4 - z_1}{L^{(3)}} \tag{3.7.12}$$

where the notation x_i, y_i, and z_i is used to denote the coordinates of each node, and $L^{(e)}$ denotes the element length. From the coordinate information given in Figure 3–16, we obtain the length and the direction cosines as

$$L^{(3)} = [(-72.0)^2 + (-48.0)^2]^{1/2} = 86.5 \text{ in.}$$

$$C_x = \frac{-72.0}{86.5} = -0.833 \qquad C_y = 0 \qquad C_z = \frac{-48.0}{86.5} = -0.550 \tag{3.7.13}$$

Using the results of Eqs. (3.7.13) in Eq. (3.7.11) yields

$$\underline{\lambda} = \begin{bmatrix} 0.69 & 0 & 0.46 \\ 0 & 0 & 0 \\ 0.46 & 0 & 0.30 \end{bmatrix} \tag{3.7.14}$$

and, from Eq. (3.7.10),

$$k^{(3)} = \frac{(0.187)(1.2 \times 10^6)}{86.5} \overset{\displaystyle d_{1x}d_{1y}d_{1z} \quad d_{4x}d_{4y}d_{4z}}{\begin{bmatrix} \underline{\lambda} & -\underline{\lambda} \\ -\underline{\lambda} & \underline{\lambda} \end{bmatrix}} \tag{3.7.15}$$

Element 1

Similarly, for element 1, we obtain

$$L^{(1)} = 80.5 \text{ in.}$$

$$C_x = -0.89 \qquad C_y = 0.45 \qquad C_z = 0$$

$$\underline{\lambda} = \begin{bmatrix} 0.79 & -0.40 & 0 \\ -0.40 & 0.20 & 0 \\ 0 & 0 & 0 \end{bmatrix}$$

and

$$\underline{k}^{(1)} = \frac{(0.302)(1.2 \times 10^6)}{80.5} \begin{array}{c} {\scriptstyle d_{1x}d_{1y}d_{1z} \quad d_{2x}d_{2y}d_{2z}} \\ \begin{bmatrix} \underline{\lambda} & -\underline{\lambda} \\ -\underline{\lambda} & \underline{\lambda} \end{bmatrix} \end{array} \qquad (3.7.16)$$

Element 2

Finally, for element 2, we obtain

$$L^{(2)} = 108 \text{ in.}$$

$$C_x = -0.667 \qquad C_y = 0.33 \qquad C_z = 0.667$$

$$\underline{\lambda} = \begin{bmatrix} 0.45 & -0.22 & -0.45 \\ -0.22 & 0.11 & 0.45 \\ -0.45 & 0.45 & 0.45 \end{bmatrix}$$

and

$$\underline{k}^{(2)} = \frac{(0.729)(1.2 \times 10^6)}{108} \begin{array}{c} {\scriptstyle d_{1x}d_{1y}d_{1z} \quad d_{3x}d_{3y}d_{3z}} \\ \begin{bmatrix} \underline{\lambda} & -\underline{\lambda} \\ -\underline{\lambda} & \underline{\lambda} \end{bmatrix} \end{array} \qquad (3.7.17)$$

Using the zero-displacement boundary conditions $d_{1y} = 0, d_{2x} = d_{2y} = d_{2z} = 0, d_{3x} = d_{3y} = d_{3z} = 0$, and $d_{4x} = d_{4y} = d_{4z} = 0$, we can cancel the corresponding rows and columns of each element stiffness matrix. After canceling appropriate rows and columns in Eqs. (3.7.15)–(3.7.17) and then superimposing the resulting element stiffness matrices, we have the total stiffness matrix for the truss as

$$\underline{K} = \begin{array}{c} {\scriptstyle d_{1x} \qquad d_{1z}} \\ \begin{bmatrix} 9000 & -2450 \\ -2450 & 4450 \end{bmatrix} \end{array} \qquad (3.7.18)$$

The global stiffness equations are then expressed by

$$\left\{ \begin{array}{c} 0 \\ -1000 \end{array} \right\} = \begin{bmatrix} 9000 & -2450 \\ -2450 & 4450 \end{bmatrix} \left\{ \begin{array}{c} d_{1x} \\ d_{1z} \end{array} \right\} \qquad (3.7.19)$$

Solving Eq. (3.7.19) for the displacements, we obtain

$$d_{1x} = -0.072 \text{ in.} \qquad (3.7.20)$$

$$d_{1z} = -0.264 \text{ in.}$$

where the minus signs in the displacements indicate these displacements to be in the negative x and z directions.

We will now determine the stress in each element. The stresses are determined by using Eq. (3.5.6) expanded to three dimensions. Thus, for an element with nodes i and j, Eq. (3.5.6) expanded to three dimensions becomes

$$\underline{\sigma} = \frac{E}{L}[-C_x \quad -C_y \quad -C_z \quad C_x \quad C_y \quad C_z] \begin{Bmatrix} d_{ix} \\ d_{iy} \\ d_{iz} \\ d_{jx} \\ d_{jy} \\ d_{jz} \end{Bmatrix} \qquad (3.7.21)$$

Derive Eq. (3.7.21) in a manner similar to that used to derive Eq. (3.5.6) (see Problem 3.45, for instance). For element 3, using Eqs. (3.7.13) for the direction cosines, along with the proper length and modulus of elasticity, we obtain the stress as

$$\underline{\sigma}^{(3)} = \frac{1.2 \times 10^6}{86.5}[0.83 \quad 0 \quad 0.55 \quad -0.83 \quad 0 \quad -0.55] \begin{Bmatrix} -0.072 \\ 0 \\ -0.264 \\ 0 \\ 0 \\ 0 \end{Bmatrix} \qquad (3.7.22)$$

Simplifying Eq. (3.7.22), we find that the result is

$$\sigma^{(3)} = -2850 \text{ psi}$$

where the negative sign in the answer indicates a compressive stress. The stresses in the other elements can be determined in a manner similar to that used for element 3. For brevity's sake, we will not show the calculations but will merely list these stresses:

$$\sigma^{(1)} = -945 \text{ psi} \qquad \sigma^{(2)} = 1440 \text{ psi} \qquad \blacksquare$$

▲ 3.8 Use of Symmetry in Structure

Different types of symmetry may exist in a structure. These include reflective or mirror, skew, axial, and cyclic. Here we introduce the most common type of symmetry, reflective symmetry. Axial symmetry occurs when a solid of revolution is generated by rotating a plane shape about an axis in the plane. These axisymmetric bodies are common, and hence their analysis is considered in Chapter 10.

In many instances, we can use reflective symmetry to facilitate the solution of a problem. **Reflective symmetry** *means correspondence in size, shape, and position of loads; material properties; and boundary conditions that are on opposite sides of a dividing line or plane.* The use of symmetry allows us to consider a reduced problem instead of the actual problem. Thus, the order of the total stiffness matrix and total set of stiffness equations can be reduced. Longhand solution time is then reduced, and computer solution time for large-scale problems is substantially decreased. Example 3.8 will be used to illustrate reflective symmetry. Additional examples of the use of symmetry are presented in Chapter 5 for beams and in Chapter 8 for plane problems.

Example 3.8

Solve the plane truss problem shown in Figure 3–17. The truss is composed of eight elements and five nodes as shown. A vertical load of $2P$ is applied at node 4. Nodes 1 and 5 are pin supports. Bar elements 1, 2, 7, and 8 have axial stiffnesses of $\sqrt{2}AE$, and bars 3–6 have axial stiffness of AE. Here again, A and E represent the cross-sectional area and modulus of elasticity of a bar.

In this problem, we will use a plane of symmetry. The vertical plane perpendicular to the plane truss passing through nodes 2, 4, and 3 is the plane of reflective symmetry because identical geometry, material, loading, and boundary conditions occur at the corresponding locations on opposite sides of this plane. For loads such as $2P$, occurring in the plane of symmetry, half of the total load must be applied to the reduced structure. For elements occurring in the plane of symmetry, half of the cross-sectional area must be used in the reduced structure. Furthermore, for nodes in the plane of symmetry, the displacement components normal to the plane of symmetry must be set to zero in the reduced structure; that is, we set $d_{2x} = 0, d_{3x} = 0$, and $d_{4x} = 0$. Figure 3–18 shows the reduced structure to be used to analyze the plane truss of Figure 3–17.

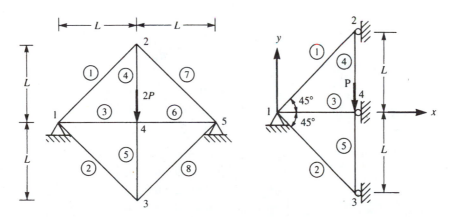

Figure 3–17 Plane truss

Figure 3–18 Truss of Figure 3–17 reduced by symmetry

We begin the solution of the problem by determining the angles θ for each bar element. For instance, for element 1, assuming $\hat{x}$ to be directed from node 1 to node 2, we obtain $\theta^{(1)} = 45°$. Table 3–2 is used in determining each element stiffness matrix.

Table 3–2 Data for the truss of Figure 3–17

Element	$\theta°$	C	S	C^2	S^2	CS
1	45°	$\sqrt{2}/2$	$\sqrt{2}/2$	1/2	1/2	1/2
2	315°	$\sqrt{2}/2$	$-\sqrt{2}/2$	1/2	1/2	−1/2
3	0°	1	0	1	0	0
4	90°	0	1	0	1	0
5	90°	0	1	0	1	0

There are a total of eight nodal components of displacement for the truss before boundary constraints are imposed. Therefore, $\underline{K}$ must be of order 8×8. For element 1, using Eq. (3.4.23) along with Table 3–2 for the direction cosines, we obtain

$$
\underline{k}^{(1)} = \frac{\sqrt{2}AE}{\sqrt{2}L}
\begin{array}{cccc}
d_{1x} & d_{1y} & d_{2x} & d_{2y}
\end{array}
\begin{bmatrix}
\frac{1}{2} & \frac{1}{2} & -\frac{1}{2} & -\frac{1}{2} \\
\frac{1}{2} & \frac{1}{2} & -\frac{1}{2} & -\frac{1}{2} \\
-\frac{1}{2} & -\frac{1}{2} & \frac{1}{2} & \frac{1}{2} \\
-\frac{1}{2} & -\frac{1}{2} & \frac{1}{2} & \frac{1}{2}
\end{bmatrix}
\tag{3.8.1}
$$

Similarly, for elements 2–5, we obtain

$$
\underline{k}^{(2)} = \frac{\sqrt{2}AE}{\sqrt{2}L}
\begin{array}{cccc}
d_{1x} & d_{1y} & d_{3x} & d_{3y}
\end{array}
\begin{bmatrix}
\frac{1}{2} & -\frac{1}{2} & -\frac{1}{2} & \frac{1}{2} \\
-\frac{1}{2} & \frac{1}{2} & \frac{1}{2} & -\frac{1}{2} \\
-\frac{1}{2} & \frac{1}{2} & \frac{1}{2} & -\frac{1}{2} \\
\frac{1}{2} & -\frac{1}{2} & -\frac{1}{2} & \frac{1}{2}
\end{bmatrix}
\tag{3.8.2}
$$

$$
\underline{k}^{(3)} = \frac{AE}{L}
\begin{array}{cccc}
d_{1x} & d_{1y} & d_{4x} & d_{4y}
\end{array}
\begin{bmatrix}
1 & 0 & -1 & 0 \\
0 & 0 & 0 & 0 \\
-1 & 0 & 1 & 0 \\
0 & 0 & 0 & 0
\end{bmatrix}
\tag{3.8.3}
$$

$$
\underline{k}^{(4)} = \frac{AE}{L}
\begin{array}{cccc}
d_{4x} & d_{4y} & d_{2x} & d_{2y}
\end{array}
\begin{bmatrix}
0 & 0 & 0 & 0 \\
0 & \frac{1}{2} & 0 & -\frac{1}{2} \\
0 & 0 & 0 & 0 \\
0 & -\frac{1}{2} & 0 & \frac{1}{2}
\end{bmatrix}
\tag{3.8.4}
$$

$$
\underline{k}^{(5)} = \frac{AE}{L}
\begin{array}{cccc}
d_{3x} & d_{3y} & d_{4x} & d_{4y}
\end{array}
\begin{bmatrix}
0 & 0 & 0 & 0 \\
0 & \frac{1}{2} & 0 & -\frac{1}{2} \\
0 & 0 & 0 & 0 \\
0 & -\frac{1}{2} & 0 & \frac{1}{2}
\end{bmatrix}
\tag{3.8.5}
$$

where, in Eqs. (3.8.1)–(3.8.5), the column labels indicate the degrees of freedom associated with each element. Also, because elements 4 and 5 lie in the plane of symmetry, half of their original areas have been used in Eqs. (3.8.4) and (3.8.5).

We will limit the solution to determining the displacement components. Therefore, considering the boundary constraints that result in zero-displacement components, we can immediately obtain the reduced set of equations by eliminating rows and columns in each element stiffness matrix corresponding to a zero-displacement component. That is, because $d_{1x} = 0$ and $d_{1y} = 0$ (owing to the pin support at node 1 in Figure 3–18) and $d_{2x} = 0, d_{3x} = 0$, and $d_{4x} = 0$ (owing to the symmetry condition), we can cancel rows and columns corresponding to these displacement components in each element stiffness matrix before assembling the total stiffness matrix. The resulting set of stiffness equations is

$$\frac{AE}{L} \begin{bmatrix} 1 & 0 & -\frac{1}{2} \\ 0 & 1 & -\frac{1}{2} \\ -\frac{1}{2} & -\frac{1}{2} & 1 \end{bmatrix} \begin{Bmatrix} d_{2y} \\ d_{3y} \\ d_{4y} \end{Bmatrix} = \begin{Bmatrix} 0 \\ 0 \\ -P \end{Bmatrix} \tag{3.8.6}$$

On solving Eq. (3.8.6) for the displacements, we obtain

$$d_{2y} = \frac{-PL}{AE} \qquad d_{3y} = \frac{-PL}{AE} \qquad d_{4y} = \frac{-2PL}{AE} \tag{3.8.7}$$

∎

The ideas presented regarding the use of symmetry should be used sparingly and cautiously in problems of vibration and buckling. For instance, a structure such as a simply supported beam has symmetry about its center but has antisymmetric vibration modes as well as symmetric vibration modes. This will be shown in Chapter 16. If only half the beam were modeled using reflective symmetry conditions, the support conditions would permit only the symmetric vibration modes.

▲ 3.9 Inclined, or Skewed, Supports ▲

In the preceding sections, the supports were oriented such that the resulting boundary conditions on the displacements were in the global directions.

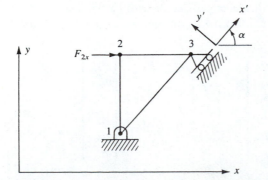

Figure 3–19 Plane truss with inclined boundary conditions at node 3

However, if a support is inclined, or skewed, at an angle α from the global x axis, as shown at node 3 in the plane truss of Figure 3–19, the resulting boundary conditions on the displacements are not in the global x-y directions but are in the local x'-y' directions. We will now describe two methods used to handle inclined supports.

In the first method, to account for inclined boundary conditions, we must perform a transformation of the global displacements at node 3 only into the local nodal coordinate system x'-y', while keeping all other displacements in the x-y global system. We can then enforce the zero-displacement boundary condition d'_{3y} in the force/displacement equations and, finally, solve the equations in the usual manner.

The transformation used is analogous to that for transforming a vector from local to global coordinates. For the plane truss, we use Eq. (3.3.16) applied to node 3 as follows:

$$\left\{ \begin{array}{c} d'_{3x} \\ d'_{3y} \end{array} \right\} = \left[\begin{array}{cc} \cos \alpha & \sin \alpha \\ -\sin \alpha & \cos \alpha \end{array} \right] \left\{ \begin{array}{c} d_{3x} \\ d_{3y} \end{array} \right\} \tag{3.9.1}$$

Rewriting Eq. (3.9.1), we have

$$\{d'_3\} = [t_3]\{d_3\} \tag{3.9.2}$$

where

$$[t_3] = \left[\begin{array}{cc} \cos \alpha & \sin \alpha \\ -\sin \alpha & \cos \alpha \end{array} \right] \tag{3.9.3}$$

We now write the transformation for the entire nodal displacement vector as

$$\{d'\} = [T_1]\{d\} \tag{3.9.4}$$

or

$$\{d\} = [T_1]^T\{d'\} \tag{3.9.5}$$

where the transformation matrix for the entire truss is the 6×6 matrix

$$[T_1] = \left[\begin{array}{ccc} [I] & [0] & [0] \\ [0] & [I] & [0] \\ [0] & [0] & [t_3] \end{array} \right] \tag{3.9.6}$$

Each submatrix in Eq. (3.9.6) (the identity matrix $[I]$, the null matrix $[0]$, and matrix $[t_3]$ has the same 2×2 order, that order in general being equal to the number of degrees of freedom at each node.

To obtain the desired displacement vector with global displacement components at nodes 1 and 2 and local displacement components at node 3, we use Eq. (3.9.5) to obtain

$$\left\{ \begin{array}{c} d_{1x} \\ d_{1y} \\ d_{2x} \\ d_{2y} \\ d_{3x} \\ d_{3y} \end{array} \right\} = \left[\begin{array}{ccc} [I] & [0] & [0] \\ [0] & [I] & [0] \\ [0] & [0] & [t_3]^T \end{array} \right] \left\{ \begin{array}{c} d'_{1x} \\ d'_{1y} \\ d'_{2x} \\ d'_{2y} \\ d'_{3x} \\ d'_{3y} \end{array} \right\} \tag{3.9.7}$$

In Eq. (3.9.7), we observe that only the node 3 global components are transformed, as indicated by the placement of the $[t_3]^T$ matrix. We denote the square matrix in Eq. (3.9.7) by $[T_1]^T$. In general, we place a 2×2 $[t]$ matrix in $[T_1]$ wherever the transformation from global to local displacements is needed (where skewed supports exist).

Upon considering Eqs. (3.9.5) and (3.9.6), we observe that only node 3 components of $\{d\}$ are really transformed to local (skewed) axes components. This transformation is indeed necessary whenever the local axes x'-y' fixity directions are known.

Furthermore, the global force vector can also be transformed by using the same transformation as for $\{d'\}$:

$$\{f'\} = [T_1]\{f\} \tag{3.9.8}$$

In global coordinates, we then have

$$\{f\} = [K]\{d\} \tag{3.9.9}$$

Premultiplying Eq. (3.9.9) by $[T_1]$, we have

$$[T_1]\{f\} = [T_1][K]\{d\} \tag{3.9.10}$$

For the truss in Figure 3–19, the left side of Eq. (3.9.10) is

$$\begin{bmatrix} [I] & [0] & [0] \\ [0] & [I] & [0] \\ [0] & [0] & [t_3] \end{bmatrix} \begin{Bmatrix} f_{1x} \\ f_{1y} \\ f_{2x} \\ f_{2y} \\ f_{3x} \\ f_{3y} \end{Bmatrix} = \begin{Bmatrix} f_{1x} \\ f_{1y} \\ f_{2x} \\ f_{2y} \\ f'_{3x} \\ f'_{3y} \end{Bmatrix} \tag{3.9.11}$$

where the fact that local forces transform similarly to Eq. (3.9.2) as

$$\{f'_3\} = [t_3]\{f_3\} \tag{3.9.12}$$

has been used in Eq. (3.9.11). From Eq. (3.9.11), we see that only the node 3 components of $\{f\}$ have been transformed to the local axes components, as desired.

Using Eq. (3.9.5) in Eq. (3.9.10), we have

$$[T_1]\{f\} = [T_1][K][T_1]^T\{d'\} \tag{3.9.13}$$

Using Eq. (3.9.11), we find that the form of Eq. (3.9.13) becomes

$$\begin{Bmatrix} F_{1x} \\ F_{1y} \\ F_{2x} \\ F_{2y} \\ F'_{3x} \\ F'_{3y} \end{Bmatrix} = [T_1][K][T_1]^T \begin{Bmatrix} d_{1x} \\ d_{1y} \\ d_{2x} \\ d_{2y} \\ d'_{3x} \\ d'_{3y} \end{Bmatrix} \tag{3.9.14}$$

as $d_{1x} = d'_{1x}, d_{1y} = d'_{1y}, d_{2x} = d'_{2x}$, and $d_{2y} = d'_{2y}$ from Eq. (3.9.7). Equation (3.9.14) is the desired form that allows all known global and inclined boundary conditions to be enforced. The global forces now result in the left side of Eq. (3.9.14). To solve Eq. (3.9.14), first perform the matrix triple product $[T_1][K][T_1]^T$. Then invoke the following boundary conditions (for the truss in Figure 3–19):

$$d_{1x} = 0 \qquad d_{1y} = 0 \qquad d'_{3y} = 0 \qquad (3.9.15)$$

Then substitute the known value of the applied force F_{2x} along with $F_{2y} = 0$ and $F'_{3x} = 0$ into Eq. (3.9.14). Finally, partition the equations with known displacements —here equations 1, 2, and 6 of Eq. (3.9.14)—and then simultaneously solve those associated with the unknown displacements d_{2x}, d_{2y}, and d'_{3x}.

After solving for the displacements, return to Eq. (3.9.14) to obtain the global reactions F_{1x} and F_{1y} and the inclined roller reaction F'_{3y}.

Example 3.9

For the plane truss shown in Figure 3–20, determine the displacements and reactions. Let $E = 210$ GPa, $A = 6.00 \times 10^{-4}$ m^2 for elements 1 and 2, and $A = 6\sqrt{2} \times 10^{-4}$ m^2 for element 3.

We begin by using Eq. (3.4.23) to determine each element stiffness matrix.

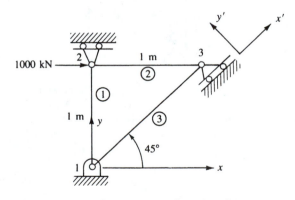

Figure 3–20 Plane truss with inclined support

Element 1

$$\cos \theta = 0 \qquad \sin \theta = 1$$

$$\underline{k}^{(1)} = \frac{(6.0 \times 10^{-4} \text{ m}^2)(210 \times 10^9 \text{ N/m}^2)}{1 \text{ m}} \begin{array}{c} \begin{array}{cccc} d_{1x} & d_{1y} & d_{2x} & d_{2y} \end{array} \\ \begin{bmatrix} 0 & 0 & 0 & 0 \\ & 1 & 0 & -1 \\ & & 0 & 0 \\ \text{Symmetry} & & & 1 \end{bmatrix} \end{array} \qquad (3.9.16)$$

Element 2

$$\cos \theta = 1 \qquad \sin \theta = 0$$

$$k^{(2)} = \frac{(6.0 \times 10^{-4} \text{ m}^2)(210 \times 10^9 \text{ N/m}^2)}{1 \text{ m}} \begin{array}{c} \begin{matrix} d_{2x} & d_{2y} & d_{3x} & d_{3y} \end{matrix} \\ \begin{bmatrix} 1 & 0 & -1 & 0 \\ & 0 & 0 & 0 \\ & & 1 & 0 \\ \text{Symmetry} & & & 0 \end{bmatrix} \end{array} \qquad (3.9.17)$$

Element 3

$$\cos \theta = \frac{\sqrt{2}}{2} \qquad \sin \theta = \frac{\sqrt{2}}{2}$$

$$k^{(3)} = \frac{(6\sqrt{2} \times 10^{-4} \text{ m}^2)(210 \times 10^9 \text{ N/m}^2)}{\sqrt{2} \text{ m}} \begin{array}{c} \begin{matrix} d_{1x} & d_{1y} & d_{3x} & d_{3y} \end{matrix} \\ \begin{bmatrix} 0.5 & 0.5 & -0.5 & -0.5 \\ & 0.5 & -0.5 & -0.5 \\ & & 0.5 & 0.5 \\ \text{Symmetry} & & & 0.5 \end{bmatrix} \end{array} \qquad (3.9.18)$$

Using the direct stiffness method on Eqs. (3.9.16)–(3.9.18), we obtain the global $\underline{K}$ matrix as

$$\underline{K} = 1260 \times 10^5 \text{ N/m} \begin{bmatrix} 0.5 & 0.5 & 0 & 0 & -0.5 & -0.5 \\ & 1.5 & 0 & -1 & -0.5 & -0.5 \\ & & 1 & 0 & -1 & 0 \\ & & & 1 & 0 & 0 \\ & & & & 1.5 & 0.5 \\ \text{Symmetry} & & & & & 0.5 \end{bmatrix} \qquad (3.9.19)$$

Next we obtain the transformation matrix $\underline{T}_1$ using Eq. (3.9.6) to transform the global displacements at node 3 into local nodal coordinates x'-y'. In using Eq. (3.9.6), the angle α is 45°.

$$[T_1] = \begin{bmatrix} 1 & 0 & 0 & 0 & 0 & 0 \\ 0 & 1 & 0 & 0 & 0 & 0 \\ 0 & 0 & 1 & 0 & 0 & 0 \\ 0 & 0 & 0 & 1 & 0 & 0 \\ 0 & 0 & 0 & 0 & \sqrt{2}/2 & \sqrt{2}/2 \\ 0 & 0 & 0 & 0 & -\sqrt{2}/2 & \sqrt{2}/2 \end{bmatrix} \qquad (3.9.20)$$

Next we use Eq. (3.9.14) (in general, we would use Eq. (3.9.13)) to express the assembled equations. First define $\underline{K}^* = \underline{T_1}\underline{K}\underline{T_1}^T$ and evaluate in steps as follows:

$$\underline{T_1}\underline{K} = 1260 \times 10^5 \begin{bmatrix} 0.5 & 0.5 & 0 & 0 & -0.5 & -0.5 \\ 0.5 & 1.5 & 0 & -1 & -0.5 & -0.5 \\ 0 & 0 & 1 & 0 & -1 & 0 \\ 0 & -1 & 0 & 1 & 0 & 0 \\ -0.707 & -0.707 & -0.707 & 0 & 1.414 & 0.707 \\ 0 & 0 & 0.707 & 0 & -0.707 & 0 \end{bmatrix} \quad (3.9.21)$$

and

$$\underline{T_1}\underline{K}\underline{T_1}^T = 1260 \times 10^5 \text{ N/m} \begin{matrix} \begin{matrix} d_{1x} & d_{1y} & d_{2x} & d_{2y} & d'_{3x} & d'_{3y} \end{matrix} \\ \begin{bmatrix} 0.5 & 0.5 & 0 & 0 & -0.707 & 0 \\ 0.5 & 1.5 & 0 & -1 & -0.707 & 0 \\ 0 & 0 & 1 & 0 & -0.707 & 0.707 \\ 0 & -1 & 0 & 1 & 0 & 0 \\ -0.707 & -0.707 & -0.707 & 0 & 1.500 & -0.500 \\ 0 & 0 & 0.707 & 0 & -0.500 & 0.500 \end{bmatrix} \end{matrix}$$
$$(3.9.22)$$

Applying the boundary conditions, $d_{1x} = d_{1y} = d_{2y} = d'_{3y} = 0$, to Eq. (3.9.22), we obtain

$$\begin{Bmatrix} F_{2x} = 1000 \text{ kN} \\ F'_{3x} = 0 \end{Bmatrix} = (126 \times 10^3 \text{ kN/m}) \begin{bmatrix} 1 & -0.707 \\ -0.707 & 1.50 \end{bmatrix} \begin{Bmatrix} d_{2x} \\ d'_{3x} \end{Bmatrix} \quad (3.9.23)$$

Solving Eq. (3.9.23) for the displacements yields

$$d_{2x} = 11.91 \times 10^{-3} \text{ m} \quad (3.9.24)$$

$$d'_{3x} = 5.613 \times 10^{-3} \text{ m}$$

Postmultiplying the known displacement vector times Eq. (3.9.22) (see Eq. (3.9.14), we obtain the reactions as

$$F_{1x} = -500 \text{ kN}$$
$$F_{1y} = -500 \text{ kN}$$
$$F_{2y} = 0 \quad (3.9.25)$$
$$F'_{3y} = 707 \text{ kN}$$

The free-body diagram of the truss with the reactions is shown in Figure 3–21. You can easily verify that the truss is in equilibrium. ∎

In the second method used to handle skewed boundary conditions, we use a boundary element of large stiffness to constrain the desired displacement. This is the method used in the Algor program.

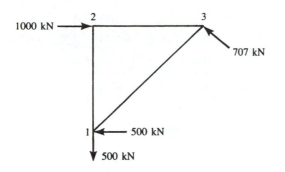

Figure 3–21 Free-body diagram of the truss of Figure 3–20

Boundary elements are used to specify nonzero displacements (as illustrated in Example 4.3) and rotations to nodes. They are also used to evaluate reactions at rigid and flexible supports. Boundary elements are two-node elements. The line defined by the two nodes specifies the direction along which the force reaction is evaluated or the displacement is specified. In the case of moment reaction, the line specifies the axis about which the moment is evaluated and the rotation is specified.

We consider boundary elements that are used to obtain reaction forces (rigid boundary elements) or specify translational displacements (displacement boundary elements) as truss elements with only one nonzero translational stiffness. Boundary elements used to either evaluate reaction moments or specify rotations behave like beam elements with only one nonzero stiffness corresponding to the rotational stiffness about the specified axis.

The elastic boundary elements are used to model flexible supports and to calculate reactions at skewed or inclined boundaries. Elastic boundary elements are described in more detail in Example 4.5, where we use the boundary element to solve a truss with a skewed support. Also consult Reference [9] for more details about using boundary elements.

▲ 3.10 Potential Energy Approach to Derive Bar Element Equations ▲

We now present the principle of minimum potential energy to derive the bar element equations. Recall from Section 2.6 that the total potential energy π_p was defined as the sum of the internal strain energy U and the potential energy of the external forces Ω:

$$\pi_p = U + \Omega \qquad (3.10.1)$$

To evaluate the strain energy for a bar, we consider only the work done by the internal forces during deformation. Because we are dealing with a one-dimensional bar, the internal force doing work is given in Figure 3–22 as $\sigma_x(\Delta y)(\Delta z)$, due only to normal stress σ_x. The displacement of the x face of the element is $\Delta x(\varepsilon_x)$; the displacement of the $x + \Delta x$ face is $\Delta x(\varepsilon_x + d\varepsilon_x)$. The change in displacement is then

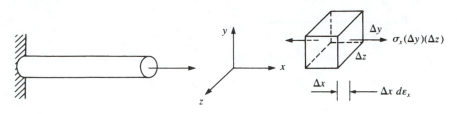

Figure 3–22   Internal force in a one-dimensional bar

$\Delta x \, d\varepsilon_x$, where $d\varepsilon_x$ is the differential change in strain occurring over length Δx. The differential internal work (or strain energy) dU is the internal force multiplied by the displacement through which the force moves, given by

$$dU = \sigma_x (\Delta y)(\Delta z)(\Delta x) \, d\varepsilon_x \tag{3.10.2}$$

Rearranging and letting the volume of the element approach zero, we obtain, from Eq. (3.10.2),

$$dU = \sigma_x \, d\varepsilon_x \, dV \tag{3.10.3}$$

For the whole bar, we then have

$$U = \iiint_V \left\{ \int_0^{\varepsilon_x} \sigma_x \, d\varepsilon_x \right\} dV \tag{3.10.4}$$

Now, for a linear-elastic (Hooke's law) material as shown in Figure 3–23, we see that $\sigma_x = E\varepsilon_x$. Hence substituting this relationship into Eq. (3.10.4), integrating with respect to ε_x, and then resubstituting σ_x for $E\varepsilon_x$, we have

$$U = \frac{1}{2} \iiint_V \sigma_x \varepsilon_x \, dV \tag{3.10.5}$$

as the expression for the strain energy for one-dimensional stress.

The potential energy of the external forces, being opposite in sign from the external work expression because the potential energy of external forces is lost when the

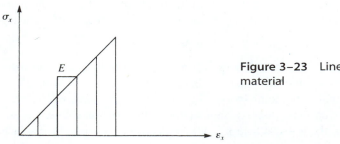

Figure 3–23   Linear-elastic (Hooke's law) material

work is done by the external forces, is given by

$$\Omega = -\iiint\limits_{V} \hat{X}_b \hat{u}\, dV - \iint\limits_{S_1} \hat{T}_x \hat{u}_s\, dS - \sum_{i=1}^{M} \hat{f}_{ix} \hat{d}_{ix} \tag{3.10.6}$$

where the first, second, and third terms on the right side of Eq. (3.10.6) represent the potential energy of (1) body forces $\hat{X}_b$, typically from the self-weight of the bar (in units of force per unit volume) moving through displacement function $\hat{u}$, (2) surface loading or traction $\hat{T}_x$, typically from distributed loading acting along the surface of the element (in units of force per unit surface area) moving through displacements $\hat{u}_s$, where $\hat{u}_s$ are the displacements occurring over surface S_1, and (3) nodal concentrated forces $\hat{f}_{ix}$ moving through nodal displacements $\hat{d}_{ix}$. The forces $\hat{X}_b$, $\hat{T}_x$, and $\hat{f}_{ix}$ are considered to act in the local $\hat{x}$ direction of the bar as shown in Figure 3–24. In Eqs. (3.10.5) and (3.10.6), V is the volume of the body and S_1 is the part of the surface S on which surface loading acts. For a bar element with two nodes and one degree of freedom per node, $M = 2$.

We are now ready to describe the finite element formulation of the bar element equations by using the principle of minimum potential energy.

The finite element process seeks a minimum in the potential energy within the constraint of an assumed displacement pattern within each element. The greater the number of degrees of freedom associated with the element (usually meaning increasing the number of nodes), the more closely will the solution approximate the true one and ensure complete equilibrium (provided the true displacement can, in the limit, be approximated). An approximate finite element solution found by using the stiffness method will always provide an approximate value of potential energy greater than or equal to the correct one. This method also results in a structure behavior that is predicted to be physically stiffer than, or at best to have the same stiffness as, the actual one. This is explained by the fact that the structure model is allowed to displace only into shapes defined by the terms of the assumed displacement field within each element of the structure. The correct shape is usually only approximated by the assumed field, although the correct shape can be the same as the assumed field. The assumed field effectively constrains the structure from deforming in its natural manner. This constraint effect stiffens the predicted behavior of the structure.

Apply the following steps when using the principle of minimum potential energy to derive the finite element equations.

1. Formulate an expression for the total potential energy.
2. Assume the displacement pattern to vary with a finite set of undetermined parameters (here these are the nodal displacements d_{ix}), which are substituted into the expression for total potential energy.
3. Obtain a set of simultaneous equations minimizing the total potential energy with respect to these nodal parameters. These resulting equations represent the element equations.

The resulting equations are the approximate (or possibly exact) equilibrium equations whose solution for the nodal parameters seeks to minimize the potential energy when back-substituted into the potential energy expression. The preceding

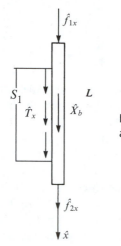

Figure 3–24 General forces acting on a one-dimensional bar

three steps will now be followed to derive the bar element equations and stiffness matrix.

Consider the bar element of length L, with constant cross-sectional area A, shown in Figure 3–24. Using Eqs. (3.10.5) and (3.10.6), we find that the total potential energy, Eq. (3.10.1), becomes

$$\pi_p = \frac{A}{2} \int_0^L \sigma_x \varepsilon_x \, d\hat{x} - \hat{f}_{1x}\hat{d}_{1x} - \hat{f}_{2x}\hat{d}_{2x} - \iint_{S_1} \hat{u}_s \hat{T}_x \, dS - \iiint_V \hat{u} \hat{X}_b \, dV \qquad (3.10.7)$$

because A is a constant and variables σ_x and ε_x at most vary with $\hat{x}$.

From Eqs. (3.1.3) and (3.1.4), we have the axial displacement function expressed in terms of the shape functions and nodal displacements by

$$\hat{u} = [N]\{\hat{d}\} \qquad \hat{u}_s = [N_S]\{\hat{d}\} \qquad (3.10.8)$$

where

$$[N] = \left[1 - \frac{\hat{x}}{L} \quad \frac{\hat{x}}{L} \right] \qquad (3.10.9)$$

$[N_S]$ is the shape function matrix evaluated over the surface that the distributed surface traction acts and

$$\{\hat{d}\} = \begin{Bmatrix} \hat{d}_{1x} \\ \hat{d}_{2x} \end{Bmatrix} \qquad (3.10.10)$$

Then, using the strain/displacement relationship $\varepsilon_x = d\hat{u}/d\hat{x}$, we can write the axial strain as

$$\{\varepsilon_x\} = \left[-\frac{1}{L} \quad \frac{1}{L} \right]\{\hat{d}\} \qquad (3.10.11)$$

or

$$\{\varepsilon_x\} = [B]\{\hat{d}\} \qquad (3.10.12)$$

where we define

$$[B] = \left[-\frac{1}{L} \quad \frac{1}{L} \right] \tag{3.10.13}$$

The axial stress/strain relationship is given by

$$\{\sigma_x\} = [D]\{\varepsilon_x\} \tag{3.10.14}$$

where

$$[D] = [E] \tag{3.10.15}$$

for the one-dimensional stress/strain relationship and E is the modulus of elasticity. Now, by Eq. (3.10.12), we can express Eq. (3.10.14) as

$$\{\sigma_x\} = [D][B]\{\hat{d}\} \tag{3.10.16}$$

Using Eq. (3.10.7) expressed in matrix notation form, we have the total potential energy given by

$$\pi_p = \frac{A}{2} \int_0^L \{\sigma_x\}^T \{\varepsilon_x\} \, d\hat{x} - \{\hat{d}\}^T \{P\} - \iint_{S_1} \{\hat{u}_s\}^T \{\hat{T}_x\} \, dS - \iiint_V \{\hat{u}\}^T \{\hat{X}_b\} \, dV \tag{3.10.17}$$

where $\{P\}$ now represents the concentrated nodal loads and where in general both σ_x and ε_x are column matrices. For proper matrix multiplication, we must place the transpose on $\{\sigma_x\}$. Similarly, $\{\hat{u}\}$ and $\{\hat{T}_x\}$ in general are column matrices, so for proper matrix multiplication, $\{\hat{u}\}$ is transposed in Eq. (3.10.17).

Using Eqs. (3.10.8), (3.10.12), and (3.10.16) in Eq. (3.10.17), we obtain

$$\pi_p = \frac{A}{2} \int_0^L \{\hat{d}\}^T [B]^T [D]^T [B]\{\hat{d}\} \, d\hat{x} - \{\hat{d}\}^T \{P\}$$
$$- \iint_{S_1} \{\hat{d}\}^T [N_S]^T \{\hat{T}_x\} \, dS - \iiint_V \{\hat{d}\}^T [N]^T \{\hat{X}_b\} \, dV \tag{3.10.18}$$

In Eq. (3.10.18), π_p is seen to be a function of $\{\hat{d}\}$; that is, $\pi_p = \pi_p(\hat{d}_{1x}, \hat{d}_{2x})$. However, $[B]$ and $[D]$, Eqs. (3.10.13) and (3.10.15), and the nodal degrees of freedom $\hat{d}_{1x}$ and $\hat{d}_{2x}$ are not functions of $\hat{x}$. Therefore, integrating Eq. (3.10.18) with respect to $\hat{x}$ yields

$$\pi_p = \frac{AL}{2} \{\hat{d}\}^T [B]^T [D]^T [B]\{\hat{d}\} - \{\hat{d}\}^T \{\hat{f}\} \tag{3.10.19}$$

where

$$\{\hat{f}\} = \{P\} + \iint_{S_1} [N_S]^T \{\hat{T}_x\} \, dS + \iiint_V [N]^T \{\hat{X}_b\} \, dV \tag{3.10.20}$$

From Eq. (3.10.20), we observe three separate types of load contributions from concentrated nodal forces, surface tractions, and body forces, respectively. We define these surface tractions and body-force matrices as

$$\{\hat{f}_s\} = \iint\limits_{S_1} [N_S]^T \{\hat{T}_x\}\, dS \qquad (3.10.20a)$$

$$\{\hat{f}_b\} = \iiint\limits_{V} [N]^T \{\hat{X}_b\}\, dV \qquad (3.10.20b)$$

The expression for $\{\hat{f}\}$ given by Eq. (3.10.20) then describes how certain loads can be considered to best advantage.

Loads calculated by Eqs. (3.10.20a) and (3.10.20b) are called consistent because they are based on the same shape functions $[N]$ used to calculate the element stiffness matrix. The loads calculated by Eq. (3.10.20a) and (3.10.20b) are also statically equivalent to the original loading; that is, both $\{\hat{f}_s\}$ and $\{\hat{f}_b\}$ and the original loads yield the same resultant force and same moment about an arbitrarily chosen point.

The minimization of π_p with respect to each nodal displacement requires that

$$\frac{\partial \pi_p}{\partial \hat{d}_{1x}} = 0 \qquad \text{and} \qquad \frac{\partial \pi_p}{\partial \hat{d}_{2x}} = 0 \qquad (3.10.21)$$

Now we explicitly evaluate π_p given by Eq. (3.10.19) to apply Eq. (3.10.21). We define the following for convenience:

$$\{U^*\} = \{\hat{d}\}^T [B]^T [D]^T [B]\{\hat{d}\} \qquad (3.10.22)$$

Using Eqs. (3.10.10), (3.10.13), and (3.10.15) in Eq. (3.10.22) yields

$$\{U^*\} = [\hat{d}_{1x} \quad \hat{d}_{2x}] \left\{ \begin{matrix} -\dfrac{1}{L} \\ \dfrac{1}{L} \end{matrix} \right\} [E] \left[-\dfrac{1}{L} \quad \dfrac{1}{L} \right] \left\{ \begin{matrix} \hat{d}_{1x} \\ \hat{d}_{2x} \end{matrix} \right\} \qquad (3.10.23)$$

Simplifying Eq. (3.10.23), we obtain

$$U^* = \frac{E}{L^2}(\hat{d}_{1x}^2 - 2\hat{d}_{1x}\hat{d}_{2x} + \hat{d}_{2x}^2) \qquad (3.10.24)$$

Also, the explicit expression for $\{\hat{d}\}^T\{\hat{f}\}$ is

$$\{\hat{d}\}^T\{\hat{f}\} = \hat{d}_{1x}\hat{f}_{1x} + \hat{d}_{2x}\hat{f}_{2x} \qquad (3.10.25)$$

Therefore, using Eqs. (3.10.24) and (3.10.25) in Eq. (3.10.19) and then applying Eqs. (3.10.21), we obtain

$$\frac{\partial \pi_p}{\partial \hat{d}_{1x}} = \frac{AL}{2}\left[\frac{E}{L^2}(2\hat{d}_{1x} - 2\hat{d}_{2x})\right] - \hat{f}_{1x} = 0 \qquad (3.10.26)$$

and
$$\frac{\partial \pi_p}{\partial \hat{d}_{2x}} = \frac{AL}{2}\left[\frac{E}{L^2}(-2\hat{d}_{1x} + 2\hat{d}_{2x})\right] - \hat{f}_{2x} = 0$$

In matrix form, we express Eqs. (3.10.26) as

$$\frac{\partial \pi_p}{\partial \{\hat{d}\}} = \frac{AE}{L}\begin{bmatrix} 1 & -1 \\ -1 & 1 \end{bmatrix}\begin{Bmatrix} \hat{d}_{1x} \\ \hat{d}_{2x} \end{Bmatrix} - \begin{Bmatrix} \hat{f}_{1x} \\ \hat{f}_{2x} \end{Bmatrix} = \begin{Bmatrix} 0 \\ 0 \end{Bmatrix} \tag{3.10.27}$$

or, because $\{\hat{f}\} = [\hat{k}]\{\hat{d}\}$, we have the stiffness matrix for the bar element obtained from Eq. (3.10.27) as

$$[\hat{k}] = \frac{AE}{L}\begin{bmatrix} 1 & -1 \\ -1 & 1 \end{bmatrix} \tag{3.10.28}$$

As expected, Eq. (3.10.28) is identical to the stiffness matrix obtained in Section 3.1.

Finally, instead of the cumbersome process of explicitly evaluating π_p, we can use the matrix differentiation as given by Eq. (2.6.12) and apply it directly to Eq. (3.10.19) to obtain

$$\frac{\partial \pi_p}{\partial \{\hat{d}\}} = AL[B]^T[D][B]\{\hat{d}\} - \{\hat{f}\} = 0 \tag{3.10.29}$$

where $[D]^T = [D]$ has been used in writing Eq. (3.10.29). The result of the evaluation of $AL[B]^T[D][B]$ is then equal to $[\hat{k}]$ given by Eq. (3.10.28). Throughout this text, we will use this matrix differentiation concept (also see Appendix A), which greatly simplifies the task of evaluating $[\hat{k}]$.

To illustrate the use of Eq. (3.10.20a) to evaluate the equivalent nodal loads for a bar subjected to axial loading traction $\hat{T}_x$, we now solve Example 3.10.

Example 3.10

A bar of length L is subjected to a linearly distributed axial loading that varies from zero at node 1 to a maximum at node 2 (Figure 3–25). Determine the energy equivalent nodal loads.

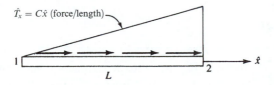

$\hat{T}_x = C\hat{x}$ (force/length)

Figure 3–25 Element subjected to linearly varying axial load

Using Eq. (3.10.20a) and shape functions from Eq. (3.10.9), we solve for the energy equivalent nodal forces of the distributed loading as follows:

$$\{\hat{f}_0\} = \begin{Bmatrix} \hat{f}_{1x} \\ \hat{f}_{2x} \end{Bmatrix} = \iint_{S_1} [N]^T \{\hat{T}_x\}\, dS = \int_0^L \begin{Bmatrix} 1 - \dfrac{\hat{x}}{L} \\[2mm] \dfrac{\hat{x}}{L} \end{Bmatrix} \{C\hat{x}\}\, d\hat{x} \qquad (3.10.30)$$

$$= \begin{Bmatrix} \dfrac{C\hat{x}^2}{2} - \dfrac{C\hat{x}^3}{3L} \\[3mm] \dfrac{C\hat{x}^3}{3L} \end{Bmatrix}_0^L$$

$$= \begin{Bmatrix} \dfrac{CL^2}{6} \\[3mm] \dfrac{CL^2}{3} \end{Bmatrix} \qquad\qquad (3.10.31)$$

where the integration was carried out over the length of the bar, because $\hat{T}_x$ is in units of force/length.

Note that the total load is the area under the load distribution given by

$$F = \frac{1}{2}(L)(CL) = \frac{CL^2}{2} \qquad (3.10.32)$$

Therefore, comparing Eq. (3.10.31) with (3.10.32), we find that the equivalent nodal loads for a linearly varying load are

$$\hat{f}_{1x} = \tfrac{1}{3}F = \text{one-third of the total load}$$
$$\hat{f}_{2x} = \tfrac{2}{3}F = \text{two-thirds of the total load} \qquad (3.10.33)$$

In summary, for the simple two-noded bar element subjected to a linearly varying load (triangular loading), place one-third of the total load at the node where the distributed loading begins (zero end of the load) and two-thirds of the total load at the node where the peak value of the distributed load ends.   ■

We now illustrate (Example 3.11) a complete solution for a bar subjected to a surface traction loading.

Example 3.11

For the rod loaded axially as shown in Figure 3–26, determine the axial displacement and axial stress. Let $E = 30 \times 10^6$ psi, $A = 2$ in.^{2}, and $L = 60$ in. Use (a) one and (b) two elements in the finite element solutions. (In Section 3.11 one-, two-, four-, and eight-element solutions will be presented from the computer program Algor. (The use of Algor for solving problems with axially loaded members only is described in Section 4.2.)

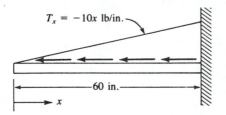

Figure 3–26   Rod subjected to triangular load distribution

(a) One-element solution (Figure 3–27).

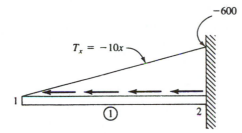

Figure 3–27   One-element model

From Eq. (3.10.20a), the distributed load matrix is evaluated as follows:

$$\{F_0\} = \int_0^L [N]^T \{T_x\} \, dx \tag{3.10.34}$$

where T_x is a line load in units of pounds per inch and $\hat{f}_0 = F_0$ as $\underline{x} = \hat{x}$. Therefore, using Eq. (3.1.4) for $[N]$ in Eq. (3.10.34), we obtain

$$\{F_0\} = \int_0^L \left\{ \begin{array}{c} 1 - \dfrac{x}{L} \\[2mm] \dfrac{x}{L} \end{array} \right\} \{-10x\} \, dx \tag{3.10.35}$$

or
$$\left\{ \begin{array}{c} F_{1x} \\ F_{2x} \end{array} \right\} = \left\{ \begin{array}{c} \dfrac{-10L^2}{2} + \dfrac{10L^2}{3} \\[3mm] \dfrac{-10L^2}{3} \end{array} \right\} = \left\{ \begin{array}{c} \dfrac{-10L^2}{6} \\[3mm] \dfrac{-10L^2}{3} \end{array} \right\} = \left\{ \begin{array}{c} \dfrac{-10(60)^2}{6} \\[3mm] \dfrac{-10(60)^2}{3} \end{array} \right\}$$

or
$$F_{1x} = -6000 \text{ lb} \qquad F_{2x} = -12{,}000 \text{ lb} \tag{3.10.36}$$

Using Eq. (3.10.33), we could have determined the same forces at nodes 1 and 2—that is, one-third of the total load is at node 1 and two-thirds of the total load is at node 2.

Using Eq. (3.10.28), we find that the stiffness matrix is given by

$$k^{(1)} = 10^6 \begin{bmatrix} 1 & -1 \\ -1 & 1 \end{bmatrix}$$

The element equations are then

$$10^6 \begin{bmatrix} 1 & -1 \\ -1 & 1 \end{bmatrix} \begin{Bmatrix} d_{1x} \\ 0 \end{Bmatrix} = \begin{Bmatrix} -6000 \\ R_{2x} - 12,000 \end{Bmatrix} \tag{3.10.37}$$

Solving Eq. 1 of Eq. (3.10.37), we obtain

$$d_{1x} = -0.006 \text{ in.} \tag{3.10.38}$$

The stress is obtained from Eq. (3.10.14) as

$$\begin{aligned}
\{\sigma_x\} &= [D]\{\varepsilon_x\} \\
&= E[B]\{d\} \\
&= E\begin{bmatrix} -\dfrac{1}{L} & \dfrac{1}{L} \end{bmatrix} \begin{Bmatrix} d_{1x} \\ d_{2x} \end{Bmatrix} \\
&= E\left(\dfrac{d_{2x} - d_{1x}}{L}\right) \\
&= 30 \times 10^6 \left(\dfrac{0 + 0.006}{60}\right) \\
&= 3000 \text{ psi } (T) \tag{3.10.39}
\end{aligned}$$

(b) Two-element solution (Figure 3–28).

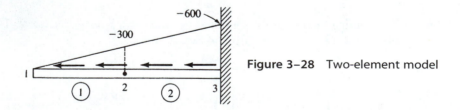

Figure 3–28 Two-element model

We first obtain the element forces. For element 2, we divide the load into a uniform part and a triangular part. For the uniform part, half the total uniform load is placed at each node associated with the element. Therefore, the total uniform part is

$$(30 \text{ in.})(-300 \text{ lb/in.}) = -9000 \text{ lb}$$

and using Eq. (3.10.33) for the triangular part of the load, we have, for element 2,

$$\begin{Bmatrix} f_{2x}^{(2)} \\ f_{3x}^{(2)} \end{Bmatrix} = \begin{Bmatrix} -[\frac{1}{2}(9000) + \frac{1}{3}(4500)] \\ -[\frac{1}{2}(9000) + \frac{2}{3}(4500)] \end{Bmatrix} = \begin{Bmatrix} -6000 \text{ lb} \\ -7500 \text{ lb} \end{Bmatrix} \tag{3.10.40}$$

For element 1, the total force is from the triangle-shaped distributed load only and is given by

$$\tfrac{1}{2}(30 \text{ in.})(-300 \text{ lb/in.}) = -4500 \text{ lb}$$

On the basis of Eq. (3.10.33), this load is separated into nodal forces as shown:

$$\begin{Bmatrix} f_{1x}^{(1)} \\ f_{2x}^{(1)} \end{Bmatrix} = \begin{Bmatrix} \tfrac{1}{3}(-4500) \\ \tfrac{2}{3}(-4500) \end{Bmatrix} = \begin{Bmatrix} -1500 \text{ lb} \\ -3000 \text{ lb} \end{Bmatrix} \tag{3.10.41}$$

The final nodal force matrix is then

$$\begin{Bmatrix} F_{1x} \\ F_{2x} \\ F_{3x} \end{Bmatrix} = \begin{Bmatrix} -1500 \\ -6000 - 3000 \\ R_{3x} - 7500 \end{Bmatrix} \tag{3.10.42}$$

The element stiffness matrices are now

$$k^{(1)} = k^{(2)} = \frac{AE}{L/2} \begin{array}{cc} 1 & 2 \\ 2 & 3 \end{array} \begin{bmatrix} 1 & -1 \\ -1 & 1 \end{bmatrix} = (2 \times 10^6) \begin{array}{cc} 1 & 2 \\ 2 & 3 \end{array} \begin{bmatrix} 1 & -1 \\ -1 & 1 \end{bmatrix} \tag{3.10.43}$$

The assembled global stiffness matrix is

$$K = (2 \times 10^6) \begin{bmatrix} 1 & -1 & 0 \\ -1 & 2 & -1 \\ 0 & -1 & 1 \end{bmatrix} \frac{\text{lb}}{\text{in.}} \tag{3.10.44}$$

The assembled global equations are then

$$(2 \times 10^6) \begin{bmatrix} 1 & -1 & 0 \\ -1 & 2 & -1 \\ 0 & -1 & 1 \end{bmatrix} \begin{Bmatrix} d_{1x} \\ d_{2x} \\ d_{3x} = 0 \end{Bmatrix} = \begin{Bmatrix} -1500 \\ -9000 \\ R_{3x} - 7500 \end{Bmatrix} \tag{3.10.45}$$

where the boundary condition $d_{3x} = 0$ has been substituted into Eq. (3.10.45). Now, solving equations 1 and 2 of Eq. (3.10.45), we obtain

$$d_{1x} = -0.006 \text{ in.}$$
$$d_{2x} = -0.00525 \text{ in.} \tag{3.10.46}$$

The element stresses are as follows:

Element 1

$$\sigma_x = E \begin{bmatrix} -\dfrac{1}{30} & \dfrac{1}{30} \end{bmatrix} \begin{Bmatrix} d_{1x} = -0.006 \\ d_{2x} = -0.00525 \end{Bmatrix}$$

$$= 750 \text{ psi } (T) \tag{3.10.47}$$

Element 2

$$\sigma_x = E\left[-\frac{1}{30} \quad \frac{1}{30}\right]\left\{\begin{array}{l} d_{2x} = -0.00525 \\ d_{3x} = 0 \end{array}\right\}$$

$$= 5250 \text{ psi } (T) \tag{3.10.48}$$

■

▲　## 3.11 Comparison of Finite Element Solution to Exact Solution for Bar　▲

We will now compare the finite element solutions for Example 3.11 using one, two, four, and eight elements to model the bar element and the exact solution. The exact solution for displacement is obtained by solving the equation

$$\delta = \frac{1}{AE}\int_0^x P(x)\, dx \tag{3.11.1}$$

where, using the following free-body diagram,

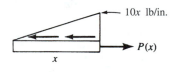

we have

$$P(x) = \tfrac{1}{2}x(10x) = 5x^2 \text{ lb} \tag{3.11.2}$$

Therefore, substituting Eq. (3.11.2) into Eq. (3.11.1), we have

$$\delta = \frac{1}{AE}\int_0^x 5x^2\, dx$$

$$= \frac{5x^3}{3AE} + C_1 \tag{3.11.3}$$

Now, applying the boundary condition at $x = L$, we obtain

$$\delta(L) = 0 = \frac{5L^3}{3AE} + C_1$$

or

$$C_1 = -\frac{5L^3}{3AE} \tag{3.11.4}$$

Substituting Eq. (3.11.4) into Eq. (3.11.3) makes the final expression for displacement

$$\delta = \frac{5}{3AE}(x^3 - L^3) \tag{3.11.5}$$

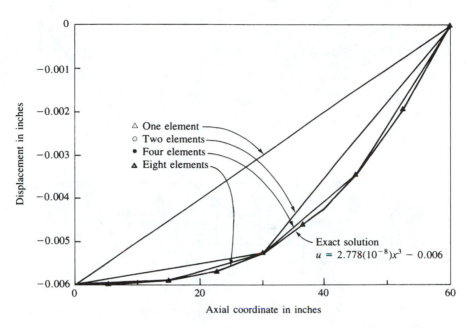

Figure 3–29 Comparison of exact and finite element solutions for axial displacement (along length of bar)

The exact solution for axial stress is obtained by solving the equation

$$\sigma(x) = \frac{P(x)}{A} = \frac{5x^2}{2 \text{ in}^2} = 2.5x^2 \text{ psi} \tag{3.11.6}$$

Figure 3–29 shows a plot of Eq. (3.11.5) along with the finite element solutions (part of which were obtained in Example 3.11). Some conclusions from these results follow.

1. The finite element solutions match the exact solution at the node points. The reason why these nodal values are correct is that the element nodal forces were calculated on the basis of being energy-equivalent to the distributed load based on the assumed linear displacement field within each element. (For uniform cross-sectional bars and beams, the nodal degrees of freedom are exact. In general, computed nodal degrees of freedom are not exact.)
2. Although the node values for displacement match the exact solution, the values at locations between the nodes are poor using few elements (see one- and two-element solutions) because we used a linear displacement function within each element, whereas the exact solution, Eq. (3.11.5), is a cubic function. However, because we use increasing numbers of elements, the finite element solution converges to the exact solution (see the four- and eight-element solutions in Figure 3–29).

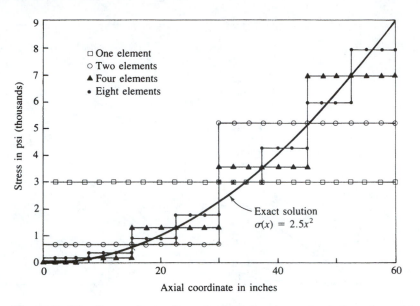

Figure 3-30 Comparison of exact and finite element solutions for axial stress (along length of bar)

3. The stress is derived from the slope of the displacement curve as $\sigma = E\varepsilon = E(du/dx)$. Therefore, by the finite element solution, because u is a linear function in each element, axial stress is constant in each element. It then takes even more elements to model the first derivative of the displacement function or, equivalently, the axial stress. This is shown in Figure 3-30, where the best results occur for the eight-element solution.

4. The best approximation of the stress occurs at the midpoint of the element, not at the nodes (Figure 3-30). This is because the derivative of displacement is better predicted between the nodes than at the nodes.

5. The stress is not continuous across element boundaries. Therefore, equilibrium is not satisfied across element boundaries. Also, equilibrium within each element is, in general, not satisfied. This is shown in Figure 3-31 for element 1 in the two-element solution and element 1 in the eight-element solution [in the eight-element solution the forces are obtained from the Algor computer code (Chapter 4) solution]. As the number of elements used increases, the discontinuity in the stress decreases across element boundaries, and the approximation of equilibrium improves.

Finally, in Figure 3-32, we show the convergence of axial stress at the fixed end $(x = L)$ as the number of elements increases.

However, if we formulate the problem in a customary general way, as described in detail in Chapter 5 for beams subjected to distributed loading, we can obtain the

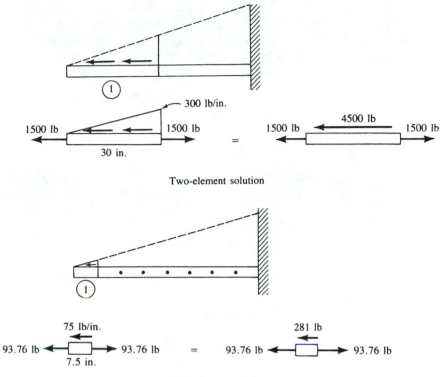

Two-element solution

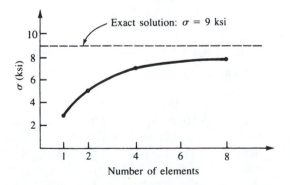

Eight-element solution

Figure 3–31 Free-body diagram of element 1 in both two- and eight-element models, showing that equilibrium is not satisfied

Figure 3–32 Axial stress at fixed end as number of elements increases

exact stress distribution with any of the models used. That is, letting $\hat{f} = \hat{k}\hat{d} - \hat{f}_0$, where $\hat{f}_0$ is the initial nodal replacement force system of the distributed load on each element, we subtract the initial replacement force system from the $\hat{k}\hat{d}$ result. This yields the nodal forces in each element. For example, considering element 1 of the two-element model, we have [see also Eqs. (3.10.33) and (3.10.41)]

$$\hat{f}_0 = \left\{ \begin{array}{c} -1500 \text{ lb} \\ -3000 \text{ lb} \end{array} \right\}$$

Using $\hat{f} = \hat{k}\hat{d} - \hat{f}_0$, we obtain

$$\hat{f} = \frac{2(30 \times 10^6)}{(30 \text{ in.})} \begin{bmatrix} 1 & -1 \\ -1 & 1 \end{bmatrix} \left\{ \begin{array}{c} -0.006 \text{ in.} \\ -0.00525 \text{ in.} \end{array} \right\} - \left\{ \begin{array}{c} -1500 \text{ lb} \\ -3000 \text{ lb} \end{array} \right\}$$

$$= \left\{ \begin{array}{c} -1500 + 1500 \\ 1500 + 3000 \end{array} \right\} = \left\{ \begin{array}{c} 0 \\ 4500 \end{array} \right\}$$

as the actual nodal forces. Drawing a free-body diagram of element 1, we have

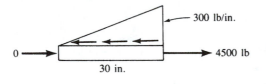

$$\sum F_x = 0: \quad -\tfrac{1}{2}(300 \text{ lb/in.})(30 \text{ in.}) + 4500 \text{ lb} = 0$$

For other kinds of elements (other than beams), this adjustment is ignored in practice. The adjustment is less important for plane and solid elements than for beams. Also, these adjustments are more difficult to formulate for an element of general shape.

▲ 3.12 Galerkin's Residual Method and Its Application to a One-Dimensional Bar

General Formulation

We developed the bar finite element equations by the direct method in Section 3.1 and by the potential energy method (one of a number of variational methods) in Section 3.10. In fields other than structural/solid mechanics, it is quite probable that a variational principle, analogous to the principle of minimum potential energy, for instance, may not be known or even exist. In some flow problems in fluid mechanics and in mass transport problems (Chapter 13), we often have only the differential equation and boundary conditions available. However, the finite element method can still be applied.

The methods of weighted residuals applied directly to the differential equation can be used to develop the finite element equations. In this section, we describe

Galerkin's residual method in general and then apply it to the bar element. This development provides the basis for later applications of Galerkin's method to the beam element in Chapter 5 and to the nonstructural heat-transfer element (specifically, the one-dimensional combined conduction, convection, and mass transport element described in Chapter 13). Because of the mass transport phenomena, the variational formulation is not known (or certainly is difficult to obtain), so Galerkin's method is necessarily applied to develop the finite element equations.

There are a number of other residual methods. Among them are collocation, least squares, and least squares collocation. (For more on these methods, see Reference [5].) However, because Galerkin's method is better known than the other residual methods, it is the only one described in this text.

In weighted residual methods, a trial or approximate function is chosen to approximate the independent variable, such as a displacement or a temperature, in a problem defined by a differential equation. This trial function will not, in general, satisfy the governing differential equation. Thus substituting the trial function into the differential equation results in a residual over the whole region of the problem as follows:

$$\iiint_V R \, dV = \text{minimum} \tag{3.12.1}$$

In the residual method, we require that a weighted value of the residual be a minimum over the whole region. The weighting functions allow the weighted integral of residuals to go to zero. If we denote the weighting function by W, the general form of the weighted residual integral is

$$\iiint_V RW \, dV = 0 \tag{3.12.2}$$

Using Galerkin's method, we choose the interpolation function, such as Eq. (3.1.3), in terms of N_i shape functions for the independent variable in the differential equation. In general, this substitution yields the residual $R \neq 0$. By the Galerkin criterion, the shape functions N_i are chosen to play the role of the weighting functions W. Thus for each i, we have

$$\iiint_V R N_i \, dV = 0 \qquad (i = 1, 2, \ldots, n) \tag{3.12.3}$$

Equation (3.12.3) results in a total of n equations. Equation (3.12.3) applies to points within the region of a body without reference to boundary conditions such as specified applied loads or displacements. To obtain boundary conditions, we apply integration by parts to Eq. (3.12.3), which yields integrals applicable for the region and its boundary.

Bar Element Formulation

We now illustrate Galerkin's method to formulate the bar element stiffness equations. We begin with the basic differential equation, without distributed load, derived in

Section 3.1 as

$$\frac{d}{d\hat{x}}\left(AE\frac{d\hat{u}}{d\hat{x}}\right) = 0 \tag{3.12.4}$$

where constants A and E are now assumed. The residual R is now defined to be Eq. (3.12.4). Applying Galerkin's criterion [Eq. (3.12.3)] to Eq. (3.12.4), we have

$$\int_0^L \frac{d}{d\hat{x}}\left(AE\frac{d\hat{u}}{d\hat{x}}\right)N_i \, d\hat{x} = 0 \qquad (i = 1, 2) \tag{3.12.5}$$

We now apply integration by parts to Eq. (3.12.5). Integration by parts is given in general by

$$\int u \, dv = uv - \int v \, du \tag{3.12.6}$$

where u and v are simply variables in the general equation. Letting

$$u = N_i \qquad du = \frac{dN_i}{d\hat{x}} \, d\hat{x}$$

$$dv = \frac{d}{d\hat{x}}\left(AE\frac{d\hat{u}}{d\hat{x}}\right) d\hat{x} \qquad v = AE\frac{d\hat{u}}{d\hat{x}} \tag{3.12.7}$$

in Eq. (3.12.5) and integrating by parts according to Eq. (3.12.6), we find that Eq. (3.12.5) becomes

$$\left(N_i AE\frac{d\hat{u}}{d\hat{x}}\right)\Bigg|_0^L - \int_0^L AE\frac{d\hat{u}}{d\hat{x}}\frac{dN_i}{d\hat{x}} \, d\hat{x} = 0 \tag{3.12.8}$$

where the integration by parts introduces the boundary conditions.

Recall that, because $\hat{u} = [N]\{\hat{d}\}$, we have

$$\frac{d\hat{u}}{d\hat{x}} = \frac{dN_1}{d\hat{x}}\hat{d}_{1x} + \frac{dN_2}{d\hat{x}}\hat{d}_{2x} \tag{3.12.9}$$

or, when Eqs. (3.1.4) are used for $N_1 = 1 - \hat{x}/L$ and $N_2 = \hat{x}/L$,

$$\frac{d\hat{u}}{d\hat{x}} = \left[-\frac{1}{L} \quad \frac{1}{L}\right]\begin{Bmatrix} \hat{d}_{1x} \\ \hat{d}_{2x} \end{Bmatrix} \tag{3.12.10}$$

Using Eq. (3.12.10) in Eq. (3.12.8), we then express Eq. (3.12.8) as

$$AE\int_0^L \frac{dN_i}{d\hat{x}}\left[-\frac{1}{L} \quad \frac{1}{L}\right]d\hat{x}\begin{Bmatrix} \hat{d}_{1x} \\ \hat{d}_{2x} \end{Bmatrix} = \left(N_i AE\frac{d\hat{u}}{d\hat{x}}\right)\Bigg|_0^L \qquad (i = 1, 2) \tag{3.12.11}$$

Equation (3.12.11) is really two equations (one for $N_i = N_1$ and one for $N_i = N_2$). First, using the weighting function $N_i = N_1$, we have

$$AE\int_0^L \frac{dN_1}{d\hat{x}}\left[-\frac{1}{L} \quad \frac{1}{L}\right]d\hat{x}\begin{Bmatrix} \hat{d}_{1x} \\ \hat{d}_{2x} \end{Bmatrix} = \left(N_1 AE\frac{d\hat{u}}{d\hat{x}}\right)\Bigg|_0^L \tag{3.12.12}$$

Substituting for $dN_1/d\hat{x}$, we obtain

$$AE \int_0^L \begin{bmatrix} -\dfrac{1}{L} \end{bmatrix} \begin{bmatrix} -\dfrac{1}{L} & \dfrac{1}{L} \end{bmatrix} d\hat{x} \begin{Bmatrix} \hat{d}_{1x} \\ \hat{d}_{2x} \end{Bmatrix} = \hat{f}_{1x} \tag{3.12.13}$$

where $\hat{f}_{1x} = AE(d\hat{u}/d\hat{x})$ because $N_1 = 1$ at $x = 0$ and $N_1 = 0$ at $x = L$. Evaluating Eq. (3.12.13) yields

$$\frac{AE}{L}(\hat{d}_{1x} - \hat{d}_{2x}) = \hat{f}_{1x} \tag{3.12.14}$$

Similarly, using $N_i = N_2$, we obtain

$$AE \int_0^L \begin{bmatrix} \dfrac{1}{L} \end{bmatrix} \begin{bmatrix} -\dfrac{1}{L} & \dfrac{1}{L} \end{bmatrix} d\hat{x} \begin{Bmatrix} \hat{d}_{1x} \\ \hat{d}_{2x} \end{Bmatrix} = \left(N_2 AE \frac{d\hat{u}}{d\hat{x}} \right)\Bigg|_0^L \tag{3.12.15}$$

Simplifying Eq. (3.12.15) yields

$$\frac{AE}{L}(\hat{d}_{2x} - \hat{d}_{1x}) = \hat{f}_{2x} \tag{3.12.16}$$

where $\hat{f}_{2x} = AE(d\hat{u}/d\hat{x})$ because $N_2 = 1$ at $x = L$ and $N_2 = 0$ at $x = 0$. Equations (3.12.14) and (3.12.16) are then seen to be the same as Eqs. (3.1.13) and (3.10.27) derived, respectively, by the direct and the variational method.

▲ References

[1] Turner, M. J., Clough, R. W., Martin, H. C., and Topp, L. J., "Stiffness and Deflection Analysis of Complex Structures," *Journal of the Aeronautical Sciences*, Vol. 23, No. 9, Sept. 1956, pp. 805–824.

[2] Martin, H. C., "Plane Elasticity Problems and the Direct Stiffness Method," *The Trend in Engineering*, Vol. 13, Jan. 1961, pp. 5–19.

[3] Melosh, R. J., "Basis for Derivation of Matrices for the Direct Stiffness Method," *Journal of the American Institute of Aeronautics and Astronautics*, Vol. 1, No. 7, July 1963, pp. 1631–1637.

[4] Oden, J. T., and Ripperger, E. A., *Mechanics of Elastic Structures*, 2nd ed., McGraw-Hill, New York, 1981.

[5] Finlayson, B. A., *The Method of Weighted Residuals and Variational Principles*, Academic Press, New York, 1972.

[6] Zienkiewicz, O. C., *The Finite Element Method*, 3rd ed., McGraw-Hill, London, 1977.

[7] Cook, R. D., Malkus, D. S., and Plesha, M. E., *Concepts and Applications of Finite Element Analysis*, 3rd ed., Wiley, New York, 1989.

[8] Forray, M. J., *Variational Calculus in Science and Engineering*, McGraw-Hill, New York, 1968.

[9] Linear Stress and Dynamics Reference Division, Docutech On-Line Documentation, Algor Interactive Systems, Pittsburgh, PA.

▲ Problems

3.1 **a.** Compute the total stiffness matrix $\underline{K}$ of the assemblage shown in Figure P3–1 by superimposing the stiffness matrices of the individual bars. Note that $\underline{K}$ should be in terms of $A_1, A_2, A_3, E_1, E_2, E_3, L_1, L_2$, and L_3. Here A, E, and L are generic symbols used for cross-sectional area, modulus of elasticity, and length, respectively.

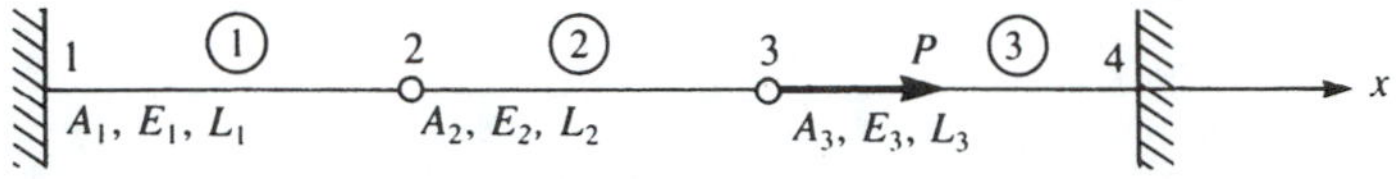

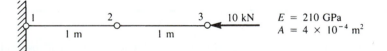

Figure P3–1

b. Now let $A_1 = A_2 = A_3 = A, E_1 = E_2 = E_3 = E$, and $L_1 = L_2 = L_3 = L$. If nodes 1 and 4 are fixed and a force P acts at node 3 in the positive x direction, find expressions for the displacement of nodes 2 and 3 in terms of A, E, L, and P.

c. Now let $A = 1$ in^2, $E = 10 \times 10^6$ psi, $L = 10$ in., and $P = 1000$ lb.
  i. Determine the numerical values of the displacements of nodes 2 and 3.
  ii. Determine the numerical values of the reactions at nodes 1 and 4.
  iii. Determine the stresses in elements 1–3.

3.2–3.11 For the bar assemblages shown in Figures P3–2–P3–11, determine the nodal displacements, the forces in each element, and the reactions. Use the direct stiffness method for these problems.

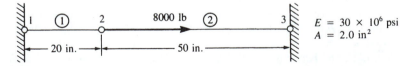

Figure P3–2

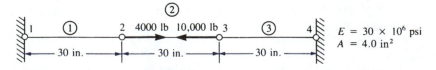

Figure P3–3

Figure P3–4

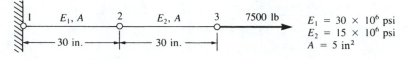

Figure P3–5

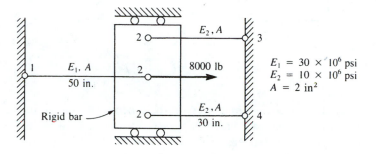

Figure P3–6

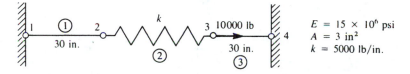

Figure P3–7

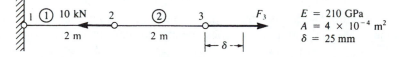

Figure P3–8

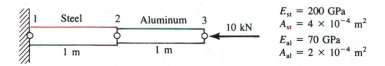

Figure P3–9

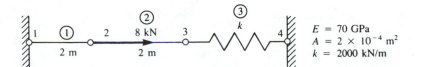

Figure P3–10

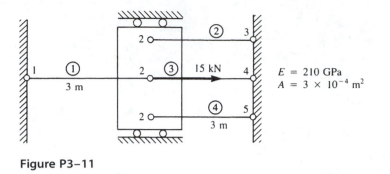

Figure P3–11

3.12 Solve for the axial displacement and stress in the tapered bar shown in Figure P3–12 using one and then two constant-area elements. Evaluate the area at the center of each element length. Use that area for each element. Let $A_0 = 2$ in^2, $L = 20$ in., $E = 10 \times 10^6$ psi, and $P = 1000$ lb. Compare your finite element solutions with the exact solution.

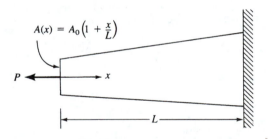

$$A(x) = A_0\left(1 + \frac{x}{L}\right)$$

Figure P3–12

3.13 Determine the stiffness matrix for the bar element with end nodes and midlength node shown in Figure P3–13. Let axial displacement $u = a_1 + a_2x + a_3x^2$. (This is a higher-order element in that strain now varies linearly through the element.)

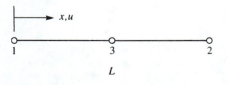

Figure P3–13

3.14 Consider the following displacement function for the two-noded bar element:

$$u = a + bx^2$$

Is this a valid displacement function? Discuss why or why not.

3.15 For each of the bar elements shown in Figure P3–15, evaluate the global x-y stiffness matrix.

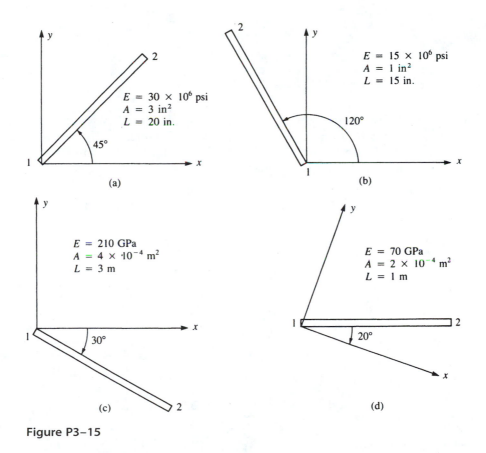

Figure P3–15

3.16 For the bar elements shown in Figure P3–16, the global displacements have been determined to be $d_{1x} = 0.5$ in., $d_{1y} = 0.0$, $d_{2x} = 0.25$ in., and $d_{2y} = 0.75$ in. Determine the local $\hat{x}$ displacements at each end of the bars. Let $E = 12 \times 10^6$ psi, $A = 0.5$ in^2, and $L = 60$ in. for each element.

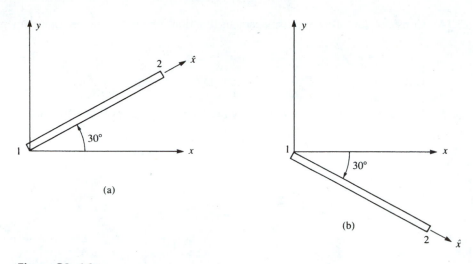

Figure P3–16

3.17 For the bar elements shown in Figure P3–17, the global displacements have been determined to be $d_{1x} = 0.0, d_{1y} = 2.5$ mm, $d_{2x} = 5.0$ mm, and $d_{2y} = 3.0$ mm. Determine the local $\hat{x}$ displacements at the ends of each bar. Let $E = 210$ GPa, $A = 10 \times 10^{-4}$ m^2, and $L = 3$ m for each element.

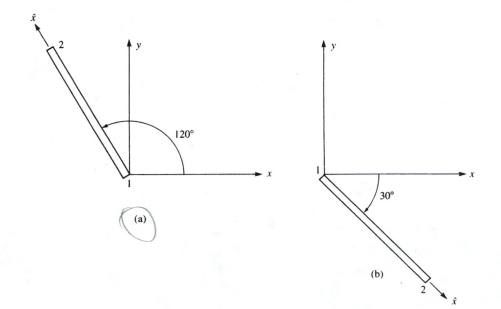

Figure P3–17

3.18 Using the method of Section 3.5, determine the axial stress in each of the bar elements shown in Figure P3–18.

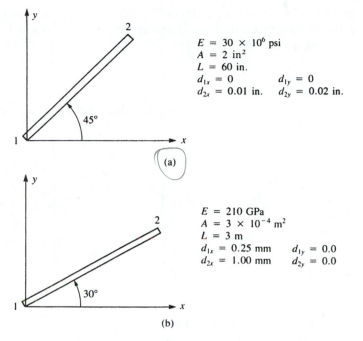

$E = 30 \times 10^6$ psi
$A = 2$ in^2
$L = 60$ in.
$d_{1x} = 0$ $d_{1y} = 0$
$d_{2x} = 0.01$ in. $d_{2y} = 0.02$ in.

(a)

$E = 210$ GPa
$A = 3 \times 10^{-4}$ m^2
$L = 3$ m
$d_{1x} = 0.25$ mm $d_{1y} = 0.0$
$d_{2x} = 1.00$ mm $d_{2y} = 0.0$

(b)

Figure P3–18

3.19 **a.** Assemble the stiffness matrix for the assemblage shown in Figure P3–19 by super-imposing the stiffness matrices of the springs. Here k is the stiffness of each spring.
b. Find the x and y components of deflection of node 1.

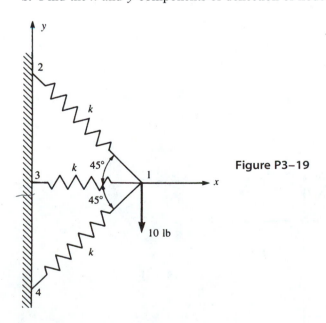

Figure P3–19

10 lb

3.20 For the plane truss structure shown in Figure P3–20, determine the displacement of node 2 using the stiffness method. Also determine the stress in element 1. Let $A = 5$ in^2, $E = 1 \times 10^6$ psi, and $L = 100$ in.

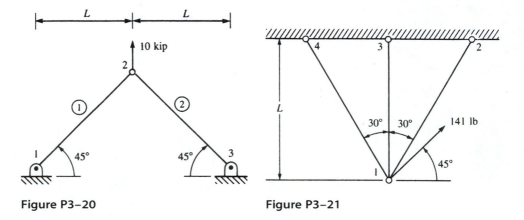

Figure P3–20   **Figure P3–21**

3.21 Find the horizontal and vertical displacements of node 1 for the truss shown in Figure P3–21. Assume AE is the same for each element.

3.22 For the truss shown in Figure P3–22 solve for the horizontal and vertical components of displacement at node 1 and determine the stress in each element. Also verify force equilibrium at node 1. All elements have $A_1 = 1$ in.^{2} and $E = 10 \times 10^6$ psi. Let $L = 100$ in.

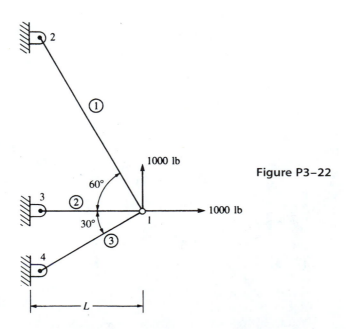

Figure P3–22

3.23 For the truss shown in Figure P3–23, solve for the horizontal and vertical components of displacement at node 1. Also determine the stress in element 1. Let $A = 1$ in^2, $E = 10.0 \times 10^6$ psi, and $L = 100$ in.

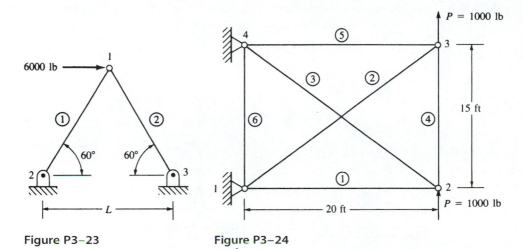

Figure P3–23   **Figure P3–24**

3.24 Determine the nodal displacements and the element forces for the truss shown in Figure P3–24. Assume all elements have the same AE.

3.25 Now remove the element connecting nodes 2 and 4 in Figure P3–24. Then determine the nodal displacements and element forces.

3.26 Now remove *both* cross elements in Figure P3–24. Can you determine the nodal displacements? If not, why?

3.27 Determine the displacement components at node 3 and the element forces for the plane truss shown in Figure P3–27. Let $A = 3$ in^2 and $E = 30 \times 10^6$ psi for all elements. Verify force equilibrium at node 3.

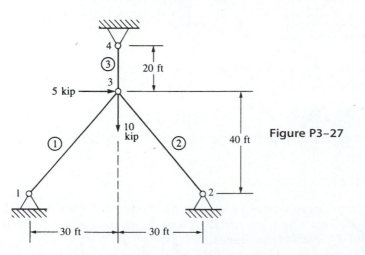

Figure P3–27

3.28 Show that for the transformation matrix $\underline{T}$ of Eq. (3.4.15), $\underline{T}^T = \underline{T}^{-1}$ and hence Eq. (3.4.21) is indeed correct, thus also illustrating that $\underline{k} = \underline{T}^T \underline{\hat{k}} \underline{T}$ is the expression for the global stiffness matrix for an element.

3.29–3.30 For the plane trusses shown in Figures P3–29 and P3–30, determine the horizontal and vertical displacements of node 1 and the stresses in each element. All elements have $E = 210$ GPa and $A = 4.0 \times 10^{-4}$ m^2.

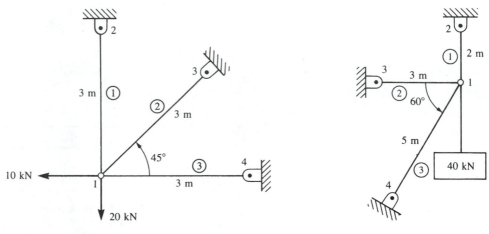

Figure P3–29   Figure P3–30

3.31 Remove element 1 from Figure P3–30 and solve the problem. Compare the displacements and stresses to the results for Problem 3.30.

3.32 For the plane truss shown in Figure P3–32, determine the nodal displacements, the element forces and stresses, and the support reactions. All elements have $E = 70$ GPa and $A = 3.0 \times 10^{-4}$ m^2. Verify force equilibrium at nodes 2 and 4. Use symmetry in your model.

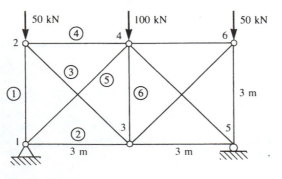

Figure P3–32

3.33 For the plane truss supported by the spring at node 1 in Figure P3–33, determine the nodal displacements and the stresses in each element. Let $E = 210$ GPa and $A = 5.0 \times 10^{-4}$ m^2 for both truss elements.

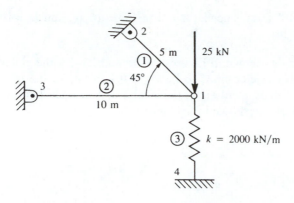

Figure P3–33

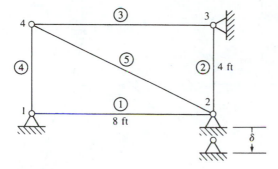

Figure P3–34

3.34 For the plane truss shown in Figure P3–34, node 2 settles an amount $\delta = 0.05$ in. Determine the forces and stresses in each element due to this settlement. Let $E = 30 \times 10^6$ psi and $A = 2$ in^2 for each element.

3.35 For the symmetric plane truss shown in Figure P3–35, determine (a) the deflection of node 1 and (b) the stress in element 1. AE/L for element 3 is twice AE/L for the other

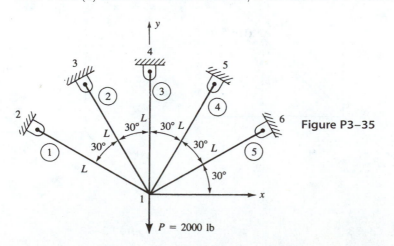

Figure P3–35

elements. Let $AE/L = 10^6$ lb/in. Then let $A = 1$ in^2, $L = 10$ in., and $E = 10 \times 10^6$ psi to obtain numerical results.

3.36–3.37 For the space truss elements shown in Figures P3–36 and P3–37, the global displacements at node 1 have been determined to be $d_{1x} = 0.1$ in., $d_{1y} = 0.2$ in., and $d_{1z} = 0.15$ in. Determine the displacement along the local $\hat{x}$ axis at node 1 of the elements. The coordinates, in inches, are shown in the figures.

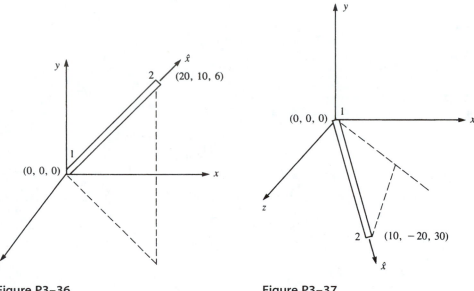

Figure P3–36   Figure P3–37

3.38–3.39 For the space truss elements shown in Figures P3–38 and P3–39, the global displacements at node 2 have been determined to be $d_{2x} = 5$ mm, $d_{2y} = 10$ mm, and $d_{2z} = 15$

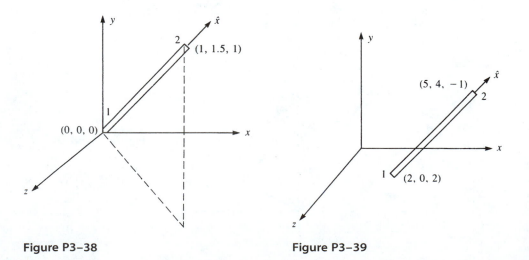

Figure P3–38   Figure P3–39

mm. Determine the displacement along the local $\hat{x}$ axis at node 2 of the elements. The coordinates, in meters, are shown in the figures.

3.40–3.41   For the space trusses shown in Figures P3–40 and P3–41, determine the nodal displacements and the stresses in each element. Let $E = 210$ GPa and $A = 10 \times 10^{-4}$ m^2 for all elements. Verify force equilibrium at node 1. The coordinates of each node, in meters, are shown in the figure. All supports are ball-and-socket joints.

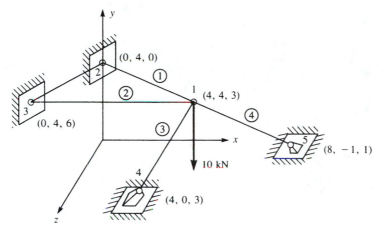

Figure P3–40

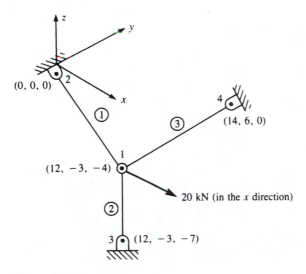

Figure P3–41

3.42   For the space truss subjected to a 1000-lb load in the x direction, as shown in Figure P3–42, determine the displacement of node 5. Also determine the stresses in each element. Let $A = 4$ in^2 and $E = 30 \times 10^6$ psi for all elements. The coordinates of each node, in inches, are shown in the figure. Nodes 1–4 are supported by ball-and-socket joints (fixed supports).

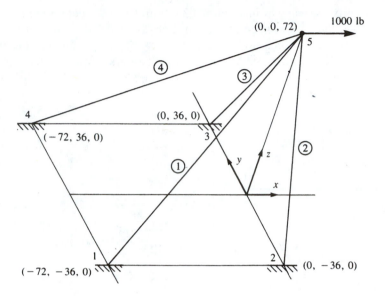

Figure P3–42

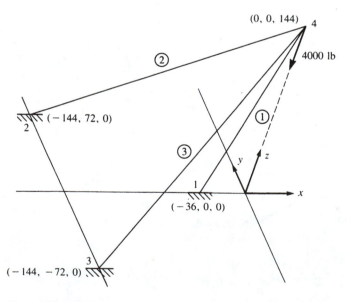

Figure P3–43

3.43 For the space truss subjected to the 4000-lb load acting as shown in Figure P3–43, determine the displacement of node 4. Also determine the stresses in each element. Let

$A = 6$ in^2 and $E = 30 \times 10^6$ psi for all elements. The coordinates of each node, in inches, are shown in the figure. Nodes 1–3 are supported by ball-and-socket joints (fixed supports).

3.44 Verify Eq. (3.7.9) for $\underline{k}$ by first expanding $\underline{T}^*$, given by Eq. (3.7.7), to a 6×6 square matrix in a manner similar to that done in Section 3.4 for the two-dimensional case. Then expand $\hat{\underline{k}}$ to a 6×6 matrix by adding appropriate rows and columns of zeros (for the $\hat{d}_z$ terms) to Eq. (3.4.17). Finally, perform the matrix triple product $\underline{k} = \underline{T}^T \hat{\underline{k}} \underline{T}$.

3.45 Derive Eq. (3.7.21) for stress in space truss elements by a process similar to that used to derive Eq. (3.5.6) for stress in a plane truss element.

3.46 For the truss shown in Figure P3–46, use symmetry to determine the displacements of the nodes and the stresses in each element. All elements have $E = 30 \times 10^6$ psi. Elements 1, 2, 4, and 5 have $A = 10$ in^2 and element 3 has $A = 20$ in^2.

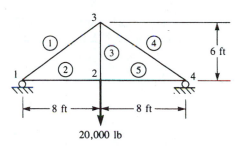

Figure P3–46

3.47 All elements of the structure in Figure P3–47 have the same AE except element 1, which has an axial stiffness of $2AE$. Find the displacements of the nodes and the stresses in elements 2, 3, and 4 by using symmetry. Check equilibrium at node 4. You might want to use the results obtained from the stiffness matrix of Problem 3.24.

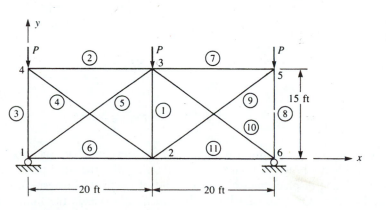

Figure P3–47

3.48 For the roof truss shown in Figure P3–48, use symmetry to determine the displacements of the nodes and the stresses in each element. All elements have $E = 210$ GPa and $A = 10 \times 10^{-4}$ m^2.

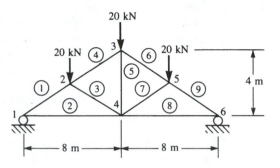

Figure P3–48

3.49–3.51 For the plane trusses with inclined supports shown in Figures P3–49–P3–51, solve for the nodal displacements and element stresses in the bars. Let $A = 2$ in^2, $E = 30 \times 10^6$ psi, and $L = 30$ in. for each truss.

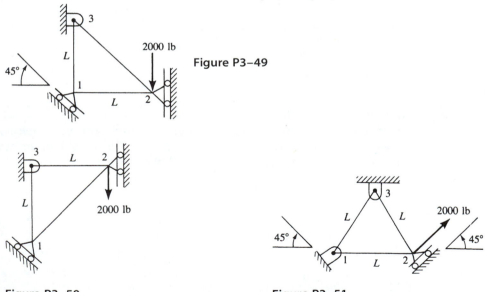

Figure P3–49

Figure P3–50

Figure P3–51

3.52 Use the principle of minimum potential energy developed in Section 3.10 to solve the bar problems shown in Figure P3–52. That is, plot the total potential energy for variations in the displacement of the free end of the bar to determine the minimum potential energy. Observe that the displacement that yields the minimum potential energy also yields the stable equilibrium position. Use displacement increments of

0.002 in., beginning with $x = -0.004$. Let $E = 30 \times 10^6$ psi and $A = 2$ in^2 for the bars.

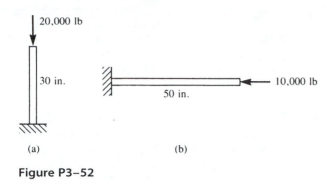

(a)   (b)

Figure P3–52

3.53 Derive the stiffness matrix for the nonprismatic bar shown in Figure P3–53 using the principle of minimum potential energy. Let E be constant.

$A(x) = A_0 + A_0 \dfrac{x}{L}$

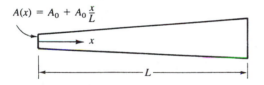

Figure P3–53

3.54 For the bar subjected to the linear varying axial load shown in Figure P3–54, determine the nodal displacements and axial stress distribution using (a) two equal-length elements and (b) four equal-length elements. Let $A = 2$ in.^{2} and $E = 30 \times 10^6$ psi. Compare the finite element solution with an exact solution.

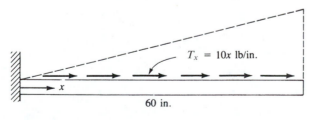

$T_x = 10x$ lb/in.

60 in.

Figure P3–54

3.55 For the bar subjected to the uniform line load in the axial direction shown in Figure P3–55, determine the nodal displacements and axial stress distribution using (a) two equal-length elements and (b) four equal-length elements. Compare the finite element results with an exact solution. Let $A = 2$ in^2 and $E = 30 \times 10^6$ psi.

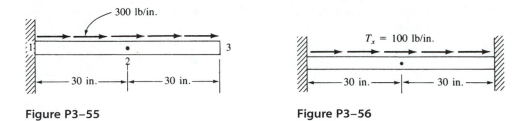

Figure P3–55   Figure P3–56

3.56 For the bar fixed at both ends and subjected to the uniformly distributed loading shown in Figure P3–56, determine the displacement at the middle of the bar and the stress in the bar. Let $A = 2$ in^2 and $E = 30 \times 10^6$ psi.

3.57 For the bar hanging under its own weight shown in Figure P3–57, determine the nodal displacements using (a) two equal-length elements and (b) four equal-length elements. Let $A = 2$ in^2, $E = 30 \times 10^6$ psi, and weight density $\rho_w = 0.283$ lb/in^3. (*Hint:* The internal force is a function of x. Use the potential energy approach.)

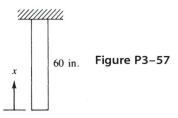

60 in.   **Figure P3–57**

3.58 Determine the energy equivalent nodal forces for the axial distributed loading shown acting on the bar elements in Figure P3–58.

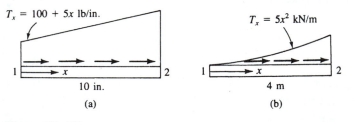

Figure P3–58

Algor Program for Truss Analysis

Introduction

In this chapter, we will give a general overview of the Algor computer program [1] and the computer flowchart for solution of both two- and three-dimensional truss structures. This computer program is based on the stiffness method. We will then discuss the manner in which the Algor program is used to model a truss and will illustrate step-by-step solutions of trusses, including two- and three-dimensional trusses with different material properties, different cross-sectional areas, inclined or skewed boundary supports, and nonzero boundary conditions that might occur during a settlement problem. These step-by-step solutions are based on the Algor computer program.

▲ 4.1 Overview of the Algor System and Flowcharts for the Solution of a Truss Problem Using Algor ▲

In Figure 4–1(a), we present a flowchart for a typical finite element process used for the analysis of two- and three-dimensional trusses. The Algor program for truss analysis is based on this flowchart. The basic steps that are used in the development of a finite element analysis of a truss are summarized as follows:

1. Draw the geometry of the model truss.
2. Apply the nodal forces and boundary conditions.
3. Define the element type (truss) and element properties, such as cross-sectional area and modulus of elasticity.
4. Review the model. (This step is optional but is recommended.)
5. Compute the nodal displacements and element stresses.
6. View and interpret the results.

In Figure 4–1(b), we present a flow diagram based on the preceding steps that includes the specific Algor programs [1] used in these steps.

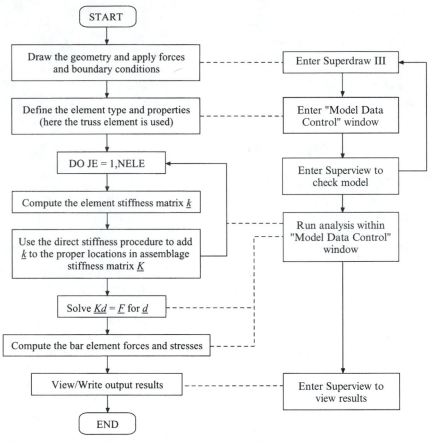

(a) Flowchart of truss finite element program
(NELE represents the number of elements)

(b) Flowchart of Algor process to perform
linear stress analysis, including truss analysis

Figure 4–1 Flowcharts for a typical finite element process used for the analysis of two- and three-dimensional trusses

Step 1 Draw the Geometry of the Model

The geometry of the model is normally drawn or defined by using the Algor program called **Superdraw III** [2], although other drawing programs such as AutoCAD and CADKEY may be used and imported into the Algor system. When we draw the geometry, we are actually defining both the description of the overall dimensions of the model truss and the individual elements that make up the model. We also define a global coordinate system and input the nodal coordinates during this process. When the individual elements, with their associated nodes, are created, we will have defined the topology or connectivity (which nodes are connected to which elements). The element numbering and the node numbering are done internally within Algor. This drawing process is normally the most time-consuming part of finite element analysis. We will later use the automatic mesh-generating capabilities for two- and three-dimensional bodies to reduce the error involved with modeling. The development of the model must then be done with great care to increase the accuracy of our results.

The Algor system also includes a number of other algorithms that facilitate error checking during the defining of the model.

For finite element models, group and surface properties are used by the Algor program to indicate various properties. Groups in your model are used to assign different material properties and different element types to specific elements. You will have to assign different group numbers to elements with different material properties, such as steel and aluminum parts in your model. You can assign one group to steel (say, group 1) and one group to aluminum (say, group 2). If you have different thicknesses of elements, which might occur in a plane stress problem, then different groups must be assigned to each element with a different thickness. Also, if your model has both truss and beam elements in it, you will need two groups in the model, one for truss elements and one for beam elements.

The status region spanning the bottom of the Superdraw III screen consists of "Surface", "Layer", and "Group" status buttons. These status buttons show the current setting for these parameters and give you the option to change them with a single mouse click. When you click on one of these buttons, a dialog box appears and prompts you to select or specify a new object parameter. To change current surface, layer, or group numbers from the pull-down menus, select "Options", then select "Current Object Parameters", and then choose "Surface Number", "Layer Number", or "Group Number". A dialog box will appear and prompt you to select or specify a new object parameter. After "Options" is selected, the "Current Object Parameters" pull-down menu also contains the option "Set All Based on Selected." This option sets the current surface, group, and layer from a selected object.

If you change the current surface, layer, or group, the new object parameters will take effect only when you create a new object. In order to change the surface, layer, or group of an existing object, first select the object or objects you want to change by going to the "Select" menu and clicking on the lines (objects) you want to change. Examples 4.4 and 4.5 illustrate this for changing groups. Then, select "Modify", select "Update Object Parameters", and choose "Surface Number", "Layer Number", or "Group Number". A dialog box will appear and prompt you to select or specify a new object parameter, such as a new group number, say, 2. The Superdraw III Reference Division [2] gives more details on using surfaces and groups in your model. The "Layers" option is used to group sections of a model together as you construct it, if you are building a model in parts.

Step 2 Apply the Nodal Forces and Boundary Conditions

Next we specify the nodal forces and boundary conditions using the Superdraw III "FEA Add" menu. In general, the forces either are nodal applied values or are distributed over some side or edge of one or more elements. For trusses, the forces are nodal applied values expressed in global-coordinate components at the nodes. The boundary or support conditions are the degrees of freedom that are specified at certain nodes and in the global-coordinate directions. They are often zero in value but may be any specified nonzero value. For instance, if the support is a pin connection in a two-dimensional truss, then the x and y displacements are both specified as zero at that nodal support. The forces and boundary conditions are applied as text in Superdraw III.

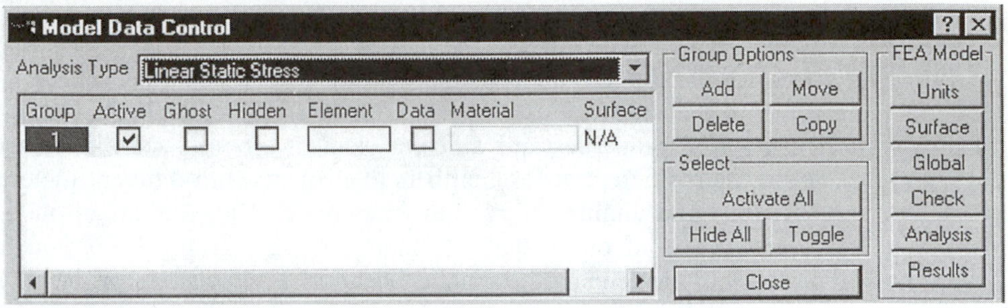

Figure 4–2 "Model Data Control" window

Step 3   Define the Element Type and Properties

The element type (truss) is defined by selecting the "Model Data Control" button located toward the bottom of the computer screen, or by selecting "Tools: Model Data Control" from the Superdraw III menu screen. Next, click on the box under the "Element" column in the "Model Data Control" window (Figure 4–2). Then click on "Truss". For each group in a model you must define the element type of the lines of that group. This is done within Superdraw III as discussed in step 1. You must do this for each group of different element types and for elements with different material or geometric properties. For instance, if there are two different cross-sectional areas for truss elements in your model, two groups must be assigned, one for each different cross-sectional area. Also, if you have two different materials, such as steel and aluminum truss elements, you will need two groups in the model, one for steel and one for aluminum. The cross-sectional area is input on the "Element Definition" screen (Figure 4–3) by clicking on the box in the "Data" column. The modulus of elasticity E of the various truss elements is input by clicking on the box under the "Material" column. This opens the "Element Material Specification" screen (Figure 4–4). On this screen, you can select a material from the Algor material library or from a user-

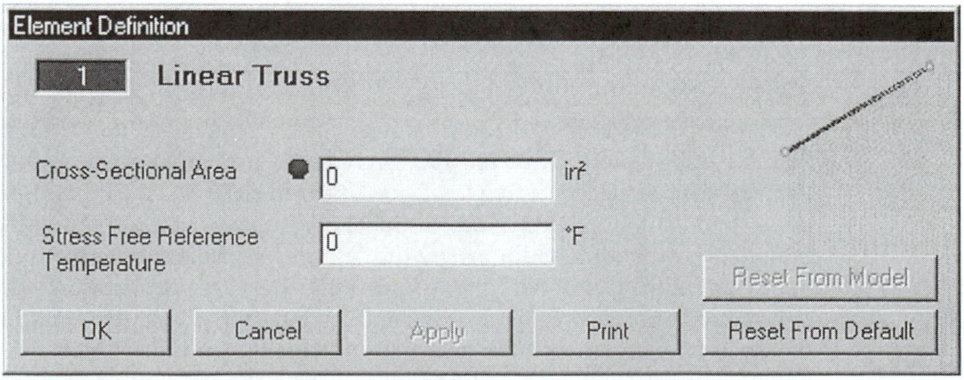

Figure 4–3 "Element Definition" screen

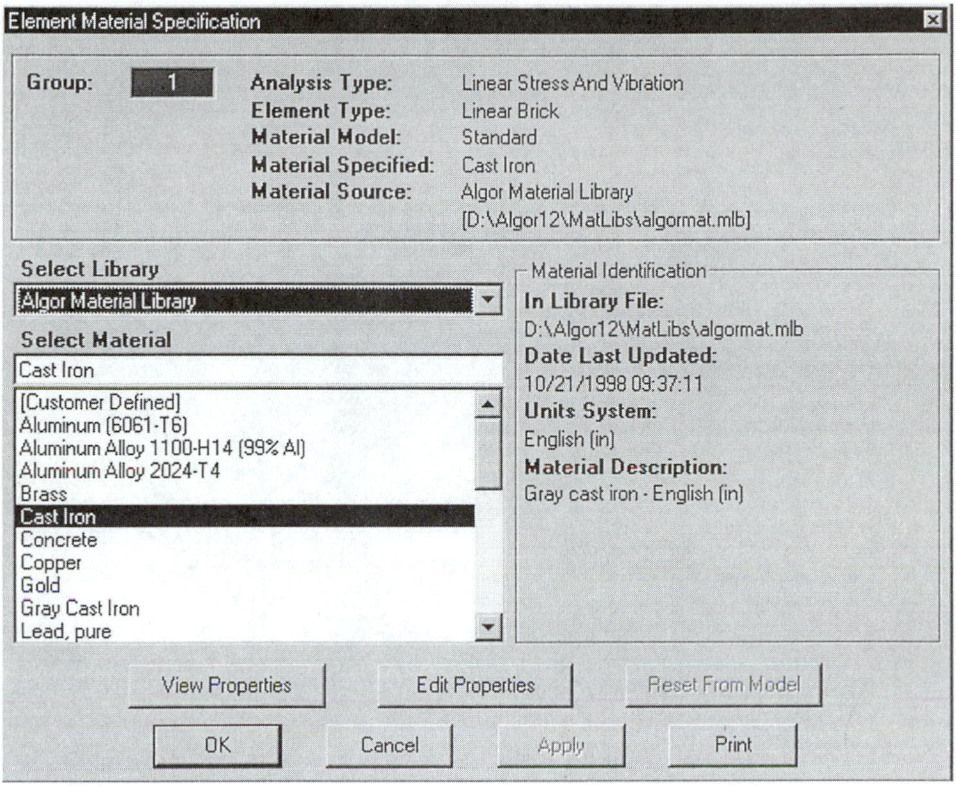

Figure 4–4 "Element Material Specification" screen

defined library, or you can select "Customer Defined" to input your own material properties. Examples 4.4 and 4.5 illustrate the use of different groups.

After the element properties are entered through the "Model Data Control" window, the decoder (transparent to the user) then reads the geometry information created in Superdraw III and forms the elements based on the type selected (in this chapter, we select the truss). This process of forming the elements, numbering the nodes, and recording the nodes associated with each element is all done in a transparent manner in the stress decoder.

There is also some general global data that must be entered on the "Global Data" screen for every model before running the analysis. You can activate the "Global Data" screen by clicking the "Global" button on the right side of the "Model Data Control" window shown in Figure 4–2. For each analysis type, a different "Global Data" screen appears. For the linear stress analysis, a "Multipliers" tab screen appears as shown in Figure 4–5. If you have applied loads or surface pressures acting on your model, then you will want to enter a value in the "Pressure" column in the "Load Case Multipliers" section. This value is 1 if the actual load magnitudes are input. The data to be entered may vary depending on the analysis type. For instance, a different screen appears for a truss analysis than for a plane stress analysis.

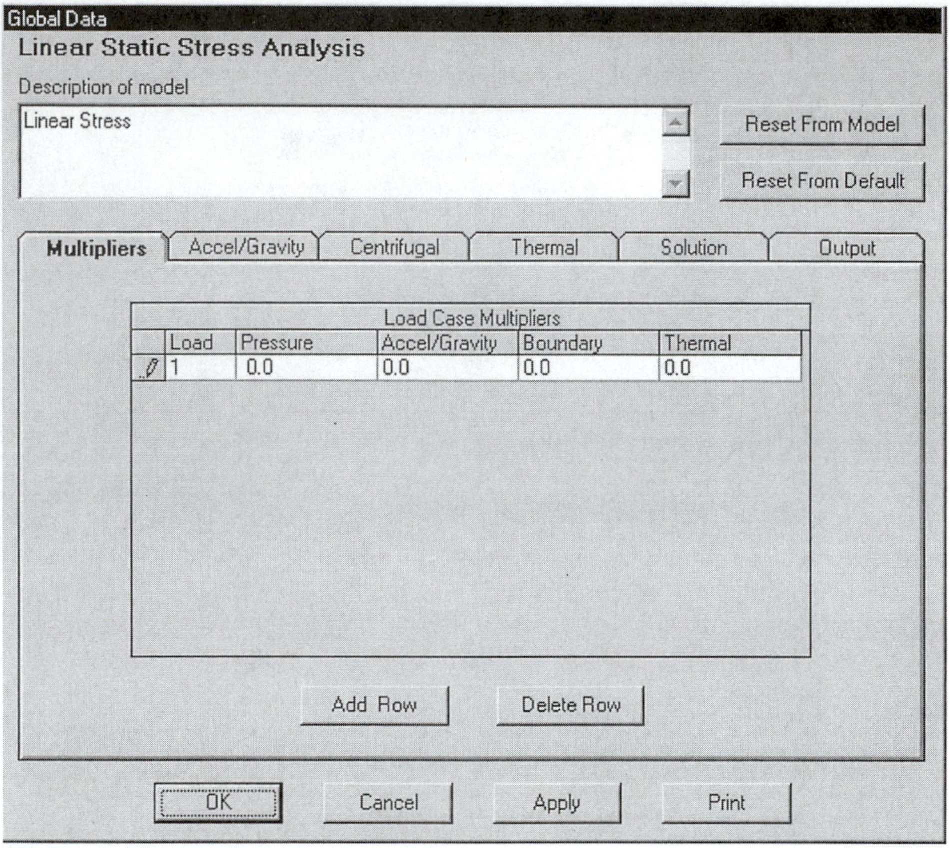

Global Data

Linear Static Stress Analysis

Description of model

| Linear Stress | ▲ | Reset From Model |
| | ▼ | Reset From Default |

| Multipliers | Accel/Gravity | Centrifugal | Thermal | Solution | Output |

Load Case Multipliers

	Load	Pressure	Accel/Gravity	Boundary	Thermal
✎	1	0.0	0.0	0.0	0.0

Add Row   Delete Row

OK   Cancel   Apply   Print

Figure 4–5 "Multipliers" tab screen

Step 4 Review or Check the Model

After the model has been constructed and decoded, it should be viewed and checked to verify that the generated model is the model we wanted. In Algor, select the "Check" button from the "Model Data Control" window; this will verify your elements and then activate the **Superview** [3] screen. Using the various display and shading options for the model increases our confidence in its geometric integrity. Using the *Inquiry* options in Superview enables us to verify the dimensions of the model and the forces, pressures, and boundary conditions. As indicated in Figure 4–1(b), if the model is not satisfactory, we can loop back to Superdraw III and/or the "Model Data Control" window to correct modeling errors that show up during the Superview examination. Although this step is optional, if there are errors in the model, we can often save time by reviewing the model before running it through the analysis processor and having the processor indicate errors and then having to return to Superview and Superdraw III.

Step 5 Run the Analysis to Compute the Nodal Displacements and Element Stresses

After the model is complete, it can be run through the analysis processor to compute the nodal displacements, element forces, and element stresses. Select the "Analysis" button from the "Model Data Control" window. This brings up the "Analysis" screen. Then select the "Analyze" button to start the processor. On this screen, you can also set the amount of memory you want the analysis to access. By default this value is set to 40% of your computer memory. The processor for running the linear stress analysis of trusses under static loading conditions is called *SSAP0* and is transparent to the user. Other processors are used for problems such as heat transfer or dynamic stress analysis. For an overview of the numerous processors in the Algor system, consult Appendix F or the "Docutech" on-line "Technical User Documentation" of Algor.

To look at some solution options, click the "Solution" tab (Figure 4–6) on the "Global Data" screen. For instance, you can click on the box next to "Avoid band-

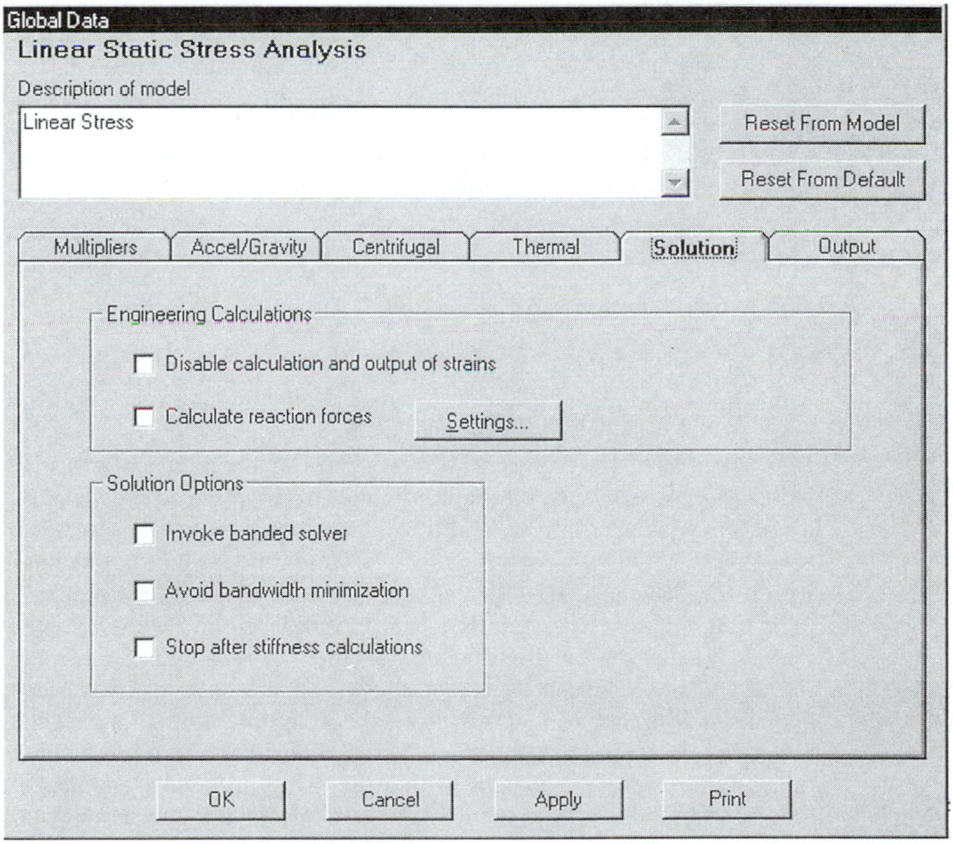

Figure 4–6 "Solution" tab screen

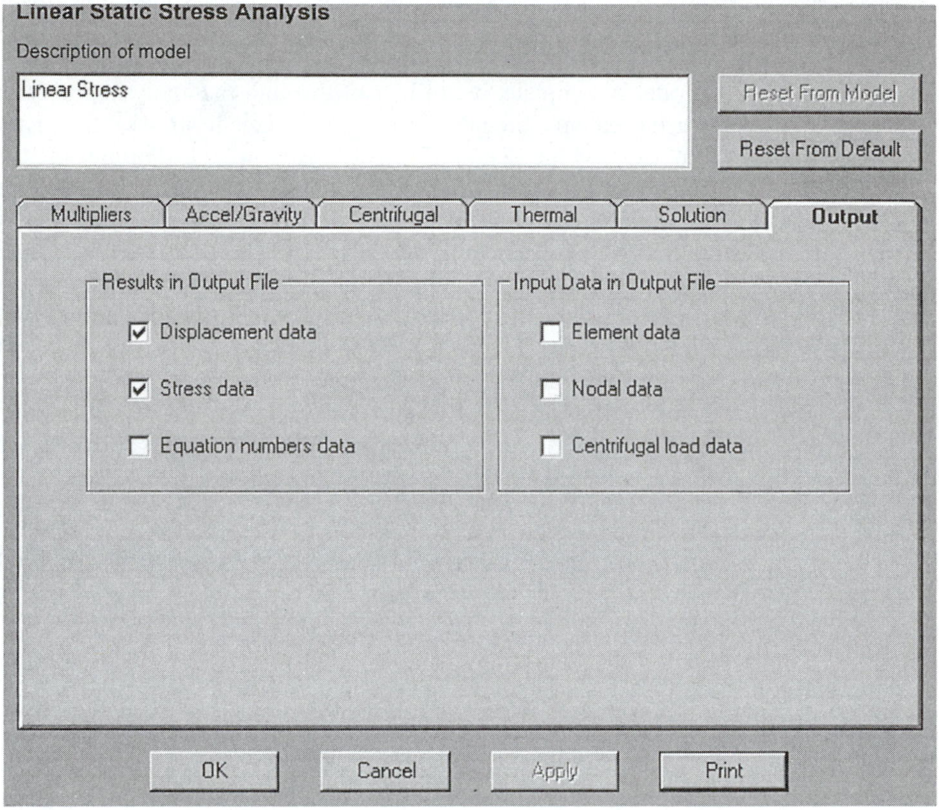

Linear Static Stress Analysis

Description of model

Linear Stress

| Reset From Model |
| Reset From Default |

| Multipliers | Accel/Gravity | Centrifugal | Thermal | Solution | **Output** |

┌─ Results in Output File ──────────┐   ┌─ Input Data in Output File ──────────┐

☑ Displacement data   ☐ Element data

☑ Stress data   ☐ Nodal data

☐ Equation numbers data   ☐ Centrifugal load data

| OK | Cancel | Apply | Print |

Figure 4–7 "Output" tab screen

width minimization" if you do not want to minimize the bandwidth of the stiffness matrix. This usually makes the analysis take longer.

As indicated in Figure 4–1(a), the primary or initial unknowns that are solved for in the processor are the nodal displacements $\underline{d}$. This process of determining the displacements for both two- and three-dimensional trusses was detailed in Chapter 3. Selecting the "Output" tab (Figure 4–7) while still viewing the "Global Data" screen within the "Model Data Control" window and clicking the box next to "Displacement data" places the nodal displacement output in the filename you have been using with a .L extension on the filename. Another program is automatically called within the processor to compute element stresses on the basis of the nodal displacements associated with each element. A detailed description of how these element stresses are computed was presented in Section 3.5 for plane trusses, and in Section 3.7 for three-dimensional trusses. While still within the "Output" menu, clicking the box next to "Stress data" places the element stresses and forces in the filename with a .S extension on the filename.

Clicking the box next to "Element data" writes the element data to the .L file. Element data consists of the nodes connecting the element, element loads, and element material properties. Finally, clicking the box next to "Nodal data" writes the nodal data to the .L file. This nodal data includes the node numbers, boundary conditions, and nodal loads.

Step 6   View and Interpret the Results

The results, such as nodal displacements and element stresses, can now be viewed by using the graphical review program Superview. Select the "Results" button in the "Model Data Control" window to get into Superview and view your results. This is the same program we used to view the model before running or processing it. However, the displacements and stresses are now available because we have run the model through the processor *SSAP0*. The displacements, stresses, and reaction forces can be viewed and displayed in graphical form in Superview. We can also obtain hard copy of the displacements and stresses by printing files that were created during the modeling, decoding, and processing phases. Each of these files has the model name assigned by the user and an extension used to distinguish one file from another. Two files of interest are the name.L and name.S files, where "name" is the name of the model created by the user. The name.L file stores such information as the echo check of the input data and the nodal displacement output. The name.S file stores the element stresses. Additional files (with their extensions) generated by Algor programs during creation and analysis of the model are listed in Appendix F.6.

▲ 4.2 Algor Example Solutions for Truss Analysis ▲

We now solve a series of example problems wherein we use the Algor computer program for truss analysis. These examples illustrate numerous concepts, such as how to include different element properties in each bar element, support settlement, and inclined or skewed boundary conditions. Detailed steps will be listed, including specific keystrokes and descriptions of the purpose of these keystrokes.

The following notation is used in the tutorials:

1. Keyboard keys with specific names are shown in angle brackets, for example, ⟨Esc⟩, ⟨Enter⟩, ⟨Ctrl⟩, ⟨Tab⟩.
2. Keyboard keys to be pressed at the same time are shown with a hyphen separating them, for example, ⟨Ctrl⟩-C.
3. Command names are enclosed in quotation marks, for example, the "Add" command.
4. Colons are sometimes used to separate commands in a command sequence; for example, "View:Enclose" means to select the "Enclose" command from the "View" menu. (Often we just list these commands line by line.)
5. Numeric data that is to be entered is typed in Courier type, for example, −10000.
6. Filenames that are user-supplied are shown in italic, for example, *ex45*.

7. Option or button names are enclosed in quotation marks and shown as they appear on-screen, for example, the "Single" option on the menu or the "Model Data" control button.

Example 4.1

Solve the simple two-dimensional truss shown in Figure 4–8 to become familiar with some of the basic commands used in the Algor program. The truss members have a constant cross-sectional area $A = 1$ in^2 and a modulus of elasticity $E = 30e6$ psi. The commands and keystrokes used to create and solve the plane truss, along with explanations, follow.

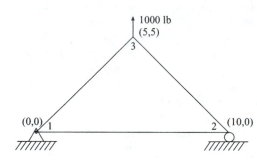

Figure 4–8 Two-dimensional truss (coordinates are in inches)

Step 1 Start Superdraw III

Double-click on the "Algor FEA" icon. This brings up the Superdraw III program, the Algor main graphics user interface. All your work will be done from this program. You will construct, analyze, and review your results from Superdraw III. Each menu option and a short example are available in the Superdraw III Reference Division (accessed through "Docutech", the on-line "Technical User Documentation").

Step 2 Create the Truss Model

Select the "Add" menu from the top main menu bar.

Click "Line". A "Line" menu appears on the upper right part of the screen. Then enter the first data point of the line and proceed by entering successive points until the truss is created.

0<Tab>0	Enter the first data point of line 1.
<Enter>	Press the ⟨Enter⟩ key to actually enter the data point.
10<Tab>0	Enter the second point of line 1; this is also the first point of line 2.
<Enter>	
5<Tab>5	Enter the second point of line 2; this is also the first point of line 3.
<Enter>	
0<Tab>0	Enter the second point of line 3; this also takes you back to the starting point.
<Enter>	

Click "Done" on the "Line" menu.

Select "View". This takes you to the draw options menu.

Click "Enclose". This encloses the model. You should now see the line drawing of the truss enclosed on the screen. It should look like Figure 4–8.

Step 3   Eliminate Duplicate Lines

Although this step is optional, during the process of building a model, some lines may be duplicated. To clean up the model, use the following commands.

Click "Modify" on the main menu bar.

Select "Clean:Duplicate". This notation means to select "Clean" and then "Duplicate" in succession.

Click "Perform Cleaning" on the "Duplicate" menu. You will see the following message on the lower part of the screen:

"3 Kept 0 Deleted. done."

Click "Done" on the "Duplicate" menu located in the upper right corner of the screen.

Step 4   Add in Boundary Conditions and Applied Force

Select the "FEA Add" menu from the main menu bar and highlight "Stress and Vibration Analysis".

Click "Boundary Conditions". The "Boundary Conditions" menu appears. The default boundary condition is "Use @ Symbol for Full". "Full" means full constraint or prevention of translation and rotations (when relevant for the element you choose). Move the cursor to node 1 and

Right-click (snap-to) on node 1. The "@" sign appears next to node 1, indicating complete fixity of node 1.

Return to the "Boundary Condition" menu.

Select "Value".

Click "*1)tx" This removes the asterisk in front of the "tx" to release the x direction translational degree of freedom (a roller effect is created). Move the cursor to node 2 and

Right-click on node 2. "TyzRxyz" appears next to node 2.

Click "Done".

Select the "FEA Add" menu again and move to highlight "Stress and Vibration Analysis".

Click "Y Direction". Note that this is the default direction for the force, but for illustration we will use it. You will see "DX 0 DY 0.97 DZ 0" at the bottom of the screen.

Click "Done". This takes you to the "Nodal Forces" menu.

Click "Magnitude". A prompt "Magnitude" appears at the bottom of the screen asking you to enter the magnitude of the force.

`1000`   Enter the magnitude of the force as positive 1000 lb.

`<Enter>`   You should see the message "Click on node to apply the current values".

Right-click on node 3. (The upper node is node 3.) The 1000-lb force appears as an arrow at node 3.

Click "Done".

Select "View:Zoom:In" to zoom in and create a box around node 3 to see "1000;L1;lbf" next to node 3.

Click "Done".

Click "View:Enclose" to see the whole truss again.

Step 5 Add the Material Properties Using the "Model Data Control" Window

After the mesh has been created and nodal boundary conditions and forces applied, you can select "Tools" and "Model Data Control" from Superdraw III to complete the data input for analysis. The "Model Data Control" window will appear. Alternatively, you can

Click on the "Model Data" button located on the lower part of the screen and highlighted in red. You will see the "Analysis Type" line with "Linear Static Stress" highlighted as the default analysis type.

Click on the box below the "Element" column. A table of group 1 element types appears.

Click "Truss". You now have created a truss. All lines are truss elements.

Click "OK".

Click on the box below the "Data" column. It will respond with "Please enter the model name first".

Click "OK" and then enter the model name in the "File_name" area that appears.

ex41   The name of the file is *ex41*.

Click "Save". By default the "Units Definition" menu appears.

Click "OK" to accept the default "English (in)" units system. Otherwise, you can scroll down to select another choice of units. The "Element Definition" menu appears next. Enter the cross-sectional area of the truss elements. All elements have the same area in this example.

Hold the left click of the mouse over the default value of 0 until a blue flashing box appears on the 0.

1.   Enter the area 1.0 square inch.
        Leave "Stress Free Reference Temperature" as 0.

Click "OK". This sends you back to the "Material Data Control" window.

Click on the box below the "Material" column. Select one of the materials or select "Customer Defined". Assume we select "Customer Defined".

Select "View Properties:Unlock Properties" and enter the modulus of elasticity of steel as you define its value.

30e6   Other properties are not needed, so just enter the modulus of steel as 30e6.

Click "OK".

Click "OK". A second click takes you out of the "Element Material Selection" menu and back to the "Model Data Control" window.

Step 6   Enter the Global Data

Click on the "Global" button on the right-side menu of the "Model Data Control" window. Under "Load Case Multipliers" and under "Pressure", type a 1.

1   Type 1 or the program will not run. This indicates one load case is defined and it is an applied force. All forces will be multiplied by this multiplier of 1. While still within the "Global" menu,

Click on the "Output" tab from the right side of the menu. Then

Click on the box next to "Displacement data" and next to "Stress data". This creates a nodal displacement output in the .L file, and element forces and stresses in the .S file.

Click on the box next to "Nodal data" and "Element data" if you want a list of the nodal data and element data information.

Click "OK". You are now out of the "Global" menu.

Step 7   Check the Model

Click "Check" under "FEA Model" on the right side of the "Model Data Control" window. This takes you to the Superview main menu, and you can check your model. The truss model appears automatically, showing the boundary conditions and load arrow.

Click "Done". This takes you out of the Superview menu and back to the "Model Data Control" menu.

Step 8   Run the Analysis

Click on the "Analysis" button under the "FEA Model" column in the "Model Data Control" window.

Click "Analyze" in the lower left corner of the "Linear Stress and Vibration-Static Stress" menu. (See the green box next to "Analyze".) Wait a short time for the analysis to be performed. A warning message will appear.

Click "OK".

Click "Done" on the lower right side of the screen.

Step 9   Review the Results

Click on the "Results" button in the "Model Data Control" window. This takes you to Superview again, and you can review the results. If you want to view the displaced truss,

Select "Displaced" from the main menu.

Click "Displ on" so that an asterisk appears in front of it (you should see "*Displ on"). The asterisk means the displaced mode is active.

Click "With undi" so that the undisplaced model also appears.

Click "Calc scal" if you want the program to enlarge or scale the displaced model. You should now see the displaced and undisplaced truss appear on the screen as shown in Figure 4–9.

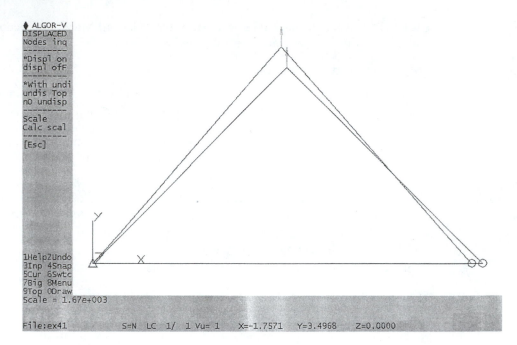

Figure 4–9 Displaced truss superimposed on undisplaced truss of Example 4–1

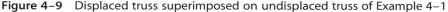

Click "Nodes inq" on the "Displaced" menu.

Click "Get" to get a node.

Click on the top node to see the x and y displacements appear at the bottom of the screen. You will see "DX $= -8.333e-05$ DY $= 3.1904e-04$".

<Esc>

<Esc>   You are now back in the main menu of Superview. If you want to view the stresses,

Click "Stress di". You are now going to post a truss stress.

Click "Post".

Click "Beam-trus".

Click "P/A". The axial stress will be posted. You will now see the P/A stress plot for the model with colored boxes representing different stresses on each element.

<Esc>

Click "Done". This takes you out of Superview. You should be able to find the *ex41.L* file (storing the echo check of your data and the nodal displacements) and the *ex41.S* file (storing the element forces and stresses) to print hard copies of each. ∎

Tables 4–1 and 4–2 are the hard copies of the *ex41.L* and *ex41.S* files. In the *ex41.L* file you will see information on the number of nodes, number of element types, nodal coordinates and their boundary conditions, number of elements created, material properties, element connectivity data, and loads on the nodes. The displacements

Table 4–1   *Ex41.L* file from Algor program solution

1**** Algor (c) Linear Stress Analysis—Version 12.00—WIN 15—FEB—1999

          DATE: November 3, 1999
          TIME: 4:47 PM
          INPUT FILE............C:\WINNT\Profiles\Logan\Desktop\ex41

1**** CONTROL INFORMATION

          number of node points   (NUMNP) =   3
          number of element types   (NELTYP) =   1
          number of load cases   (LL)   =   1
          number of frequencies   (NF)   =   0
          geometric stiffness flag   (GEOSTF) =   0
          analysis type code   (NDYN)   =   0
          solution mode   (MODEX) =   0
          equations per block   (KEQB)   =   0
          weight and c.g. flag   (IWTCG) =   0
          bandwidth minimization flag (MINBND) =   0
          gravitational constant   (GRAV)   = 3.8640E+02

          bandwidth minimization specified

1**** NODAL DATA

| NODE | BOUNDARY CONDITION CODES | | | | | | NODAL POINT COORDINATES | | | |
NO.	DX	DY	DZ	RX	RY	RZ	X	Y	Z	T
1	1	1	1	1	1	1	0.000E+00	0.000E+00	0.000E+00	0.000E+00
2	0	1	1	1	1	1	1.000E+01	0.000E+00	0.000E+00	0.000E+00
3	0	0	0	1	1	1	5.000E+00	5.000E+00	0.000E+00	0.000E+00

**** PRINT OF EQUATION NUMBERS SUPPRESSED

1**** TRUSS ELEMENTS

          number of truss members =   3
          number of diff members =   1

1**** ELEMENT LOAD FACTORS

	CASE A	CASE B	CASE C	CASE D
X-DIR	0.000E+00	0.000E+00	0.000E+00	0.000E+00
Y-DIR	0.000E+00	0.000E+00	0.000E+00	0.000E+00
Z-DIR	0.000E+00	-1.000E+00	0.000E+00	0.000E+00
TEMP	0.000E+00	0.000E+00	0.000E+00	1.000E+00

1**** ELEMENT CONNECTIVITY DATA

ELEMENT NUMBER	NODE I	NODE J	MAT'L INDEX	TEMP
1	1	2	1	1.00
2	3	1	1	1.00
3	2	3	1	1.00

Table 4–1 *(Continued)*

1**** MATERIAL/AREA DATA

INDEX	E	ALPHA	MASS DENSITY	AREA	WEIGHT DENSITY
1	3.0000E+07	0.0000E+00	7.3240E-04	1.0000E+00	2.8300E-01

1**** BANDWIDTH MINIMIZATION

    minbnd (bandwidth control parameter) =   1
**** MINIMIZER DID NOT NEED TO REDUCE BANDWIDTH
**** Hard disk file size information for processor:

    Available hard disk space on drive = 741.605 megabytes

1**** NODAL LOADS (STATIC) OR MASSES (DYNAMIC)

NODE NUMBER	LOAD CASE	X-AXIS FORCE	Y-AXIS FORCE	Z-AXIS FORCE	X-AXIS MOMENT	Y-AXIS MOMENT	Z-AXIS MOMENT
3	1	0.000E+00	1.000E+03	0.000E+00	0.000E+00	0.000E+00	0.000E+00

1**** ELEMENT LOAD MULTIPLIERS

load case	case A	case B	case C	case D
1	1.000E+00	0.000E+00	0.000E+00	0.000E+00

**** EQUATION PARAMETERS
    Number of equations   =   4
    Minimum bandwidth   =   1
    Maximum bandwidth   =   4
    Average bandwidth   =   1
    Storage required   (KB) =   0
    Total memory allocated (KB) = 12567
    Total memory free   (KB) = 12567

**** Proceeding with in-core fast solver ...
    Warning: 0 stiffness replaced by 1e20 for DOF   4

1**** STIFFNESS MATRIX PARAMETERS

    minimum non-zero diagonal element = 4.2426E+06
    maximum diagonal element   = 5.1213E+06
    maximum/minimum   = 1.2071E+00
    average diagonal element   = 4.5355E+06

**** Begin in-core solution
**** load case #1
**** End in-core solution

1**** STATIC ANALYSIS

    LOAD CASE =   1

Table 4–1   (*Continued*)

NODE number	Displacements/Rotations (degrees) of nodes					
	X-translation	Y-translation	Z-translation	X-rotation	Y-rotation	Z-rotation
1	0.0000E+00	0.0000E+00	0.0000E+00	0.0000E+00	0.0000E+00	0.0000E+00
2	-1.6667E-04	0.0000E+00	0.0000E+00	0.0000E+00	0.0000E+00	0.0000E+00
3	-8.3333E-05	3.1904E-04	0.0000E+00	0.0000E+00	0.0000E+00	0.0000E+00

```
1**** TEMPORARY FILE STORAGE (MEGABYTES)

        UNIT NO.   7 :    .000
        UNIT NO.   8 :    .000
        UNIT NO.   9 :    .000
        UNIT NO.  10 :    .000
        UNIT NO.  11 :    .000
        UNIT NO.  12 :    .000
        UNIT NO.  13 :    .000
        UNIT NO.  14 :    .000
        UNIT NO.  15 :    .000
        UNIT NO.  17 :    .000

        TOTAL        :    .000

1**** End of file                                            ■
```

Table 4–2   *Ex41.S* file from Algor program solution

```
Algor (R) FEA Stress Processor
Version 12.00-WIN 15-FEB-1999
Copyright (c) 1989-1999 Algor, Inc. All rights reserved.

        DATE: NOVEMBER 3, 1999
        TIME: 04:48 PM
        INPUT......C:\WINNT\Profiles\Logan\Desktop\ex41
```

```
**** 3-D truss elements:

     Number of elements  = 3
     Number of materials = 1

**** Nodal stresses for 3-D truss elements:

     El. #   LC   ND     Stress        Force

        1     1    I     5.000E+02     5.000E+02
        1     1    J    -5.000E+02    -5.000E+02

        2     1    I    -7.071E+02    -7.071E+02
        2     1    J     7.071E+02     7.071E+02

        3     1    I    -7.071E+02    -7.071E+02
        3     1    J     7.071E+02     7.071E+02

1**** End of file                                            ■
```

of the nodes are then listed. The *ex41.S* file lists the stresses and forces in each element of the truss. This problem being quite simple, a quick hand check of the results can be used to verify the element forces.

Example 4.2

Solve the plane truss problem from Example 3.5 (Figure 4–10) using the Algor program. Compare the results to the longhand solution. Let the cross-sectional area of all members be $A = 2$ in^2 and let $E = 30 \times 10^6$ psi.

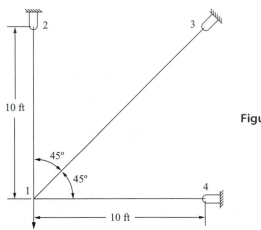

Figure 4–10　Three-element plane truss

Step 1　Start Superdraw III

Select the "Start" button in Windows NT/95/98 and then proceed to "Programs: Algor Software:Algor FEA".

Then click on "Algor FEA" to start the program.

Or double-click the "Algor FEA" icon on your desktop. Superdraw III will appear. This is the Algor main graphics user interface. All your work will be done from this program. You will construct, analyze, and review your results from Superdraw III. Each menu option and a short example are available in the Superdraw III Reference Division (accessed through "Docutech", the on-line "Technical User documentation").

Step 2　Create the Truss Model

Select the "Add" menu from the top main menu bar.

Click "Line". A "Line" menu appears in the upper right part of the screen.

Click "Single" on the menu on the upper right side of the screen. Then enter the data points (coordinates of the horizontal line 1) in succession.

0<Tab>0	This is point one of element 1.
<Enter>	Press the ⟨Enter⟩ key to actually enter the data.
120<Tab>0	This is the other endpoint of element 1.
<Enter>	
0<Tab>0	Repeat the process for element 2.
<Enter>	
120<Tab>120	
<Enter>	
0<Tab>0	Repeat the process for element 3.
<Enter>	
0<Tab>120	
<Enter>	You have finished entering points.

Click "Done" on the menu in the upper right corner.

Select "View". This takes you to the draw options menu.

Click on "Enclose". This encloses the model. You now see the line drawing of the truss enclosed on the screen. It should look like Figure 4–10.

Step 3   Eliminate Duplicate Lines

Click "Modify" on the main menu bar.

Select "Clean:Duplicate". This notation means to select "Clean" and then "Duplicate" in succession.

Click "Perform Cleaning" on the "Duplicate" menu. You will see the following message on the lower part of the screen:

"3 Kept 0 Deleted. done"

Click "Done" on the "Duplicate" menu located in the upper right corner of the screen.

Step 4   Add in the Boundary Conditions and Applied Force

Select the "FEA Add" menu from the main menu bar and highlight "Stress and Vibration Analysis".

Click "Boundary Conditions". The "Boundary Conditions" menu appears. The default boundary condition is "Use @ Symbol for Full". "Full" means full constraint or prevention of translations and rotations (when relevant for the element you choose). You may want to left-click and hold on the "Boundary Conditions" label and move it out of the way to see all the nodes of the model. Move the cursor to node 2 and

Right-click on node 2. The "@" sign appears next to the node, indicating node 2 is completely fixed. Move the cursor to node 3 and

Right-click on node 3. Then move the cursor to node 4 and

Right-click on node 4. It may be difficult to see the "@" sign at this time. Therefore, Click "Done". Then

Select "View" and "Zoom" and "In" and create a box around node 2, for instance, to see the "@" sign. You may have to repeat this process of zooming one or more times to see the "@" sign.

Click "Done".

Note: If you have created one or more spurious or misplaced boundary condition "@" signs, go to "Select" from the top menu; then choose "Box" and create a box around the object (say, around an "@" sign that you want to delete). Click "Modify" on the top menu and then click "Delete". The "@" sign should disappear.

Select "View" and "Enclose" again to see the whole truss model.

Click on the "FEA Add" menu again and move to highlight "Stress and Vibration Analysis".

Click "Nodal Forces". You will now add the nodal force at the free node.

Click "Vector". You will need to create the direction of the force.

Click "Y Direction". Note that this is the default direction for the force, but for illustration we will use it. You will see "DX 0 DY 0.97 DZ 0" appear at the bottom of the screen.

Click "Done". This takes you back to the "Nodal Forces" menu.

Click "Magnitude". A prompt ("Magnitude") will appear at the bottom of the screen asking you to enter the magnitude of the force.

`-10000`   Enter the magnitude of the force as negative 10,000 lb.

`<Enter>`   You might have to move the screen to see the message "Click on node to apply the current values".

Move the mouse cursor to node 1 and right-click on node 1. The −10000-lb force appears.

Click "Done".

Select "View" and "Zoom".

Click "In" to zoom in and create a box around node 1 to see "−10000;L1;Lbf". Repeat this process as necessary.

Click "Done".

Click "View" and "Enclose" to see the whole truss again.

Step 5   Add the Material Properties Using the "Model Data Control" Window

After the mesh has been created and nodal boundary conditions and forces applied, you can select "Tools" and "Model Data Control" from Superdraw III to complete the data input for analysis. The "Model Data Control" window will appear. Alternatively, you can

Click on the "Model Data" control button located on the lower part of the screen highlighted in red.

For the "Analysis Type", "Linear Static Stress" appears highlighted as the default. You can scroll down for other analysis types as appropriate.

Click on the box below the "Element" column. A table of group 1 element types appears.

Click on "Truss". You now have created a truss. All lines are truss elements.

Click "OK".

Click on the box below the "Data" column. It will respond with "Please enter the model name first".

Click "OK" and then enter the model name in the "File_name" area that appears.

ex42   The name of the file is *ex42*.

Click "Save". By default the "Units Definition" menu appears.

Click "OK" to accept the default "English (in)" units system. Otherwise, scroll down to select another choice of units.

The "Element Definition" menu appears next.

Enter the cross-sectional area of the truss elements. All elements have the same area in this example. Hold the left click of the mouse over the default value of 0 until a blue flashing box appears on the 0.

2.0   Enter a 2. This is the cross-sectional area of the truss elements.
              Leave the "Stress Free Reference Temperature" as 0.

Click "OK". This sends you back to the "Material Data Control" window.

Click on the box below the "Material" column. Select one of the materials or select "Customer Defined". Assume we select "Customer Defined".

Then select "View Properties:Unlock Properties" and enter the modulus of elasticity of steel as you define its value.

30e6   Other properties are not needed, so just enter the modulus of steel as 30e6.

Click "OK".

Click "OK". A second click takes you out of the "Element Material Selection" menu and back to the "Model Data Control" window.

Step 6   Enter the Global Data

Click on the "Global" button on the right-side menu of the "Model Data Control" window. Under "Load Case Multipliers" and under "Pressure", type a 1.

1   Type a 1 under the "Pressure" column as you have applied mechanical loads to the truss. The program will not run without the load multiplier. The load case multiplier is what all the loads are multiplied by. Usually this value is 1. While still within the "Global" menu,

Click the "Solution" tab to view the solution options. For instance, you can ask for the banded solver to be used instead of the default skyline solver. (Refer to Appendix B.4.)

Click "Output" on the right side of the menu and then

Click the boxes next to "Displacement data" and next to "Stress data". This creates nodal displacement output in the .L file and element forces and stresses in the .S file. You can also click "Element data" to have element data written to the .L file as well. The element data consists of the nodes assigned to the elements, element loads, and material properties. You can click "Nodal data" to have nodal data written to the .L file. The nodal data consists of node numbers, boundary conditions, locations of nodes, and nodal forces.

Click "OK". You are now out of the "Global" menu.

Step 7   Check the Model

Click "Check" under the "FEA Model" column on the right side of the "Model Data Control" window. This takes you to Superview's main menu, and you can check your model as explained in Section 4.1 and detailed by the example in Appendix F. The truss model appears automatically, showing the boundary conditions as red arrows and the load arrow in yellow.

Click "Done". This takes you out of the Superview menu and back to the "Model Data Control" window.

Step 8   Run the Analysis

Click on the "Analysis" button under the "FEA Model" column in the "Model Data Control" window.

Click "Analyze" in the lower left corner of the "Linear Stress and Vibration-Static Stress" menu. (See the green box next to "Analyze".) Wait a short time for the analysis to be performed. A warning message appears. Just click "OK".

Click "Done" on the lower right of the screen.

Step 9   Review the Results

Click on the "Results" button in the "Model Data Control" window. This takes you to Superview again, and you can view the results.

Select "Displaced" from the main menu of Superview.

Click "Displ on" so that "*Displ on" appears. The asterisk means the displaced mode is active.

Click "With undi" so that "*With undi" appears. This allows the displaced and undisplaced models to be superimposed as shown in Figure 4–11(a).

Click "Calc scal" if you want the program to enlarge or scale the displaced model.

Click "Nodes inq" on the "Displaced" menu.

Click "Get" to get a node.

Click on node 1. Move the cursor to node 1 and click on it to see the x and y displacements appear at the bottom of the screen. You will see "DX = 4.142e−003 DY = 1.586e−002 . . .".

<Esc>

<Esc>   You are now back in the main menu of Superview.

Click "Stress di". You are now going to post a truss stress.

Click "Post".

Click "Beam-trus". You must select "Beam-trus" because you have a truss model.

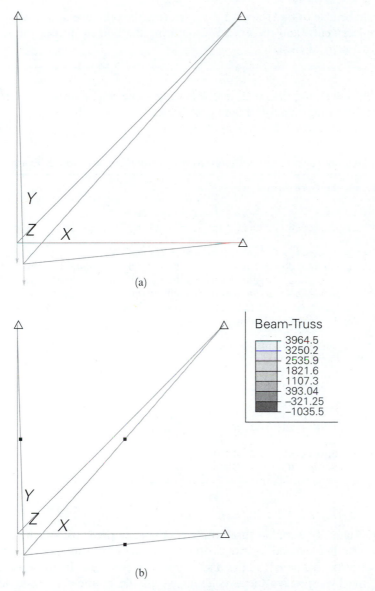

(a)

(b)

Figure 4–11 (a) Displaced truss and (b) stress plot of truss in Figure 4–10

Click "1)P/A". The axial stress will be posted. You now see the P/A stress plot [Figure 4–11(b)] for the model, with colored boxes representing different stresses on each element. For instance, red on the vertical element indicates a stress of 3964.5 psi.

<Esc>

Click "Done". This takes you out of Superview.

Click "Close" to close the "Model Data Control" window.

You should be able to find the *ex45.L* file (storing the echo check of your data and the nodal displacements) and the *ex45.S* file (storing the element forces and stresses) and print a hard copy of each. ■

Table 4–3 shows the nodal displacements and element stresses and forces taken from the Algor program ex45.L and ex45.S files.

Table 4–3 Nodal displacements and element stresses from the Algor solution of Example 4.2

Displacements/Rotations(degrees) of nodes

NODE number	X- translation	Y- translation	Z- translation	X- rotation	Y- rotation	Z- rotation
1	4.1421E-03	-1.5858E-02	0.0000E+00	0.0000E+00	0.0000E+00	0.0000E+00
2	0.0000E+00	0.0000E+00	0.0000E+00	0.0000E+00	0.0000E+00	0.0000E+00
3	0.0000E+00	0.0000E+00	0.0000E+00	0.0000E+00	0.0000E+00	0.0000E+00
4	0.0000E+00	0.0000E+00	0.0000E+00	0.0000E+00	0.0000E+00	0.0000E+00

**** Nodal stresses for 3-D truss elements:

El. #	LC	ND	Stress	Force
1	1	I	1.036E+03	2.071E+03
1	1	J	-1.036E+03	-2.071E+03
2	1	I	-3.964E+03	-7.929E+03
2	1	J	3.964E+03	7.929E+03
3	1	I	-1.464E+03	-2.929E+03
3	1	J	1.464E+03	2.929E+03

■

The Algor program has three kinds of boundary elements, which are two-noded elements. The line defined by the two nodes specifies the direction along which the force reaction is evaluated or the displacement is specified. In the case of moment reaction, the line specifies the axis about which the moment is evaluated and the rotation is specified. The boundary elements are:

1. Apply specified translational or rotational displacements.
   This boundary element has a large spring stiffness and is used to specify nonzero displacements (as illustrated in Example 4.3) and rotations to nodes. This element must be oriented in the global x, y, z directions.
2. Rigidly support nodes and calculate support reactions.
   This boundary element has a large spring stiffness and also must be aligned with the global x, y, z axes. The number and direction of the

elements is determined by the boundary element axes you selected in Superdraw. The number of axes you select is the number of boundary elements you will apply to your model (you can define up to three rigid boundary elements with one command).

3. Apply elastic support nodes.

This element has normal stiffness and can be inclined or skewed to the global x, y, z axes. The element is used to model flexible supports and to calculate reactions at these skewed supports. Elastic boundary elements are described in more detail in Example 4.5. Consult Reference [4] for more details about using boundary elements.

Example 4.3

Solve Example 3.6 (see Figure 4–12) using the Algor program. This problem illustrates how the Algor program handles support settlement using the *boundary element*. Let $A = 6.0 \times 10^{-4}$ m^2 for both members and let $E = 210$ GPa. The support settlement is 50 mm down at node 1.

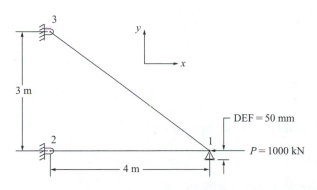

Figure 4–12 Two-bar truss with support settlement

Step 1 Model the Truss Using Superdraw III

Click "Add".

Go to "Line". Begin to enter the coordinates.

(0,0,0)   Enter the first coordinate of the truss.

<Enter>

(4,0,0)   Enter the second coordinate of the truss.

<Enter>

(0,3,0)   Enter the third coordinate of the truss.

<Enter>

Click "Done".
Go to "View" and click "Enclose". Finally, to apply the correct units to the model,
Click on the "Model Data" button on the lower portion of the screen.
Click "Units" in the "Model Data Control" window.
Select "Metric mks" for the unit system and click "OK".

Step 2 Add Boundary Conditions and Apply Force

Click "FEA Add" and go to "Stress and Vibration Analysis".
Click "Boundary Conditions".
Make sure there is a check next to "Use @ symbol for full" only.
Click "Box Apply".
With the cursor make a box around nodes 2 and 3.
An "@" symbol should appear at these nodes.
Click "Done".
Again, click "FEA Add" and go to "Stress and Vibration Analysis".
Click "Nodal Forces".
Click "Vector".
Click "X direction".
Click "Negate".
Click "Done".
Click "Magnitude".
Type "1000000".
<Enter>
Click "Box Apply" and make a box around node 1.
Click "Done".

Step 3 Add in the Boundary Element

Click "FEA Add" and go to "Stress and Vibration Analysis".
Select "Linear extension" and
Click "Boundary elements".
Make sure there is a check next to "Translate".
Place a Check next to "Displ".
Click "Magnitude".
Type 0.05 and ⟨Enter⟩.
Click "Done".
Click "Vector".
Click "Y direction".
Click "Negate".

Click "Done".
Click "Box Apply" and make a box around node 1.
Click "Done".

Step 4   Add Element and Material Type

Click on the "Model Data" button to open the "Model Data Control" window.
Click on the box below "Element".
Select "Truss".
Click "OK".
Click on the box below "Data".
Type "0.0006" next to area.
Click "OK".
Click next to the box under "Material".
Click "Customer Defined".
Click "Edit Properties".
Type "210e9" for "Modulus of elasticity".
Click "OK".
Again click "OK" to exit the "Material Data Control" window.

Step 5   Enter Global Data

In the "Model Data Control" window, click "Global".
Type the value "1" under the words "Pressure" and "Boundary".
Click on the "Output" tab.
Place checks next to "Nodal data", "Element data", "Stress data", and "Displacement data".
Click "OK".

Step 6   Perform Analysis on the Model

In the "Model Data Control" window, click "Analysis".
Click "Analyze".
Click "OK".
Click "Done".

Step 7   Review the Results

Follow the same steps as in Example 4.1 to view the results. The results for the nodal displacements and element stresses are shown in Table 4–4 and Figures 4–13(a) and (b). The comparison to the longhand solution (Example 3.6) is very satisfactory.   ■

Table 4–4 Nodal displacements and element stresses from the Algor solution of Example 4.3

```
Displacements/Rotations(degrees) of nodes
```

NODE number	X- translation	Y- translation	Z- translation	X- rotation	Y- rotation	Z- rotation
1	0.0000E+00	0.0000E+00	0.0000E+00	0.0000E+00	0.0000E+00	0.0000E+00
2	-3.3694E-02	-5.0000E-02	0.0000E+00	0.0000E+00	0.0000E+00	0.0000E+00
3	0.0000E+00	0.0000E+00	0.0000E+00	0.0000E+00	0.0000E+00	0.0000E+00
4	0.0000E+00	0.0000E+00	0.0000E+00	0.0000E+00	0.0000E+00	0.0000E+00

```
**** 3-D truss elements:

     Number of elements  = 2
     Number of materials = 1


**** Nodal stresses for 3-D truss elements:
```

El. #	LC	ND	Stress	Force
1	1	I	1.769E+09	1.061E+06
1	1	J	-1.769E+09	-1.061E+06
2	1	I	-1.279E+08	-7.672E+04
2	1	J	1.279E+08	7.672E+04

```
1**** BOUNDARY ELEMENTS

     number of elements =                    1

1**** BOUNDARY ELEMENT FORCES/MOMENTS
```

ELEMENT NO.	CASE (MODE)	FORCE	MOMENT
1	1	-4.6031E+04	0.0000E+00

```
Note: A boundary element load or displacement is defined to be positive in the
      direction from the structure attachment node (NP) to the reference node
      (NI). Therefore, a positive output force means that the element is in
      compression. Please refer to REFERENCE MANUAL for more information.


1**** End of file                                                        ▪
```

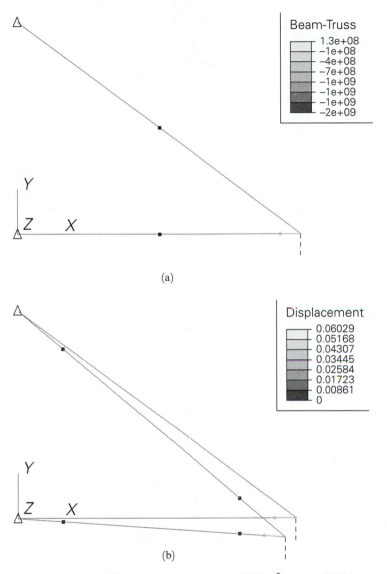

Beam-Truss

	1.3e+08
	−1e+08
	−4e+08
	−7e+08
	−1e+09
	−1e+09
	−1e+09
	−2e+09

(a)

Displacement

	0.06029
	0.05168
	0.04307
	0.03445
	0.02584
	0.01723
	0.00861
	0

(b)

Figure 4–13 (a) Truss element stresses in N/m^2 units and (b) truss displacement plot in meters

Example 4.4

Solve the three-dimensional truss problem shown in Figure 4–14 using the Algor program. This is Example 3.7, which was solved longhand. This example illustrates how to include elements with different cross-sectional areas using the *Group* command.

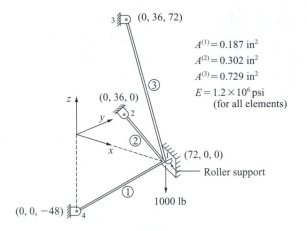

$A^{(1)} = 0.187 \text{ in}^2$
$A^{(2)} = 0.302 \text{ in}^2$
$A^{(3)} = 0.729 \text{ in}^2$
$E = 1.2 \times 10^6 \text{ psi}$
(for all elements)

Figure 4–14 Space truss

Step 1 Start Superdraw III

Click on the "Algor FEA" icon to start Superdraw III.

Step 2 Create the Truss Model

Select the "Add" menu from the top main menu bar.

Click "Line". A menu appears on the upper right part of the screen.

Click "Single" on the main menu on the upper right part of the screen. Then enter the data points (coordinates of lines 1, 2, and 3 in succession).

0<Tab>0<Tab>-48	Coordinates of the first point of line 1.
<Enter>	Press the ⟨Enter⟩ key to actually enter the data.
72<Tab>0<Tab>0	This is the endpoint of line 1.
<Enter>	
0<Tab>36<Tab>0	This is the start point of line 2.
<Enter>	
72<Tab>0<Tab>0	This is the endpoint of line 2.
<Enter>	
0<Tab>36<Tab>72	This is the start point of line 3.
<Enter>	
72<Tab>0<Tab>0	This is the endpoint of line 3.

Click "Done" on the menu on the upper right side of the screen.

Select "View" from the main menu bar. Under "View:Predefined Views", click the "Isometric" option.

Select "View" and click on "Enclose". The model should appear on the screen as an isometric drawing.

Step 3 Eliminate Duplicate Lines

Click "Modify" on main menu.

Select "Clean:Duplicate" from the "Duplicate" menu. You will see the following message on the lower part of the screen:

3 Kept 0 Deleted. done

Click "Done" on the "Duplicate" menu.

Step 4 Add Boundary Conditions and Force

Click "FEA Add" on the main menu and highlight "Stress and Vibration Analysis".

Click "Boundary Conditions". The "Boundary Condition" menu appears. The default boundary condition is "Use @ Symbol for Full". "Full" means full constraint or prevention of translations and rotations (when relevant).

Right-click on nodes labeled 2, 3, and 4 in the figure. These nodes are then fixed to the wall. After fixing the three nodes,

Click "Change Values" on the "Boundary Condition" menu. This takes you to the "BC Values" menu.

Click "Tx constrained" to remove the x-displacement constraint.

Click "Tz constrained" to remove the z-displacement constraint.

Click "Done" to exit this menu.

Right-click on node 1. You now have a roller effect at node 1.

Click "Done".

Go to "View:Zoom:In" and create a box around node 1. Repeat this process two times to clearly see the label "TyRxyz" appear next to the node.

Click "Done".

Select "View:Enclose". The model should appear.

Click on the "FEA Add" menu again and move to highlight "Stress and Vibration Analysis".

Click "Nodal Force" to create a force.

Click "Vector".

Click "Z Direction" because the force is in the z direction. You will see "DX 0 DY 0 DZ 0.97" at the bottom of the screen. This means the force will now be in the z direction.

Click "Done". This takes you back to the "Nodal Forces" menu.

Click "Magnitude" and enter the magnitude.

-1000 This is the magnitude of the force.

\<Enter\>

Move the cursor to node 1 and right-click on node 1.

Click "Done".

You have to select "View: Zoom: In" two times to see "$-1000;L1;LBS$" appear next to node 1.

Click "Done".

Click "View:Enclose" to see the whole truss again.

Step 5 Create New Groups to Change the Area of Elements 1, 2, and 3

(No element lines are presently highlighted as we begin this part).

Line (element) 1 (see figure of truss) will by default have group property 1.

Click "Select" from the main menu bar.

Click "Point" and then move the cursor to line 2 and left-click on line 2 (labeled on truss, see Figure 4–14). A red highlight box will appear on the line.

Click "Modify: Update Object Parameters:Group Number".

2   Enter a 2 for the second group of properties. The line will turn red. These properties will be defined in the "Model Data Control" window in step 6.

<Enter>

Click "Select" from the main menu bar.

Click "None" to unhighlight all lines.

Click "Select".

Select "Point" and move the cursor to line 3 and click on line 3. A red highlight box will appear on line 3.

Click "Done".

Select "Modify:Update Object Parameters:Group Number".

3   Enter a 3 for the third group of properties to be added in the "Model Data Control" window. The line will become yellowish.

<Enter>

Step 6 Add the Material Properties Using the "Model Data Control" Window

Click on the "Model Data" control button located on the lower part of the screen and highlighted in red.

The "Model Data Control" window appears.

For "Analysis Type", "Linear Static Stress" appears highlighted as the default.

In the "Group" column, you will see three groups—1, 2, and 3 with green, red, and yellow boxes around each number, respectively.

Click on the box below "Element".

Click "Truss".

Click "OK". Repeat this process two more times so that each element is defined to be a truss.

Click on the box below "Data". The next response will be "Please enter the model name first".

Click "OK" and then enter a model name in the "File_name" area that appears.

ex44   The name of the file is *ex44*.

Click "Save". By default the "Units Definition" menu appears.

Click "OK" if you accept the default "English (in)" units system. Otherwise scroll down to select another choice of units.

The "Element Definition" menu appears next. (Or click the box under "Data" to get to the "Element Definition" menu).

Enter the cross-sectional areas of each element on the proper lines. You will see a column of group numbers, "Gr 1", "Gr 2", and "Gr 3". Enter the areas as follows:

0.187　　　　Cross-sectional area of element 1.

Click "OK". You are now back in the "Data" window. Click the box next to "Data" again. This takes you to the "Element Definition" menu again.

0.302　　　　Cross-sectional area of element 2.

Click "OK". Repeat the above steps for the next element area.

0.729　　　　Cross-sectional area of element 3.

Click "OK". This sends you back to the "Model Data Control" window.

Click on the box below "Material". You are now in the "Element Material Selection" window.

Select "Customer Defined".

Then click "View Properties:Unlock Properties" and enter a modulus of elasticity of 1.2e6 psi for element 1 in box 1.

1.2e6

Click "OK". This takes you back to the "Element Material Selection" window.

Select "Customer Defined" again.

Click "View Properties:Unlock Properties" again.

Then enter the modulus of elasticity for element 2 in box 2.

1.2e6

Click "OK". This takes you to the "Element Material Selection" window again. You need to enter the modulus for the last element in box 3.

Select "Customer Defined" again.

Click "View Properties:Unlock Properties" again.

Enter the modulus of elasticity.

1.2e6

Click "OK".

Click "OK" again to go back to the "Model Data Control" window.

Step 7　Enter the Global Data

Click on the "Global" button on the right-side menu of the "Model Data Control" window.

Under "Load Case Multipliers" and under "Pressure", type a 1.

1　　　　　　　Type a 1 or the program will not run. This indicates the one load case is defined as an applied force (denoted by the 1 under the "Pressure" column).

While still within the "Global" menu, click "Output" on the right side of the menu;

then click on the box next to "Displacement data" and on the box next to "Stress data". This creates nodal displacement output in the .L file, and element forces and stresses in the .S file.

Click "OK". You are now out of the "Global" menu.

Step 8 Check the Model

Click "Check" under "FEA Model" on the right side of the "Model Data Control" window.

This takes you to Superview's main menu, and you can check your model.

The model appears with three different colored lines.

Click "Done" when you are ready to exit this menu.

Step 9 Run the Analysis

Click the "Analysis" button under the "FEA Model" column.

Click "Analyze" in the lower left corner of the window. Wait a short time for the analysis to be performed. Then click the "OK" button.

Click "Done".

Step 10 Review the Results

Follow the same steps as in Example 4.1 to view the displaced truss and the stresses in each element. ■

The results for the nodal displacements and element stresses shown in Table 4–5 compare closely with the longhand solution of Example 3.7.

Example 4.5

Solve Example 3.9 (see Figure 4–15) using the Algor program. This problem illustrates how Algor handles skewed or inclined boundary conditions using the *boundary element* and again shows how to define different cross-sectional areas using the *Group* command. Let $A^{(1)} = A^{(2)} = 6 \times 10^{-4}$ m^2, $A^{(3)} = 8.485 \times 10^{-4}$ m^2, and $E = 210$ GPa.

Elastic boundary elements [4] are used to model flexible supports and to calculate reactions at skewed or inclined boundaries. The stiffness of elastic boundary elements is defined by the user. When we model skewed supports, we must select the proper stiffness of the element. For the case of an unyielding support, as exists at the inclined roller support in Figure 4–15, a two-step approach is suggested:

1. Model the structure using fixed boundary conditions at the skewed support. Perform the static analysis to obtain the maximum value of the structure's stiffness matrix $k_{\max}$. This value is the "maximum diagonal element" under the heading "STIFFNESS MATRIX PARAMETERS" from the .L file. For instance, see Table 4–1, *ex41.L* file of Example 4.1 for a value of 5.1213 E+6.
2. Replace the fixed boundary support conditions at the skewed support with a translational elastic boundary element (ET in Algor) oriented

Table 4–5 Nodal displacements and element stresses from the Algor solution of Example 4.4

```
Displacements/Rotations(degrees) of nodes

NODE      X-            Y-            Z-            X-          Y-          Z-
number  translation  translation  translation  rotation    rotation    rotation

   1    0.0000E+00   0.0000E+00    0.0000E+00  0.0000E+00  0.0000E+00  0.0000E+00
   2   -7.1114E-02   0.0000E+00   -2.6624E-01  0.0000E+00  0.0000E+00  0.0000E+00
   3    0.0000E+00   0.0000E+00    0.0000E+00  0.0000E+00  0.0000E+00  0.0000E+00
   4    0.0000E+00   0.0000E+00    0.0000E+00  0.0000E+00  0.0000E+00  0.0000E+00


**** 3-D truss elements:

     Number of elements  = 3
     Number of materials = 3


**** Nodal stresses for 3-D truss elements:

     El. #   LC   ND     Stress         Force

        1    1    I     2.869E+03      5.364E+02
        1    1    J    -2.869E+03     -5.364E+02

        2    1    I     9.482E+02      2.864E+02
        2    1    J    -9.482E+02     -2.864E+02

        3    1    I    -1.445E+03     -1.054E+03
        3    1    J     1.445E+03      1.054E+03


1**** End of file                                              ■
```

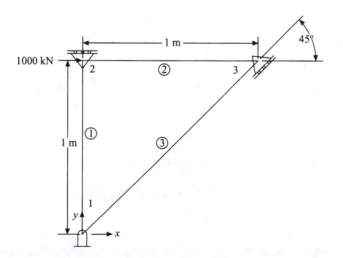

Figure 4–15 Plane truss with inclined support

along the direction of the restrained displacement. Also assign a stiffness of order 5 to 10 times k_{max} in order to provide sufficient stiffness in the spring to restrain the displacement. In step 4 of Example 4.5, we illustrate the step-by-step process to create a boundary element at node 3 of the truss in Figure 4–15.

That is, the boundary element is selected and made elastic. Then a vector perpendicular to the 45° line at node 3 is created. The stiffness is selected, and a magnitude of 1.E+12, which happens to be the default value assigned by Algor and works satisfactorily in this example, is selected. (We could have used 5 to 10 times k_{max} taken from an initial run with the support fixed.)

We will refer to the previous tutorials for the steps that are the same as in those tutorials. Details will be provided only for new concepts.

Step 1 Start Superdraw III

Click on the "Algor FEA" icon to start Superdraw III.

Step 2 Create the Truss Model

Select the "Add" menu from the top main menu bar.

Click "Line". A menu appears on the upper right side of the screen. Now enter the data points in succession starting at the origin and finishing at the origin.

0<Tab>0   This is the first point for element (line) 1 of the truss.

<Enter>

0<Tab>1   This is the second point of element 1 and the first point of element 2.

<Enter>

1<Tab>1   This is the second point of element 2 and the first point of element 3.

<Enter>

0<Tab>0   This is the second point of element 3.

<Enter>

Click "Done".

Select "View" from the main menu.

Click "Enclose". You now see the three lines forming the truss. The figure should look like Figure 4–15.

Step 3 Eliminate Duplicate Lines

Follow the same steps as in the previous example to eliminate duplicate lines.

Step 4 Add Boundary Conditions and Apply Force

Click "FEA Add" on the main menu and highlight "Stress and Vibration Analysis".

Click "Boundary Conditions". The "Boundary Condition" menu appears. The default boundary condition of complete fixity can be used at the bottom node (node 1 in Figure 4–15).

Right-click on node 1. You should see the "@" sign next to node 1.

Click "Change Values" on the "Boundary Condition" menu.

Left-click "Tx Constrained" to remove the x translational boundary condition.

Click "Done".

Right-click on top node 2. "TyzRxyz" will appear next to node 2. You have now created a roller at node 2.

Click "Done".

Click "View:Enclose". You should now see the model with boundary conditions indicated.

Click "FEA Add:Stress and Vibration Analysis".

Click "Force". You will now add the force.

Click "Vector". This allows you to direct the force in the x direction.

Click "X- direction". You should see "DX 0.97 DY 0 DZ 0" at the bottom of the screen.

Click "Done".

Click "Magnitude". This allows you to enter the magnitude of the force.

1000   A 1000-kN force is created.

<Enter>   This actually enters the force magnitude.

Right-click (snap-to) on node 2. A 1000 appears with an arrow.

Click "Done".

Step 5   Add Boundary Element

For more on the boundary element, see the beginning of Example 4.5.

Click "FEA Add:Stress and Vibration Analysis".

Select "Linear Extensions".

Click "Boundary Elements"

Click "Elastic" and make sure the default "Translate" is checked.

Click "Vector". You need to define a vector that gives the direction of the boundary element. See the beginning of Example 4.5 in this tutorial and page 160 of the text for more details on boundary elements.

1<Tab>1   Create the start point for the unit vector of the boundary element.

<Enter>

2<Tab>0   Input the endpoint of the vector for the boundary element. This element will then be perpendicular to a 45° angle as shown in Figure 4–15.

<Enter>

Click "Done".

Click "Stiffness" on the "Boundary Elements" menu.

1e12   Type the stiffness of the boundary element. A value about 10^4 times the largest diagonal stiffness of the structure is recommended. You can run the analysis and check the output for this value. If need be, rerun the problem with the recommended stiffness. A stiffness of 1e12 should work in this problem.

`<Enter>`

Right-click on node 3. Then "ET"; 1e+012" appears with an arrow perpendicular to the 45° line 1–3.

Click "Done".

Click "View:Enclose" to view the whole model.

Step 6 Create a New Group to Change the Area of Elements

Warning: In Release 12.0, you must remove the default check mark in front of "Maintain Group Boundary". In Release 12.02, this default no longer exists.

Click "Select".

Click "Point".

Click on element 3, the diagonal element. Make sure no other elements are highlighted. If necessary, click "Select:None" first. Then select element 3.

Click "Done".

Click "Modify" on the main menu.

Select "Update Object Parameters" and click "Maintain Group Boundary" to remove the check mark. This takes away the maintaining of group boundaries, as we want to change these boundaries for this truss. We want to make the diagonal member have a different group.

Click "Modify".

Select "Update Object Parameters".

Click "Group Number".

2   Enter a 2 so that the diagonal element will have a different group number than the other elements. The diagonal element line is now red. The cross-sectional area of the diagonal element is different than that of the other two elements. The cross-sectional areas will be entered in the "Model Data Control" window.

Click "Inquire:Properties of Object" to inquire about the group status of each line.

Click on the diagonal line, and you should see "... Group = 2".

Step 7 Add the Material Properties Using the "Model Data Control" Window

Click on the "Model Data" button at the bottom of the screen. The "Model Data Control" window appears.

For "Analysis Type", "Linear Static Stress" appears highlighted as the default type of analysis.

In the "Group" column, you see two groups—1 with a green box around it and 2 with a red box around it

Click on the box below "Element".

Click "Truss". You now have truss elements.

Click "OK". Repeat, clicking on the second box below "Element".

Click on the box below "Data". The next response is "Please enter the model name first".

Click "OK" and enter the model name in the "File_name" area.

Ex45 The name of this file is "*Ex45*".

Click "Save". The "Units Definition" menu now appears.

Click "OK" if you accept the units or scroll down to select other units.

The "Element Definition" menu appears next. Or click on the box under "Data" to get the "Element Definition" menu.

Enter the cross-sectional areas of each element on the proper lines. You will see a column of group numbers, "Gr 1" and "Gr 2". Enter the areas as follows:

6e-4   Cross-sectional area of the vertical and horizontal elements.

Click "OK".

Click on the second box below the "Data" box.

8.485e-4   Cross-sectional area of diagonal element.

Click "OK".

Click the box under "Material".

Select "Customer Defined".

Click "View Properties" and then click "Unlock Properties" and enter the modulus of elasticity for the group 1 material.

210e6   The value of $E = 210e6$ kPa.

Click "OK". This takes you back to the "Element Material Selection" window.

Select "Customer Defined" again.

Click "View Properties" and click " Unlock Properties" again.

Enter the modulus of elasticity of the group 2 material.

210e6

Click "OK".

Click "OK" again. You should be back at the "Model Data Control" window.

Step 8 Enter the Global Data

Follow the steps in the previous example.

Step 9 Check the Model

Follow the steps in the previous example.

Step 10 Run the Analysis

Follow the steps in the previous example.

Step 11 Review the Results

Follow the steps in the previous example.   ■

Table 4–6 Nodal displacements, element stresses, and boundary element reaction from the Algor solution of Example 4.5

NODE number	X- translation	Y- translation	Z- translation	X- rotation	Y- rotation	Z- rotation
1	0.0000E+00	0.0000E+00	0.0000E+00	0.0000E+00	0.0000E+00	0.0000E+00
2	1.1905E-02	0.0000E+00	0.0000E+00	0.0000E+00	0.0000E+00	0.0000E+00
3	3.9684E-03	3.9683E-03	0.0000E+00	0.0000E+00	0.0000E+00	0.0000E+00
4	0.0000E+00	0.0000E+00	0.0000E+00	0.0000E+00	0.0000E+00	0.0000E+00

```
**** 3-D truss elements:

     Number of elements  = 3
     Number of materials = 2

**** Nodal stresses for 3-D truss elements:
```

El. #	LC	ND	Stress	Force
1	1	I	0.000E+00	0.000E+00
1	1	J	0.000E+00	0.000E+00
2	1	I	-8.334E+05	-7.071E+02
2	1	J	8.334E+05	7.071E+02
3	1	I	1.667E+06	1.000E+03
3	1	J	-1.667E+06	-1.000E+03

```
1**** BOUNDARY ELEMENTS

     number of elements =                    1

1**** BOUNDARY ELEMENT FORCES/MOMENTS
```

ELEMENT NO.	CASE (MODE)	FORCE	MOMENT
1	1	7.0711E+02	0.0000E+00

```
Note: A boundary element load or displacement is defined to be positive in the
      direction from the structure attachment node (NP) to the reference node
      (NI). Therefore, a positive output force means that the element is in
      compression. Please refer to REFERENCE MANUAL for more information.   ■
```

The results for the nodal displacements compare closely with those of Example 3.9. The displacement of node two in the x direction is the same in Table 4–6 as in Eq. (3.9.24). The displacements in Algor are all global ones. To obtain the displacement along the incline, the global x and y components from Algor must both be separated into components along and perpendicular to the incline. The results then compare closely with the displacement along the incline, x' axis; see Eq. (3.9.24). The boundary element force shown in Table 4–6 compares closely with the longhand solution for F'_{3y} of Eq. (3.9.25).

References

[1] Docutech On-Line Technical User Documentation, Algor, Inc., Pittsburgh, PA.
[2] Superdraw Reference Division, Docutech On-Line Technical User Documentation, Algor, Inc., Pittsburgh, PA.
[3] Superview Reference Division, Docutech On-Line Technical User Documentation, Algor, Inc., Pittsburgh, PA.
[4] Linear Stress and Vibration Reference Division, Docutech On-Line Technical User Documentation, Algor, Inc., Pittsburgh, PA.

Problems

Use the Algor computer program to solve Problems 4.1–4.19

4.1 For the bridge truss shown in Figure P4–1, use symmetry to determine the displacements of the nodes and the stresses in each element. All elements have $E = 210$ GPa and $A = 30 \times 10^{-4}$ m^2 except element 7, which has $A = 60 \times 10^{-4}$ m^2.

$$210 \times 10^9 \ N/m^2 \qquad 2101/0^6 N/m^2$$

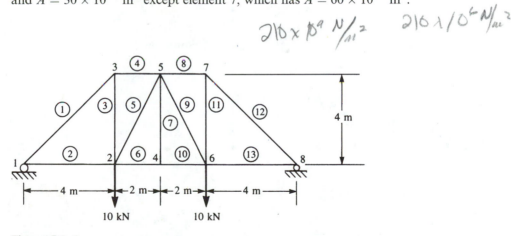

Figure P4–1

4.2 A typical truss made of structural steel for a mill building is shown in Figure P4–2. Design the truss members such that the maximum stress in any member is 15 ksi. Assume all elements have the same cross-sectional area.

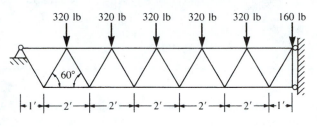

320 lb   320 lb   320 lb   320 lb   320 lb   160 lb

60°

|←1'→|←——2'——→|←——2'——→|←——2'——→|←——2'——→|←——2'——→|←1'→|

Figure P4–2

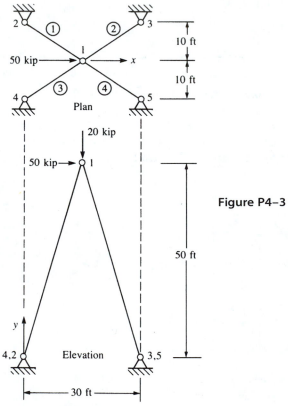

10 ft

10 ft

50 kip →   x

Plan

20 kip

50 kip → ○ 1

Figure P4–3

50 ft

y ↑

4,2   Elevation   3,5

|←——— 30 ft ———→|

4.3 For the three-dimensional truss shown in Figure P4–3, determine the displacement components at node 1 and the stresses in each element. Let $A = 4$ in^2 and $E = 30 \times 10^6$ psi.

4.4 If the space truss shown in Figure P4–4 is to be constructed of 30,000 psi yield-strength steel and a safety factor of 2.0 is to be used, determine the cross-sectional area required for the elements. All elements must have the same final cross-sectional area. The loading on node 7 is a vertical force of 4500 lb and a horizontal force of 2800 lb. Nodes 1 and 2 are completely fixed, whereas node 3 is constrained from vertical displacement.

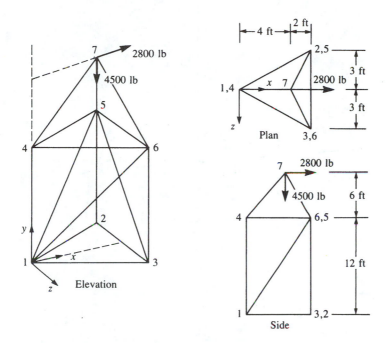

Figure P4–4

4.5 For the space truss shown in Figure P4–5, all nodes other than node 1 are fixed (ball-and-socket joints). Determine the displacement components at node 1 and the element stresses. Let the cross-sectional areas of elements 2 and 5 be 0.729 in², that of elements 1 and 4 be 0.302 in², and that of element 3 be 0.374 in². Let $E = 1.2 \times 10^6$ psi for all elements. A force of 2000 lb is applied at node 1 in the negative z direction. The coordinates of each node (in units of inches) are shown in the figure.

4.6 For the space truss shown in Figure P4–6, nodes 3, 4, 5, and 6 are ball-and-socket joints. Determine the displacement components at nodes 1 and 2 and the element stresses. Let $E = 70$ GPa and $A = 4 \times 10^{-4}$ m² for all elements. A force of 50 kN is applied at node 1 in the negative y direction.

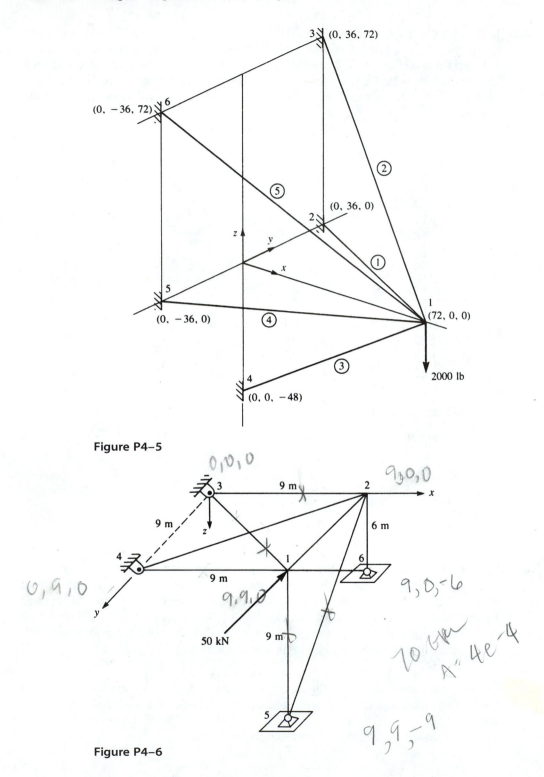

Figure P4–5

Figure P4–6

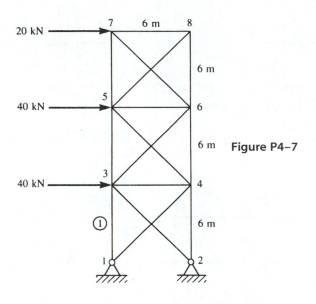

Figure P4–7

4.7 For the plane truss shown in Figure P4–7, determine the nodal displacements and the element stresses. Nodes 1 and 2 are pin joints. Let $E = 210$ GPa and $A = 3 \times 10^{-4}$ m² for all elements.

4.8 For the space truss shown in Figure P4–8, determine the nodal displacements and the element stresses. Let $E = 210$ GPa and $A = 1 \times 10^{-4}$ m² for all elements. Nodes 1–3 are ball-and-socket joints.

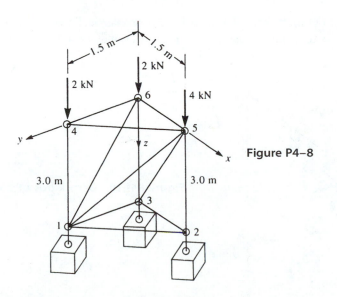

Figure P4–8

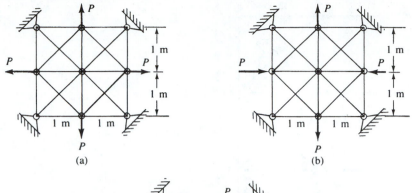

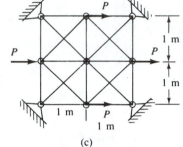

Figure P4–9

4.9 For the plane truss shown in Figure P4–9, determine the nodal displacements and element stresses. Let $A = 1 \times 10^{-4}$ m^2 and $E = 200$ GPa for each element and let $P = 10$ kN.

4.10–4.19 Use the Algor program to solve the truss design problems shown in Figures P4.10–P4.19. Determine the single most critical cross-sectional area based on maximum

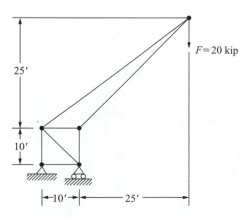

Figure P4–10   Derrick truss (FS = 4.0)

allowable yield strength or buckling strength (based on either Euler's or Johnson's formula as relevant) using the factor of safety (FS) listed next to each truss. Recommend a common structural shape and size for each truss. List the largest three nodal displacements and their locations. Also include both a plot of the deflected shape of the truss and a stress plot.

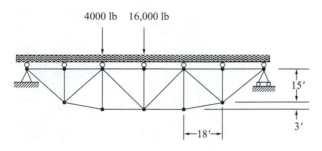

Figure P4–11 Truss bridge (FS = 3.0)

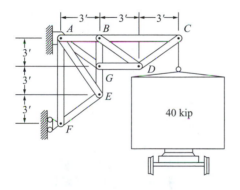

Figure P4–12 Boxcar lift (FS = 3.0)

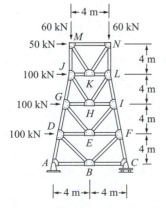

Figure P4–13 Tower (FS = 2.5)

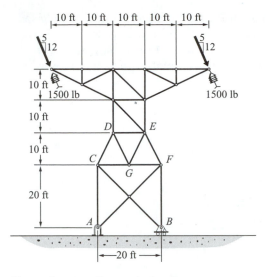

Figure P4–14 Transmission line truss (FS = 2.5)

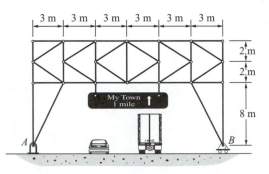

Figure P4–15 Signboard truss (FS = 2.0). Sign weight = 1000 N

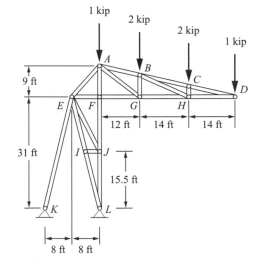

Figure P4–16 Stadium roof truss (FS = 3.0)

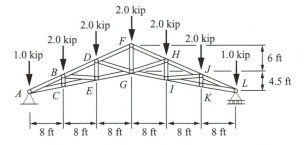

Figure P4–17 Howe scissors roof truss (FS = 2.0)

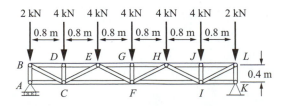

Figure P4–18 Floor truss (FS = 2.0)

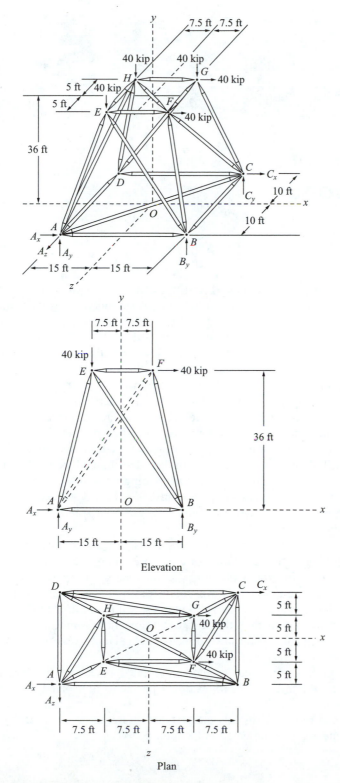

Figure P4–19 Space truss (FS = 2.0)

5

Development of Beam Equations

Introduction

We begin this chapter by developing the stiffness matrix for the bending of a beam element, the most common of all structural elements as evidenced by its prominence in buildings, bridges, towers, and many other structures. The beam element is considered to be straight and to have constant cross-sectional area. We will first derive the beam element stiffness matrix by using the principles developed for simple beam theory.

We will then present simple examples to illustrate the assemblage of beam element stiffness matrices and the solution of beam problems by the direct stiffness method presented in Chapter 2. The solution of a beam problem illustrates that the degrees of freedom associated with a node are a transverse displacement and a rotation. We will include the nodal shear forces and bending moments and the resulting shear force and bending moment diagrams as part of the total solution.

Next, we will discuss procedures for handling distributed loading, because beams and frames are often subjected to distributed loading as well as concentrated nodal loading. We will follow the discussion with solutions of beams subjected to distributed loading and compare a finite element solution to an exact solution for a beam subjected to a distributed loading.

We will then develop the beam element stiffness matrix for a beam element with a nodal hinge and illustrate the solution of a beam with an internal hinge.

To further acquaint you with the potential energy approach for developing stiffness matrices and equations, we will again develop the beam bending element equations using this approach. We hope to increase your confidence in this approach. It will be used throughout much of this text to develop stiffness matrices and equations for more complex elements, such as two-dimensional (plane) stress, axisymmetric, and three-dimensional stress.

Next, the Galerkin residual method is applied to derive the beam element equations. Finally, a number of beam example problems are solved using the Algor program.

The concepts presented in this chapter are prerequisite to understanding the concepts for frame analysis presented in Chapter 6.

▲ 5.1 Beam Stiffness ▲

In this section, we will derive the stiffness matrix for a simple beam element. A **beam** *is a long, slender structural member generally subjected to transverse loading that produces significant bending effects as opposed to twisting or axial effects.* This bending deformation is measured as a transverse displacement and a rotation. Hence, the degrees of freedom considered per node are a transverse displacement and a rotation (as opposed to only an axial displacement for the bar element of Chapter 3).

Consider the beam element shown in Figure 5–1. The beam is of length L with axial local coordinate $\hat{x}$ and transverse local coordinate $\hat{y}$. The local transverse nodal displacements are given by $\hat{d}_{iy}$'s and the rotations by $\hat{\phi}_i$'s. The local nodal forces are given by $\hat{f}_{iy}$'s and the bending moments by $\hat{m}_i$'s as shown. We initially neglect all axial effects.

At all nodes, the following sign conventions are used:

1. Moments are positive in the counterclockwise direction.
2. Rotations are positive in the counterclockwise direction.
3. Forces are positive in the positive $\hat{y}$ direction.
4. Displacements are positive in the positive $\hat{y}$ direction.

Figure 5–2 indicates the sign conventions used in simple beam theory for positive shear forces $\hat{V}$ and bending moments $\hat{m}$.

The differential equation governing elementary linear-elastic beam behavior [1] is derived as follows. Consider the beam shown in Figure 5–3 subjected to a distributed loading $w(\hat{x})$ (force/length). From force and moment equilibrium of a differ-

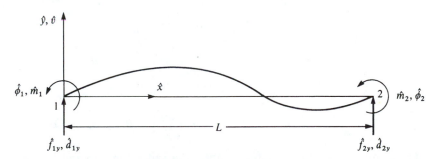

Figure 5–1 Beam element with positive nodal displacements, rotations, forces, and moments

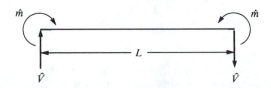

Figure 5–2 Beam theory sign conventions for shear forces and bending moments

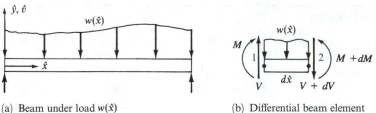

(a) Beam under load $w(\hat{x})$

(b) Differential beam element

Figure 5–3 Beam under distributed load

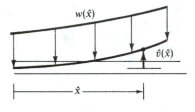

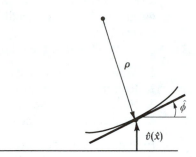

(a) Portion of deflected curve of beam

(b) Radius of deflected curve at $\hat{v}(\hat{x})$

Figure 5–4 Deflected curve of beam

ential element of the beam, we have

$$\Sigma F_y = 0: \quad V - (V + dV) - w(\hat{x})\,dx = 0 \tag{5.1.1a}$$

Or, simplifying Eq. (5.1.1a), we obtain

$$-w\,d\hat{x} - dV = 0 \quad \text{or} \quad w = -\frac{dV}{d\hat{x}} \tag{5.1.1b}$$

$$\Sigma M_2 = 0: \quad -V\,dx + dM + w(\hat{x})\,d\hat{x}\left(\frac{d\hat{x}}{2}\right) = 0 \quad \text{or} \quad V = \frac{dM}{d\hat{x}} \tag{5.1.1c}$$

The final form of Eq. (5.1.1c), relating the shear force to the bending moment, is obtained by dividing the left equation by $d\hat{x}$ and then taking the limit of the equation as $d\hat{x}$ approaches 0. The $w(\hat{x})$ term then disappears.

Also, the curvature κ of the beam is related to the moment by

$$\kappa = \frac{1}{\rho} = \frac{M}{EI} \tag{5.1.1d}$$

where ρ is the radius of the deflected curve shown in Figure 5–4b, $\hat{v}$ is the transverse displacement function in the $\hat{y}$ direction (see Figure 5–4a), E is the modulus of elasticity, and I is the principal moment of inertia about the $\hat{z}$ axis (where the $\hat{z}$ axis is perpendicular to the $\hat{x}$ and $\hat{y}$ axes).

The curvature for small slopes $\hat{\phi} = d\hat{v}/d\hat{x}$ is given by

$$\kappa = \frac{d^2\hat{v}}{d\hat{x}^2} \tag{5.1.1e}$$

Using Eq. (5.1.1e) in (5.1.1d), we obtain

$$\frac{d^2\hat{v}}{d\hat{x}^2} = \frac{M}{EI} \tag{5.1.1f}$$

Solving Eq. (5.1.1f) for M and substituting this result into (5.1.1c) and (5.1.1b), we obtain

$$\frac{d^2}{d\hat{x}^2}\left(EI\frac{d^2\hat{v}}{d\hat{x}^2}\right) = -w(\hat{x}) \tag{5.1.1g}$$

For constant EI and only nodal forces and moments, Eq. (5.1.1g) becomes

$$EI\frac{d^4\hat{v}}{d\hat{x}^4} = 0 \tag{5.1.1h}$$

We will now follow the steps outlined in Chapter 1 to develop the stiffness matrix and equations for a beam element and then to illustrate complete solutions for beams.

Step 1 Select the Element Type

Represent the beam by labeling nodes at each end and in general by labeling the element number (Figure 5–1).

Step 2 Select a Displacement Function

Assume the transverse displacement variation through the element length to be

$$\hat{v}(\hat{x}) = a_1\hat{x}^3 + a_2\hat{x}^2 + a_3\hat{x} + a_4 \tag{5.1.2}$$

The complete cubic displacement function Eq. (5.1.2) is appropriate because there are four total degrees of freedom (a transverse displacement $\hat{d}_{iy}$ and a small rotation $\hat{\phi}_i$ at each node). The cubic function also satisfies the basic beam differential equation— further justifying its selection. In addition, the cubic function also satisfies the conditions of displacement and slope continuity at nodes shared by two elements.

Using the same procedure as described in Section 2.2, we express $\hat{v}$ as a function of the nodal degrees of freedom $\hat{d}_{1y}$, $\hat{d}_{2y}$, $\hat{\phi}_1$, and $\hat{\phi}_2$ as follows:

$$
\begin{aligned}
\hat{v}(0) &= \hat{d}_{1y} = a_4 \\[4pt]
\frac{d\hat{v}(0)}{d\hat{x}} &= \hat{\phi}_1 = a_3 \\[4pt]
\hat{v}(L) &= \hat{d}_{2y} = a_1L^3 + a_2L^2 + a_3L + a_4 \\[4pt]
\frac{d\hat{v}(L)}{d\hat{x}} &= \hat{\phi}_2 = 3a_1L^2 + 2a_2L + a_3
\end{aligned}
\tag{5.1.3}
$$

where $\hat{\phi} = d\hat{v}/d\hat{x}$ for the assumed small rotation $\hat{\phi}$. Solving Eqs. (5.1.3) for a_1 through a_4 in terms of the nodal degrees of freedom and substituting into Eq. (5.1.2), we have

$$\hat{v} = \left[\frac{2}{L^3}(\hat{d}_{1y} - \hat{d}_{2y}) + \frac{1}{L^2}(\hat{\phi}_1 + \hat{\phi}_2)\right]\hat{x}^3$$

$$+ \left[-\frac{3}{L^2}(\hat{d}_{1y} - \hat{d}_{2y}) - \frac{1}{L}(2\hat{\phi}_1 + \hat{\phi}_2)\right]\hat{x}^2 + \hat{\phi}_1\hat{x} + \hat{d}_{1y} \qquad (5.1.4)$$

In matrix form, we express Eq. (5.1.4) as

$$\hat{v} = [N]\{\hat{d}\} \qquad (5.1.5)$$

where

$$\{\hat{d}\} = \begin{Bmatrix} \hat{d}_{1y} \\ \hat{\phi}_1 \\ \hat{d}_{2y} \\ \hat{\phi}_2 \end{Bmatrix} \qquad (5.1.6a)$$

and where

$$[N] = [N_1 \quad N_2 \quad N_3 \quad N_4] \qquad (5.1.6b)$$

and

$$N_1 = \frac{1}{L^3}(2\hat{x}^3 - 3\hat{x}^2L + L^3) \qquad N_2 = \frac{1}{L^3}(\hat{x}^3L - 2\hat{x}^2L^2 + \hat{x}L^3)$$

$$(5.1.7)$$

$$N_3 = \frac{1}{L^3}(-2\hat{x}^3 + 3\hat{x}^2L) \qquad N_4 = \frac{1}{L^3}(\hat{x}^3L - \hat{x}^2L^2)$$

N_1, N_2, N_3, and N_4 are called the **shape functions** for a beam element. For the beam element, $N_1 = 1$ when evaluated at node 1 and $N_1 = 0$ when evaluated at node 2. Because N_2 is associated with $\hat{\phi}_1$, we have, from the second of Eqs. (5.1.7), $(dN_2/d\hat{x}) = 1$ when evaluated at node 1. Shape functions N_3 and N_4 have analogous results for node 2.

Step 3 Define the Strain/Displacement and Stress/Strain Relationships

Assume the following axial strain/displacement relationship to be valid:

$$\varepsilon_x(\hat{x}, \hat{y}) = \frac{d\hat{u}}{d\hat{x}} \qquad (5.1.8)$$

where $\hat{u}$ is the axial displacement function. From the deformed configuration of the beam shown in Figure 5–5, we relate the axial displacement to the transverse displacement by

$$\hat{u} = -\hat{y}\frac{d\hat{v}}{d\hat{x}} \qquad (5.1.9)$$

where we should recall from elementary beam theory [1] the basic assumption that cross sections of the beam (such as cross section $ABCD$) that are planar before bending deformation remain planar after deformation and, in general, rotate through a

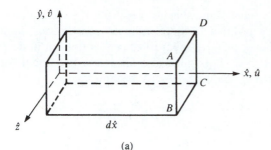

(a)

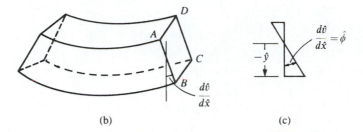

(b)   (c)

Figure 5–5 Beam segment (a) before deformation and (b) after deformation; (c) angle of rotation of cross section *ABCD*

small angle $(d\hat{v}/d\hat{x})$. Using Eq. (5.1.9) in Eq. (5.1.8), we obtain

$$\varepsilon_x(\hat{x}, \hat{y}) = -\hat{y}\frac{d^2\hat{v}}{d\hat{x}^2} \tag{5.1.10}$$

From elementary beam theory, the bending moment and shear force are related to the transverse displacement function. Because we will use these relationships in the derivation of the beam element stiffness matrix, we now present them as

$$\hat{m}(\hat{x}) = EI\frac{d^2\hat{v}}{d\hat{x}^2} \qquad \hat{V} = EI\frac{d^3\hat{v}}{d\hat{x}^3} \tag{5.1.11}$$

Step 4   Derive the Element Stiffness Matrix and Equations

First, derive the element stiffness matrix and equations using a direct equilibrium approach. We now relate the nodal and beam theory sign conventions for shear forces and bending moments (Figures 5–1 and 5–2), along with Eqs. (5.1.4) and (5.1.11), to obtain

$$\hat{f}_{1y} = \hat{V} = EI\frac{d^3\hat{v}(0)}{d\hat{x}^3} = \frac{EI}{L^3}(12\hat{d}_{1y} + 6L\hat{\phi}_1 - 12\hat{d}_{2y} + 6L\hat{\phi}_2)$$

$$\hat{m}_1 = -\hat{m} = -EI\frac{d^2\hat{v}(0)}{d\hat{x}^2} = \frac{EI}{L^3}(6L\hat{d}_{1y} + 4L^2\hat{\phi}_1 - 6L\hat{d}_{2y} + 2L^2\hat{\phi}_2)$$

$$\hat{f}_{2y} = -\hat{V} = -EI\frac{d^3\hat{v}(L)}{d\hat{x}^3} = \frac{EI}{L^3}(-12\hat{d}_{1y} - 6L\hat{\phi}_1 + 12\hat{d}_{2y} - 6L\hat{\phi}_2)$$

$$\hat{m}_2 = \hat{m} = EI\frac{d^2\hat{v}(L)}{d\hat{x}^2} = \frac{EI}{L^3}(6L\hat{d}_{1y} + 2L^2\hat{\phi}_1 - 6L\hat{d}_{2y} + 4L^2\hat{\phi}_2)$$

$$(5.1.12)$$

where the minus signs in the second and third of Eqs. (5.1.12) are the result of opposite nodal and beam theory positive bending moment conventions at node 1 and opposite nodal and beam theory positive shear force conventions at node 2 as seen by comparing Figures 5–1 and 5–2. Equations (5.1.12) relate the nodal forces to the nodal displacements. In matrix form, Eqs. (5.1.12) become

$$
\begin{Bmatrix} \hat{f}_{1y} \\ \hat{m}_1 \\ \hat{f}_{2y} \\ \hat{m}_2 \end{Bmatrix} = \frac{EI}{L^3} \begin{bmatrix} 12 & 6L & -12 & 6L \\ 6L & 4L^2 & -6L & 2L^2 \\ -12 & -6L & 12 & -6L \\ 6L & 2L^2 & -6L & 4L^2 \end{bmatrix} \begin{Bmatrix} \hat{d}_{1y} \\ \hat{\phi}_1 \\ \hat{d}_{2y} \\ \hat{\phi}_2 \end{Bmatrix}
\tag{5.1.13}
$$

where the stiffness matrix is then

$$
\hat{k} = \frac{EI}{L^3} \begin{bmatrix} 12 & 6L & -12 & 6L \\ 6L & 4L^2 & -6L & 2L^2 \\ -12 & -6L & 12 & -6L \\ 6L & 2L^2 & -6L & 4L^2 \end{bmatrix}
\tag{5.1.14}
$$

Equation (5.1.13) indicates that $\hat{k}$ relates transverse forces and bending moments to transverse displacements and rotations, whereas axial effects have been neglected.

▲ 5.2 Example of Assemblage of Beam Stiffness Matrices

Step 5 Assemble the Element Equations to Obtain the Global
Equations and Introduce Boundary Conditions

Consider the beam in Figure 5–6 as an example to illustrate the procedure for assemblage of beam element stiffness matrices. Assume EI to be constant throughout the beam. A force of 1000 lb and a moment of 1000 lb-ft are applied to the beam at midlength. The left end is a fixed support and the right end is a pin support.

First, we discretize the beam into two elements with nodes 1–3 as shown. We include a node at midlength because applied force and moment exist at midlength

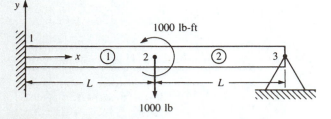

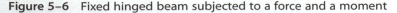

Figure 5–6 Fixed hinged beam subjected to a force and a moment

and, at this time, loads are assumed to be applied only at nodes. (Another procedure for handling loads applied on elements will be discussed in Section 5.4.)

Using Eq. (5.1.14), we find that the global stiffness matrices for the two elements are now given by

$$
\underline{k}^{(1)} = \frac{EI}{L^3}
\begin{array}{cccc}
d_{1y} & \phi_1 & d_{2y} & \phi_2
\end{array}
\begin{bmatrix}
12 & 6L & -12 & 6L \\
6L & 4L^2 & -6L & 2L^2 \\
-12 & -6L & 12 & -6L \\
6L & 2L^2 & -6L & 4L^2
\end{bmatrix}
\tag{5.2.1}
$$

and

$$
\underline{k}^{(2)} = \frac{EI}{L^3}
\begin{array}{cccc}
d_{2y} & \phi_2 & d_{3y} & \phi_3
\end{array}
\begin{bmatrix}
12 & 6L & -12 & 6L \\
6L & 4L^2 & -6L & 2L^2 \\
-12 & -6L & 12 & -6L \\
6L & 2L^2 & -6L & 4L^2
\end{bmatrix}
\tag{5.2.2}
$$

where the degrees of freedom associated with each beam element are indicated by the usual labels above the columns in each element stiffness matrix. Here the local coordinate axes for each element coincide with the global x and y axes of the whole beam. Consequently, the local and global stiffness matrices are identical, so hats (^) are not needed in Eqs. (5.2.1) and (5.2.2).

The total stiffness matrix can now be assembled for the beam by using the direct stiffness method. When the total (global) stiffness matrix has been assembled, the external global nodal forces are related to the global nodal displacements. Through direct superposition and Eqs. (5.2.1) and (5.2.2), the governing equations for the beam are thus given by

$$
\begin{Bmatrix}
F_{1y} \\
M_1 \\
F_{2y} \\
M_2 \\
F_{3y} \\
M_3
\end{Bmatrix}
= \frac{EI}{L^3}
\begin{bmatrix}
12 & 6L & -12 & 6L & 0 & 0 \\
6L & 4L^2 & -6L & 2L^2 & 0 & 0 \\
-12 & -6L & 12+12 & -6L+6L & -12 & 6L \\
6L & 2L^2 & -6L+6L & 4L^2+4L^2 & -6L & 2L^2 \\
0 & 0 & -12 & -6L & 12 & -6L \\
0 & 0 & 6L & 2L^2 & -6L & 4L^2
\end{bmatrix}
\begin{Bmatrix}
d_{1y} \\
\phi_1 \\
d_{2y} \\
\phi_2 \\
d_{3y} \\
\phi_3
\end{Bmatrix}
\tag{5.2.3}
$$

Now considering the boundary conditions, or constraints, of the fixed support at node 1 and the hinge (pinned) support at node 3, we have

$$
\phi_1 = 0 \qquad d_{1y} = 0 \qquad d_{3y} = 0
\tag{5.2.4}
$$

On considering the third, fourth, and sixth equations of Eqs. (5.2.3) corresponding to the rows with unknown degrees of freedom and using Eqs. (5.2.4), we obtain

$$
\begin{Bmatrix}
-1000 \\
1000 \\
0
\end{Bmatrix}
= \frac{EI}{L^3}
\begin{bmatrix}
24 & 0 & 6L \\
0 & 8L^2 & 2L^2 \\
6L & 2L^2 & 4L^2
\end{bmatrix}
\begin{Bmatrix}
d_{2y} \\
\phi_2 \\
\phi_3
\end{Bmatrix}
\tag{5.2.5}
$$

where $F_{2y} = -1000$ lb, $M_2 = 1000$ lb-ft, and $M_3 = 0$ have been substituted into the reduced set of equations. We could now solve Eq. (5.2.5) simultaneously for the unknown nodal displacement d_{2y} and the unknown nodal rotations ϕ_2 and ϕ_3. We leave the final solution for you to obtain. Section 5.3 provides complete solutions to beam problems.

▲ 5.3 Examples of Beam Analysis Using the Direct Stiffness Method ▲

We will now perform complete solutions for beams with various boundary supports and loads to illustrate further the use of the equations developed in Section 5.1.

Example 5.1

Using the direct stiffness method, solve the problem of the propped cantilever beam subjected to end load P in Figure 5–7. The beam is assumed to have constant EI and length $2L$. It is supported by a roller at midlength and is built in at the right end.

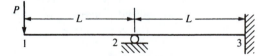

Figure 5–7 Propped cantilever beam

We have discretized the beam and established global coordinate axes as shown in Figure 5–7. We will determine the nodal displacements and rotations, the reactions, and the complete shear force and bending moment diagrams.

Using Eq. (5.1.14) for each element, along with superposition, we obtain the structure total stiffness matrix by the same method as described in Section 5.2 for obtaining the stiffness matrix in Eq. (5.2.3). The $\underline{K}$ is

$$
\underline{K} = \frac{EI}{L^3}
\begin{array}{c}
\begin{array}{cccccc} d_{1y} & \phi_1 & d_{2y} & \phi_2 & d_{3y} & \phi_3 \end{array} \\
\begin{bmatrix}
12 & 6L & -12 & 6L & 0 & 0 \\
 & 4L^2 & -6L & 2L^2 & 0 & 0 \\
 & & 12+12 & -6L+6L & -12 & 6L \\
 & & & 4L^2+4L^2 & -6L & 2L^2 \\
 & & & & 12 & -6L \\
\text{Symmetry} & & & & & 4L^2
\end{bmatrix}
\end{array}
\qquad (5.3.1)
$$

The governing equations for the beam are then given by

$$\begin{Bmatrix} F_{1y} \\ M_1 \\ F_{2y} \\ M_2 \\ F_{3y} \\ M_3 \end{Bmatrix} = \frac{EI}{L^3} \begin{bmatrix} 12 & 6L & -12 & 6L & 0 & 0 \\ 6L & 4L^2 & -6L & 2L^2 & 0 & 0 \\ -12 & -6L & 24 & 0 & -12 & 6L \\ 6L & 2L^2 & 0 & 8L^2 & -6L & 2L^2 \\ 0 & 0 & -12 & -6L & 12 & -6L \\ 0 & 0 & 6L & 2L^2 & -6L & 4L^2 \end{bmatrix} \begin{Bmatrix} d_{1y} \\ \phi_1 \\ d_{2y} \\ \phi_2 \\ d_{3y} \\ \phi_3 \end{Bmatrix} \tag{5.3.2}$$

On applying the boundary conditions

$$d_{2y} = 0 \qquad d_{3y} = 0 \qquad \phi_3 = 0 \tag{5.3.3}$$

and partitioning the equations associated with unknown displacements [the first, second, and fourth equations of Eqs. (5.3.2)] from those equations associated with known displacements in the usual manner, we obtain the final set of equations for a longhand solution as

$$\begin{Bmatrix} -P \\ 0 \\ 0 \end{Bmatrix} = \frac{EI}{L^3} \begin{bmatrix} 12 & 6L & 6L \\ 6L & 4L^2 & 2L^2 \\ 6L & 2L^2 & 8L^2 \end{bmatrix} \begin{Bmatrix} d_{1y} \\ \phi_1 \\ \phi_2 \end{Bmatrix} \tag{5.3.4}$$

where $F_{1y} = -P$, $M_1 = 0$, and $M_2 = 0$ have been used in Eq. (5.3.4). We will now solve Eq. (5.3.4) for the nodal displacement and nodal slopes. We obtain the transverse displacement at node 1 as

$$d_{1y} = -\frac{7PL^3}{12EI} \tag{5.3.5}$$

where the minus sign indicates that the displacement of node 1 is downward.

The slopes are

$$\phi_1 = \frac{3PL^2}{4EI} \qquad \phi_2 = \frac{PL^2}{4EI} \tag{5.3.6}$$

where the positive signs indicate counterclockwise rotations at nodes 1 and 2.

We will now determine the global nodal forces. To do this, we substitute the known global nodal displacements and rotations, Eqs. (5.3.5) and (5.3.6), into Eq. (5.3.2). The resulting equations are

$$\begin{Bmatrix} F_{1y} \\ M_1 \\ F_{2y} \\ M_2 \\ F_{3y} \\ M_3 \end{Bmatrix} = \frac{EI}{L^3} \begin{bmatrix} 12 & 6L & -12 & 6L & 0 & 0 \\ 6L & 4L^2 & -6L & 2L^2 & 0 & 0 \\ -12 & -6L & 24 & 0 & -12 & 6L \\ 6L & 2L^2 & 0 & 8L^2 & -6L & 2L^2 \\ 0 & 0 & -12 & -6L & 12 & -6L \\ 0 & 0 & 6L & 2L^2 & -6L & 4L^2 \end{bmatrix} \begin{Bmatrix} -\dfrac{7PL^3}{12EI} \\[2mm] \dfrac{3PL^2}{4EI} \\[2mm] 0 \\[2mm] \dfrac{PL^2}{4EI} \\[2mm] 0 \\[2mm] 0 \end{Bmatrix} \tag{5.3.7}$$

Multiplying the matrices on the right-hand side of Eq. (5.3.7), we obtain the global nodal forces and moments as

$$F_{1y} = -P \qquad M_1 = 0 \qquad F_{2y} = \tfrac{5}{2}P$$
$$M_2 = 0 \qquad F_{3y} = -\tfrac{3}{2}P \qquad M_3 = \tfrac{1}{2}PL \tag{5.3.8}$$

The results of Eqs. (5.3.8) can be interpreted as follows: The value of $F_{1y} = -P$ is the applied force at node 1, as it must be. The values of F_{2y}, F_{3y}, and M_3 are the reactions from the supports as felt by the beam. The moments M_1 and M_2 are zero because no applied or reactive moments are present on the beam at node 1 or node 2.

It is generally necessary to determine the local nodal forces associated with each element of a large structure to perform a stress analysis of the entire structure. We will thus consider the forces in element 1 of this example to illustrate this concept (element 2 can be treated similarly). Using Eqs. (5.3.5) and (5.3.6) in the $\hat{f} = \hat{k}\hat{d}$ equation for element 1 [also see Eq. (5.1.13)], we have

$$\begin{Bmatrix} \hat{f}_{1y} \\ \hat{m}_1 \\ \hat{f}_{2y} \\ \hat{m}_2 \end{Bmatrix} = \frac{EI}{L^3} \begin{bmatrix} 12 & 6L & -12 & 6L \\ 6L & 4L^2 & -6L & 2L^2 \\ -12 & -6L & 12 & -6L \\ 6L & 2L^2 & -6L & 4L^2 \end{bmatrix} \begin{Bmatrix} -\dfrac{7PL^3}{12EI} \\[2ex] \dfrac{3PL^2}{4EI} \\[2ex] 0 \\[2ex] \dfrac{PL^2}{4EI} \end{Bmatrix} \tag{5.3.9}$$

where, again, because the local coordinate axes of the element coincide with the global axes of the whole beam, we have used the relationships $\underline{d} = \hat{\underline{d}}$ and $\underline{k} = \hat{\underline{k}}$ (that is, the local nodal displacements are also the global nodal displacements, and so forth). Equation (5.3.9) yields

$$\hat{f}_{1y} = -P \qquad \hat{m}_1 = 0 \qquad \hat{f}_{2y} = P \qquad \hat{m}_2 = -PL \tag{5.3.10}$$

A free-body diagram of element 1, shown in Figure 5–8(a), should help you to understand the results of Eqs. (5.3.10). The figure shows a nodal transverse force of

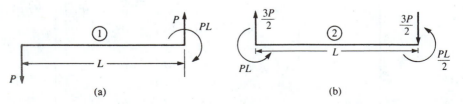

(a) (b)

Figure 5–8 Free-body diagrams showing forces and moments on (a) element 1 and (b) element 2

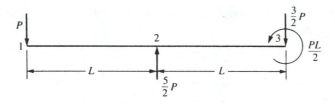

Figure 5–9 Nodal forces and moment on the beam

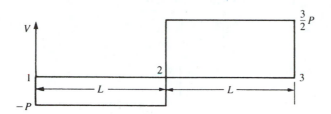

Figure 5–10 Shear force diagram for the beam of Figure 5–9

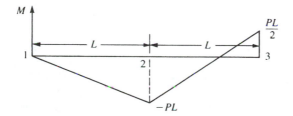

Figure 5–11 Bending moment diagram for the beam of Figure 5–9

---

negative P at node 1 and of positive P and negative moment PL at node 2. These values are consistent with the results given by Eqs. (5.3.10). For completeness, the free-body diagram of element 2 is shown in Figure 5–8(b). We can easily verify the element nodal forces by writing an equation similar to Eq. (5.3.9). From the results of Eqs. (5.3.8), the nodal forces and moments for the whole beam are shown on the beam in Figure 5–9. Using the beam sign conventions established in Section 5.1, we obtain the shear force V and bending moment M diagrams shown in Figures 5–10 and 5–11. ■

In general, for complex beam structures, we will use the element local forces to determine the shear force and bending moment diagrams for each element. We can then use these values for design purposes. Chapter 6 will further discuss this concept as used in computer codes.

Example 5.2

Determine the nodal displacements and rotations, global nodal forces, and element forces for the beam shown in Figure 5–12. We have discretized the beam as indicated by the node numbering. The beam is fixed at nodes 1 and 5 and has a roller support at node 3. Vertical loads of 10,000 lb each are applied at nodes 2 and 4. Let $E = 30 \times 10^6$ psi and $I = 500$ in^4 throughout the beam.

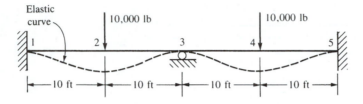

Figure 5–12 Beam example

We must have consistent units; therefore, the 10-ft lengths in Figure 5–12 will be converted to 120 in. during the solution. Using Eq. (5.1.10), along with superposition of the four beam element stiffness matrices, we obtain the global stiffness matrix and the global equations as given by Eq. (5.3.11) on page 199. Here the lengths of each element are the same. Thus, we can factor an L out of the superimposed stiffness matrix.

For a longhand solution, we reduce Eq. (5.3.11) in the usual manner by application of the boundary conditions

$$d_{1y} = \phi_1 = d_{3y} = d_{5y} = \phi_5 = 0$$

The resulting equation is

$$
\begin{Bmatrix} -10,000 \\ 0 \\ 0 \\ -10,000 \\ 0 \end{Bmatrix} = \frac{EI}{L^3}
\begin{bmatrix}
24 & 0 & 6L & 0 & 0 \\
0 & 8L^2 & 2L^2 & 0 & 0 \\
6L & 2L^2 & 8L^2 & -6L & 2L^2 \\
0 & 0 & -6L^2 & 24 & 0 \\
0 & 0 & 2L^2 & 0 & 8L^2
\end{bmatrix}
\begin{Bmatrix} d_{2y} \\ \phi_2 \\ \phi_3 \\ d_{4y} \\ \phi_4 \end{Bmatrix}
\qquad (5.3.12)
$$

The rotations (slopes) at nodes 2–4 are equal to zero because of symmetry in loading, geometry, and material properties about a plane perpendicular to the beam length and

$$
\begin{Bmatrix} F_{1y} \\ M_1 \\ F_{2y} \\ M_2 \\ F_{3y} \\ M_3 \\ F_{4y} \\ M_4 \\ F_{5y} \\ M_5 \end{Bmatrix}
= \frac{EI}{L^3}
\begin{bmatrix}
12 & 6L & -12 & 6L & 0 & 0 & 0 & 0 & 0 & 0 \\
6L & 4L^2 & -6L & 2L^2 & 0 & 0 & 0 & 0 & 0 & 0 \\
-12 & -6L & 12+12 & -6L+6L & -12 & 6L & 0 & 0 & 0 & 0 \\
6L & 2L^2 & -6L+6L & 4L^2+4L^2 & -6L & 2L^2 & 0 & 0 & 0 & 0 \\
0 & 0 & -12 & -6L & 12+12 & -6L+6L & -12 & 6L & 0 & 0 \\
0 & 0 & 6L & 2L^2 & -6L+6L & 4L^2+4L^2 & -6L & 2L^2 & 0 & 0 \\
0 & 0 & 0 & 0 & -12 & -6L & 12+12 & -6L+6L & -12 & 6L \\
0 & 0 & 0 & 0 & 6L & 2L^2 & -6L+6L & 4L^2+4L^2 & -6L & 2L^2 \\
0 & 0 & 0 & 0 & 0 & 0 & -12 & -6L & 12 & -6L \\
0 & 0 & 0 & 0 & 0 & 0 & 6L & 2L^2 & -6L & 4L^2
\end{bmatrix}
\begin{Bmatrix} d_{1y} \\ \phi_1 \\ d_{2y} \\ \phi_2 \\ d_{3y} \\ \phi_3 \\ d_{4y} \\ \phi_4 \\ d_{5y} \\ \phi_5 \end{Bmatrix}
$$

$$(5.3.11)$$

passing through node 3. Therefore, $\phi_2 = \phi_3 = \phi_4 = 0$, and we can further reduce Eq. (5.3.12) to

$$\begin{Bmatrix} -10{,}000 \\ -10{,}000 \end{Bmatrix} = \frac{EI}{L^3} \begin{bmatrix} 24 & 0 \\ 0 & 24 \end{bmatrix} \begin{Bmatrix} d_{2y} \\ d_{4y} \end{Bmatrix} \tag{5.3.13}$$

Solving for the displacements using $L = 120$ in., $E = 30 \times 10^6$ psi, and $I = 500$ in^4 in Eq. (5.3.13), we obtain

$$d_{2y} = d_{4y} = -0.048 \text{ in.} \tag{5.3.14}$$

as expected because of symmetry.

As observed from the solution of this problem, the greater the static redundancy (degrees of static indeterminacy or number of unknown forces and moments that cannot be determined by equations of statics), the smaller the kinematic redundancy (unknown nodal degrees of freedom, such as displacements or slopes)—hence, the fewer the number of unknown degrees of freedom to be solved for. Moreover, the use of symmetry, when applicable, reduces the number of unknown degrees of freedom even further. We can now back-substitute the results from Eq. (5.3.14), along with the numerical values for E, I, and L, into Eq. (5.3.12) to determine the global nodal forces as

$$
\begin{aligned}
F_{1y} &= 5000 \text{ lb} & M_1 &= 25{,}000 \text{ lb-ft} \\
F_{2y} &= 10{,}000 \text{ lb} & M_2 &= 0 \\
F_{3y} &= 10{,}000 \text{ lb} & M_3 &= 0 \\
F_{4y} &= 10{,}000 \text{ lb} & M_4 &= 0 \\
F_{5y} &= 5000 \text{ lb} & M_5 &= -25{,}000 \text{ lb-ft}
\end{aligned}
\tag{5.3.15}
$$

Once again, the global nodal forces (and moments) at the support nodes (nodes 1, 3, and 5) can be interpreted as the reaction forces, and the global nodal forces at nodes 2 and 4 are the applied nodal forces.

However, for large structures we must obtain the local element shear force and bending moment at each node end of the element because these values are used in the design/analysis process. We will again illustrate this concept for the element connecting nodes 1 and 2 in Figure 5–12. Using the local equations for this element, for which all nodal displacements have now been determined, we obtain

$$\begin{Bmatrix} \hat{f}_{1y} \\ \hat{m}_1 \\ \hat{f}_{2y} \\ \hat{m}_2 \end{Bmatrix} = \frac{EI}{L^3} \begin{bmatrix} 12 & 6L & -12 & 6L \\ 6L & 4L^2 & -6L & 2L^2 \\ -12 & -6L & 12 & -6L \\ 6L & 2L^2 & -6L & 4L^2 \end{bmatrix} \begin{Bmatrix} \hat{d}_{1y} = 0 \\ \hat{\phi}_1 = 0 \\ \hat{d}_{2y} = -0.048 \\ \hat{\phi}_2 = 0 \end{Bmatrix} \tag{5.3.16}$$

Simplifying Eq. (5.3.16), we have

$$\begin{Bmatrix} \hat{f}_{1y} \\ \hat{m}_1 \\ \hat{f}_{2y} \\ \hat{m}_2 \end{Bmatrix} = \begin{Bmatrix} 5000 \text{ lb} \\ 25{,}000 \text{ lb-ft} \\ -5000 \text{ lb} \\ 25{,}000 \text{ lb-ft} \end{Bmatrix} \tag{5.3.17}$$

If you wish, you can draw a free-body diagram to confirm the equilibrium of the element. ▪

Finally, you should note that because of reflective symmetry about a vertical plane passing through node 3, we could have initially considered one-half of this beam and used the following model. The fixed support at node 3 is due to the slope being zero at node 3 because of the symmetry in the loading and support conditions.

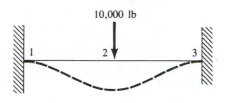

10,000 lb

Example 5.3

Determine the nodal displacements and rotations and the global and element forces for the beam shown in Figure 5–13. We have discretized the beam as shown by the node numbering. The beam is fixed at node 1, has a roller support at node 2, and has an elastic spring support at node 3. A downward vertical force of $P = 50$ kN is applied at node 3. Let $E = 210$ GPa and $I = 2 \times 10^{-4}$ m^4 throughout the beam, and let $k = 200$ kN/m.

Using Eq. (5.1.14) for each beam element and Eq. (2.2.18) for the spring element as well as the direct stiffness method, we obtain the structure stiffness matrix as

$$
\underline{K} = \frac{EI}{L^3}
\begin{array}{c}
\begin{array}{ccccccc}
d_{1y} & \phi_1 & d_{2y} & \phi_2 & d_{3y} & \phi_3 & d_{4y}
\end{array} \\
\begin{bmatrix}
12 & 6L & -12 & 6L & 0 & 0 & 0 \\
 & 4L^2 & -6L & 2L^2 & 0 & 0 & 0 \\
 & & 24 & 0 & -12 & 6L & 0 \\
 & & & 8L^2 & -6L & 2L^2 & 0 \\
 & & & & 12 + \dfrac{kL^3}{EI} & -6L & -\dfrac{kL^3}{EI} \\
 & & & & & 4L^2 & 0 \\
\text{Symmetry} & & & & & & \dfrac{kL^3}{EI}
\end{bmatrix}
\end{array}
\tag{5.3.18a}
$$

where the spring stiffness matrix $\underline{k}_s$ given below by Eq. (5.3.18b) has been directly added into the global stiffness matrix corresponding to its degrees of freedom at nodes 3 and 4.

$$
\underline{k}_s =
\begin{array}{c}
\begin{array}{cc} d_{3y} & d_{4y} \end{array} \\
\begin{bmatrix} k & -k \\ -k & k \end{bmatrix}
\end{array}
\tag{5.3.18b}
$$

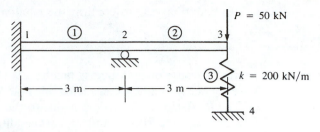

Figure 5–13 Beam example

It is easier to solve the problem using the general variables, later making numerical substitutions into the final displacement expressions. The governing equations for the beam are then given by

$$
\begin{Bmatrix}
F_{1y} \\
M_1 \\
F_{2y} \\
M_2 \\
F_{3y} \\
M_3 \\
F_{4y}
\end{Bmatrix}
=
\frac{EI}{L^3}
\begin{bmatrix}
12 & 6L & -12 & 6L & 0 & 0 & 0 \\
 & 4L^2 & -6L & 2L^2 & 0 & 0 & 0 \\
 & & 24 & 0 & -12 & 6L & 0 \\
 & & & 8L^2 & -6L & 2L^2 & 0 \\
 & & & & 12+k' & -6L & -k' \\
 & & & & & 4L^2 & 0 \\
\text{Symmetry} & & & & & & k'
\end{bmatrix}
\begin{Bmatrix}
d_{1y} \\
\phi_1 \\
d_{2y} \\
\phi_2 \\
d_{3y} \\
\phi_3 \\
d_{4y}
\end{Bmatrix}
\tag{5.3.19}
$$

where $k' = kL^3/(EI)$ is used to simplify the notation. We now apply the boundary conditions

$$
d_{1y} = 0 \qquad \phi_1 = 0 \qquad d_{2y} = 0 \qquad d_{4y} = 0
\tag{5.3.20}
$$

We delete the first three equations and the seventh equation of Eqs. (5.3.19). The remaining three equations are

$$
\begin{Bmatrix}
0 \\
-P \\
0
\end{Bmatrix}
=
\frac{EI}{L^3}
\begin{bmatrix}
8L^2 & -6L & 2L^2 \\
-6L & 12+k' & -6L \\
2L^2 & -6L & 4L^2
\end{bmatrix}
\begin{Bmatrix}
\phi_2 \\
d_{3y} \\
\phi_3
\end{Bmatrix}
\tag{5.3.21}
$$

Solving Eqs. (5.3.21) simultaneously for the displacement at node 3 and the rotations at nodes 2 and 3, we obtain

$$
d_{3y} = -\frac{7PL^3}{EI}\left(\frac{1}{12+7k'}\right) \qquad \phi_2 = -\frac{3PL^2}{EI}\left(\frac{1}{12+7k'}\right)
$$

$$
\phi_3 = -\frac{9PL^2}{EI}\left(\frac{1}{12+7k'}\right)
\tag{5.3.22}
$$

The influence of the spring stiffness on the displacements is easily seen in Eq. (5.3.22). Solving for the numerical displacements using $P = 50$ kN, $L = 3$ m, $E = 210$ GPa

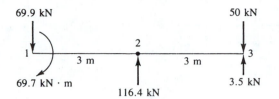

Figure 5–14 Free-body diagram of beam of Figure 5–13

$(= 210 \times 10^6 \text{ kN/m}^2)$, $I = 2 \times 10^{-4} \text{ m}^4$, and $k' = 0.129$ in Eq. (5.3.22), we obtain

$$d_{3y} = \frac{-7(50 \text{ kN})(3 \text{ m})^3}{(210 \times 10^6 \text{ kN/m}^2)(2 \times 10^{-4} \text{ m}^4)} \left(\frac{1}{12 + 7(0.129)} \right) = -0.0174 \text{ m} \quad (5.3.23)$$

Similar substitutions into Eq. (5.3.26) yield

$$\phi_2 = -0.00249 \text{ rad} \qquad \phi_3 = -0.00747 \text{ rad} \qquad (5.3.24)$$

We now back-substitute the results from Eqs. (5.3.23) and (5.3.24), along with numerical values for P, E, I, L, and k', into Eq. (5.3.19) to obtain the global nodal forces as

$$F_{1y} = -69.9 \text{ kN} \qquad M_1 = -69.7 \text{ kN} \cdot \text{m}$$

$$F_{2y} = 116.4 \text{ kN} \qquad M_2 = 0.0 \text{ kN} \cdot \text{m} \qquad (5.3.25)$$

$$F_{3y} = -50.0 \text{ kN} \qquad M_3 = 0.0 \text{ kN} \cdot \text{m}$$

For the beam-spring structure, an additional global force F_{4y} is determined at the base of the spring as follows:

$$F_{4y} = -d_{3y}k = (0.0174)200 = 3.5 \text{ kN} \qquad (5.3.26)$$

This force provides the additional global y force for equilibrium of the structure.

A free-body diagram, including the forces and moments from Eqs. (5.3.25) and (5.3.26) acting on the beam, is shown in Figure 5–14. ∎

▲ 5.4 Distributed Loading ▲

Beam members can support distributed loading as well as concentrated nodal loading. Therefore, we must be able to account for distributed loading. Consider the fixed-fixed beam subjected to a uniformly distributed loading w shown in Figure 5–15. The reactions, determined from structural analysis theory [2], are shown in Figure 5–16. These reactions are called *fixed-end reactions*. In general, **fixed-end reactions** are those reactions at the ends of an element if the ends of the element are assumed to be fixed—that is, if displacements and rotations are prevented. (Those of you who are

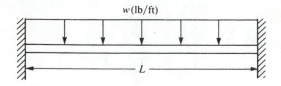

Figure 5–15 Fixed-fixed beam subjected to a uniformly distributed load

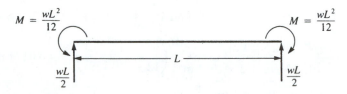

Figure 5–16 Fixed-end reactions for the beam of Figure 5–15

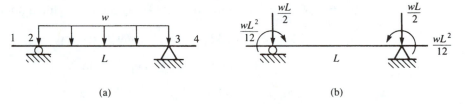

(a)   (b)

Figure 5–17 (a) Beam with a distributed load and (b) the equivalent nodal force system

unfamiliar with the analysis of indeterminate structures should assume these reactions as given and proceed with the rest of the discussion; we will develop these results in a subsequent presentation of the work-equivalence method.) Therefore, guided by the results from structural analysis for the case of a uniformly distributed load, we replace the load by concentrated nodal forces and moments tending to have the same effect on the beam as the actual distributed load. Figure 5–17 illustrates this idea for a beam. We have replaced the uniformly distributed load by a statically equivalent force system consisting of a concentrated nodal force and moment at each end of the member carrying the distributed load. That is, both the statically equivalent concentrated nodal forces and moments and the original distributed load have the same resultant force and same moment about an arbitrarily chosen point. These statically equivalent forces are always of opposite sign from the fixed-end reactions. If we want to analyze the behavior of loaded member 2–3 in better detail, we can place a node at midspan and use the same procedure just described for each of the two elements representing the horizontal member. That is, to determine the maximum deflection and maximum moment in the beam span, a node is needed at midspan of beam segment 2–3.

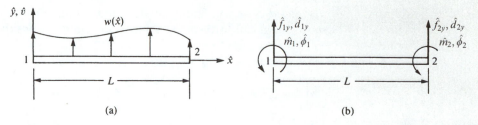

Figure 5–18 (a) Beam element subjected to a general load and (b) the statically equivalent nodal force system

Work-Equivalence Method

We can use the work-equivalence method to replace a distributed load by a set of discrete loads. This method is based on the concept that the work of the distributed load $w(\hat{x})$ in going through the displacement field $\hat{v}(\hat{x})$ is equal to the work done by nodal loads $\hat{f}_{iy}$ and $\hat{m}_i$ in going through nodal displacements $\hat{d}_{iy}$ and $\hat{\phi}_i$ for arbitrary nodal displacements. To illustrate the method, we consider the example shown in Figure 5–18. The work due to the distributed load is given by

$$W_{\text{distributed}} = \int_0^L w(\hat{x})\hat{v}(\hat{x}) \, d\hat{x} \tag{5.4.1}$$

where $\hat{v}(\hat{x})$ is the transverse displacement given by Eq. (5.1.4). The work due to the discrete nodal forces is given by

$$W_{\text{discrete}} = \hat{m}_1\hat{\phi}_1 + \hat{m}_2\hat{\phi}_2 + \hat{f}_{1y}\hat{d}_{1y} + \hat{f}_{2y}\hat{d}_{2y} \tag{5.4.2}$$

We can then determine the nodal moments and forces $\hat{m}_1, \hat{m}_2, \hat{f}_{1y}$, and $\hat{f}_{2y}$ used to replace the distributed load by using the concept of work equivalence—that is, by setting $W_{\text{distributed}} = W_{\text{discrete}}$ for arbitrary displacements $\hat{\phi}_1, \hat{\phi}_2, \hat{d}_{1y}$, and $\hat{d}_{2y}$.

Example of Load Replacement

To illustrate more clearly the concept of work equivalence, we will now consider a beam subjected to a specified distributed load. Consider the uniformly loaded beam shown in Figure 5–19(a). The support conditions are not shown because they are not relevant to the replacement scheme. By letting $W_{\text{discrete}} = W_{\text{distributed}}$ and by assuming

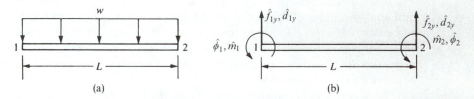

Figure 5–19 (a) Beam subjected to a uniformly distributed loading and (b) the equivalent nodal forces to be determined

arbitrary $\hat{\phi}_1, \hat{\phi}_2, \hat{d}_{1y},$ and $\hat{d}_{2y}$, we will find equivalent nodal forces $\hat{m}_1, \hat{m}_2, \hat{f}_{1y},$ and $\hat{f}_{2y}$. Figure 5–19(b) shows the nodal forces and moments directions as positive based on Figure 5–1.

Using Eqs. (5.4.1) and (5.4.2) for $W_{\text{distributed}} = W_{\text{discrete}}$, we have

$$\int_0^L w(\hat{x})\hat{v}(\hat{x})\, d\hat{x} = \hat{m}_1\hat{\phi}_1 + \hat{m}_2\hat{\phi}_2 + \hat{f}_{1y}\hat{d}_{1y} + \hat{f}_{2y}\hat{d}_{2y} \tag{5.4.3}$$

where $\hat{m}_1\hat{\phi}_1$ and $\hat{m}_2\hat{\phi}_2$ are the work due to concentrated nodal moments moving through their respective nodal rotations and $\hat{f}_{1y}\hat{d}_{1y}$ and $\hat{f}_{2y}\hat{d}_{2y}$ are the work due to the nodal forces moving through nodal displacements. Evaluating the left-hand side of Eq. (5.4.3) by substituting $w(\hat{x}) = -w$ and $\hat{v}(\hat{x})$ from Eq. (5.1.4), we obtain the work due to the distributed load as

$$\int_0^L w(\hat{x})\hat{v}(\hat{x})\, d\hat{x} = -\frac{Lw}{2}(\hat{d}_{1y} - \hat{d}_{2y}) - \frac{L^2 w}{4}(\hat{\phi}_1 + \hat{\phi}_2) - Lw(\hat{d}_{2y} - \hat{d}_{1y})$$

$$+ \frac{L^2 w}{3}(2\hat{\phi}_1 + \hat{\phi}_2) - \hat{\phi}_1\left(\frac{L^2 w}{2}\right) - \hat{d}_{1y}(wL) \tag{5.4.4}$$

Now using Eqs. (5.4.3) and (5.4.4) for arbitrary nodal displacements, we let $\hat{\phi}_1 = 1,$ $\hat{\phi}_2 = 0, \hat{d}_{1y} = 0,$ and $\hat{d}_{2y} = 0$ and then obtain

$$\hat{m}_1(1) = -\left(\frac{L^2 w}{4} - \frac{2}{3}L^2 w + \frac{L^2}{2}w\right) = -\frac{wL^2}{12} \tag{5.4.5}$$

Similarly, letting $\hat{\phi}_1 = 0, \hat{\phi}_2 = 1, \hat{d}_{1y} = 0,$ and $\hat{d}_{2y} = 0$ yields

$$\hat{m}_2(1) = -\left(\frac{L^2 w}{4} - \frac{L^2 w}{3}\right) = \frac{wL^2}{12} \tag{5.4.6}$$

Finally, letting all nodal displacements equal zero except first $\hat{d}_{1y}$ and then $\hat{d}_{2y}$, we obtain

$$\hat{f}_{1y}(1) = -\frac{Lw}{2} + Lw - Lw = -\frac{Lw}{2}$$

$$\hat{f}_{2y}(1) = \frac{Lw}{2} - Lw = -\frac{Lw}{2} \tag{5.4.7}$$

We can conclude that, in general, for any given load function $w(\hat{x})$, we can multiply by $\hat{v}(\hat{x})$ and then integrate according to Eq. (5.4.3) to obtain the concentrated nodal forces (and/or moments) used to replace the distributed load. Moreover, we can obtain the load replacement by using the concept of fixed-end reactions from structural analysis theory. Tables of fixed-end reactions have been generated for numerous load cases and can be found in texts on structural analysis such as Reference [2]. A table of equivalent nodal forces has been generated in Appendix D of this text, guided by the fact that fixed-end reaction forces are of opposite sign from those obtained by the work equivalence method.

Hence, if a concentrated load is applied other than at the natural intersection of two elements, we can use the concept of equivalent nodal forces to replace the concentrated load by nodal concentrated values acting at the beam ends, instead of creating a node on the beam at the location where the load is applied. We provide examples of this procedure for handling concentrated loads on elements in beam Example 5.5 and in plane frame Example 6.3.

General Formulation

In general, we can account for distributed loads or concentrated loads acting on beam elements by starting with the following formulation application for a general structure:

$$\underline{F} = \underline{K}\underline{d} - \underline{F}_o \tag{5.4.8}$$

where $\underline{F}_o$ are called the *equivalent nodal forces*, now expressed in terms of global-coordinate components, which are of such magnitude that they yield the same displacements at the nodes as would the distributed load. Using the table in Appendix D of equivalent nodal forces $\hat{f}_o$ expressed in terms of local-coordinate components, we can express $\underline{F}_o$ in terms of global-coordinate components.

Recall from Section 3.10 the derivation of the element equations by the principle of minimum potential energy. Starting with Eqs. (3.10.19) and (3.10.20), the minimization of the total potential energy resulted in the same form of equation as Eq. (5.4.8) where $\underline{F}_o$ now represents the same work-equivalent force replacement system as given by Eq. (3.10.20a) for surface traction replacement. Also, $\underline{F} = \underline{P}$ [$\underline{P}$ from Eq. (3.10.20)] represents the global nodal concentrated forces. Because we now assume that concentrated nodal forces are not present ($\underline{F} = 0$), as we are solving beam problems with distributed loading only in this section, we can write Eq. (5.4.8) as

$$\underline{F}_o = \underline{K}\underline{d} \tag{5.4.9}$$

On solving for $\underline{d}$ in Eq. (5.4.9) and then substituting the global displacements $\underline{d}$ and equivalent nodal forces $\underline{F}_o$ into Eq. (5.4.8), we obtain the actual global nodal forces $\underline{F}$. For example, using the definition of $\hat{f}_o$ and Eqs. (5.4.5)–(5.4.7) (or using load case 4 in Appendix D) for a uniformly distributed load w acting over a one-element beam, we have

$$\underline{F}_o = \left\{ \begin{array}{c} \dfrac{-wL}{2} \\[2ex] \dfrac{-wL^2}{12} \\[2ex] \dfrac{-wL}{2} \\[2ex] \dfrac{wL^2}{12} \end{array} \right\} \tag{5.4.10}$$

This concept can be applied on a local basis to obtain the local nodal forces $\hat{f}$ in individual elements of structures by applying Eq. (5.4.8) locally as

$$\hat{\underline{f}} = \hat{\underline{k}}\hat{\underline{d}} - \hat{\underline{f}}_o \tag{5.4.11}$$

where $\hat{f}_o$ are the equivalent local nodal forces.

Examples 5.4–5.6 illustrate the method of equivalent nodal forces for solving beams subjected to distributed and concentrated loadings. We will use global-coordinate notation in Examples 5.4–5.6—treating the beam as a general structure rather than as an element.

Example 5.4

For the cantilever beam subjected to the uniform load w in Figure 5–20, solve for the right-end vertical displacement and rotation and then for the nodal forces. Assume the beam to have constant EI throughout its length.

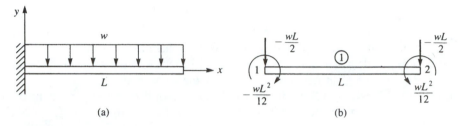

(a)   (b)

Figure 5–20 (a) Cantilever beam subjected to a uniformly distributed load and (b) the work equivalent nodal force system

We begin by discretizing the beam. Here only one element will be used to represent the whole beam. Next, the distributed load is replaced by its work-equivalent nodal forces as shown in Figure 5–20(b). The work-equivalent nodal forces are those that result from the uniformly distributed load acting over the whole beam given by Eq. (5.4.10). (Or see appropriate load case 4 in Appendix D.) Using Eq. (5.4.9) and the beam element stiffness matrix, and realizing $\hat{k} = k$ as the local $\hat{x}$ axis is coincident with the global x axis, we obtain

$$\frac{EI}{L^3}\begin{bmatrix} 12 & -6L \\ -6L & 4L^2 \end{bmatrix}\begin{Bmatrix} d_{2y} \\ \phi_2 \end{Bmatrix} = \begin{Bmatrix} \dfrac{-wL}{2} \\[2mm] \dfrac{wL^2}{12} \end{Bmatrix} \tag{5.4.12}$$

where we have applied the nodal forces from Figure 5–20(b) and the boundary conditions $d_{1y} = 0$ and $\phi_1 = 0$ to reduce the number of matrix equations for the normal longhand solution. Solving Eq. (5.4.12) for the displacements, we obtain

$$\begin{Bmatrix} d_{2y} \\ \phi_2 \end{Bmatrix} = \frac{L}{6EI}\begin{bmatrix} 2L^2 & 3L \\ 3L & 6 \end{bmatrix}\begin{Bmatrix} \dfrac{-wL}{2} \\[2mm] \dfrac{wL^2}{12} \end{Bmatrix} \tag{5.4.13}$$

Simplifying Eq. (5.4.13), we obtain the displacement and rotation as

$$\begin{Bmatrix} d_{2y} \\ \phi_2 \end{Bmatrix} = \begin{Bmatrix} \dfrac{-wL^4}{8EI} \\[2mm] \dfrac{-wL^3}{6EI} \end{Bmatrix} \tag{5.4.14}$$

The negative signs in the answers indicate that d_{2y} is downward and ϕ_2 is clockwise. In this case, the method of replacing the distributed load by discrete concentrated loads gives exact solutions for the displacement and rotation as could be obtained by classical methods, such as double integration [1]. This is expected, as the work-equivalence method ensures that the nodal displacement and rotation from the finite element method match those from an exact solution. ∎

We will now illustrate the procedure for obtaining the global nodal forces. For convenience, we first define the product $\underline{K}\underline{d}$ to be $\underline{F}^{(e)}$, where $\underline{F}^{(e)}$ are called the *effective global nodal forces*. On using Eq. (5.4.14) for $\underline{d}$, we then have

$$
\begin{Bmatrix} F_{1y}^{(e)} \\ M_1^{(e)} \\ F_{2y}^{(e)} \\ M_2^{(e)} \end{Bmatrix} = \frac{EI}{L^3} \begin{bmatrix} 12 & 6L & -12 & 6L \\ 6L & 4L^2 & -6L & 2L^2 \\ -12 & -6L & 12 & -6L \\ 6L & 2L^2 & -6L & 4L^2 \end{bmatrix} \begin{Bmatrix} 0 \\ 0 \\ \dfrac{-wL^4}{8EI} \\ \dfrac{-wL^3}{6EI} \end{Bmatrix} \tag{5.4.15}
$$

Simplifying Eq. (5.4.15), we obtain

$$
\begin{Bmatrix} F_{1y}^{(e)} \\ M_1^{(e)} \\ F_{2y}^{(e)} \\ M_2^{(e)} \end{Bmatrix} = \begin{Bmatrix} \dfrac{wL}{2} \\ \dfrac{5wL^2}{12} \\ \dfrac{-wL}{2} \\ \dfrac{wL^2}{12} \end{Bmatrix} \tag{5.4.16}
$$

We then use Eqs. (5.4.10) and (5.4.16) in Eq. (5.4.8) to obtain the correct global nodal forces as

$$
\begin{Bmatrix} F_{1y} \\ M_1 \\ F_{2y} \\ M_2 \end{Bmatrix} = \begin{Bmatrix} \dfrac{wL}{2} \\ \dfrac{5wL^2}{12} \\ \dfrac{-wL}{2} \\ \dfrac{wL^2}{12} \end{Bmatrix} - \begin{Bmatrix} \dfrac{-wL}{2} \\ \dfrac{-wL^2}{12} \\ \dfrac{-wL}{2} \\ \dfrac{wL^2}{12} \end{Bmatrix} = \begin{Bmatrix} wL \\ \dfrac{wL^2}{2} \\ 0 \\ 0 \end{Bmatrix} \tag{5.4.17}
$$

In Eq. (5.4.17), F_{1y} is the vertical force reaction and M_1 is the moment reaction as applied by the clamped support at node 1. The results for displacement given by

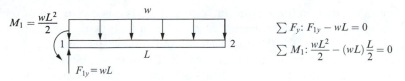

$$M_1 = \frac{wL^2}{2}$$

$$\sum F_y: F_{1y} - wL = 0$$

$$\sum M_1: \frac{wL^2}{2} - (wL)\frac{L}{2} = 0$$

$$F_{1y} = wL$$

Figure 5–20(c) Free-body diagram and equations of equilibrium for beam of Figure 5–(20)a.

Eq. (5.4.14) and the global nodal forces given by Eq. (5.4.17) are sufficient to complete the solution of the cantilever beam problem.

A free-body diagram of the beam using the reactions from Eq. (5.4.17) verifies both force and moment equilibrium as shown in Figure 5–20(c).

We will solve the following example to illustrate the procedure for handling concentrated loads acting on beam elements at locations other than nodes.

Example 5.5

For the cantilever beam subjected to the concentrated load P in Figure 5–21, solve for the right-end vertical displacement and rotation and the nodal forces, including reactions, by replacing the concentrated load with equivalent nodal forces acting at each end of the beam. Assume EI constant throughout the beam.

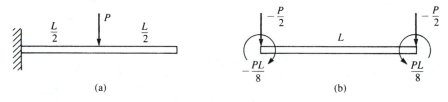

(a)   (b)

Figure 5–21 (a) Cantilever beam subjected to a concentrated load and (b) the equivalent nodal force replacement system

We begin by discretizing the beam. Here only one element is used with nodes at each end of the beam. We then replace the concentrated load as shown in Figure 5–21(b) by using appropriate loading case 1 in Appendix D. Using Eq. (5.4.9) and the beam element stiffness matrix Eq. (5.1.14), we obtain

$$\frac{EI}{L^3}\begin{bmatrix} 12 & -6L \\ -6L & 4L^2 \end{bmatrix}\begin{Bmatrix} d_{2y} \\ \phi_2 \end{Bmatrix} = \begin{Bmatrix} \dfrac{-P}{2} \\ \dfrac{PL}{8} \end{Bmatrix} \qquad (5.4.18)$$

where we have applied the nodal forces from Figure 5–21(b) and the boundary conditions $d_{1y} = 0$ and $\phi_1 = 0$ to reduce the number of matrix equations for the usual longhand solution. Solving Eq. (5.4.18) for the displacements, we obtain

$$
\left\{ \begin{array}{c} d_{2y} \\ \phi_2 \end{array} \right\} = \frac{L}{6EI} \begin{bmatrix} 2L^2 & 3L \\ 3L & 6 \end{bmatrix} \left\{ \begin{array}{c} \dfrac{-P}{2} \\[2mm] \dfrac{PL}{8} \end{array} \right\} \tag{5.4.19}
$$

Simplifying Eq. (5.4.19), we obtain the displacement and rotation as

$$
\left\{ \begin{array}{c} d_{2y} \\ \phi_2 \end{array} \right\} = \left\{ \begin{array}{c} \dfrac{-5PL^3}{48EI} \quad \downarrow \\[4mm] \dfrac{-PL^2}{8EI} \quad \curvearrowright \end{array} \right\} \tag{5.4.20}
$$

To obtain the unknown nodal forces, we begin by evaluating the effective nodal forces $\underline{F}^{(e)} = \underline{K}\underline{d}$ as

$$
\left\{ \begin{array}{c} F_{1y}^{(e)} \\[1mm] M_1^{(e)} \\[1mm] F_{2y}^{(e)} \\[1mm] M_2^{(e)} \end{array} \right\} = \frac{EI}{L^3} \begin{bmatrix} 12 & 6L & -12 & 6L \\ 6L & 4L^2 & -6L & 2L^2 \\ -12 & -6L & 12 & -6L \\ 6L & 2L^2 & -6L & 4L^2 \end{bmatrix} \left\{ \begin{array}{c} 0 \\[1mm] 0 \\[1mm] \dfrac{-5PL^3}{48EI} \\[2mm] \dfrac{-PL^2}{8EI} \end{array} \right\} \tag{5.4.21}
$$

Simplifying Eq. (5.4.21), we obtain

$$
\left\{ \begin{array}{c} F_{1y}^{(e)} \\[1mm] M_1^{(e)} \\[1mm] F_{2y}^{(e)} \\[1mm] M_2^{(e)} \end{array} \right\} = \left\{ \begin{array}{c} \dfrac{P}{2} \\[2mm] \dfrac{3PL}{8} \\[2mm] \dfrac{-P}{2} \\[2mm] \dfrac{PL}{8} \end{array} \right\} \tag{5.4.22}
$$

Then using Eq. (5.4.22) and the equivalent nodal forces from Figure 5–21(b) in Eq. (5.4.8), we obtain the correct nodal forces as

$$
\left\{ \begin{array}{c} F_{1y} \\ M_1 \\ F_{2y} \\ M_2 \end{array} \right\} = \left\{ \begin{array}{c} \dfrac{P}{2} \\[2mm] \dfrac{3PL}{8} \\[2mm] \dfrac{-P}{2} \\[2mm] \dfrac{PL}{8} \end{array} \right\} - \left\{ \begin{array}{c} \dfrac{-P}{2} \\[2mm] \dfrac{-PL}{8} \\[2mm] \dfrac{-P}{2} \\[2mm] \dfrac{PL}{8} \end{array} \right\} = \left\{ \begin{array}{c} P \\[2mm] \dfrac{PL}{2} \\[2mm] 0 \\[2mm] 0 \end{array} \right\} \tag{5.4.23}
$$

We can see from Eq. (5.4.23) that F_{1y} is equivalent to the vertical reaction force and M_1 is the reaction moment as applied by the clamped support at node 1. ∎

Finally, to illustrate the procedure for handling concentrated nodal forces and distributed loads acting simultaneously on beam elements, we will solve the following example.

Example 5.6

For the cantilever beam subjected to the concentrated free-end load P and the uniformly distributed load w acting over the whole beam as shown in Figure 5–22, determine the free-end displacements and the nodal forces.

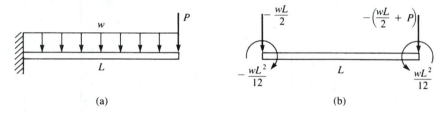

(a)   (b)

Figure 5–22 (a) Cantilever beam subjected to a concentrated load and a distributed load and (b) the equivalent nodal force replacement system

Once again, the beam is modeled using one element with nodes 1 and 2, and the distributed load is replaced as shown in Figure 5–22(b) using appropriate loading case 4 in Appendix D. Using the beam element stiffness Eq. (5.1.14), we obtain

$$\frac{EI}{L^3}\begin{bmatrix} 12 & -6L \\ -6L & 4L^2 \end{bmatrix}\begin{Bmatrix} d_{2y} \\ \phi_2 \end{Bmatrix} = \begin{Bmatrix} \dfrac{-wL}{2} - P \\ \dfrac{wL^2}{12} \end{Bmatrix} \tag{5.4.24}$$

where we have applied the nodal forces from Figure 5–22(b) and the boundary conditions $d_{1y} = 0$ and $\phi_1 = 0$ to reduce the number of matrix equations for the usual longhand solution. Solving Eq. (5.4.24) for the displacements, we obtain

$$\begin{Bmatrix} d_{2y} \\ \phi_2 \end{Bmatrix} = \begin{Bmatrix} \dfrac{-wL^4}{8EI} - \dfrac{PL^3}{3EI} \\ \dfrac{-wL^3}{6EI} - \dfrac{PL^2}{2EI} \end{Bmatrix} \begin{matrix} \downarrow \\ \curvearrowright \end{matrix} \tag{5.4.25}$$

Next, we obtain the effective nodal forces using $\underline{F}^{(e)} = \underline{K}\underline{d}$ as

$$
\left\{
\begin{array}{c}
F_{1y}^{(e)} \\
M_1^{(e)} \\
F_{2y}^{(e)} \\
M_2^{(e)}
\end{array}
\right\}
= \frac{EI}{L^3}
\left[
\begin{array}{cccc}
12 & 6L & -12 & 6L \\
6L & 4L^2 & -6L & 2L^2 \\
-12 & -6L & 12 & -6L \\
6L & 2L^2 & -6L & 4L^2
\end{array}
\right]
\left\{
\begin{array}{c}
0 \\
0 \\
-\dfrac{wL^4}{8EI} - \dfrac{PL^3}{3EI} \\
-\dfrac{wL^3}{6EI} - \dfrac{PL^2}{2EI}
\end{array}
\right\}
\tag{5.4.26}
$$

Simplifying Eq. (5.4.26), we obtain

$$
\left\{
\begin{array}{c}
F_{1y}^{(e)} \\
M_1^{(e)} \\
F_{2y}^{(e)} \\
M_2^{(e)}
\end{array}
\right\}
=
\left\{
\begin{array}{c}
P + \dfrac{wL}{2} \\
PL + \dfrac{5wL^2}{12} \\
-P - \dfrac{wL}{2} \\
\dfrac{wL^2}{12}
\end{array}
\right\}
\tag{5.4.27}
$$

Finally, subtracting the equivalent nodal force matrix [see Figure 5–22(b)] from the effective force matrix of Eq. (5.4.27), we obtain the correct nodal forces as

$$
\left\{
\begin{array}{c}
F_{1y} \\
M_1 \\
F_{2y} \\
M_2
\end{array}
\right\}
=
\left\{
\begin{array}{c}
P + \dfrac{wL}{2} \\
PL + \dfrac{5wL^2}{12} \\
-P - \dfrac{wL}{2} \\
\dfrac{wL^2}{12}
\end{array}
\right\}
-
\left\{
\begin{array}{c}
\dfrac{-wL}{2} \\
\dfrac{-wL^2}{12} \\
\dfrac{-wL}{2} \\
\dfrac{wL^2}{12}
\end{array}
\right\}
=
\left\{
\begin{array}{c}
P + wL \\
PL + \dfrac{wL^2}{2} \\
-P \\
0
\end{array}
\right\}
\tag{5.4.28}
$$

From Eq. (5.4.28), we see that F_{1y} is equivalent to the vertical reaction force, M_1 is the reaction moment at node 1, and F_{2y} is equal to the applied downward force P at node 2. (Remember that only the equivalent nodal force matrix is subtracted, not the original concentrated load matrix. This is based on the general formulation, Eq. (5.4.8).] ■

In general, for any structure in which an equivalent nodal force replacement is made, the actual nodal forces acting on the structure are determined by first evaluating the effective nodal forces $\underline{F}^{(e)}$ for the structure and then subtracting the equivalent nodal forces $\underline{F}_o$ for the structure, as indicated in Eq. (5.4.8). Similarly, for any element of a structure in which equivalent nodal force replacement is made, the actual local nodal forces acting on the element are determined by first evaluating the effective local nodal forces $\hat{f}^{(e)}$ for the element and then subtracting the equivalent local nodal forces $\hat{f}_o$ associated only with the element, as indicated in Eq. (5.4.11). We provide other examples of this procedure in plane frame Examples 6.2 and 6.3.

5.5 Comparison of the Finite Element Solution to the Exact Solution for a Beam

We will now compare the finite element solution to the exact classical beam theory solution for the cantilever beam shown in Figure 5–23 subjected to a uniformly distributed load. Both one- and two-element finite element solutions will be presented and compared to the exact solution obtained by the direct double-integration method. Let $E = 30 \times 10^6$ psi, $I = 100$ in^4, $L = 100$ in., and uniform load $w = 20$ lb/in.

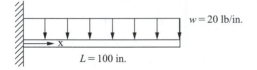

Figure 5–23 Cantilever beam subjected to uniformly distributed load

To obtain the solution from classical beam theory, we use the double-integration method [1]. Therefore, we begin with the moment-curvature equation

$$y'' = \frac{M(x)}{EI} \tag{5.5.1}$$

where the double prime superscript indicates differentiation with respect to x and M is expressed as a function of x by using a section of the beam as shown:

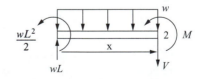

$$\Sigma F_y = 0: \quad V(x) = wL - wx \tag{5.5.2}$$

$$\Sigma M_2 = 0: \quad M(x) = \frac{-wL^2}{2} + wLx - (wx)\left(\frac{x}{2}\right)$$

Using Eq. (5.5.2) in Eq. (5.5.1), we have

$$y'' = \frac{1}{EI}\left(\frac{-wL^2}{2} + wLx - \frac{wx^2}{2}\right) \tag{5.5.3}$$

On integrating Eq. (5.5.3) with respect to x, we obtain an expression for the slope of the beam as

$$y' = \frac{1}{EI}\left(\frac{-wL^2x}{2} + \frac{wLx^2}{2} - \frac{wx^3}{6}\right) + C_1 \tag{5.5.4}$$

Integrating Eq. (5.5.4) with respect to x, we obtain the deflection expression for the beam as

$$y = \frac{1}{EI}\left(\frac{-wL^2x^2}{4} + \frac{wLx^3}{6} - \frac{wx^4}{24}\right) + C_1x + C_2 \tag{5.5.5}$$

Applying the boundary conditions $y = 0$ and $y' = 0$ at $x = 0$, we obtain

$$y'(0) = 0 = C_1 \qquad y(0) = 0 = C_2 \tag{5.5.6}$$

Using Eq. (5.5.6) in Eqs. (5.5.4) and (5.5.5), the final beam theory solution expressions for y' and y are then

$$y' = \frac{1}{EI}\left(\frac{-wx^3}{6} + \frac{wLx^2}{2} - \frac{wL^2x}{2}\right) \tag{5.5.7}$$

and

$$y = \frac{1}{EI}\left(\frac{-wx^4}{24} + \frac{wLx^3}{6} - \frac{wL^2x^2}{4}\right) \tag{5.5.8}$$

The one-element finite element solution for slope and displacement is given in variable form by Eqs. (5.4.14). Using the numerical values of this problem in Eqs. (5.4.14), we obtain the slope and displacement at the free end (node 2) as

$$\hat{\phi}_2 = \frac{-wL^3}{6EI} = \frac{-(20 \text{ lb/in.})(100 \text{ in.})^3}{6(30 \times 10^6 \text{ psi})(100 \text{ in.}^4)} = -0.00111 \text{ rad} \tag{5.5.9}$$

$$\hat{d}_{2y} = \frac{-wL^4}{8EI} = \frac{-(20 \text{ lb/in.})(100 \text{ in.})^4}{8(30 \times 10^6 \text{ psi})(100 \text{ in.}^4)} = -0.0833 \text{ in.}$$

The slope and displacement given by Eq. (5.5.9) identically match the beam theory values, as Eqs. (5.5.7) and (5.5.8) evaluated at $x = L$ are identical to the variable form of the finite element solution given by Eqs. (5.4.14). The reason why these nodal values from the finite element solution are correct is that the element nodal forces were calculated on the basis of being energy or work equivalent to the distributed load based on the assumed cubic displacement field within each beam element.

Values of displacement and slope at other locations along the beam for the finite element solution are obtained by using the assumed cubic displacement function [Eq. (5.1.4)] as

$$\hat{v}(x) = \frac{1}{L^3}(-2x^3 + 3x^2L)\hat{d}_{2y} + \frac{1}{L^3}(x^3L - x^2L^2)\hat{\phi}_2 \tag{5.5.10}$$

where the boundary conditions $\hat{d}_{1y} = \hat{\phi}_1 = 0$ have been used in Eq. (5.5.10). Using the numerical values in Eq. (5.5.10), we obtain the displacement at the midlength of the beam as

$$\hat{v}(x = 50 \text{ in.}) = \frac{1}{(100 \text{ in.})^3}[-2(50 \text{ in.})^3 + 3(50 \text{ in.})^2(100 \text{ in.})](-0.0833 \text{ in.})$$

$$+ \frac{1}{(100 \text{ in.})^3}[(50 \text{ in.})^3(100 \text{ in.}) - (50 \text{ in.})^2(100 \text{ in.})^2]$$

$$\times (-0.00111 \text{ rad}) = -0.0278 \text{ in.} \qquad (5.5.11)$$

Using the beam theory [Eq. (5.5.8)], the deflection is

$$y(x = 50 \text{ in.}) = \frac{20 \text{ lb/in.}}{30 \times 10^6 \text{ psi}(100 \text{ in.}^4)}$$

$$\times \left[\frac{-(50 \text{ in.})^4}{24} + \frac{(100 \text{ in.})(50 \text{ in.})^3}{6} - \frac{(100 \text{ in.})^2(50 \text{ in.})^2}{4} \right]$$

$$= -0.0295 \text{ in.} \qquad (5.5.12)$$

We conclude that the beam theory solution for midlength displacement, $y = -0.0295$ in., is greater than the finite element solution for displacement, $\hat{v} = -0.0278$ in. In general, the displacements evaluated using the cubic function for $\hat{v}$ are lower as predicted by the finite element method than by the beam theory except at the nodes. This is always true for beams subjected to some form of distributed load that are modeled using the cubic displacement function. The exception to this result is at the nodes, where the beam theory and finite element results are identical because of the work-equivalence concept used to replace the distributed load by work-equivalent discrete loads at the nodes.

The beam theory solution predicts a quartic (fourth-order) polynomial expression for y [Eq. (5.5.5)] for a beam subjected to uniformly distributed loading, while the finite element solution $\hat{v}(x)$ assumes a cubic displacement behavior in each beam element under all load conditions. The finite element solution predicts a stiffer structure than the actual one. This is expected, as the finite element model forces the beam into specific modes of displacement and effectively yields a stiffer model than the actual structure. However, as more and more elements are used in the model, the finite element solution converges to the beam theory solution.

For the special case of a beam subjected to only nodal concentrated loads, the beam theory predicts a cubic displacement behavior, as the moment is a linear function and is integrated twice to obtain the resulting cubic displacement function. A simple verification of this cubic displacement behavior would be to solve the cantilevered beam subjected to an end load. In this special case, the finite element solution for displacement matches the beam theory solution for all locations along the beam length, as both functions $y(x)$ and $\hat{v}(x)$ are then cubic functions.

Monotonic convergence of the solution of a particular problem is discussed in Reference [8], and proof that compatible and complete displacement functions (as described in Section 3.2) used in the displacement formulation of the finite element method yield an upper bound on the true stiffness, hence a lower bound on the displacement of the problem, is discussed in Reference [8].

Under uniformly distributed loading, the beam theory solution predicts a quadratic moment and a linear shear force in the beam. However, the finite element solution using the cubic displacement function predicts a linear bending moment and a constant shear force within each beam element used in the model.

We will now determine the bending moment and shear force in the present problem based on the finite element method. The bending moment is given by

$$M = EIv'' = EI\frac{d^2(N\underline{d})}{dx^2} = EI\frac{(d^2N)}{dx^2}\underline{d} \tag{5.5.13}$$

as $\underline{d}$ is not a function of x. Or in terms of the gradient matrix $\underline{B}$ we have

$$M = EI\underline{B}\underline{d} \tag{5.5.14}$$

where

$$\underline{B} = \frac{d^2N}{dx^2} = \left[\left(-\frac{6}{L^2} + \frac{12x}{L^3}\right)\left(-\frac{4}{L} + \frac{6x}{L^2}\right)\left(\frac{6}{L^2} - \frac{12x}{L^3}\right)\left(-\frac{2}{L} + \frac{6x}{L^2}\right)\right] \tag{5.5.15}$$

The shape functions given by Eq. (5.1.7) are used to obtain Eq. (5.5.15) for the $\underline{B}$ matrix. For the single-element solution, the bending moment is then evaluated by substituting Eq. (5.5.15) for $\underline{B}$ into Eq. (5.5.14) and multiplying $\underline{B}$ by $\underline{d}$ to obtain

$$M = EI\left[\left(-\frac{6}{L^2} + \frac{12x}{L^3}\right)\hat{d}_{1x} + \left(-\frac{4}{L} + \frac{6x}{L^2}\right)\hat{\phi}_1 + \left(\frac{6}{L^2} - \frac{12x}{L^3}\right)\hat{d}_{2x} + \left(-\frac{2}{L} + \frac{6x}{L^2}\right)\hat{\phi}_2\right] \tag{5.5.16}$$

Evaluating the moment at the wall, $x = 0$, with $\hat{d}_{1x} = \hat{\phi}_1 = 0$, and $\hat{d}_{2x}$ and $\hat{\phi}_2$ given by Eq. (5.4.14) in Eq. (5.5.16), we have

$$M(x = 0) = -\frac{10wL^2}{24} = -83{,}333 \text{ lb-in.} \tag{5.5.17}$$

Using Eq. (5.5.16) to evaluate the moment at $x = 50$ in., we have

$$M(x = 50 \text{ in.}) = -33{,}333 \text{ lb-in.} \tag{5.5.18}$$

Evaluating the moment at $x = 100$ in. by using Eq. (5.5.16) again, we obtain

$$M(x = 100 \text{ in.}) = 16{,}667 \text{ lb-in.} \tag{5.5.19}$$

The beam theory solution using Eq. (5.5.2) predicts

$$M(x = 0) = \frac{-wL^2}{2} = -100{,}000 \text{ lb-in.} \tag{5.5.20}$$

$$M(x = 50 \text{ in.}) = -25{,}000 \text{ lb-in.}$$

and
$$M(x = 100 \text{ in.}) = 0$$

Figure 5–24(a)–(c) show the plots of the displacement variation, bending moment variation, and shear force variation through the beam length for the beam theory and the one-element finite element solutions. Again, the finite element solution for dis-

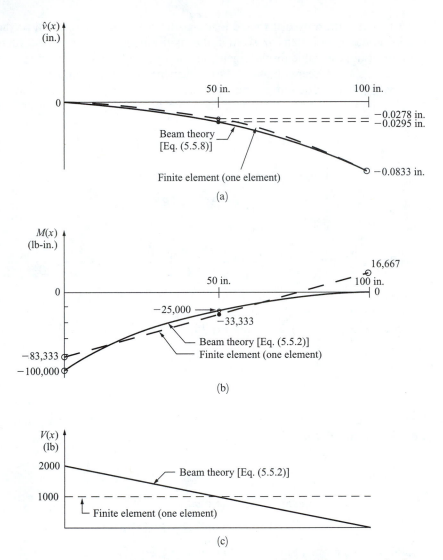

(a)

(b)

(c)

Figure 5–24 Comparison of beam theory and finite element results for a cantilever beam subjected to a uniformly distributed load: (a) displacement diagrams, (b) bending moment diagrams, and (c) shear force diagrams

placement matches the beam theory solution at the nodes but predicts smaller displacements (less deflection) at other locations along the beam length.

The bending moment is derived by taking two derivatives on the displacement function. It then takes more elements to model the second derivative of the displacement function. Therefore, the finite element solution does not predict the bending moment as well as it does the displacement. For the uniformly loaded beam, the finite element model predicts a linear bending moment variation as shown in Figure

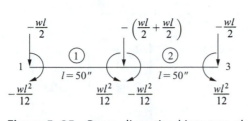

Figure 5–25 Beam discretized into two elements and work-equivalent load replacement for each element

5–24(b). The best approximation for bending moment appears at the midpoint of the element.

The shear force is derived by taking three derivatives on the displacement function. For the uniformly loaded beam, the resulting shear force shown in Figure 5–24(c) is a constant throughout the single-element model. Again, the best approximation for shear force is at the midpoint of the element.

It should be noted that if we use Eq. (5.4.11), that is, $f = \underline{k}\underline{d} - \underline{f}_o$, and subtract off the $\underline{f}_o$ matrix, we also obtain the correct nodal forces and moments in each element. For instance, from the one-element finite element solution we have for the bending moment at node 1

$$m_1^{(1)} = \frac{EI}{L^3}\left[-6L\left(\frac{-wL^4}{8EI}\right) + 2L^2\left(\frac{-wL^3}{6EI}\right)\right] - \left(\frac{-wL^2}{12}\right) = \frac{wL^2}{2}$$

and

$$m_2^{(1)} = 0$$

To improve the finite element solution we need to use more elements in the model (refine the mesh) or use a higher-order element, such as a fifth-order approximation for the displacement function, that is, $\hat{v}(x) = a_1 + a_2 x + a_3 x^2 + a_4 x^3 + a_5 x^4 + a_6 x^5$, with three nodes (with an extra node at the middle of the element).

We now present the two-element finite element solution for the cantilever beam subjected to a uniformly distributed load. Figure 5–25 shows the beam discretized into two elements of equal length and the work-equivalent load replacement for each element. Using the beam element stiffness matrix [Eq. (5.1.13)], we obtain the element stiffness matrices as follows:

$$
\underline{k}^{(1)} = \underline{k}^{(2)} = \frac{EI}{l^3}
\begin{array}{cc}
\begin{array}{cccc} 1 & & 2 & \\ 2 & & 3 & \end{array} \\
\begin{bmatrix}
12 & 6l & -12 & 6l \\
6l & 4l^2 & -6l & 2l^2 \\
-12 & -6l & 12 & -6l \\
6l & 2l^2 & -6l & 4l^2
\end{bmatrix}
\end{array}
\tag{5.5.21}
$$

where $l = 50$ in. is the length of each element and the numbers above the columns indicate the degrees of freedom associated with each element.

Applying the boundary conditions $\hat{d}_{1y} = 0$ and $\hat{\phi}_1 = 0$ to reduce the number of equations for a normal longhand solution, we obtain the global equations for solution as

$$\frac{EI}{l^3}\begin{bmatrix} 24 & 0 & -12 & 6l \\ 0 & 8l^2 & -6l & 2l^2 \\ -12 & -6l & 12 & -6l \\ 6l & 2l^2 & -6l & 4l^2 \end{bmatrix}\begin{Bmatrix} \hat{d}_{2y} \\ \hat{\phi}_2 \\ \hat{d}_{3y} \\ \hat{\phi}_3 \end{Bmatrix} = \begin{Bmatrix} -wl \\ 0 \\ -wl/2 \\ wl^2/12 \end{Bmatrix} \tag{5.5.22}$$

Solving Eq. (5.5.22) for the displacements and slopes, we obtain

$$\hat{d}_{2y} = \frac{-17wl^4}{24EI} \qquad \hat{d}_{3y} = \frac{-2wl^4}{EI} \qquad \hat{\phi}_2 = \frac{-7wl^3}{6EI} \qquad \hat{\phi}_3 = \frac{-4wl^3}{3EI} \tag{5.5.23}$$

Substituting the numerical values $w = 20$ lb/in., $l = 50$ in., $E = 30 \times 10^6$ psi, and $I = 100$ in^4 into Eq. (5.5.23), we obtain

$$\hat{d}_{2y} = -0.02951 \text{ in.} \qquad \hat{d}_{3y} = -0.0833 \text{ in.} \qquad \hat{\phi}_2 = -9.722 \times 10^{-4} \text{ rad}$$

$$\hat{\phi}_3 = -11.11 \times 10^{-4} \text{ rad}$$

The two-element solution yields nodal displacements that match the beam theory results exactly [see Eqs. (5.5.9) and (5.5.12)]. A plot of the two-element displacement throughout the length of the beam would be a cubic displacement within each element. Within element 1, the plot would start at a displacement of 0 at node 1 and finish at a displacement of −0.0295 at node 2. A cubic function would connect these values. Similarly, within element 2, the plot would start at a displacement of −0.0295 and finish at a displacement of −0.0833 in. at node 2. A cubic function would again connect these values.

▲ 5.6 Beam Element with Nodal Hinge ▲

In some beams an internal hinge may be present. In general, this internal hinge causes a discontinuity in the slope of the deflection curve at the hinge.

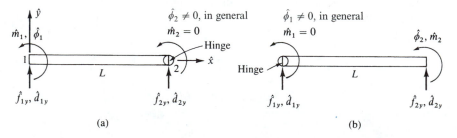

Figure 5–26 Beam element with (a) hinge at right end and (b) hinge at left end

Also, the bending moment is zero at the hinge. We could construct other types of connections that release other generalized end forces; that is, connections can be designed to make the shear force or axial force zero at the connection. These special conditions can be treated by starting with the generalized unreleased beam stiffness matrix [Eq. (5.1.14)] and eliminating the known zero force or moment. This yields a modified stiffness matrix with the desired force or moment equal to zero and the corresponding displacement or slope eliminated.

We now consider the most common cases of a beam element with a nodal hinge at the right end or left end, as shown in Figure 5–26. For the beam element with a hinge at its right end, the moment $\hat{m}_2$ is zero and we partition the $\hat{\underline{k}}$ matrix [Eq. (5.1.14)] to eliminate the degree of freedom $\hat{\phi}_2$ (which is not zero, in general) associated with $\hat{m}_2 = 0$ as follows:

$$\hat{\underline{k}} = \frac{EI}{L^3} \begin{bmatrix} 12 & 6L & -12 & 6L \\ 6L & 4L^2 & -6L & 2L^2 \\ -12 & -6L & 12 & -6L \\ 6L & 2L^2 & -6L & 4L^2 \end{bmatrix} \tag{5.6.1}$$

We condense out the degree of freedom $\hat{\phi}_2$ associated with $\hat{m}_2 = 0$. Partitioning allows us to condense out the degree of freedom $\hat{\phi}_2$ associated with $\hat{m}_2 = 0$. That is, Eq. (5.6.1) is partitioned as shown below:

$$\hat{\underline{k}} = \begin{bmatrix} \underline{K}_{11} & \underline{K}_{12} \\ 3 \times 3 & 3 \times 1 \\ \underline{K}_{21} & \underline{K}_{22} \\ 1 \times 3 & 1 \times 1 \end{bmatrix} \tag{5.6.2}$$

The condensed stiffness matrix is then found by using the equation $\hat{\underline{f}} = \hat{\underline{k}}\hat{\underline{d}}$ partitioned as follows:

$$\left\{ \begin{matrix} \underline{f}_1 \\ 3 \times 1 \\ \underline{f}_2 \\ 1 \times 1 \end{matrix} \right\} = \begin{bmatrix} \underline{K}_{11} & \underline{K}_{12} \\ 3 \times 3 & 3 \times 1 \\ \underline{K}_{21} & \underline{K}_{22} \\ 1 \times 3 & 1 \times 1 \end{bmatrix} \left\{ \begin{matrix} \underline{d}_1 \\ 3 \times 1 \\ \underline{d}_2 \\ 1 \times 1 \end{matrix} \right\} \tag{5.6.3}$$

where
$$\underline{d}_1 = \left\{ \begin{matrix} \hat{d}_{1y} \\ \hat{\phi}_1 \\ \hat{d}_{2y} \end{matrix} \right\} \qquad \underline{d}_2 = \{\hat{\phi}_2\} \tag{5.6.4}$$

Equations (5.6.3) in expanded form are

$$\underline{f}_1 = \underline{K}_{11}\underline{d}_1 + \underline{K}_{12}\underline{d}_2 \tag{5.6.5}$$

$$\underline{f}_2 = \underline{K}_{21}\underline{d}_1 + \underline{K}_{22}\underline{d}_2$$

Solving for $\underline{d}_2$ in the second of Eqs. (5.6.5), we obtain

$$\underline{d}_2 = \underline{K}_{22}^{-1}(\underline{f}_2 - \underline{K}_{21}\underline{d}_1) \tag{5.6.6}$$

Substituting Eq. (5.6.6) into the first of Eqs. (5.6.5), we obtain

$$\underline{f}_1 = (\underline{K}_{11} - \underline{K}_{12}\underline{K}_{22}^{-1}\underline{K}_{21})\underline{d}_1 + \underline{K}_{12}\underline{K}_{22}^{-1}\underline{f}_2 \tag{5.6.7}$$

Combining the second term on the right side of Eq. (5.6.7) with $\underline{f}_1$, we obtain

$$\underline{f}_c = \underline{K}_c\underline{d}_1 \tag{5.6.8}$$

where the condensed stiffness matrix is

$$\underline{K}_c = \underline{K}_{11} - \underline{K}_{12}\underline{K}_{22}^{-1}\underline{K}_{21} \tag{5.6.9}$$

and the condensed force matrix is

$$\underline{f}_c = \underline{f}_1 - \underline{K}_{12}\underline{K}_{22}^{-1}\underline{f}_2 \tag{5.6.10}$$

Substituting the partitioned parts of $\hat{\underline{k}}$ from Eq. (5.6.1) into Eq. (5.6.9), we obtain the condensed stiffness matrix as

$$\underline{K}_c = [K_{11}] - [K_{12}][K_{22}]^{-1}[K_{21}]$$

$$= \frac{EI}{L^3}\begin{bmatrix} 12 & 6L & -12 \\ 6L & 4L^2 & -6L \\ -12 & -6L & 12 \end{bmatrix} - \frac{EI}{L^3}\begin{Bmatrix} 6L \\ 2L^2 \\ -6L \end{Bmatrix}\frac{1}{4L^2}[6L \quad 2L^2 \quad -6L]$$

$$= \frac{3EI}{L^3}\begin{bmatrix} 1 & L & -1 \\ L & L^2 & -L \\ -1 & -L & 1 \end{bmatrix} \tag{5.6.11}$$

and the element equations (force/displacement equations) with the hinge at node 2 are

$$\begin{Bmatrix} \hat{f}_{1y} \\ \hat{m}_1 \\ \hat{f}_{2y} \end{Bmatrix} = \frac{3EI}{L^3}\begin{bmatrix} 1 & L & -1 \\ L & L^2 & -L \\ -1 & -L & 1 \end{bmatrix}\begin{Bmatrix} \hat{d}_{1y} \\ \hat{\phi}_1 \\ \hat{d}_{2y} \end{Bmatrix} \tag{5.6.12}$$

The generalized rotation $\hat{\phi}_2$ has been eliminated from the equation and will not be calculated using this scheme. However, $\hat{\phi}_2$ is not zero in general. We can expand Eq. (5.6.12) to include $\hat{\phi}_2$ by adding zeros in the fourth row and column of the $\hat{\underline{k}}$ matrix to maintain $\hat{m}_2 = 0$, as follows:

$$\begin{Bmatrix} \hat{f}_{1y} \\ \hat{m}_1 \\ \hat{f}_{2y} \\ \hat{m}_2 \end{Bmatrix} = \frac{3EI}{L^3}\begin{bmatrix} 1 & L & -1 & 0 \\ L & L^2 & -L & 0 \\ -1 & -L & 1 & 0 \\ 0 & 0 & 0 & 0 \end{bmatrix}\begin{Bmatrix} \hat{d}_{1y} \\ \hat{\phi}_1 \\ \hat{d}_{2y} \\ \hat{\phi}_2 \end{Bmatrix} \tag{5.6.13}$$

For the beam element with a hinge at its left end, the moment $\hat{m}_1$ is zero, and we partition the $\hat{\underline{k}}$ matrix [Eq. (5.1.14)] to eliminate the zero moment $\hat{m}_1$ and its corresponding rotation $\hat{\phi}_1$ to obtain

$$\begin{Bmatrix} \hat{f}_{1y} \\ \hat{f}_{2y} \\ \hat{m}_2 \end{Bmatrix} = \frac{3EI}{L^3} \begin{bmatrix} 1 & -1 & L \\ -1 & 1 & -L \\ L & -L & L^2 \end{bmatrix} \begin{Bmatrix} \hat{d}_{1y} \\ \hat{d}_{2y} \\ \hat{\phi}_2 \end{Bmatrix} \qquad (5.6.14)$$

The expanded form of Eq. (5.6.14) including $\hat{\phi}_1$ is

$$\begin{Bmatrix} \hat{f}_{1y} \\ \hat{m}_1 \\ \hat{f}_{2y} \\ \hat{m}_2 \end{Bmatrix} = \frac{3EI}{L^3} \begin{bmatrix} 1 & 0 & -1 & L \\ 0 & 0 & 0 & 0 \\ -1 & 0 & 1 & -L \\ L & 0 & -L & L^2 \end{bmatrix} \begin{Bmatrix} \hat{d}_{1y} \\ \hat{\phi}_1 \\ \hat{d}_{2y} \\ \hat{\phi}_2 \end{Bmatrix} \qquad (5.6.15)$$

Example 5.7

Determine the displacement and rotation at node 2 and the element forces for the uniform beam with an internal hinge at node 2 shown in Figure 5–27. Let EI be a constant.

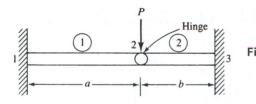

Figure 5–27   Beam with internal hinge

We can assume the hinge is part of element 1. Therefore, using Eq. (5.6.13), the stiffness matrix of element 1 is

$$\begin{matrix} d_{1y} & \phi_1 & d_{2y} & \phi_2 \end{matrix}$$
$$\underline{k}^{(1)} = \frac{3EI}{a^3} \begin{bmatrix} 1 & a & -1 & 0 \\ a & a^2 & -a & 0 \\ -1 & -a & 1 & 0 \\ 0 & 0 & 0 & 0 \end{bmatrix} \qquad (5.6.16)$$

The stiffness matrix of element 2 is obtained from Eq. (5.1.14) as

$$\begin{matrix} d_{2y} & \phi_2 & d_{3y} & \phi_3 \end{matrix}$$
$$\underline{k}^{(2)} = \frac{EI}{b^3} \begin{bmatrix} 12 & 6b & -12 & 6b \\ 6b & 4b^2 & -6b & 2b^2 \\ -12 & -6b & 12 & -6b \\ 6b & 2b^2 & -6b & 4b^2 \end{bmatrix} \qquad (5.6.17)$$

Superimposing Eqs. (5.6.16) and (5.6.17) and applying the boundary conditions

$$d_{1y} = 0, \qquad \phi_1 = 0, \qquad d_{3y} = 0, \qquad \phi_3 = 0$$

we obtain the total stiffness matrix and total set of equations as

$$
EI \begin{bmatrix} \dfrac{3}{a^3} + \dfrac{12}{b^3} & \dfrac{6}{b^2} \\[2mm] \dfrac{6}{b^2} & \dfrac{4}{b} \end{bmatrix} \begin{Bmatrix} d_{2y} \\ \phi_2 \end{Bmatrix} = \begin{Bmatrix} -P \\ 0 \end{Bmatrix} \tag{5.6.18}
$$

Solving Eq. (5.6.18), we obtain

$$
d_{2y} = \frac{-a^3 b^3 P}{3(b^3 + a^3)EI}
$$

$$
\phi_2 = \frac{a^3 b^2 P}{2(b^3 + a^3)EI} \tag{5.6.19}
$$

The value ϕ_2 is actually that associated with element 2—that is, ϕ_2 in Eq. (5.6.19) is actually $\phi_2^{(2)}$. The value of ϕ_2 at the right end of element 1 ($\phi_2^{(1)}$) is, in general, not equal to $\phi_2^{(2)}$. If we had chosen to assume the hinge to be part of element 2, then we would have used Eq. (5.1.14) for the stiffness matrix of element 1 and Eq. (5.6.15) for the stiffness matrix of element 2. This would have enabled us to obtain $\phi_2^{(1)}$, which is different from $\phi_2^{(2)}$.

Using Eq. (5.6.12) for element 1, we obtain the element forces as

$$
\begin{Bmatrix} \hat{f}_{1y} \\ \hat{m}_1 \\ \hat{f}_{2y} \end{Bmatrix} = \frac{3EI}{a^3} \begin{bmatrix} 1 & a & -1 \\ a & a^2 & -a \\ -1 & -a & 1 \end{bmatrix} \begin{Bmatrix} 0 \\ 0 \\ \dfrac{-a^3 b^3 P}{3(b^3 + a^3)EI} \end{Bmatrix} \tag{5.6.20}
$$

Simplifying Eq. (5.6.20), we obtain the forces as

$$
\hat{f}_{1y} = \frac{b^3 P}{b^3 + a^3}
$$

$$
\hat{m}_1 = \frac{ab^3 P}{b^3 + a^3} \tag{5.6.21}
$$

$$
\hat{f}_{2y} = -\frac{b^3 P}{b^3 + a^3}
$$

Using Eq. (5.6.17) and the results from Eq. (5.6.19), we obtain the element 2 forces as

$$
\begin{Bmatrix} \hat{f}_{2y} \\ \hat{m}_2 \\ \hat{f}_{3y} \\ \hat{m}_3 \end{Bmatrix} = \frac{EI}{b^3} \begin{bmatrix} 12 & 6b & -12 & 6b \\ 6b & 4b^2 & -6b & 2b^2 \\ -12 & -6b & 12 & -6b \\ 6b & 2b^2 & -6b & 4b^2 \end{bmatrix} \begin{Bmatrix} -\dfrac{a^3 b^3 P}{3(b^3 + a^3)EI} \\[2mm] \dfrac{a^3 b^2 P}{2(b^3 + a^3)EI} \\[2mm] 0 \\[2mm] 0 \end{Bmatrix} \tag{5.6.22}
$$

Simplifying Eq. (5.6.22), we obtain the element forces as

$$\hat{f}_{2y} = -\frac{a^3 P}{b^3 + a^3}$$

$$\hat{m}_2 = 0$$

$$\hat{f}_{3y} = \frac{a^3 P}{b^3 + a^3} \tag{5.6.23}$$

$$\hat{m}_3 = -\frac{ba^3 P}{b^3 + a^3} \qquad ■$$

▲ 5.7 Potential Energy Approach to Derive Beam Element Equations

We will now derive the beam element equations using the principle of minimum potential energy. The procedure is similar to that used in Section 3.10 in deriving the bar element equations. Again, our primary purpose in applying the principle of minimum potential energy is to enhance your understanding of the principle. It will be used routinely in subsequent chapters to develop element stiffness equations. We use the same notation here as in Section 3.10.

The total potential energy for a beam is

$$\pi_p = U + \Omega \tag{5.7.1}$$

where the general one-dimensional expression for the strain energy U for a beam is given by

$$U = \iiint_V \frac{1}{2} \sigma_x \varepsilon_x \, dV \tag{5.7.2}$$

and for a single beam element subjected to both distributed and concentrated nodal loads, the potential energy of forces is given by

$$\Omega = -\iint_{S_1} \hat{T}_y \hat{v} \, dS - \sum_{i=1}^{2} \hat{P}_{iy} \hat{d}_{iy} - \sum_{i=1}^{2} \hat{m}_i \hat{\phi}_i \tag{5.7.3}$$

where body forces are now neglected. The terms on the right-hand side of Eq. (5.7.3) represent the potential energy of (1) transverse surface loading $\hat{T}_y$ (in units of force per unit surface area, acting over surface S_1 and moving through displacements over which $\hat{T}_y$ act); (2) nodal concentrated force $\hat{P}_{iy}$ moving through displacements $\hat{d}_{iy}$; and (3) moments $\hat{m}_i$ moving through rotations $\hat{\phi}_i$. Again, $\hat{v}$ is the transverse displacement function for the beam element of length L shown in Figure 5–28.

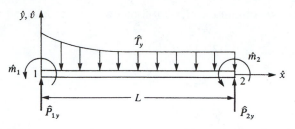

Figure 5–28 Beam element subjected to surface loading and concentrated nodal forces

Consider the beam element to have constant cross-sectional area A. The differential volume for the beam element can then be expressed as

$$dV = dA \, d\hat{x} \tag{5.7.4}$$

and the differential area over which the surface loading acts is

$$dS = b \, d\hat{x} \tag{5.7.5}$$

where b is the constant width. Using Eqs. (5.7.4) and (5.7.5) in Eqs. (5.7.1)–(5.7.3), the total potential energy becomes

$$\pi_p = \iiint_{\hat{x}\,A} \frac{1}{2}\sigma_x \varepsilon_x \, dA \, d\hat{x} - \int_0^L b\hat{T}_y \hat{v} \, d\hat{x} - \sum_{i=1}^{2}(\hat{P}_{iy}\hat{d}_{iy} + \hat{m}_i\hat{\phi}_i) \tag{5.7.6}$$

Substituting Eq. (5.1.4) for $\hat{v}$ into the strain/displacement relationship Eq. (5.1.10), repeated here for convenience as

$$\varepsilon_x = -\hat{y}\frac{d^2\hat{v}}{d\hat{x}^2} \tag{5.7.7}$$

we express the strain in terms of nodal displacements and rotations as

$$\{\varepsilon_x\} = -\hat{y}\left[\frac{12\hat{x} - 6L}{L^3} \quad \frac{6\hat{x}L - 4L^2}{L^3} \quad \frac{-12\hat{x} + 6L}{L^3} \quad \frac{6\hat{x}L - 2L^2}{L^3}\right]\{\hat{d}\} \tag{5.7.8}$$

or

$$\{\varepsilon_x\} = -\hat{y}[B]\{\hat{d}\} \tag{5.7.9}$$

where we define

$$[B] = \left[\frac{12\hat{x} - 6L}{L^3} \quad \frac{6\hat{x}L - 4L^2}{L^3} \quad \frac{-12\hat{x} + 6L}{L^3} \quad \frac{6\hat{x}L - 2L^2}{L^3}\right] \tag{5.7.10}$$

The stress/strain relationship is given by

$$\{\sigma_x\} = [D]\{\varepsilon_x\} \tag{5.7.11}$$

where

$$[D] = [E] \tag{5.7.12}$$

and E is the modulus of elasticity. Using Eq. (5.7.9) in Eq. (5.7.11), we obtain

$$\{\sigma_x\} = -\hat{y}[D][B]\{\hat{d}\} \tag{5.7.13}$$

Next, the total potential energy Eq. (5.7.6) is expressed in matrix notation as

$$\pi_p = \iiint_{\hat{x}\ A} \frac{1}{2}\{\sigma_x\}^T\{\varepsilon_x\}\,dA\,d\hat{x} - \int_0^L b\hat{T}_y[\hat{v}]^T\,d\hat{x} - \{\hat{d}\}^T\{\hat{P}\} \qquad (5.7.14)$$

Using Eqs. (5.1.5), (5.7.9), (5.7.12), and (5.7.13), and defining $w = b\hat{T}_y$ as the line load (load per unit length) in the $\hat{y}$ direction, we express the total potential energy, Eq. (5.7.14), in matrix form as

$$\pi_p = \int_0^L \frac{EI}{2}\{\hat{d}\}^T[B]^T[B]\{\hat{d}\}\,d\hat{x} - \int_0^L w\{\hat{d}\}^T[N]^T\,d\hat{x} - \{\hat{d}\}^T\{\hat{P}\} \qquad (5.7.15)$$

where we have used the definition of the moment of inertia

$$I = \iint_A y^2\,dA \qquad (5.7.16)$$

to obtain the first term on the right-hand side of Eq. (5.7.15). In Eq. (5.7.15), π_p is now expressed as a function of $\{\hat{d}\}$.

Differentiating π_p in Eq. (5.7.15) with respect to $\hat{d}_{1y}, \hat{\phi}_1, \hat{d}_{2y}$, and $\hat{\phi}_2$ and equating each term to zero to minimize π_p, we obtain four element equations, which are written in matrix form as

$$EI\int_0^L [B]^T[B]\,d\hat{x}\{\hat{d}\} - \int_0^L [N]^T w\,d\hat{x} - \{\hat{P}\} = 0 \qquad (5.7.17)$$

The derivation of the four element equations is left as an exercise (see Problem 5.32). Representing the nodal force matrix as the sum of those nodal forces resulting from distributed loading and concentrated loading, we have

$$\{\hat{f}\} = \int_0^L [N]^T w\,d\hat{x} + \{\hat{P}\} \qquad (5.7.18)$$

Using Eq. (5.7.18), the four element equations given by explicitly evaluating Eq. (5.7.17) are then identical to Eq. (5.1.13). The integral term on the right side of Eq. (5.7.18) also represents the work-equivalent replacement of a distributed load by nodal concentrated loads. For instance, letting $w(\hat{x}) = -w$ (constant), substituting shape functions from Eq. (5.1.7) into the integral, and then performing the integration result in the same nodal equivalent loads as given by Eqs. (5.4.5)–(5.4.7).

Because $\{\hat{f}\} = [\hat{k}]\{\hat{d}\}$, we have, from Eq. (5.7.17),

$$[\hat{k}] = EI\int_0^L [B]^T[B]\,d\hat{x} \qquad (5.7.19)$$

Using Eq. (5.7.10) in Eq. (5.7.19) and integrating, $[\hat{k}]$ is evaluated in explicit form as

$$[\hat{k}] = \frac{EI}{L^3}\begin{bmatrix} 12 & 6L & -12 & 6L \\ & 4L^2 & -6L & 2L^2 \\ & & 12 & -6L \\ \text{Symmetry} & & & 4L^2 \end{bmatrix} \qquad (5.7.20)$$

Equation (5.7.20) represents the local stiffness matrix for a beam element. As expected, Eq. (5.7.20) is identical to Eq. (5.1.14) developed previously.

▲ ## 5.8 Galerkin's Method for Deriving Beam Element Equations ▲

We will now illustrate Galerkin's method to formulate the beam element stiffness equations. We begin with the basic differential Eq. (5.1.1a) with transverse loading w now included; that is,

$$EI \frac{d^4 \hat{v}}{d\hat{x}^4} + w = 0 \tag{5.8.1}$$

We now define the residual R to be Eq. (5.8.1). Applying Galerkin's criterion [Eq. (3.12.3)] to Eq. (5.8.1), we have

$$\int_0^L \left(EI \frac{d^4 \hat{v}}{d\hat{x}^4} + w \right) N_i \, d\hat{x} = 0 \qquad (i = 1, 2, 3, 4) \tag{5.8.2}$$

where the shape functions N_i are defined by Eqs. (5.1.7).

We now apply integration by parts twice to the first term in Eq. (5.8.2) to yield

$$\int_0^L EI(\hat{v}_{,\hat{x}\hat{x}\hat{x}\hat{x}}) N_i \, d\hat{x} = \int_0^L EI(\hat{v}_{,\hat{x}\hat{x}})(N_{i,\hat{x}\hat{x}}) \, d\hat{x} + EI[N_i(\hat{v}_{,\hat{x}\hat{x}\hat{x}}) - (N_{i,\hat{x}})(\hat{v}_{,\hat{x}\hat{x}})]_0^L \tag{5.8.3}$$

where the notation of the comma followed by the subscript $\hat{x}$ indicates differentiation with respect to $\hat{x}$. Again, integration by parts introduces the boundary conditions.

Because $\hat{v} = [N]\{\hat{d}\}$ as given by Eq. (5.1.5), we have

$$\hat{v}_{,\hat{x}\hat{x}} = \left[\frac{12\hat{x} - 6L}{L^3} \frac{6\hat{x}L - 4L^2}{L^3} \frac{-12\hat{x} + 6L}{L^3} \frac{6\hat{x}L - 2L^2}{L^3} \right] \{\hat{d}\} \tag{5.8.4}$$

or, using Eq. (5.7.10),

$$\hat{v}_{,\hat{x}\hat{x}} = [B]\{\hat{d}\} \tag{5.8.5}$$

Substituting Eq. (5.8.5) into Eq. (5.8.3), and then Eq. (5.8.3) into Eq. (5.8.2), we obtain

$$\int_0^L (N_{i,\hat{x}\hat{x}}) EI[B] \, d\hat{x}\{\hat{d}\} + \int_0^L N_i w \, d\hat{x} + [N_i \hat{V} - (N_{i,\hat{x}})\hat{m}]|_0^L = 0 \qquad (i = 1, 2, 3, 4) \tag{5.8.6}$$

where Eqs. (5.1.11) have been used in the boundary terms. Equation (5.8.6) is really four equations (one each for $N_i = N_1, N_2, N_3$, and N_4). Instead of directly evaluating Eq. (5.8.6) for each N_i, as was done in Section 3.12, we can express the four equations of Eq. (5.8.6) in matrix form as

$$\int_0^L [B]^T EI[B]\, d\hat{x}\{\hat{d}\} = \int_0^L -[N]^T w\, d\hat{x} + ([N]^T,_{\hat{x}}\hat{m} - [N]^T \hat{V})|_0^L \qquad (5.8.7)$$

where we have used the relationship $[N],_{xx} = [B]$ in Eq. (5.8.7).

Observe that the integral term on the left side of Eq. (5.8.7) is identical to the stiffness matrix previously given by Eq. (5.7.19) and that the first term on the right side of Eq. (5.8.7) represents the equivalent nodal forces due to distributed loading [also given in Eq. (5.7.18)]. The two terms in parentheses on the right side of Eq. (5.8.7) are the same as the concentrated force matrix $\{\hat{P}\}$ of Eq. (5.7.18). We explain this by evaluating $[N],_{\hat{x}}$ and $[N]$, where $[N]$ is defined by Eq. (5.1.6), at the ends of the element as follows:

$$[N],_{\hat{x}}|_0 = [0 \quad 1 \quad 0 \quad 0] \qquad [N],_{\hat{x}}|_L = [0 \quad 0 \quad 0 \quad 1]$$
$$[N]|_0 = [1 \quad 0 \quad 0 \quad 0] \qquad [N]|_L = [0 \quad 0 \quad 1 \quad 0] \qquad (5.8.8)$$

Therefore, when we use Eqs. (5.8.8) in Eq. (5.8.7), the following terms result:

$$\begin{Bmatrix} 0 \\ 0 \\ 0 \\ 1 \end{Bmatrix} \hat{m}(L) - \begin{Bmatrix} 0 \\ 1 \\ 0 \\ 0 \end{Bmatrix} \hat{m}(0) - \begin{Bmatrix} 0 \\ 0 \\ 1 \\ 0 \end{Bmatrix} \hat{V}(L) + \begin{Bmatrix} 1 \\ 0 \\ 0 \\ 0 \end{Bmatrix} \hat{V}(0) \qquad (5.8.9)$$

These nodal shear forces and moments are illustrated in Figure 5–29.

Note that when element matrices are assembled, two shear forces and two moments from adjacent elements contribute to the concentrated force and concentrated moment at the node common to the adjacent elements as shown in Figure 5–30. These

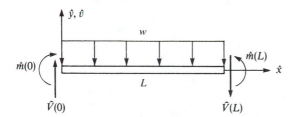

Figure 5–29 Beam element with shear forces, moments, and a distributed load

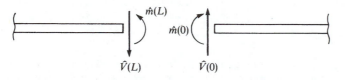

Figure 5–30 Shear forces and moments acting on adjacent elements meeting at a node

concentrated shear forces $\hat{V}(0) - \hat{V}(L)$ and moments $\hat{m}(L) - \hat{m}(0)$ are often zero; that is, $\hat{V}(0) = \hat{V}(L)$ and $\hat{m}(L) = \hat{m}(0)$ occur except when a concentrated nodal force or moment exists at the node. In the actual computations, we handle the expressions given by Eq. (5.8.9) by including them as concentrated nodal values making up the matrix $\{P\}$.

▲ 5.9 Algor Example Solutions for Beam Analysis ▲

From the presentation in this chapter, we observe that the beam element is represented by a line drawn between two end nodes, i and j. In this chapter, we have considered the beam in the plane only. In Chapter 6, we will develop the stiffness matrix that allows us to analyze three-dimensional frames. The Algor program includes this generalized beam element that allows each node to translate and rotate about all three axes. For additional details on the beam element and beam design editor (BEdit), consult References [3–4].

Algor uses the *Surface* command to designate the orientation of a beam element. For instance, consider the beam element shown in Figure 5–31. Using the Algor convention for a rectangular cross section, the local axes, 2 and 3, are in the plane of the cross section, and the local 1 axis is always along the beam length directed from the i node to the j node. A plane is created through the i and j nodes and a third node called the k node. The local 1 axis lies in this plane, while the local 2 axis also lies in this i-j-k plane, perpendicular to the local 1 axis and pointing toward the k node. The local 3 axis is then perpendicular to the 1-2 plane (or i-j-k plane), and its direction is established by the right-hand rule of 1 cross 2 equal to 3. The location of the k node is defined by the *Surface* command. By setting a certain surface number to a beam element, the location of the k node and hence the orientation of beam local axes 2 and 3 are established. For instance, in Figure 5–31, the global X-Y-Z axes are established

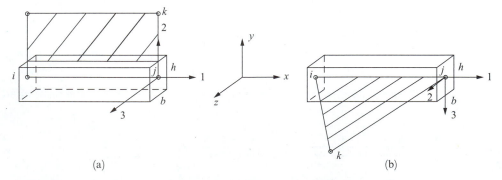

(a) (b)

Figure 5–31 Beam element illustrating the orientation of local 1-2-3 axes based on the *Surface* command: (a) surface number = 1 defines the k node to be in the positive y direction and (b) Surface number = 2 defines the k node to be in the positive z direction. (In both cases the local 2 axis always points toward the k node.)

Table 5–1 Correlation of surface number and k node (axis 2 is oriented toward the k node)

Superdraw Surface Number	First-Choice k-node Location	Second-Choice k-node Location
0 or 1	1e14 in $+y$	1e14 in $-x$
2	1e14 in $+z$	1e14 in $+y$
3	1e14 in $+x$	1e14 in $+z$
4	1e14 in $-y$	1e14 in $+x$
5	1e14 in $-z$	1e14 in $-y$
6	1e14 in $-x$	1e14 in $-z$
7–255	Use the center of a circle with the same surface number	Not applicable

by the user input into Algor as shown. Then for nodes i and j at the left and right ends of the beam, if we assign a specific surface number of 1 (the default number) in Superdraw III to a beam element, we see from Table 5–1 that the k node is in the positive Y direction as its first choice, and since the local 2 axis points to the k node, the local 2 axis is also in the global Y direction.

The first-choice location is where the k node is created provided the i, j, and k nodes form a plane as shown in Figure 5–31(a). If the beam is colinear with the k node, then a unique plane cannot be found. Then the second choice in Table 5–1 is used for that element.

If instead we assign the surface number 2 to the element, then based on Table 5–1, the k node is in the positive Z direction as its first choice, and since the local 2 axis points to the k node, the local 2 axis now points to the global Z as shown in Figure 5–31(b).

Six k nodes are created by default in the Algor program. These six k nodes are the last nodes listed in the nodal data table and can be viewed in the .L file if the "Displacement data" is checked in the "Output" tab of the "Global" window in Algor. Based on Table 5–1, these six nodes are located at 1e14 in the positive and negative X directions, at 1e14 in the positive and negative Y directions, and at 1e14 in the positive and negative Z directions, respectively.

The orientation of the elements can be checked in the BEdit processor using the "F5:Draw sw:shOw" icon. The top of the beam symbol points in the local 2-axis direction. Again, I2 may correspond to the strong or weak axis depending on how you enter the data in BEdit. Example 5.13 illustrates the use of the *Surface* (number) command to properly orient a beam element in a desired orientation.

In Superview, the orientation of the beam elements can be checked by using the "Options:Ele opt:2) beam:Orient" succession of commands and choosing a color. A line is added at the midpoint of the beam element pointing in the direction of axis 2.

Now that the element axes have been determined using the *Surface* command, the cross-sectional properties can be entered properly in the BEdit processor as explained in the following. In Algor, the cross-sectional or geometric properties used to define the beam element include

A Cross-sectional area

$Sa2$ Shear area in local 2 direction

$Sa3$ Shear area in local 3 direction

$J1$ Torsional constant about the 1 axis

$I2$ Moment of inertia about the 2 axis

$I3$ Moment of inertia about the 3 axis

$S2$ Section modulus about the 2 axis

$S3$ Section modulus about the 3 axis

Remember that the torsional constant $J1$ is not equal to $I2$ plus $I3$ except for circular cross sections. Torsional constants are listed for various cross-sectional shapes in Table 6–1 and in Reference [9] and for the circle, tube and rectangle in BEdit by clicking "Add/Mod:Sectional" and doing a "Help" on "Value".

Because we consider only beams in the plane in this chapter, we are only concerned with properties A, $I2$, and/or $I3$, $S2$, and/or $S3$. For cross sections that have a different $I2$ from $I3$, we must be careful to input the proper moment of inertia based on the axis about which bending is taking place. This was explained in the preceding paragraphs that described the *Surface* command. Shear areas can normally be set to zero except for very short beams. Shear areas for various cross sections can be found in References [5–6]. For instance, for rectangular cross sections we use the shear area as $Sa2 = Sa3 = 5/6$ of the base times the height of the cross section; for wide-flange sections with bending occurring about the strong axis we use $Sa2 =$ thickness of web times height of the cross section; for thin-walled box sections we use 2 times the product of the thickness times height; and for a circular cross section we use 0.9 times the cross-sectional area.

Algor uses the *Layer* command to designate cross-sectional properties of the beam elements. Example 6.8 illustrates the use of the *Layer* command for a frame composed of members with different cross-sectional properties.

Algor uses the *Group* command to designate different material properties of beam elements. Example 5.9 illustrates use of the *Group* command to designate different material properties, such as Young's modulus E.

To use the Algor program for beam analysis, we now solve a series of example problems. The program is based on the flowchart of Figure 4–1(a) with the beam element replacing the truss element and with BEdit used with the "Model Data Control" window to enter beam sectional properties, such as cross-sectional area and moment of inertia, member end releases, such as internal hinges, distributed loading, and lumped masses. These examples illustrate numerous concepts, such as basic input commands used in the BEdit processor for beam analysis (Example 5.2 reworked using Algor), use of the *Group* command to differentiate between beam elements with different material properties (Example 5.3 reworked using Algor), how to handle distributed loading (Problem 5.22 worked using Algor), internal hinges or member end releases (Example 5.7 reworked using Algor), use of the gap element when there is a gap between a support and the beam (see Problem 5.6), and use of the *Surface* command to properly orient a beam (Example 5.13). Detailed steps will be listed, including specific keystrokes and descriptions of the purpose of these keystrokes.

Example 5.8

For the beam shown in Figure 5–32, solve for the displacements at node 2 and the bending stress about the local 3 axis ($M3/S3$) using Algor. Let $I = 500$ in^4 and $E = 30 \times 10^6$ psi. This is Example 5.2, which was solved longhand.

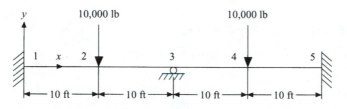

Figure 5–32   Beam example

Step 1   Start Superdraw III

Click on the "Algor" icon on your desktop computer screen or click the start button at the bottom left of the screen and go to "Programs-Algor Software-Algor FEA 12" to start Superdraw III.

Step 2   Create the Beam Model

Select the "Add" menu from the top main menu bar.

Click "Line". A menu appears on the upper right side of the screen. Now enter the data points in succession starting at the origin and finishing at the end of the beam. To simplify the problem we will convert all measurements to inches.

0<Tab>0   This is the first point of the beam.

<Enter>

120<Tab>0   This is the second point of the beam.

<Enter>

240<Tab>0   This is the third point of the beam.

<Enter>

360<Tab>0   This is the fourth point of the beam.

<Enter>

480<Tab>0   This is the fifth and final point of the beam.

<Enter>

Click "Done".

Select "View" from the main menu.

Click "Enclose". You now can see a line that represents the beam in Figure 5–32.

Step 3   Eliminate Duplicate Lines

Click "Modify" on the main menu.

Select "Clean:Duplicate". This notation means to select "Clean" and then "Duplicate" in succession.

Click "Perform Cleaning" on the "Duplicate" menu. You will see the following message at the lower part of the screen:

"4 Kept 0 Deleted. done"

Click "Done" on the "Duplicate" menu located in the upper right corner of the screen.

Step 4   Add in the Boundary Conditions

Click on the "FEA Add" menu from the main menu.

Click "Stress and Vibration:Boundary Conditions", and the on-screen "Boundary Conditions" menu will appear. The default boundary condition is "Use @ symbol for Full". "Full" means full constraints on translation and rotation when this condition is applied to a particular node. This particular model needs to be constrained fully at the far end nodes, while it is constrained in only Y-translational and X- and Y-rotational at the middle node-roller support.

First we will fix the two far ends by simply right-clicking on the two farthest end nodes. An "@" symbol will appear next to the node telling you the node is completely constrained. You may have to zoom in to see this node. To do this,

Click on the "View" menu and select "Zoom" and then "In". Make a box around the node, and you now should be able to see the "@" symbol.

To make the changes to apply the next constraint, the defaults must be reset.

Click "Change Values" in the "Boundary Conditions" window. This will allow you to take away constraints from particular translations and rotations. Tx, Ty, and Tz are the translational constraints, while Rx, Ry, and Rz are the rotational constraints. For this constraint simply unselect Tx, Tz, and Rz.

Click "Done".

Right-click at the center of the beam; a Ty, Tz, Rx, and Ry term appears at the node. As before, you may have to zoom in to see this term.

Step 5   Apply Forces to the Beam

Click "FEA Add" on the main menu.

Click "Stress and Vibration Analysis".

Click "Nodal Forces". A "Nodal Forces" window appears.

Click "Vector". You will need to create the direction of the force.

Click "Y direction". Note that this is the default direction for the force, but for illustration we will use it. You will see "DX 0 DY 0.97 DZ 0" appear at the bottom of the screen.

Click "Done". This takes you back to the "Nodal Forces" window.

Click "Magnitude". A prompt "Magnitude"appears at the bottom of the screen asking you to enter the magnitude of the force.

–10000   Enter the magnitude of the force.

<Enter>   You might have to move the screen to see the message "Click on node to apply the current values".

Move the mouse cursor to nodes 2 and 4 and right-click on the nodes. The $-10{,}000$-lb force appears.

Click "Done". You may have to "Zoom" in again to see these forces.

Step 6   Add the Material Properties Using the "Model Data Control" Window

Click on the "Model Data" control button located at the lower part of the screen and highlighted in red.

For the "Analysis Type" line, "Linear Static Stress" appears highlighted as the default. You can scroll down for other analysis types as appropriate.

Click on the box below the "Element" column. A table of group 1 element types appears.

Click "Beam". You now have created a beam. All lines are beam elements.

Click "OK".

Click on the box below the "Data" column. It will respond with "Please enter model name first".

Click "OK" and then enter the model name in the "File_name" area that appears.

Ex58   The name of the file is now *Ex58*.

Click "Save". By default, the "Units Definition" menu appears.

Click "OK" to accept the default "English(in)" units system. Otherwise, scroll down to select another choice of units.

The "Beam Design Editor" window now appears.

Click "Add/Mod" to bring up the "Add/Modify" window.

Click "Sectional" to go to the "Sectional" window. This is where you will enter the sectional properties of the beam.

Click "Value". A cursor now appears at the bottom of the screen next to the property "A". You can scroll over to the other properties by hitting the ⟨Tab⟩ key. Do this until you come to the "I" (moment of inertia) terms. The ⟨Esc⟩ key allows you to delete the current terms.

Under these "I" terms enter the value of 500.

500   These are the values of the moments of inertia that resist the bending.

<Enter>

<Esc>

<Esc>   You are now in the main menu of the "Beam Design Editor" again.

Click "Quit". You are then asked "Save Current Work?"

Click "Yes". This takes you back to the "Model Data Control" window.

Click the box below "Material". This brings up the on-screen menu "Element Material Selection".

We can select one of the predefined materials or the "Customer Defined" materials. In this case we will select "Customer Defined".

Click "Customer Defined".

Click "View Properties:Edit Properties" and enter the modulus of elasticity.

30e6   Modulus of elasticity.

Click "OK" to apply the properties to the "Customer Defined" material.

Click "OK" to return to the "Model Data Control" window.

Step 7   Enter the Global Data

Click on the "Global" button on the right-side menu of the "Model Data Control" window.

Under "Load Case Multipliers" and under "Pressure", type in 1.

Click on the "Output" tab to view the output options.

Click on the box next to "Displacement data" and on the box next to "Stress data".

This creates nodal displacement output in the .L file, and element forces and stresses in the .S file. These files can be viewed later with a word processor.

Step 8   Check the Model

Click "Check" under the "FEA Model" column in the "Model Data Control" window. This takes you to Superview's main menu, and you can check your model as explained in Section 4.1 and detailed by example in Appendix F. The truss model appears automatically, showing the boundary conditions in red and any forces that may be applied in yellow.

Click "Done". This takes you back to the "Model Data Control" menu.

Step 9   Run the Analysis

Click on the "Analysis" button under the "FEA Model" column in the "Model Data Control" window.

Click "Analyze" in lower left corner of the "Linear Stress and Vibration-Static Stress" menu.

A warning message appears after the analysis has completed but has no bearing on the outcome of the analysis.

Click "OK" to close the warning message.

Click "Done" to return to the "Model Data Control" menu.

Step 10   Review the Results

Click on the "Results" button in the "Model Data Control" window. This takes you to Superview once again where you can view the results.

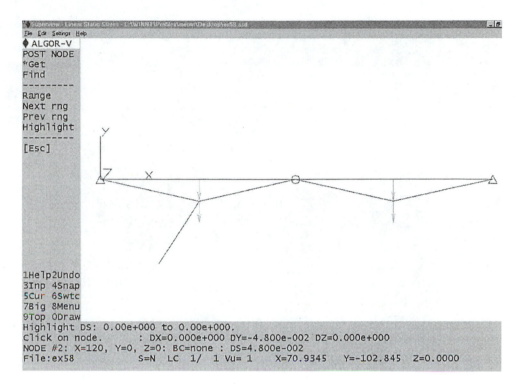

Figure 5–33   Displaced and undisplaced beams of Example 5.8

Click "Stress-di".

Select "Post".

Click "Beam and truss stress".

For bending stress about the local 3 ($M3/S3$):

Click "M3/S3".

You will see purple boxes on the elements. These represent the values of the largest and smallest stresses in each element. Here positive and negative values should appear at the bottom of the screen. These values should be $+/-6.9e6$. These values do not mean anything as we have assigned section modules as $S3 = 0.4333$ (its default). Hence, $M3/S3 = 3e5/0.4333 = 6.923e6$ psi in the beam element stress table in Figure 5–34.

Select "Esc".

<Esc>

<Esc>

You should now be back at the main menu screen.

Click "Displaced".

Click "Displaced On", "With Undi", and "Calc Scal".

You now see the displaced and undisplaced beams superimposed (Figure 5–33).

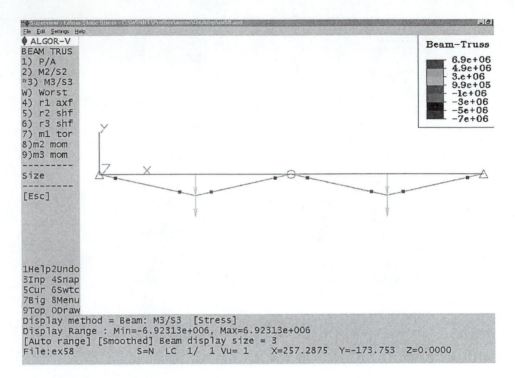

Figure 5–34 *M3/S3* values for each element of Example 5.8

Click "Nod Inq" for node inquiry.

Click "Get".

Right-click on node 2 of the deformed beam and you will see the magnitude of the displacement appear at the bottom of the screen as

$$\text{"DY} = -4.8e{-}02.\text{"}$$

Click ⟨Esc⟩ and then "Quit" to leave Superview.

You should be able to find the *ex58.L* file (storing the echo check of your data and the nodal displacements) and the *ex58.S* file (storing the element forces and stresses) where you saved your *ex58 file*. You may want to print a hard copy of these files.

Figure 5–33 shows the symmetric scaled deflections as expected for this symmetric beam (symmetry about a plane perpendicular to the beam and passing through the roller at node 3). The basic elements have straight sides, and since deflection plots depict only the translations (deflections) and not the rotations of the nodes, the sides remain straight in the deflected plot.

Table 5–2 shows the nodal displacements and element stresses and forces taken from the Algor program *ex58.L* and *ex58.S* files. The last six nodes are the "*k* nodes". They have zero displacement. The Algor solution for the displacements (shown in Table 5–2) compares exactly with the longhand solution [Eq. (5.3.14)].

Table 5–2   ex58.L and ex58.S files

```
Algor (R) Linear Static Stress
Version 12.00-WIN 15-FEB-1999
Copyright (c) 1984—1999 Algor, Inc. All rights reserved.

       DATE: DECEMBER 17,1999
       TIME: 02:00 PM
       INPUT FILE............C:\WINNT\Profiles\mermt\Desktop\ex58
```

1**** CONTROL INFORMATION

```
        number of node points        (NUMNP)  =          11
        number of element types      (NELTYP) =           1
        number of load cases         (LL)     =           1
        number of frequencies        (NF)     =           0
        geometric stiffness flag     (GEOSTF) =           0
        analysis type code           (NDYN)   =           0
        solution mode                (MODEX)  =           0
        equations per block          (KEQB)   =           0
        weight and c.g. flag         (IWTCG)  =           0
        bandwidth minimization flag  (MINBND) =           0
        gravitational constant       (GRAV)   = 3.8640E+02
```

```
        bandwidth minimization specified
 **** PRINT OF NODAL DATA SUPPRESSED
 **** PRINT OF EQUATION NUMBERS SUPPRESSED
 **** PRINT OF TYPE-2 ELEMENT DATA SUPPRESSED
```

1**** BANDWIDTH MINIMIZATION

```
     minbnd (bandwidth control parameter) = 1
 **** MINIMIZER DID NOT NEED TO REDUCE BANDWIDTH
 **** Hard disk file size information for processor:

     Available hard disk space on current drive = 445.844 megabytes
```

1**** NODAL LOADS (STATIC) OR MASSES (DYNAMIC)

NODE NUMBER	LOAD CASE	X-AXIS FORCE	Y-AXIS FORCE	Z-AXIS FORCE	X-AXIS MOMENT	Y-AXIS MOMENT	Z-AXIS MOMENT
2	1	0.000E+00	−1.000E+04	0.000E+00	0.000E+00	0.000E+00	0.000E+00
4	1	0.000E+00	−1.000E+04	0.000E+00	0.000E+00	0.000E+00	0.000E+00

1**** ELEMENT LOAD MULTIPLIERS

load case	case A	case B	case C	case D
1	1.000E+00	0.000E+00	0.000E+00	0.000E+00

```
 **** EQUATION PARAMETERS
      Number of equations        =     15
      Minimum bandwidth          =      1
      Maximum bandwidth          =      9
```

Table 5–2 (*Continued*)

```
        Average bandwidth        =      5
        Storage required       (KB) =      1
        Total memory allocated (KB) = 45608
        Total memory free      (KB) = 45608

 ****  Proceeding with in-core fast solver ...

1****  STIFFNESS MATRIX PARAMETERS

        minimum non-zero diagonal element = 3.5000E+04
        maximum diagonal element          = 1.0000E+09
        maximum/minimum                   = 2.8571E+04
        average diagonal element          = 3.3351E+08

 ****  BEGIN IN-CORE SOLUTION
 ****  load case #      1
 ****  END IN-CORE SOLUTION

1****  STATIC ANALYSIS

        LOAD CASE =      1

        Displacements/Rotations(degrees) of nodes
```

NODE number	X-translation	Y-translation	Z-translation	X-rotation	Y-rotation	Z-rotation
1	0.0000E+00	0.0000E+00	0.0000E+00	0.0000E+00	0.0000E+00	0.0000E+00
2	0.0000E+00	−4.8000E−02	0.0000E+00	0.0000E+00	0.0000E+00	0.0000E+00
3	0.0000E+00	0.0000E+00	0.0000E+00	0.0000E+00	0.0000E+00	0.0000E+00
4	0.0000E+00	−4.8000E−02	0.0000E+00	0.0000E+00	0.0000E+00	0.0000E+00
5	0.0000E+00	0.0000E+00	0.0000E+00	0.0000E+00	0.0000E+00	0.0000E+00
6	0.0000E+00	0.0000E+00	0.0000E+00	0.0000E+00	0.0000E+00	0.0000E+00
7	0.0000E+00	0.0000E+00	0.0000E+00	0.0000E+00	0.0000E+00	0.0000E+00
8	0.0000E+00	0.0000E+00	0.0000E+00	0.0000E+00	0.0000E+00	0.0000E+00
9	0.0000E+00	0.0000E+00	0.0000E+00	0.0000E+00	0.0000E+00	0.0000E+00
10	0.0000E+00	0.0000E+00	0.0000E+00	0.0000E+00	0.0000E+00	0.0000E+00
11	0.0000E+00	0.0000E+00	0.0000E+00	0.0000E+00	0.0000E+00	0.0000E+00

```
1****  TEMPORARY FILE STORAGE (MEGABYTES)

        UNIT NO.   7:   0.000
        UNIT NO.   8:   0.001
        UNIT NO.   9:   0.000
        UNIT NO.  10:   0.000
        UNIT NO.  11:   0.000
        UNIT NO.  12:   0.000
        UNIT NO.  13:   0.000
        UNIT NO.  14:   0.000
        UNIT NO.  15:   0.000
        UNIT NO.  17:   0.000

        TOTAL      :   0.001 Megabytes
```

Table 5–2 (*Continued*)

```
Algor (R) FEA Stress Processor
Version 12.00-WIN 15-FEB-1999
Copyright (c) 1989—1999 Algor, Inc. All rights reserved.
DATE: DECEMBER 17,1999
TIME: 02:00 PM
INPUT......C:\WINNT\Profiles\mermt\Desktop\ex58
```

1**** BEAM ELEMENTS

```
    number of beam elements         = 4
    number of area property sets    = 2
    number of fixed end force sets   = 4
    number of materials             = 1
    number of intermediate load sets = 4
```

1**** BEAM ELEMENT FORCES AND MOMENTS

ELEMENT NO.	CASE (MODE)	AXIAL FORCE R1	SHEAR FORCE R2	SHEAR FORCE R3	TORSION MOMENT M1	BENDING MOMENT M2	BENDING MOMENT M3
1	1	0.000E+00	5.000E+03	0.000E+00	0.000E+00	0.000E+00	3.000E+05
		0.000E+00	−5.000E+03	0.000E+00	0.000E+00	0.000E+00	3.000E+05
2	1	0.000E+00	−5.000E+03	0.000E+00	0.000E+00	0.000E+00	−3.000E+05
		0.000E+00	5.000E+03	0.000E+00	0.000E+00	0.000E+00	−3.000E+05
3	1	0.000E+00	5.000E+03	0.000E+00	0.000E+00	0.000E+00	3.000E+05
		0.000E+00	−5.000E+03	0.000E+00	0.000E+00	0.000E+00	3.000E+05
4	1	0.000E+00	−5.000E+03	0.000E+00	0.000E+00	0.000E+00	−3.000E+05
		0.000E+00	5.000E+03	0.000E+00	0.000E+00	0.000E+00	−3.000E+05

1**** BEAM ELEMENT STRESSES

ELEMENT NO.	CASE (MODE)	P/A	P/A+M2/S2	P/A−M2/S2	P/A+M3/S3	P/A−M3/S3	WORST SUM
1	1	0.000E+00	0.000E+00	0.000E+00	−6.923E+06	6.923E+06	6.923E+06
		0.000E+00	0.000E+00	0.000E+00	6.923E+06	−6.923E+06	6.923E+06
2	1	0.000E+00	0.000E+00	0.000E+00	6.923E+06	−6.923E+06	6.923E+06
		0.000E+00	0.000E+00	0.000E+00	−6.923E+06	6.923E+06	6.923E+06
3	1	0.000E+00	0.000E+00	0.000E+00	−6.923E+06	6.923E+06	6.923E+06
		0.000E+00	0.000E+00	0.000E+00	6.923E+06	−6.923E+06	6.923E+06
4	1	0.000E+00	0.000E+00	0.000E+00	6.923E+06	−6.923E+06	6.923E+06
		0.000E+00	0.000E+00	0.000E+00	−6.923E+06	6.923E+06	6.923E+06

1**** End of file ■

Example 5.9

We now solve Example 5.3 (Figure 5–35) to become familiar with the *Group* command to enable the model to have different material (mechanical) properties such as E and G. Let $A = 1$ m^2, $I = 2 \times 10^{-4}$ m^4, and $E = 210$ GPa for the horizontal beam, and let $A = 1$ m^2, $I = 2 \times 10^{-4}$ m^4, $L = 1$ m, and $E = 200$ kPa for the vertical member. (Note that the *Layer* command is used for different geometric properties, such as cross-sectional areas or moments of inertia.)

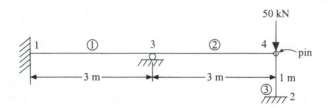

Figure 5–35 Beam supported by vertical member with different modulus of elasticity

Step 1 Start Superdraw III

Click on the "Algor" icon on your desktop computer screen or click on the "Start" button at the bottom left of the screen and go to "Programs-Algor Software-Algor FEA 12" to start Superdraw III.

Step 2 Create the Beam Model

Since the unit system is different from the default (English (in)), we must first change the unit system in Algor.

Click "Model Data" on the bottom of the screen.

Click "Units" under the "FEA Model" heading. It will respond with "Please enter model name first".

Click "OK" and then enter the model name in the "File_name" area that appears.

ex59   The name of the file is now *ex59*.

Click "OK". Now the units window opens.

Select "Metric mks (SI)" from the units pull-down menu.

Click "OK". You are now ready to draw the beam.

Click on the "Add" menu on the top main menu bar.

Click "Line". A menu appears on the upper right side of the screen. Now enter the data points in succession, starting at the origin and finishing at the end of the beam.

0<Tab>0   This is the first point of the beam.

<Enter>

3<Tab>0   This is the second point of the beam.

<Enter>

6<Tab>0   This is the point at which the load will be added.

<Enter>

In order to analyze the problem with different material properties, we will have to draw the vertical beam in a different group. The group we are currently in is group 1. To switch to a different group, simply

Click on the "Hidden" box in the "Model Data Control" window. You are prompted to choose a new group.

Click "Yes". This takes you to the "Group" window. Either type in a 2 or select the number 2 from above.

<Enter>

Click "Add" on the top main menu bar.

Click "Line". A menu appears on the upper right side of the screen. Now enter the data points in succession, starting at the top of the vertical beam.

6<Tab>0   This is the third point of the beam.

<Enter>

6<Tab>-1   This is the fourth and final point of the beam.

<Enter>

Click "Done".

Select "View" from the main menu.

Click "Enclose".

Now go back to the "Model Data Control" window.

Click on the box under "Active" to reactivate group 1.

You will see a line that represents the beam in Figure 5–35.

Step 3   Eliminate Duplicate Lines

Click "Modify" on the main menu.

Select "Clean:Duplicate". This notation means to select "Clean" and then "Duplicate" in succession.

Click "Perform Cleaning" on the "Duplicate" menu. You will see the following message at the lower part of the screen:

"3 Kept 0 Deleted. done"

Click "Done" on the "Duplicate" menu located in the upper right corner of the screen.

Step 4   Add in the Boundary Conditions

Click "FEA Add" on the main menu.

Click "Stress and Vibration:Boundary Conditions", and the on-screen "Boundary Conditions" menu will appear. The default boundary condition is "Use @ symbol for Full". "Full" means full constraints on translation and rotation when this condition is applied to a particular node. This particular model needs to be constrained fully at the far end nodes while it is constrained in only y-translational and x-and y-rotational at the middle node roller support.

First we will fix the two far ends by simply right-clicking on the two farthest end nodes. An "@" symbol will appear next to the node telling you the node is completely constrained. You may have to zoom in to see this node. To do this,

Click on the "View" menu and select "Zoom" and then "In". Make a box around the node, and you should be able to see the "@" symbol.

To make the changes to apply the next constraint, the defaults must be reset.

Click "Change Values" in the "Boundary Conditions" window. This will allow you to take away constraints from particular translations and rotations. Tx, Ty, Tz are the translational constraints, while Rx, Ry, Rz are the rotational constraints. For this constraint simply unselect the terms Tx and Rz.

"Click Done". Now right-click at the center of the beam; Ty, Tz, Rx, Ry terms should appear at the node. As before, you may have to zoom in to see these terms.

Step 5   Apply Forces to the Beam

Click "FEA Add" on the main menu.

Click "Stress and Vibration Analysis".

Click "Nodal Forces"; a "Nodal Forces" window appears.

Click "Vector". You will need to create the direction of the force.

Click "Ydirection". Note that this is the default direction for the force, but for illustration we will use it. You will see "DX 0 DY 0.97 DZ 0" at the bottom of the screen.

Click "Done". This takes you back to the "Nodal Forces" window.

Click "Magnitude" A prompt "Magnitude" appears at the bottom of the screen asking you to enter the magnitude of the force.

−50000   Enter the magnitude of the force.

<Enter>   You might have to move the screen to see the message "Click on node to apply the current values".

Move the mouse cursor to the top of the vertical member and right-click on the node. The −50000-N force appears.

Click "Done". You may have to "Zoom" in again to see these forces.

Step 6   Add the Material Properties Using the "Model Data Control" Window

Click on the "Model Data" control button located at the lower part of the screen and highlighted in red.

For the "Analysis Type" line, "Linear Static Stress" appears highlighted as the default.

Click on the box below the "Element" column. A table of group 1 element types appears.

Click "Beam". You now have created a beam. All lines are beam elements.

Click "OK".

Click on the box below the "Data" column. The "Beam Design Editor" window now appears. Now enter the "File_name" as *ex59*.

Click "Add/Mod" to go to the "Add/Modify" window.

Click "Sectional" to take you to the "Sectional" window. You will enter the sectional properties of the beam here.

Click "Value". A cursor is now at the bottom of the screen next to the property "A". You can scroll over to the other properties by pressing the ⟨Tab⟩ key. Do this until you come to the "I" terms. The ⟨Esc⟩ key allows you to delete the current terms. Under these "I" terms enter the value 2e-4.

`2e-4`   These are the values of the moments of inertia that resist the bending.

`<Enter>`

`<Esc>`

`<Esc>`   You are now in the "Beam Design Editor" main menu again.

Click "donE". You are asked "Save Current Work?"

Click "Yes". This takes you back to the "Model Data Control" window. Repeat the process for group 2.

Click on the box below the "Element" column. A table of group 2 element types appears.

Click "Beam". You now have created a beam. All lines are beam elements.

Click "OK".

Click on the box below the "Data" column. The "Beam Design Editor" window now appears.

Click "Add/Mod" to bring up the "Add/Modify" window.

Click "Sectional" to go to the "Sectional" window. You will enter the sectional properties of the beam here.

Click "Value". A cursor is now at the bottom of the screen next to the property "A". You can scroll over to the other properties by hitting the ⟨Tab⟩ key. Do this until you come to the "I" terms. The ⟨Esc⟩ key allows you to delete the current terms. Under these "I" terms enter the value of 2e-4.

`2e-4`   These are the values of the moments of inertia that resist the bending.

`<Enter>`

`<Esc>`

`<Esc>`   You are now in the "Beam Design Editor" main menu again.

Click "Add/Mod".

Click "End rel" for the end release.

Select "Value".

Click on r1, r2, and r3 so that an asterisk appears in front of each. For example, *r1, *r2, *r3.

Left-click on the top portion of the vertical member. A brown circle will appear at the top of the vertical member. Return to the main menu.

Click "donE". You are asked "Save Current Work?"

Click "Yes". This takes you back to the "Model Data Control" window.

Click on the box below "Material". This brings up the on-screen menu "Element Material Selection".

We can select one of the predefined materials or "Customer Defined" materials. In this case we will select "Customer Defined".

Click "Customer Defined".

Click "View Properties:Edit Properties" and enter the modulus of elasticity.

200e3   Modulus of elasticity.

Click "OK" to apply the properties to the "Customer Defined" material.

Click "OK" to return to the "Model Data Control" window.

Repeat this process for group 1.

Step 7 Enter the Global Data

Click on the "Global" button on the right-side menu of the "Model Data Control" window.

Under "Load Case Multipliers" and under "Pressure", type in 1.

Click on the "Output" tab to view the output options.

Click on the box next to "Displacement data" and the box next to "Stress data".

This creates nodal displacement output in the .L file and element forces and stresses in the .S file. These files can be viewed later with a word processor.

Step 8 Check the Model

Click "Check" under the "FEA Model" column in the "Model Data Control" window. This takes you to Superview's main menu, and you can check your model as explained in Section 4.1 and detailed by example in Appendix F. The beam model appears automatically, showing the boundary conditions in red and any forces that may be applied in yellow.

Click "Done". This takes you back to the "Model Data Control" menu.

Step 9 Run the Analysis

Click on the "Analysis" button under the "FEA Model" column in the "Model Data Control" window.

Click "Analyze" in lower left corner of the "Linear Stress and Vibration-Static Stress" menu.

A warning message appears after the analysis has been completed but has no bearing on the outcome of the analysis.

Click "OK" to close the warning message.

Click "Done" to return to the "Model Data Control" menu.

Step 10 Review the Results

Click on the "Results" button in the "Model Data Control" window. This takes you to Superview once again, where you can view the results.

Click "Results".

Click "Beam and truss stress".

This gives you the worst-case bending stress.

This gives a maximum of 32e6 Pa and a minimum of 3488 Pa.

Select ⟨Esc⟩.

<Esc>

Select "Disp vec" on the "Post" menu.

The maximum displacement is −0.01744 m.

<Esc>

<Esc>

<Esc>

You should now be back at the main menu screen.

Click "Done". ■

You should be able to find the *ex59.L* file (storing the echo check of your data and the nodal displacements) and the *ex59.S* file (storing the element forces and stresses) where you saved your *ex59 file*. You may want to print a hard copy of these files.

Table 5–3 shows the nodal displacements taken from the Algor program *ex59.L* and *ex59.S* files. The nodal displacement at node 4 in Table 5–3 is identical to the

Table 5–3 Displacements/rotations from the *ex59.L* file from Algor program solution

Displacements/Rotations(degrees) of nodes

NODE number	X- translation	Y- translation	Z- translation	X- rotation	Y- rotation	Z- rotation
1	0.0000E+00	0.0000E+00	0.0000E+00	0.0000E+00	0.0000E+00	0.0000E+00
2	0.0000E+00	0.0000E+00	0.0000E+00	0.0000E+00	0.0000E+00	0.0000E+00
3	0.0000E+00	0.0000E+00	0.0000E+00	0.0000E+00	0.0000E+00	-1.4276E-01
4	0.0000E+00	-1.7442E-02	0.0000E+00	0.0000E+00	0.0000E+00	-4.2829E-01
5	0.0000E+00	0.0000E+00	0.0000E+00	0.0000E+00	0.0000E+00	0.0000E+00
6	0.0000E+00	0.0000E+00	0.0000E+00	0.0000E+00	0.0000E+00	0.0000E+00
7	0.0000E+00	0.0000E+00	0.0000E+00	0.0000E+00	0.0000E+00	0.0000E+00
8	0.0000E+00	0.0000E+00	0.0000E+00	0.0000E+00	0.0000E+00	0.0000E+00
9	0.0000E+00	0.0000E+00	0.0000E+00	0.0000E+00	0.0000E+00	0.0000E+00
10	0.0000E+00	0.0000E+00	0.0000E+00	0.0000E+00	0.0000E+00	0.0000E+00

1**** MATERIAL PROPERTIES

INDEX	E	MU	MASS DENSITY	WEIGHT DENSITY	THERMAL EXPANSION X	Y	Z	REFERENCE TEMPERATURE
1	2.10E+11	.300	7.32E-04	2.83E-01	0.00E+00	0.00E+00	0.00E+00	0.000E+00
2	2.00E+05	.300	7.32E-04	2.83E-01	0.00E+00	0.00E+00	0.00E+00	0.000E+00 ■

longhand displacement [Eq. (5.3.23)]. (Note that node 4 in the Algor model corresponds to node 3 in the longhand solution.) Also, the rotations at nodes 3 and 4 in Algor are expressed in units of degrees, whereas the corresponding longhand solutions for rotations at nodes 2 and 3 are in radians. On conversion from radians to degrees, the answers are identical; see Eq. (5.3.24).

The output file also includes the echo check of all data as in Example 5.8. We list below the material properties part of this only, because this is the only new information showing the two different material properties for Young's modulus.

Example 5.10

We now solve Problem 5.22 (Figure 5–36) to illustrate how to analyze a beam subjected to distributed loading. Let $I = 150$ in^4 and $E = 29 \times 10^6$ psi.

Applying distributed loads to beams is described in [3]. Figure 5–37 shows some of the terms that are input to create a distributed load on a beam. Note that the load does not have to be uniform and does not have to act over the whole length of the beam. The values of A and B are expressed as a ratio of the beam length. The range of

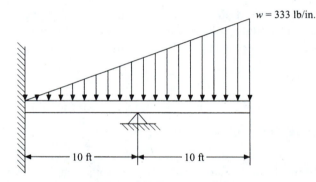

Figure 5–36 Beam subjected to distributed load

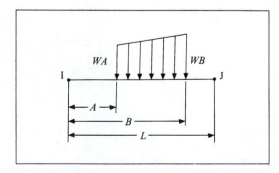

Figure 5–37 Sample distributed loads

these values is from 0.0 to 1.0. The values for *WA* and *WB* are the magnitudes of the loads at locations *A* and *B*, respectively. The load is then a straight line connecting these values.

Step 1   Start Superdraw III

Select the "Start" button of Windows NT/95/98 and then proceed to "Programs: Algor FEA".

Then click "Algor FEA" to start the program.

Or double-click on the "Algor FEA" icon on your desktop computer screen. The Superdraw III program will appear. This is the Algor main graphics user interface. All your work will be done from this program. You will construct, analyze, and review your results from Superdraw III.

Each menu option and a short example are available in the Superdraw III Reference Division (accessed through "Docutech", the on-line "Technical User Documentation").

Step 2   Create the Beam Model

Select the "Add" menu from the top main menu bar.

Click "Line". A "Line" menu appears in the upper right part of the screen. Then enter the data points (coordinates of the horizontal line 1) in succession.

`0<Tab>0`   This point is the left side of the beam element.

`<Enter>`   Press the ⟨Enter⟩ key to actually enter the data.

`120<Tab>0`   Enter the second point of the model.

`<Enter>`

`240<Tab>0`   This is the right end of the model.

`<Enter>`

Click "Done"

Select "View:Enclose". This encloses the model. You should see two highlighted lines.

Step 3   Add in the Boundary Conditions

Click on the "FEA Add" menu on the main menu bar and highlight "Stress and Vibration Analysis".

Click "Boundary Conditions". The default is "Use @ Symbol for Full".

Click "Text Attributes". This changes the size of the boundary condition text, making it easy to see that you have placed your boundary conditions in the right place.

Click "Height."

`10`   This is the height of the text.

`<Enter>`

`<Done>`   This takes you back to the "Boundary Conditions" menu.

Right-click on the left side of the model. An "@" symbol appears next to the node.

Click "Change Values".

Click "Tx Constrained". This releases translation in the x direction.

Click "Rz Constrained". This allows rotation about the z axis.

Click "Done".

Right-click on the middle of the model. "TyzRxy" should appear next to the node.

Click "Done". This closes the "Boundary Conditions" menu.

Step 4 Eliminate Duplicate Lines

Click "Modify" on the main menu.

Click "Clean:Duplicate". This notation means click "Clean" and then click "Duplicate" in succession. This command brings up the on-screen "Duplicate" menu.

Click "Perform Cleaning" on the "Duplicate" menu. You will see the following message within the text bar near the bottom of the screen.

"4 Kept 0 Deleted"

Click "Done" on the "Duplicate" menu located on the screen.

Step 5 Enter Data in the "Model Data Control" Window

Click on the "Model Data" control button located on the lower part of the screen highlighted in red. This brings up the on-screen "Model Data Control" menu.

Click on the "Analysis Type" button (▼).

Click "Linear Static Stress".

Click on the box below "Element". This brings up the "Element Type" menu.

Click "Beam".

Click "OK".

Click on the box below "Data". The "Save As" menu appears. Enter the filename and click "Save". The "Units Definition menu appears. Click "OK" if "English (in)" is in the "Unit System" box.

The "Beam Design Editor" opens.

Click "Add/Mod":"Sectional":"Value". Then enter the beam data.

⟨Tab⟩, ⟨Tab⟩, ⟨Tab⟩, ⟨Tab⟩, ⟨Esc⟩, Type 150, ⟨Tab⟩, ⟨Esc⟩, Type 150, ⟨Enter⟩

You have entered the beam's moment of inertia data.

Click on ⟨Esc⟩, ⟨Esc⟩.

Click "Add/Mod", "Loading", "Distribut", "Select-b".

You are selecting the left half of the beam to apply the distributed force.

Click on the left side of the model.

Click ⟨Esc⟩. This takes you back to the "Distribut" menu.

Click "Value". This is where you enter the load for the left half of the model.

⟨Tab⟩, ⟨Esc⟩, 0, ⟨Tab⟩, ⟨Tab⟩, ⟨Tab⟩, ⟨Esc⟩, −165.5, ⟨Enter⟩

You have entered data for a load that starts at 0 lb at the left end and ends at −165 lb at the right end.

Click "Add". The loading should appear on your screen.

Click "Select-b", "None".

Click on the right half of the model.

Click ⟨Esc⟩. This takes you to the "Distribut" menu.

Click "Value".

⟨Tab⟩, ⟨Esc⟩, −165.5, ⟨Tab⟩, ⟨Tab⟩, ⟨Tab⟩, ⟨Esc⟩, −333, ⟨Enter⟩

This is data for a loading that starts at −165 lb at the left end and ends at −333 lb at the right end.

Click "Add". The load should appear on the model.

Click ⟨Esc⟩, ⟨Esc⟩, ⟨Esc⟩, "Quit".

Click "Yes" to save current work.

You should be back in Superdraw III with the "Model Data Control" window open.

Click on the box below "Material".

The "Element Material" menu appears

Click on "[Customer Defined]"

Click "View Properties".

Click "Unlock Properties".

Enter 29e6 in the "Modulus of Elasticity" box.

Click "OK".

Click "OK".

Step 6   Enter Global Data

Click on the "Global" button on the right side of the "Model Data Control" window. Under "Multipliers" and "Pressure", type in 1.

Click on the "Output" tab to view the output options.

Click on the boxes next to "Displacement data" and "Stress data".

Click on the boxes next to "Element data" and "Nodal data".

This creates output files that will be viewed later.

Click "OK". You are out of the "Global" menu and back at the "Model Data Control" menu.

Step 7   Check the Model

Click "Check" under the "FEA Model" column in the "Model Data Control" window. This takes you to Superview's main menu, and you can check your model as explained in Section 4.1 and detailed by example in Appendix F. The beam model appears automatically, showing the boundary conditions in red.

Click "Done". This takes you back to the "Model Data Control" menu.

Step 8   Run the Analysis

Click on the "Analysis" button under the "FEA Model" column in the "Model Data Control" window.

Table 5–4 Displacements/rotations from the ex510.L file

	Displacements/Rotations(degrees) of nodes					
NODE number	X– translation	Y– translation	Z– translation	X– rotation	Y– rotation	Z– rotation
1	0.0000E+00	0.0000E+00	0.0000E+00	0.0000E+00	0.0000E+00	0.0000E+00
2	0.0000E+00	0.0000E+00	0.0000E+00	0.0000E+00	0.0000E+00	-7.4146E-01
3	0.0000E+00	-3.2710E+00	0.0000E+00	0.0000E+00	0.0000E+00	-1.8458E+00
4	0.0000E+00	0.0000E+00	0.0000E+00	0.0000E+00	0.0000E+00	0.0000E+00
5	0.0000E+00	0.0000E+00	0.0000E+00	0.0000E+00	0.0000E+00	0.0000E+00
6	0.0000E+00	0.0000E+00	0.0000E+00	0.0000E+00	0.0000E+00	0.0000E+00
7	0.0000E+00	0.0000E+00	0.0000E+00	0.0000E+00	0.0000E+00	0.0000E+00
8	0.0000E+00	0.0000E+00	0.0000E+00	0.0000E+00	0.0000E+00	0.0000E+00
9	0.0000E+00	0.0000E+00	0.0000E+00	0.0000E+00	0.0000E+00	0.0000E+00 ∎

Click "Analyze" in the lower left corner of the "Linear Stress and Vibration-Static Stress" menu.

Click "Done" to return to the "Model Data Control" menu.

Step 9 Review the Results

Click on the "Results" button in the "Model Data Control" window.

This takes you to Superview once again, where you can view the results. You can now plot the displaced model.

Click "Displaced".

Click "With undi".

Click "Disp on".

Click "Calc scal".

Click "Nodes inq".

Click "Get".

Right-click on the right side of beam.

You will see "DY=−3.27" which represents the y displacement of 3.277 in. downward. This value matches the value from the longhand solution for Problem 5.22.

The nodal displacement and rotations listed in Table 5–4 compare exactly with those from the longhand solution for Problem 5.22. ∎

Example 5.11

We now solve Example 5.7 (Figure 5–38) to become familiar with how to model a beam with an internal hinge. This will enhance our understanding of member end releases. Let $a = 100$ in., $b = 100$ in., $P = 600$ lb, $I = 10$ in^4, and $E = 10^6$ psi.

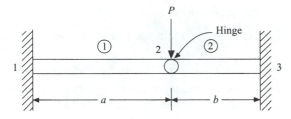

Figure 5–38 Beam with internal hinge

End releases are applied to beam nodes. These end releases have the effect of removing the specified translation or rotation response as described in Section 5.6. For example, if an *R3* end release (rotation about the local 3 axis) were applied to one end of a beam element (say, at node 2 of element 1 in Figure 5–38), this would effectively create a hinge at the end of the beam element and allow rotation about the local 3 axis. We do not want to create the same release *R3* at the left end of member 2 in Figure 5–38. Only the right end of member one or the left end of member two should be released. This is illustrated in this example. Translational end releases can be applied as well. They are considered sliders in the direction of application.

Step 1 Start Superdraw III

Select the "Start" button of Windows NT/95/98 and then proceed to "Programs: Algor FEA".

Then click "Algor FEA" to start the program.

Or double-click the "Algor FEA" icon on your desktop computer screen. The Superdraw III program will appear. This is the Algor main graphics user interface. All your work will be done from this program. You will construct, analyze, and review your results from Superdraw III.

Each menu option and a short example are available in the Superdraw III Reference Division (accessed through "Docutech", the on-line "Technical User Documentation").

Step 2 Create the Beam Model

Select the "Add" menu from the top main menu bar.

Click "Line". A "Line" menu appears in the upper right part of the screen. Then enter the data points (coordinates of the horizontal line 1) in succession.

`0<Tab>0`   This point is the left side of the beam element.

`<Enter>`   Press the ⟨Enter⟩ key to actually enter the data.

`100<Tab>0`   Enter the second point of the model.

`<Enter>`

`200<Tab>0`   This is the right end of model.

`<Enter>`

Click "Done".

Select "View:Enclose". This encloses the model. You should see two highlighted lines.

Step 3 Add in the Boundary Conditions and Forces

Click on the "FEA Add" menu on the main menu bar and highlight "Stress and Vibration Analysis".

Click "Boundary Conditions". The default is "Use @ Symbol for Full".

Click "Text Attributes". This changes the size of the boundary condition text, making it easy to see that you have placed your boundary conditions in the right place.

Click on "Height".

10   This is the height of the text.

<Enter>

<Done>   This takes you back to the "Boundary Conditions" menu.

Right-click on the left side of the model. An "@" symbol appears next to the node.

Right-click on the right side of the model.

Click "Done". This closes the "Boundary Conditions" menu.

Click "FEA Add", "Stress and Vibration Analysis", "Nodal Forces".

Click "Vector", "Y Direction", "Length", 10, "Enter", "Done", "Magnitude".

Enter −600 ⟨Enter⟩

Click "Arrow Size". Enter 3 ⟨Enter⟩.

Right-click on the middle of the beam.

The load shows up on the screen.

Click "Done".

Step 4 Eliminate Duplicate Lines

Click "Modify" on the main menu.

Click "Clean:Duplicate". This notation means click "Clean" and then click "Duplicate" in succession. This command brings up the on-screen "Duplicate" menu.

Click "Perform Cleaning" on the "Duplicate" menu. You will see the following message in the text bar near the bottom of the screen.

"6 Kept, 0 Deleted"

Click "Done" on the "Duplicate" menu located on the screen.

Step 5 Enter Data in the "Model Data Control" Window

Click on the "Model Data" control button in the lower part of the screen and highlighted in red. This brings up the on-screen menu "Model Data Control".

Click on the "Analysis Type" button (▼).

Click "Linear Static Stress".

Click on the box below "Element". This brings up the "Element Type" menu.

Click "Beam".

Click "OK".

Click on the box below "Data". The "Save As" menu appears. Enter the "File_name".

ex511

Click "Save". The "Units Definition" menu appears. Click "OK" if "English (in)" is in the "Unit System" box.

The "Beam Design Editor" opens.

Click "Add/Mod":"Sectional":"Value". Now enter beam data.

⟨Tab⟩, ⟨Tab⟩, ⟨Tab⟩, ⟨Tab⟩, ⟨Esc⟩, Type 10, ⟨Tab⟩, ⟨Esc⟩, Type 10, ⟨Enter⟩

You have entered the beam's moment of inertia data.

Click "⟨Esc⟩", "⟨Esc⟩"

Click "Add/Mod", "End Rel", "Value", "r3". This makes the hinge in the beam.

Click "⟨Esc⟩".

Left-click on the middle of the beam. This applies the release. A small circle should show up.

Click "⟨Esc⟩", "⟨Esc⟩":"Quit".

Click "Yes" to save current work.

You should be back in Superdraw III with the "Model Data Control" window open.

Click on the box below "Material".

The "Element Material" menu appears.

Click on "[Customer Defined]".

Click "View Properties".

Click "Unlock Properties".

Enter 10e6 in the "Modulus of Elasticity" box.

Click "OK".

Click "OK".

Step 6   Enter Global Data

Click on the "Global" button on the right side of the "Model Data Control" window. Under "Multipliers" and "Pressure", type 1.

Click on the "Output" tab to view the output options.

Click on the boxes next to "Displacement data" and "Stress data".

Click on the boxes next to "Element data" and "Nodal data".

This creates output files that will be viewed later.

Click "OK". You are out of the "Global" menu and back in the "Model Data Control" menu.

Step 7   Check the Model

Click "Check" under the "FEA Model" column in the "Model Data Control" window. This takes you to Superview's main menu, and you can check your model as explained in Section 4.1 and detailed by example in Appendix F. The beam model appears automatically, showing the boundary conditions in red.

Click "Done". This takes you back to the "Model Data Control" menu.

Step 8 Run the Analysis

Click on the "Analysis" button under the "FEA Model" column in the "Model Data Control" window.

Click "Analyze" on the lower left corner of the "Linear Stress and Vibration-Static Stress" menu.

Click "Done" to return to the "Model Data Control" menu.

Step 9 Review the Results

Click on the "Results" button in the "Model Data Control" window.

This takes you to Superview once again, where you can view the results.

You can now plot the displaced model.

Click "Displaced".

Click "With undi".

Click "Disp on".

Click "Calc scal".

Click "Nodes inq".

Click "Get".

Right-click on the middle of the beam.

You will see "DY = −10", which represents the y displacement of 10 in. downward.

■

The displacements and the member end forces are listed in Table 5–5. On substituting numbers from the Algor file into the longhand Eq. (5.6.19), we find that the longhand and Algor solutions are identical for the displacement at the hinge (node 2). The nodal forces from Algor and the longhand solution [Eq. (5.6.21)] are identical.

Gap elements are described in detail in Reference [3]. There are four types of gap elements used in Algor. The gap element must be identified by its own *Group* number. We use the compression with nonzero gap element in Example 5.12. To explain this element, refer to Figure 5–39. In this figure, X_{orig} is the original distance between the endpoints (nodes) A and B of the gap/cable element in Figure 5–39, and X_{new} is the distance between the gap/cable element nodes after loading has been applied. In the figure, line AB is the original gap orientation, line $A'D'$ is parallel to line AB, and line $A'C$ is the projection of line $A'B'$ onto line $A'D$. This projection is considered to be X_{new}.

Four types of gap elements are used to indicate the desired compression or tension with or without a gap. They are (1) a compression gap with a distance between the two endpoints (used in Example 5.12), (2) a tension gap with a distance between the two endpoints, (3) a compression gap with a zero gap, and (4) a tension gap with a zero gap. A compression gap is not activated until the gap is closed. In Example 5.12, the gap element is then not activated until the 0.5-in. gap is closed. Likewise, a tension gap is not activated until the gap is opened.

See Figure 5–40 for the "Gap Element Definition" screen. The information you need to supply in this screen is (1) the type of gap/cable you want applied (in Example

Table 5–5 Displacements and member end forces from Example 5.11

Displacements/Rotations(degrees) of nodes

NODE number	X- translation	Y- translation	Z- translation	X- rotation	Y- rotation	Z- rotation
1	0.0000E+00	0.0000E+00	0.0000E+00	0.0000E+00	0.0000E+00	0.0000E+00
2	0.0000E+00	-1.0000E+01	0.0000E+00	0.0000E+00	0.0000E+00	8.5944E+00
3	0.0000E+00	0.0000E+00	0.0000E+00	0.0000E+00	0.0000E+00	0.0000E+00
4	0.0000E+00	0.0000E+00	0.0000E+00	0.0000E+00	0.0000E+00	0.0000E+00
5	0.0000E+00	0.0000E+00	0.0000E+00	0.0000E+00	0.0000E+00	0.0000E+00
6	0.0000E+00	0.0000E+00	0.0000E+00	0.0000E+00	0.0000E+00	0.0000E+00
7	0.0000E+00	0.0000E+00	0.0000E+00	0.0000E+00	0.0000E+00	0.0000E+00
8	0.0000E+00	0.0000E+00	0.0000E+00	0.0000E+00	0.0000E+00	0.0000E+00
9	0.0000E+00	0.0000E+00	0.0000E+00	0.0000E+00	0.0000E+00	0.0000E+00

1**** BEAM ELEMENT FORCES AND MOMENTS

ELEMENT NO.	CASE (MODE)	AXIAL FORCE R1	SHEAR FORCE R2	SHEAR FORCE R3	TORSION MOMENT M1	BENDING MOMENT M2	BENDING MOMENT M3
1	1	0.000E+00	3.000E+02	0.000E+00	0.000E+00	0.000E+00	3.000E+04
		0.000E+00	-3.000E+02	0.000E+00	0.000E+00	0.000E+00	0.000E+00
2	1	0.000E+00	-3.000E+02	0.000E+00	0.000E+00	0.000E+00	0.000E+00
		0.000E+00	3.000E+02	0.000E+00	0.000E+00	0.000E+00	-3.000E+04 ■

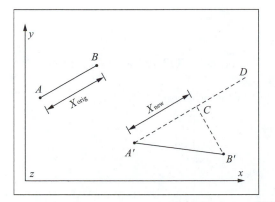

Figure 5–39 How X_{new} is determined

Element Definition

1 Linear Gap/Cable

Type	Compression with Gap ▼
Stiffness ●	0	lbf/in
Maximum Iterations	200

Reset From Model

| OK | Cancel | Apply | Print | Reset From Default |

Figure 5–40 "Gap Element Definition" screen

5.12 "Compression with Gap" is selected), (2) the gap/cable stiffness (in Example 5.12 a value of 10 times 10^{10} is appropriate, as we have a rigid support at the structure boundary to calculate the support reaction), and (3) the maximum number of gap iterations, or use 0 for the default of 200 iterations.

Example 5.12

Consider a cantilever beam with a gap as shown in Figure 5–41. This example will make use of combining beam and gap elements for analysis. For the beam use 4130 steel that has $A = 1$ in^2, $I_2 = 100$ in^4, and $I_3 = 100$ in^4. (This is Problem 5.6 at the end of the chapter.)

Step 1 Start Superdraw III

Select the "Start" button of Windows NT/95/98 and then proceed to "Programs: Algor Software:Algor FEA".

Then click "Algor FEA" to start the program.

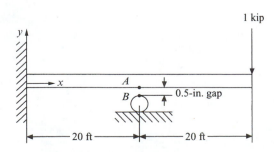

Figure 5–41 Beam with gap

Or double-click on the "Algor FEA" icon on your desktop computer screen. The Superdraw III program will appear. This is the Algor main graphics user interface. All your work will be done from this program. You will construct, analyze, and review your results from Superdraw III. Each menu option and a short example are available in the Superdraw III Reference Division (accessed through "Docutech", the on-line "Technical User Documentation").

Step 2 Create the Beam Model

Select the "Add" menu from the top main menu bar.

Click "Line". A "Line" menu appears on the screen.

Click "Single" on the "Line" menu and begin entering the data points (coordinates of the elements) in succession.

`0<Tab>0`	This is the location of the first node.
`<Enter>`	Press the ⟨Enter⟩ key to actually enter the data.
`240<Tab>0`	This is the endpoint of element 1.
`<Enter>`	
`240<Tab>0`	Repeat the process for element 2.
`<Enter>`	
`480<Tab>0`	
`<Enter>`	All the data points have been entered for the beam. Now enter the gap element data points *A* and *B* as shown in Figure 5–41.
`240<Tab>0`	
`<Enter>`	
`240<Tab>-0.5`	

Click "Done" on the "Line" menu on the screen.

Click "View" on the main menu.

Click "Enclose" on the "View" menu to see the full view of the model. It should look like Figure 5–41.

Step 3 Eliminate Duplicate Lines

Click "Modify" on the main menu.

Click "Clean:Duplicate". This notation means click "Clean" and then click "Duplicate" in succession. This command brings up the on-screen "Duplicate" menu.

Click "Perform Cleaning" on the "Duplicate" menu. You will see the following message within the text bar near the bottom of the screen.

"2 Kept 0 Deleted. done"

Click "Done" on the "Duplicate" menu located on the screen.

Step 4 Apply a Group to the Gap Element

A new group needs to be created in order to change the vertical line element into a different group to be defined as a gap element in the "Model Data Control" window in step 6. The horizontal element will become a beam element.

Click "Select:Point" on the main menu. You may need to zoom in on the gap element.

Click on the short gap element. It should now be highlighted with a small red box.

Click "Modify:Update Object Parameters:Group Number". This brings up a menu on the screen called "Group". Type 2 in the box with the cursor in it and enter.

2

`<Enter>`   The on-screen menu should disappear, and the bar element turns to a red color.

Step 5   Apply Boundary Conditions and Nodal Forces

Click on the "FEA Add" menu on the main menu.

Click "Stress and Vibration Analysis:Boundary Conditions", and the on-screen menu appears. The default boundary condition is "Use @ symbol for Full". "Full" means full constraints on translation and rotation when this condition is applied to a particular node. The model needs full constraints at the left node of the first element.

Right-click on the left node of the first horizontal element. The "@" sign appears on the screen next to the node, indicating that the node is fully constrained.

Click "Done" within the on-screen "Boundary Conditions" menu.

If there is difficulty in seeing the "@" symbol on-screen you may wish to zoom in closer to each node to view the boundary conditions.

Click "View:Zoom:In" and create a box around node 2 with the mouse and the left mouse button. You may have to repeat this process of zooming in one or more times to see the "@" sign.

Click "Done".

Note: If you have created one or more spurious or misplaced boundary condition "@" signs, click "Select:Box" on the main menu and draw a box around the object that you wish to delete (in this case a boundary condition). After the object is highlighted, click "Modify:Delete". At this time the "@" symbol should disappear.

Click "View:Enclose" to view the entire model again.

Click "FEA Add:Stress and Vibration Analysis:Nodal Forces". This brings up the on-screen menu "Nodal Forces". We need to confirm the direction of the nodal forces.

Click "Vector".

Click "Y Direction". Nodal forces will be applied in the y direction.

Click "Done". Now add the magnitude of the force.

Click "Magnitude". Now enter the value.

−1000

`<Enter>`   Now apply the force.

Right-click on the far right node of the beam. The force should now show on-screen. Zoom in if necessary.

Click "View:Enclose" to view the entire model again if necessary.

Step 6 Add Material Properties and Define the Type of Gap
Element Using the "Model Data Control" Window

Click on the "Model Data" control button located on the lower part of the screen and highlighted in red. This brings up the on-screen menu "Model Data Control". Alternatively, you can access the "Model Data Control" menu by clicking "Tools:Model Data Control".

On the "Analysis Type" line, "Linear Static Stress" appears as the default. You can select another analysis type by scrolling down the menu to an appropriate analysis algorithm, but in our case, "Linear Static Stress" will do nicely.

In the "Modal Data Control" window there should be two groups listed. The first group is the beam element, and the second group is the truss element.

Click on the group 1 row box below the "Element" column. An on-screen menu appears called "Group 1" elements. These are the different types of elements we can choose from. For this analysis both a beam and gap element are needed.

Click "Beam".

Click "OK".

Click on the group 2 row box below the "Element" column. An on-screen menu appears called "Group 2" elements.

Click "Gap".

Click "OK".

Click on the group 1 row box below "Data". It responds with "Please enter the model name first".

Click "OK". Then enter the model name in the "File_name" area that appears.

ex5_12   The name of the file is "5_12".

Click "Save". By default, the "Units Definition" menu appears on the screen.

For this analysis we need "English (in)" units.

Click "OK".

The program takes you to the "Beam Design Editor". We need to add area and cross-sectional properties.

Click "Add/Mod" on the main menu.

Click "Sectional" on the "Add/Mod" menu.

Click "Value" on the "Sectional" menu.

The area and "I" values are shown at the bottom. We need to delete some of the current values and add new ones. The current "A" value is correct as 1.

<Tab>

<Tab>

<Tab>

<Tab>   You should now be at the "*I2*" location.

<Esc>   This deletes the current "*I2*" value

100

```
<Tab>
<Esc>          This deletes the current "I3" value
100
<ENTER>        The "I2" and "I3" should now be $I_2 = 100$ in$^4$ and $I_3 = 100$ in$^4$.
```

Now leave the "Beam Design Editor" and return to the Superdraw III screen.

```
<Esc>
<Esc>
```

Click "Quit". The program asks if you wish to save.

Click "YES". The program should return you to the "Modal Data Control" on-screen menu.

The "Data" box within the group 1 row should be satisfied with a check mark.

Click on the group 2 row box below "Data".

The "Element Definition" menu appears next on the screen.

We must specify what type of gap element must be used. "Compression with Gap" is the default, and the one that we will be using. Now there must be a stiffness entered into the program. A large stiffness of 10×10^{10} is an appropriate stiffness for this model. Enter this number in the required area for stiffness.

```
10e10
<Enter>
```

Click "OK". This should satisfy "Data" in the group 2 row box on the "Modal Data Control" menu with a check mark.

Click on the group 1 row box below the "Material" column. This brings up the on-screen menu "Element Material Selection". We can select one of the predefined materials for a steel. The steel that we will choose is standard steel (steel 4130).

Click "Steel (4130)".

Click "OK".

The group 1 row "Material" box should be satisfied with "Steel (4130)". A material is not necessary for the gap element.

Step 7 Enter the Global Data

Click on the "Global" button on the right-side menu of the "Modal Data Control" window. Under "Load Case Multipliers" and under "Pressures", type a 1.

```
1              This enters values of 1 under "Pressure" on the "Load Case Multi-
               pliers" menu.
```

Under "Boundary" enter a multiplier of 1.

```
1
```

Click "Output" tab to view the output options.

Click on the box next to "Displacement data".

Click on the box next to "Stress data".

This creates nodal displacement output in the .L file and element forces and stresses in the .S file. These files can be viewed later with a word processor.

Click "OK". You are out of the "Global" menu and back in the "Model Data Control" menu.

Step 8   Check the Model

Click "Check" under the "FEA Model" column in the "Model Data Control" window. This takes you to Superview's main menu, and you can check your model as explained in Section 4.1 and detailed by example in Appendix F. The model appears automatically, showing the boundary conditions in red and any forces that may be applied in yellow.

Click "Done". This takes you back to the "Model Data Control" menu.

Step 9   Run the Analysis

Click on "Analysis" button under the "FEA Model" column in the "Model Data Control" window.

Click "Analyze" in lower left corner of the "Linear Stress and Vibration-Static Stress" menu. A warning message appears after the analysis has been completed but has no bearing on the outcome of the analysis.

Click "OK" to close the warning message.

Click "Done" to return to the "Model Data Control" menu.

Step 10   Review the Results

Click on the "Results" button in the "Model Data Control" window. This takes you to Superview once again where you can view the results.

Click "Displaced" on the main menu of Superview.

Click "Displ on" so that "*With undi" appears. This allows the displaced and undisplaced models to be superimposed (Figure 5–42).

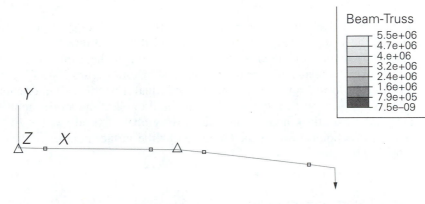

Figure 5–42   Plot of displaced beam and worst stress

Click "Calc scal" to enlarge the scale of the displacement for easier viewing.

Click "Nodes inq" on the "Displaced" menu.

Click "Get" to get a node.

Move the cursor to the free node and click on it to see the x and y displacements appear at the bottom of the screen. You will see "DX = 0, DY = 3.938e0, DZ = 0".

<Esc>

<Esc>   You are now back in the main menu of Superview.

Click "Results". You are now going to post a true stress.

Click "Beam and truss stress". You must select "Beam and truss stress" because you have a truss model.

Click "W)Worst". The worst-case stress will be posted. You now see the worst-case stresses. See the Figure 5–42 plot of the model with colored boxes representing different stresses on each element. For instance, red on the vertical element indicates a stress of 5.5×10^6 psi.

<Esc>

Click "Done" This takes you out of Superview.

Click "Close" to close the "Model Data Control" window.

You should be able to find the *ex5_12.L* file (storing the echo check of your data and the nodal displacements) and the *ex5_12.S* file (storing the element forces and stresses) and print a hard copy of each.   ■

Table 5–6 shows the nodal displacements and element stresses and forces from the Algor program *ex5_12.L* and *ex5_12.S* files. The results compare closely with the longhand solution of Problem 5.6.

Example 5.13

We now solve Problem 5.5 (see Figure 5–43). In the previous examples in this chapter, we used the same moments of inertia in the local 2 and 3 directions without concern for the orientation of the cross section. However, in this example we consider the beam to have a rectangular cross section as shown in Figure 5–43. The moments of inertia in the local 2 and 3 directions are not equal. For the desired orientation of the beam shown, we must use the Surface command to select the proper number (2 in this case) that makes the k node located along the global z axis at $z = 1 \times 10^{14}$. The local 2 axis always points towards the k node set by the surface number of the element (see Table 5–1).

Step 1   Start Superdraw III

Click on the "Algor FEA" icon on your desktop computer screen to start Algor.

Table 5–6   Displacements/Rotations, Beam element Forces and Stresses

Displacements/Rotations(degrees) of nodes

NODE number	X- translation	Y- translation	Z- translation	X- rotation	Y- rotation	Z- rotation
1	0.0000E+00	0.0000E+00	0.0000E+00	0.0000E+00	0.0000E+00	0.0000E+00
2	0.0000E+00	0.0000E+00	0.0000E+00	0.0000E+00	0.0000E+00	0.0000E+00
3	0.0000E+00	-5.0000E-01	0.0000E+00	0.0000E+00	0.0000E+00	-4.5407E-01
4	0.0000E+00	-3.9380E+00	0.0000E+00	0.0000E+00	0.0000E+00	-1.0041E+00
5	0.0000E+00	0.0000E+00	0.0000E+00	0.0000E+00	0.0000E+00	0.0000E+00
6	0.0000E+00	0.0000E+00	0.0000E+00	0.0000E+00	0.0000E+00	0.0000E+00
7	0.0000E+00	0.0000E+00	0.0000E+00	0.0000E+00	0.0000E+00	0.0000E+00
8	0.0000E+00	0.0000E+00	0.0000E+00	0.0000E+00	0.0000E+00	0.0000E+00
9	0.0000E+00	0.0000E+00	0.0000E+00	0.0000E+00	0.0000E+00	0.0000E+00
10	0.0000E+00	0.0000E+00	0.0000E+00	0.0000E+00	0.0000E+00	0.0000E+00

1**** BEAM ELEMENT FORCES AND MOMENTS

ELEMENT NO.	CASE (MODE)	AXIAL FORCE R1	SHEAR FORCE R2	SHEAR FORCE R3	TORSION MOMENT M1	BENDING MOMENT M2	BENDING MOMENT M3
1	1	0.000E+00	-1.174E+03	0.000E+00	0.000E+00	0.000E+00	-4.187E+04
		0.000E+00	1.174E+03	0.000E+00	0.000E+00	0.000E+00	-2.400E+05
2	1	0.000E+00	1.000E+03	0.000E+00	0.000E+00	0.000E+00	2.400E+05
		0.000E+00	-1.000E+03	0.000E+00	0.000E+00	0.000E+00	0.000E+00

1**** BEAM ELEMENT STRESSES

ELEMENT NO.	CASE (MODE)	P/A	P/A+M2/S2	P/A-M2/S2	P/A+M3/S3	P/A-M3/S3	WORST SUM
1	1	0.000E+00	0.000E+00	0.000E+00	9.664E+05	-9.664E+05	9.664E+05
		0.000E+00	0.000E+00	0.000E+00	-5.539E+06	5.539E+06	5.539E+06
2	1	0.000E+00	0.000E+00	0.000E+00	-5.539E+06	5.539E+06	5.539E+06
		0.000E+00	0.000E+00	0.000E+00	0.000E+00	0.000E+00	0.000E+00

Figure 5–43   Beam with cross section having different moments of inertia

Step 2 Create the Beam Model

Select the "Add" menu.

Click "Line" and enter the nodal data points.

0<Tab>0   This is the left node of the beam.

<Enter>

240<Tab>0   This is the middle point of the beam.

<Enter>

480<Tab>0   This is the right endpoint of the beam.

<Enter>

Click "Done".

Select "View" from the main menu and click "Enclose". The line drawing should appear.

Step 3 Eliminate Duplicate Lines

Do this as in previous examples.

Step 4 Add in the Boundary Conditions and Nodal Force

Do this as in previous examples. *Note*: Remove the "Tx" and "Rz" constraints at the middle support.

Step 5 Set the Surface Number Property for Proper Orientation of
Beam Cross Section

Choose "Select" from the main menu bar.

Click "All" to highlight all the elements. This step is not necessary if the beam elements are already highlighted.

Select "Modify".

Select "Update Object Parameters".

Click "Surface Number".

2   Enter a 2. This defines the elements to have a surface number that locates the k node at $z = 1e14$ as shown in Table 5–1. The local 2 axis of the beam cross section always points toward the k node. This is the desired orientation as we want the moment of inertia $I_2 = 100$. We want bending to take place about the local 2 axis of the cross section.

<Enter>

Select "Inquire" from the main menu.

Select "Properties of Object . . ." and click on each element one at a time and note the status of each. You will see a status line at the bottom of the screen with "Surface = 2 layer = 1 group = 1 . . . = LINE length=240.00".

Click "Done".

Step 6　Add the Material Properties Using the "Model Data Control" Window

Click on the "Model Data" control button at the lower part of the screen.

Follow the usual steps in this window.

Click on the box below "Element".

Click "Beam". You now have identified the elements as beams.

Click on the box below "Data" and click "OK". You should now be in the "Beam Design Editor" window. Now follow the usual steps in this window.

Important: On the "Add/Mod:Sectional" menu make $I_2 = 100$ and $I_3 = 15$.

On the "Sectional" menu enter the following properties:

Click "Sectional" to enter the "Sectional" menu.

Click "Value".

<Esc>	Clear the default value throughout the following data input.
21.56	Cross-sectional area A.
<Tab>	
<Esc>	
17.97	Shear area $Sa2$ (5/6 of A)
<Tab>	
<Esc>	
17.97	Shear area $Sa3$ (5/6 of A)
<Tab>	
<Esc>	
360	Torsional constant $J1$ (0.3 times the base times the height cubed [7]).
<Tab>	
<Esc>	
100	The moment of inertia $I2$ is 100.
<Tab>	
<Esc>	
15	The moment of inertia $I3$ is 15.
<Tab>	
<Esc>	
26.81	The section modulus $S2$ ($S2 = I2/$(half the depth $= 7.46/2)$].
<Tab>	
<Esc>	
10.38	The section modulus $S3$ ($S3 = I3/(2.89/2)$].
<Enter>	
<Esc>	
<Esc>	You should now be out of the "Beam Design Editor".

Table 5–7 Ex513.L file from the Algor solution

```
                    Displacements/Rotations(degrees) of nodes

  NODE      X-           Y-           Z-           X-           Y-           Z-
 number  translation  translation  translation   rotation     rotation     rotation

    1     0.0000E+00  -2.6894E+00   0.0000E+00   0.0000E+00   0.0000E+00   8.2528E-01
    2     0.0000E+00   0.0000E+00   0.0000E+00   0.0000E+00   0.0000E+00   0.0000E+00
    3     0.0000E+00   0.0000E+00   0.0000E+00   0.0000E+00   0.0000E+00   0.0000E+00
    4     0.0000E+00   0.0000E+00   0.0000E+00   0.0000E+00   0.0000E+00   0.0000E+00
    5     0.0000E+00   0.0000E+00   0.0000E+00   0.0000E+00   0.0000E+00   2.7524E-01
    6     0.0000E+00   0.0000E+00   0.0000E+00   0.0000E+00   0.0000E+00   0.0000E+00
    7     0.0000E+00   0.0000E+00   0.0000E+00   0.0000E+00   0.0000E+00   0.0000E+00
    8     0.0000E+00   0.0000E+00   0.0000E+00   0.0000E+00   0.0000E+00   0.0000E+00
    9     0.0000E+00   0.0000E+00   0.0000E+00   0.0000E+00   0.0000E+00   0.0000E+00   ■
```

Step 7 Enter the Global Data

Enter the global information in the usual manner.

Step 8 Check the Model

Check the model in the usual manner.

Step 9 Run the Analysis

Run the analysis in the usual manner.

Step 10 Review the Results

Review the results in the usual manner and then exit the program.

Table 5–7 shows the displacements from the .L file. The displacements and rotations from Table 5–7 compare very closely to those from the longhand solution of Problem 5.5. ■

▲ **References**

[1] Gere, J. M., and Timoshenko, S. P., *Mechanics of Materials*, 5th ed., Brooks/Cole Publishers, Pacific Grove, CA, 2001.

[2] Hsieh, Y. Y., *Elementary Theory of Structures*, 2nd ed., Prentice-Hall, Englewood Cliffs, NJ, 1982.

[3] *Linear Stress and Dynamics Reference Division*, Docutech On-Line Documentation, Algor, Inc., Pittsburgh, PA, 1999.

[4] *Superdraw Reference Division*, Docutech On-Line Documentation, Algor, Inc., Pittsburgh, PA, 1999.

[5] Spyrakos, Constantine C., *Finite Element Modeling in Engineering Practice*, West Virginia University Press, Morgantown, WV, 1994.

[6] Boresi, A. P., Schmidt, R. J., and Sidebottom, O. M., *Advanced Mechanics of Materials*, 5th Ed., Wiley, New York, 1993.

[7] Melosh, R. J., "Basis for Derivation of Matrices for the Direct Stiffness Method," *Journal of the American Institute of Aeronautics and Astronautics*, Vol. 1, No. 7, pp. 1631–1637, July 1963.

[8] Fraeijes de Veubeke, B., "Upper and Lower Bounds in Matrix Structural Analysis," *Matrix Methods of Structural Analysis*, AGAR Dograph 72, B. Fraeijes de Veubeke, ed., Macmillan, New York, 1964.

[9] Roark, R. J., and Young, W. C., *Formulas for Stress and Strain*, 6th ed., McGraw-Hill, New York, 1989.

▲ Problems

5.1 Use Eqs. (5.1.7) to plot the shape functions N_1 and N_3 and the derivatives $(dN_2/d\hat{x})$ and $(dN_4/d\hat{x})$, which represent the shapes (variations) of the slopes $\hat{\phi}_1$ and $\hat{\phi}_2$ over the length of the beam element.

5.2 Derive the element stiffness matrix for the beam element in Figure 5–1 if the rotational degrees of freedom are assumed positive clockwise instead of counterclockwise. Compare the two different nodal sign conventions and discuss. Compare the resulting stiffness matrix to Eq. (5.1.14).

Solve all problems using the finite element stiffness method.

5.3 For the beam shown in Figure P5–3, determine the rotation at pin support A and the rotation and displacement under the load P. Determine the reactions. Draw the shear force and bending moment diagrams. Let EI be constant throughout the beam.

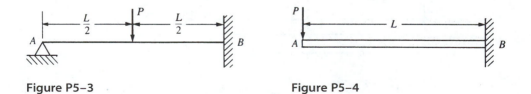

Figure P5–3   Figure P5–4

5.4 For the cantilever beam subjected to the free-end load P shown in Figure P5–4, determine the maximum deflection and the reactions. Let EI be constant throughout the beam.

5.5–5.11 For the beams shown in Figures P5–5–P5–11, determine the displacements and the slopes at the nodes, the forces in each element, and the reactions. Also, draw the shear force and bending moment diagrams.

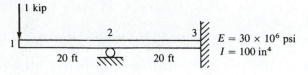

Figure P5–5

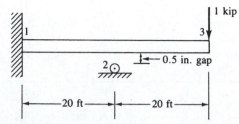

Figure P5–6

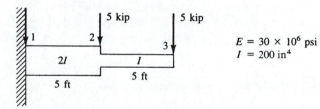

Figure P5–7

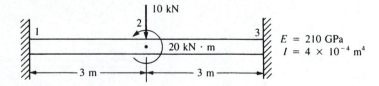

Figure P5–8

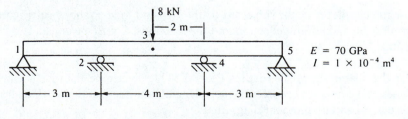

Figure P5–9

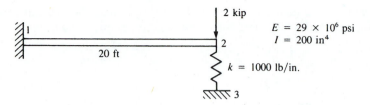

Figure P5–10

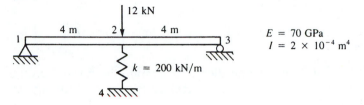

Figure P5–11

5.12   For the fixed-fixed beam subjected to the uniform load w shown in Figure P5–12, determine the midspan deflection and the reactions. Draw the shear force and bending moment diagrams. The middle section of the beam has a bending stiffness of $2EI$; the other sections have bending stiffnesses of EI.

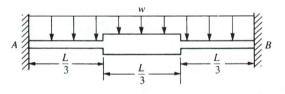

Figure P5–12

5.13   Determine the midspan deflection and the reactions and draw the shear force and bending moment diagrams for the fixed-fixed beam subjected to uniformly distributed load w shown in Figure P5–13. Assume EI constant throughout the beam. Compare your answers with the classical solution (that is, with the appropriate equivalent joint forces given in Appendix D).

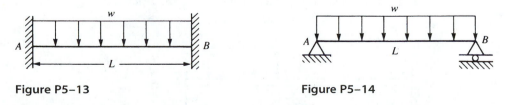

Figure P5–13   **Figure P5–14**

5.14   Determine the midspan deflection and the reactions and draw the shear force and bending moment diagrams for the simply supported beam subjected to the uniformly distributed load w shown in Figure P5–14. Assume EI constant throughout the beam.

5.15 For the beam loaded as shown in Figure P5–15, determine the free-end deflection and the reactions and draw the shear force and bending moment diagrams.

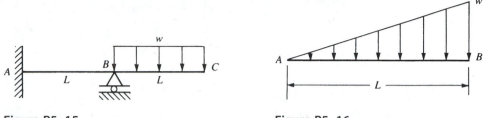

Figure P5–15   **Figure P5–16**

5.16 Using the concept of work equivalence, determine the nodal forces and moments (called *equivalent nodal forces*) used to replace the linearly varying distributed load shown in Figure P5–16.

5.17 Assume the beam of Figure P5–16 to be fixed at each end. Determine the reactions and draw the shear force and bending moment diagrams. Assume EI constant throughout the beam. Use two beam elements to discretize the beam. Compare your answers for the reactions with the appropriate equivalent joint forces given in Appendix D.

5.18 For the beam subjected to the linearly varying line load w shown in Figure P5–18, determine the right-end rotation and the reactions. Assume EI constant throughout the beam.

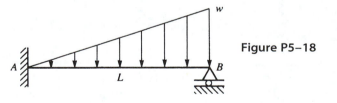

Figure P5–18

5.19–5.24 For the beams shown in Figures P5–19–P5–24, determine the nodal displacements and slopes, the forces in each element, and the reactions.

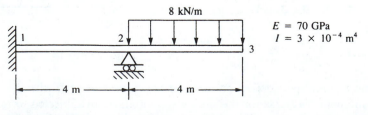

8 kN/m

$E = 70$ GPa
$I = 3 \times 10^{-4}$ m^4

4 m   4 m

Figure P5–19

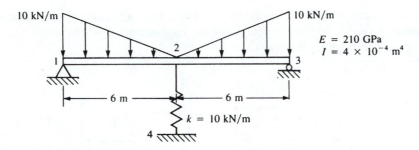

$$E = 210 \text{ GPa}$$
$$I = 4 \times 10^{-4} \text{ m}^4$$

$k = 10 \text{ kN/m}$

Figure P5–20

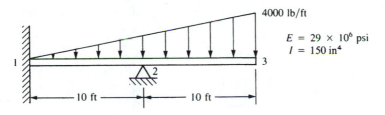

$$E = 29 \times 10^6 \text{ psi}$$
$$I = 200 \text{ in}^4$$

Figure P5–21

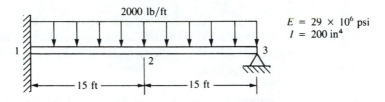

$$E = 29 \times 10^6 \text{ psi}$$
$$I = 150 \text{ in}^4$$

Figure P5–22

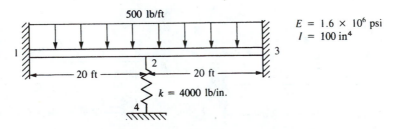

$$E = 1.6 \times 10^6 \text{ psi}$$
$$I = 100 \text{ in}^4$$

$k = 4000 \text{ lb/in.}$

Figure P5–23

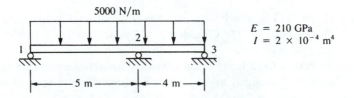

$$E = 210 \text{ GPa}$$
$$I = 2 \times 10^{-4} \text{ m}^4$$

Figure P5–24

5.25 For the beam shown in Figure P5–25 subjected to the concentrated load P and distributed load w, determine the midspan displacement and the reactions. Let EI be constant throughout the beam.

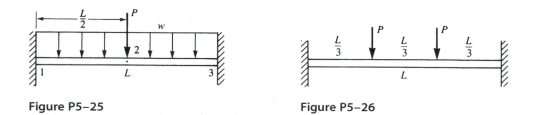

Figure P5–25   **Figure P5–26**

5.26 For the beam shown in Figure P5–26 subjected to the two concentrated loads P, determine the deflection at the midspan. Use the equivalent load replacement method. Let EI be constant throughout the beam.

5.27 For the beam shown in Figure P5–27 subjected to the concentrated load P and the linearly varying line load w, determine the free-end deflection and rotation and the reactions. Use the equivalent load replacement method. Let EI be constant throughout the beam.

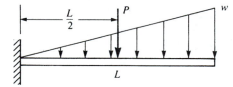

Figure P5–27

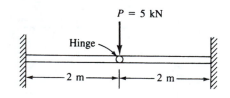

Figure P5–28

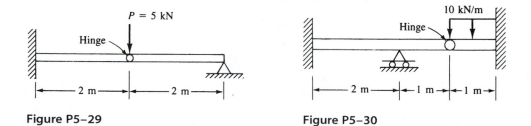

Figure P5–29   **Figure P5–30**

5.28–5.30 For the beams shown in Figures P5–28–P5–30, with internal hinge, determine the deflection at the hinge. Let $E = 210$ GPa and $I = 2 \times 10^{-4}$ m⁴.

5.31 Derive the stiffness matrix for a beam element with a nodal linkage—that is, the shear is 0 at node i, but the usual shear and moment resistance are present at node j (see Figure P5–31).

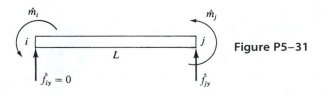

Figure P5–31

5.32 Explicitly evaluate π_p of Eq. (5.7.15); then differentiate π_p with respect to $\hat{d}_{1y}, \hat{\phi}_1, \hat{d}_{2y}$, and $\hat{\phi}_2$ and set each of these equations to zero (that is, minimize π_p) to obtain the four element equations for the beam element. Then express these equations in matrix form.

5.33 Determine the free-end deflection for the tapered beam shown in Figure P5–33. Here $I(x) = I_0(1 + nx/L)$ where I_0 is the moment of inertia at $x = 0$. Compare the exact beam theory solution with a 2-element finite element solution for $n = 2$.

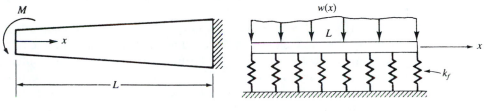

Figure P5–33   Figure P5–34

5.34 Derive the equations for the beam element on an elastic foundation (Figure P5–34) using the principle of minimum potential energy. Here k_f is the subgrade spring constant per unit length. The potential energy of the beam is

$$\pi_p = \int_0^L \frac{1}{2} EI(v'')^2 \, dx + \int_0^L \frac{k_f v^2}{2} \, dx - \int_0^L wv \, dx$$

5.35 Derive the equations for the beam element on an elastic foundation (see Figure P5–34) using Galerkin's method. The basic differential equation for the beam on an elastic foundation is

$$(EIv'')'' = -w + k_f v$$

5.36–51 Solve problems 5.5–5.11, 5.19–5.24, and 5.28–5.30 using the Algor program or other suitable computer program.

6

Frame and Grid Equations

Introduction

Many structures, such as buildings (Figure 6–1) and bridges, are composed of frames and/or grids. This chapter develops the equations and methods for solution of plane frames and grids.

First, we will develop the stiffness matrix for a beam element arbitrarily oriented in a plane. We will then include the axial nodal displacement degree of freedom in the local beam element stiffness matrix. Then we will combine these results to develop the stiffness matrix, including axial deformation effects, for an arbitrarily oriented beam element, thus making it possible to analyze plane frames. Specific examples of plane frame analysis follow. We will then consider frames with inclined or skewed supports.

Next, we will develop the grid element stiffness matrix. We will present the solution of a grid deck system to illustrate the application of the grid equations. We will then develop the stiffness matrix for a beam element arbitrarily oriented in space. We will also consider the concept of substructure analysis.

Finally, numerous rigid frame (including a space frame) and grid example problems are solved using the Algor computer program.

▲ 6.1 Two-Dimensional Arbitrarily Oriented Beam Element ▲

We can derive the stiffness matrix for an arbitrarily oriented beam element, as shown in Figure 6–2, in a manner similar to that used for the bar element in Chapter 3. The local axes $\hat{x}$ and $\hat{y}$ are located along the beam element and transverse to the beam element, respectively, and the global axes x and y are located to be convenient for the total structure.

Recall that we can relate local displacements to global displacements by using Eq. (3.3.16), repeated here for convenience as

$$\left\{ \begin{array}{c} \hat{d}_x \\ \hat{d}_y \end{array} \right\} = \left[\begin{array}{cc} C & S \\ -S & C \end{array} \right] \left\{ \begin{array}{c} d_x \\ d_y \end{array} \right\} \tag{6.1.1}$$

Figure 6–1 Rigid building frame (© Bill Bloom/Courtesy Midwest Steel, Inc.)

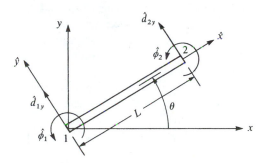

Figure 6–2 Arbitrarily oriented beam element

Using the second equation of Eqs. (6.1.1) for the beam element, we relate local nodal degrees of freedom to global degrees of freedom by

$$\begin{Bmatrix} \hat{d}_{1y} \\ \hat{\phi}_1 \\ \hat{d}_{2y} \\ \hat{\phi}_2 \end{Bmatrix} = \begin{bmatrix} -S & C & 0 & 0 & 0 & 0 \\ 0 & 0 & 1 & 0 & 0 & 0 \\ 0 & 0 & 0 & -S & C & 0 \\ 0 & 0 & 0 & 0 & 0 & 1 \end{bmatrix} \begin{Bmatrix} d_{1x} \\ d_{1y} \\ \phi_1 \\ d_{2x} \\ d_{2y} \\ \phi_2 \end{Bmatrix} \qquad (6.1.2)$$

where, for a beam element, we define

$$\underline{T} = \begin{bmatrix} -S & C & 0 & 0 & 0 & 0 \\ 0 & 0 & 1 & 0 & 0 & 0 \\ 0 & 0 & 0 & -S & C & 0 \\ 0 & 0 & 0 & 0 & 0 & 1 \end{bmatrix} \qquad (6.1.3)$$

as the *transformation matrix*. The axial effects are not yet included. Equation (6.1.2) indicates that rotation is invariant with respect to either coordinate system. For example, $\hat{\phi}_1 = \phi_1$, and moment $\hat{m}_1 = m_1$ can be considered to be a vector pointing normal to the $\hat{x}$-$\hat{y}$ plane or to the x-y plane by the usual right-hand rule. From either viewpoint, the moment is in the $\hat{z} = z$ direction. Therefore, moment is unaffected as the element changes orientation in the x-y plane.

Substituting Eq. (6.1.3) for $\underline{T}$ and Eq. (5.1.14) for $\hat{\underline{k}}$ into Eq. (3.4.22), $\underline{k} = \underline{T}^T \hat{\underline{k}} \underline{T}$, we obtain the global element stiffness matrix as

$$\underline{k} = \frac{EI}{L^3}\begin{array}{c}\begin{matrix} d_{1x} & \quad d_{1y} & \quad \phi_1 & \quad d_{2x} & \quad d_{2y} & \quad \phi_2 \end{matrix} \\ \begin{bmatrix} 12S^2 & -12SC & -6LS & -12S^2 & 12SC & -6LS \\ & 12C^2 & 6LC & 12SC & -12C^2 & 6LC \\ & & 4L^2 & 6LS & -6LC & 2L^2 \\ & & & 12S^2 & -12SC & 6LS \\ & & & & 12C^2 & -6LC \\ \text{Symmetry} & & & & & 4L^2 \end{bmatrix}\end{array} \quad (6.1.4)$$

where, again, $C = \cos\theta$ and $S = \sin\theta$. It is not necessary here to expand $\underline{T}$ given by Eq. (6.1.3) to make it a square matrix to be able to use Eq. (3.4.22). Because Eq. (3.4.22) is a generally applicable equation, the matrices used must merely be of the correct order for matrix multiplication (see Appendix A for more on matrix multiplication). The stiffness matrix Eq. (6.1.4) is the global element stiffness matrix for a beam element that includes shear and bending resistance. Local axial effects are not yet included. The transformation from local to global stiffness by multiplying matrices $\underline{T}^T \hat{\underline{k}} \underline{T}$, as done in Eq. (6.1.4), is usually done on the computer.

We will now include the axial effects in the element, as shown in Figure 6–3. The element now has three degrees of freedom per node $(\hat{d}_{ix}, \hat{d}_{iy}, \hat{\phi}_i)$. For axial effects, we recall from Eq. (3.1.13),

$$\left\{ \begin{matrix} \hat{f}_{1x} \\ \hat{f}_{2x} \end{matrix} \right\} = \frac{AE}{L}\begin{bmatrix} 1 & -1 \\ -1 & 1 \end{bmatrix}\left\{ \begin{matrix} \hat{d}_{1x} \\ \hat{d}_{2x} \end{matrix} \right\} \quad (6.1.5)$$

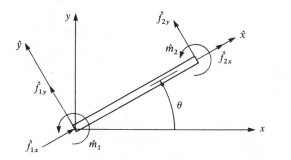

Figure 6–3   Local forces acting on a beam element

Combining the axial effects of Eq. (6.1.5) with the shear and principal bending moment effects of Eq. (5.1.13), we have, in local coordinates,

$$
\begin{Bmatrix} \hat{f}_{1x} \\ \hat{f}_{1y} \\ \hat{m}_1 \\ \hat{f}_{2x} \\ \hat{f}_{2y} \\ \hat{m}_2 \end{Bmatrix} = \begin{bmatrix} C_1 & 0 & 0 & -C_1 & 0 & 0 \\ 0 & 12C_2 & 6C_2L & 0 & -12C_2 & 6C_2L \\ 0 & 6C_2L & 4C_2L^2 & 0 & -6C_2L & 2C_2L^2 \\ -C_1 & 0 & 0 & C_1 & 0 & 0 \\ 0 & -12C_2 & -6C_2L & 0 & 12C_2 & -6C_2L \\ 0 & 6C_2L & 2C_2L^2 & 0 & -6C_2L & 4C_2L^2 \end{bmatrix} \begin{Bmatrix} \hat{d}_{1x} \\ \hat{d}_{1y} \\ \hat{\phi}_1 \\ \hat{d}_{2x} \\ \hat{d}_{2y} \\ \hat{\phi}_2 \end{Bmatrix} \quad (6.1.6)
$$

where

$$
C_1 = \frac{AE}{L} \quad \text{and} \quad C_2 = \frac{EI}{L^3} \quad (6.1.7)
$$

and, therefore,

$$
\underline{\hat{k}} = \begin{bmatrix} C_1 & 0 & 0 & -C_1 & 0 & 0 \\ 0 & 12C_2 & 6C_2L & 0 & -12C_2 & 6C_2L \\ 0 & 6C_2L & 4C_2L^2 & 0 & -6C_2L & 2C_2L^2 \\ -C_1 & 0 & 0 & C_1 & 0 & 0 \\ 0 & -12C_2 & -6C_2L & 0 & 12C_2 & -6C_2L \\ 0 & 6C_2L & 2C_2L^2 & 0 & -6C_2L & 4C_2L^2 \end{bmatrix} \quad (6.1.8)
$$

The $\hat{k}$ matrix in Eq. (6.1.8) now has three degrees of freedom per node and now includes axial effects (in the $\hat{x}$ direction), as well as shear force effects (in the $\hat{y}$ direction) and principal bending moment effects (about the $\hat{z} = z$ axis). Using Eqs. (6.1.1) and (6.1.2), we now relate the local to the global displacements by

$$
\begin{Bmatrix} \hat{d}_{1x} \\ \hat{d}_{1y} \\ \hat{\phi}_1 \\ \hat{d}_{2x} \\ \hat{d}_{2y} \\ \hat{\phi}_2 \end{Bmatrix} = \begin{bmatrix} C & S & 0 & 0 & 0 & 0 \\ -S & C & 0 & 0 & 0 & 0 \\ 0 & 0 & 1 & 0 & 0 & 0 \\ 0 & 0 & 0 & C & S & 0 \\ 0 & 0 & 0 & -S & C & 0 \\ 0 & 0 & 0 & 0 & 0 & 1 \end{bmatrix} \begin{Bmatrix} d_{1x} \\ d_{1y} \\ \phi_1 \\ d_{2x} \\ d_{2y} \\ \phi_2 \end{Bmatrix} \quad (6.1.9)
$$

where $\underline{T}$ has now been expanded to include local axial deformation effects as

$$
\underline{T} = \begin{bmatrix} C & S & 0 & 0 & 0 & 0 \\ -S & C & 0 & 0 & 0 & 0 \\ 0 & 0 & 1 & 0 & 0 & 0 \\ 0 & 0 & 0 & C & S & 0 \\ 0 & 0 & 0 & -S & C & 0 \\ 0 & 0 & 0 & 0 & 0 & 1 \end{bmatrix} \quad (6.1.10)
$$

Substituting $\underline{T}$ from Eq. (6.1.10) and $\hat{k}$ from Eq. (6.1.8) into Eq. (3.4.22), we obtain the general transformed global stiffness matrix for a beam element that includes axial force, shear force, and bending moment effects as follows:

$$\underline{k} = \frac{E}{L} \times$$

$$\begin{bmatrix} AC^2 + \frac{12I}{L^2}S^2 & \left(A - \frac{12I}{L^2}\right)CS & -\frac{6I}{L}S & -\left(AC^2 + \frac{12I}{L^2}S^2\right) & -\left(A - \frac{12I}{L^2}\right)CS & -\frac{6I}{L}S \\ & AS^2 + \frac{12I}{L^2}C^2 & \frac{6I}{L}C & -\left(A - \frac{12I}{L^2}\right)CS & -\left(AS^2 + \frac{12I}{L^2}C^2\right) & \frac{6I}{L}C \\ & & 4I & \frac{6I}{L}S & -\frac{6I}{L}C & 2I \\ & & & AC^2 + \frac{12I}{L^2}S^2 & \left(A - \frac{12I}{L^2}\right)CS & \frac{6I}{L}S \\ & & & & AS^2 + \frac{12I}{L^2}C^2 & -\frac{6I}{L}C \\ \text{Symmetry} & & & & & 4I \end{bmatrix}$$

$$(6.1.11)$$

The analysis of a rigid plane frame can be undertaken by applying stiffness matrix Eq. (6.1.11). *A **rigid plane frame** is defined here as a series of beam elements rigidly connected to each other*; that is, the original angles made between elements at their joints remain unchanged after the deformation due to applied loads or applied displacements.

Furthermore, moments are transmitted from one element to another at the joints. Hence, moment continuity exists at the rigid joints. In addition, the element centroids, as well as the applied loads, lie in a common plane (*x-y* plane). From Eq. (6.1.11), we observe that the element stiffnesses of a frame are functions of E, A, L, I, and the angle of orientation θ of the element with respect to the global-coordinate axes. It should be noted that computer programs often refer to the frame element as a beam element, with the understanding that the program is using the stiffness matrix in Eq. (6.1.11) for plane frame analysis.

▲ 6.2 Rigid Plane Frame Examples ▲

To illustrate the use of the equations developed in Section 6.1, we will now perform complete solutions for the following rigid plane frames.

Example 6.1

As the first example of rigid plane frame analysis, solve the simple "bent" shown in Figure 6–4.

The frame is fixed at nodes 1 and 4 and subjected to a positive horizontal force of 10,000 lb applied at node 2 and to a positive moment of 5000 lb-in. applied at node 3. The global-coordinate axes and the element lengths are shown in Figure 6–4. Let

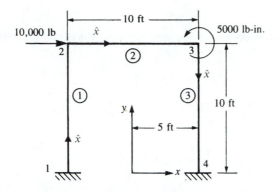

Figure 6–4 Plane frame for analysis, also showing local $\hat{x}$ axis for each element

$E = 30 \times 10^6$ psi and $A = 10$ in² for all elements, and let $I = 200$ in⁴ for elements 1 and 3, and $I = 100$ in⁴ for element 2.

Using Eq. (6.1.11), we obtain the global stiffness matrices for each element.

Element 1

For element 1, the angle between the global x and the local $\hat{x}$ axes is 90° (counterclockwise) because $\hat{x}$ is assumed to be directed from node 1 to node 2. Therefore,

$$C = \cos 90° = \frac{x_2 - x_1}{L^{(1)}} = \frac{-5 - (-5)}{120} = 0$$

$$S = \sin 90° = \frac{y_2 - y_1}{L^{(1)}} = \frac{120 - 0}{120} = 1$$

Also,

$$\frac{12I}{L^2} = \frac{12(200)}{(10 \times 12)^2} = 0.167 \text{ in}^2 \tag{6.2.1}$$

$$\frac{6I}{L} = \frac{6(200)}{10 \times 12} = 10.0 \text{ in}^3$$

$$\frac{E}{L} = \frac{30 \times 10^6}{10 \times 12} = 250,000 \text{ lb/in}^3$$

Then, using Eqs. (6.2.1) to help in evaluating Eq. (6.1.11) for element 1, we obtain the element global stiffness matrix as

$$
k^{(1)} = 250,000
\begin{array}{c}
\begin{array}{cccccc}
d_{1x} & d_{1y} & \phi_1 & d_{2x} & d_{2y} & \phi_2
\end{array} \\
\left[
\begin{array}{cccccc}
0.167 & 0 & -10 & -0.167 & 0 & -10 \\
0 & 10 & 0 & 0 & -10 & 0 \\
-10 & 0 & 800 & 10 & 0 & 400 \\
-0.167 & 0 & 10 & 0.167 & 0 & 10 \\
0 & -10 & 0 & 0 & 10 & 0 \\
-10 & 0 & 400 & 10 & 0 & 800
\end{array}
\right]
\begin{array}{c}
\\ \\ \frac{\text{lb}}{\text{in.}} \\ \\ \\
\end{array}
\end{array}
\tag{6.2.2}
$$

where all diagonal terms are positive.

Element 2

For element 2, the angle between x and $\hat{x}$ is zero because $\hat{x}$ is directed from node 2 to node 3. Therefore,

$$C = 1 \qquad S = 0$$

Also,

$$\frac{12I}{L^2} = \frac{12(100)}{120^2} = 0.0835 \text{ in}^2$$

$$\frac{6I}{L} = \frac{6(100)}{120} = 5.0 \text{ in}^3 \tag{6.2.3}$$

$$\frac{E}{L} = 250{,}000 \text{ lb/in}^3$$

Using the quantities obtained in Eqs. (6.2.3) in evaluating Eq. (6.1.11) for element 2, we obtain

$$
k^{(2)} = 250{,}000
\begin{matrix}
d_{2x} & d_{2y} & \phi_2 & d_{3x} & d_{3y} & \phi_3 \\
\begin{bmatrix}
10 & 0 & 0 & -10 & 0 & 0 \\
0 & 0.0835 & 5 & 0 & 0.0835 & 5 \\
0 & 5 & 400 & 0 & -5 & 200 \\
-10 & 0 & 0 & 10 & 0 & 0 \\
0 & 0.0835 & -5 & 0 & 0.0835 & -5 \\
0 & 5 & 200 & 0 & -5 & 400
\end{bmatrix}
\end{matrix}
\begin{matrix} \\ \\ \frac{\text{lb}}{\text{in.}} \\ \\ \end{matrix}
\tag{6.2.4}
$$

Element 3

For element 3, the angle between x and $\hat{x}$ is $270°$ (or $-90°$) because $\hat{x}$ is directed from node 3 to node 4. Therefore,

$$C = 0 \qquad S = -1$$

Therefore, evaluating Eq. (6.1.11) for element 3, we obtain

$$
k^{(3)} = 250{,}000
\begin{matrix}
d_{3x} & d_{3y} & \phi_3 & d_{4x} & d_{4y} & \phi_4 \\
\begin{bmatrix}
0.167 & 0 & 10 & -0.167 & 0 & 10 \\
0 & 10 & 0 & 0 & -10 & 0 \\
10 & 0 & 800 & -10 & 0 & 400 \\
-0.167 & 0 & -10 & 0.167 & 0 & -10 \\
0 & -10 & 0 & 0 & 10 & 0 \\
10 & 0 & 400 & -10 & 0 & 800
\end{bmatrix}
\end{matrix}
\begin{matrix} \\ \\ \frac{\text{lb}}{\text{in.}} \\ \\ \end{matrix}
\tag{6.2.5}
$$

Superposition of Eqs. (6.2.2), (6.2.4), and (6.2.5) and application of the boundary conditions $d_{1x} = d_{1y} = \phi_1 = 0$ and $d_{4x} = d_{4y} = \phi_4 = 0$ at nodes 1 and 4 yield the reduced set of equations for a longhand solution as

$$
\begin{Bmatrix} 10{,}000 \\ 0 \\ 0 \\ 0 \\ 0 \\ 5000 \end{Bmatrix} = 250{,}000 \begin{bmatrix} 10.167 & 0 & 10 & -10 & 0 & 0 \\ 0 & 10.0835 & 5 & 0 & 0.0835 & 5 \\ 10 & 5 & 1200 & 0 & -5 & 200 \\ -10 & 0 & 0 & 10.167 & 0 & 10 \\ 0 & 0.0835 & -5 & 0 & 10.0835 & -5 \\ 0 & 5 & 200 & 10 & -5 & 1200 \end{bmatrix} \begin{Bmatrix} d_{2x} \\ d_{2y} \\ \phi_2 \\ d_{3x} \\ d_{3y} \\ \phi_3 \end{Bmatrix}
$$
(6.2.6)

Solving Eq. (6.2.6) for the displacements and rotations, we have

$$
\begin{Bmatrix} d_{2x} \\ d_{2y} \\ \phi_2 \\ d_{3x} \\ d_{3y} \\ \phi_3 \end{Bmatrix} = \begin{Bmatrix} 0.211 \text{ in.} \\ 0.00148 \text{ in.} \\ -0.00153 \text{ rad} \\ 0.209 \text{ in.} \\ -0.00148 \text{ in.} \\ -0.00149 \text{ rad} \end{Bmatrix}
$$
(6.2.7)

The results indicate that the top of the frame moves to the right with negligible vertical displacement and small rotations of elements at nodes 2 and 3.

The element forces can now be obtained using $\hat{f} = \hat{k}\underline{T}\underline{d}$ for each element, as was previously done in solving truss and beam problems. We will illustrate this procedure only for element 1. For element 1, on using Eq. (6.1.10) for $\underline{T}$ and Eq. (6.2.7) for the displacements at node 2, we have

$$
\underline{T}\underline{d} = \begin{bmatrix} 0 & 1 & 0 & 0 & 0 & 0 \\ -1 & 0 & 0 & 0 & 0 & 0 \\ 0 & 0 & 1 & 0 & 0 & 0 \\ 0 & 0 & 0 & 0 & 1 & 0 \\ 0 & 0 & 0 & -1 & 0 & 0 \\ 0 & 0 & 0 & 0 & 0 & 1 \end{bmatrix} \begin{Bmatrix} d_{1x} = 0 \\ d_{1y} = 0 \\ \phi_1 = 0 \\ d_{2x} = 0.211 \\ d_{2y} = 0.00148 \\ \phi_2 = -0.00153 \end{Bmatrix}
$$
(6.2.8)

On multiplying the matrices in Eq. (6.2.8), we obtain

$$
\underline{T}\underline{d} = \begin{Bmatrix} 0 \\ 0 \\ 0 \\ 0.00148 \\ -0.211 \\ -0.00153 \end{Bmatrix}
$$
(6.2.9)

Then

$$\hat{f} = \hat{\underline{k}}\underline{T}\underline{d} = 250{,}000 \begin{bmatrix} 10 & 0 & 0 & -10 & 0 & 0 \\ 0 & 0.167 & 10 & 0 & -0.167 & 10 \\ 0 & 10 & 800 & 0 & -10 & 400 \\ -10 & 0 & 0 & 10 & 0 & 0 \\ 0 & -0.167 & -10 & 0 & 0.167 & -10 \\ 0 & 10 & 400 & 0 & -10 & 800 \end{bmatrix} \begin{Bmatrix} 0 \\ 0 \\ 0 \\ 0.00148 \\ -0.211 \\ -0.00153 \end{Bmatrix}$$

(6.2.10)

Simplifying Eq. (6.2.10), we obtain the local forces acting on element 1 as

$$\begin{Bmatrix} \hat{f}_{1x} \\ \hat{f}_{1y} \\ \hat{m}_1 \\ \hat{f}_{2x} \\ \hat{f}_{2y} \\ \hat{m}_2 \end{Bmatrix} = \begin{Bmatrix} -3700 \text{ lb} \\ 4990 \text{ lb} \\ 376{,}000 \text{ lb-in.} \\ 3700 \text{ lb} \\ -4990 \text{ lb} \\ 223{,}000 \text{ lb-in.} \end{Bmatrix}$$

(6.2.11)

A free-body diagram of each element is shown in Figure 6–5 along with equilibrium verification. In Figure 6–5, the $\hat{x}$ axis is directed from node 1 to node 2—consistent with the order of the nodal degrees of freedom used in developing the stiffness matrix for the element. Since the x-y plane was initially established as shown in Figure 6–4, the z axis is directed outward—consequently, so is the $\hat{z}$ axis (recall $\hat{z} = z$). The $\hat{y}$ axis is then established such that $\hat{x}$ cross $\hat{y}$ yields the direction of $\hat{z}$. The signs on the resulting element forces in Eq. (6.2.11) are thus consistently shown in Figure 6–5. The forces in elements 2 and 3 can be obtained in a manner similar to that used to obtain Eq. (6.2.11) for the nodal forces in element 1. Here we report only the final results for the forces in elements 2 and 3 and leave it to your discretion to perform the detailed calculations. The element forces (shown in Figure 6–5(b) and (c)) are as follows:

Element 2

$$\hat{f}_{2x} = 5010 \text{ lb} \qquad \hat{f}_{2y} = -3700 \text{ lb} \qquad \hat{m}_2 = -223{,}000 \text{ lb-in.}$$
$$\hat{f}_{3x} = -5010 \text{ lb} \qquad \hat{f}_{3y} = 3700 \text{ lb} \qquad \hat{m}_3 = -221{,}000 \text{ lb-in.}$$

(6.2.12a)

Element 3

$$\hat{f}_{3x} = 3700 \text{ lb} \qquad \hat{f}_{3y} = 5010 \text{ lb} \qquad \hat{m}_3 = 226{,}000 \text{ lb-in.}$$
$$\hat{f}_{4x} = -3700 \text{ lb} \qquad \hat{f}_{4y} = -5010 \text{ lb} \qquad \hat{m}_4 = 375{,}000 \text{ lb-in.}$$

(6.2.12b)

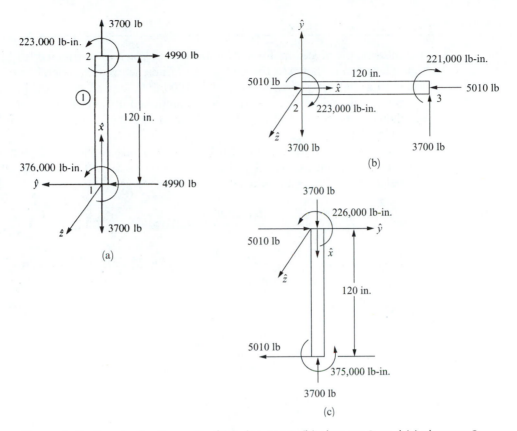

Figure 6–5 Free-body diagrams of (a) element 1, (b) element 2, and (c) element 3

Considering the free body of element 1, the equilibrium equations are

$$\sum F_{\hat{x}}: \ -4990 + 4990 = 0$$

$$\sum F_{\hat{y}}: \ -3700 + 3700 = 0$$

$$\sum M_2: \ 376{,}000 + 223{,}000 - 4990(120 \text{ in.}) \cong 0$$

Considering moment equilibrium at node 2, we see from Eqs. (6.2.12a) and (6.2.12b) that on element 1, $\hat{m}_2 = 223{,}000$ lb-in., and the opposite value, $-223{,}000$ lb-in., occurs on element 2. Similarly, moment equilibrium is satisfied at node 3, as $\hat{m}_3$ from elements 2 and 3 add to the 5000 lb-in. applied moment. That is, from Eqs. (6.2.12a) and (6.2.12b) we have

$$-221{,}000 + 226{,}000 = 5000 \text{ lb-in.} \qquad \blacksquare$$

Example 6.2

To illustrate the procedure for solving frames subjected to distributed loads, solve the rigid plane frame shown in Figure 6–6. The frame is fixed at nodes 1 and 3 and subjected to a uniformly distributed load of 1000 lb/ft applied downward over element 2. The global-coordinate axes have been established at node 1. The element lengths are shown in the figure. Let $E = 30 \times 10^6$ psi, $A = 100$ in², and $I = 1000$ in⁴ for both elements of the frame.

We begin by replacing the distributed load acting on element 2 by nodal forces and moments acting at nodes 2 and 3. Using Eqs. (5.4.5)–(5.4.7) (or Appendix D), the equivalent nodal forces and moments are calculated as

$$f_{2y} = -\frac{wL}{2} = -\frac{(1000)40}{2} = -20,000 \text{ lb} = -20 \text{ kip}$$

$$(6.2.13)$$

$$m_2 = -\frac{wL^2}{12} = -\frac{(1000)40^2}{12} = -133,333 \text{ lb-ft} = -1600 \text{ k-in.}$$

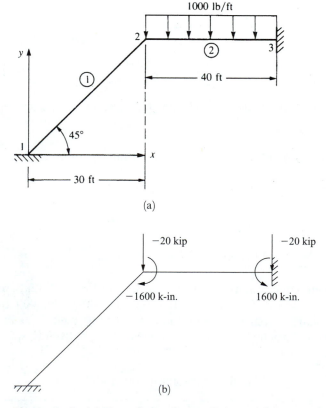

Figure 6–6 (a) Plane frame for analysis and (b) equivalent nodal forces on frame

$$f_{3y} = -\frac{wL}{2} = -\frac{(1000)40}{2} = -20,000 \text{ lb} = -20 \text{ kip}$$

$$m_3 = \frac{wL^2}{12} = \frac{(1000)40^2}{12} = 133,333 \text{ lb-ft} = 1600 \text{ k-in.}$$

We then use Eq. (6.1.11), to determine each element stiffness matrix:

Element 1

$$\theta^{(1)} = 45° \qquad C = 0.707 \qquad S = 0.707 \qquad L^{(1)} = 42.4 \text{ ft} = 509.0 \text{ in.}$$

$$\frac{E}{L} = \frac{30 \times 10^3}{509} = 58.93$$

$$\underline{k}^{(1)} = 58.93 \begin{bmatrix} 50.02 & 49.98 & 8.33 \\ 49.98 & 50.02 & -8.33 \\ 8.33 & -8.33 & 4000 \end{bmatrix} \frac{\text{lb}}{\text{in.}} \tag{6.2.14}$$

Simplifying Eq. (6.2.14), we obtain

$$\begin{array}{ccc} d_{2x} & d_{2y} & \phi_2 \end{array}$$
$$\underline{k}^{(1)} = \begin{bmatrix} 2948 & 2945 & 491 \\ 2945 & 2948 & -491 \\ 491 & -491 & 235,700 \end{bmatrix} \frac{\text{lb}}{\text{in.}} \tag{6.2.15}$$

where only the parts of the stiffness matrix associated with degrees of freedom at node 2 are included because node 1 is fixed.

Element 2

$$\theta^{(2)} = 0° \qquad C = 1 \qquad S = 0 \qquad L^{(2)} = 40 \text{ ft} = 480 \text{ in.}$$

$$\frac{E}{L} = \frac{30 \times 10^3}{480} = 62.50$$

$$\underline{k}^{(2)} = 62.50 \begin{bmatrix} 100 & 0 & 0 \\ 0 & 0.052 & 12.5 \\ 0 & 12.5 & 4000 \end{bmatrix} \frac{\text{lb}}{\text{in.}} \tag{6.2.16}$$

Simplifying Eq. (6.2.16), we obtain

$$\begin{array}{ccc} d_{2x} & d_{2y} & \phi_2 \end{array}$$
$$\underline{k}^{(2)} = \begin{bmatrix} 6250 & 0 & 0 \\ 0 & 3.25 & 781.25 \\ 0 & 781.25 & 250,000 \end{bmatrix} \frac{\text{lb}}{\text{in.}} \tag{6.2.17}$$

where, again, only the parts of the stiffness matrix associated with degrees of freedom at node 2 are included because node 3 is fixed. On superimposing the stiffness matrices of the elements, using Eqs. (6.2.15) and (6.2.17), and using Eq. (6.2.13) for the nodal forces and moments only at node 2 (because the structure is fixed at node 3), we have

$$
\begin{Bmatrix} F_{2x} = 0 \\ F_{2y} = -20 \\ M_2 = -1600 \end{Bmatrix} = \begin{bmatrix} 9198 & 2945 & 491 \\ 2945 & 2951 & 290 \\ 491 & 290 & 485{,}700 \end{bmatrix} \begin{Bmatrix} d_{2x} \\ d_{2y} \\ \phi_2 \end{Bmatrix}
\tag{6.2.18}
$$

Solving Eq. (6.2.18) for the displacements and the rotation at node 2, we obtain

$$
\begin{Bmatrix} d_{2x} \\ d_{2y} \\ \phi_2 \end{Bmatrix} = \begin{Bmatrix} 0.0033 \text{ in.} \\ -0.0097 \text{ in.} \\ -0.0033 \text{ rad} \end{Bmatrix}
\tag{6.2.19}
$$

The results indicate that node 2 moves to the right ($d_{2x} = 0.0033$ in.) and down ($d_{2y} = -0.0097$ in.) and the rotation of the joint is clockwise ($\phi_2 = -0.0033$ rad).

The local forces in each element can now be determined. The procedure for elements that are subjected to a distributed load must be applied to element 2. Recall that the local forces are given by $\hat{f} = \hat{k} \underline{T} \underline{d}$. For element 1, we then have

$$
\underline{T}\underline{d} = \begin{bmatrix} 0.707 & 0.707 & 0 & 0 & 0 & 0 \\ -0.707 & 0.707 & 0 & 0 & 0 & 0 \\ 0 & 0 & 1 & 0 & 0 & 0 \\ 0 & 0 & 0 & 0.707 & 0.707 & 0 \\ 0 & 0 & 0 & -0.707 & 0.707 & 0 \\ 0 & 0 & 0 & 0 & 0 & 1 \end{bmatrix} \begin{Bmatrix} 0 \\ 0 \\ 0 \\ 0.0033 \\ -0.0097 \\ -0.0033 \end{Bmatrix}
\tag{6.2.20}
$$

Simplifying Eq. (6.2.20) yields

$$
\underline{T}\underline{d} = \begin{Bmatrix} 0 \\ 0 \\ 0 \\ -0.00452 \\ -0.0092 \\ -0.0033 \end{Bmatrix}
\tag{6.2.21}
$$

Using Eq. (6.2.21) and Eq. (6.1.8) for $\hat{k}$, we obtain

$$
\begin{Bmatrix} \hat{f}_{1x} \\ \hat{f}_{1y} \\ \hat{m}_1 \\ \hat{f}_{2x} \\ \hat{f}_{2y} \\ \hat{m}_2 \end{Bmatrix} = \begin{bmatrix} 5893 & 0 & 0 & -5893 & 0 & 0 \\ & 2.730 & 694.8 & 0 & -2.730 & 694.8 \\ & & 117{,}900 & 0 & -694.8 & 117{,}900 \\ & & & 5893 & 0 & 0 \\ & & & & 2.730 & -694.8 \\ \text{Symmetry} & & & & & 235{,}800 \end{bmatrix} \begin{Bmatrix} 0 \\ 0 \\ 0 \\ -0.00452 \\ -0.0092 \\ -0.0033 \end{Bmatrix}
\tag{6.2.22}
$$

Simplifying Eq. (6.2.22) yields the local forces in element 1 as

$$\hat{f}_{1x} = 26.64 \text{ kip} \qquad \hat{f}_{1y} = -2.268 \text{ kip} \qquad \hat{m}_{1x} = -389.1 \text{ k-in.}$$
$$\hat{f}_{2x} = -26.64 \text{ kip} \qquad \hat{f}_{2y} = 2.268 \text{ kip} \qquad \hat{m}_{2x} = -778.2 \text{ k-in.} \tag{6.2.23}$$

For element 2, the local forces are given by Eq. (5.4.11) because a distributed load is acting on the element. From Eqs. (6.1.10) and (6.2.19), we then have

$$\underline{T}\underline{d} = \begin{bmatrix} 1 & 0 & 0 & 0 & 0 & 0 \\ 0 & 1 & 0 & 0 & 0 & 0 \\ 0 & 0 & 1 & 0 & 0 & 0 \\ 0 & 0 & 0 & 1 & 0 & 0 \\ 0 & 0 & 0 & 0 & 1 & 0 \\ 0 & 0 & 0 & 0 & 0 & 1 \end{bmatrix} \begin{Bmatrix} 0.0033 \\ -0.0097 \\ -0.0033 \\ 0 \\ 0 \\ 0 \end{Bmatrix} \tag{6.2.24}$$

Simplifying Eq. (6.2.24), we obtain

$$\begin{Bmatrix} 0.0033 \\ -0.0097 \\ -0.0033 \\ 0 \\ 0 \\ 0 \end{Bmatrix} \tag{6.2.25}$$

Using Eq. (6.2.25) and Eq. (6.1.8) for $\hat{\underline{k}}$, we have

$$\hat{\underline{k}}\hat{\underline{d}} = \hat{\underline{k}}\underline{T}\underline{d} = \begin{bmatrix} 6250 & 0 & 0 & -6250 & 0 & 0 \\ & 3.25 & 781.1 & 0 & -3.25 & 781.1 \\ & & 250{,}000 & 0 & -781.1 & 125{,}000 \\ & & & 6250 & 0 & 0 \\ & & & & 3.25 & -781.1 \\ \text{Symmetry} & & & & & 250{,}000 \end{bmatrix} \begin{Bmatrix} 0.0033 \\ -0.0097 \\ -0.0033 \\ 0 \\ 0 \\ 0 \end{Bmatrix} \tag{6.2.26}$$

Simplifying Eq. (6.2.26) yields

$$\hat{\underline{k}}\hat{\underline{d}} = \begin{Bmatrix} 20.63 \\ -2.58 \\ -832.57 \\ -20.63 \\ 2.58 \\ -412.50 \end{Bmatrix} \tag{6.2.27}$$

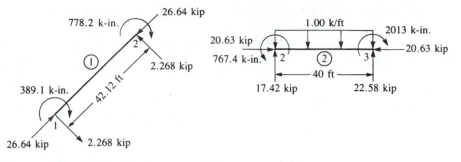

Figure 6–7 Free-body diagrams of elements 1 and 2

To obtain the actual element local nodal forces, we apply Eq. (5.4.11); that is, we must subtract the equivalent nodal forces [Eqs. (6.2.13)] from Eq. (6.2.27) to yield

$$
\begin{Bmatrix} \hat{f}_{2x} \\ \hat{f}_{2y} \\ \hat{m}_2 \\ \hat{f}_{3x} \\ \hat{f}_{3y} \\ \hat{m}_3 \end{Bmatrix} = \begin{Bmatrix} 20.63 \\ -2.58 \\ -832.57 \\ -20.63 \\ 2.58 \\ -412.50 \end{Bmatrix} - \begin{Bmatrix} 0 \\ -20 \\ -1600 \\ 0 \\ -20 \\ 1600 \end{Bmatrix}
\tag{6.2.28}
$$

Simplifying Eq. (6.2.28), we obtain

$$
\hat{f}_{2x} = 20.63 \text{ kip} \qquad \hat{f}_{2y} = 17.42 \text{ kip} \qquad \hat{m}_2 = 767.4 \text{ k-in.}
\tag{6.2.29}
$$

$$
\hat{f}_{3x} = -20.63 \text{ kip} \qquad \hat{f}_{3y} = 22.58 \text{ kip} \qquad \hat{m}_3 = -2013 \text{ k-in.}
$$

Using Eqs. (6.2.23) and (6.2.29) for the local forces in each element, we can construct the free-body diagram for each element, as shown in Figure 6–7. From the free-body diagrams, one can confirm the equilibrium of each element, the total frame, and joint 2 as desired. ∎

In Example 6.3, we will illustrate the equivalent joint force replacement method for a frame subjected to a load acting on an element instead of at one of the joints of the structure. Since no distributed loads are present, the point of application of the concentrated load could be treated as an extra joint in the analysis, and we could solve the problem in the same manner as Example 6.1.

This approach has the disadvantage of increasing the total number of joints, as well as the size of the total structure stiffness matrix $\underline{K}$. For small structures solved by computer, this does not pose a problem. However, for very large structures, this might reduce the maximum size of the structure that could be analyzed. Certainly, this additional node greatly increases the long-hand solution time for the structure. Hence, we will illustrate a standard procedure based on the concept of equivalent joint forces applied to the case of concentrated loads. We will again use Appendix D.

Example 6.3

Solve the frame shown in Figure 6–8(a). The frame consists of the three elements shown and is subjected to a 15-kip horizontal load applied at midlength of element 1. Nodes 1, 2, and 3 are fixed, and the dimensions are shown in the figure. Let $E = 30 \times 10^6$ psi, $I = 800$ in^4, and $A = 8$ in^2 for all elements.

1. We first express the applied load in the element 1 local coordinate system (here $\hat{x}$ is directed from node 1 to node 4). This is shown in Figure 6–8(b).

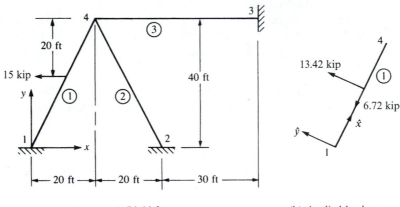

(a) Rigid frame

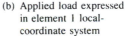

(b) Applied load expressed in element 1 local-coordinate system

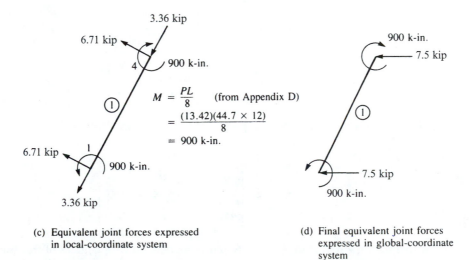

(c) Equivalent joint forces expressed in local-coordinate system

$$M = \frac{PL}{8} \quad \text{(from Appendix D)}$$

$$= \frac{(13.42)(44.7 \times 12)}{8}$$

$$= 900 \text{ k-in.}$$

(d) Final equivalent joint forces expressed in global-coordinate system

Figure 6–8 Rigid frame with a load applied on an element

2. Next, we determine the equivalent joint forces at each end of element 1, using the table in Appendix D. (These forces are of opposite sign from what are traditionally known as *fixed-end forces* in classical structural analysis theory [1].) These equivalent forces (and moments) are shown in Figure 6–8(c).

3. We then transform the equivalent joint forces from the present local-coordinate-system forces into the global-coordinate-system forces, using the equation $f = \underline{T}^T \hat{f}$, where $\underline{T}$ is defined by Eq. (6.1.10). These global joint forces are shown in Figure 6–8(d).

4. Then we analyze the structure in Figure 6–8(d), using the equivalent joint forces (plus actual joint forces, if any) in the usual manner.

5. We obtain the final internal forces developed at the ends of each element that has an applied load (here element 1 only) by subtracting step 2 joint forces from step 4 joint forces; that is, Eq. (5.4.11) is applied locally to all elements that originally had loads acting on them.

The solution of the structure as shown in Figure 6–8(d) now follows. Using Eq. (6.1.11), we obtain the global stiffness matrix for each element.

Element 1

For element 1, the angle between the global x and the local $\hat{x}$ axes is 63.43° because $\hat{x}$ is assumed to be directed from node 1 to node 4. Therefore,

$$C = \cos 63.43° = \frac{x_4 - x_1}{L^{(1)}} = \frac{20 - 0}{44.7} = 0.447$$

$$S = \sin 63.43° = \frac{y_4 - y_1}{L^{(1)}} = \frac{40 - 0}{44.7} = 0.895$$

$$\frac{12I}{L^2} = \frac{12(800)}{(44.7 \times 12)^2} = 0.0334 \qquad \frac{6I}{L} = \frac{6(800)}{44.7 \times 12} = 8.95$$

$$\frac{E}{L} = \frac{30 \times 10^3}{44.7 \times 12} = 55.9$$

Using the preceding results in Eq. (6.1.11) for $\underline{k}$, we obtain

$$
\underline{k}^{(1)} = \begin{matrix} d_{4x} & d_{4y} & \phi_4 \end{matrix} \\
\begin{bmatrix} 90.9 & 178 & 448 \\ 178 & 359 & -224 \\ 448 & -224 & 179{,}000 \end{bmatrix} \qquad (6.2.30)
$$

where only the parts of the stiffness matrix associated with degrees of freedom at node 4 are included because node 1 is fixed and, hence, not needed in the solution for the nodal displacements.

Element 3

For element 3, the angle between x and $\hat{x}$ is zero because $\hat{x}$ is directed from node 4 to node 3. Therefore,

$$C = 1 \qquad S = 0 \qquad \frac{12I}{L^2} = \frac{12(800)}{(50 \times 12)^2} = 0.0267$$

$$\frac{6I}{L} = \frac{6(800)}{50 \times 12} = 8.00 \qquad \frac{E}{L} = \frac{30 \times 10^3}{50 \times 12} = 50$$

Substituting these results into $\underline{k}$, we obtain

$$\underline{k}^{(3)} = \begin{matrix} d_{4x} & d_{4y} & \phi_4 \\ \begin{bmatrix} 400 & 0 & 0 \\ 0 & 1.334 & 400 \\ 0 & 400 & 160{,}000 \end{bmatrix} \end{matrix} \tag{6.2.31}$$

since node 3 is fixed.

Element 2

For element 2, the angle between x and $\hat{x}$ is $116.57°$ because $\hat{x}$ is directed from node 2 to node 4. Therefore,

$$C = \frac{20 - 40}{44.7} = -0.447 \qquad S = \frac{40 - 0}{44.7} = 0.895$$

$$\frac{12I}{L^2} = 0.0334 \qquad \frac{6I}{L} = 8.95 \qquad \frac{E}{L} = 55.9$$

since element 2 has the same properties as element 1. Substituting these results into $\underline{k}$, we obtain

$$\underline{k}^{(2)} = \begin{matrix} d_{4x} & d_{4y} & \phi_4 \\ \begin{bmatrix} 90.9 & -178 & 448 \\ -178 & 359 & 224 \\ 448 & 224 & 179{,}000 \end{bmatrix} \end{matrix} \tag{6.2.32}$$

since node 2 is fixed. On superimposing the stiffness matrices given by Eqs. (6.2.30), (6.2.31), and (6.2.32), and using the nodal forces given in Figure 6–8(d) at node 4 only, we have

$$\begin{Bmatrix} -7.50 \text{ kip} \\ 0 \\ -900 \text{ k-in.} \end{Bmatrix} = \begin{bmatrix} 582 & 0 & 896 \\ 0 & 719 & 400 \\ 896 & 400 & 518{,}000 \end{bmatrix} \begin{Bmatrix} d_{4x} \\ d_{4y} \\ \phi_4 \end{Bmatrix} \tag{6.2.33}$$

Simultaneously solving the three equations in Eq. (6.2.33), we obtain

$$d_{4x} = -0.0103 \text{ in.}$$

$$d_{4y} = 0.000956 \text{ in.} \tag{6.2.34}$$

$$\phi_4 = -0.00172 \text{ rad}$$

Next, we determine the element forces by again using $\hat{f} = \hat{k}\underline{T}\underline{d}$. In general, we have

$$
\underline{T}\underline{d} =
\begin{bmatrix}
C & S & 0 & 0 & 0 & 0 \\
-S & C & 0 & 0 & 0 & 0 \\
0 & 0 & 1 & 0 & 0 & 0 \\
0 & 0 & 0 & C & S & 0 \\
0 & 0 & 0 & -S & C & 0 \\
0 & 0 & 0 & 0 & 0 & 1
\end{bmatrix}
\begin{Bmatrix}
d_{ix} \\
d_{iy} \\
\phi_i \\
d_{jx} \\
d_{jy} \\
\phi_j
\end{Bmatrix}
$$

Thus, the preceding matrix multiplication yields

$$
\underline{T}\underline{d} =
\begin{Bmatrix}
Cd_{ix} + Sd_{iy} \\
-Sd_{ix} + Cd_{iy} \\
\phi_i \\
Cd_{jx} + Sd_{jy} \\
-Sd_{jx} + Cd_{jy} \\
\phi_j
\end{Bmatrix}
\tag{6.2.35}
$$

Element 1

$$
\underline{T}\underline{d} =
\begin{Bmatrix}
0 \\
0 \\
0 \\
(0.447)(-0.0103) + (0.895)(0.000956) \\
(-0.895)(-0.0103) + (0.447)(0.000956) \\
-0.00172
\end{Bmatrix}
=
\begin{Bmatrix}
0 \\
0 \\
0 \\
-0.00374 \\
0.00963 \\
-0.00172
\end{Bmatrix}
\tag{6.2.36}
$$

Using Eq. (6.1.8) for $\hat{k}$ and Eq. (6.2.36), we obtain

$$
\hat{k}\underline{T}\underline{d} =
\begin{bmatrix}
447 & 0 & 0 & -447 & 0 & 0 \\
0 & 1.868 & 500.5 & 0 & -1.868 & 500.5 \\
0 & 500.5 & 179{,}000 & 0 & -500.5 & 89{,}490 \\
-447 & 0 & 0 & 447 & 0 & 0 \\
0 & -1.868 & -500.5 & 0 & 1.868 & -500.5 \\
0 & 500.5 & 89{,}490 & 0 & -500.5 & 179{,}000
\end{bmatrix}
\times
\begin{Bmatrix}
0 \\
0 \\
0 \\
-0.00374 \\
0.00963 \\
-0.00172
\end{Bmatrix}
\tag{6.2.37}
$$

These values are now called *effective nodal forces*. Multiplying the matrices of Eq. (6.2.37) and using Eq. (5.4.11) to subtract the equivalent nodal forces in local coordinates for the element shown in Figure 6–8(c), we obtain the final nodal forces in

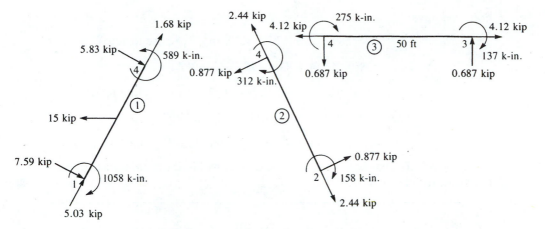

Figure 6–9 Free-body diagrams of all elements of the frame in Figure 6–8(a)

element 1 as

$$
\hat{f}^{(1)} = \left\{ \begin{array}{r} 1.67 \\ -0.88 \\ -158 \\ -1.67 \\ 0.88 \\ -311 \end{array} \right\} - \left\{ \begin{array}{r} -3.36 \\ 6.71 \\ 900 \\ -3.36 \\ 6.71 \\ -900 \end{array} \right\} = \left\{ \begin{array}{l} 5.03 \text{ kip} \\ -7.59 \text{ kip} \\ -1058 \text{ k-in.} \\ 1.68 \text{ kip} \\ -5.83 \text{ kip} \\ 589 \text{ k-in.} \end{array} \right\} \tag{6.2.38}
$$

Similarly, we can use Eqs. (6.2.35) and (6.1.8) for elements 3 and 2 to obtain the local nodal forces in these elements. Since these elements do not have any applied loads on them, the final nodal forces in local coordinates associated with each element are given by $\hat{f} = \hat{k}\underline{T}\underline{d}$. These forces have been determined as follows.

Element 3

$$\hat{f}_{4x} = -4.12 \text{ kip} \qquad \hat{f}_{4y} = -0.687 \text{ kip} \qquad \hat{m}_4 = -275 \text{ k-in.}$$
$$\hat{f}_{3x} = 4.12 \text{ kip} \qquad \hat{f}_{3y} = 0.687 \text{ kip} \qquad \hat{m}_3 = -137 \text{ k-in.}$$

$$\tag{6.2.39}$$

Element 2

$$\hat{f}_{2x} = -2.44 \text{ kip} \qquad \hat{f}_{2y} = -0.877 \text{ kip} \qquad \hat{m}_2 = -158 \text{ k-in.}$$
$$\hat{f}_{4x} = 2.44 \text{ kip} \qquad \hat{f}_{4y} = 0.877 \text{ kip} \qquad \hat{m}_4 = -312 \text{ k-in.}$$

$$\tag{6.2.40}$$

Free-body diagrams of all elements are shown in Figure 6–9. Each element has been determined to be in equilibrium, as often occurs even if errors are made in the long-hand calculations. However, equilibrium at node 4 and equilibrium of the whole frame are also satisfied. For instance, using the results of Eqs. (6.2.38)–(6.2.40) to check equilibrium at node 4, which is implicit in the formulation of the global equations, we have

$$\sum M_4 = 589 - 275 - 312 = 2 \text{ k-in.} \quad \text{(close to zero)}$$

$$\sum F_x = 1.68(0.447) + 5.83(0.895) - 2.44(0.447)$$
$$- 0.877(0.895) - 4.12 = -0.027 \text{ kip} \quad \text{(close to zero)}$$

$$\sum F_y = 1.68(0.895) - 5.83(0.447) + 2.44(0.895)$$
$$- 0.877(0.447) - 0.687 = 0.004 \text{ kip} \quad \text{(close to zero)}$$

Thus, the solution has been verified to be correct within the accuracy associated with a longhand solution. ∎

To illustrate the solution of a problem involving both bar and frame elements, we will solve the following example.

Example 6.4

The bar element 2 is used to stiffen the cantilever beam element 1, as shown in Figure 6–10. Determine the displacements at node 1 and the element forces. For the bar, let $A = 1.0 \times 10^{-3}$ m^2. For the beam, let $A = 2 \times 10^{-3}$ m^2, $I = 5 \times 10^{-5}$ m^4, and $L = 3$ m. For both the bar and the beam elements, let $E = 210$ GPa. Let the angle between the beam and the bar be 45°. A downward force of 500 kN is applied at node 1.

For brevity's sake, since nodes 2 and 3 are fixed, we keep only the parts of $\underline{k}$ for each element that are needed to obtain the global $\underline{K}$ matrix necessary for solution of the nodal degrees of freedom. Using Eq. (3.4.23), we obtain $\underline{k}$ for the bar as

$$\underline{k}^{(2)} = \frac{(1 \times 10^{-3})(210 \times 10^6)}{(3/\cos 45°)} \begin{bmatrix} 0.5 & 0.5 \\ 0.5 & 0.5 \end{bmatrix}$$

or, simplifying this equation, we obtain

$$\underline{k}^{(2)} = 70 \times 10^3 \begin{matrix} d_{1x} & d_{1y} \\ \begin{bmatrix} 0.354 & 0.354 \\ 0.354 & 0.354 \end{bmatrix} \end{matrix} \frac{\text{kN}}{\text{m}} \qquad (6.2.41)$$

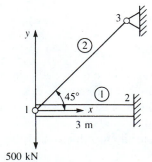

Figure 6–10 Cantilever beam with a bar element support

Using Eq. (6.1.11), we obtain $\underline{k}$ for the beam (including axial effects) as

$$\underline{k}^{(1)} = 70 \times 10^3 \begin{array}{c} \\ \\ \\ \end{array} \overset{\displaystyle d_{1x} \quad d_{1y} \quad \phi_1}{\begin{bmatrix} 2 & 0 & 0 \\ 0 & 0.067 & 0.10 \\ 0 & 0.10 & 0.20 \end{bmatrix}} \frac{\text{kN}}{\text{m}} \tag{6.2.42}$$

where $(E/L) \times 10^{-3}$ has been factored out in evaluating Eq. (6.2.42).

We assemble Eqs. (6.2.41) and (6.2.42) in the usual manner to obtain the global stiffness matrix as

$$\underline{K} = 70 \times 10^3 \begin{bmatrix} 2.354 & 0.354 & 0 \\ 0.354 & 0.421 & 0.10 \\ 0 & 0.10 & 0.20 \end{bmatrix} \frac{\text{kN}}{\text{m}} \tag{6.2.43}$$

The global equations are then written for node 1 as

$$\begin{Bmatrix} F_{1x} \\ F_{1y} \\ M_1 \end{Bmatrix} = \begin{Bmatrix} 0 \\ -500 \\ 0 \end{Bmatrix} = 70 \times 10^3 \begin{bmatrix} 2.354 & 0.354 & 0 \\ 0.354 & 0.421 & 0.10 \\ 0 & 0.10 & 0.20 \end{bmatrix} \begin{Bmatrix} d_{1x} \\ d_{1y} \\ \phi_1 \end{Bmatrix} \tag{6.2.44}$$

Solving Eq. (6.2.44), we obtain

$$d_{1x} = 0.00338 \text{ m} \qquad d_{1y} = -0.0225 \text{ m} \qquad \phi_1 = 0.0113 \text{ rad} \tag{6.2.45}$$

In general, the local element forces are obtained using $\hat{f} = \underline{\hat{k}}\,\underline{T}\,\underline{d}$. For the bar element, we then have

$$\begin{Bmatrix} \hat{f}_{1x} \\ \hat{f}_{3x} \end{Bmatrix} = \frac{AE}{L} \begin{bmatrix} 1 & -1 \\ -1 & 1 \end{bmatrix} \begin{bmatrix} C & S & 0 & 0 \\ 0 & 0 & C & S \end{bmatrix} \begin{Bmatrix} d_{1x} \\ d_{1y} \\ d_{3x} \\ d_{3y} \end{Bmatrix} \tag{6.2.46}$$

The matrix triple product of Eq. (6.2.46) yields (as one equation)

$$\hat{f}_{1x} = \frac{AE}{L}(Cd_{1x} + Sd_{1y}) \tag{6.2.47}$$

Substituting the numerical values into Eq. (6.2.47), we obtain

$$\hat{f}_{1x} = \frac{(1 \times 10^{-3} \text{ m}^2)(210 \times 10^6 \text{ kN/m}^2)}{4.24 \text{ m}} \left[\frac{\sqrt{2}}{2}(0.00338 - 0.0225) \right] \tag{6.2.48}$$

Simplifying Eq. (6.2.48), we obtain the axial force in the bar (element 2) as

$$\hat{f}_{1x} = -670 \text{ kN} \tag{6.2.49}$$

where the negative sign means $\hat{f}_{1x}$ is in the direction opposite $\hat{x}$ for element 2. Similarly, we obtain

$$\hat{f}_{3x} = 670 \text{ kN} \tag{6.2.50}$$

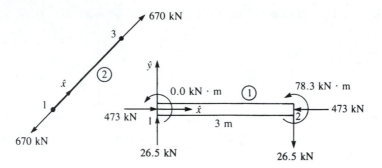

Figure 6–11 Free-body diagrams of the bar (element 2) and beam (element 1) elements of Figure 6–10

which means the bar is in tension as shown in Figure 6–11. Since the local and global axes are coincident for the beam element, we have $\hat{f} = f$ and $\hat{d} = d$. Therefore, from Eq. (6.1.6), we have at node 1

$$
\begin{Bmatrix} \hat{f}_{1x} \\ \hat{f}_{1y} \\ \hat{m}_1 \end{Bmatrix} = \begin{bmatrix} C_1 & 0 & 0 \\ 0 & 12C_2 & 6C_2L \\ 0 & 6C_2L & 4C_2L^2 \end{bmatrix} \begin{Bmatrix} d_{1x} \\ d_{1y} \\ \phi_1 \end{Bmatrix}
\tag{6.2.51}
$$

where only the upper part of the stiffness matrix is needed because the displacements at node 2 are equal to zero. Substituting numerical values into Eq. (6.2.51), we obtain

$$
\begin{Bmatrix} \hat{f}_{1x} \\ \hat{f}_{1y} \\ \hat{m}_1 \end{Bmatrix} = 70 \times 10^3 \begin{bmatrix} 2 & 0 & 0 \\ 0 & 0.067 & 0.10 \\ 0 & 0.10 & 0.20 \end{bmatrix} \begin{Bmatrix} 0.00338 \\ -0.0225 \\ 0.0113 \end{Bmatrix}
$$

The matrix product then yields

$$
\hat{f}_{1x} = 473 \text{ kN} \qquad \hat{f}_{1y} = -26.5 \text{ kN} \qquad \hat{m}_1 = 0.0 \text{ kN} \cdot \text{m}
\tag{6.2.52}
$$

Similarly, using Eq. (6.1.6), we have at node 2,

$$
\begin{Bmatrix} \hat{f}_{2x} \\ \hat{f}_{2y} \\ \hat{m}_2 \end{Bmatrix} = 70 \times 10^3 \begin{bmatrix} -2 & 0 & 0 \\ 0 & -0.067 & -0.10 \\ 0 & 0.10 & 0.10 \end{bmatrix} \begin{Bmatrix} 0.00338 \\ -0.0225 \\ 0.0113 \end{Bmatrix}
$$

The matrix product then yields

$$
\hat{f}_{2x} = -473 \text{ kN} \qquad \hat{f}_{2y} = 26.5 \text{ kN} \qquad \hat{m}_2 = -78.3 \text{ kN} \cdot \text{m}
\tag{6.2.53}
$$

To help interpret the results of Eqs. (6.2.49), (6.2.50), (6.2.52), and (6.2.53), free-body diagrams of the bar and beam elements are shown in Figure 6–11. To further verify the results, we can show a check on equilibrium of node 1 to be satisfied. You should also verify that moment equilibrium is satisfied in the beam. ■

▲ 6.3 Inclined or Skewed Supports— Frame Element

For the frame element with inclined support at node 3 in Figure 6–12, the transformation matrix $\underline{T}$ used to transform global to local nodal displacements is given by Eq. (6.1.10).

In the example shown in Figure 6–12, we use $\underline{T}$ applied to node 3 as follows:

$$
\begin{Bmatrix} d'_{3x} \\ d'_{3y} \\ \phi'_3 \end{Bmatrix} = \begin{bmatrix} \cos\alpha & \sin\alpha & 0 \\ -\sin\alpha & \cos\alpha & 0 \\ 0 & 0 & 1 \end{bmatrix} \begin{Bmatrix} d_{3x} \\ d_{3y} \\ \phi_3 \end{Bmatrix}
$$

The same steps as given in Section 3.9 then follow for the plane frame. The resulting equations for the plane frame in Figure 6–12 are

$$
[T_i]\{f\} = [T_i][K][T_i]^T\{d\}
$$

or

$$
\begin{Bmatrix} F_{1x} \\ F_{1y} \\ M_1 \\ F_{2x} \\ F_{2y} \\ M_2 \\ F'_{3x} \\ F'_{3y} \\ M_3 \end{Bmatrix} = [T_i][K][T_i]^T \begin{Bmatrix} d_{1x}=0 \\ d_{1y}=0 \\ \phi_1=0 \\ d_{2x} \\ d_{2y} \\ \phi_2 \\ d'_{3x} \\ d'_{3y}=0 \\ \phi'_3=\phi_3 \end{Bmatrix}
$$

where

$$
[T_i] = \begin{bmatrix} [I] & [0] & [0] \\ [0] & [I] & [0] \\ [0] & [0] & [t_3] \end{bmatrix}
$$

and

$$
[t_3] = \begin{bmatrix} \cos\alpha & \sin\alpha & 0 \\ -\sin\alpha & \cos\alpha & 0 \\ 0 & 0 & 1 \end{bmatrix}
$$

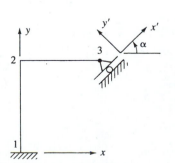

Figure 6–12 Frame with inclined support

▲ 6.4 Grid Equations

A **grid** *is a structure on which loads are applied perpendicular to the plane of the structure, as opposed to a plane frame, where loads are applied in the plane of the structure.* We will now develop the grid element stiffness matrix. The elements of a grid are assumed to be rigidly connected, so that the original angles between elements connected together at a node remain unchanged. Both torsional and bending moment continuity then exist at the node point of a grid. Examples of grids include floor and bridge deck systems. A typical grid structure subjected to loads F_1, F_2, F_3, and F_4 is shown in Figure 6–13.

We will now consider the development of the grid element stiffness matrix and element equations. A representative grid element with the nodal degrees of freedom and nodal forces is shown in Figure 6–14. The degrees of freedom at each node for a grid are a vertical deflection $\hat{d}_{iy}$ (normal to the grid), a torsional rotation $\hat{\phi}_{ix}$ about the $\hat{x}$ axis, and a bending rotation $\hat{\phi}_{iz}$ about the $\hat{z}$ axis. Any effect of axial displacement is ignored; that is, $\hat{d}_{ix} = 0$. The nodal forces consist of a transverse force $\hat{f}_{iy}$, a torsional moment $\hat{m}_{ix}$ about the $\hat{x}$ axis, and a bending moment $\hat{m}_{iz}$ about the $\hat{z}$ axis. Grid elements do not resist axial loading; that is $\hat{f}_{ix} = 0$.

To develop the local stiffness matrix for a grid element, we need to include the torsional effects in the basic beam element stiffness matrix Eq. (5.1.14). Recall that Eq. (5.1.14) already accounts for the bending and shear effects.

We can derive the torsional bar element stiffness matrix in a manner analogous to that used for the axial bar element stiffness matrix in Chapter 3. In the derivation, we simply replace $\hat{f}_{ix}$ with $\hat{m}_{ix}$, $\hat{d}_{ix}$ with $\hat{\phi}_{ix}$, E with G (the shear modulus), A with J (the torsional constant, or stiffness factor), σ with τ (shear stress), and ε with γ (shear strain).

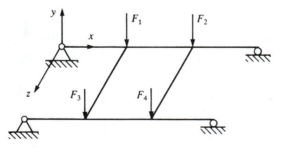

Figure 6–13 Typical grid structure

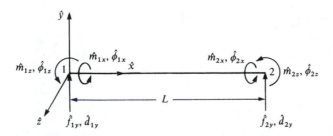

Figure 6–14 Grid element with nodal degrees of freedom and nodal forces

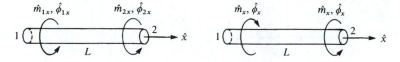

Figure 6–15 Nodal and element torque sign conventions

The actual derivation is briefly presented as follows. We assume a circular cross section with radius R for simplicity but without loss of generalization.

Step 1

Figure 6–15 shows the sign conventions for nodal torque and angle of twist and for element torque.

Step 2

We assume a linear angle-of-twist variation along the $\hat{x}$ axis of the bar such that

$$\hat{\phi} = a_1 + a_2\hat{x} \tag{6.4.1}$$

Using the usual procedure of expressing a_1 and a_2 in terms of unknown nodal angles of twist $\hat{\phi}_{1x}$ and $\hat{\phi}_{2x}$, we obtain

$$\hat{\phi} = \left(\frac{\hat{\phi}_{2x} - \hat{\phi}_{1x}}{L}\right)\hat{x} + \hat{\phi}_{1x} \tag{6.4.2}$$

or, in matrix form, Eq. (6.4.2) becomes

$$\hat{\phi} = [N_1 \quad N_2]\begin{Bmatrix} \hat{\phi}_{1x} \\ \hat{\phi}_{2x} \end{Bmatrix} \tag{6.4.3}$$

with the shape functions given by

$$N_1 = 1 - \frac{\hat{x}}{L} \qquad N_2 = \frac{\hat{x}}{L} \tag{6.4.4}$$

Step 3

We obtain the shear strain γ/angle of twist $\hat{\phi}$ relationship by considering the torsional deformation of the bar segment shown in Figure 6–16. Assuming that all radial lines, such as OA, remain straight during twisting or torsional deformation, we observe that the arc length $\widehat{AB}$ is given by

$$\widehat{AB} = \gamma_{max}\,d\hat{x} = R\,d\hat{\phi}$$

Solving for the maximum shear strain γ_{max}, we obtain

$$\gamma_{max} = \frac{R\,d\hat{\phi}}{d\hat{x}}$$

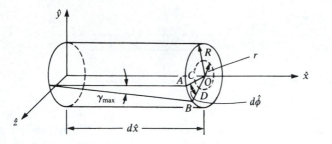

Figure 6–16 Torsional deformation of a bar segment

Similarly, at any radial position r, we then have, from similar triangles OAB and OCD,

$$\gamma = r\frac{d\hat{\phi}}{d\hat{x}} = \frac{r}{L}(\hat{\phi}_{2x} - \hat{\phi}_{1x}) \tag{6.4.5}$$

where we have used Eq. (6.4.2) to derive the final expression in Eq. (6.4.5).

The shear stress τ/shear strain γ relationship for linear-elastic isotropic materials is given by

$$\tau = G\gamma \tag{6.4.6}$$

where G is the shear modulus of the material.

Step 4

We derive the element stiffness matrix in the following manner. From elementary mechanics, we have the shear stress related to the applied torque by

$$\hat{m}_x = \frac{\tau J}{R} \tag{6.4.7}$$

where J is called the *polar moment of inertia* for the circular cross section or, generally, the *torsional constant* for noncircular cross sections. Using Eqs. (6.4.5) and (6.4.6) in Eq. (6.4.7), we obtain

$$\hat{m}_x = \frac{GJ}{L}(\hat{\phi}_{2x} - \hat{\phi}_{1x}) \tag{6.4.8}$$

By the nodal torque sign convention of Figure 6–15,

$$\hat{m}_{1x} = -\hat{m}_x \tag{6.4.9}$$

or, by using Eq. (6.4.8) in Eq. (6.4.9), we obtain

$$\hat{m}_{1x} = \frac{GJ}{L}(\hat{\phi}_{1x} - \hat{\phi}_{2x}) \tag{6.4.10}$$

Similarly,

$$\hat{m}_{2x} = \hat{m}_x \tag{6.4.11}$$

or

$$\hat{m}_{2x} = \frac{GJ}{L}(\hat{\phi}_{2x} - \hat{\phi}_{1x}) \tag{6.4.12}$$

Expressing Eqs. (6.4.10) and (6.4.12) together in matrix form, we have the resulting torsion bar stiffness matrix equation:

$$\left\{\begin{array}{c} \hat{m}_{1x} \\ \hat{m}_{2x} \end{array}\right\} = \frac{GJ}{L}\begin{bmatrix} 1 & -1 \\ -1 & 1 \end{bmatrix}\left\{\begin{array}{c} \hat{\phi}_{1x} \\ \hat{\phi}_{2x} \end{array}\right\} \tag{6.4.13}$$

Hence, the stiffness matrix for the torsion bar is

$$\hat{\underline{k}} = \frac{GJ}{L}\begin{bmatrix} 1 & -1 \\ -1 & 1 \end{bmatrix} \tag{6.4.14}$$

The cross sections of various structures, such as bridge decks, are often not circular. However, Eqs. (6.4.13) and (6.4.14) are still general; to apply them to other cross sections, we simply evaluate the torsional constant J for the particular cross section. For instance, for cross sections made up of thin rectangular shapes such as channels, angles, or I shapes, we approximate J by

$$J = \sum \frac{1}{3}b_i t_i^3 \tag{6.4.15}$$

where b_i is the length of any element of the cross section and t_i is the thickness of any element of the cross section. In Table 6–1, we list values of J for various common cross sections. The first four cross sections are called *open sections*. Equation (6.4.15) applies only to these open cross sections. (For more information on the J concept, consult References [2] and [3], and for an extensive table of torsional constants for various cross-sectional shapes, consult Reference [4].) We assume the loading to go through the shear center of these open cross sections in order to prevent twisting of the cross section. For more on the shear center consult References [2] and [5].

On combining the torsional effects of Eq. (6.4.13) with the shear and bending effects of Eq. (5.1.13), we obtain the local stiffness matrix equation for a grid element as

$$\left\{\begin{array}{c} \hat{f}_{1y} \\ \hat{m}_{1x} \\ \hat{m}_{1z} \\ \hat{f}_{2y} \\ \hat{m}_{2x} \\ \hat{m}_{2z} \end{array}\right\} = \begin{bmatrix} \dfrac{12EI}{L^3} & 0 & \dfrac{6EI}{L^2} & \dfrac{-12EI}{L^3} & 0 & \dfrac{6EI}{L^2} \\[2mm] & \dfrac{GJ}{L} & 0 & 0 & \dfrac{-GJ}{L} & 0 \\[2mm] & & \dfrac{4EI}{L} & \dfrac{-6EI}{L^2} & 0 & \dfrac{2EI}{L} \\[2mm] & & & \dfrac{12EI}{L^3} & 0 & \dfrac{-6EI}{L^2} \\[2mm] & & & & \dfrac{GJ}{L} & 0 \\[2mm] \text{Symmetry} & & & & & \dfrac{4EI}{L} \end{bmatrix} \left\{\begin{array}{c} \hat{d}_{1y} \\ \hat{\phi}_{1x} \\ \hat{\phi}_{1z} \\ \hat{d}_{2y} \\ \hat{\phi}_{2x} \\ \hat{\phi}_{2z} \end{array}\right\} \tag{6.4.16}$$

Table 6–1 Torsional constants J and shear centers SC for various cross sections

Cross Section	Torsional Constant

1. Channel

$$J = \frac{t^3}{3}(h + 2b)$$

$$e = \frac{h^2 b^2 t}{4I}$$

2. Angle

$$J = \tfrac{1}{3}(b_1 t_1^3 + b_2 t_2^3)$$

3. Z section

$$J = \frac{t^3}{3}(2b + h)$$

4. Wide-flanged beam with unequal flanges

$$J = \tfrac{1}{3}(b_1 t_1^3 + b_2 t_2^3 + h t_w^3)$$

5. Solid circular

$$J = \frac{\pi}{2} r^4$$

6. Closed hollow rectangular

$$J = \frac{2t t_1 (a - t)^2 (b - t_1)^2}{a t + b t_1 - t^2 - t_1^2}$$

where, from Eq. (6.4.16), the local stiffness matrix for a grid element is

$$
\hat{k}_G =
\begin{array}{cccccc}
\hat{d}_{1y} & \hat{\phi}_{1x} & \hat{\phi}_{1z} & \hat{d}_{2y} & \hat{\phi}_{2x} & \hat{\phi}_{2z}
\end{array}
$$

$$
\hat{k}_G =
\begin{bmatrix}
\dfrac{12EI}{L^3} & 0 & \dfrac{6EI}{L^2} & \dfrac{-12EI}{L^3} & 0 & \dfrac{6EI}{L^2} \\[2ex]
0 & \dfrac{GJ}{L} & 0 & 0 & \dfrac{-GJ}{L} & 0 \\[2ex]
\dfrac{6EI}{L^2} & 0 & \dfrac{4EI}{L} & \dfrac{-6EI}{L^2} & 0 & \dfrac{2EI}{L} \\[2ex]
\dfrac{-12EI}{L^3} & 0 & \dfrac{-6EI}{L^2} & \dfrac{12EI}{L^3} & 0 & \dfrac{-6EI}{L^2} \\[2ex]
0 & \dfrac{-GJ}{L} & 0 & 0 & \dfrac{GJ}{L} & 0 \\[2ex]
\dfrac{6EI}{L^2} & 0 & \dfrac{2EI}{L} & \dfrac{-6EI}{L^2} & 0 & \dfrac{4EI}{L}
\end{bmatrix}
\tag{6.4.17}
$$

and the degrees of freedom are in the order (1) vertical deflection, (2) torsional rotation, and (3) bending rotation, as indicated by the notation used above the columns of Eq. (6.4.17).

The transformation matrix relating local to global degrees of freedom for a grid is given by

$$
T_G =
\begin{bmatrix}
1 & 0 & 0 & 0 & 0 & 0 \\
0 & C & S & 0 & 0 & 0 \\
0 & -S & C & 0 & 0 & 0 \\
0 & 0 & 0 & 1 & 0 & 0 \\
0 & 0 & 0 & 0 & C & S \\
0 & 0 & 0 & 0 & -S & C
\end{bmatrix}
\tag{6.4.18}
$$

where θ is now positive, taken counterclockwise from x to $\hat{x}$ in the x-z plane (Figure 6–17) and

$$
C = \cos\theta = \frac{x_j - x_i}{L} \qquad S = \sin\theta = \frac{z_j - z_i}{L}
$$

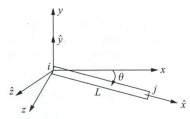

Figure 6–17 Grid element arbitrarily oriented in the x-z plane

where L is the length of the element from node i to node j. As indicated by Eq. (6.4.18) for a grid, the vertical deflection $\hat{d}_y$ is invariant with respect to a coordinate transformation (that is, $y = \hat{y}$) (Figure 6–17).

The global stiffness matrix for a grid element arbitrarily oriented in the x-z plane is then given by using Eqs. (6.4.17) and (6.4.18) in

$$\underline{k}_G = \underline{T}_G^T \hat{\underline{k}}_G \underline{T}_G \tag{6.4.19}$$

Now that we have formulated the global stiffness matrix for the grid element, the procedure for solution then follows in the same manner as that for the plane frame.

To illustrate the use of the equations developed in Section 6.4, we will now solve the following grid structures.

Example 6.5

Analyze the grid shown in Figure 6–18. The grid consists of three elements, is fixed at nodes 2, 3, and 4, and is subjected to a downward vertical force (perpendicular to the x-z plane passing through the grid elements) of 100 kip. The global-coordinate axes have been established at node 3, and the element lengths are shown in the figure. Let $E = 30 \times 10^3$ ksi, $G = 12 \times 10^3$ ksi, $I = 400$ in^4, and $J = 110$ in^4 for all elements of the grid.

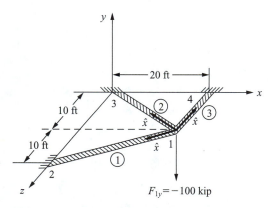

Figure 6–18 Grid for analysis showing local $\hat{x}$ axis for each element

Substituting Eq. (6.4.17) for the local stiffness matrix and Eq. (6.4.18) for the transformation matrix into Eq. (6.4.19), we can obtain each element global stiffness matrix. To expedite the longhand solution, the boundary conditions at nodes 2, 3, and 4,

$$d_{2y} = \phi_{2x} = \phi_{2z} = 0 \qquad d_{3y} = \phi_{3x} = \phi_{3z} = 0 \qquad d_{4y} = \phi_{4x} = \phi_{4z} = 0 \tag{6.4.20}$$

make it possible to use only the upper left-hand 3×3 partitioned part of the local stiffness and transformation matrices associated with the degrees of freedom at node 1. Therefore, the global stiffness matrices for each element are as follows:

Element 1

For element 1, we assume the local $\hat{x}$ axis to be directed from node 1 to node 2 for the formulation of the element stiffness matrix. We need the following expressions to evaluate the element stiffness matrix:

$$C = \cos \theta = \frac{x_2 - x_1}{L^{(1)}} = \frac{-20 - 0}{22.36} = -0.894$$

$$S = \sin \theta = \frac{z_2 - z_1}{L^{(1)}} = \frac{10 - 0}{22.36} = 0.447$$

$$\frac{12EI}{L^3} = \frac{12(30 \times 10^3)(400)}{(22.36 \times 12)^3} = 7.45$$

$$\frac{6EI}{L^2} = \frac{6(30 \times 10^3)(400)}{(22.36 \times 12)^2} = 1000 \tag{6.4.21}$$

$$\frac{GJ}{L} = \frac{(12 \times 10^3)(110)}{(22.36 \times 12)} = 4920$$

$$\frac{4EI}{L} = \frac{4(30 \times 10^3)(400)}{(22.36 \times 12)} = 179{,}000$$

Considering the boundary condition Eqs. (6.4.20), using the results of Eqs. (6.4.21) in Eq. (6.4.17) for $\hat{k}_G$ and Eq. (6.4.18) for $\underline{T}_G$, and then applying Eq. (6.4.19), we obtain the upper left-hand 3×3 partitioned part of the global stiffness matrix for element 1 as

$$\underline{k}^{(1)} = \begin{bmatrix} 1 & 0 & 0 \\ 0 & -0.894 & -0.447 \\ 0 & 0.447 & -0.894 \end{bmatrix} \begin{bmatrix} 7.45 & 0 & 1000 \\ 0 & 4920 & 0 \\ 1000 & 0 & 179{,}000 \end{bmatrix} \begin{bmatrix} 1 & 0 & 0 \\ 0 & -0.894 & 0.447 \\ 0 & -0.447 & -0.894 \end{bmatrix}$$

Performing the matrix multiplications, we obtain the global element grid stiffness matrix

$$\underline{k}^{(1)} = \begin{array}{ccc} d_{1y} & \phi_1 & \phi_2 \\ \begin{bmatrix} 7.45 & -447 & -894 \\ -447 & 39{,}700 & 69{,}600 \\ -894 & 69{,}600 & 144{,}000 \end{bmatrix} \end{array} \frac{\text{kip}}{\text{in.}} \tag{6.4.22}$$

where the labels next to the columns indicate the degrees of freedom.

Element 2

For element 2, we assume the local $\hat{x}$ axis to be directed from node 1 to node 3 for the formulation of the element stiffness matrix. We need the following expressions to evaluate the element stiffness matrix:

$$C = \frac{x_3 - x_1}{L^{(2)}} = \frac{-20 - 0}{22.36} = -0.894$$

$$S = \frac{z_3 - z_1}{L^{(2)}} = \frac{-10 - 0}{22.36} = -0.447$$

(6.4.23)

Other expressions used in Eq. (6.4.17) are identical to those in Eqs. (6.4.21) for element 1 because E, G, I, J, and L are identical. Evaluating Eq. (6.4.19) for the global stiffness matrix for element 2, we obtain

$$\underline{k}^{(2)} = \begin{bmatrix} 1 & 0 & 0 \\ 0 & -0.894 & 0.447 \\ 0 & -0.447 & -0.894 \end{bmatrix} \begin{bmatrix} 7.45 & 0 & 1000 \\ 0 & 4920 & 0 \\ 1000 & 0 & 179,000 \end{bmatrix} \begin{bmatrix} 1 & 0 & 0 \\ 0 & -0.894 & -0.447 \\ 0 & 0.447 & -0.894 \end{bmatrix}$$

Simplifying, we obtain

$$\underline{k}^{(2)} = \begin{bmatrix} d_{1y} & \phi_{1x} & \phi_{1z} \\ 7.45 & 447 & -894 \\ 447 & 39,700 & -69,600 \\ -894 & -69,600 & 144,000 \end{bmatrix} \frac{\text{kip}}{\text{in.}}$$

(6.4.24)

Element 3

For element 3, we assume the local $\hat{x}$ axis to be directed from node 1 to node 4. We need the following expressions to evaluate the element stiffness matrix:

$$C = \frac{x_4 - x_1}{L^{(3)}} = \frac{20 - 20}{10} = 0$$

$$S = \frac{z_4 - z_1}{L^{(3)}} = \frac{0 - 10}{10} = -1$$

(6.4.25)

$$\frac{12EI}{L^3} = \frac{12(30 \times 10^3)(400)}{(10 \times 12)^3} = 83.3$$

$$\frac{6EI}{L^2} = \frac{6(30 \times 10^3)(400)}{(10 \times 12)^2} = 5000$$

$$\frac{GJ}{L} = \frac{(12 \times 10^3)(110)}{(10 \times 12)} = 11,000$$

$$\frac{4EI}{L} = \frac{4(30 \times 10^3)(400)}{(10 \times 12)} = 400,000$$

Using Eqs. (6.4.25), we can obtain the upper part of the global stiffness matrix for element 3 as

$$\underline{k}^{(3)} = \begin{array}{c} \begin{array}{ccc} d_{1y} & \phi_{1x} & \phi_{1z} \end{array} \\ \begin{bmatrix} 83.3 & 5000 & 0 \\ 5000 & 400,000 & 0 \\ 0 & 0 & 11,000 \end{bmatrix} \end{array} \frac{\text{kip}}{\text{in.}} \qquad (6.4.26)$$

Superimposing the global stiffness matrices from Eqs. (6.4.22), (6.4.24), and (6.4.26), we obtain the total stiffness matrix of the grid (with boundary conditions applied) as

$$\underline{K}_G = \begin{array}{c} \begin{array}{ccc} d_{1y} & \phi_{1x} & \phi_{1z} \end{array} \\ \begin{bmatrix} 98.2 & 5000 & -1790 \\ 5000 & 479,000 & 0 \\ -1790 & 0 & 299,000 \end{bmatrix} \end{array} \frac{\text{kip}}{\text{in.}} \qquad (6.4.27)$$

The grid matrix equation then becomes

$$\begin{Bmatrix} F_{1y} = -100 \\ M_{1x} = 0 \\ M_{1z} = 0 \end{Bmatrix} = \begin{bmatrix} 98.2 & 5000 & -1790 \\ 5000 & 479,000 & 0 \\ -1790 & 0 & 299,000 \end{bmatrix} \begin{Bmatrix} d_{1y} \\ \phi_{1x} \\ \phi_{1z} \end{Bmatrix} \qquad (6.4.28)$$

The force F_{1y} is negative because the load is applied in the negative y direction. Solving for the displacement and the rotations in Eq. (6.4.28), we obtain

$$d_{1y} = -2.83 \text{ in.}$$

$$\phi_{1x} = 0.0295 \text{ rad} \qquad (6.4.29)$$

$$\phi_{1z} = -0.0169 \text{ rad}$$

The results indicate that the y displacement at node 1 is downward as indicated by the minus sign, the rotation about the x axis is positive, and the rotation about the z axis is negative. Based on the downward loading location with respect to the supports, these results are expected.

Having solved for the unknown displacement and the rotations, we can obtain the local element forces on formulating the element equations in a manner similar to that for the beam and the plane frame. The local forces (which are needed in the design/analysis stage) are found by applying the equation $\hat{f} = \hat{k}_G \underline{T}_G \underline{d}$ for each element as follows:

Element 1

Using Eqs. (6.4.17) and (6.4.18) for $\hat{k}_G$ and $\underline{T}_G$ and Eq. (6.4.29), we obtain

$$\underline{T}_G\underline{d} = \begin{bmatrix} 1 & 0 & 0 & 0 & 0 & 0 \\ 0 & -0.894 & 0.447 & 0 & 0 & 0 \\ 0 & -0.447 & -0.894 & 0 & 0 & 0 \\ 0 & 0 & 0 & 1 & 0 & 0 \\ 0 & 0 & 0 & 0 & -0.894 & 0.447 \\ 0 & 0 & 0 & 0 & -0.447 & -0.894 \end{bmatrix} \begin{Bmatrix} -2.83 \\ 0.0295 \\ -0.0169 \\ 0 \\ 0 \\ 0 \end{Bmatrix}$$

Multiplying the matrices, we obtain

$$\underline{T}_G\underline{d} = \begin{Bmatrix} -2.83 \\ -0.0339 \\ 0.00192 \\ 0 \\ 0 \\ 0 \end{Bmatrix} \tag{6.4.30}$$

Then $\hat{f} = \hat{k}_G\underline{T}_G\underline{d}$ becomes

$$\begin{Bmatrix} \hat{f}_{1y} \\ \hat{m}_{1x} \\ \hat{m}_{1z} \\ \hat{f}_{2y} \\ \hat{m}_{2x} \\ \hat{m}_{2z} \end{Bmatrix} = \begin{bmatrix} 7.45 & 0 & 1000 & -7.45 & 0 & 1000 \\ 0 & 4920 & 0 & 0 & -4920 & 0 \\ 1000 & 0 & 179,000 & -1000 & 0 & 89,500 \\ -7.45 & 0 & -1000 & 7.45 & 0 & -1000 \\ 0 & -4920 & 0 & 0 & 4920 & 0 \\ 1000 & 0 & 89,500 & -1000 & 0 & 179,000 \end{bmatrix} \begin{Bmatrix} -2.83 \\ -0.0339 \\ 0.00192 \\ 0 \\ 0 \\ 0 \end{Bmatrix}$$

$$\tag{6.4.31}$$

Multiplying the matrices in Eq. (6.4.31), we obtain the local element forces as

$$\begin{Bmatrix} \hat{f}_{1y} \\ \hat{m}_{1x} \\ \hat{m}_{1z} \\ \hat{f}_{2y} \\ \hat{m}_{2x} \\ \hat{m}_{2z} \end{Bmatrix} = \begin{Bmatrix} -19.2 \text{ kip} \\ -167 \text{ k-in.} \\ -2480 \text{ k-in.} \\ 19.2 \text{ kip} \\ 167 \text{ k-in.} \\ -2660 \text{ k-in.} \end{Bmatrix} \tag{6.4.32}$$

The directions of the forces acting on element 1 are shown in the free-body diagram of element 1 in Figure 6–19.

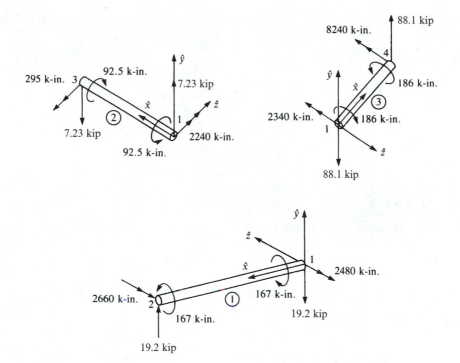

Figure 6–19 Free-body diagrams of the elements of Figure 6–18 showing local-coordinate systems for each

Element 2

Similarly, using $\hat{\underline{f}} = \hat{\underline{k}}_G \underline{T}_G \underline{d}$ for element 2, with the direction cosines in Eqs. (6.4.23), we obtain

$$
\begin{Bmatrix} \hat{f}_{1y} \\ \hat{m}_{1x} \\ \hat{m}_{1z} \\ \hat{f}_{3y} \\ \hat{m}_{3x} \\ \hat{m}_{3z} \end{Bmatrix} =
\begin{bmatrix}
7.45 & 0 & 1000 & -7.45 & 0 & 1000 \\
0 & 4920 & 0 & 0 & -4920 & 0 \\
1000 & 0 & 179,000 & -1000 & 0 & 89,500 \\
-7.45 & 0 & -1000 & 7.45 & 0 & -1000 \\
0 & -4920 & 0 & 0 & 4920 & 0 \\
1000 & 0 & 89,500 & -1000 & 0 & 179,000
\end{bmatrix}
$$

$$
\times
\begin{bmatrix}
1 & 0 & 0 & 0 & 0 & 0 \\
0 & -0.894 & -0.447 & 0 & 0 & 0 \\
0 & 0.447 & -0.894 & 0 & 0 & 0 \\
0 & 0 & 0 & 1 & 0 & 0 \\
0 & 0 & 0 & 0 & -0.894 & -0.447 \\
0 & 0 & 0 & 0 & 0.447 & -0.894
\end{bmatrix}
\begin{Bmatrix} -2.83 \\ 0.0295 \\ -0.0169 \\ 0 \\ 0 \\ 0 \end{Bmatrix}
$$

$$(6.4.33)$$

Multiplying the matrices in Eq. (6.4.33), we obtain the local element forces as

$$\hat{f}_{1y} = 7.23 \text{ kip}$$

$$\hat{m}_{1x} = -92.5 \text{ k-in.}$$

$$\hat{m}_{1z} = 2240 \text{ k-in.}$$

$$\hat{f}_{3y} = -7.23 \text{ kip}$$ (6.4.34)

$$\hat{m}_{3x} = 92.5 \text{ k-in.}$$

$$\hat{m}_{3z} = -295 \text{ k-in.}$$

Element 3

Finally, using the direction cosines in Eqs. (6.4.25), we obtain the local element forces as

$$
\begin{Bmatrix} \hat{f}_{1y} \\ \hat{m}_{1x} \\ \hat{m}_{1z} \\ \hat{f}_{3y} \\ \hat{m}_{3x} \\ \hat{m}_{3z} \end{Bmatrix} =
\begin{bmatrix}
83.3 & 0 & 5000 & -83.3 & 0 & 5000 \\
0 & 11{,}000 & 0 & 0 & -11{,}000 & 0 \\
5000 & 0 & 400{,}000 & -5000 & 0 & 200{,}000 \\
-83.3 & 0 & -5000 & 83.33 & 0 & -5000 \\
0 & -11{,}000 & 0 & 0 & 11{,}000 & 0 \\
5000 & 0 & 200{,}000 & -5000 & 0 & 400{,}000
\end{bmatrix}
$$

$$
\times
\begin{bmatrix}
1 & 0 & 0 & 0 & 0 & 0 \\
0 & 0 & -1 & 0 & 0 & 0 \\
0 & 1 & 0 & 0 & 0 & 0 \\
0 & 0 & 0 & 1 & 0 & 0 \\
0 & 0 & 0 & 0 & 0 & -1 \\
0 & 0 & 0 & 0 & 1 & 0
\end{bmatrix}
\begin{Bmatrix}
-2.83 \\
0.0295 \\
-0.0169 \\
0 \\
0 \\
0
\end{Bmatrix}
$$ (6.4.35)

Multiplying the matrices in Eq. (6.4.35), we obtain the local element forces as

$$\hat{f}_{1y} = -88.1 \text{ kip}$$

$$\hat{m}_{1x} = 186 \text{ k-in.}$$

$$\hat{m}_{1z} = -2340 \text{ k-in.}$$

$$\hat{f}_{4y} = 88.1 \text{ kip}$$ (6.4.36)

$$\hat{m}_{4x} = -186 \text{ k-in.}$$

$$\hat{m}_{4z} = -8240 \text{ k-in.}$$

Free-body diagrams for all elements are shown in Figure 6–19. Each element is in equilibrium. For each element, the $\hat{x}$ axis is shown directed from the first node to the

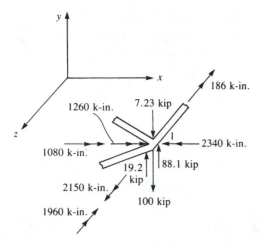

Figure 6–20 Free-body diagram of node 1 of Figure 6–18

second node, the $\hat{y}$ axis coincides with the global y axis, and the $\hat{z}$ axis is perpendicular to the $\hat{x}$-$\hat{y}$ plane with its direction given by the right-hand rule.

To verify equilibrium of node 1, we draw a free-body diagram of the node showing all forces and moments transferred from node 1 of each element, as in Figure 6–20. In Figure 6–20, the local forces and moments from each element have been transformed to global components, and any applied nodal forces have been included. To perform this transformation, recall that, in general, $\hat{f} = \underline{T}f$, and therefore $f = \underline{T}^T\hat{f}$ because $\underline{T}^T = \underline{T}^{-1}$. Since we are transforming forces at node 1 of each element, only the upper 3×3 part of Eq. (6.4.18) for $\underline{T}_G$ need be applied. Therefore, by premultiplying the local element forces and moments at node 1 by the transpose of the transformation matrix for each element, we obtain the global nodal forces and moments as follows:

Element 1

$$\left\{ \begin{array}{c} f_{1y} \\ m_{1x} \\ m_{1z} \end{array} \right\} = \left[\begin{array}{ccc} 1 & 0 & 0 \\ 0 & -0.894 & -0.447 \\ 0 & 0.447 & -0.894 \end{array} \right] \left\{ \begin{array}{c} -19.2 \\ -167 \\ -2480 \end{array} \right\}$$

Simplifying, we obtain the global-coordinate force and moments as

$$f_{1y} = -19.2 \text{ kip} \qquad m_{1x} = 1260 \text{ k-in.} \qquad m_{1z} = 2150 \text{ k-in.} \qquad (6.4.37)$$

where $f_{1y} = \hat{f}_{1y}$ because $y = \hat{y}$.

Element 2

$$\left\{ \begin{array}{c} f_{1y} \\ m_{1x} \\ m_{1z} \end{array} \right\} = \left[\begin{array}{ccc} 1 & 0 & 0 \\ 0 & -0.894 & 0.447 \\ 0 & -0.447 & -0.894 \end{array} \right] \left\{ \begin{array}{c} 7.23 \\ -92.5 \\ 2240 \end{array} \right\}$$

Simplifying, we obtain the global-coordinate force and moments as

$$f_{1y} = 7.23 \text{ kip} \qquad m_{1x} = 1080 \text{ k-in.} \qquad m_{1z} = -1960 \text{ k-in.} \qquad (6.4.38)$$

Element 3

$$\begin{Bmatrix} f_{1y} \\ m_{1x} \\ m_{1z} \end{Bmatrix} = \begin{bmatrix} 1 & 0 & 0 \\ 0 & 0 & 1 \\ 0 & -1 & 0 \end{bmatrix} \begin{Bmatrix} -88.1 \\ 186 \\ -2340 \end{Bmatrix}$$

Simplifying, we obtain the global-coordinate force and moments as

$$f_{1y} = -88.1 \text{ kip} \qquad m_{1x} = -2340 \text{ k-in.} \qquad m_{1z} = -186 \text{ k-in.} \qquad (6.4.39)$$

Then forces and moments from each element that are equal in magnitude but opposite in sign will be applied to node 1. Hence, the free-body diagram of node 1 is shown in Figure 6–20. Force and moment equilibrium are verified as follows:

$$\sum F_{1y} = -100 - 7.23 + 19.2 + 88.1 = 0.07 \text{ kip} \qquad \text{(close to zero)}$$

$$\sum M_{1x} = -1260 - 1080 + 2340 = 0.0 \text{ k-in.}$$

$$\sum M_{1z} = -2150 + 1960 + 186 = -4.00 \text{ k-in.} \qquad \text{(close to zero)}$$

Thus, we have verified the solution to be correct within the accuracy associated with a longhand solution. ∎

Example 6.6

Analyze the grid shown in Figure 6–21. The grid consists of two elements, is fixed at nodes 1 and 3, and is subjected to a downward vertical load of 22 kN. The global-coordinate axes and element lengths are shown in the figure. Let $E = 210$ GPa, $G = 84$ GPa, $I = 16.6 \times 10^{-5}$ m^4, and $J = 4.6 \times 10^{-5}$ m^4.

As in Example 6.5, we use the boundary conditions and express only the part of the stiffness matrix associated with the degrees of freedom at node 2. The boundary conditions at nodes 1 and 3 are

$$d_{1y} = \phi_{1x} = \phi_{1z} = 0 \qquad d_{3y} = \phi_{3x} = \phi_{3z} = 0 \qquad (6.4.40)$$

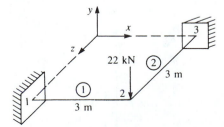

Figure 6–21 Grid example

The global stiffness matrices for each element are obtained as follows:

Element 1

For element 1, we have the local $\hat{x}$ axis coincident with the global x axis. Therefore, we obtain

$$C = \frac{x_2 - x_1}{L^{(1)}} = \frac{3}{3} = 1 \qquad S = \frac{z_2 - z_1}{L^{(1)}} = \frac{3 - 3}{3} = 0$$

Other expressions needed to evaluate the stiffness matrix are

$$\frac{12EI}{L^3} = \frac{12(210 \times 10^6 \text{ kN/m}^2)(16.6 \times 10^{-5} \text{ m}^4)}{(3 \text{ m})^3} = 1.55 \times 10^4$$

$$\frac{6EI}{L^2} = \frac{6(210 \times 10^6)(16.6 \times 10^{-5})}{(3)^2} = 2.32 \times 10^4$$

$$\frac{GJ}{L} = \frac{(84 \times 10^6)(4.6 \times 10^{-5})}{3} = 1.28 \times 10^3 \qquad (6.4.41)$$

$$\frac{4EI}{L} = \frac{4(210 \times 10^6)(16.6 \times 10^{-5})}{3} = 4.65 \times 10^4$$

Considering the boundary condition Eqs. (6.4.40), using the results of Eqs. (6.4.41) in Eq. (6.4.17) for $\hat{\underline{k}}_G$ and Eq. (6.4.18) for $\underline{T}_G$, and then applying Eq. (6.4.19), we obtain the reduced part of the global stiffness matrix associated only with the degrees of freedom at node 2 as

$$\underline{k}^{(1)} = \begin{bmatrix} 1 & 0 & 0 \\ 0 & 1 & 0 \\ 0 & 0 & 1 \end{bmatrix} \begin{bmatrix} 1.55 & 0 & -2.32 \\ 0 & 0.128 & 0 \\ -2.32 & 0 & 4.65 \end{bmatrix} (10^4) \begin{bmatrix} 1 & 0 & 0 \\ 0 & 1 & 0 \\ 0 & 0 & 1 \end{bmatrix}$$

Since the local axes associated with element 1 are parallel to the global axes, we observe that $\underline{T}_G$ is merely the identity matrix; therefore, $\underline{k}_G = \hat{\underline{k}}_G$. Performing the matrix multiplications, we obtain

$$\underline{k}^{(1)} = \begin{bmatrix} 1.55 & 0 & -2.32 \\ 0 & 0.128 & 0 \\ -2.32 & 0 & 4.65 \end{bmatrix} (10^4) \; \frac{\text{kN}}{\text{m}} \qquad (6.4.42)$$

Element 2

For element 2, we assume the local $\hat{x}$ axis to be directed from node 2 to node 3 for the formulation of $\underline{k}$. Therefore,

$$C = \frac{x_3 - x_2}{L^{(2)}} = \frac{0 - 0}{3} = 0 \qquad S = \frac{z_3 - z_2}{L^{(2)}} = \frac{0 - 3}{3} = -1 \qquad (6.4.43)$$

Other expressions used in Eq. (6.4.17) are identical to those obtained in Eqs. (6.4.41) for element 1. Evaluating Eq. (6.4.19) for the global stiffness matrix, we obtain

$$
\underline{k}^{(2)} = \begin{bmatrix} 1 & 0 & 0 \\ 0 & 0 & 1 \\ 0 & -1 & 0 \end{bmatrix} \begin{bmatrix} 1.55 & 0 & 2.32 \\ 0 & 0.128 & 0 \\ 2.32 & 0 & 4.65 \end{bmatrix} (10^4) \begin{bmatrix} 1 & 0 & 0 \\ 0 & 0 & -1 \\ 0 & 1 & 0 \end{bmatrix}
$$

where the reduced part of $\underline{k}$ is now associated with node 2 for element 2. Again performing the matrix multiplications, we have

$$
\underline{k}^{(2)} = \begin{bmatrix} 1.55 & 2.32 & 0 \\ 2.32 & 4.65 & 0 \\ 0 & 0 & 0.128 \end{bmatrix} (10^4) \ \frac{\text{kN}}{\text{m}} \tag{6.4.44}
$$

Superimposing the global stiffness matrices from Eqs. (6.4.42) and (6.4.44), we obtain the total global stiffness matrix (with boundary conditions applied) as

$$
\underline{K}_G = \begin{bmatrix} 3.10 & 2.32 & -2.32 \\ 2.32 & 4.78 & 0 \\ -2.32 & 0 & 4.78 \end{bmatrix} (10^4) \ \frac{\text{kN}}{\text{m}} \tag{6.4.45}
$$

The grid matrix equation becomes

$$
\begin{Bmatrix} F_{2y} = -22 \\ M_{2x} = 0 \\ M_{2z} = 0 \end{Bmatrix} = \begin{bmatrix} 3.10 & 2.32 & -2.32 \\ 2.32 & 4.78 & 0 \\ -2.32 & 0 & 4.78 \end{bmatrix} \begin{Bmatrix} d_{2y} \\ \phi_{2x} \\ \phi_{2z} \end{Bmatrix} (10^4) \tag{6.4.46}
$$

Solving for the displacement and the rotations in Eq. (6.4.46), we obtain

$$
d_{2y} = -0.259 \times 10^{-2} \text{ m}
$$

$$
\phi_{2x} = 0.126 \times 10^{-2} \text{ rad} \tag{6.4.47}
$$

$$
\phi_{2z} = -0.126 \times 10^{-2} \text{ rad}
$$

We determine the local element forces by applying the local equation $\hat{\underline{f}} = \hat{\underline{k}}_G \underline{T}_G \underline{d}$ for each element as follows:

Element 1

Using Eq. (6.4.17) for $\hat{\underline{k}}_G$, Eq. (6.4.18) for $\underline{T}_G$, and Eqs. (6.4.47), we obtain

$$
\underline{T}_G \underline{d} = \begin{bmatrix} 1 & 0 & 0 & 0 & 0 & 0 \\ 0 & 1 & 0 & 0 & 0 & 0 \\ 0 & 0 & 1 & 0 & 0 & 0 \\ 0 & 0 & 0 & 1 & 0 & 0 \\ 0 & 0 & 0 & 0 & 1 & 0 \\ 0 & 0 & 0 & 0 & 0 & 1 \end{bmatrix} \begin{Bmatrix} 0 \\ 0 \\ 0 \\ -0.259 \times 10^{-2} \\ 0.126 \times 10^{-2} \\ -0.126 \times 10^{-2} \end{Bmatrix}
$$

Multiplying the matrices, we have

$$
\underline{T_G}\underline{d} = \left\{ \begin{array}{c} 0 \\ 0 \\ 0 \\ -0.259 \times 10^{-2} \\ 0.126 \times 10^{-2} \\ -0.126 \times 10^{-2} \end{array} \right\} \tag{6.4.48}
$$

Using Eqs. (6.4.17), (6.4.41), and (6.4.48), we obtain the local element forces as

$$
\left\{ \begin{array}{c} \hat{f}_{1y} \\ \hat{m}_{1x} \\ \hat{m}_{1z} \\ \hat{f}_{2y} \\ \hat{m}_{2x} \\ \hat{m}_{2z} \end{array} \right\} = (10^4) \begin{bmatrix} 1.55 & 0 & 2.32 & -1.55 & 0 & 2.32 \\ & 0.128 & 0 & 0 & -0.128 & 0 \\ & & 4.65 & -2.32 & 0 & 2.33 \\ & & & 1.55 & 0 & -2.32 \\ & & & & 0.128 & 0 \\ \text{Symmetry} & & & & & 4.65 \end{bmatrix} \left\{ \begin{array}{c} 0 \\ 0 \\ 0 \\ -0.259 \times 10^{-2} \\ 0.126 \times 10^{-2} \\ -0.126 \times 10^{-2} \end{array} \right\}
$$

$$\tag{6.4.49}$$

Multiplying the matrices in Eq. (6.4.49), we obtain

$$
\begin{array}{lll}
\hat{f}_{1y} = 11.0 \text{ kN} & \hat{m}_{1x} = -1.50 \text{ kN} \cdot \text{m} & \hat{m}_{1z} = 31.0 \text{ kN} \cdot \text{m} \\
\hat{f}_{2y} = -11.0 \text{ kN} & \hat{m}_{2x} = 1.50 \text{ kN} \cdot \text{m} & \hat{m}_{2z} = 1.50 \text{ kN} \cdot \text{m}
\end{array} \tag{6.4.50}
$$

Element 2

We can obtain the local element forces for element 2 in a similar manner. Because the procedure is the same as that used to obtain the element 1 local forces, we will not show the details but will only list the final results:

$$
\begin{array}{lll}
\hat{f}_{2y} = -11.0 \text{ kN} & \hat{m}_{2x} = 1.50 \text{ kN} \cdot \text{m} & \hat{m}_{2z} = -1.50 \text{ kN} \cdot \text{m} \\
\hat{f}_{3y} = 11.0 \text{ kN} & \hat{m}_{3x} = -1.50 \text{ kN} \cdot \text{m} & \hat{m}_{3z} = -31.0 \text{ kN} \cdot \text{m}
\end{array} \tag{6.4.51}
$$

Free-body diagrams showing the local element forces are shown in Figure 6–22. ■

▲ 6.5 Beam Element Arbitrarily Oriented in Space ▲

In this section, we develop the stiffness matrix for the beam element arbitrarily oriented in space, or three dimensions. This element can then be used to analyze frames in three-dimensional space.

First we consider bending about two axes, as shown in Figure 6–23.

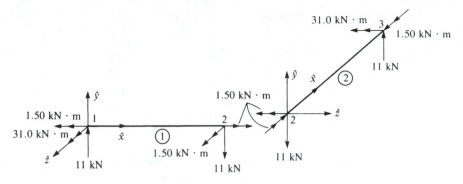

Figure 6–22   Free-body diagram of each element of Figure 6–21

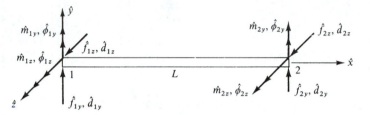

Figure 6–23   Bending about two axes $\hat{y}$ and $\hat{z}$

We establish the following sign convention for the axes. Now we choose positive $\hat{x}$ from node 1 to 2. Then $\hat{y}$ is the principal axis for which the moment of inertia is minimum, I_y. By the right-hand rule we establish $\hat{z}$, and the maximum moment of inertia is I_z.

Bending in $\hat{x}$-$\hat{z}$ Plane

First consider bending in the x-z plane due to $\hat{m}_y$. Then clockwise rotation $\hat{\phi}_y$ is in the same sense as before for single bending. The stiffness matrix due to bending in the x-z plane is then

$$
\underline{\hat{k}}_y = \frac{EI_y}{L^4}
\begin{bmatrix}
12L & -6L^2 & -12L & -6L^2 \\
 & 4L^3 & 6L^2 & 2L^3 \\
 & & 12L & 6L^2 \\
\text{Symmetry} & & & 4L^3
\end{bmatrix}
\tag{6.5.1}
$$

where I_y is the moment of inertia of the cross section about the principal axis $\hat{y}$, the weak axis; that is, $I_y < I_z$.

Bending in the $\hat{z}$-$\hat{y}$ Plane

Now we consider bending in the $\hat{x}$-$\hat{y}$ plane due to $\hat{m}_z$. Now positive rotation $\hat{\phi}_z$ is counterclockwise instead of clockwise. Therefore, some signs change in the stiffness matrix for bending in the $\hat{x}$-$\hat{y}$ plane. The resulting stiffness matrix is

$$\hat{\underline{k}}_z = \frac{EI_z}{L^4}\begin{bmatrix} 12L & 6L^2 & -12L & 6L^2 \\ & 4L^3 & -6L^2 & 2L^3 \\ & & 12L & -6L^2 \\ \text{Symmetry} & & & 4L^3 \end{bmatrix} \tag{6.5.2}$$

Direct superposition of Eqs. (6.5.1) and (6.5.2) with the axial stiffness matrix Eq. (3.1.14) and the torsional stiffness matrix Eq. (6.4.14) yields the element stiffness matrix for the beam or frame element in three-dimensional space as

$$\hat{\underline{k}} =$$

	$\hat{d}_{1x}$	$\hat{d}_{1y}$	$\hat{d}_{1z}$	$\hat{\phi}_{1x}$	$\hat{\phi}_{1y}$	$\hat{\phi}_{1z}$	$\hat{d}_{2x}$	$\hat{d}_{2y}$	$\hat{d}_{2z}$	$\hat{\phi}_{2x}$	$\hat{\phi}_{2y}$	$\hat{\phi}_{2z}$
	$\dfrac{AE}{L}$	0	0	0	0	0	$-\dfrac{AE}{L}$	0	0	0	0	0
	0	$\dfrac{12EI_z}{L^3}$	0	0	0	$\dfrac{6EI_z}{L^2}$	0	$-\dfrac{12EI_z}{L^3}$	0	0	0	$\dfrac{6EI_z}{L^2}$
	0	0	$\dfrac{12EI_y}{L^3}$	0	$-\dfrac{6EI_y}{L^2}$	0	0	0	$-\dfrac{12EI_y}{L^3}$	0	$-\dfrac{6EI_y}{L^2}$	0
	0	0	0	$\dfrac{GJ}{L}$	0	0	0	0	0	$-\dfrac{GJ}{L}$	0	0
	0	0	$-\dfrac{6EI_y}{L^2}$	0	$\dfrac{4EI_y}{L}$	0	0	0	$\dfrac{6EI_y}{L^2}$	0	$\dfrac{2EI_y}{L}$	0
	0	$\dfrac{6EI_z}{L^2}$	0	0	0	$\dfrac{4EI_z}{L}$	0	$-\dfrac{6EI_z}{L^2}$	0	0	0	$\dfrac{2EI_z}{L}$
	$-\dfrac{AE}{L}$	0	0	0	0	0	$\dfrac{AE}{L}$	0	0	0	0	0
	0	$-\dfrac{12EI_z}{L^3}$	0	0	0	$-\dfrac{6EI_z}{L^2}$	0	$\dfrac{12EI_z}{L^3}$	0	0	0	$\dfrac{6EI_z}{L^2}$
	0	0	$-\dfrac{12EI_y}{L^3}$	0	$\dfrac{6EI_y}{L^2}$	0	0	0	$\dfrac{12EI_y}{L^3}$	0	$\dfrac{6EI_y}{L^2}$	0
	0	0	0	$-\dfrac{GJ}{L}$	0	0	0	0	0	$\dfrac{GJ}{L}$	0	0
	0	0	$-\dfrac{6EI_y}{L^2}$	0	$\dfrac{2EI_y}{L}$	0	0	0	$\dfrac{6EI_y}{L^2}$	0	$\dfrac{4EI_y}{L}$	0
	0	$\dfrac{6EI_z}{L^2}$	0	0	0	$\dfrac{2EI_z}{L}$	0	$-\dfrac{6EI_z}{L^2}$	0	0	0	$\dfrac{4EI_z}{L}$

$$\tag{6.5.3}$$

The transformation from local to global axis system is accomplished as follows:

$$\underline{k} = \underline{T}^T \hat{\underline{k}} \underline{T} \tag{6.5.4}$$

where $\hat{\underline{k}}$ is given by Eq. (6.5.3) and $\underline{T}$ is given by

$$\underline{T} = \begin{bmatrix} \underline{\lambda}_{3\times3} & & & \\ & \underline{\lambda}_{3\times3} & & \\ & & \underline{\lambda}_{3\times3} & \\ & & & \underline{\lambda}_{3\times3} \end{bmatrix} \tag{6.5.5}$$

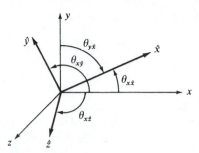

Figure 6–24 Direction cosines associated with the x axis

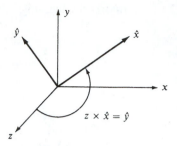

Figure 6–25 Illustration showing how local $\hat{y}$ axis is determined

where

$$\underline{\lambda} = \begin{bmatrix} C_{x\hat{x}} & C_{y\hat{x}} & C_{z\hat{x}} \\ C_{x\hat{y}} & C_{y\hat{y}} & C_{z\hat{y}} \\ C_{x\hat{z}} & C_{y\hat{z}} & C_{z\hat{z}} \end{bmatrix} \tag{6.5.6}$$

Here $C_{y\hat{x}}$ and $C_{x\hat{y}}$ are not necessarily equal. The direction cosines are shown in part in Figure 6–24.

Remember that direction cosines of the $\hat{x}$ axis member are

$$\hat{x} = \cos \theta_{x\hat{x}} \bar{i} + \cos \theta_{y\hat{x}} \bar{j} + \cos \theta_{z\hat{x}} \bar{k} \tag{6.5.7}$$

where

$$\cos \theta_{x\hat{x}} = \frac{x_2 - x_1}{L} = l$$

$$\cos \theta_{y\hat{x}} = \frac{y_2 - y_1}{L} = m \tag{6.5.8}$$

$$\cos \theta_{z\hat{x}} = \frac{z_2 - z_1}{L} = n$$

The $\hat{y}$ axis is selected to be perpendicular to the $\hat{x}$ and z axes in such a way that the cross product of global z with $\hat{x}$ results in the $\hat{y}$ axis, as shown in Figure 6–25. Therefore,

$$z \times \hat{x} = \hat{y} = \frac{1}{D} \begin{vmatrix} \bar{i} & \bar{j} & \bar{k} \\ 0 & 0 & 1 \\ l & m & n \end{vmatrix} \tag{6.5.9}$$

$$\hat{y} = -\frac{m}{D} \bar{i} + \frac{l}{D} \bar{j} \tag{6.5.10}$$

and

$$D = (l^2 + m^2)^{1/2}$$

The $\hat{z}$ axis will be determined by the orthogonality condition $\hat{z} = \hat{x} \times \hat{y}$ as follows:

$$\hat{z} = \hat{x} \times \hat{y} = \frac{1}{D} \begin{vmatrix} \bar{i} & \bar{j} & \bar{k} \\ l & m & n \\ -m & l & 0 \end{vmatrix} \tag{6.5.11}$$

or
$$\hat{z} = -\frac{ln}{D}\bar{i} - \frac{mn}{D}\bar{j} + D\bar{k} \qquad (6.5.12)$$

Combining Eqs. (6.5.7), (6.5.10), and (6.5.12), the 3×3 transformation matrix becomes

$$\underline{\lambda}_{3\times3} = \begin{bmatrix} l & m & n \\ -\dfrac{m}{D} & \dfrac{l}{D} & 0 \\ -\dfrac{ln}{D} & -\dfrac{mn}{D} & D \end{bmatrix} \qquad (6.5.13)$$

This vector $\underline{\lambda}$ rotates a vector from the local coordinate system into the global one. This is the $\underline{\lambda}$ used in the $\underline{T}$ matrix. In summary, we have

$$\cos\theta_{x\hat{y}} = -\frac{m}{D}$$

$$\cos\theta_{y\hat{y}} = \frac{l}{D}$$

$$\cos\theta_{z\hat{y}} = 0$$

$$\cos\theta_{x\hat{z}} = -\frac{ln}{D} \qquad (6.5.14)$$

$$\cos\theta_{y\hat{z}} = -\frac{mn}{D}$$

$$\cos\theta_{z\hat{z}} = D$$

Two exceptions arise when local and global axes have special orientations with respect to each other. If the local $\hat{x}$ axis coincides with the global z axis, then the member is parallel to the global z axis and the $\hat{y}$ axis becomes uncertain, as shown in Figure 6–26(a). In this case the local $\hat{y}$ axis is selected as the global y axis. Then, for

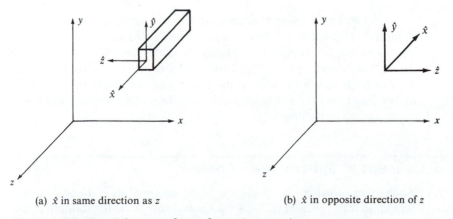

(a) $\hat{x}$ in same direction as z (b) $\hat{x}$ in opposite direction of z

Figure 6–26 Special cases of transformation matrices

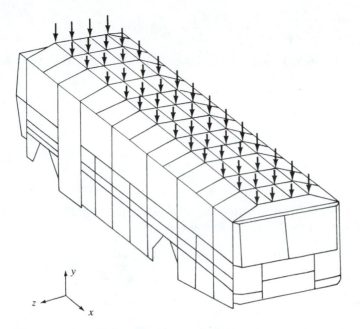

Figure 6–27 Finite element model of bus frame subjected to roof load

the positive $\hat{x}$ axis in the same direction as the global z, $\underline{\lambda}$ becomes

$$\underline{\lambda} = \begin{bmatrix} 0 & 0 & 1 \\ 0 & 1 & 0 \\ -1 & 0 & 0 \end{bmatrix} \tag{6.5.15}$$

For the positive $\hat{x}$ axis opposite the global z [Figure 6–26(b)], $\underline{\lambda}$ becomes

$$\underline{\lambda} = \begin{bmatrix} 0 & 0 & -1 \\ 0 & 1 & 0 \\ 1 & 0 & 0 \end{bmatrix} \tag{6.5.16}$$

An example using the frame element in three-dimensional space is shown in Figure 6–27. Figure 6–27 shows a bus frame subjected to a static roof-crush analysis. In this model, 599 frame elements and 357 nodes were used. A total downward load of 100 kN was uniformly spread over the 56 nodes of the roof portion of the frame. Figure 6–28 shows the rear of the frame and the displaced view of the rear frame. Other frame models with additional loads simulating rollover and front-end collisions were studied in Reference [6].

▲ 6.6 Concept of Substructure Analysis ▲

Sometimes structures are too large to be analyzed as a single system or treated as a whole; that is, the final stiffness matrix and equations for solution exceed the memory capacity of the computer. A procedure to overcome this problem is to separate the

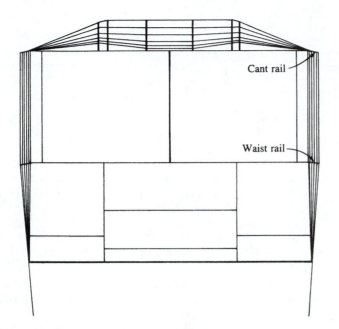

Cant rail

Waist rail

Figure 6–28 Displaced view of the frame of Figure 6–27 made of square section members

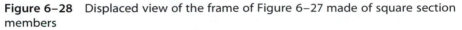

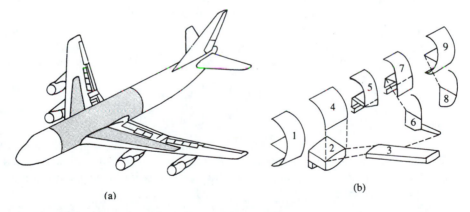

(a)

(b)

Figure 6–29 Airplane frame showing substructuring. (a) Boeing 747 aircraft (shaded area indicates portion of the airframe analyzed by finite element method). (b) Substructures for finite element analysis of shaded region

whole structure into smaller units called **substructures**. For example, the space frame of an airplane, as shown in Figure 6–29(a), may require thousands of nodes and elements to model and describe completely the response of the whole structure. If we separate the aircraft into substructures, such as parts of the fuselage or body, wing sections, and so on, as shown in Figure 6–29(b), then we can solve the problem more readily and on computers with limited memory.

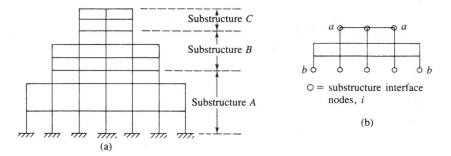

Figure 6–30 (a) Rigid frame for substructure analysis and (b) substructure B

The analysis of the airplane frame is performed by treating each substructure separately while ensuring force and displacement compatibility at the intersections where partitioning occurs.

To describe the procedure of substructuring, consider the rigid frame shown in Figure 6–30 (even though this frame could be analyzed as a whole). First we define individual separate substructures. Normally, we make these substructures of similar size, and to reduce computations, we make as few cuts as possible. We then separate the frame into three parts, A, B, and C.

We now analyze a typical substructure B shown in Figure 6–30(b). This substructure includes the beams at the top (a-a), but the beams at the bottom (b-b) are included in substructure A, although the beams at top could be included in substructure C and the beams at the bottom could be included in substructure B.

The force/displacement equations for substructure B are partitioned with the interface displacements and forces separated from the interior ones as follows:

$$\left\{ \begin{array}{c} \underline{F}_i^B \\ \underline{F}_e^B \end{array} \right\} = \left[\begin{array}{c|c} \underline{K}_{ii}^B & \underline{K}_{ie}^B \\ \hline \underline{K}_{ei}^B & \underline{K}_{ee}^B \end{array} \right] \left\{ \begin{array}{c} \underline{d}_i^B \\ \underline{d}_e^B \end{array} \right\} \tag{6.6.1}$$

where the superscript B denotes the substructure B, subscript i denotes the interface nodal forces and displacements, and subscript e denotes the interior nodal forces and displacements to be eliminated by static condensation. Using static condensation, Eq. (6.6.1) becomes

$$\underline{F}_i^B = \underline{K}_{ii}^B \underline{d}_i^B + \underline{K}_{ie}^B \underline{d}_e^B \tag{6.6.2}$$

$$\underline{F}_e^B = \underline{K}_{ei}^B \underline{d}_i^B + \underline{K}_{ee}^B \underline{d}_e^B \tag{6.6.3}$$

We eliminate the interior displacements $\underline{d}_e$ by solving Eq. (6.6.3) for $\underline{d}_e^B$, as follows:

$$\underline{d}_e^B = [\underline{K}_{ee}^B]^{-1}[\underline{F}_e^B - \underline{K}_{ei}^B \underline{d}_i^B] \tag{6.6.4}$$

Then we substitute Eq. (6.6.4) for $\underline{d}_e^B$ into Eq. (6.6.2) to obtain

$$\underline{F}_i^B - \underline{K}_{ie}^B[\underline{K}_{ee}^B]^{-1}\underline{F}_e^B = (\underline{K}_{ii}^B - \underline{K}_{ie}^B[\underline{K}_{ee}^B]^{-1}\underline{K}_{ei}^B)\underline{d}_i^B \tag{6.6.5}$$

We define

$$\bar{F}_i^B = \underline{K}_{ie}^B [\underline{K}_{ee}^B]^{-1} \underline{F}_e^B \quad \text{and} \quad \bar{K}_{ii}^B = \underline{K}_{ii}^B - \underline{K}_{ie}^B [\underline{K}_{ee}^B]^{-1} \underline{K}_{ei}^B \tag{6.6.6}$$

Substituting Eq. (6.6.6) into (6.6.5), we obtain

$$\underline{F}_i^B - \bar{F}_i^B = \bar{K}_{ii}^B \underline{d}_i^B \tag{6.6.7}$$

Similarly, we can write force/displacement equations for substructures A and C. These equations can be partitioned in a manner similar to Eq. (6.6.1) to obtain

$$\left\{ \begin{array}{c} \underline{F}_i^A \\ \hline \underline{F}_e^A \end{array} \right\} = \left[\begin{array}{c|c} \underline{K}_{ii}^A & \underline{K}_{ie}^A \\ \hline \underline{K}_{ei}^A & \underline{K}_{ee}^A \end{array} \right] \left\{ \begin{array}{c} \underline{d}_i^A \\ \hline \underline{d}_e^A \end{array} \right\} \tag{6.6.8}$$

Eliminating $\underline{d}_e^A$, we obtain

$$\underline{F}_i^A - \bar{F}_i^A = \bar{K}_{ii}^A \underline{d}_i^A \tag{6.6.9}$$

Similarly, for substructure C, we have

$$\underline{F}_i^C - \bar{F}_i^C = \bar{K}_{ii}^C \underline{d}_i^C \tag{6.6.10}$$

The whole frame is now considered to be made of superelements A, B, and C connected at interface nodal points (each superelement being made up of a collection of individual smaller elements). Using compatibility, we have

$$\underline{d}_{i\,\text{top}}^A = \underline{d}_{i\,\text{bottom}}^B \quad \text{and} \quad \underline{d}_{i\,\text{top}}^B = \underline{d}_{i\,\text{bottom}}^C \tag{6.6.11}$$

That is, the interface displacements at the common locations where cuts were made must be the same.

The response of the whole structure can now be obtained by direct superposition of Eqs. (6.6.7), (6.6.9), and (6.6.10), where now the final equations are expressed in terms of the interface displacements at the eight interface nodes only [Figure 6–30(b)] as

$$\underline{F}_i - \bar{F}_i = \bar{K}_{ii} \underline{d}_i \tag{6.6.12}$$

The solution of Eq. (6.6.12) gives the displacements at the interface nodes. To obtain the displacements within each substructure, we use the force-displacement Eqs. (6.6.4) for $\underline{d}_e^B$ with similar equations for substructures A and C. Example 6.7 illustrates the concept of substructure analysis. In order to solve by hand, a relatively simple structure is used.

Example 6.7

Solve for the displacement and rotation at node 3 for the beam in Figure 6–31 by using substructuring. Let $E = 29 \times 10^3$ ksi and $I = 1000$ in^4.

To illustrate the substructuring concept, we divide the beam into two substructures, labeled 1 and 2 in Figure 6–32. The 10-kip force has been assigned to node 3 of substructure 2, although it could have been assigned to either substructure or a fraction of it assigned to each substructure.

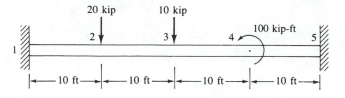

Figure 6–31 Beam analyzed by substructuring

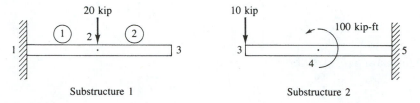

Substructure 1   Substructure 2

Figure 6–32 Beam of Figure 6–31 separated into substructures

The stiffness matrix for each beam element is given by Eq. (5.1.14) as

$$
\underset{\sim}{k}^{(1)} = \underset{\sim}{k}^{(2)} = \underset{\sim}{k}^{(3)} = \underset{\sim}{k}^{(4)} = \frac{29 \times 10^6}{(120)^3}
\begin{bmatrix}
12 & 6(120) & -12 & 6(120) \\
6(120) & 4(120)^2 & -6(120) & 2(120)^2 \\
-12 & -6(120) & 12 & -6(120) \\
6(120) & 2(120)^2 & -6(120) & 4(120)^2
\end{bmatrix}
$$

$$\begin{array}{cc} 1 & 2 \\ 2 & 3 \\ 3 & 4 \\ 4 & 5 \end{array}$$

(6.6.13)

$$
= 16.78
\begin{bmatrix}
12 & 720 & -12 & 720 \\
720 & 57{,}600 & -720 & 28{,}800 \\
-12 & -720 & 12 & -720 \\
720 & 28{,}800 & -720 & 57{,}600
\end{bmatrix}
$$

(6.6.14)

For substructure 1, we add the stiffness matrices of elements 1 and 2 together. The equations are

$$
16.78
\begin{bmatrix}
12 + 12 & -720 + 720 & -12 & 720 \\
-720 + 720 & 57{,}600 + 57{,}600 & -720 & 28{,}800 \\
-12 & -720 & 12 & -720 \\
720 & 28{,}800 & -720 & 57{,}600
\end{bmatrix}
\begin{Bmatrix}
d_{2y} \\ \phi_2 \\ d_{3y} \\ \phi_3
\end{Bmatrix}
=
\begin{Bmatrix}
-20 \\ 0 \\ 0 \\ 0
\end{Bmatrix}
$$

(6.6.15)

where the boundary conditions $d_{1y} = \phi_1 = 0$ were used to reduce the equations.

Rewriting Eq. (6.6.15) with the interface displacements first allows us to use Eq. (6.6.6) to condense out, or eliminate, the interior degrees of freedom, d_{2y} and ϕ_2. These reordered equations are

$$16.78(12d_{3y} - 720\phi_3 - 12d_{2y} - 720\phi_2) = 0$$

$$16.78(-720d_{3y} + 57{,}600\phi_3 + 720d_{2y} + 28{,}800\phi_2) = 0$$

$$16.78(-12d_{3y} + 720\phi_3 + 24d_{2y} + \phi_2) = -20 \qquad (6.6.16)$$

$$16.78(-720d_{2y} + 28{,}800\phi_3 + 0d_{2y} + 115{,}200\phi_2) = 0$$

Using Eq. (6.6.6), we obtain equations for the interface degrees of freedom as

$$16.78\left\{\begin{bmatrix} 12 & -720 \\ -720 & 57{,}600 \end{bmatrix} - \begin{bmatrix} -12 & -720 \\ 720 & 28{,}800 \end{bmatrix}\begin{bmatrix} 24 & 0 \\ 0 & 115{,}200 \end{bmatrix}^{-1}\begin{bmatrix} -12 & 720 \\ -720 & 28{,}800 \end{bmatrix}\right\}\begin{Bmatrix} d_{3y} \\ \phi_3 \end{Bmatrix}$$

$$= \begin{Bmatrix} 0 \\ 0 \end{Bmatrix} - \begin{bmatrix} -12 & -720 \\ 720 & 28{,}800 \end{bmatrix}\begin{bmatrix} 24 & 0 \\ 0 & 115{,}200 \end{bmatrix}^{-1}\begin{Bmatrix} -20 \\ 0 \end{Bmatrix} \qquad (6.6.17)$$

Simplifying Eq. (6.6.17), we obtain

$$\begin{bmatrix} 25.17 & -3020 \\ -3020 & 483{,}264 \end{bmatrix}\begin{Bmatrix} d_{3y} \\ \phi_3 \end{Bmatrix} = \begin{Bmatrix} -10 \\ 600 \end{Bmatrix} \qquad (6.6.18)$$

For substructure 2, we add the stiffness matrices of elements 3 and 4 together. The equations are

$$16.78\begin{bmatrix} 12 & 720 & -12 & 720 \\ 720 & 57{,}600 & -720 & 28{,}800 \\ -12 & -720 & 12+12 & -720+720 \\ 720 & 28{,}800 & -720+720 & 57{,}600+57{,}600 \end{bmatrix}\begin{Bmatrix} d_{3y} \\ \phi_3 \\ d_{4y} \\ \phi_4 \end{Bmatrix} = \begin{Bmatrix} -10 \\ 0 \\ 0 \\ 1200 \end{Bmatrix}$$

$$(6.6.19)$$

where boundary conditions $d_{5y} = \phi_5 = 0$ were used to reduce the equations.

Using static condensation, Eq. (6.6.6), we obtain equations with only the interface displacements d_{3y} and ϕ_3. These equations are

$$16.78\left\{\begin{bmatrix} 12 & 720 \\ 720 & 57{,}600 \end{bmatrix} - \begin{bmatrix} -12 & 720 \\ -720 & 28{,}800 \end{bmatrix}\begin{bmatrix} 24 & 0 \\ 0 & 115{,}200 \end{bmatrix}^{-1}\begin{bmatrix} -12 & -720 \\ 720 & 28{,}800 \end{bmatrix}\right\}\begin{Bmatrix} d_{3y} \\ \phi_3 \end{Bmatrix}$$

$$= \begin{Bmatrix} -10 \\ 0 \end{Bmatrix} - \begin{bmatrix} -12 & 720 \\ -720 & 28{,}800 \end{bmatrix}\begin{bmatrix} 24 & 0 \\ 0 & 115{,}200 \end{bmatrix}^{-1}\begin{Bmatrix} 0 \\ 1200 \end{Bmatrix} \qquad (6.6.20)$$

Simplifying Eq. (6.6.20), we obtain

$$\begin{bmatrix} 25.17 & 3020 \\ 3020 & 483{,}264 \end{bmatrix}\begin{Bmatrix} d_{3y} \\ \phi_3 \end{Bmatrix} = \begin{Bmatrix} -17.5 \\ -300 \end{Bmatrix} \qquad (6.6.21)$$

Adding Eqs. (6.6.18) and (6.6.21), we obtain the final nodal equilibrium equations at the interface degrees of freedom as

$$\begin{bmatrix} 50.34 & 0 \\ 0 & 966{,}528 \end{bmatrix} \begin{Bmatrix} d_{3y} \\ \phi_3 \end{Bmatrix} = \begin{Bmatrix} -27.5 \\ 300 \end{Bmatrix} \tag{6.6.22}$$

Solving Eq. (6.6.22) for the displacement and rotation at node 3, we obtain

$$d_{3y} = -0.5463 \text{ in.} \tag{6.6.23}$$

$$\phi_3 = 0.0003104 \text{ rad}$$

We could now return to Eq. (6.6.15) or Eq. (6.6.16) to obtain d_{2y} and ϕ_2 and to Eq. (6.6.19) to obtain d_{4y} and ϕ_4. ∎

We emphasize that this example is used as a simple illustration of substructuring and is not typical of the size of problems where substructuring is normally performed. Generally, substructuring is used when the number of degrees of freedom is very large, as might occur, for instance, for very large structures such as the airframe in Figure 6–29.

▲ **6.7 Algor Example Solutions for Plane Frame, Grid, and Space Frame Analysis** ▲

To use the Algor computer program [10] for plane frame, grid, and space frame analysis, we now solve a series of example problems. The program is based on the flowchart of Section 4.1 with a frame or grid replacing the truss in Figure 4–1(a) and the BEdit processor accessed from the "Model Data Control" window in Figure 4–1(b). Longhand frame Examples 6.1 and 6.4, a rigid building frame, grid Example 6.6, and a space frame are solved using the Algor program. These examples illustrate such concepts as applying a moment to a rigid frame, mixing beam and truss elements together, and analyzing grid and space frame structures.

Example 6–8

We now consider the plane frame solved longhand in Example 6.1 to illustrate the use of the Algor computer program for plane frame analysis. The frame is again shown in Figure 6–33. Let $I_1 = I_3 = 200$ in^4 and $I_2 = 100$ in^4. Young's modulus is 30×10^6 psi for all members. The steps used to create and analyze the frame follow.

Step 1 Model the Frame Using SuperDraw III

Click on the "Model Data" button in the lower portion of the screen.

Click on the "Units" button. You will be asked to specify a model name. Save the model, and the "Unit System" window will appear.

Pick "English (in)" for the unit system.

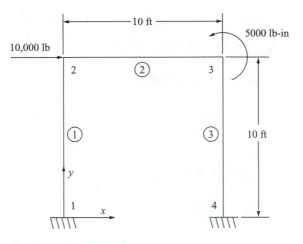

Figure 6–33 Rigid frame

Click "OK".
Click "Add".
Go to "Line".

(0,0,0)	Enter the first coordinate of the beam.
(0,120,0)	Enter the second coordinate of the beam.
(120,120,0)	Enter the third coordinate of the beam.
(120,0,0)	Enter the fourth coordinate of the beam.

Click "Done".
Go to "View" and click "Enclose".

Step 2 Create a New Layer for the Horizontal Member

Warning: In Release 12.0, we must remove the default check mark in front of "Maintain Group Boundary" to create different groups. In Release 12.02, this default no longer exists.

Click "Select" and then "None".

Click "Select" and then "Point" and right-click next to the middle of the horizontal element.

Click "Done".

Click "Modify" on the main menu.

Select "Update Object Parameters" and click "Maintain Group Boundary" to remove the check mark.

Click "Modify".

Select "Update Object Parameters".

Click "Layer Number".

Type 2 and ⟨Enter⟩. The horizontal member should now be layer 2.

Step 3 Add Boundary Conditions and Apply Force

Click "FEA Add" and go to "Stress and Vibration Analysis".

Click "Boundary Conditions".

Make sure there is a check next to "Use @ Symbol for Full" only.

Click "Box Apply".

With the cursor, make a box around nodes 1 and 4.

An "@" symbol should appear at these nodes.

Click "Done".

Again, click "FEA Add" and go to "Stress and Vibration Analysis".

Click "Nodal Forces".

Click "Vector".

Click "X direction".

Click "Done".

Click "Magnitude".

Type "10000" and ⟨Enter⟩.

Click "Box Apply" and make a box around node 2.

Click "Done".

Again, click "FEA Add" and go to "Stress and Vibration Analysis".

Click on "Nodal Moments".

Click "Vector".

Click "Z direction".

Click "Done".

Click "Magnitude".

Type 5000 and ⟨Enter⟩. You have now created a nodal moment.

Click "Box Apply" and make a box around node 3.

Click "Done".

Step 4 Add Material Properties Using the "Model Data Control"
          Window

Click on the "Model Data" button to open the "Model Data Control" window.

Click on the box below "Element" for group 1.

Select "Beam".

Click "OK".

Click on the box below "Data". This takes you into "Beam Design Editor".

Click "Add/Mod".

Click "Sectional".

Click "Value".

10   In the lower portion of the screen enter the value 10 for the area.

200   Tab over and enter the value 200 for both I_2 and I_3.

<Enter>   when done.

Click "Next". ID = 2 appears and horizontal beam is highlighted.

Click "Value" and make I2 and I3 = 100.

Click ⟨Esc⟩ twice and then "Done" to exit.

Click "Yes" to save.

Click next to the box under "Material".

Click "Customer Defined".

Click "Edit Properties".

Type 30e6 for a modulus of elasticity.

Click "OK".

Again click "OK" to exit the "Material" window.

Step 5   Enter Global Data

In the "Model Data Control" window:

Click "Global".

1   Type the value 1 under the word "Pressure".

Click on the "Output" tab.

Place checks next to "Nodal data", "Element data", and "Displacement data".

Click "OK".

Step 6   Perform Analysis on the Model

In the "Model Data Control" window:

Click "Analysis".

Click "Analyze".

Click "OK".

Click "Done".

Step 7   Review the Results

Review the results in Superview in the usual manner.   ■

The nodal displacements from the Algor *ex68.L* file are listed in Table 6–2. The horizontal displacement at the top of the frame (node 3, *X*-translation) is the same as the longhand solution from Eq. (6.2.7). Figure 6–34 shows a plot of the undisplaced and displaced frame.

You should always interpret the postprocessor displayed information carefully. For the frame (as for all beam elements), Superview provides the following stress output for each element: P/A, M_2/S_2, M_3/S_3, and worst sum. P/A is the usual axial stress from axial force P, and M_2/S_2 and M_3/S_3 are the bending stresses about the local transverse principal axes of the cross-sectional area (about the local 2 and 3 axes), respectively. The worst sum is calculated from the following formula:

$$\text{Worst sum} = (\text{sign of } P/A)[\text{ABS}(P/A) + \text{ABS}(M_2/S_2) + \text{ABS}(M_3/S_3)]$$

Table 6–2 Nodal displacements (in.) and element stresses (psi) from the Algor solution of Example 6.8

Displacements/Rotations(degrees) of nodes

NODE number	X-translation	Y-translation	Z-translation	X-rotation	Y-rotation	Z-rotation
1	0.0000E+00	0.0000E+00	0.0000E+00	0.0000E+00	0.0000E+00	0.0000E+00
2	0.0000E+00	0.0000E+00	0.0000E+00	0.0000E+00	0.0000E+00	0.0000E+00
3	2.1136E-01	1.4813E-03	0.0000E+00	0.0000E+00	0.0000E+00	-8.7435E-02
4	2.0936E-01	-1.4813E-03	0.0000E+00	0.0000E+00	0.0000E+00	-8.5142E-02
5	0.0000E+00	0.0000E+00	0.0000E+00	0.0000E+00	0.0000E+00	0.0000E+00
6	0.0000E+00	0.0000E+00	0.0000E+00	0.0000E+00	0.0000E+00	0.0000E+00
7	0.0000E+00	0.0000E+00	0.0000E+00	0.0000E+00	0.0000E+00	0.0000E+00
8	0.0000E+00	0.0000E+00	0.0000E+00	0.0000E+00	0.0000E+00	0.0000E+00
9	0.0000E+00	0.0000E+00	0.0000E+00	0.0000E+00	0.0000E+00	0.0000E+00
10	0.0000E+00	0.0000E+00	0.0000E+00	0.0000E+00	0.0000E+00	0.0000E+00

1**** BEAM ELEMENT FORCES AND MOMENTS

ELEMENT NO.	CASE (MODE)	AXIAL FORCE R1	SHEAR FORCE R2	SHEAR FORCE R3	TORSION MOMENT M1	BENDING MOMENT M2	BENDING MOMENT M3
1	1	5.008E+03	-3.703E+03	0.000E+00	0.000E+00	0.000E+00	-2.232E+05
		-5.008E+03	3.703E+03	0.000E+00	0.000E+00	0.000E+00	-2.212E+05

1**** BEAM ELEMENT STRESSES

ELEMENT NO.	CASE (MODE)	P/A	P/A+M2/S2	P/A-M2/S2	P/A+M3/S3	P/A-M3/S3	WORST SUM
1	1	-5.008E+02	-5.008E+02	-5.008E+02	5.150E+06	-5.151E+06	-5.151E+06
		-5.008E+02	-5.008E+02	-5.008E+02	-5.105E+06	5.104E+06	-5.105E+06 ■

This formula may seem appropriate to determine the maximum normal stress on a cross section of the beam. However, you should verify that this equation is not valid for circular cross sections, as the largest M_2/S_2 and M_3/S_3 will not add at any same point on the cross section. The worst sum will then overestimate the largest normal stress on the circular cross section. Therefore, you must use this worst sum result with caution.

Example 6.9

We now consider the cantilever beam with a bar element support that was previously solved longhand in Example 6.4 to illustrate how to combine the different element types (beam and truss) into one model. The model is shown in Figure 6–35. Let $E = 210$ GPa, $A = 2 \times 10^{-3}$ m^2, and $I = 5 \times 10^{-5}$ m^4 for the beam, and let $E = 210$ GPa and $A = 1 \times 10^{-3}$ m^2 for the bar.

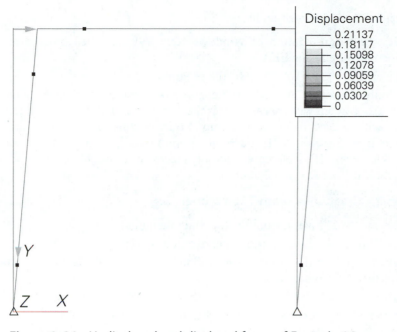

Figure 6–34 Undisplaced and displaced frame of Example 6.8

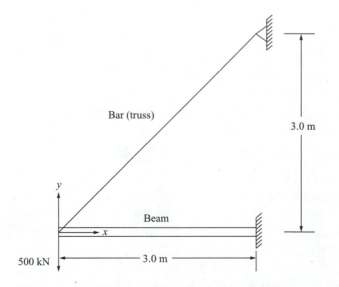

Figure 6–35 Cantilever beam with bar element

Step 1   Start Superdraw III

Select the "Start" button of Windows NT/95/98 and then proceed to "Programs: Algor Software:Algor FEA"

Then click "Algor FEA" to start the program.

Or double-click the "Algor FEA" icon on your desktop computer screen. The Superdraw III program will appear. This is the Algor main graphics user interface. All your work will be done from this program. You will construct, analyze, and review your results from Superdraw III. Each menu option and a short example are available in the Superdraw III Reference Division (accessed through "Docutech", the on-line "Technical User Documentation").

Step 2   Create the Beam-Truss Model

Select the "Add" menu from the top main menu bar.

Click "Line". A "Line" menu appears on the screen.

Click "Single" on the "Line" menu and begin entering the data points (coordinates of the elements) in succession.

0<Tab>0   This is the location of the free node.

<Enter>   Press the ⟨Enter⟩ key to actually enter the data.

3<Tab>0   This is the endpoint of element 1.

<Enter>

0<Tab>0   Repeat the process for element 2.

<Enter>

3<Tab>3

<Enter>   All data points have been entered.

Click "Done" on the "Line" menu on the screen.

Click "View" on the main menu.

Click "Enclose" within the "View" menu to see the full view of the model. It should look similar to Figure 6–35.

Step 3   Eliminate Duplicate Lines

Click "Modify" on the main menu.

Click "Clean:Duplicate". This notation means to click "Clean" and then click "Duplicate" in succession. This command brings up the on-screen "Duplicate" menu.

Click "Perform Cleaning" on the "Duplicate" menu. You will see the following message in the text bar near the bottom of the screen.

"2 Kept 0 Deleted. done"

Click "Done" on the "Duplicate" menu located on the screen.

Step 4   Apply a Group to the Bar (Truss) Element

A new group needs to be created in order to change the angled element into a truss, leaving the horizontal element to become a beam element.

Click "Select:Point" on the main menu.

Click on the bar (truss) element. It should now be highlighted with a small red box.

Click "Modify:Update Object Parameters:Group Number" This brings up a menu screen called "Group". Type 2 in the box with the cursor in it and ⟨Enter⟩.

2

⟨Enter⟩   The on-screen menu disappears, and the bar element turns to a red color.

Step 5   Apply Boundary Conditions and Nodal Forces

Click on the "FEA Add" menu on the main menu.

Click "Stress and Vibration Analysis:Boundary Conditions", and the on-screen menu appears. The default boundary condition is "Use @ symbol for Full". "Full" means full constraints on translation and rotation when this condition is applied to a particular node. The model needs full constraints at the two right nodes of the elements.

Right-click on the right node of the horizontal element. The "@" sign appears on the screen next to the node, indicating that the node is fully constrained.

Right-click on the right node of the bar element. The nodes have been fully constrained.

Click "Done" within the on-screen "Boundary Conditions" menu.

If you have difficulty seeing the "@" symbol on screen, you may wish to zoom in closer to each node to view the boundary conditions.

Click "View:Zoom:In" and create a box around node 2 with the mouse and the left mouse button. You may have to repeat this process of zooming in one or more times to see the "@" sign.

Click "Done".

Note: If you have created one or more spurious or misplaced boundary condition "@" signs, click "Select:Box" on the main menu and draw a box around the object that you wish to delete (in this case a boundary condition). After the object is highlighted, click "Modify:Delete". At this time the "@" symbol should disappear.

Click "View:Enclose" to view the entire model again.

Click "FEA Add:Stress and Vibration Analysis:Nodal Forces". This brings up the on-screen menu "Nodal Forces". We need to confirm the direction of the nodal forces.

Click "Vector".

Click "Y Direction". Nodal forces will be applied in the *y* direction.

Click "Done". Now add the magnitude of the force.

Click "Magnitude" and enter the value.

−500000

⟨Enter⟩   Now apply the force.

Right-click on the free node. The force should now show on the screen. Zoom in if necessary.

Click "View:Enclose" to view the entire model again if necessary.

Step 6   Add Material Properties Using the "Model Data Control" Window

Click on the "Model Data" control button on the lower part of the screen and highlighted in red. This brings up the on-screen menu "Model Data Control". Alternatively, you can access the "Model Data Control" menu by clicking "Tools:Model Data Control".

For the "Analysis Type" line, "Linear Static Stress" appears as the default. You could select another analysis type by scrolling down menu to an appropriate analysis algorithm, but in our case, "Linear Static Stress" will do nicely.

In the "Model Data Control" window there should be two groups listed. The first group is the beam element, and the second group is the truss element.

Click on the group 1 row box below the "Element" column. An on-screen menu appears called "Group 1" elements. These are the different types of elements we can choose from. For this analysis both a beam and a truss element are needed.

Click "Beam".

Click "OK".

Click on the group 2 row box below the "Element" column. An on-screen menu appears called "Group 2" elements.

Click "Truss".

Click "OK".

Click on the group 1 row box below "Data". It will respond with "Please enter the model name first".

Click "OK". Then enter the model name in the "File_name" area that appears.

ex6_9   The name of the file is "*ex6_9*".

Click "Save". By default, the "Units Definition" menu appears on-screen.

For this analysis we need metric MKS units.

Click "Metric MKS" on the on-screen menu.

Click "OK".

The program takes you to the "Beam Design Editor". We need to add area and cross-sectional properties.

Click "Add/Mod" on the main menu.

Click "Sectional" on the "Add/Mod" menu.

Click "Value" on the "Sectional" menu.

The area and "I" values are shown at the bottom. We need to delete some of the current values and add new ones. Delete the current "*A*" value using the escape key.

```
<Esc>
2e-3
<Tab>
<Tab>
<Tab>
<Tab>
```

<Tab>

<Esc>   This deletes the current "*I3*" value.

5e-5

<Enter>   The area and "*I3*" should now be $A = 2 \times 10^{-3}$ m^2 and $I = 5 \times 10^{-5}$ m^4.

Now leave the "Beam Design Editor" and return to the Superdraw III screen.

<Esc>

<Esc>

Click "done". The program will ask if you wish to save.

Click "YES", and the program will return you to the "Model Data Control" on-screen menu.

The "Data" box within the group 1 row should be satisfied with a check mark.

Click on the Group 2 row box below "Data".

The "Element Definition" menu appears next on the screen.

Enter the cross-sectional area of the truss elements. Hold the left click of the mouse over the default value of 0 until a blue box appears on the 0. Now enter in the cross-sectional area of the truss element.

1e-3

<Enter>

Click "OK". This should satisfy "Data" in the group 2 row box on the "Model Control Data" menu with a check mark.

Click on the group 1 row box below the "Material" column. This brings up the on-screen menu "Element Material Selection". We can select one of the predefined materials for a steel. The steel we will choose is standard steel (steel 4130).

Click "Steel (4130)".

Click "OK". Repeat this for the truss in the group 2 row.

Click on the group 2 row box below the "Material" column.

Click "Steel (4130)".

Click "OK".

Both "Material" boxes should be satisfied with "Steel (4130)".

Step 7   Enter the Global Data

Click on the "Global" button on the right-side menu of the "Model Data Control" window. Under "Load Case Multipliers" and under "Pressure", type in 1.

1   This enters values of 1 under the "Pressure" on the "Load Case Multipliers" menu.

Click on the "Output" tab to view the output options.

Click on the box next to "Displacement data".

Click on the box next to "Stress data".

This creates nodal displacement output in the .*L* file and element forces and stresses in the .*S* file. These files can be viewed later with a word processor.

Click "OK". You are out of the "Global" menu and back in the "Model Data Control" menu.

Step 8   Check the Model

Click "Check" under the "FEA Model" column in the "Model Data Control" window. This takes you to Superview's main menu, and you can check your model as explained in Section 4.1 and detailed by example in Appendix F. The model appears automatically, showing the boundary conditions in red and any forces that may be applied in yellow.

Click "Done". This takes you back to the "Model Data Control" menu.

Step 9   Run the Analysis

Click on the "Analysis" button under the "FEA Model" column in the "Model Data Control" window.

Click "Analyze" in the lower left corner of the "Linear Stress and Vibration-Static Stress" menu. A warning message will appear after the analysis has been completed but has no bearing on the outcome of the analysis.

Click "OK" to close the warning message.

Click "Done" to return to the "Model Data Control" menu.

Step 10   Review the Results

Click on the "Results" button in the "Model Data Control" window. This takes you to the Superview program once again where you can view the results.

Click "Results" on the main menu line of Superview.

Click "Displaced" on the main menu of Superview.

Click "Displ on" so that "*With undi" appears. This allows the displaced and undisplaced models to be superimposed. (Figure 6–36 shows the displaced model.)

Click "Calc scal" to enlarge the scale of the displacement for easier viewing.

Click "Nodes inq" on the "Displaced" menu.

Click "Get" to get a node.

Click on node 1. Move the cursor to the free node and click on it to see the x and y displacements appear at the bottom of the screen. You will see "DX$=-3.45*10^{-3}$, DY$=-2.287*10^{-2}$" meters.

<Esc>

<Esc>   You are now back in the main menu of Superview.

Click "Results". You are now going to post a true stress.

Click "Beam and Truss Stress". You must select "Beam and Truss Stress" because you have a beam-truss model.

Click "1)P/A". The axial stress is posted. You now see the P/A stress. See Figure 6–36 for a plot of the model with colored boxes representing different stresses on each element. For instance, red on the truss element indicates a stress of 6.7×10^8 Pa.

<Esc>

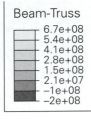

Figure 6–36 Displaced structure of Example 6.9

Click "Done". This takes you out of Superview.

Click "Close" to close the "Model Data Control" window.

You should be able to find the *ex6_9.L* file (storing the echo check of your data and the nodal displacements) and the *ex6_9.S* file (storing the element forces and stresses) and print a hard copy of each.

Table 6–3 shows the nodal displacements and element stresses and forces taken from the Algor program *ex6_9.L* and *ex6_9.S* files. The nodal displacements at the free node are the same from the Algor output (Table 6–3) as from the longhand solution [Eq. (6.2.45)]. ∎

Example 6.10

For the rigid building frame with loading as shown in Figure 6–37, determine the deflection and stress of each element. All the vertical members are W12 × 72, roof members are W18 × 55 and floor members are W24 × 68. Use ASTM A-36 for all elements.

Step 1 Start Superdraw III

Select the "Start" button of Windows NT/95/98 and then proceed to "Programs: Algor Software:Algor FEA".

Then click "Algor FEA" to start the program.

Table 6–3 Displacements and element forces and stresses for Example 6.9

Displacements/Rotations(degrees) of nodes

NODE number	X-translation	Y-translation	Z-translation	X-rotation	Y-rotation	Z-rotation
1	3.4354E-03	-2.2869E-02	0.0000E+00	0.0000E+00	0.0000E+00	6.5515E-01
2	0.0000E+00	0.0000E+00	0.0000E+00	0.0000E+00	0.0000E+00	0.0000E+00
3	0.0000E+00	0.0000E+00	0.0000E+00	0.0000E+00	0.0000E+00	0.0000E+00
4	0.0000E+00	0.0000E+00	0.0000E+00	0.0000E+00	0.0000E+00	0.0000E+00
5	0.0000E+00	0.0000E+00	0.0000E+00	0.0000E+00	0.0000E+00	0.0000E+00
6	0.0000E+00	0.0000E+00	0.0000E+00	0.0000E+00	0.0000E+00	0.0000E+00
7	0.0000E+00	0.0000E+00	0.0000E+00	0.0000E+00	0.0000E+00	0.0000E+00
8	0.0000E+00	0.0000E+00	0.0000E+00	0.0000E+00	0.0000E+00	0.0000E+00
9	0.0000E+00	0.0000E+00	0.0000E+00	0.0000E+00	0.0000E+00	0.0000E+00

1**** BEAM ELEMENT FORCES AND MOMENTS

ELEMENT NO.	CASE (MODE)	AXIAL FORCE R1	SHEAR FORCE R2	SHEAR FORCE R3	TORSION MOMENT M1	BENDING MOMENT M2	BENDING MOMENT M3
1	1	4.737E+05	-2.628E+04	0.000E+00	0.000E+00	0.000E+00	0.000E+00
		-4.737E+05	2.628E+04	0.000E+00	0.000E+00	0.000E+00	-7.884E+04

1**** BEAM ELEMENT STRESSES

ELEMENT NO.	CASE (MODE)	P/A	P/A+M2/S2	P/A-M2/S2	P/A+M3/S3	P/A-M3/S3	WORST SUM
1	1	-2.369E+08	-2.369E+08	-2.369E+08	-2.369E+08	-2.369E+08	-2.369E+08
		-2.369E+08	-2.369E+08	-2.369E+08	-2.387E+08	-2.350E+08	-2.387E+08

**** 3-D truss elements:

```
    Number of elements    =    1
    Number of materials   =    2
```

**** Nodal stresses for 3-D truss elements:

El. #	LC	ND	Stress	Force
1	1	I	-6.699E+08	-6.699E+05
1	1	J	6.699E+08	6.699E+05

■

Or double-click on the "Algor FEA" icon on your desktop computer screen. The Superdraw III program will appear. This is the Algor main graphics user interface. All your work will be done from this program. You will construct, analyze, and review your results from Superdraw III. Each menu option and a short example are available in the Superdraw III Reference Division (accessed through "Docutech", the on-line "Technical User Documentation".)

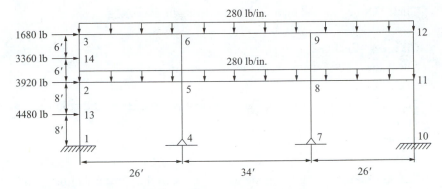

Figure 6-37 Rigid frame with wind and dead loads

Step 2 Create the Frame Model

Select the "Add" menu from the top main menu bar.

Click "Line". A "Line" menu appears in the upper right part of the screen. Then enter the data points in succession.

0<Tab>0 Enter node 1, the first point of the first column.

<Enter>

0<Tab>96 Enter node 13, the midpoint of the beam where a wind load will later be placed.

<Enter>

0<Tab>192 Enter node 2.

<Enter>

0<Tab>264 Enter node 14, the midpoint of the beam where a wind load will later be placed.

<Enter>

0<Tab>336 Enter node 3. This completes the first column.

<Enter>

Click "Done" on the menu in the upper right corner.

Select "View". This takes you to the draw options menu.

Click "Enclose". This encloses the model. Four elements appear as indicated by red highlight boxes.

Select the "Add" menu from the top main menu bar.

Click "Line".

312<Tab>0 Enter node 4, the first point of the second column.

<Enter>

312<Tab>192 Enter node 5.

<Enter>

312<Tab>336 Enter node 6. This completes the second column.

<Enter>

Click "Done" on the menu in the upper right corner.
Select "View".
Click on "Enclose".
Select the "Add" menu from the top main menu bar.
Click "Line".

720<Tab>0	Enter node 7, the first point of the third column.
<Enter>	
720<Tab>192	Enter node 8.
<Enter>	
720<Tab>336	Enter node 9. This finishes the third column.
<Enter>	

Click "Done" on the menu in the upper right corner.
Select "View".
Click "Enclose".
Select the "Add" menu from the top main menu bar.
Click "Line".

1032<Tab>0	Enter node 10, the first point of the fourth column.
<Enter>	
1032<Tab>192	Enter node 11.
<Enter>	
1032<Tab>336	Enter node 12. This finishes the fourth column.
<Enter>	

Click "Done" on the menu in the upper right corner.
Select "View".
Click "Enclose".
Select the "Select" menu from the top main menu bar.
Click "All". This selects all the columns.
Select "Modify" from the top main menu bar.
Click "Update Object Parameters; Group Number".

2	This changes the group number from 1 to 2. It is necessary to change group numbers for every set of members that will have different material properties or cross sections, or if there is a mixture of beams and trusses in the model. In this example problem, all the columns have the cross section of W12 × 72 and are of the same material.

Click "OK".
Select the "Add" menu from the top main menu bar.
Click "Line".

0<Tab>336	Enter the first point of the roof or right-click on node 3.
<Enter>	
312<Tab>336	Enter the second point of the roof or right-click on node 6.
<Enter>	

720<Tab>336   Enter the third point of the roof or right-click on node 9.

<Enter>

1032<Tab>336   Enter the last point of the roof or right-click on node 12.

<Enter>

Click "Done".

Select "Modify" from the top main menu bar.

Click "Update Object Parameters;Group Number".

3   This changes the group number from 1 to 3. It is not necessary to select the roof numbers since they are automatically selected (shown by the red squares in the middle of each member) as the members are drawn. For the roof members, it is necessary to give them a separate group number because they are W18 × 55's.

Click "OK".

Select the "Add" menu from the top main menu bar.

Click "Line".

0<Tab>192   Enter the first point of the floor or right-click on node 2.

<Enter>

312<Tab>192   Enter the second point of the floor or right-click on node 5.

<Enter>

720<Tab>192   Enter the third point of the floor or right-click on node 8.

<Enter>

1032<Tab>192   Enter the last point of the floor or right-click on node 11.

<Enter>   This completes the model. The floor members will remain in group 1.

Click "Done".

Step 3 Eliminate Duplicate Lines

Click "Modify" on the main menu.

Select "Clean:Duplicate". This notation means to select "Clean" and "Duplicate" in succession.

Click "Perform Cleaning" on the "Duplicate" menu. You will see the following message in the lower part of the screen:

"16 Kept, 0 Deleted. done."

Click "Done" on the "Duplicate" menu located in the upper right corner of the screen.

Step 4 Add the Boundary Conditions (Reactions) and Applied Forces

Select the "FEA Add" menu from the main menu bar and highlight "Stress and Vibration Analysis".

Click "Boundary Conditions". The "Boundary Conditions" menu appears.

The default boundary condition is "Use @ Symbol for Full". "Full" means full constraint or prevention of translations and rotations. For this problem, the outer reactions are fixed and the middle ones are pinned. First, however, before applying these conditions, it is necessary to change the text height so the reaction values will be visible. Therefore,

Click "Text Attributes". This takes you to the "Text Setup" window.

Click "Height".

15　　　　　　This changes the text height of reaction values to 15 in.

<Enter>

Click "Done". This takes you back to the "Boundary Conditions" window.

Click on nodes 1 and 10 to apply the full constraint boundary conditions to the outer columns, and the "@" symbol should be visible.

Click "Change Values" to make the middle reactions pinned.

Click on the box next to "Rz constrained" to remove the reaction in the z direction.

Click "Done".

Click on nodes 4 and 7 to apply the pinned boundary conditions to the inner columns, and "TxyzRxy" should be visible next to these nodes.

Click "Done"

Select the "FEA Add" menu from the main menu bar and highlight "Stress and Vibration Analysis".

Click "Nodal Forces", and the "Nodal Forces" window appears in the upper right-hand corner.

Click "Vector" to choose the direction in which the wind loads will be applied.

Click "X Direction"

Click "Done." This takes you back to the "Nodal Forces" window.

Click "Text Attributes". This takes you to the "Text Setup" window.

Click "Height".

15　　　　　　This changes the text height of applied force values to 15 in.

<Enter>

Click "Done".

Click "Arrow Size".

5　　　　　　Change the arrow size to 5 in.

<Enter>

Click "Magnitude".

1680　　　　　Enter the value of the wind load at node 3.

<Enter>

Click on node 3 to apply the force. "1680;L1; LBS" appears next to the node along with an arrow pointing in the positive x direction.

Click "Magnitude".

3360　　　　　Enter the value of the wind load at node 14.

<Enter>

Click on node 14 to apply the force. "3360;L1; LBS" appears next to the node along with an arrow pointing in the positive x direction.

Click "Magnitude".

3920   Enter the value of the wind load at node 2.

<Enter>

Click on node 2 to apply the force. "3920;L1; LBS" appears next to the node along with an arrow pointing in the positive x direction.

Click "Magnitude".

4480   Enter the value of the wind load at node 13.

<Enter>

Click on node 13 to apply the force. "4480;L1; LBS" appears next to the node along with an arrow pointing in the positive x direction.

Click "Done". All four wind loads have been applied.

Step 5   **Add the Material Properties Using the "Model Data Control" Window**

Click on the "Model Data" button located at the lower part of the screen and highlighted in red.

For the "Analysis Type" line, "Linear Static Stress" appears highlighted as the default. You can scroll down for other analysis types as appropriate.

Click on the box below the "Element" column next to group 1. A table of group 1 element types appears.

Click on "Beam".

Click "OK". You have now created a beam. All lines are beam elements in group 1. Now repeat the process for groups 2 and 3.

Click on the box below the "Element" column next to group 2. A table of group 2 element types appears.

Click on "Beam". You have now created beams for all group 2 elements.

Click "OK".

Click on the box below the "Element" column next to group 3. A table of group 3 element types appears.

Click on "Beam". You have now created beams for all group 3 elements.

Click "OK".

Click on the box below the "Data" column next to group 2. You will see "Please enter the model name first."

Click "OK" and then enter the model name in the "File_name" area that appears.

Hw3   The name of the file is *Hw3*.

Click "Save". By default, the "Units Definition" menu appears.

Click "OK" to accept the default "English (in)" unit system. For this reason all lengths were entered in inches and all forces in pounds.

The "Beam Design Editor" window appears next, and the columns and reactions are visible.

Click "Add/Mod" to modify the default values.

Click "Sectional".

Click "Value". The cursor then appears at "A=1" for you to enter the cross-sectional area. Delete the default value and enter

21.1   This is the cross-sectional area of the W12 × 72 columns.

⟨Tab⟩ ⟨Tab⟩ ⟨Tab⟩ ⟨Tab⟩ ⟨Tab⟩ until the cursor is at "I3="

597   This enters the cross-sectional value of 597 in^4.

⟨Tab⟩ ⟨Tab⟩ until the cursor is at "S3="

154   This enters the section modulus of 154.

<Enter>   The values entered are now stored.

⟨Esc⟩ ⟨Esc⟩ ⟨Quit⟩ to exit the "Beam Design Editor" window. It will respond "Save Current Work?"

Click "Yes". Now update the sectional and distributed loading properties for groups 1 and 3.

Click on the box below the "Data" column next to group 3. The "Beam Design Editor" window appears next, and the roof members appear in blue.

Click "Add/Mod" to modify the default values.

Click "Sectional".

Click "Value". The cursor then appears at "A=1" for you to enter the cross-sectional area. Delete the default value and enter

16.2   This is the moment of inertia of the W18 × 55 beams.

⟨Tab⟩ ⟨Tab⟩ ⟨Tab⟩ ⟨Tab⟩ ⟨Tab⟩ until the cursor is at "I3="

890   This enters the moment of inertia value of 890 in^4.

⟨Tab⟩ ⟨Tab⟩ until the cursor is at "S3="

98.3   This enters the section modulus of 98.3 in^3.

<Enter>   The values entered are now stored.

<Esc>

Click "Loading" to add the distributed load.

Click "Distribut".

Click "Value".

⟨Tab⟩ to "Wy:−100". Delete the default value and type

−280   This enters the distributed loading to 280 lb/in. in the negative y direction at the beginning of the beam.

⟨Tab⟩ ⟨Tab⟩ ⟨Tab⟩ to "Wy:−100". Delete the default value and type

−280   This enters the distributed loading to 280 lb/in. in the negative y direction at the end of the beam.

<Enter>

Click on all three members to apply the loads, and yellow arrows pointing down should appear to represent the distributed load.

⟨Esc⟩ ⟨Esc⟩ ⟨Esc⟩ ⟨Quit⟩ to exit the "Beam Design Editor" window. It will respond "Save Current Work?"

Click "Yes". Now update the sectional and loading properties for group 1.

Click on the box below the "Data" column next to group 1. The "Beam Design Editor" window appears next, and the floor members appear in blue.

Click "Add/Mod" to modify the default values.

Click "Sectional".

Click "Value". The cursor then appears at "A=1" for you to enter the cross-sectional area. Delete the default value and enter

20.1   This is the moment of inertia of the W24 × 68 beams.

⟨Tab⟩ ⟨Tab⟩ ⟨Tab⟩ ⟨Tab⟩ ⟨Tab⟩ until the cursor is at "I3="

1830   This enters the moment of inertia value of 1830 in^4.

⟨Tab⟩ ⟨Tab⟩ until the cursor is at "S3="

154   This enters the section modulus of 154 in^3.

<Enter><Esc>

Click "Loading".

Click "Distribut".

Click "Value".

⟨Tab⟩ to "Wy:−100". Delete the default value and type

−280   This enters the distributed loading to 280 lb/in. in the negative y direction at the beginning of the beam.

⟨Tab⟩ ⟨Tab⟩ ⟨Tab⟩ to "Wy:−100". Delete the default value and type

−280   This enters the distributed loading to 280 lb/in. in the negative y direction at the end of the beam.

<Enter>

Click on all three members to apply the loads, and yellow arrows pointing down should appear to represent the distributed load.

⟨Esc⟩ ⟨Esc⟩ ⟨Esc⟩ ⟨Quit⟩ to exit the "Beam Design Editor" window. It will respond "Save Current Work?"

Click "Yes". Now all three groups have been updated with their sectional and distributed loading values.

Click on the box under "Material" for group 1. The "Element Material Selection" window appears for Group 1.

Scroll down and select "Steel (ASTM-A36)", as this is a common structural steel.

Click "OK". Repeat this procedure for groups 2 and 3 since they are all made of the same material.

Click on the box under "Material" for group 2. The "Element Material Selection" window appears for Group 2.

Scroll down and select "Steel (ASTM-A36)" again.

Click "OK". Repeat this procedure for group 3.

Click on the box under "Material" for group 3. The "Element Material Selection" window appears for group 3.

Scroll down and select "Steel (ASTM-A36)" once more.

Click "OK". Now the material properties have been updated for all three groups.

Step 6   Enter the Global Data

Click on the "Global" button under "FEA Model" on the right-hand side of the "Model Data Control" window. Under "Load Case Multipliers" and "Pressure" type in 1.

1   Type a 1 under the "Pressure" column, as you have applied mechanical loads to the frame. The program will not run without the load multiplier. The load case multiplier is what all the loads are multiplied by, and usually this value is 1. While still within the "Global" menu,

Click "Output" on the right side of the menu and then

Click on the boxes next to "Displacement data" and "Stress data." This creates nodal displacement output in the .*L* file and element forces and stresses in the .*S* file. You can also click on "Element Data" to have element data written to the .*L* file as well. The element data consists of the nodes assigned to the elements, nodal data written to the .*L* file. The nodal data consists of node numbers, boundary conditions, locations of nodes, and nodal forces.

Click "OK". You are now out of the "Global" menu.

Step 7   Run the Analysis

Click on the "Analysis" button under the "FEA Model" column on the right side of the "Model Data Control" window.

Click "Analyze" in the lower left corner of the "Linear Stress and Vibration-Static Stress" menu. (See green box next to "Analyze.") Wait a short time for the analysis to be performed. A "Geometry Warning" comes up.

Click "No."

An "Algor Analysis" window pops up indicating that the analysis is finished.

Click "OK".

Click "Done" on the lower right of the screen.

Step 8   Review the Results

Click on the "Results" button in the "Model Data Control" window. This takes you to Superview again, and you can view the results. The "Beam Post-Processing" window comes up.

Click "OK" to accept the default "Stress and Displacement Contours".

Select "Displaced" from the main menu of Superview.

Click "Displ on" so that "*Displ on" appears. The asterisk means the displaced mode is active.

Click "With undi" so that "*With undi" appears. This allows the displaced and undisplaced models to be superimposed.

Click "Nodes Inq" on the "Displaced" menu.

Click "Get" to get a node to find the displacement of a node.

Click on node 2 (which Algor calls node 6). Move the cursor to node 2 and click on it to see the x and y displacements appear at the bottom of the screen. You will see "DX=1.806e−001 DY=−2.350e−002"

<Esc><Esc>

Click "Stress di". You are now going to post a beam stress.

Click "LgNd box"

Click "Wind Adj" to adjust the legend box so that it does not cover part of the structure.

<Esc>

Click "Post".

Click "Beam-Trus". You must select this option since you have a frame model.

Click "M3/S3". This shows the stresses at the end of each member by a highlighted box whose color represents a stress value found in the legend box.

<Esc><Esc><Esc>

Click "Done" to exit Superview.

Click on the "x" in the upper right-hand corner of the screen to exit Algor. ■

Figure 6–38 shows the bending stress M_3/S_3 and displaced frame for Example 6.10 superimposed in one figure.

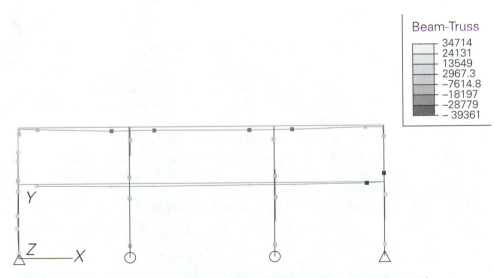

Figure 6–38 M_3/S_3 (psi) and displaced (inches) frame plots superimposed

Example 6.11

We now solve Example 6.6 (Figure 6–39) using the Algor program. The steps to create and analyze the grid follow. Let $I_y = I_z = 16.6 \times 10^{-5}$ m^4, $J = 4.6 \times 10^{-5}$ m^4, $E = 210$ GPa, and $G = 84$ GPa.

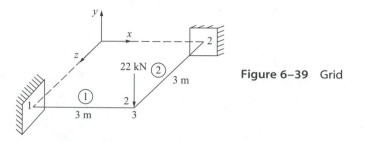

Figure 6–39 Grid

Step 1   Start Superdraw III to Construct the Basic Model

Select the "Start" button of Windows NT/95/98 and then proceed to "Pro-grams:Algor Software:Algor FEA".

Then click "Algor FEA" to start the program.

Or double-click on the "Algor FEA" icon on your desktop. The Superdraw III pro-gram will appear. This is the Algor main graphics user interface. All your work will be done from this program. You will construct, analyze, and review your results from Superdraw III. Each menu option and a short example are available in the Superdraw III Reference Division (accessed through "Docutech", the on-line "Technical User Documentation").

Step 2   Create the Two-Dimensional Model

Select "View". This takes you to the draw options menu.

Click on "Pre-defined Views:XZ Front" to view the drawing in the *x-z* plane. This is the equivalent of looking up the *y* axis from beneath; the *x*-axis is horizontal on the screen, and the *z*-axis is vertical on the screen.

Select the "Add" menu from the top main menu bar.

Click "Line". A "Line" menu appears in the upper right part of the screen. If a check mark appears in the box next to "Single", click on it to unselect. This allows end-to-end construction of lines.

3<Tab>0<Tab>0	This is node 2 of element 2.
<Enter>	Press the ⟨Enter⟩ key to actually enter the data.
3<Tab>0<Tab>3	This is the common node 3.
<Enter>	
0<Tab>0<Tab>3	This is node 1 of element 1.
<Enter>	

Click "Done" on the menu in the upper right corner.

Select "View". This takes you to the draw options menu.

Click "Enclose". This encloses the model. You now see the line drawing of the beam enclosed on the screen. It should look like Figure 6–39 shown in this example.

Step 3   Eliminate Duplicate Lines

Click "Modify" on the main menu.

Select "Clean:Duplicate". This notation means to select "Clean" and then Duplicate" in succession.

Click "Perform Cleaning" on the "Duplicate" menu. You will see the following message at the lower part of the screen:

"2 Kept 0 Deleted. done"

Click "Done" on the "Duplicate" menu located in the upper right corner of the screen.

Step 4   Add in the Boundary Conditions and Nodal Force

Click on the "FEA Add" menu on the main menu bar and highlight "Stress and Vibration Analysis".

Click "Boundary Conditions". The "Boundary Conditions" menu appears. The default boundary condition is "Use @ Symbol for Full". "Full" means full constraint or prevention of translations and rotations when relevant for the element you choose). You may want to left-click and hold on the "Boundary Conditions" label to move it out of the way to see all the nodes of the model.

Move the cursor to node 2 and

Right-click on node 2. The "@" sign appears next to the node, indicating that node 2 is completely fixed. Move the cursor to node 1 and

Right-click on node 1.

Click "Done". It may be difficult to see the "@" sign at this time. Therefore,

Click "Done".

Note: If you have created one or more spurious or misplaced boundary condition "@" signs, go to "Select" on the top menu, select "Box", and create a box around the object (say, around an "@" sign you want to delete). Then click "Modify" on the top menu and click "Delete". The "@" sign should disappear.

Select "View" and "Enclose" again to see the whole beam model.

Click on the "FEA Add" menu again and move to highlight "Stress and Vibration Analysis".

Click "Nodal Forces". You will now add the nodal forces at the free node.

Click "Vector". You will need to create the direction of the force.

Click on "Y Direction". Note this is the default direction for the force, but for illustration we will use it. You will see "DX 0 DY 0.97 DZ 0" appear at the bottom line of the screen.

Click "Done". This takes you back to the "Nodal Forces" menu.

Click "Magnitude". A prompt ("Magnitude") will appear at the bottom of the screen asking you to enter the magnitude of the force.

-22000 Enter the magnitude of the force as negative 22,000 lb.

<Enter> You might have to move the screen to see the message "Click on node to apply current values".

Move the mouse cursor to node 3 and right-click on node 3. The -22000-lb force appears.

Click "Done".

Step 5 Add the Material Properties Using the Model Data Control Window

After the mesh has been created and nodal boundary conditions and forces applied, you can select "Tools" and "Model Data Control" from Superdraw III to complete the data input for analysis. The "Model Data Control" window will appear. Or click on the "Model Data" control button located at the lower part of the screen and highlighted in red.

For the "Analysis type" line, "Linear Static Stress" appears highlighted as the default. You can scroll down for other analysis types as appropriate.

Click on the box below the "Element" column label. A table of group 1 element types appears.

Click on "Beam". You now have created a beam. All lines are beam elements.

Click "OK".

Click on the box below the "Data" column. It will respond with "Please enter the model name first".

Click "OK" and then enter the filename in the "File_name" area that appears.

ex611 The name of the file is *ex611*.

Click "Save". By default, the "Units Definition" menu appears.

Click "English (in)", and then select "Metric MKS (SI)" for the units system.

Click "Add/Mod".

Click "Sectional". This allows you to add the geometric properties of the beam.

Click "Value". This allows you to change the current values for *J1* to 4.6e-5, *I2* to 16.6e-5, and *I3* to 16.6e-5.

⟨Tab⟩ Press ⟨Tab⟩ a total of three times to get to *J1*.

Press ⟨Esc⟩ to remove the existing value of *J1*.

4.6e-5

<Tab> This gets you to *I2*.

16.6e-5

<Enter>

<Tab> This gets you to *I3*.

16.6e-5

<Enter>

<Esc>

<Esc>   You are now in the main menu of the "Beam Design Editor" again.

Click "Quit". You are then asked "Save Current Work?"

Click "Yes". This takes you back to the "Model Data Control" window.

Click on the box below "Material". This takes you to the "Element Material Selection" menu and back to the "Model Data Control" window.

Click "Customer Defined" in the scrolling box below "Select Material".

Click "Edit Properties".

The *Lock* Icon should appear with a green background. If it does not, click on it to "Unlock Properties".

Double-click on the box beside "Modulus of elasticity" and then enter

210e9

<Tab>

Click on the box beside "Poisson's ratio". Press the delete key and then enter

0.3

<Enter>

Click "OK".

Click "OK". A second click takes you out of the "Element Material Selection" menu and back to the "Model Data Control" window.

Step 6   Enter the Global Data

Click on the "Global" button on the right-side menu in the "Model Data Control" window. Under "Load case Multipliers" and under "Pressure", type a 1.

1   Type a 1 under the "Pressure" column as you have applied a mechanical load to the beam. The program will not run without the load multiplier. The load case multiplier is what all the loads are multiplied by, and usually this value is 1. While still within the "Global" menu,

Click on the "Solution" tab to view the solution options. For instance, you can ask for the banded solver to be used instead of the default skyline solver.

Click "Calculate reaction forces".

Click "Output" on the right-side menu and then

Click on the boxes next to "Displacement data" and "Stress data". This creates nodal displacement output in the .*L* file and element forces and stresses in the .*S* file.

Click "Element data" to have element data written to the .*L* file. The element data consists of the nodes assigned to the elements, element loads, and material properties.

Click "Nodal data" to have nodal data written to the .*L* file. The nodal data consists of node numbers, boundary conditions, locations of nodes, and nodal forces.

Click "Apply".

Click "OK".

Step 7   Check the Model

Click "Check" under the "FEA Model" column on the right side of the "Model Data Control" window. This takes you to Superview's main menu, and you can check your model as explained in Section 4.1 and detailed by example in Appendix F. The beam model appears automatically, showing the boundary conditions as red arrows and the load arrow in yellow.

Click "Done". This takes you out of the Superview menu and back to the "Model Data Control" window.

Step 8   Run the Analysis

Click on the "Analysis" button under the "FEA Model" column in the "Model Data Control" window.

Click "Analyze" in the lower left corner of the "Linear Stress and Vibration-Static Stress" menu. (See the green box next to "Analyze".) Wait a short time for the analysis to be performed.

Click "Done" on the lower right of the screen.

Step 9   Review the Results

Click on the "Results" button in the "Model Data Control" window. This takes you to the Superview program again, and you can view the results.

Click "Draw" on the lower left of the screen.

Click "View" on the upper left menu.

Click "Isometric".

Click "Enclose".

Click ⟨Esc⟩ to return to the Superview main menu.

Select "Displaced" from the main menu of Superview.

Click "Displ on" so that "*Displ on" appears. The asterisk means the displaced mode is active.

Click "With Undi" so that "*With Undi" appears.

Click "Calc scal" to show exaggerated displacements.

Click "Nodes inq" on the "Displaced" menu.

Click "Get" to get a node.

Move the cursor to the free node and click on it to view its respective translation data.

Click ⟨Esc⟩ to return to the main menu of Superview.

Click "Options".

Click "General".

Click "Node num" to display the node numbers as Algor labeled them.

Click "Ele num" to display the element numbers used by Algor.

Click ⟨Esc⟩ twice to return to the Superview main menu.

Click "Stress-di".

Click "*Post".

Click "disp Vec" to display the linear displacements of nodes. If the displacements are not displayed, click "Vector" and then click "Negate". Click "Esc" and then click "Magnitude".

Click ⟨Esc⟩ three times to return to the main menu.

Q   Press Q to quit the program once completed.   ∎

The nodal displacements taken from the *ex611.L* file and shown in Table 6–4 compare closely with Eq. (6.4.47) from the longhand solution. In the following data, node 3 equates to Figure 6–21 node 2.

Table 6–4   Displacements/rotations of nodes for Example 6.11

Displacements/Rotations(degrees) of nodes

NODE number	X- translation	Y- translation	Z- translation	X- rotation	Y- rotation	Z- rotation
1	0.0000E+00	0.0000E+00	0.0000E+00	0.0000E+00	0.0000E+00	0.0000E+00
2	0.0000E+00	0.0000E+00	0.0000E+00	0.0000E+00	0.0000E+00	0.0000E+00
3	0.0000E+00	-2.6481E-03	0.0000E+00	7.4031E-02	0.0000E+00	-7.4031E-02
4	0.0000E+00	0.0000E+00	0.0000E+00	0.0000E+00	0.0000E+00	0.0000E+00
5	0.0000E+00	0.0000E+00	0.0000E+00	0.0000E+00	0.0000E+00	0.0000E+00
6	0.0000E+00	0.0000E+00	0.0000E+00	0.0000E+00	0.0000E+00	0.0000E+00
7	0.0000E+00	0.0000E+00	0.0000E+00	0.0000E+00	0.0000E+00	0.0000E+00
8	0.0000E+00	0.0000E+00	0.0000E+00	0.0000E+00	0.0000E+00	0.0000E+00
9	0.0000E+00	0.0000E+00	0.0000E+00	0.0000E+00	0.0000E+00	0.0000E+00 ∎

Example 6.12

Determine the nodal displacements and rotations for the rigid space frame shown in Figure 6–40. Let $E = 30 \times 10^3$ ksi, $G = 10 \times 10^3$ ksi, $J = 50$ in^4, $I_y = 100$ in^4, $I_z = 1000$ in^4, and $A = 10$ in^2 for all members.

Step 1   Start Superdraw III

Click on the "Algor FEA" icon to start Superdraw III.

Step 2   Create the Model

The model will look like Figure 6–40 when we complete Step 2.

Click "Add" on the main menu bar.

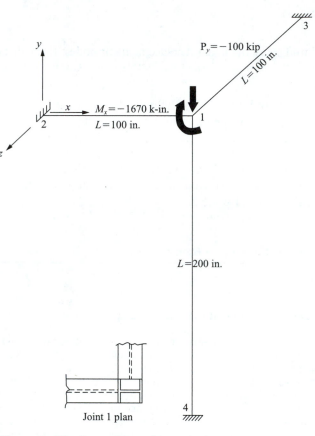

Figure 6–40 Space frame for analysis

Click "Line".
(0, 0, 0)
<Enter>
(100, 0, 0)
<Enter>
(100, -200, 0)
<Enter>
Click "Done".
Click "Add" on the main menu bar.
Click "Line".
(100, 0, 0)
<Enter>

```
(100, 0, -100)
<Enter>
```
Click "Done".
Select "View".
Select "Pre-defined".
Select "Isometric".
Select "View".
Select "Enclose". The model should look like Figure 6–40.

Step 3   Add in Boundary Conditions

Select "FEA Add:Stress and Vibration Analysis".
Click "Boundary Conditions".
Click "Change Values".
Make sure all the boxes have check marks in them.
Click "Done".
Right-click on the three outer ends of the three lines.
Click "Done".

Step 4   Add Nodal Forces and Moments

Select "FEA Add:Stress and Vibration Analysis".
Select "Nodal Forces".
Click "Vector".
Click "Y Direction".
Click "Done".
Click "Magnitude".
```
-100
<Enter>
```
Click "Done".
Right-click on the intersection of the three lines.
Click "Done".
Select "FEA Add:Stress and Vibration Analysis".
Select "Nodal Moments".
Click "Vector".
Click "X Direction".
Click "Done".
Click "Magnitude".
```
-1670
```

\<Enter\>

Right-click on the intersection of the three lines.

Click "Done".

Step 5 Go to "Model Data Control" Window to Enter Material
Properties

Click on the "Model Data" control button.

Click "Element".

Click "Beam".

Click "OK".

Select the box below "Data".

You will be asked to save the model. Save the model under an appropriate name and continue.

The "Units" window then pops up. Click "OK" to use the default units.

This brings you to the "Beam Design Editor" window.

Your model should look like the figure below.

Click "Add/mod".

Click "Sectional".

Click "Value".

\<Esc\>

Type 10.

\<Tab\>

\<Tab\>

\<Tab\>

\<Esc\>

Type 50.

\<Tab\>

\<Esc\>

Type 100.

\<Tab\>

\<Esc\>

Type 1000.

\<Enter\>

\<Esc\>

\<Esc\>

Click "Quit".

You will be asked to save. Click "Yes".

Click "Material".
Select "Customer Defined".
Click "Edit Properties".
Set the modulus of elasticity to 30e3.
Set Poisson's ratio to 0.30.
Click "OK".
Click "OK".
Select "Global".
Under "Load Case Multipliers" and under "Pressure", type a "1".
Click on the "Output" tab.
Click on the boxes next to "Displacement Data" and "Stress Data".
Click "OK".

Step 6 Run the Analysis

Click on the "Analysis" button in the "Model Data Control" window.
Click "Analyze". Wait for a message telling you the analysis is complete.
Click "OK".
Click "Done".

Step 7 Review the Results

Click on the "Results" button in the "Model Data Control" window.
Click "OK".
Click "Displaced".
Click "Displ on".
Click "With undi".
Click "Calc scal" to show exaggerated displacements.
Click "Nodes inq".
Click "Get".
Left-click on the center node.
Displacements should be: "DX=6.27E−004 DY=−4.27E−002 DZ=−1.085E−004"
<Esc>
<Esc>
Click "Stress-di".
Click "Post".

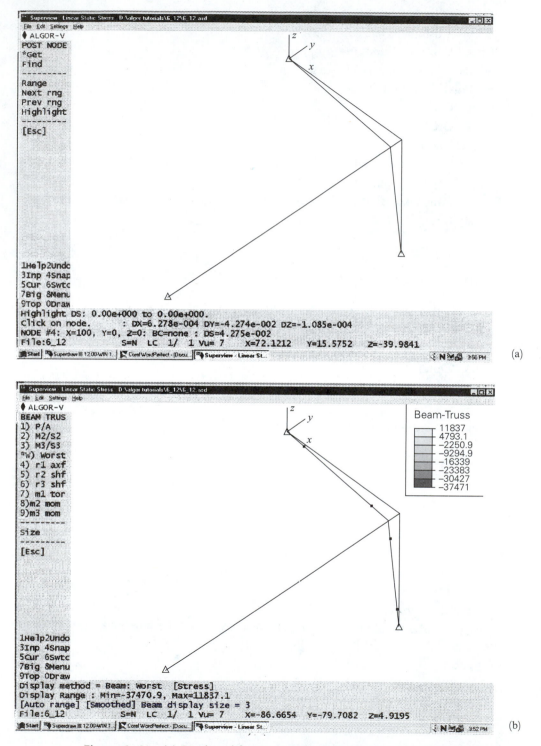

(a)

(b)

Figure 6–41 (a) Displaced frame and (b) worst stresses of Example 6.12

Click "Beam-trus".

The results should look like the plots in Figure 6–41(a) and (b).   ■

The nodal displacements/rotations shown in Table 6–5 are exactly those found in Reference [9].

The beam element forces and moments are listed in Table 6–6. To improve understanding of the output in Table 6–6, free-body diagrams of all elements are shown in Figure 6–42. (The very small forces and moments are not shown on the diagrams.) Equilibrium of each element and of the whole frame can be verified. For instance, at node 1, equilibrium of forces in the y direction and of moments about the x axis yields

$$\sum F_Y = 0 : (-100 + 64.11 + 7.69 + 28.2) \text{ kip} = 0$$

$$\sum M_X = 0 : (-1670 + 1623 + 4.105 + 42.65) \text{ k-ft} = 0$$

Table 6–5 Free node displacements/rotations for Example 6.12

Translations (inches)	Rotations (degrees)
x = 6.2777E-4	x = -4.077E-02
y = -4.2743E-02	y = 3.1265E-04
z = -1.0854E-04	z = -2.4501E-02

Table 6–6 Beam element forces and moments for Example 6.12

1**** BEAM ELEMENT FORCES AND MOMENTS

ELEMENT NO.	CASE (MODE)	AXIAL FORCE R1	SHEAR FORCE R2	SHEAR FORCE R3	TORSION MOMENT M1	BENDING MOMENT M2	BENDING MOMENT M3
1	1	−1.883E+00	7.690E+00	−5.915E-03	4.105E+00	1.320E-01	5.128E+02
		1.883E+00	−7.690E+00	5.915E-03	−4.105E+00	4.594E-01	2.562E+02
2	1	6.411E+01	−1.896E+00	−3.197E-01	−1.574E-02	2.130E+01	−1.255E+02
		−6.411E+01	1.896E+00	3.197E-01	1.574E-02	4.265E+01	−2.538E+02
3	1	3.256E-01	2.820E+01	1.278E-02	2.467E+00	−8.026E-01	1.196E+03
		−3.256E-01	−2.820E+01	−1.278E-02	−2.467E+00	−4.752E-01	1.623E+03

■

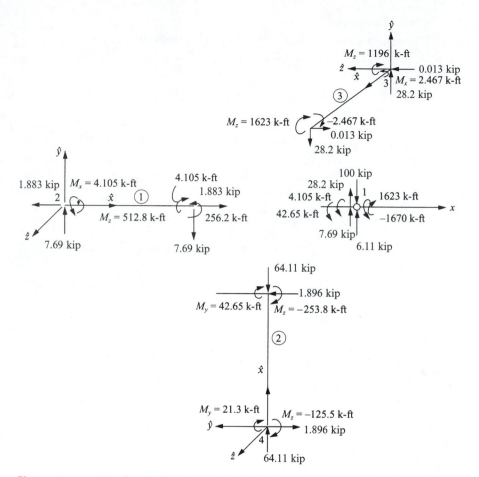

Figure 6–42 Free-body diagrams of elements of frame for Example 6.12 ∎

▲ References

[1] Kassimali, A., *Structural Analysis*, 2nd ed., Brooks/Cole Publishers, Pacific Grove, CA, 1999.

[2] Budynas, R. G., *Advanced Strength and Applied Stress Analysis*, 2nd ed., McGraw-Hill, New York, 1999.

[3] Allen, H. G., and Bulson, P. S., *Background to Buckling*, McGraw-Hill, London, 1980.

[4] Roark, R. J., and Young, W. C., *Formulas for Stress and Strain*, 6th ed., McGraw-Hill, New York, 1989.

[5] Gere, J. M., *Mechanics of Materials*, 5th ed., Brooks/Cole Publishers, Pacific Grove, CA, 2001.

[6] Parakh, Z. K., *Finite Element Analysis of Bus Frames Under Simulated Crash Loadings*, M.S. Thesis, Rose-Hulman Institute of Technology, Terre Haute, Indiana, May 1989.

[7] Martin, H. C., *Introduction to Matrix Methods of Structural Analysis*, McGraw-Hill, New York, 1966.

[8] *Superdraw Reference Division*, Docutech On-line Documentation, Algor, Inc., Pittsburgh, PA 15238.

[9] Kardestuncer, H., *Elementary Matrix Analysis of Structures*, McGraw-Hill, New York, 1974.

[10] *Linear Stress and Dynamics Reference Division*, Docutech On-line Documentation, Algor, Inc., Pittsburgh, PA 15238.

[11] Juvinall, R. C., and Marshek, K. M., *Fundamentals of Machine Component Design*, 3rd ed., p. 200, Wiley, 2000.

▲ Problems

Solve all problems using the finite element stiffness method.

6.1 For the rigid frame shown in Figure P6–1, determine (1) the displacement components and the rotation at node 2, (2) the support reactions, and (3) the forces in each element. Then check equilibrium at node 2. Let $E = 30 \times 10^6$ psi, $A = 10$ in^2, and $I = 500$ in^4 for both elements.

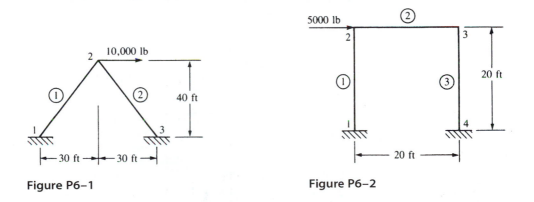

Figure P6–1   Figure P6–2

6.2 For the rigid frame shown in Figure P6–2, determine (1) the nodal displacement components and rotations, (2) the support reactions, and (3) the forces in each element. Let $E = 30 \times 10^6$ psi, $A = 10$ in^2, and $I = 200$ in^4 for all elements.

6.3 For the rigid stairway frame shown in Figure P6–3, determine (1) the displacements at node 2, (2) the support reactions, and (3) the local nodal forces acting on each element. Draw the bending moment diagram for the whole frame. Remember that the angle between elements 1 and 2 is preserved as deformation takes place; similarly for the angle between elements 2 and 3. Furthermore, owing to symmetry, $d_{2x} = -d_{3x}$, $d_{2y} = d_{3y}$, and $\phi_2 = -\phi_3$. What size A36 steel channel section would be needed to keep the allowable bending stress less than two-thirds of the yield stress? (For A36 steel, the yield stress is 36,000 psi.)

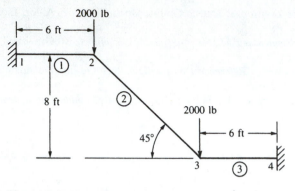

Figure P6–3

6.4 For the rigid frame shown in Figure P6–4, determine (1) the nodal displacements and rotation at node 4, (2) the reactions, and (3) the forces in each element. Then check equilibrium at node 4. Finally, draw the shear force and bending moment diagrams for each element. Let $E = 30 \times 10^3$ ksi, $A = 8$ in^2, and $I = 800$ in^4 for all elements.

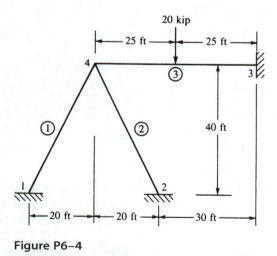

Figure P6–4

6.5–6.15 For the rigid frames shown in Figures P6–5–P6–15, determine the displacements and rotations of the nodes, the element forces, and the reactions. The values of E, A, and I to be used are listed next to each figure.

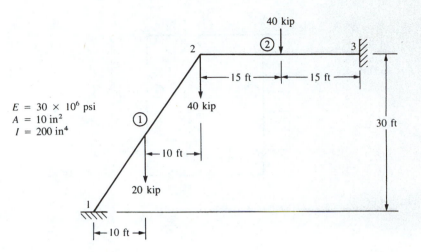

Figure P6–5

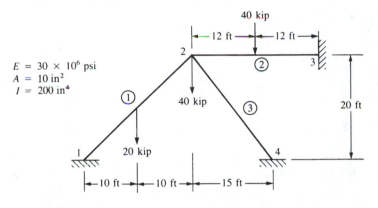

Figure P6–6

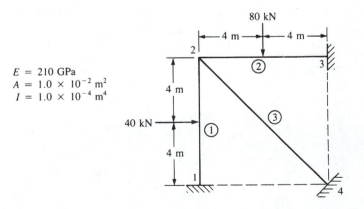

Figure P6–7

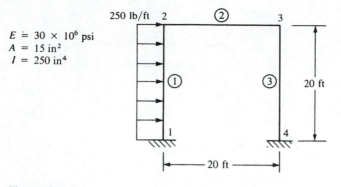

$E \doteq 30 \times 10^6$ psi
$A = 15$ in^2
$I = 250$ in^4

250 lb/ft

20 ft

20 ft

Figure P6-8

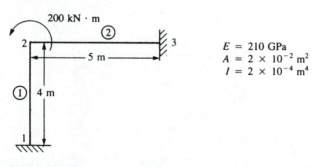

200 kN · m

5 m

4 m

$E = 210$ GPa
$A = 2 \times 10^{-2}$ m^2
$I = 2 \times 10^{-4}$ m^4

Figure P6-9

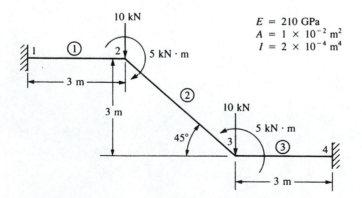

10 kN

5 kN · m

3 m

3 m

45°

10 kN

5 kN · m

3 m

$E = 210$ GPa
$A = 1 \times 10^{-2}$ m^2
$I = 2 \times 10^{-4}$ m^4

Figure P6-10

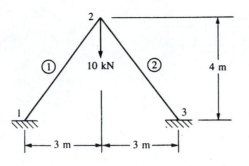

Figure P6–11

$E = 70$ GPa
$A = 3 \times 10^{-2}$ m^2
$I = 3 \times 10^{-4}$ m^4

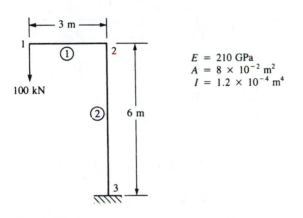

Figure P6–12

$E = 210$ GPa
$A = 8 \times 10^{-2}$ m^2
$I = 1.2 \times 10^{-4}$ m^4

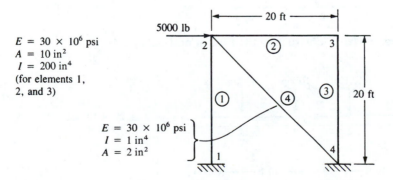

$E = 30 \times 10^6$ psi
$A = 10$ in^2
$I = 200$ in^4
(for elements 1, 2, and 3)

$E = 30 \times 10^6$ psi
$I = 1$ in^4
$A = 2$ in^2

Figure P6–13

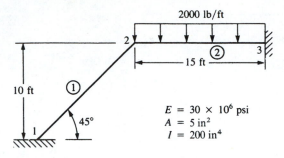

Figure P6–14

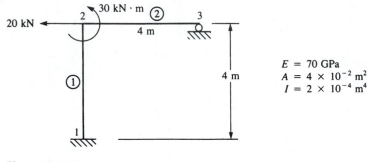

Figure P6–15

6.16–6.18 Solve the structures in Figures P6–16–P6–18 by using substructuring.

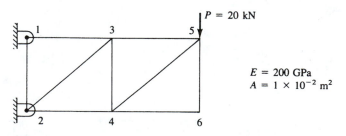

Figure P6–16 (Substructure the truss at nodes 3 and 4)

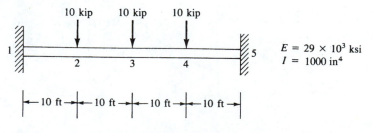

Figure P6–17 (Substructure the beam at node 3)

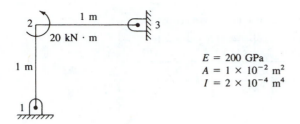

Figure P6–18 (Substructure the frame at node 2)

Solve Problems 6.19–6.39 by using the Algor computer program.

6.19 For the rigid frame shown in Figure P6–19, determine (1) the nodal displacement components and (2) the support reactions. (3) Draw the shear force and bending moment diagrams. For all elements, let $E = 30 \times 10^6$ psi, $I = 200$ in^4, and $A = 10$ in^2.

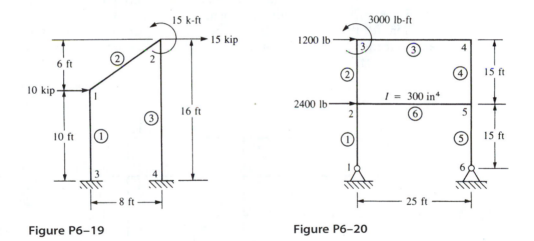

Figure P6–19   **Figure P6–20**

6.20 For the rigid frame shown in Figure P6–20, determine (1) the nodal displacement components and (2) the support reactions. (3) Draw the shear force and bending moment diagrams. Let $E = 30 \times 10^6$ psi, $I = 200$ in^4, and $A = 10$ in^2 for all elements, except as noted in the figure.

6.21 For the slant-legged rigid frame shown in Figure P6–21, size the structure for minimum weight based on a maximum bending stress of 20 ksi in the horizontal beam elements and a maximum compressive stress (due to bending and direct axial load) of 15 ksi in the slant-legged elements. Use the same element size for the two slant-legged

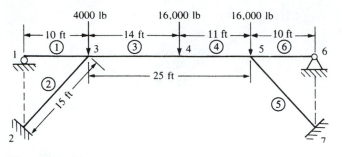

Figure P6–21

elements and the same element size for the two 10-foot sections of the horizontal element. Assume A36 steel is used.

6.22 For the rigid building frame shown in Figure P6–22, determine the forces in each element and calculate the bending stresses. Assume all the vertical elements have $A = 10$ in^2 and $I = 100$ in^4 and all horizontal elements have $A = 15$ in^2 and $I = 150$ in^4. Let $E = 29 \times 10^6$ psi for all elements. Let $c = 5$ in. for the vertical elements and $c = 6$ in. for the horizontal elements, where c denotes the distance from the neutral axis to the top or bottom of the beam cross section, as used in the bending stress formula $\sigma = (Mc/I)$.

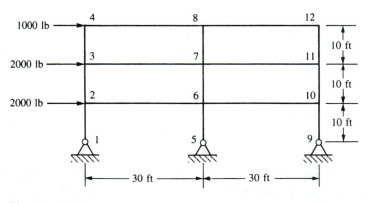

Figure P6–22

6.23–6.38 For the rigid frames or beams shown in Figures P6–23–P6–38, determine the displacements and rotations at the nodes, the element forces, and the reactions.

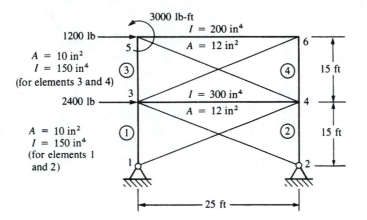

For cross members:
$I = 1.0 \text{ in}^4$
$A = 2.0 \text{ in}^2$

$E = 30 \times 10^6$ psi
(for all members)

Figure P6–23

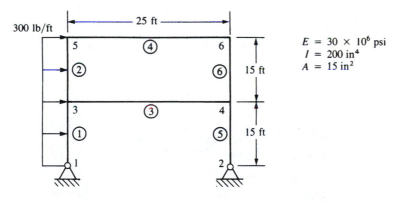

$E = 30 \times 10^6$ psi
$I = 200 \text{ in}^4$
$A = 15 \text{ in}^2$

Figure P6–24

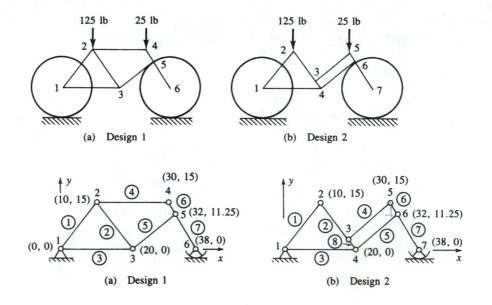

(a) Design 1 (b) Design 2

Case 1
$E = 30 \times 10^6$ psi
Case 2
$E = 10 \times 10^6$ psi

$A_1 = 0.1$ in^2
$A_2 = A_3 = A_4 = A_5 = 0.15$ in^2
$A_6 = A_7 = A_8 = 0.3$ in^2
$I_1 = 0.01$ in^4
$I_2 = I_3 = I_4 = I_5 = 0.02$ in^4
$I_6 = I_7 = I_8 = 0.1$ in^4

Figure P6–25 Two bicycle frame models (coordinates shown in inches)

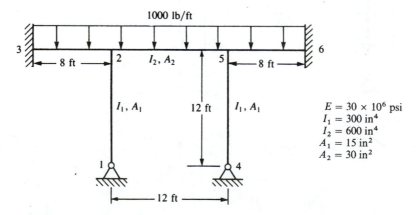

$E = 30 \times 10^6$ psi
$I_1 = 300$ in^4
$I_2 = 600$ in^4
$A_1 = 15$ in^2
$A_2 = 30$ in^2

Figure P6–26

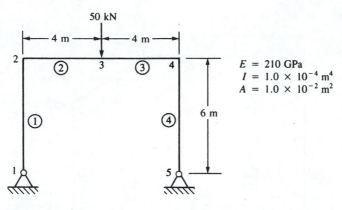

Figure P6–27

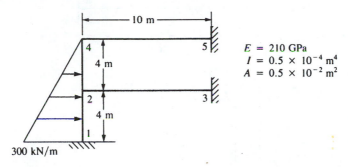

Figure P6–28

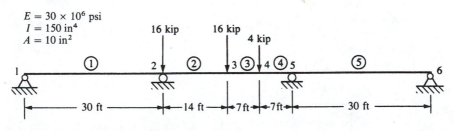

Figure P6–29

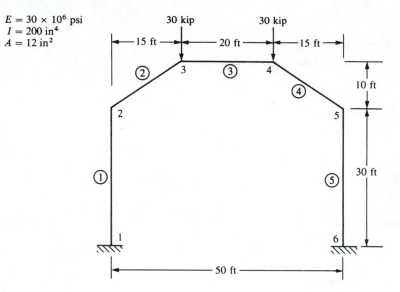

Figure P6–30

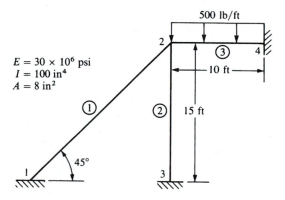

Figure P6–31

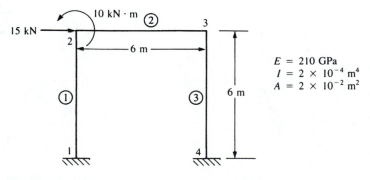

Figure P6–32

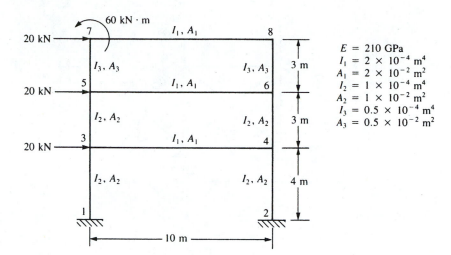

Figure P6–33

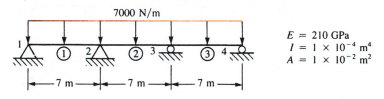

Figure P6–34

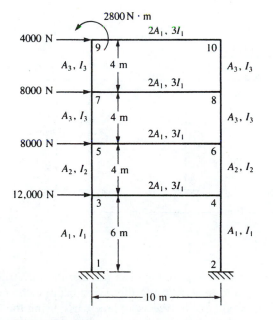

Figure P6–35

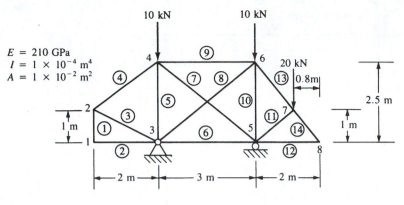

Figure P6–36

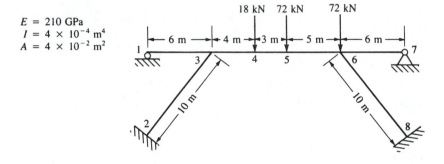

Figure P6–37

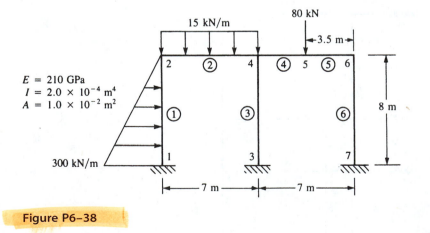

6.39 Consider the plane structure shown in Figure P6–39. First assume the structure to be a plane frame with rigid joints, and analyze using the Algor beam element. Then assume the structure to be pin-jointed and analyze as a plane truss, using the Algor truss element. If the structure is actually a truss, is it appropriate to model it as a rigid

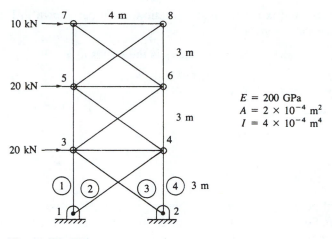

Figure P6–39

frame? How can you model the truss using the frame (or beam) element? In other words, what idealization could you make in your model to use the beam element to approximate a truss?

6.40 For the tapered beam shown in Figure P6–40, determine the maximum deflection using one, two, four, and eight elements. Calculate the moment of inertia at the mid-length station for each element. Let $E = 30 \times 10^6$ psi, $I_0 = 100$ in^4, and $L = 100$ in. Run cases where $n = 1, 3$, and 7. Use the Algor beam element. The analytical solution for $n = 7$ is given by Reference [7]:

$$v_1 = \frac{PL^3}{49EI_0}(1/7 \ln 8 + 2.5) = \frac{1}{17.55}\frac{PL^3}{EI_0}$$

$$\theta_1 = \frac{PL^2}{49EI_0}(\ln 8 - 7) = -\frac{1}{9.95}\frac{PL^2}{EI_0}$$

$$I(x) = I_0\left(1 + n\frac{x}{L}\right)$$

where n = arbitrary numerical factor and I_0 = moment of inertia of section at $x = 0$.

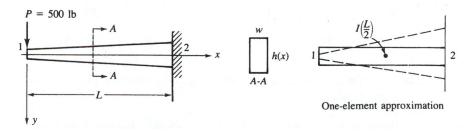

Figure P6–40 Tapered cantilever beam

6.41 Derive the stiffness matrix for the nonprismatic torsion bar shown in Figure P6–41. The radius of the shaft is given by

$$r = r_0 + (x/L)r_0, \text{ where } r_0 \text{ is the radius at } x = 0.$$

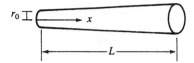

Figure P6–41

6.42 Derive the total potential energy for the prismatic circular cross-section torsion bar shown in Figure P6–42. Also determine the equivalent nodal torques for the bar subjected to uniform torque per unit length (lb-in./in.). Let G be the shear modulus and J be the polar moment of inertia of the bar.

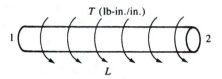

Figure P6–42

6.43 For the grid shown in Figure P6–43, determine the nodal displacements and the local element forces. Let $E = 30 \times 10^6$ psi, $G = 12 \times 10^6$ psi, $I = 200$ in^4, and $J = 100$ in^4 for both elements.

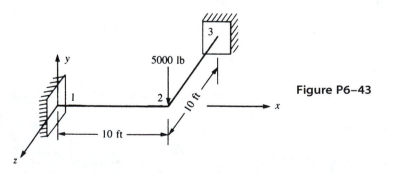

Figure P6–43

6.44 Resolve Problem 6–43 with an additional nodal moment of 1000 k-in. applied about the x axis at node 2.

6.45–6.46 For the grids shown in Figures P6–45 and P6–46, determine the nodal displacements and the local element forces. Let $E = 210$ GPa, $G = 84$ GPa, $I = 2 \times 10^{-4}$ m^4, $J = 1 \times 10^{-4}$ m^4, and $A = 1 \times 10^{-2}$ m^2.

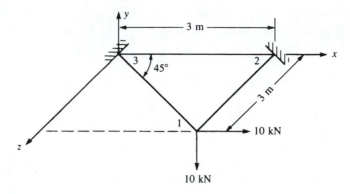

Figure P6–45

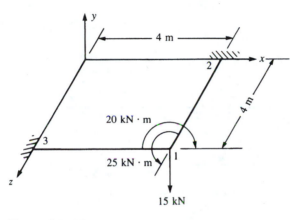

Figure P6–46

6.47–6.52 Solve the grid structures shown in Figures P6–47–P6–52 using the Algor beam element. For grids P6–47–P6–49, let $E = 30 \times 10^6$ psi, $G = 12 \times 10^6$ psi, $I = 200$ in^4, and $J = 100$ in^4, except as noted in the figures. In Figure P6–49, let the cross elements

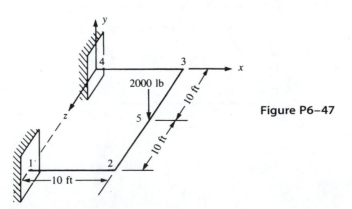

Figure P6–47

have $I = 50$ in^4 and $J = 20$ in^4, with dimensions and loads as in Figure P6–48. For grids P6–50–P6–52, let $E = 210$ GPa, $G = 84$ GPa, $I = 2 \times 10^{-4}$ m^4, $J = 1 \times 10^{-4}$ m^4, and $A = 1 \times 10^{-2}$ m^2.

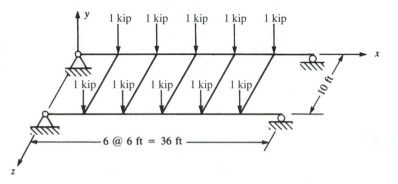

Figure P6–48

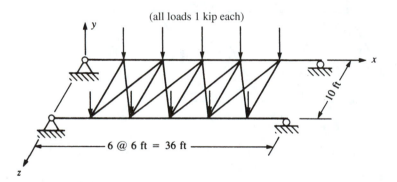

Figure P6–49

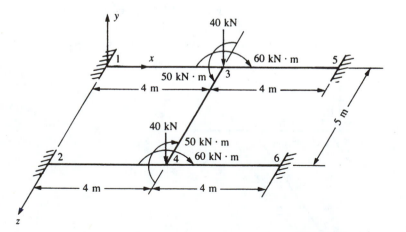

Figure P6–50

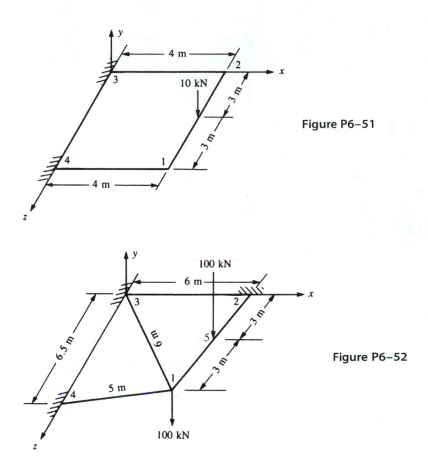

Figure P6–51

Figure P6–52

6.53–6.54 Determine the displacements and reactions for the space frames shown in Figures P6–53 and P6–54. Let $I_x = 100$ in^4, $I_y = 200$ in^4, $I_z = 1000$ in^4, $E = 30{,}000$ ksi, $G = 10{,}000$ ksi, and $A = 100$ in^2 for both frames.

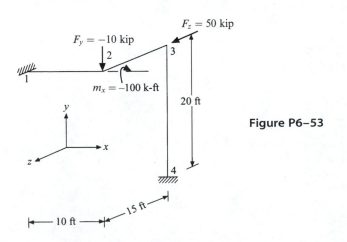

Figure P6–53

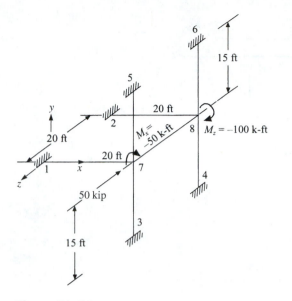

Figure P6–54

Use the Algor computer program to assist in the design problems in Problems 6.55–6.60.

6.55 Design a jib crane as shown in Figure P6–55 that will support a downward load of 6000 lb. Choose a common structural steel shape for all members. Use allowable stresses of $0.66S_y$ (S_y is the yield strength of the material) in bending, and $0.60S_y$ in

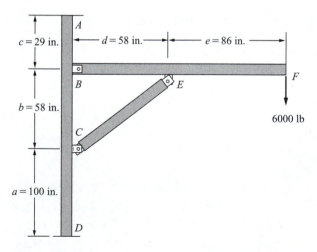

Figure P6–55

tension on gross areas. The maximum deflection should not exceed 1/360 of the length of the horizontal beam. Buckling should be checked using Euler's or Johnson's method as applicable.

6.56 Design the support members, *AB* and *CD*, for the platform lift shown in Figure P6–56. Select a mild steel and choose suitable cross-sectional shapes with no more than a 4:1 ratio of moments of inertia between the two principal directions of the cross section. You may choose two different cross sections to make up each arm to reduce weight. The actual structure has four support arms, but the loads shown are for one side of the platform with the two arms shown. The loads shown are under operating conditions. Use a factor of safety of 2 for human safety. In developing the finite element model, remove the platform and replace it with statically equivalent loads at the joints at *B* and *D*. Use truss elements or beam elements with low bending stiffness to model the arms from *B* to *D*, the intermediate connection, *E* to *F*, and the hydraulic actuator. The allowable stresses are $0.66S_y$ in bending and $0.60S_y$ in tension. Check buckling using either Euler's method or Johnson's method as appropriate. Also check maximum deflections. Any deflection greater than 1/360 of the length of member *AB* is considered too large.

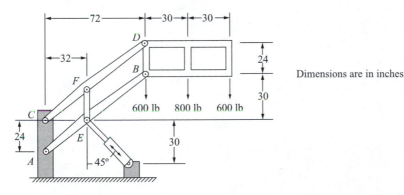

Dimensions are in inches

Figure P6–56

6.57 A two-story building frame is to be designed as shown in Figure P6–57. The members are all to be I-beams with rigid connections. We would like the floor joists beams to have a 15-in. depth and the columns to have a 10 in. width. The material is to be A36 structural steel. Two horizontal loads and vertical loads are shown. Select members such that the allowable bending in the beams is 24,000 psi. Check buckling in the columns using Euler's or Johnson's method as appropriate. The allowable deflection in the beams should not exceed 1/360 of each beam span. The overall sway of the frame should not exceed 0.5 in.

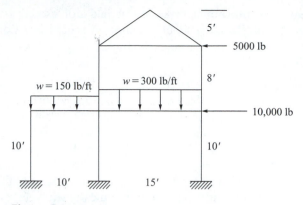

Figure P6–57

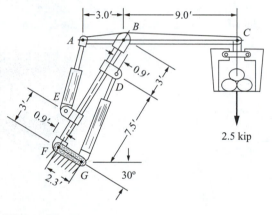

Figure P6–58

6.58 A pulpwood loader as shown in Figure P6–58 is to be designed to lift 2.5 kip. Select a steel and determine a suitable tubular cross section for the main upright member BF that has attachments for the hydraulic cylinder actuators AE and DG. Select a steel and determine a suitable box section for the horizontal load arm AC. The horizontal load arm may have two different cross sections AB and BC to reduce weight. The finite element model should use beam elements for all members except the hydraulic cylinders, which should be truss elements. The pinned joint at B between the upright and the horizontal beam is best modeled with end release of the end node of the top element on the upright member. The allowable bending stress is $0.66S_y$ in members AB and BC. Member BF should be checked for buckling. The allowable deflection at C should be less than $1/360$ of the length of BC. As a bonus, the client would like you to select the size of the hydraulic cylinders AE and DG.

6.59 A piston ring (with a split as shown in Figure P6–59) is to be expanded by a tool to facilitate its installation. The ring is sufficiently thin (0.2 in. depth) to justify using conventional straight-beam bending formulas. The ring requires a displacement of 0.1 in. at its separation for installation. Determine the force required to produce this separation. In addition, determine the largest stress in the ring. Let $E = 18 \times 10^6$ psi, $G = 7 \times 10^6$ psi, cross-sectional area $A = 0.06$ in.^{2}, and principal moment of inertia $I = 4.5 \times 10^{-4}$ in.^{4}. The inner radius is 1.85 in., and the outer radius is 2.15 in. Use models with 4, 6, 8, 10, and 20 elements in a symmetric model until convergence to the same results occurs. Plot the displacement versus the number of elements for a constant force F predicted by the conventional beam theory equation of Reference [11].

$$\delta = \frac{3\pi FR^3}{EI} + \frac{\pi FR}{EA} + \frac{6\pi FR}{5GA} \qquad \text{where } R = 2.0 \text{ in. and } \delta = 0.1 \text{ in.}$$

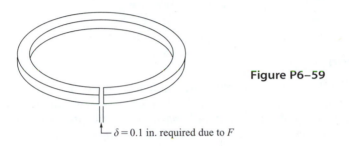

Figure P6–59

$\delta = 0.1$ in. required due to F

6.60 A small hydraulic floor crane as shown in Figure P6–60 carries a 5000-lb load. Determine the size of the beam and column needed. Select either a standard box section or a wide-flange section. Assume a rigid connection between the beam and column. The column is rigidly connected to the floor. The allowable bending stress in the beam is $0.60S_y$. The allowable deflection is $1/360$ of the beam length. Check the column for buckling.

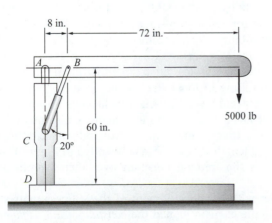

Figure P6–60

7

Development of the Plane Stress and Plane Strain Stiffness Equations

Introduction

In Chapters 2–6, we considered only line elements. Two or more line elements are connected only at common nodes, forming framed or articulated structures such as trusses, frames, and grids. Line elements have geometric properties such as cross-sectional area and moment of inertia associated with their cross sections. However, only one local coordinate $\hat{x}$ along the length of the element is required to describe a position along the element (hence, they are called *line elements* or one-dimensional elements). Nodal compatibility is then enforced during the formulation of the nodal equilibrium equations for a line element.

This chapter considers the two-dimensional finite element. Two-dimensional (planar) elements are defined by three or more nodes in a two-dimensional plane (that is, *x-y*). The elements are connected at common nodes and/or along common edges to form continuous structures such as those shown in Figures 1–3, 1–4, 1–6, and 7–6(b). Nodal displacement compatibility is then enforced during the formulation of the nodal equilibrium equations for two-dimensional elements. If proper displacement functions are chosen, compatibility along common edges is also obtained. The two-dimensional element is extremely important for (1) plane stress analysis, which includes problems such as plates with holes, fillets, or other changes in geometry that are loaded in their plane resulting in local stress concentrations, such as illustrated in Figure 7–1; and (2) plane strain analysis, which includes problems such as a long underground box culvert subjected to a uniform load acting constantly over its length, as illustrated in Figure 1–3, a long, cylindrical control rod subjected to a load that remains constant over the rod length (or depth), as illustrated in Figure 1–4, and dams and pipes subjected to loads that remain constant over their lengths as shown in Figure 7–2.

We begin this chapter with the development of the stiffness matrix for a basic two-dimensional or plane finite element, called the *constant-strain triangular element*. We consider the constant-strain triangle (CST) stiffness matrix because its derivation is the simplest among the available two-dimensional elements. The element is called a CST because it has a constant strain throughout it.

We will derive the CST stiffness matrix by using the principle of minimum potential energy because the energy formulation is the most feasible for the development of the equations for both two- and three-dimensional finite elements.

We will then present a simple, thin-plate plane stress example problem to illustrate the assemblage of the plane element stiffness matrices using the direct stiffness method as presented in Chapter 2. We will present the total solution, including the stresses within the plate.

▲ 7.1 Basic Concepts of Plane Stress and Plane Strain ▲

In this section, we will describe the concepts of plane stress and plane strain. These concepts are important because the developments in this chapter are directly applicable only to systems assumed to behave in a plane stress or plane strain manner. Therefore, we will now describe these concepts in detail.

Plane Stress

Plane stress *is defined to be a state of stress in which the normal stress and the shear stresses directed perpendicular to the plane are assumed to be zero.* For instance, in Figures 7–1(a) and 7–1(b), the plates in the *x-y* plane shown subjected to surface tractions T (pressure acting on the surface edge or face of a member in units of force/area) in the plane are under a state of plane stress; that is, the normal stress σ_z and the shear stresses τ_{xz} and τ_{yz} are assumed to be zero. Generally, members that are thin (those with a small z dimension compared to the in-plane x and y dimensions) and whose loads act only in the *x-y* plane can be considered to be under plane stress.

Plane Strain

Plane strain *is defined to be a state of strain in which the strain normal to the x-y plane ε_z and the shear strains γ_{xz} and γ_{yz} are assumed to be zero.* The assumptions of plane strain are realistic for long bodies (say, in the z direction) with constant cross-sectional area subjected to loads that act only in the x and/or y directions and do not vary in

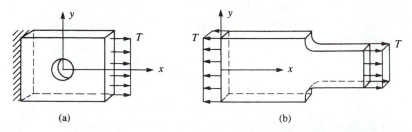

Figure 7–1 Plane stress problems: (a) plate with hole; (b) plate with fillet

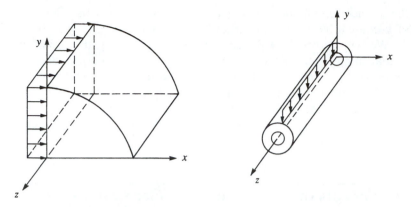

Figure 7–2 Plane strain problems: (a) dam subjected to horizontal loading; (b) pipe subjected to a vertical load

the z direction. Some plane strain examples are shown in Figure 7–2 [and in Figures 1–3 (a long underground box culvert) and 1–4 (a hydraulic cylinder rod end)]. In these examples, only a unit thickness (1 in. or 1 ft) of the structure is considered because each unit thickness behaves identically (except near the ends). The finite element models of the structures in Figure 7–2 consist of appropriately discretized cross sections in the x-y plane with the loads acting over unit thicknesses in the x and/or y directions only.

Two-Dimensional State of Stress and Strain

The concept of a two-dimensional state of stress and strain and the stress/strain relationships for plane stress and plane strain are necessary to understand fully the development and applicability of the stiffness matrix for the plane stress/plane strain triangular element. Therefore, we briefly outline the essential concepts of two-dimensional stress and strain (see References [1] and [2] and Appendix C for more details on this subject).

First, we illustrate the two-dimensional state of stress using Figure 7–3. The infinitesimal element with sides dx and dy has normal stresses σ_x and σ_y acting in the x and y directions (here on the vertical and horizontal faces), respectively. The shear stress τ_{xy} acts on the x edge (vertical face) in the y direction. The shear stress τ_{yx} acts on the y edge (horizontal face) in the x direction. Moment equilibrium of the element results in τ_{xy} being equal in magnitude to τ_{yx}. See Appendix C.1 for proof of this equality. Hence, three independent stresses exist and are represented by the vector column matrix

$$\{\sigma\} = \begin{Bmatrix} \sigma_x \\ \sigma_y \\ \tau_{xy} \end{Bmatrix} \tag{7.1.1}$$

The element equilibrium equations are derived in Appendix C.1.

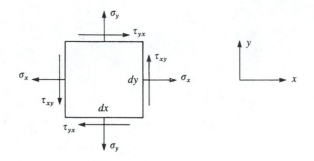

Figure 7–3 Two-dimensional state of stress

The stresses given by Eq. (7.1.1) will be expressed in terms of the nodal displacement degrees of freedom. Hence, once the nodal displacements are determined, these stresses can be evaluated directly.

Recall from strength of materials [2] that the **principal stresses**, which are the maximum and minimum normal stresses in the two-dimensional plane, can be obtained from the following expressions:

$$\sigma_1 = \frac{\sigma_x + \sigma_y}{2} + \sqrt{\left(\frac{\sigma_x - \sigma_y}{2}\right)^2 + \tau_{xy}^2} = \sigma_{\max}$$

$$\sigma_2 = \frac{\sigma_x + \sigma_y}{2} - \sqrt{\left(\frac{\sigma_x - \sigma_y}{2}\right)^2 + \tau_{xy}^2} = \sigma_{\min}$$

(7.1.2)

Also, the **principal angle** θ_p, which defines the normal whose direction is perpendicular to the plane on which the maximum or minimum principal stress acts, is defined by

$$\tan 2\theta_p = \frac{2\tau_{xy}}{\sigma_x - \sigma_y}$$

(7.1.3)

Figure 7–4 shows the principal stresses σ_1 and σ_2 and the angle θ_p. Recall (as Figure 7–4 indicates) that the shear stress is zero on the planes having principal (maximum and minimum) normal stresses.

In Figure 7–5, we show an infinitesimal element used to represent the general two-dimensional state of strain at some point in a structure. The element is shown to be displaced by amounts u and v in the x and y directions at point A, and to displace or extend an additional (incremental) amount $(\partial u / \partial x)\, dx$ along line AB, and $(\partial v / \partial y)\, dy$ along line AC in the x and y directions, respectively. Furthermore, observing lines AB and AC, we see that point B moves upward an amount $(\partial v / \partial x)\, dx$ with respect to A, and point C moves to the right an amount $(\partial u / \partial y)\, dy$ with respect to A.

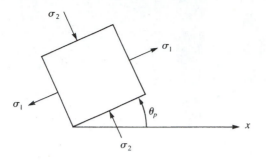

Figure 7–4 Principal stresses and their directions

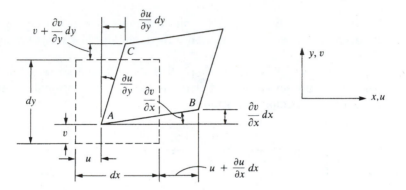

Figure 7–5 Displacements and rotations of lines of an element in the *x-y* plane

From the general definitions of normal and shear strains and the use of Figure 7–5, we obtain

$$\varepsilon_x = \frac{\partial u}{\partial x} \qquad \varepsilon_y = \frac{\partial v}{\partial y} \qquad \gamma_{xy} = \frac{\partial u}{\partial y} + \frac{\partial v}{\partial x} \tag{7.1.4}$$

Appendix C.2 shows a detailed derivation of Eqs. (7.1.4). Hence, recall that the strains ε_x and ε_y are the changes in length per unit length of material fibers originally parallel to the x and y axes, respectively, when the element undergoes deformation. These strains are then called *normal* (or *extensional* or *longitudinal*) *strains*. The strain γ_{xy} is the change in the original right angle made between dx and dy when the element undergoes deformation. The strain γ_{xy} is then called a *shear strain*.

The strains given by Eqs. (7.1.4) are generally represented by the vector column matrix

$$\{\varepsilon\} = \begin{Bmatrix} \varepsilon_x \\ \varepsilon_y \\ \gamma_{xy} \end{Bmatrix} \tag{7.1.5}$$

The relationships between strains and displacements referred to the x and y directions given by Eqs. (7.1.4) are sufficient for your understanding of subsequent material in this chapter.

We now present the stress/strain relationships for isotropic materials for both plane stress and plane strain. For plane stress, we assume the following stresses to be zero:

$$\sigma_z = \tau_{xz} = \tau_{yz} = 0 \qquad (7.1.6)$$

Applying Eq. (7.1.6) to the three-dimensional stress/strain relationship [see Appendix C, Eq. (C.3.10)], the shear strains $\gamma_{xz} = \gamma_{yz} = 0$, but $\varepsilon_z \neq 0$. For plane stress conditions, we then have

$$\{\sigma\} = [D]\{\varepsilon\} \qquad (7.1.7)$$

where
$$[D] = \frac{E}{1 - v^2} \begin{bmatrix} 1 & v & 0 \\ v & 1 & 0 \\ 0 & 0 & \dfrac{1-v}{2} \end{bmatrix} \qquad (7.1.8)$$

is called the *stress/strain matrix* (or *constitutive matrix*), E is the modulus of elasticity, and v is Poisson's ratio. In Eq. (7.1.7), $\{\sigma\}$ and $\{\varepsilon\}$ are defined by Eqs. (7.1.1) and (7.1.5), respectively.

For plane strain, we assume the following strains to be zero:

$$\varepsilon_z = \gamma_{xz} = \gamma_{yz} = 0 \qquad (7.1.9)$$

Applying Eq. (7.1.9) to the three-dimensional stress/strain relationship [Eq. (C.3.10)], the shear stresses $\tau_{xz} = \tau_{yz} = 0$, but $\sigma_z \neq 0$. The stress/strain matrix then becomes

$$[D] = \frac{E}{(1+v)(1-2v)} \begin{bmatrix} 1-v & v & 0 \\ v & 1-v & 0 \\ 0 & 0 & \dfrac{1-2v}{2} \end{bmatrix} \qquad (7.1.10)$$

The $\{\sigma\}$ and $\{\varepsilon\}$ matrices remain the same as for the plane stress case. The basic partial differential equations for plane stress, as derived in Reference [1], are

$$\frac{\partial^2 u}{\partial x^2} + \frac{\partial^2 u}{\partial y^2} = \frac{1+v}{2}\left(\frac{\partial^2 u}{\partial y^2} - \frac{\partial^2 v}{\partial x \partial y}\right)$$

$$\frac{\partial^2 v}{\partial x^2} + \frac{\partial^2 v}{\partial y^2} = \frac{1+v}{2}\left(\frac{\partial^2 v}{\partial x^2} - \frac{\partial^2 u}{\partial x \partial y}\right) \qquad (7.1.11)$$

▲ 7.2 Derivation of the Constant-Strain Triangular Element Stiffness Matrix and Equations ▲

To illustrate the steps and introduce the basic equations necessary for the plane triangular element, consider the thin plate subjected to tensile surface traction loads T_S in Figure 7–6(a).

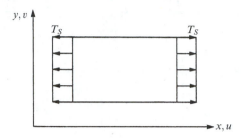

Figure 7–6(a) Thin plate in tension

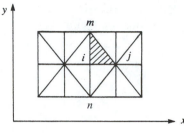

Figure 7–6(b) Discretized plate of Figure 7–6(a) using triangular elements

Step 1 Select Element Type

To analyze the plate, we consider the basic triangular element in Figure 7–7 taken from the discretized plate, as shown in Figure 7–6(b). The discretized plate has been divided into triangular elements, each with nodes such as i, j, and m. We use triangular elements because boundaries of irregularly shaped bodies can be closely approximated in this way, and because the expressions related to the triangular element are comparatively simple. This discretization is called a *coarse-mesh generation* if a few large elements are used. Each node has two degrees of freedom—an x and a y displacement. We will let u_i and v_i represent the node i displacement components in the x and y directions, respectively.

Here all formulations are based on this counterclockwise system of labeling of nodes, although a formulation based on a clockwise system of labeling could be used. Remember that a consistent labeling procedure for the whole body is necessary to avoid

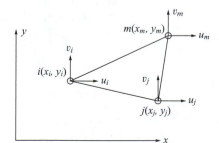

Figure 7–7 Basic triangular element showing degrees of freedom

problems in the calculations such as negative element areas. Here (x_i, y_i), (x_j, y_j), and (x_m, y_m) are the known nodal coordinates of nodes i, j, and m, respectively.

The nodal displacement matrix is given by

$$\{d\} = \begin{Bmatrix} \underline{d_i} \\ \underline{d_j} \\ \underline{d_m} \end{Bmatrix} = \begin{Bmatrix} u_i \\ v_i \\ u_j \\ v_j \\ u_m \\ v_m \end{Bmatrix} \tag{7.2.1}$$

Step 2   Select Displacement Functions

We select a linear displacement function for each element as

$$u(x, y) = a_1 + a_2 x + a_3 y$$
$$v(x, y) = a_4 + a_5 x + a_6 y \tag{7.2.2}$$

where $u(x, y)$ and $v(x, y)$ describe displacements at any interior point (x_i, y_i) of the element.

The linear function ensures that compatibility will be satisfied. A linear function with specified endpoints has only one path through which to pass—that is, through the two points. Hence, the linear function ensures that the displacements along the edge and at the nodes shared by adjacent elements, such as edge i-j of the two elements shown in Figure 7–6(b), are equal. Using Eqs. (7.2.2), the general displacement function $\{\psi\}$, which stores the functions u and v, can be expressed as

$$\{\psi\} = \begin{Bmatrix} a_1 + a_2 x + a_3 y \\ a_4 + a_5 x + a_6 y \end{Bmatrix} = \begin{bmatrix} 1 & x & y & 0 & 0 & 0 \\ 0 & 0 & 0 & 1 & x & y \end{bmatrix} \begin{Bmatrix} a_1 \\ a_2 \\ a_3 \\ a_4 \\ a_5 \\ a_6 \end{Bmatrix} \tag{7.2.3}$$

To obtain the a's in Eqs. (7.2.2), we begin by substituting the coordinates of the nodal points into Eqs. (7.2.2) to yield

$$u_i = u(x_i, y_i) = a_1 + a_2 x_i + a_3 y_i$$
$$u_j = u(x_j, y_j) = a_1 + a_2 x_j + a_3 y_j$$
$$u_m = u(x_m, y_m) = a_1 + a_2 x_m + a_3 y_m$$
$$v_i = v(x_i, y_i) = a_4 + a_5 x_i + a_6 y_i \tag{7.2.4}$$
$$v_j = v(x_j, y_j) = a_4 + a_5 x_j + a_6 y_j$$
$$v_m = v(x_m, y_m) = a_4 + a_5 x_m + a_6 y_m$$

We can solve for the a's beginning with the first three of Eqs. (7.2.4) expressed in matrix form as

$$\left\{\begin{array}{c} u_i \\ u_j \\ u_m \end{array}\right\} = \left[\begin{array}{ccc} 1 & x_i & y_i \\ 1 & x_j & y_j \\ 1 & x_m & y_m \end{array}\right] \left\{\begin{array}{c} a_1 \\ a_2 \\ a_3 \end{array}\right\} \tag{7.2.5}$$

or, solving for the a's, we have

$$\{a\} = [x]^{-1}\{u\} \tag{7.2.6}$$

where $[x]$ is the 3×3 matrix on the right side of Eq. (7.2.5). The method of cofactors (Appendix A) is one possible method for finding the inverse of $[x]$. Thus,

$$[x]^{-1} = \frac{1}{2A} \left[\begin{array}{ccc} \alpha_i & \alpha_j & \alpha_m \\ \beta_i & \beta_j & \beta_m \\ \gamma_i & \gamma_j & \gamma_m \end{array}\right] \tag{7.2.7}$$

where

$$2A = \left|\begin{array}{ccc} 1 & x_i & y_i \\ 1 & x_j & y_j \\ 1 & x_m & y_m \end{array}\right| \tag{7.2.8}$$

is the determinant of $[x]$, which on evaluation is

$$2A = x_i(y_j - y_m) + x_j(y_m - y_i) + x_m(y_i - y_j) \tag{7.2.9}$$

Here A is the area of the triangle, and

$$\begin{array}{lll} \alpha_i = x_j y_m - y_j x_m & \alpha_j = y_i x_m - x_i y_m & \alpha_m = x_i y_j - y_i x_j \\ \beta_i = y_j - y_m & \beta_j = y_m - y_i & \beta_m = y_i - y_j \\ \gamma_i = x_m - x_j & \gamma_j = x_i - x_m & \gamma_m = x_j - x_i \end{array} \tag{7.2.10}$$

Having determined $[x]^{-1}$, we can now express Eq. (7.2.6) in expanded matrix form as

$$\left\{\begin{array}{c} a_1 \\ a_2 \\ a_3 \end{array}\right\} = \frac{1}{2A} \left[\begin{array}{ccc} \alpha_i & \alpha_j & \alpha_m \\ \beta_i & \beta_j & \beta_m \\ \gamma_i & \gamma_j & \gamma_m \end{array}\right] \left\{\begin{array}{c} u_i \\ u_j \\ u_m \end{array}\right\} \tag{7.2.11}$$

Similarly, using the last three of Eqs. (7.2.4), we can obtain

$$\left\{\begin{array}{c} a_4 \\ a_5 \\ a_6 \end{array}\right\} = \frac{1}{2A} \left[\begin{array}{ccc} \alpha_i & \alpha_j & \alpha_m \\ \beta_i & \beta_j & \beta_m \\ \gamma_i & \gamma_j & \gamma_m \end{array}\right] \left\{\begin{array}{c} v_i \\ v_j \\ v_m \end{array}\right\} \tag{7.2.12}$$

We will derive the general x displacement function $u(x, y)$ of $\{\psi\}$ (v will follow analogously) in terms of the coordinate variables x and y, known coordinate variables

$\alpha_i, \alpha_j, \ldots, \gamma_m$, and unknown nodal displacements u_i, u_j, and u_m. Beginning with Eqs. (7.2.2) expressed in matrix form, we have

$$\{u\} = [1 \quad x \quad y] \begin{Bmatrix} a_1 \\ a_2 \\ a_3 \end{Bmatrix} \tag{7.2.13}$$

Substituting Eq. (7.2.11) into Eq. (7.2.13), we obtain

$$\{u\} = \frac{1}{2A} [1 \quad x \quad y] \begin{bmatrix} \alpha_i & \alpha_j & \alpha_m \\ \beta_i & \beta_j & \beta_m \\ \gamma_i & \gamma_j & \gamma_m \end{bmatrix} \begin{Bmatrix} u_i \\ u_j \\ u_m \end{Bmatrix} \tag{7.2.14}$$

Expanding Eq. (7.2.14), we have

$$\{u\} = \frac{1}{2A} [1 \quad x \quad y] \begin{Bmatrix} \alpha_i u_i + \alpha_j u_j + \alpha_m u_m \\ \beta_i u_i + \beta_j u_j + \beta_m u_m \\ \gamma_i u_i + \gamma_j u_j + \gamma_m u_m \end{Bmatrix} \tag{7.2.15}$$

Multiplying the two matrices in Eq. (7.2.15) and rearranging, we obtain

$$u(x, y) = \frac{1}{2A} \{(\alpha_i + \beta_i x + \gamma_i y)u_i + (\alpha_j + \beta_j x + \gamma_j y)u_j + (\alpha_m + \beta_m x + \gamma_m y)u_m\} \tag{7.2.16}$$

Similarly, replacing u_i by v_i, u_j by v_j, and u_m by v_m in Eq. (7.2.16), we have the y displacement given by

$$v(x, y) = \frac{1}{2A} \{(\alpha_i + \beta_i x + \gamma_i y)v_i + (\alpha_j + \beta_j x + \gamma_j y)v_j + (\alpha_m + \beta_m x + \gamma_m y)v_m\} \tag{7.2.17}$$

To express Eqs. (7.2.16) and (7.2.17) for u and v in simpler form, we define

$$N_i = \frac{1}{2A}(\alpha_i + \beta_i x + \gamma_i y)$$

$$N_j = \frac{1}{2A}(\alpha_j + \beta_j x + \gamma_j y) \tag{7.2.18}$$

$$N_m = \frac{1}{2A}(\alpha_m + \beta_m x + \gamma_m y)$$

Thus, using Eqs. (7.2.18), we can rewrite Eqs. (7.2.16) and (7.2.17) as

$$u(x, y) = N_i u_i + N_j u_j + N_m u_m$$

$$v(x, y) = N_i v_i + N_j v_j + N_m v_m \tag{7.2.19}$$

Expressing Eqs. (7.2.19) in matrix form, we obtain

$$\{\psi\} = \left\{ \begin{array}{c} u(x,y) \\ v(x,y) \end{array} \right\} = \left\{ \begin{array}{c} N_i u_i + N_j u_j + N_m u_m \\ N_i v_i + N_j v_j + N_m v_m \end{array} \right\}$$

or

$$\{\psi\} = \begin{bmatrix} N_i & 0 & N_j & 0 & N_m & 0 \\ 0 & N_i & 0 & N_j & 0 & N_m \end{bmatrix} \left\{ \begin{array}{c} u_i \\ v_i \\ u_j \\ v_j \\ u_m \\ v_m \end{array} \right\} \tag{7.2.20}$$

Finally, expressing Eq. (7.2.20) in abbreviated matrix form, we have

$$\{\psi\} = [N]\{d\} \tag{7.2.21}$$

where $[N]$ is given by

$$[N] = \begin{bmatrix} N_i & 0 & N_j & 0 & N_m & 0 \\ 0 & N_i & 0 & N_j & 0 & N_m \end{bmatrix} \tag{7.2.22}$$

We have now expressed the general displacements as functions of $\{d\}$, in terms of the shape functions $N_i, N_j,$ and N_m. The shape functions represent the shape of $\{\psi\}$ when plotted over the surface of a typical element. For instance, N_i represents the shape of the variable u when plotted over the surface of the element for $u_i = 1$ and all other degrees of freedom equal to zero; that is, $u_j = u_m = v_i = v_j = v_m = 0$. In addition, $u(x_i, y_i)$ must be equal to u_i. Therefore, we must have $N_i = 1$, $N_j = 0$, and $N_m = 0$ at (x_i, y_i). Similarly, $u(x_j, y_j) = u_j$. Therefore, $N_i = 0$, $N_j = 1$, and $N_m = 0$ at (x_j, y_j). Figure 7–8 shows the shape variation of N_i plotted over the surface of a typical element. Note that N_i does not equal zero except along a line connecting and including nodes j and m.

Finally, $N_i + N_j + N_m = 1$ for all x and y locations on the surface of the element so that u and v will yield a constant value when rigid-body displacement occurs. The

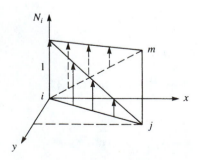

Figure 7–8 Variation of N_i over the *x-y* surface of a typical element

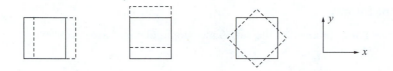

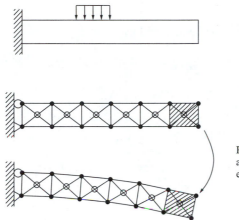

(a) Rigid-body modes of a plane stress element (from left to right, pure
    translation in x and y directions and pure rotation)

Rigid-body translation
and rotation occurs for
elements to right of load

(b) Cantilever beam modeled using constant-strain triangle elements;
    elements to the right of the loading are stress-free

Figure 7–9 Unstressed elements in a cantilever beam modeled with CST

proof of this relationship follows that given for the bar element in Section 3.2 and is
left as an exercise (Problem 7.1).

The requirement of completeness for the constant-strain triangle element used
in a two-dimensional plane stress element is illustrated in Figure 7–9. The element
must be able to translate uniformly in either the x or y direction in the plane and
to rotate without straining as shown in Figure 7–9(a). The reason that the element must be able to translate as a rigid body and to rotate stress-free is illustrated
in the example of a cantilever beam modeled with plane stress elements as shown
in Figure 7–9(b). By simple statics the beam elements beyond the loading are stress-
free. Hence these elements must be free to translate and rotate without stretching or
changing shape.

Step 3   **Define the Strain/Displacement and Stress/Strain
             Relationships**

We express the element strains and stresses in terms of the unknown nodal displacements.

Element Strains

The strains associated with the two-dimensional element are given by

$$\{\varepsilon\} = \left\{ \begin{array}{c} \varepsilon_x \\ \varepsilon_y \\ \gamma_{xy} \end{array} \right\} = \left\{ \begin{array}{c} \dfrac{\partial u}{\partial x} \\[2mm] \dfrac{\partial v}{\partial y} \\[2mm] \dfrac{\partial u}{\partial y} + \dfrac{\partial v}{\partial x} \end{array} \right\} \tag{7.2.23}$$

Using Eqs. (7.2.19) for the displacements, we have

$$\frac{\partial u}{\partial x} = u_{,x} = \frac{\partial}{\partial x}(N_i u_i + N_j u_j + N_m u_m) \tag{7.2.24}$$

or

$$u_{,x} = N_{i,x} u_i + N_{j,x} u_j + N_{m,,x} u_m \tag{7.2.25}$$

where the comma followed by a variable indicates differentiation with respect to that variable. We have used $u_{i,x} = 0$ because $u_i = u(x_i, y_i)$ is a constant value; similarly, $u_{j,x} = 0$ and $u_{m,x} = 0$.

Using Eqs. (7.2.18), we can evaluate the expressions for the derivatives of the shape functions in Eq. (7.2.25) as follows:

$$N_{i,x} = \frac{1}{2A}\frac{\partial}{\partial x}(\alpha_i + \beta_i x + \gamma_i y) = \frac{\beta_i}{2A} \tag{7.2.26}$$

Similarly,

$$N_{j,x} = \frac{\beta_j}{2A} \quad \text{and} \quad N_{m,x} = \frac{\beta_m}{2A} \tag{7.2.27}$$

Therefore, using Eqs. (7.2.26) and (7.2.27) in Eq. (7.2.25), we have

$$\frac{\partial u}{\partial x} = \frac{1}{2A}(\beta_i u_i + \beta_j u_j + \beta_m u_m) \tag{7.2.28}$$

Similarly, we can obtain

$$\frac{\partial v}{\partial y} = \frac{1}{2A}(\gamma_i v_i + \gamma_j v_j + \gamma_m v_m)$$

$$\frac{\partial u}{\partial y} + \frac{\partial v}{\partial x} = \frac{1}{2A}(\gamma_i u_i + \beta_i v_i + \gamma_j u_j + \beta_j v_j + \gamma_m u_m + \beta_m v_m) \tag{7.2.29}$$

Using Eqs. (7.2.28) and (7.2.29) in Eq. (7.2.23), we obtain

$$\{\varepsilon\} = \frac{1}{2A}\begin{bmatrix} \beta_i & 0 & \beta_j & 0 & \beta_m & 0 \\ 0 & \gamma_i & 0 & \gamma_j & 0 & \gamma_m \\ \gamma_i & \beta_i & \gamma_j & \beta_j & \gamma_m & \beta_m \end{bmatrix} \left\{ \begin{array}{c} u_i \\ v_i \\ u_j \\ v_j \\ u_m \\ v_m \end{array} \right\} \tag{7.2.30}$$

or

$$\{\varepsilon\} = [\underline{B}_i \quad \underline{B}_j \quad \underline{B}_m] \left\{ \begin{array}{c} \underline{d}_i \\ \underline{d}_j \\ \underline{d}_m \end{array} \right\}$$

(7.2.31)

where

$$[B_i] = \frac{1}{2A} \begin{bmatrix} \beta_i & 0 \\ 0 & \gamma_i \\ \gamma_i & \beta_i \end{bmatrix} \qquad [B_j] = \frac{1}{2A} \begin{bmatrix} \beta_j & 0 \\ 0 & \gamma_j \\ \gamma_j & \beta_j \end{bmatrix} \qquad [B_m] = \frac{1}{2A} \begin{bmatrix} \beta_m & 0 \\ 0 & \gamma_m \\ \gamma_m & \beta_m \end{bmatrix}$$

(7.2.32)

Finally, in simplified matrix form, Eq. (7.2.31) can be written as

$$\{\varepsilon\} = [B]\{d\}$$

(7.2.33)

where

$$[B] = [\underline{B}_i \quad \underline{B}_j \quad \underline{B}_m]$$

(7.2.34)

The $\underline{B}$ matrix is independent of the x and y coordinates. It depends solely on the element nodal coordinates, as seen from Eqs. (7.2.32) and (7.2.10). The strains in Eq. (7.2.33) will be constant; hence, the element is called a *constant-strain triangle* (CST).

Stress/Strain Relationship

In general, the in-plane stress/strain relationship is given by

$$\left\{ \begin{array}{c} \sigma_x \\ \sigma_y \\ \tau_{xy} \end{array} \right\} = [D] \left\{ \begin{array}{c} \varepsilon_x \\ \varepsilon_y \\ \gamma_{xy} \end{array} \right\}$$

(7.2.35)

where $[D]$ is given by Eq. (7.1.8) for plane stress problems and by Eq. (7.1.10) for plane strain problems. Using Eq. (7.2.33) in Eq. (7.2.35), we obtain the in-plane stresses in terms of the unknown nodal degrees of freedom as

$$\{\sigma\} = [D][B]\{d\}$$

(7.2.36)

where the stresses $\{\sigma\}$ are also constant everywhere within the element.

Step 4 Derive the Element Stiffness Matrix and Equations

Using the principle of minimum potential energy, we can generate the equations for a typical constant-strain triangular element. Keep in mind that for the basic plane stress element, the total potential energy is now a function of the nodal displacements $u_i, v_i, u_j, \ldots, v_m$ (that is, $\{d\}$) such that

$$\pi_p = \pi_p(u_i, v_i, u_j, \ldots, v_m)$$

(7.2.37)

Here the total potential energy is given by

$$\pi_p = U + \Omega_b + \Omega_p + \Omega_s$$

(7.2.38)

where the strain energy is given by

$$U = \frac{1}{2} \iiint\limits_V \{\varepsilon\}^T \{\sigma\} \, dV \tag{7.2.39}$$

or, using Eq. (7.2.35), we have

$$U = \frac{1}{2} \iiint\limits_V \{\varepsilon\}^T [D] \{\varepsilon\} \, dV \tag{7.2.40}$$

where we have used $[D]^T = [D]$ in Eq. (7.2.40).

The potential energy of the body forces is given by

$$\Omega_b = - \iiint\limits_V \{\psi\}^T \{X\} \, dV \tag{7.2.41}$$

where $\{\psi\}$ is again the general displacement function, and $\{X\}$ is the body weight/unit volume or weight density matrix (typically, in units of pounds per cubic inch or kilonewtons per cubic meter).

The potential energy of concentrated loads is given by

$$\Omega_p = -\{d\}^T \{P\} \tag{7.2.42}$$

where $\{d\}$ represents the usual nodal displacements, and $\{P\}$ now represents the concentrated external loads.

The potential energy of distributed loads (or surface tractions) moving through respective surface displacements is given by

$$\Omega_s = - \iint\limits_S \{\psi_S\}^T \{T_S\} \, dS \tag{7.2.43}$$

where $\{T_S\}$ represents the surface tractions (typically in units of pounds per square inch or kilonewtons per square meter), $\{\psi_S\}$ represents the field of surface displacements through which the surface tractions act, and S represents the surfaces over which the tractions $\{T_S\}$ act. Similar to Eq. (7.2.21), we express $\{\psi_S\}$ as $\{\psi_S\} = [N_S]\{d\}$, where $[N_S]$ represents the shape function matrix evaluated along the surface where the surface traction acts.

Using Eq. (7.2.21) for $\{\psi\}$ and Eq. (7.2.33) for the strains in Eqs. (7.2.40)–(7.2.43), we have

$$\pi_p = \frac{1}{2} \iiint\limits_V \{d\}^T [B]^T [D][B]\{d\} \, dV - \iiint\limits_V \{d\}^T [N]^T \{X\} \, dV$$

$$- \{d\}^T \{P\} - \iint\limits_S \{d\}^T [N_S]^T \{T_S\} \, dS \tag{7.2.44}$$

The nodal displacements $\{d\}$ are independent of the general x-y coordinates, so $\{d\}$ can be taken out of the integrals of Eq. (7.2.44). Therefore,

$$\pi_p = \frac{1}{2}\{d\}^T \iiint_V [B]^T [D][B]\, dV \{d\} - \{d\}^T \iiint_V [N]^T \{X\}\, dV$$

$$- \{d\}^T \{P\} - \{d\}^T \iint_S [N_S]^T \{T_S\}\, dS \tag{7.2.45}$$

From Eqs. (7.2.41)–(7.2.43) we can see that the last three terms of Eq. (7.2.45) represent the total load system $\{f\}$ on an element; that is,

$$\{f\} = \iiint_V [N]^T \{X\}\, dV + \{P\} + \iint_S [N_S]^T \{T_S\}\, dS \tag{7.2.46}$$

where the first, second, and third terms on the right side of Eq. (7.2.46) represent the body forces, the concentrated nodal forces, and the surface tractions, respectively. Using Eq. (7.2.46) in Eq. (7.2.45), we obtain

$$\pi_p = \frac{1}{2}\{d\}^T \iiint_V [B]^T [D][B]\, dV \{d\} - \{d\}^T \{f\} \tag{7.2.47}$$

Taking the first variation, or equivalently, as shown in Chapters 2 and 3, the partial derivative of π_p with respect to the nodal displacements since $\pi_p = \pi_p(\underline{d})$ (as was previously done for the bar and beam elements in Chapters 3 and 5, respectively), we obtain

$$\frac{\partial \pi_p}{\partial \{d\}} = \left[\iiint_V [B]^T [D][B]\, dV \right] \{d\} - \{f\} = 0 \tag{7.2.48}$$

Rewriting Eq. (7.2.48), we have

$$\iiint_V [B]^T [D][B]\, dV \{d\} = \{f\} \tag{7.2.49}$$

where the partial derivative with respect to matrix $\{d\}$ was previously defined by Eq. (2.6.12). From Eq. (7.2.49) we can see that

$$[k] = \iiint_V [B]^T [D][B]\, dV \tag{7.2.50}$$

For an element with constant thickness, t, Eq. (7.2.50) becomes

$$[k] = t \iint_A [B]^T [D][B]\, dx\, dy \tag{7.2.51}$$

where the integrand is not a function of x or y for the constant-strain triangular element and thus can be taken out of the integral to yield

$$[k] = tA[B]^T [D][B] \tag{7.2.52}$$

where A is given by Eq. (7.2.9), $[B]$ is given by Eq. (7.2.34), and $[D]$ is given by Eq. (7.1.8) or Eq. (7.1.10). We will assume elements of constant thickness. (This assumption is convergent to the actual situation as the element size is decreased.)

From Eq. (7.2.52) we see that $[k]$ is a function of the nodal coordinates (because $[B]$ and A are defined in terms of them) and of the mechanical properties E and v (of which $[D]$ is a function). The expansion of Eq. (7.2.52) for an element is

$$[k] = \begin{bmatrix} [k_{ii}] & [k_{ij}] & [k_{im}] \\ [k_{ji}] & [k_{jj}] & [k_{jm}] \\ [k_{mi}] & [k_{mj}] & [k_{mm}] \end{bmatrix} \tag{7.2.53}$$

where the 2×2 submatrices are given by

$$[k_{ii}] = [B_i]^T [D][B_i] tA$$

$$[k_{ij}] = [B_i]^T [D][B_j] tA \tag{7.2.54}$$

$$[k_{im}] = [B_i]^T [D][B_m] tA$$

and so forth. In Eqs. (7.2.54), $[B_i]$, $[B_j]$, and $[B_m]$ are defined by Eqs. (7.2.32). The $[k]$ matrix is seen to be a 6×6 matrix (equal in order to the number of degrees of freedom per node, two, times the total number of nodes per element, three).

In general, Eq. (7.2.46) must be used to evaluate the surface and body forces. When Eq. (7.2.46) is used to evaluate the surface and body forces, these forces are called *consistent loads* because they are derived from the consistent (energy) approach. For higher-order elements, typically with quadratic or cubic displacement functions, Eq. (7.2.46) should be used. However, for the CST element, the body and surface forces can be lumped at the nodes with equivalent results (this is illustrated in Section 7.3) and added to any concentrated nodal forces to obtain the element force matrix. The element equations are then given by

$$\begin{Bmatrix} f_{1x} \\ f_{1y} \\ f_{2x} \\ f_{2y} \\ f_{3x} \\ f_{3y} \end{Bmatrix} = \begin{bmatrix} k_{11} & k_{12} & \dots & k_{16} \\ k_{21} & k_{22} & \dots & k_{26} \\ \vdots & \vdots & & \vdots \\ k_{61} & k_{62} & \dots & k_{66} \end{bmatrix} \begin{Bmatrix} u_1 \\ v_1 \\ u_2 \\ v_2 \\ u_3 \\ v_3 \end{Bmatrix} \tag{7.2.55}$$

Step 5 Assemble the Element Equations to Obtain the Global Equations and Introduce Boundary Conditions

We obtain the global structure stiffness matrix and equations by using the direct stiffness method as

$$[K] = \sum_{e=1}^{N} [k^{(e)}] \tag{7.2.56}$$

and
$$\{F\} = [K]\{d\} \tag{7.2.57}$$

where, in Eq. (7.2.56), all element stiffness matrices are defined in terms of the global x-y coordinate system, $\{d\}$ is now the total structure displacement matrix, and

$$\{F\} = \sum_{e=1}^{N}\{f^{(e)}\} \tag{7.2.58}$$

is the column of equivalent global nodal loads obtained by lumping body forces and distributed loads at the proper nodes (as well as including concentrated nodal loads) or by consistently using Eq. (7.2.46). (Further details regarding the treatment of body forces and surface tractions will be given in Section 7.3.)

In the formulation of the element stiffness matrix Eq. (7.2.52), the matrix has been derived for a general orientation in global coordinates. Equation (7.2.52) then applies for all elements. All element matrices are expressed in the global-coordinate orientation. Therefore, no transformation from local to global equations is necessary. However, for completeness, we will now describe the method to use if the local axes for the constant-strain triangular element are not parallel to the global axes for the whole structure.

If the local axes for the constant-strain triangular element are not parallel to the global axes for the whole structure, we must apply rotation-of-axes transformations similar to those introduced in Chapter 3 by Eq. (3.3.16) to the element stiffness matrix, as well as to the element nodal force and displacement matrices. We illustrate the transformation of axes for the triangular element shown in Figure 7–10, considering the element to have local axes $\hat{x}$-$\hat{y}$ not parallel to global axes x-y. Local nodal forces are shown in the figure. The transformation from local to global equations follows the procedure outlined in Section 3.4. We have the same general expressions, Eqs. (3.4.14), (3.4.16), and (3.4.22), to relate local to global displacements, forces, and stiffness matrices, respectively; that is,

$$\hat{\underline{d}} = \underline{T}\underline{d} \qquad \hat{\underline{f}} = \underline{T}\underline{f} \qquad \underline{k} = \underline{T}^{T}\hat{\underline{k}}\underline{T} \tag{7.2.59}$$

where Eq. (3.4.15) for the transformation matrix $\underline{T}$ used in Eqs. (7.2.59) must be expanded because two additional degrees of freedom are present in the constant-strain

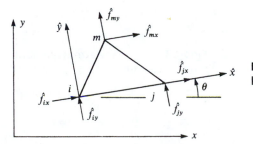

Figure 7–10 Triangular element with local axes not parallel to global axes

triangular element. Thus, Eq. (3.4.15) is expanded to

$$
\underline{T} =
\begin{bmatrix}
C & S & 0 & 0 & 0 & 0 \\
-S & C & 0 & 0 & 0 & 0 \\
0 & 0 & C & S & 0 & 0 \\
0 & 0 & -S & C & 0 & 0 \\
0 & 0 & 0 & 0 & C & S \\
0 & 0 & 0 & 0 & -S & C
\end{bmatrix}
\begin{matrix}
u_i \\ v_i \\ u_j \\ v_j \\ u_m \\ v_m
\end{matrix}
\tag{7.2.60}
$$

where $C = \cos\theta$, $S = \sin\theta$, and θ is shown in Figure 7–10.

Step 6   Solve for the Nodal Displacements

We determine the unknown global structure nodal displacements by solving the system of algebraic equations given by Eq. (7.2.57).

Step 7   Solve for the Element Forces (Stresses)

Having solved for the nodal displacements, we obtain the strains and stresses in the global x and y directions in the elements by using Eqs. (7.2.33) and (7.2.36). Finally, we determine the maximum and minimum in-plane principal stresses σ_1 and σ_2 by using the transformation Eqs. (7.1.2), where these stresses are usually assumed to act at the centroid of the element. The angle that one of the principal stresses makes with the x axis is given by Eq. (7.1.3).

Example 7.1

Evaluate the stiffness matrix for the element shown in Figure 7–11. The coordinates are shown in units of inches. Assume plane stress conditions. Let $E = 30 \times 10^6$ psi, $v = 0.25$, and thickness $t = 1$ in. Assume the element nodal displacements have been determined to be $u_1 = 0.0$, $v_1 = 0.0025$ in., $u_2 = 0.0012$ in., $v_2 = 0.0$, $u_3 = 0.0$, and $v_3 = 0.0025$ in. Determine the element stresses.

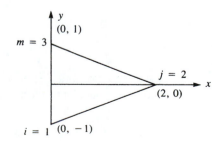

Figure 7–11   Plane stress element for stiffness matrix evaluation

We use Eq. (7.2.52) to obtain the element stiffness matrix. To evaluate $\underline{k}$, we first use Eqs. (7.2.10) to obtain the β's and γ's as follows:

$$\beta_i = y_j - y_m = 0 - 1 = -1 \qquad \gamma_i = x_m - x_j = 0 - 2 = -2$$
$$\beta_j = y_m - y_i = 1 - (-1) = 2 \qquad \gamma_j = x_i - x_m = 0 - 0 = 0 \qquad (7.2.61)$$
$$\beta_m = y_i - y_j = -1 - 0 = -1 \qquad \gamma_m = x_j - x_i = 2 - 0 = 2$$

Using Eqs. (7.2.32) and (7.2.34), we obtain matrix $\underline{B}$ as

$$\underline{B} = \frac{1}{2(2)}\begin{bmatrix} -1 & 0 & 2 & 0 & -1 & 0 \\ 0 & -2 & 0 & 0 & 0 & 2 \\ -2 & -1 & 0 & 2 & 2 & -1 \end{bmatrix} \qquad (7.2.62)$$

where we have used $A = 2$ in^2 in Eq. (7.2.62).

Using Eq. (7.1.8) for plane stress conditions,

$$\underline{D} = \frac{30 \times 10^6}{1 - (0.25)^2}\begin{bmatrix} 1 & 0.25 & 0 \\ 0.25 & 1 & 0 \\ 0 & 0 & \dfrac{1 - 0.25}{2} \end{bmatrix} \text{psi} \qquad (7.2.63)$$

Substituting Eqs. (7.2.62) and (7.2.63) into Eq. (7.2.52), we obtain

$$\underline{k} = \frac{(2)30 \times 10^6}{4(0.9375)}\begin{bmatrix} -1 & 0 & -2 \\ 0 & -2 & -1 \\ 2 & 0 & 0 \\ 0 & 0 & 2 \\ -1 & 0 & 2 \\ 0 & 2 & -1 \end{bmatrix}$$

$$\times \begin{bmatrix} 1 & 0.25 & 0 \\ 0.25 & 1 & 0 \\ 0 & 0 & 0.375 \end{bmatrix} \frac{1}{2(2)}\begin{bmatrix} -1 & 0 & 2 & 0 & -1 & 0 \\ 0 & -2 & 0 & 0 & 0 & 2 \\ -2 & -1 & 0 & 2 & 2 & -1 \end{bmatrix}$$

Performing the matrix triple product, we have

$$\underline{k} = 4.0 \times 10^6 \begin{bmatrix} 2.5 & 1.25 & -2 & -1.5 & -0.5 & 0.25 \\ 1.25 & 4.375 & -1 & -0.75 & -0.25 & -3.625 \\ -2 & -1 & 4 & 0 & -2 & 1 \\ -1.5 & -0.75 & 0 & 1.5 & 1.5 & -0.75 \\ -0.5 & -0.25 & -2 & 1.5 & 2.5 & -1.25 \\ 0.25 & -3.625 & 1 & -0.75 & -1.25 & 4.375 \end{bmatrix} \frac{\text{lb}}{\text{in.}} \qquad (7.2.64)$$

To evaluate the stresses, we use Eq. (7.2.36). Substituting Eqs. (7.2.62) and (7.2.63), along with the given nodal displacements, into Eq. (7.2.36), we obtain

$$
\begin{Bmatrix} \sigma_x \\ \sigma_y \\ \tau_{xy} \end{Bmatrix} = \frac{30 \times 10^6}{1 - (0.25)^2} \begin{bmatrix} 1 & 0.25 & 0 \\ 0.25 & 1 & 0 \\ 0 & 0 & 0.375 \end{bmatrix}
$$

$$
\times \frac{1}{2(2)} \begin{bmatrix} -1 & 0 & 2 & 0 & -1 & 0 \\ 0 & -2 & 0 & 0 & 0 & 2 \\ -2 & -1 & 0 & 2 & 2 & -1 \end{bmatrix} \begin{Bmatrix} 0.0 \\ 0.0025 \\ 0.0012 \\ 0.0 \\ 0.0 \\ 0.0025 \end{Bmatrix} \tag{7.2.65}
$$

Performing the matrix triple product in Eq. (7.2.65), we have

$$
\sigma_x = 19{,}200 \text{ psi} \qquad \sigma_y = 4800 \text{ psi} \qquad \tau_{xy} = -15{,}000 \text{ psi} \tag{7.2.66}
$$

Finally, the principal stresses and principal angle are obtained by substituting the results from Eqs. (7.2.66) into Eqs. (7.1.2) and (7.1.3) as follows:

$$
\sigma_1 = \frac{19{,}200 + 4800}{2} + \left[\left(\frac{19{,}200 - 4800}{2} \right)^2 + (-15{,}000)^2 \right]^{1/2}
$$

$$
= 28{,}639 \text{ psi}
$$

$$
\sigma_2 = \frac{19{,}200 + 4800}{2} - \left[\left(\frac{19{,}200 - 4800}{2} \right)^2 + (-15{,}000)^2 \right]^{1/2} \tag{7.2.67}
$$

$$
= -4639 \text{ psi}
$$

$$
\theta_p = \frac{1}{2} \tan^{-1} \left[\frac{2(-15{,}000)}{19{,}200 - 4800} \right] = -32.2° \qquad \blacksquare
$$

▲ 7.3 Treatment of Body and Surface Forces

Body Forces

Using the first term on the right side of Eq. (7.2.46), we can evaluate the body forces at the nodes as

$$
\{f_b\} = \iiint\limits_V [N]^T \{X\} \, dV \tag{7.3.1}
$$

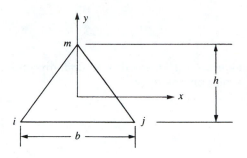

Figure 7–12   Element with centroidal coordinate axes

where

$$\{X\} = \begin{Bmatrix} X_b \\ Y_b \end{Bmatrix} \tag{7.3.2}$$

and X_b and Y_b are the weight densities in the x and y directions in units of force/unit volume, respectively. These forces may arise, for instance, because of actual body weight (gravitational forces), angular velocity (called *centrifugal body forces*, as described in Chapter 10), or inertial forces in dynamics.

In Eq. (7.3.1), $[N]$ is a linear function of x and y; therefore, the integration must be carried out. Without lack of generality, the integration is simplified if the origin of the coordinates is chosen at the centroid of the element. For example, consider the element with coordinates shown in Figure 7–12. With the origin of the coordinate placed at the centroid of the element, we have, from the definition of the centroid, $\iint x\, dA = \iint y\, dA = 0$ and therefore,

$$\iint \beta_i x\, dA = \iint \gamma_i y\, dA = 0 \tag{7.3.3}$$

and

$$\alpha_i = \alpha_j = \alpha_m = \frac{2A}{3} \tag{7.3.4}$$

Using Eqs. (7.3.2)–(7.3.4) in Eq. (7.3.1), the body force at node i is then represented by

$$\{f_{bi}\} = \begin{Bmatrix} X_b \\ Y_b \end{Bmatrix} \frac{tA}{3} \tag{7.3.5}$$

Similarly, considering the j and m node body forces, we obtain the same results as in Eq. (7.3.5). In matrix form, the element body forces are

$$\{f_b\} = \begin{Bmatrix} f_{bix} \\ f_{biy} \\ f_{bjx} \\ f_{bjy} \\ f_{bmx} \\ f_{bmy} \end{Bmatrix} = \begin{Bmatrix} X_b \\ Y_b \\ X_b \\ Y_b \\ X_b \\ Y_b \end{Bmatrix} \frac{At}{3} \tag{7.3.6}$$

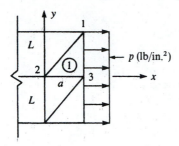

Figure 7-13 Element with uniform surface traction acting on one edge

From the results of Eq. (7.3.6), we can conclude that the body forces are distributed to the nodes in three equal parts. The signs depend on the directions of X_b and Y_b with respect to the positive x and y global coordinates. For the case of body weight only, because of the gravitational force associated with the y direction, we have only Y_b $(X_b = 0)$.

Surface Forces

Using the third term on the right side of Eq. (7.2.46), we can evaluate the surface forces at the nodes as

$$\{f_s\} = \iint_S [N_S]^T \{T_S\}\, dS \tag{7.3.7}$$

We will now illustrate the use of Eq. (7.3.7) by considering the example of a uniform stress p (say, in pounds per square inch) acting between nodes 1 and 3 on the edge of element 1 in Figure 7–13. In Eq. (7.3.7), the surface traction now becomes

$$\{T_S\} = \begin{Bmatrix} p_x \\ p_y \end{Bmatrix} = \begin{Bmatrix} p \\ 0 \end{Bmatrix} \tag{7.3.8}$$

and

$$[N_S]^T = \begin{bmatrix} N_1 & 0 \\ 0 & N_1 \\ N_2 & 0 \\ 0 & N_2 \\ N_3 & 0 \\ 0 & N_3 \end{bmatrix} \text{ evaluated at } x = a,\ y = y \tag{7.3.9}$$

Using Eqs. (7.3.8) and (7.3.9), we express Eq. (7.3.7) as

$$\{f_s\} = \int_0^t \int_0^L \begin{bmatrix} N_1 & 0 \\ 0 & N_1 \\ N_2 & 0 \\ 0 & N_2 \\ N_3 & 0 \\ 0 & N_3 \end{bmatrix} \begin{Bmatrix} p \\ 0 \end{Bmatrix} dz\, dy \quad \text{evaluated at } x = a,\ y = y \tag{7.3.10}$$

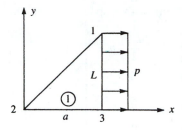

Figure 7-14 Representative element subjected to edge surface traction p

where the N's are evaluated at $x = a$ and $y = y$ because the load acts on that edge. Simplifying Eq. (7.3.10), we obtain

$$\{f_s\} = t \int_0^L \begin{bmatrix} N_1 p \\ 0 \\ N_2 p \\ 0 \\ N_3 p \\ 0 \end{bmatrix} dy \quad \text{evaluated at } x = a, \ y = y \tag{7.3.11}$$

Now, by Eqs. (7.2.18) (with $i = 1$), we have

$$N_1 = \frac{1}{2A}(\alpha_1 + \beta_1 x + \gamma_1 y) \tag{7.3.12}$$

For convenience, we choose the coordinate system for the element as shown in Figure 7-14. Using the definition Eqs. (7.2.10), we obtain

$$\alpha_i = x_j y_m - y_j x_m$$

or, with $i = 1, j = 2$, and $m = 3$,

$$\alpha_1 = x_2 y_3 - y_2 x_3 \tag{7.3.13}$$

Substituting the coordinates into Eq. (7.3.13), we obtain

$$\alpha_1 = 0 \tag{7.3.14}$$

Similarly, again using Eqs. (7.2.10), we obtain

$$\beta_1 = 0 \qquad \gamma_1 = a \tag{7.3.15}$$

Therefore, substituting Eqs. (7.3.14) and (7.3.15) into Eq. (7.3.12), we obtain

$$N_1 = \frac{ay}{2A} \tag{7.3.16}$$

Similarly, using Eqs. (7.2.18), we can show that

$$N_2 = \frac{L(a - x)}{2A} \quad \text{and} \quad N_3 = \frac{Lx - ay}{2A} \tag{7.3.17}$$

On substituting Eqs. (7.3.16) and (7.3.17) for N_1, N_2, and N_3 into Eq. (7.3.11), evaluating N_1, N_2, and N_3 at $x = a$ and $y = y$ (the coordinates corresponding to the location of the surface load p), and then integrating with respect to y, we obtain

$$\{f_s\} = \frac{t}{2(aL/2)} \begin{Bmatrix} a\left(\dfrac{L^2}{2}\right)p \\ 0 \\ 0 \\ 0 \\ \left(L^2 - \dfrac{L^2}{2}\right)ap \\ 0 \end{Bmatrix} \tag{7.3.18}$$

where the shape function $N_2 = 0$ between nodes 1 and 3, as should be the case according to the definitions of the shape functions. Simplifying Eq. (7.3.18), we finally obtain

$$\{f_s\} = \begin{Bmatrix} f_{s1x} \\ f_{s1y} \\ f_{s2x} \\ f_{s2y} \\ f_{s3x} \\ f_{s3y} \end{Bmatrix} = \begin{Bmatrix} pLt/2 \\ 0 \\ 0 \\ 0 \\ pLt/2 \\ 0 \end{Bmatrix} \tag{7.3.19}$$

Figure 7–15 illustrates the results for the surface load equivalent nodal forces for both elements 1 and 2.

We can conclude that for a constant-strain triangle, a distributed load on an element edge can be treated as concentrated loads acting at the nodes associated with the loaded edge by making the two kinds of load statically equivalent [which is equivalent to applying Eq. (7.3.7)]. However, for higher-order elements such as the linear-strain triangle (discussed in Chapter 9), the load replacement should be made by using Eq. (7.3.7), which was derived by the principle of minimum potential energy.

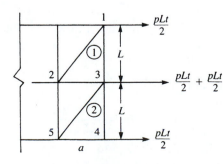

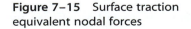

Figure 7–15 Surface traction equivalent nodal forces

For higher-order elements, this load replacement by use of Eq. (7.3.7) is generally not equal to the apparent statically equivalent one; however, it is consistent in that this replacement results directly from the energy approach.

We now recognize the force matrix $\{f_s\}$ defined by Eq. (7.3.7), and based on the principle of minimum potential energy, to be equivalent to that based on work equivalence, which we previously used in Chapter 5 when discussing distributed loads acting on beams.

▲ 7.4 Explicit Expression for the Constant-Strain Triangle Stiffness Matrix

Although the stiffness matrix is generally formulated internally in most computer programs by performing the matrix triple product indicated by Eq. (7.4.1), it is still a valuable learning experience to evaluate the stiffness matrix explicitly for the constant-strain triangular element. Hence, we will consider the plane strain case specifically in this development.

First, recall that the stiffness matrix is given by

$$[k] = tA[B]^T[D][B] \tag{7.4.1}$$

where, for the plane strain case, $[D]$ is given by Eq. (7.1.10) and $[B]$ is given by Eq. (7.2.34). On substituting the matrices $[D]$ and $[B]$ into Eq. (7.4.1), we obtain

$$[k] = \frac{tE}{4A(1+v)(1-2v)}
\begin{bmatrix}
\beta_i & 0 & \gamma_i \\
0 & \gamma_i & \beta_i \\
\beta_j & 0 & \gamma_j \\
0 & \gamma_j & \beta_j \\
\beta_m & 0 & \gamma_m \\
0 & \gamma_m & \beta_m
\end{bmatrix}$$

$$\times
\begin{bmatrix}
1-v & v & 0 \\
v & 1-v & 0 \\
0 & 0 & \dfrac{1-2v}{2}
\end{bmatrix}
\begin{bmatrix}
\beta_i & 0 & \beta_j & 0 & \beta_m & 0 \\
0 & \gamma_i & 0 & \gamma_j & 0 & \gamma_m \\
\gamma_i & \beta_i & \gamma_j & \beta_j & \gamma_m & \beta_m
\end{bmatrix}
\tag{7.4.2}$$

On multiplying the matrices in Eq. (7.4.2), we obtain Eq. (7.4.3), the explicit constant-strain triangle stiffness matrix for the plane strain case. Note that $[k]$ is a function of the difference in the x and y nodal coordinates, as indicated by the γ's and β's, of the material properties E and v, and of the thickness t and surface area A of the element.

$$\underline{k} = \frac{tE}{4A(1+v)(1-2v)}$$

$$\times \begin{bmatrix}
\beta_i^2(1-v) + \gamma_i^2\left(\frac{1-2v}{2}\right) & \beta_i\gamma_i v + \beta_i\gamma_i\left(\frac{1-2v}{2}\right) & \beta_i\beta_j(1-v) + \gamma_i\gamma_j\left(\frac{1-2v}{2}\right) \\[2mm]
 & \gamma_i^2(1-v) + \beta_i^2\left(\frac{1-2v}{2}\right) & \beta_j\gamma_i v + \beta_i\gamma_j\left(\frac{1-2v}{2}\right) \\[2mm]
 & & \beta_j^2(1-v) + \gamma_j^2\left(\frac{1-2v}{2}\right) \\[2mm]
\text{Symmetry}
\end{bmatrix}$$

$$\begin{bmatrix}
\beta_i\gamma_j v + \beta_j\gamma_i\left(\frac{1-2v}{2}\right) & \beta_i\beta_m(1-v) + \gamma_i\gamma_m\left(\frac{1-2v}{2}\right) & \beta_i\gamma_m v + \beta_m\gamma_i\left(\frac{1-2v}{2}\right) \\[2mm]
\gamma_i\gamma_j(1-v) + \beta_i\beta_j\left(\frac{1-2v}{2}\right) & \beta_m\gamma_i v + \beta_i\gamma_m\left(\frac{1-2v}{2}\right) & \gamma_i\gamma_m(1-v) + \beta_i\beta_m\left(\frac{1-2v}{2}\right) \\[2mm]
\beta_j\gamma_j v + \beta_j\gamma_j\left(\frac{1-2v}{2}\right) & \beta_j\beta_m(1-v) + \gamma_j\gamma_m\left(\frac{1-2v}{2}\right) & \beta_j\gamma_m v + \gamma_j\beta_m\left(\frac{1-2v}{2}\right) \\[2mm]
\gamma_j^2(1-v) + \beta_j^2\left(\frac{1-2v}{2}\right) & \beta_m\gamma_j v + \beta_j\gamma_m\left(\frac{1-2v}{2}\right) & \gamma_j\gamma_m(1-v) + \beta_j\beta_m\left(\frac{1-2v}{2}\right) \\[2mm]
 & \beta_m^2(1-v) + \gamma_m^2\left(\frac{1-2v}{2}\right) & \gamma_m\beta_m v + \beta_m\gamma_m\left(\frac{1-2v}{2}\right) \\[2mm]
 & & \gamma_m^2(1-v) + \beta_m^2\left(\frac{1-2v}{2}\right)
\end{bmatrix}$$

$$(7.4.3)$$

For the plane stress case, we need only replace $1-v$ by 1, $(1-2v)/2$ by $(1-v)/2$, and $(1+v)(1-2v)$ outside the brackets by $1-v^2$ in Eq. (7.4.3).

Finally, it should be noted that for Poisson's ratio v approaching 0.5, as in rubberlike materials and plastic solids, for instance, a material becomes incompressible [2]. For plane strain, as v approaches 0.5, the denominator becomes zero in the material property matrix [see Eq. (7.1.10)] and hence in the stiffness matrix, Eq. (7.4.3). A value of v near 0.5 can cause ill-conditioned structural equations. A special formulation (called a *penalty formulation* [3]) has been used in this case.

▲ 7.5 Finite Element Solution of a Plane Stress Problem ▲

To illustrate the finite element method for a plane stress problem, we now present a detailed solution.

Example 7.2

For a thin plate subjected to the surface traction shown in Figure 7–16, determine the nodal displacements and the element stresses. The plate thickness $t = 1$ in., $E = 30 \times 10^6$ psi, and $v = 0.30$.

Discretization

To illustrate the finite element method solution for the plate, we first discretize the plate into two elements, as shown in Figure 7–17. It should be understood that the coarseness of the mesh will not yield as true a predicted behavior of the plate as would a finer mesh, particularly near the fixed edge. However, since we are performing a longhand solution, we will use a coarse discretization for simplicity (but without loss of generality of the method).

In Figure 7–17, the original tensile surface traction in Figure 7–16 has been converted to nodal forces as follows:

$$F = \tfrac{1}{2}TA$$

$$F = \tfrac{1}{2}(1000 \text{ psi})(1 \text{ in.} \times 10 \text{ in.})$$

$$F = 5000 \text{ lb}$$

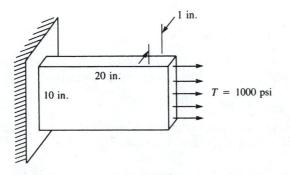

Figure 7–16 Thin plate subjected to tensile stress

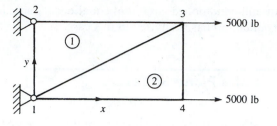

Figure 7–17 Discretized plate

In general, for higher-order elements, Eq. (7.3.7) should be used to convert distributed surface tractions to nodal forces. However, for the CST element, we have shown in Section 7.3 that a statically equivalent force replacement can be used directly, as has been done here.

The governing global matrix equation is

$$\{F\} = [K]\{d\} \tag{7.5.1}$$

Expanding matrices in Eq. (7.5.1), we obtain

$$
\begin{Bmatrix} F_{1x} \\ F_{1y} \\ F_{2x} \\ F_{2y} \\ F_{3x} \\ F_{3y} \\ F_{4x} \\ F_{4y} \end{Bmatrix}
=
\begin{Bmatrix} R_{1x} \\ R_{1y} \\ R_{2x} \\ R_{2y} \\ 5000 \\ 0 \\ 5000 \\ 0 \end{Bmatrix}
= [K]
\begin{Bmatrix} d_{1x} \\ d_{1y} \\ d_{2x} \\ d_{2y} \\ d_{3x} \\ d_{3y} \\ d_{4x} \\ d_{4y} \end{Bmatrix}
= [K]
\begin{Bmatrix} 0 \\ 0 \\ 0 \\ 0 \\ d_{3x} \\ d_{3y} \\ d_{4x} \\ d_{4y} \end{Bmatrix}
\tag{7.5.2}
$$

where $[K]$ is an 8×8 matrix (two degrees of freedom per node with four nodes) before deleting rows and columns to account for the fixed boundary support conditions at nodes 1 and 2.

Assemblage of the Stiffness Matrix

We assemble the global stiffness matrix by superposition of the individual element stiffness matrices. By Eq. (7.2.52), the stiffness matrix for an element is

$$[k] = tA[B]^T[D][B] \tag{7.5.3}$$

In Figure 7–18 for element 1, we have coordinates $x_i = 0$, $y_i = 0$, $x_j = 20$, $y_j = 10$, $x_m = 0$, and $y_m = 10$, since the global coordinate axes are set up at node 1, and

$$A = \tfrac{1}{2}bh$$

$$A = (\tfrac{1}{2})(20)(10) = 100 \text{ in}^2$$

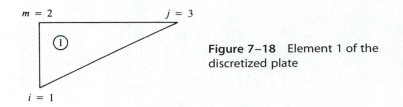

Figure 7–18 Element 1 of the discretized plate

or, in general, A can be obtained equivalently by the nodal coordinate formula of Eq. (7.2.9).

We will now evaluate $[B]$, where $[B]$ is given by Eq. (7.2.34), expanded here as

$$[B] = \frac{1}{2A} \begin{bmatrix} \beta_i & 0 & \beta_j & 0 & \beta_m & 0 \\ 0 & \gamma_i & 0 & \gamma_j & 0 & \gamma_m \\ \gamma_i & \beta_i & \gamma_j & \beta_j & \gamma_m & \beta_m \end{bmatrix} \qquad (7.5.4)$$

and, from Eqs. (7.2.10),

$$\beta_i = y_j - y_m = 10 - 10 = 0$$
$$\beta_j = y_m - y_i = 10 - 0 = 10$$
$$\beta_m = y_i - y_j = 0 - 10 = -10$$
$$\gamma_i = x_m - x_j = 0 - 20 = -20 \qquad (7.5.5)$$
$$\gamma_j = x_i - x_m = 0 - 0 = 0$$
$$\gamma_m = x_j - x_i = 20 - 0 = 20$$

Therefore, substituting Eqs. (7.5.5) into Eq. (7.5.4), we obtain

$$[B] = \frac{1}{200} \begin{bmatrix} 0 & 0 & 10 & 0 & -10 & 0 \\ 0 & -20 & 0 & 0 & 0 & 20 \\ -20 & 0 & 0 & 10 & 20 & -10 \end{bmatrix} \frac{1}{\text{in.}} \qquad (7.5.6)$$

For plane stress, the $[D]$ matrix is conveniently expressed here as

$$[D] = \frac{E}{(1 - \nu^2)} \begin{bmatrix} 1 & \nu & 0 \\ \nu & 1 & 0 \\ 0 & 0 & \dfrac{1 - \nu}{2} \end{bmatrix} \qquad (7.5.7)$$

With $\nu = 0.3$ and $E = 30 \times 10^6$ psi, we obtain

$$[D] = \frac{30(10^6)}{0.91} \begin{bmatrix} 1 & 0.3 & 0 \\ 0.3 & 1 & 0 \\ 0 & 0 & 0.35 \end{bmatrix} \text{ psi} \qquad (7.5.8)$$

Then
$$[B]^T[D] = \frac{30(10^6)}{200(0.91)} \begin{bmatrix} 0 & 0 & -20 \\ 0 & -20 & 0 \\ 10 & 0 & 0 \\ 0 & 0 & 10 \\ -10 & 0 & 20 \\ 0 & 20 & -10 \end{bmatrix} \begin{bmatrix} 1 & 0.3 & 0 \\ 0.3 & 1 & 0 \\ 0 & 0 & 0.35 \end{bmatrix} \tag{7.5.9}$$

Simplifying Eq. (7.5.9) yields

$$[B]^T[D] = \frac{(0.15)(10^6)}{0.91} \begin{bmatrix} 0 & 0 & -7 \\ -6 & -20 & 0 \\ 10 & 3 & 0 \\ 0 & 0 & 3.5 \\ -10 & -3 & 7 \\ 6 & 20 & -3.5 \end{bmatrix} \tag{7.5.10}$$

Using Eqs. (7.5.10) and (7.5.6) in Eq. (7.5.3), we have the stiffness matrix for element 1 as

$$[k] = (1)(100) \frac{(0.15)(10^6)}{0.91} \begin{bmatrix} 0 & 0 & -7 \\ -6 & -20 & 0 \\ 10 & 3 & 0 \\ 0 & 0 & 3.5 \\ -10 & -3 & 7 \\ 6 & 20 & -3.5 \end{bmatrix}$$

$$\times \frac{1}{2(100)} \begin{bmatrix} 0 & 0 & 10 & 0 & -10 & 0 \\ 0 & -20 & 0 & 0 & 0 & 20 \\ -20 & 0 & 0 & 10 & 20 & -10 \end{bmatrix} \tag{7.5.11}$$

Finally, simplifying Eq. (7.5.11) yields

$$[k] = \frac{75{,}000}{0.91} \begin{matrix} \begin{matrix} u_1 & v_1 & u_3 & v_3 & u_2 & v_2 \end{matrix} \\ \begin{bmatrix} 140 & 0 & 0 & -70 & -140 & 70 \\ 0 & 400 & -60 & 0 & 60 & -400 \\ 0 & -60 & 100 & 0 & -100 & 60 \\ -70 & 0 & 0 & 35 & 70 & -35 \\ -140 & 60 & -100 & 70 & 240 & -130 \\ 70 & -400 & 60 & -35 & -130 & 435 \end{bmatrix} \end{matrix} \frac{\text{lb}}{\text{in.}} \tag{7.5.12}$$

where the labels above the columns indicate the nodal order of the degrees of freedom in the element 1 stiffness matrix.

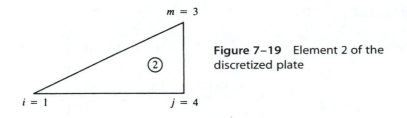

Figure 7–19 Element 2 of the discretized plate

In Figure 7–19 for element 2, we have $x_i = 0$, $y_i = 0$, $x_j = 20$, $y_j = 0$, $x_m = 20$, and $y_m = 10$. Then, from Eqs. (7.2.10), we have

$$\beta_i = y_j - y_m = 0 - 10 = -10$$

$$\beta_j = y_m - y_i = 10 - 0 = 10$$

$$\beta_m = y_i - y_j = 0 - 0 = 0$$

$$\gamma_i = x_m - x_j = 20 - 20 = 0 \qquad (7.5.13)$$

$$\gamma_j = x_i - x_m = 0 - 20 = -20$$

$$\gamma_m = x_j - x_i = 20 - 0 = 20$$

Therefore, using Eqs. (7.5.13) in Eq. (7.5.4) yields

$$[B] = \frac{1}{200} \begin{bmatrix} -10 & 0 & 10 & 0 & 0 & 0 \\ 0 & 0 & 0 & -20 & 0 & 20 \\ 0 & -10 & -20 & 10 & 20 & 0 \end{bmatrix} \frac{1}{\text{in.}} \qquad (7.5.14)$$

The $[D]$ matrix is again given by

$$[D] = \frac{30(10^6)}{0.91} \begin{bmatrix} 1 & 0.3 & 0 \\ 0.3 & 1 & 0 \\ 0 & 0 & 0.35 \end{bmatrix} \text{psi} \qquad (7.5.15)$$

Then, using Eqs. (7.5.14) and (7.5.15), we obtain

$$[B]^T[D] = \frac{30(10^6)}{200(0.91)} \begin{bmatrix} -10 & 0 & 0 \\ 0 & 0 & -10 \\ 10 & 0 & -20 \\ 0 & -20 & 10 \\ 0 & 0 & 20 \\ 0 & 20 & 0 \end{bmatrix} \begin{bmatrix} 1 & 0.3 & 0 \\ 0.3 & 1 & 0 \\ 0 & 0 & 0.35 \end{bmatrix} \qquad (7.5.16)$$

Simplifying Eq. (7.5.16) yields

$$[B]^T[D] = \frac{(0.15)(10^6)}{0.91} \begin{bmatrix} -10 & -3 & 0 \\ 0 & 0 & -3.5 \\ 10 & 3 & -7 \\ -6 & -20 & 3.5 \\ 0 & 0 & 7 \\ 6 & 20 & 0 \end{bmatrix} \qquad (7.5.17)$$

Finally, substituting Eqs. (7.5.17) and (7.5.14) into Eq. (7.5.3), we obtain the stiffness matrix for element 2 as

$$[k] = (1)(100)\frac{(0.15)(10^6)}{0.91} \begin{bmatrix} -10 & -3 & 0 \\ 0 & 0 & -3.5 \\ 10 & 3 & -7 \\ -6 & -20 & 3.5 \\ 0 & 0 & 7 \\ 6 & 20 & 0 \end{bmatrix}$$

$$\times \frac{1}{2(100)} \begin{bmatrix} -10 & 0 & 10 & 0 & 0 & 0 \\ 0 & 0 & 0 & -20 & 0 & 20 \\ 0 & -10 & -20 & 10 & 20 & 0 \end{bmatrix} \qquad (7.5.18)$$

Equation (7.5.18) simplifies to

$$[k] = \frac{75{,}000}{0.91} \begin{array}{cccccc} u_1 & v_1 & u_4 & v_4 & u_3 & v_3 \end{array} \begin{bmatrix} 100 & 0 & -100 & 60 & 0 & -60 \\ 0 & 35 & 70 & -35 & -70 & 0 \\ -100 & 70 & 240 & -130 & -140 & 60 \\ 60 & -35 & -130 & 435 & 70 & -400 \\ 0 & -70 & -140 & 70 & 140 & 0 \\ -60 & 0 & 60 & -400 & 0 & 400 \end{bmatrix} \frac{\text{lb}}{\text{in.}} \qquad (7.5.19)$$

where the degrees of freedom in the element 2 stiffness matrix are shown above the columns in Eq. (7.5.19). Rewriting the element stiffness matrices, Eqs. (7.5.12) and (7.5.19), expanded to the order of, and rearranged according to, increasing nodal degrees of freedom of the total $\underline{K}$ matrix (where we have factored out a constant 5), we obtain

Element 1

$$[k] = \frac{375{,}000}{0.91}
\begin{array}{c}
\begin{array}{cccccccc} u_1 & v_1 & u_2 & v_2 & u_3 & v_3 & u_4 & v_4 \end{array} \\
\begin{bmatrix}
28 & 0 & -28 & 14 & 0 & -14 & 0 & 0 \\
0 & 80 & 12 & -80 & -12 & 0 & 0 & 0 \\
-28 & 12 & 48 & -26 & -20 & 14 & 0 & 0 \\
14 & -80 & -26 & 87 & 12 & -7 & 0 & 0 \\
0 & -12 & -20 & 12 & 20 & 0 & 0 & 0 \\
-14 & 0 & 14 & -7 & 0 & 7 & 0 & 0 \\
0 & 0 & 0 & 0 & 0 & 0 & 0 & 0 \\
0 & 0 & 0 & 0 & 0 & 0 & 0 & 0
\end{bmatrix} \frac{\text{lb}}{\text{in.}}
\end{array}
\qquad (7.5.20)$$

Element 2

$$[k] = \frac{375{,}000}{0.91}
\begin{array}{c}
\begin{array}{cccccccc} u_1 & v_1 & u_2 & v_2 & u_3 & v_3 & u_4 & v_4 \end{array} \\
\begin{bmatrix}
20 & 0 & 0 & 0 & 0 & -12 & -20 & 12 \\
0 & 7 & 0 & 0 & -14 & 0 & 14 & -7 \\
0 & 0 & 0 & 0 & 0 & 0 & 0 & 0 \\
0 & 0 & 0 & 0 & 0 & 0 & 0 & 0 \\
0 & -14 & 0 & 0 & 28 & 0 & -28 & 14 \\
-12 & 0 & 0 & 0 & 0 & 80 & 12 & -80 \\
-20 & 14 & 0 & 0 & -28 & 12 & 48 & -26 \\
12 & -7 & 0 & 0 & 14 & -80 & -26 & 87
\end{bmatrix} \frac{\text{lb}}{\text{in.}}
\end{array}
\qquad (7.5.21)$$

Using superposition of the element stiffness matrices, Eqs. (7.5.20) and (7.5.21), now that the orders of the degrees of freedom are the same, we obtain the total global stiffness matrix as

$$[K] = \frac{375{,}000}{0.91}
\begin{array}{c}
\begin{array}{cccccccc} u_1 & v_1 & u_2 & v_2 & u_3 & v_3 & u_4 & v_4 \end{array} \\
\begin{bmatrix}
48 & 0 & -28 & 14 & 0 & -26 & -20 & 12 \\
0 & 87 & 12 & -80 & -26 & 0 & 14 & -7 \\
-28 & 12 & 48 & -26 & -20 & 14 & 0 & 0 \\
14 & -80 & -26 & 87 & 12 & -7 & 0 & 0 \\
0 & -26 & -20 & 12 & 48 & 0 & -28 & 14 \\
-26 & 0 & 14 & -7 & 0 & 87 & 12 & -80 \\
-20 & 14 & 0 & 0 & -28 & 12 & 48 & -26 \\
12 & -7 & 0 & 0 & 14 & -80 & -26 & 87
\end{bmatrix} \frac{\text{lb}}{\text{in.}}
\end{array}
\qquad (7.5.22)$$

[Alternatively, we could have applied the direct stiffness method to Eqs. (7.5.12) and (7.5.19) to obtain Eq. (7.5.22).] Substituting $[K]$ into $\{F\} = [K]\{d\}$ of Eq. (7.5.2), we have

$$
\begin{Bmatrix} R_{1x} \\ R_{1y} \\ R_{2x} \\ R_{2y} \\ 5000 \\ 0 \\ 5000 \\ 0 \end{Bmatrix} = \frac{375{,}000}{0.91} \begin{bmatrix} 48 & 0 & -28 & 14 & 0 & -26 & -20 & 12 \\ 0 & 87 & 12 & -80 & -26 & 0 & 14 & -7 \\ -28 & 12 & 48 & -26 & -20 & 14 & 0 & 0 \\ 14 & -80 & -26 & 87 & 12 & -7 & 0 & 0 \\ 0 & -26 & -20 & 12 & 48 & 0 & -28 & 14 \\ -26 & 0 & 14 & -7 & 0 & 87 & 12 & -80 \\ -20 & 14 & 0 & 0 & -28 & 12 & 48 & -26 \\ 12 & -7 & 0 & 0 & 14 & -80 & -26 & 87 \end{bmatrix} \begin{Bmatrix} 0 \\ 0 \\ 0 \\ 0 \\ d_{3x} \\ d_{3y} \\ d_{4x} \\ d_{4y} \end{Bmatrix}
$$

$$(7.5.23)$$

Applying the support or boundary conditions by eliminating rows and columns corresponding to displacement matrix rows and columns equal to zero [namely, rows and columns 1–4 in Eq. (7.5.23)], we obtain

$$
\begin{Bmatrix} 5000 \\ 0 \\ 5000 \\ 0 \end{Bmatrix} = \frac{375{,}000}{0.91} \begin{bmatrix} 48 & 0 & -28 & 14 \\ 0 & 87 & 12 & -80 \\ -28 & 12 & 48 & -26 \\ 14 & -80 & -26 & 87 \end{bmatrix} \begin{Bmatrix} d_{3x} \\ d_{3y} \\ d_{4x} \\ d_{4y} \end{Bmatrix}
$$

$$(7.5.24)$$

Transposing the displacement matrix to the left side, we have

$$
\begin{Bmatrix} d_{3x} \\ d_{3y} \\ d_{4x} \\ d_{4y} \end{Bmatrix} = \frac{0.91}{375{,}000} \begin{bmatrix} 48 & 0 & -28 & 14 \\ 0 & 87 & 12 & -80 \\ -28 & 12 & 48 & -26 \\ 14 & -80 & -26 & 87 \end{bmatrix}^{-1} \begin{Bmatrix} 5000 \\ 0 \\ 5000 \\ 0 \end{Bmatrix}
$$

$$(7.5.25)$$

Solving for the displacements in Eq. (7.5.25), we obtain

$$
\begin{Bmatrix} d_{3x} \\ d_{3y} \\ d_{4x} \\ d_{4y} \end{Bmatrix} = \frac{0.91}{75} \begin{Bmatrix} 0.05024 \\ 0.00034 \\ 0.05470 \\ 0.00878 \end{Bmatrix}
$$

$$(7.5.26)$$

Simplifying Eq. (7.5.26), the final displacements are given by

$$
\begin{Bmatrix} d_{3x} \\ d_{3y} \\ d_{4x} \\ d_{4y} \end{Bmatrix} = \begin{Bmatrix} 609.6 \\ 4.2 \\ 663.7 \\ 104.1 \end{Bmatrix} \times 10^{-6} \text{ in.}
$$

$$(7.5.27)$$

Comparing the finite element solution to an analytical solution, as a first approximation, we have the axial displacement given by

$$
\delta = \frac{PL}{AE} = \frac{(10{,}000)20}{10(30 \times 10^6)} = 670 \times 10^{-6} \text{ in.}
$$

for a one-dimensional bar subjected to tensile force. Hence, the nodal x displacement components of Eq. (7.5.27) for the two-dimensional plate appear to be reasonably correct, considering the coarseness of the mesh and the directional stiffness bias of the model. (For more on this subject see Section 8.5.) The y displacement would be expected to be downward at the top (node 3) and upward at the bottom (node 4) as a result of the Poisson effect. However, the directional stiffness bias due to the coarse mesh accounts for this unexpected poor result.

We now determine the stresses in each element by using Eq. (7.2.36):

$$\{\sigma\} = [D][B]\{d\} \tag{7.5.28}$$

In general, for element 1, we then have

$$\{\sigma\} = \frac{E}{(1-v^2)} \begin{bmatrix} 1 & v & 0 \\ v & 1 & 0 \\ 0 & 0 & \frac{1-v}{2} \end{bmatrix} \times \left(\frac{1}{2A}\right) \begin{bmatrix} \beta_1 & 0 & \beta_3 & 0 & \beta_2 & 0 \\ 0 & \gamma_1 & 0 & \gamma_3 & 0 & \gamma_2 \\ \gamma_1 & \beta_1 & \gamma_3 & \beta_3 & \gamma_2 & \beta_2 \end{bmatrix} \begin{Bmatrix} d_{1x} \\ d_{1y} \\ d_{3x} \\ d_{3y} \\ d_{2x} \\ d_{2y} \end{Bmatrix} \tag{7.5.29}$$

Substituting numerical values for $[B]$, given by Eq. (7.5.6); for $[D]$, given by Eq. (7.5.8); and the appropriate part of $\{d\}$, given by Eq. (7.5.27), we obtain

$$\{\sigma\} = \frac{30(10^6)(10^{-6})}{0.91(200)} \begin{bmatrix} 1 & 0.3 & 0 \\ 0.3 & 1 & 0 \\ 0 & 0 & 0.35 \end{bmatrix}$$

$$\times \begin{bmatrix} 0 & 0 & 10 & 0 & -10 & 0 \\ 0 & -20 & 0 & 0 & 0 & 20 \\ -20 & 0 & 0 & 10 & 20 & -10 \end{bmatrix} \begin{Bmatrix} 0 \\ 0 \\ 609.6 \\ 4.2 \\ 0 \\ 0 \end{Bmatrix} \tag{7.5.30}$$

Simplifying Eq. (7.5.30), we obtain

$$\begin{Bmatrix} \sigma_x \\ \sigma_y \\ \tau_{xy} \end{Bmatrix} = \begin{Bmatrix} 1005 \\ 301 \\ 2.4 \end{Bmatrix} \text{psi} \tag{7.5.31}$$

In general, for element 2, we have

$$\{\sigma\} = \frac{E}{(1-v^2)}\left(\frac{1}{2A}\right)\begin{bmatrix} 1 & v & 0 \\ v & 1 & 0 \\ 0 & 0 & \frac{1-v}{2} \end{bmatrix} \times \begin{bmatrix} \beta_1 & 0 & \beta_4 & 0 & \beta_3 & 0 \\ 0 & \gamma_1 & 0 & \gamma_4 & 0 & \gamma_3 \\ \gamma_1 & \beta_1 & \gamma_4 & \beta_4 & \gamma_3 & \beta_3 \end{bmatrix}\begin{Bmatrix} d_{1x} \\ d_{1y} \\ d_{4x} \\ d_{4y} \\ d_{3x} \\ d_{3y} \end{Bmatrix}$$

(7.5.32)

Substituting numerical values into Eq. (7.5.32), we obtain

$$\{\sigma\} = \frac{30(10^6)(10^{-6})}{0.91(200)}\begin{bmatrix} 1 & 0.3 & 0 \\ 0.3 & 1 & 0 \\ 0 & 0 & 0.35 \end{bmatrix}$$

$$\times \begin{bmatrix} -10 & 0 & 10 & 0 & 0 & 0 \\ 0 & 0 & 0 & -20 & 0 & 20 \\ 0 & -10 & -20 & 10 & 20 & 0 \end{bmatrix}\begin{Bmatrix} 0 \\ 0 \\ 663.7 \\ 104.1 \\ 609.6 \\ 4.2 \end{Bmatrix}$$

(7.5.33)

Simplifying Eq. (7.5.33), we obtain

$$\begin{Bmatrix} \sigma_x \\ \sigma_y \\ \tau_{xy} \end{Bmatrix} = \begin{Bmatrix} 995 \\ -1.2 \\ -2.4 \end{Bmatrix}\text{psi}$$

(7.5.34)

The principal stresses can now be determined from Eq. (7.1.2), and the principal angle made by one of the principal stresses can be determined from Eq. (7.1.3). (The other principal stress will be directed 90° from the first.) We determine these principal stresses for element 2 (those for element 1 will be similar) as

$$\sigma_1 = \frac{\sigma_x + \sigma_y}{2} + \left[\left(\frac{\sigma_x - \sigma_y}{2}\right)^2 + \tau_{xy}^2\right]^{1/2}$$

$$\sigma_1 = \frac{995 + (-1.2)}{2} + \left[\left(\frac{995 - (-1.2)}{2}\right)^2 + (-2.4)^2\right]^{1/2}$$

$$\sigma_1 = 497 + 498 = 995 \text{ psi}$$

$$\sigma_2 = \frac{995 + (-1.2)}{2} - 498 = -1.1 \text{ psi}$$

The principal angle is then

$$\theta_p = \frac{1}{2}\tan^{-1}\left[\frac{2\tau_{xy}}{\sigma_x - \sigma_y}\right]$$

or

$$\theta_p = \frac{1}{2}\tan^{-1}\left[\frac{2(-2.4)}{995 - (-1.2)}\right] = 0°$$

Owing to the uniform stress of 1000 psi acting only in the x direction on the edge of the plate, we would expect the stress $\sigma_x(= \sigma_1)$ to be near 1000 psi in each element. Thus, the results from Eqs. (7.5.31) and (7.5.34) for σ_x are quite good. We would expect the stress σ_y to be very small (at least near the free edge). The restraint of element 1 at nodes 1 and 2 causes a relatively large element stress σ_y, whereas the restraint of element 2 at only one node causes a very small stress σ_y. The shear stresses τ_{xy} remain close to zero, as expected. Had the number of elements been increased, with smaller ones used near the support edge, even more realistic results would have been obtained. However, a finer discretization would result in a cumbersome longhand solution and thus was not used here. Use of a computer program is recommended for a detailed solution to this plate problem and certainly for solving more complex stress/strain problems. ∎

In Chapter 8, we illustrate how to use the Algor system computer program to solve plane stress/plane strain problems.

▲ References

[1] Timoshenko, S., and Goodier, J., *Theory of Elasticity*, 3rd ed., McGraw-Hill, New York 1970.
[2] Gere, J. M., *Mechanics of Materials*, 5th ed., Brooks/Cole Publishers, Pacific Grove, CA, 2001.
[3] Cook, R. D., Malkus, D. S., and Plesha, M. E., *Concepts and Applications of Finite Element Analysis*, 3rd ed., Wiley, New York, 1989.

▲ Problems

7.1 Sketch the variations of the shape functions N_j and N_m, given by Eqs. (7.2.18), over the surface of the triangular element with nodes i, j, and m. Check that $N_i + N_j + N_m = 1$ anywhere on the element.

7.2 For a simple three-noded triangular element, show explicitly that differentiation of Eq. (7.2.47) indeed results in Eq. (7.2.48); that is, substitute the expression for $[B]$ and the plane stress condition for $[D]$ into Eq. (7.2.47), and then differentiate π_p with respect to each nodal degree of freedom in Eq. (7.2.47) to obtain Eq. (7.2.48).

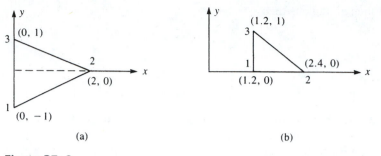

(a)   (b)

Figure P7–3

7.3 Evaluate the stiffness matrix for the elements shown in Figure P7–3. The coordinates are in units of inches. Assume plane stress conditions. Let $E = 30 \times 10^6$ psi, $v = 0.25$, and thickness $t = 1$ in.

7.4 For the elements given in Problem 7.3, the nodal displacements are given as

$$u_1 = 0.0 \qquad v_1 = 0.0025 \text{ in.} \qquad u_2 = 0.0012 \text{ in.}$$

$$v_2 = 0.0 \qquad u_3 = 0.0 \qquad v_3 = 0.0025 \text{ in.}$$

Determine the element stresses $\sigma_x, \sigma_y, \tau_{xy}, \sigma_1$, and σ_2 and the principal angle θ_p. Use the values of E, v, and t given in Problem 7.3.

7.5 Evaluate the stiffness matrix for the elements shown in Figure P7–5. The coordinates are given in units of millimeters. Assume plane stress conditions. Let $E = 210$ GPa, $v = 0.25$, and $t = 10$ mm.

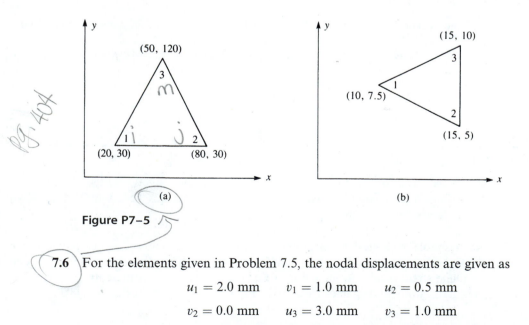

(a)   (b)

Figure P7–5

7.6 For the elements given in Problem 7.5, the nodal displacements are given as

$$u_1 = 2.0 \text{ mm} \qquad v_1 = 1.0 \text{ mm} \qquad u_2 = 0.5 \text{ mm}$$

$$v_2 = 0.0 \text{ mm} \qquad u_3 = 3.0 \text{ mm} \qquad v_3 = 1.0 \text{ mm}$$

Determine the element stresses $\sigma_x, \sigma_y, \tau_{xy}, \sigma_1$, and σ_2 and the principal angle θ_p. Use the values of E, v, and t given in Problem 7.5.

7.7 For the plane strain elements shown in Figure P7–7, the nodal displacements are given as

$$u_1 = 0.001 \text{ in.} \qquad v_1 = 0.005 \text{ in.} \qquad u_2 = 0.001 \text{ in.}$$

$$v_2 = 0.0025 \text{ in.} \qquad u_3 = 0.0 \text{ in.} \qquad v_3 = 0.0 \text{ in.}$$

Determine the element stresses $\sigma_x, \sigma_y, \tau_{xy}, \sigma_1$, and σ_2 and the principal angle θ_p. Let $E = 30 \times 10^6$ psi and $v = 0.25$, and use unit thickness for plane strain. All coordinates are in inches.

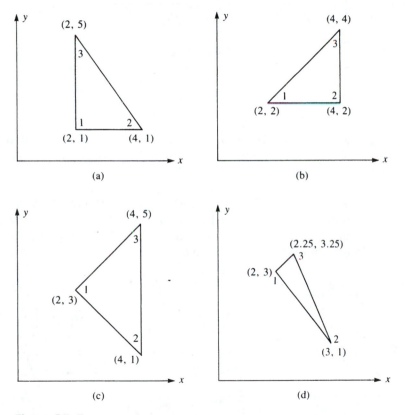

Figure P7–7

7.8 For the plane strain elements shown in Figure P7–8, the nodal displacements are given as

$$u_1 = 0.005 \text{ mm} \qquad v_1 = 0.002 \text{ mm} \qquad u_2 = 0.0 \text{ mm}$$

$$v_2 = 0.0 \text{ mm} \qquad u_3 = 0.005 \text{ mm} \qquad v_3 = 0.0 \text{ mm}$$

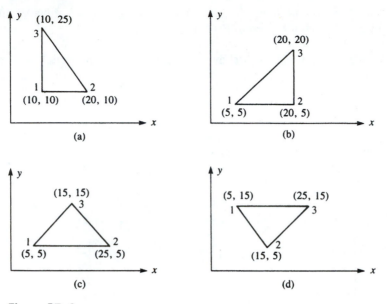

Figure P7-8

Determine the element stresses $\sigma_x, \sigma_y, \tau_{xy}, \sigma_1$, and σ_2 and the principal angle θ_p. Let $E = 70$ GPa and $v = 0.3$, and use unit thickness for plane strain. All coordinates are in millimeters.

7.9 Determine the nodal forces for (a) a linearly varying pressure p_x on the edge of the triangular element shown in Figure P7–9(a); and (b) the quadratic varying pressure shown in Figure P7–9(b) by evaluating the surface integral given by Eq. (7.3.7). Assume the element thickness is equal to t.

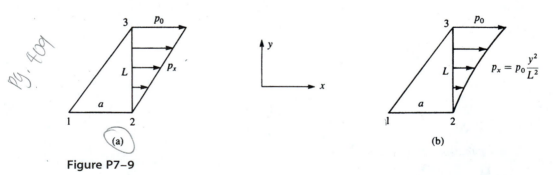

Figure P7-9

7.10 Determine the nodal displacements and the element stresses, including principal stresses, for the thin plate of Section 7.5 with a uniform shear load (instead of a tensile load) acting on the right edge, as shown in Figure P7–10. Use $E = 30 \times 10^6$ psi, $v = 0.30$, and $t = 1$ in. (*Hint:* The $[K]$ matrix derived in Section 7.5 and given by Eq. (7.5.22) can be used to solve the problem.)

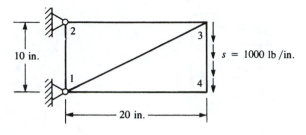

Figure P7–10

7.11 Determine the nodal displacements and the element stresses, including principal stresses, due to the loads shown for the thin plates in Figure P7–11. Use $E = 210$ GPa, $v = 0.30$, and $t = 5$ mm. Assume plane stress conditions apply. The recommended discretized plates are shown in the figures.

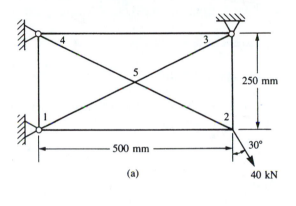

(a)

40 kN

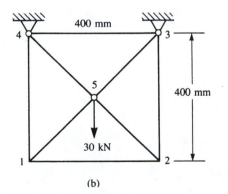

(b)

Figure P7–11

7.12 Evaluate the body force matrix for the plate shown in Figure P7–11(b). Assume the weight density to be 77.1 kN/m^3.

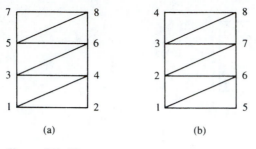

(a)   (b)

Figure P7–13

7.13 For the plane structures modeled by triangular elements shown in Figure P7–13, show that numbering in the direction that has fewer nodes, as in Figure P7–13(a) (as opposed to numbering in the direction that has more nodes), results in a reduced bandwidth. Illustrate this fact by filling in, with X's, the occupied elements in $\underline{K}$ for each mesh, as was done in Appendix B.4. Compare the bandwidths for each case.

7.14 Go through the detailed steps to evaluate Eq. (7.3.6).

7.15 How would you treat a linearly varying thickness for a three-noded triangle?

7.16 Compute the stiffness matrix of element 1 of the two-triangle element model of the rectangular plate in plane stress shown in Figure P7–16. Then use it to compute the stiffness matrix of element 2.

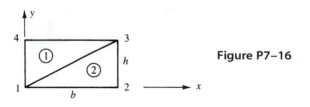

Figure P7–16

Practical Considerations in Modeling; Interpreting Results; and Use of the Algor Program for Plane Stress/Strain Analysis

Introduction

In this chapter, we will describe some modeling guidelines, including generally recommended mesh size, natural subdivisions, and more on use of symmetry and associated boundary conditions. This is followed by discussion of equilibrium, compatibility, and convergence of solution. We will then consider interpretation of stress results.

Next, we introduce the concept of static condensation, which enables us to apply the concept of the basic constant-strain triangle stiffness matrix to a quadrilateral element. Thus, both three-sided and four-sided two-dimensional elements can be used in the finite element models of actual bodies. This concept is incorporated in the Algor computer program.

We then describe the use of the Algor program. This program facilitates the solution of complex, large-number-of-degrees-of-freedom plane stress/plane strain problems that generally cannot be solved longhand because of the larger number of equations involved. Also, problems for which longhand solutions do not exist (such as those involving complex geometries and complex loads or where unrealistic, often gross, assumptions were previously made to simplify the problem to allow it to be described via a classical differential equation approach) can now be solved with a higher degree of confidence in the results by using the finite element approach (with its resulting system of algebraic equations).

A number of problems are solved using the Algor program. These step-by-step solutions provide the information we need to understand how Algor is used to solve plane stress/strain problems.

▲ 8.1 Finite Element Modeling   ▲

We will now discuss various concepts that should be considered when modeling any problem for solution by the finite element method.

General Considerations

Finite element modeling is partly an art guided by visualizing physical interactions taking place within the body. One appears to acquire good modeling techniques through experience and by working with experienced people. General-purpose programs provide some guidelines for specific types of problems [12]. In subsequent parts of this section, some significant concepts that should be considered are described.

In modeling, the user is first confronted with the sometimes difficult task of understanding the physical behavior taking place and understanding the physical behavior of the various elements available for use. Choosing the proper type of element or elements to match as closely as possible the physical behavior of the problem is one of the numerous decisions that must be made by the user. Understanding the boundary conditions imposed on the problem can, at times, be a difficult task. Also, it is often difficult to determine the kinds of loads that must be applied to a body and their magnitudes and locations. Again, working with more experienced users and searching the literature can help overcome these difficulties.

Aspect Ratio and Element Shapes

The **aspect ratio** *is defined as the ratio of the longest dimension to the shortest dimension of a quadrilateral element.* In many cases, as the aspect ratio increases, the inaccuracy of the solution increases. To illustrate this point, Figure 8–1(a) shows five different finite element models used to analyze a beam subjected to bending. Figure 8–1(b) is a plot of the resulting error in the displacement at point *A* of the beam versus the aspect ratio. Table 8–1 reports a comparison of results for the displacements at points *A* and *B* for the five models, and the exact solution [2].

There are exceptions for which aspect ratios approaching 50 still produce satisfactory results; for example, if the stress gradient is close to zero at some location of the actual problem, then large aspect ratios at that location still produce reasonable results.

In general, an element yields best results if its shape is compact and regular. Although different elements have different sensitivities to shape distortions, try to maintain (1) aspect ratios low as in Figure 8–1, cases 1 and 2, and (2) corner angles of quadrilaterals near 90°. Figure 8–2 shows elements with poor shapes that tend to promote poor results. If few of these poor element shapes exist in a model, then usually only results near these elements are poor. In the Algor program, when $\alpha \geqslant 170°$ in Figure 8–2(c), the program automatically divides the quadrilateral into two triangles.

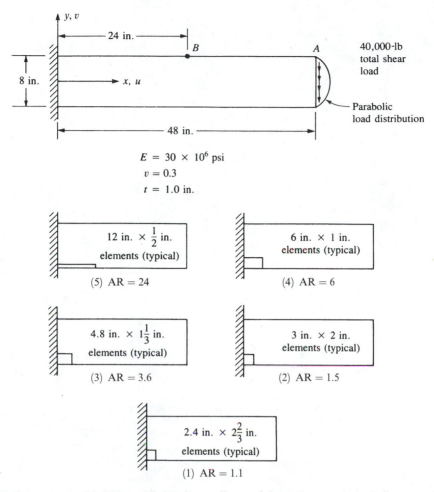

Figure 8–1 (a) Beam with loading; effects of the aspect ratio (AR) illustrated by five cases with different aspect ratios

Use of Symmetry

The appropriate use of symmetry* will often expedite the modeling of a problem. Use of symmetry allows us to consider a reduced problem instead of the actual problem. Thus, we can use a finer subdivision of elements with less labor and computer costs. For another discussion on the use of symmetry, see Reference [3].

Figures 8–3–8–5 illustrate the use of symmetry in modeling (1) a soil mass subjected to foundation loading, (2) a uniaxially loaded member with a fillet, and (3) a plate with a hole subjected to internal pressure. Note that at the plane of symmetry the

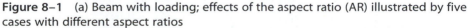

* Again, reflective *symmetry* means correspondence in size, shape, and position of loads; material properties; and boundary conditions that are on opposite sides of a dividing line or plane.

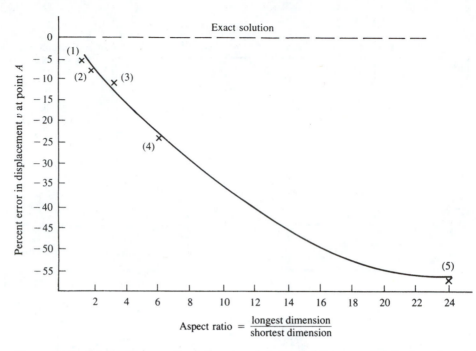

Figure 8–1 (b) Inaccuracy of solution as a function of the aspect ratio (numbers in parentheses correspond to the cases listed in Table 8–1)

Table 8–1 Comparison of results for various aspect ratios

Case	Aspect Ratio	Number of Nodes	Number of Elements	Vertical Displacement, v (in.) Point A	Point B	Percent Error in Displacement at A
1	1.1	84	60	−1.093	−0.346	5.2
2	1.5	85	64	−1.078	−0.339	6.4
3	3.6	77	60	−1.014	−0.328	11.9
4	6.0	81	64	−0.886	−0.280	23.0
5	24.0	85	64	−0.500	−0.158	56.0
Exact solution [2]				−1.152	−0.360	

displacement in the direction perpendicular to the plane must be equal to zero. This is modeled by the rollers at nodes 2–6 in Figure 8–3, where the plane of symmetry is the vertical plane passing through nodes 1–6, perpendicular to the plane of the model. In Figures 8–4(a) and 8–5(a), there are two planes of symmetry. Thus, we need model only one-fourth of the actual members, as shown in Figures 8–4(b) and 8–5(b). Therefore, rollers are used at nodes along both the vertical and horizontal planes of symmetry.

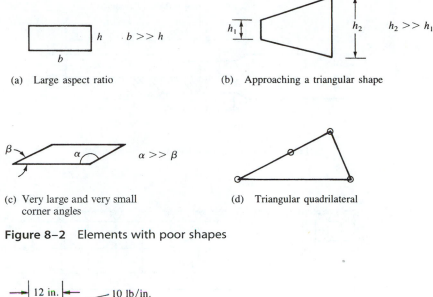

(a) Large aspect ratio

(b) Approaching a triangular shape

(c) Very large and very small corner angles

(d) Triangular quadrilateral

Figure 8–2 Elements with poor shapes

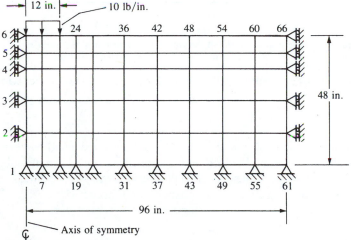

Figure 8–3 Use of symmetry applied to a soil mass subjected to foundation loading (number of nodes = 66, number of elements = 50) (2.54 cm = 1 in., 4.445 N = 1 lb)

As previously indicated in Chapter 3, in vibration and buckling problems, symmetry must be used with caution since symmetry in geometry does not imply symmetry in all vibration or buckling modes.

Natural Subdivisions at Discontinuities

Figure 8–6 illustrates various natural subdivisions for finite element discretization. For instance, nodes are required at locations of concentrated loads or discontinuity in

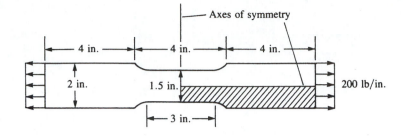

(a) Plane stress uniaxially loaded member with fillet

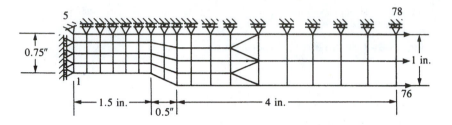

(b) Enlarged finite element model of the cross-hatched quarter of the member (number of nodes = 78, number of elements = 60) (2.54 cm = 1 in.)

Figure 8–4 Use of symmetry applied to a uniaxially loaded member with a fillet

loads, as shown in Figure 8–6(a) and (b). Nodal lines are defined by abrupt changes of plate thickness, as in Figure 8–6(c), and by abrupt changes of material properties, as in Figure 8–6(d) and (e). Other natural subdivisions occur at re-entrant corners, as in Figure 8–6(f), and along holes in members, as in Figure 8–5.

Sizing of Elements and Mesh Refinement

The discretization depends on the geometry of the structure, the loading pattern, and the boundary conditions. For instance, regions of stress concentration or high stress gradient due to fillets, holes, or re-entrant corners require a finer mesh near those regions, as indicated in Figures 8–4, 8–5, and 8–6(f).

Finally, Figure 8–4 illustrates the use of triangular elements for transitions from smaller quadrilaterals to larger quadrilaterals. This transition is necessary because for simple CST elements, intermediate nodes along element edges are inconsistent with the energy formulation of the CST equations. If intermediate nodes were used, no assurance of compatibility would be possible, and resulting holes could occur in the deformed model. Using higher-order elements, such as the linear-strain triangle described in Chapter 9, allows us to use intermediate nodes along element edges and maintain compatibility.

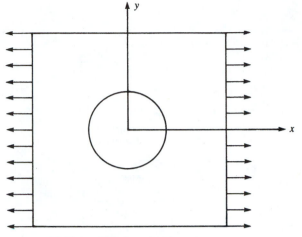

(a) Plate with hole under plane stress

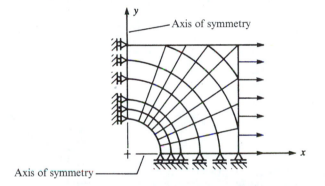

(b) Finite element model of one-quarter of the plate

Figure 8–5 Problem reduction using axes of symmetry applied to a plate with a hole subjected to tensile force

Concentrated or Point Loads

Concentrated or point loads can be applied to nodes of an element provided the element supports the degree of freedom associated with the load. For instance, truss elements and two- and three-dimensional elements support only translational degrees of freedom, and therefore concentrated nodal moments cannot be applied to these elements; only concentrated forces can be applied. However, we should realize that physically concentrated forces are usually an idealization and mathematical convenience that represent a distributed load of high intensity acting over a small area.

According to classical linear theories of elasticity for beams, plates, and solid bodies [2, 16, 17], at a point loaded by a concentrated normal force there is finite displacement and stress in a beam, finite displacement but infinite stress in a plate, and both infinite displacement and stress in a two- or three-dimensional solid body. These

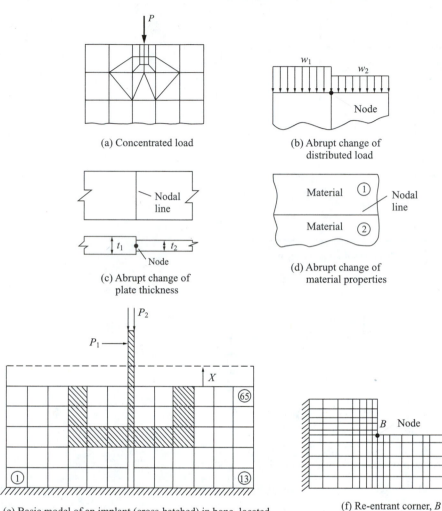

(a) Concentrated load

(b) Abrupt change of
distributed load

(c) Abrupt change of
plate thickness

(d) Abrupt change of
material properties

(e) Basic model of an implant (cross-hatched) in bone, located
at various depths X beneath the bony surface, using
rectangular elements

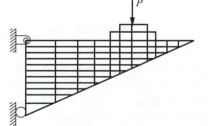

(f) Re-entrant corner, B

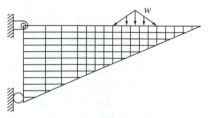

(g) Structure with a distributed load

(h) Using elements to distribute the loading
and spread the concentrated load

Figure 8–6 Natural subdivisions at discontinuities

results are the consequences of the differing assumptions about the stress fields in standard linear theories of beams, plates, and solid elastic bodies. A truly concentrated force would cause material under the load to yield, and linear elastic theories do not predict yielding.

In a finite element analysis, when a concentrated force is applied to a node of a finite element model, infinite displacement and stress are never computed. A concentrated force on a plane stress or strain model has a number of equivalent distributed loadings, which would not be expected to produce infinite displacements or infinite stresses. Infinite displacements and stresses can be approached only as the mesh around the load is highly refined. The best we can hope for is that we can highly refine the mesh in the vicinity of the concentrated load as shown in Figure 8–6(a), with the understanding that the deformations and stresses will be approximate around the load, or that these stresses near the concentrated force are not the object of study, while stresses near another point away from the force, such as *B* in Figure 8–6(f), are of concern. The preceding remarks about concentrated forces apply to concentrated reactions as well.

Finally, another way to model with a concentrated force is to use additional elements and a single concentrated load as shown in Figures 8–6(g) and (h). The shape of the distribution used to simulate a distributed load can be controlled by the relative stiffness of the elements above the loading plane to the actual structure by changing the modulus of elasticity of these elements. This method spreads the concentrated load over a number of elements of the actual structure.

Infinite Medium

Figure 8–3 shows a typical model used to represent an infinite medium (a soil mass subjected to a foundation load). The guideline for the finite element model is that enough material must be included such that the displacements at nodes and stresses within the elements become negligibly small at locations far from the foundation load. Just how much of the medium should be modeled can be determined by a trial-and-error procedure in which the horizontal and vertical distances from the load are varied and the resulting effects on the displacements and stresses are observed. Alternatively, the experiences of other investigators working on similar problems may prove helpful. For a homogeneous soil mass, experience has shown that the influence of the footing becomes insignificant if the horizontal distance of the model is taken as approximately four to six times the width of the footing and the vertical distance is taken as approximately four to ten times the width of the footing [4–6]. Also, the use of infinite elements is described in Reference [13].

After choosing the horizontal and vertical dimensions of the model, we must idealize the boundary conditions. Usually, the horizontal displacement becomes negligible far from the load, and we restrain the horizontal movement of all the nodal points on that boundary (the right-side boundary in Figure 8–3). Hence, rollers are used to restrain the horizontal motion along the right side. The bottom boundary can be completely fixed, as is modeled in Figure 8–3 by using pin supports at each nodal point along the bottom edge. Alternatively, the bottom can be constrained only against vertical movement. The choice depends on the soil conditions at the bottom of the model. Usually, complete fixity is assumed if the lower boundary is taken as bedrock.

In Figure 8–3, the left-side vertical boundary is taken to be directly under the center of the load because symmetry has been assumed. As we said before when discussing symmetry, all nodal points along the line of symmetry are restrained against horizontal displacement.

Finally, Reference [11] is recommended for additional discussion regarding guidelines in modeling with different element types, such as beams, plane stress/plane strain, and three-dimensional solids.

Checking the Model

The discretized finite element model should be checked carefully before results are computed. Ideally, a model should be checked by an analyst not involved in the preparation of the model, who is then more likely to be objective.

Preprocessors with their detailed graphical display capabilities (Figure 8–7) now make it comparatively easy to find errors, particularly the more obvious ones involved with a misplaced node or missing element or a misplaced load or boundary support. Preprocessors include such niceties as color, shrink plots, rotated views, sectioning, exploded views, and removal of hidden lines to aid in error detection.

Most commercial codes also include warnings regarding overly distorted element shapes and checking for sufficient supports. However, the user must still select the proper element types, place supports and forces in proper locations, use consistent units, etc., to obtain a successful analysis.

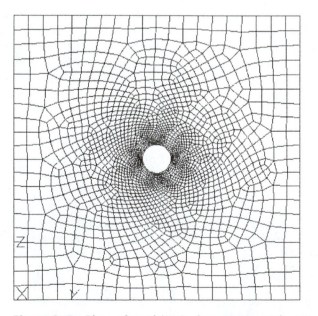

Figure 8–7 Plate of steel (20 in. long, 20 in. wide, 1 in. thick, and with a 1-in.-radius hole) discretized using an Algor preprocessor program with automatic mesh generation

Checking the Results and Typical Postprocessor Results

The results should be checked for consistency by making sure that intended support nodes have zero displacement, as required. If symmetry exists, then stresses and displacements should exhibit this symmetry. Computed results from the finite element program should be compared with results from other available techniques, even if these techniques may be cruder than the finite element results. For instance, approximate mechanics of material formulas, experimental data, and numerical analysis of simpler but similar problems may be used for comparison, particularly if you have no real idea of the magnitude of the answers. Remember to use all results with some degree of caution, as errors can crop up in such sources as textbook or handbook comparison solutions and experimental results.

In the end, the analyst should probably spend as much time processing, checking, and analyzing results as is spent in data preparation.

Finally, we present some typical postprocessor results for the plane stress problem of Figure 8–7 (Figures 8–8 and 8–9). Other examples with results are shown in Section 8.7.

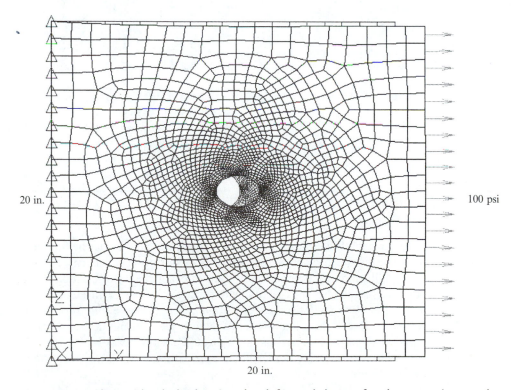

20 in.   100 psi

20 in.

Figure 8–8 Plate with a hole showing the deformed shape of a plate superimposed over an undeformed shape. Plate is fixed on the left edge and subjected to 100-psi tensile stress along the right edge. Maximum horizontal displacement is 7.046 × 10⁻⁵ in. at the center of the right edge

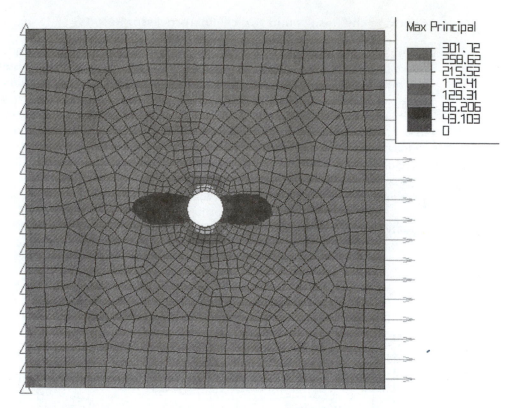

Figure 8–9 Maximum principal stress contour plot for a plate with a hole (original plot is in color). Largest principal stresses of 301 psi occur at the top and bottom of the hole, which indicates a stress concentration of 3. Stresses were obtained by using an average of the nodal values (called smoothing)

▲ 8.2 Equilibrium and Compatibility of Finite Element Results ▲

An approximate solution for a stress analysis problem using the finite element method based on assumed displacement fields does not generally satisfy all the requirements for equilibrium and compatibility that an exact theory-of-elasticity solution satisfies. However, remember that relatively few exact solutions exist. Hence, the finite element method is a very practical one for obtaining reasonable, but approximate, numerical solutions. Recall the advantages of the finite element method as described in Chapter 1 and as illustrated numerous times throughout this text.

We now describe some of the approximations generally inherent in finite element solutions.

1. Equilibrium of nodal forces and moments is satisfied. This is true because the global equation $\underline{F} = \underline{K}\,\underline{d}$ is a nodal equilibrium equation whose solution for $\underline{d}$ is such that the sums of all forces and moments

applied to each node are zero. Equilibrium of the whole structure is also satisfied because the structure reactions are included in the global forces and hence in the nodal equilibrium equations. Numerous example problems, particularly involving truss and frame analysis in Chapter 3 and 6, respectively, have illustrated the equilibrium of nodes and of total structures.

2. Equilibrium within an element is not always satisfied. However, for the constant-strain bar of Chapter 3 and the constant-strain triangle of Chapter 7, element equilibrium is satisfied. Also the cubic displacement function is shown to satisfy the basic beam equilibrium differential equation in Chapter 5 and hence to satisfy element force and moment equilibrium. However, elements such as the linear-strain triangle of Chapter 9, the axisymmetric element of Chapter 10, and the rectangular element of Chapter 11 usually only approximately satisfy the element equilibrium equations.

3. Equilibrium is not usually satisfied between elements. A differential element including parts of two adjacent finite elements is usually not in equilibrium (Figure 8–10). For line elements, such as used for truss

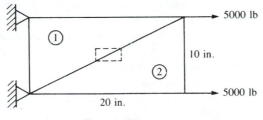

Example 7.2

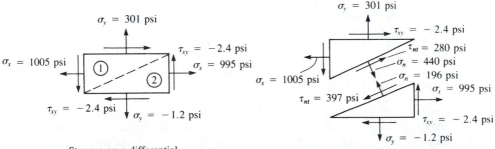

Stresses on a differential element common to both finite elements, illustrating violation of equilibrium

Stress along the diagonal between elements, showing normal and shear stresses, σ_n and τ_{nt}. *Note:* σ_n and τ_{nt} are not equal in magnitude but are opposite in sign for the two elements, and so interelement equilibrium is not satisfied

Figure 8–10 Example 7.2, illustrating violation of equilibrium of a differential element and along the diagonal edge between two elements (the coarseness of the mesh amplifies the violation of equilibrium)

and frame analysis, interelement equilibrium is satisfied, as shown in example problems in Chapters 3–6. However, for two- and three-dimensional elements, interelement equilibrium is not usually satisfied. For instance, the results of Example 7.2 indicate that the normal stress along the diagonal edge between the two elements is different in the two elements. Also, the coarseness of the mesh causes this lack of interelement equilibrium to be even more pronounced. The normal and shear stresses at a free edge usually are not zero even though theory predicts them to be. Again, Example 7.2 illustrates this, with free-edge stresses σ_y and τ_{xy} not equal to zero. However, as more elements are used (refined mesh) the σ_y and τ_{xy} stresses on the stress-free edges will approach zero.

4. Compatibility is satisfied within an element as long as the element displacement field is continuous. Hence, individual elements do not tear apart.

5. In the formulation of the element equations, compatibility is invoked at the nodes. Hence, elements remain connected at their common nodes. Similarly, the structure remains connected to its support nodes because boundary conditions are invoked at these nodes.

6. Compatibility may or may not be satisfied along interelement boundaries. For line elements such as bars and beams, interelement boundaries are merely nodes. Therefore, the preceding statement 5 applies for these line elements. The constant-strain triangle of Chapter 7 and the rectangular element of Chapter 11 remain straight-sided when deformed. Therefore, interelement compatibility exists for these elements; that is, these plane elements deform along common lines without openings, overlaps, or discontinuities. Incompatible elements, those that allow gaps or overlaps between elements, can be acceptable and even desirable. Incompatible element formulations, in some cases, have been shown to converge more rapidly to the exact solution [1]. (For more on this special topic, consult References [7] and [8].)

▲ 8.3 Convergence of Solution ▲

In Section 3.2, we presented guidelines for the selection of so-called compatible and complete displacement functions as they related to the bar element. Those four guidelines are generally applicable, and satisfaction of them has been shown to ensure monotonic convergence of the solution of a particular problem [9]. Furthermore, it has been shown [10] that these compatible and complete displacement functions used in the displacement formulation of the finite element method yield an upper bound on the true stiffness, and hence a lower bound on the displacement of the problem, as shown in Figure 8–11.

Hence, as the mesh size is reduced—that is, as the number of elements is increased—we are ensured of monotonic convergence of the solution when compatible and complete displacement functions are used. Examples of this convergence are

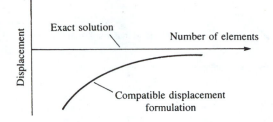

Figure 8–11 Convergence of a finite element solution based on the compatible displacement formulation

Table 8–2 Comparison of results for different numbers of elements

Case	Number of Nodes	Number of Elements	Aspect Ratio	Vertical Displacement, v (in.) Point A
1	21	12	2	−0.740
2	39	24	1	−0.980
3	45	32	3	−0.875
4	85	64	1.5	−1.078
5	105	80	1.2	−1.100
Exact solution [2]				−1.152

given in References [1] and [11], and in Table 8–2 for the beam with loading shown in Figure 8–1(a). All elements in the table are rectangular. The results in Table 8–2 indicate the influence of the number of elements (or the number of degrees of freedom as measured by the number of nodes) on the convergence toward a common solution, in this case the exact one. We again observe the influence of the aspect ratio. The higher the aspect ratio, even with a larger number of degrees of freedom, the worse the answer, as indicated by comparing cases 2 and 3.

8.4 Interpretation of Stresses

In the stiffness or displacement formulation of the finite element method used throughout this text, the primary quantities determined are the interelement nodal displacements of the assemblage. The secondary quantities, such as strain and stress in an element, are then obtained through use of $\{\varepsilon\} = [B]\{d\}$ and $\{\sigma\} = [D][B]\{d\}$. For elements using linear-displacement models, such as the bar and the constant-strain triangle, $[B]$ is constant, and since we assume $[D]$ to be constant, the stresses are constant over the element. In this case, it is common practice to assign the stress to the centroid of the element with acceptable results.

However, as illustrated in Section 3.11 for the axial member, stresses are not predicted as accurately as the displacements. For example, remember the constant-

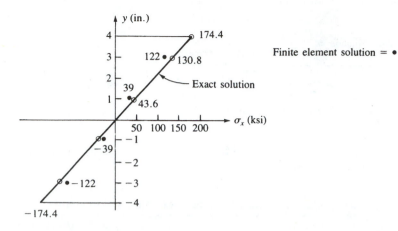

Figure 8–12 Comparison of the finite element solution and the exact solution of bending stress through a beam cross section

strain or constant-stress element has been used in modeling the beam in Figure 8–1. Therefore, the stress in each element is assumed constant. Figure 8–12 compares the exact beam theory solution for bending stress through the beam depth at the centroidal location of the elements next to the wall with the finite element solution of case 4 in Table 8–2. This finite element model consists of four elements through the beam depth. Therefore, only four stress values are obtained through the depth. Again, the best approximation of the stress appears to occur at the midpoint of each element, since the derivative of displacement is better predicted between the nodes than at the nodes.

For higher-order elements, such as the linear-strain triangle of Chapter 9, $[B]$, and hence the stresses, are functions of the coordinates. The common practice is then to evaluate directly the stresses at the centroid of the element.

An alternative procedure sometimes is to use an average (possibly weighted) value of the stresses evaluated at each node of the element. This averaging method is often based on evaluating the stresses at the Gauss points located within the element (described in Chapter 11) and then interpolating to the element nodes using the shape functions of the specific element. Then these stresses in all elements at a common node are averaged to represent the stress at the node. This averaging process in called *smoothing*. Figure 8–9 shows a maximum principal stress "fringe carpet" (dithered) contour plot obtained by smoothing.

Smoothing results in a pleasing, continuous plot which may not indicate some serious problems with the model and the results. You should always view the unsmoothed contour plots as well. Highly discontinuous contours between elements in a region of an unsmoothed plot indicate modeling problems and typically require additional refinement of the element mesh in the suspect region.

If the discontinuities in an unsmoothed contour plot are small or are in regions of little consequence, a smoothed contour plot can normally be used with a high degree of confidence in the results. There are, however, exceptions when smoothing leads to erroneous results. For instance, if the thickness or material stiffness changes significantly between adjacent elements, the stresses will normally be different from one element to the next. Smoothing will likely hide the actual results. Also, for shrink-fit problems involving one cylinder being expanded enough by heating to slip over the smaller one, the circumferential stress between the mating cylinders is normally quite different [16].

The Algor examples in Section 8.7 show additional results, such as displaced models, along with line contour stress plots and smoothed stress plots. The stresses to be plotted can be Von Mises (used in the maximum distortion energy theory to predict failure of ductile materials subjected to static loading), Tresca (used in the Tresca or maximum shear stress theory also to predict failure of ductile materials subjected to static loading) [14, 16], and maximum and minimum principal stresses.

▲ 8.5 Static Condensation

We will now consider the concept of static condensation because this concept is used in developing the stiffness matrix of a quadrilateral element in the Algor computer program.

Consider the basic quadrilateral element with external nodes 1–4 shown in Figure 8–13. An imaginary node 5 is temporarily introduced at the intersection of the diagonals of the quadrilateral to create four triangles. We then superimpose the stiffness matrices of the four triangles to create the stiffness matrix of the quadrilateral element, where the internal imaginary node 5 degrees of freedom are said to be *condensed out* so as never to enter the final equations. Hence, only the degrees of freedom associated with the four *actual* external corner nodes enter the equations.

We begin the static condensation procedure by partitioning the equilibrium equations as

$$\left[\begin{array}{c|c} k_{11} & k_{12} \\ \hline k_{21} & k_{22} \end{array}\right] \left\{\begin{array}{c} d_a \\ \hline d_i \end{array}\right\} = \left\{\begin{array}{c} F_a \\ \hline F_i \end{array}\right\} \tag{8.5.1}$$

where d_i is the vector of internal displacements corresponding to the imaginary internal node (node 5 in Figure 8–13), F_i is the vector of loads at the internal node, and d_a and F_a are the actual nodal degrees of freedom and loads, respectively, at the actual nodes. Rewriting Eq. (8.5.1), we have

$$[k_{11}]\{d_a\} + [k_{12}]\{d_i\} = \{F_a\} \tag{8.5.2}$$

$$[k_{21}]\{d_a\} + [k_{22}]\{d_i\} = \{F_i\} \tag{8.5.3}$$

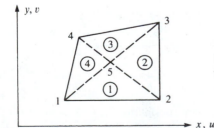

Figure 8–13   Quadrilateral element with an internal node

Solving for $\{d_i\}$ in Eq. (8.5.3), we obtain

$$\{d_i\} = -[k_{22}]^{-1}[k_{21}]\{d_a\} + [k_{22}]^{-1}\{F_i\} \tag{8.5.4}$$

Substituting Eq. (8.5.4) into Eq. (8.5.2), we obtain the condensed equilibrium equation

$$[k_c]\{d_a\} = \{F_c\} \tag{8.5.5}$$

where

$$[k_c] = [k_{11}] - [k_{12}][k_{22}]^{-1}[k_{21}] \tag{8.5.6}$$

$$\{F_c\} = \{F_a\} - [k_{12}][k_{22}]^{-1}\{F_i\} \tag{8.5.7}$$

and $[k_c]$ and $\{F_c\}$ are called the *condensed stiffness matrix* and the *condensed load vector*, respectively. Equation (8.5.5) can now be solved for the actual corner node displacements in the usual manner of solving simultaneous linear equations.

Both constant-strain triangular (CST) and constant-strain quadrilateral elements are used to analyze plane stress/plane strain problems. The quadrilateral element has the stiffness of four CST elements. An advantage of the four-CST quadrilateral is that the solution becomes less dependent on the *skew* of the subdivision mesh, as shown in Figure 8–14. Here **skew** means the *directional stiffness bias* that can be built into a model through certain discretization patterns, since the stiffness matrix of an element is a function of its nodal coordinates, as indicted by Eq. (7.2.52). The four-CST mesh of Figure 8–14(c) represents a reduction in the skew effect over the meshes of Figure 8–14(a) and (b). Figure 8–14(b) is generally worse than Figure 8–14(a) because the use of long, narrow triangles results in an element stiffness matrix that is stiffer along the narrow direction of the triangle.

The resulting stiffness matrix of the quadrilateral element will be an 8×8 matrix consisting of the stiffnesses of four triangles, as was shown in Figure 8–13. The stiffness matrix is first assembled according to the usual direct stiffness method. Then we apply static condensation as outlined in Eqs. (8.5.1)–(8.5.7) to remove the internal node 5 degrees of freedom.

The stiffness matrix of a typical triangular element (labeled element 1 in Figure 8–13) with nodes 1, 2, and 5 is given in general form by

$$[k^{(1)}] = \begin{bmatrix} k_{11}^{(1)} & k_{12}^{(1)} & k_{15}^{(1)} \\ k_{21}^{(1)} & k_{22}^{(1)} & k_{25}^{(1)} \\ k_{51}^{(1)} & k_{52}^{(1)} & k_{55}^{(1)} \end{bmatrix} \tag{8.5.8}$$

where the superscript in parentheses again refers to the element number, and each submatrix $[k_{ij}^{(1)}]$ is of order 2×2. The stiffness matrix of the quadrilateral, assembled

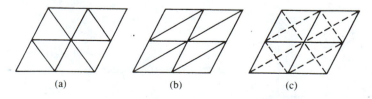

(a)   (b)   (c)

Figure 8–14   Skew effects in finite element modeling

using Eq. (8.5.8) along with similar stiffness matrices for elements 2–4 of Figure 8–12, is given by the following (before static condensation is used):

$$
[k] =
\begin{array}{ccccc}
(u_1,v_1) & (u_2,v_2) & (u_3,v_3) & (u_4,v_4) & (u_5,v_5)
\end{array}
$$

$$
[k] =
\left[
\begin{array}{cccc:c}
[k_{11}^{(1)}] + [k_{11}^{(4)}] & [k_{12}^{(1)}] & [0] & [k_{14}^{(4)}] & [k_{15}^{(1)}] + [k_{15}^{(4)}] \\
[k_{21}^{(1)}] & [k_{22}^{(1)}] + [k_{22}^{(2)}] & [k_{23}^{(2)}] & [0] & [k_{25}^{(1)}] + [k_{25}^{(2)}] \\
[0] & [k_{32}^{(2)}] & [k_{33}^{(2)}] + [k_{33}^{(3)}] & [k_{34}^{(3)}] & [k_{35}^{(2)}] + [k_{35}^{(3)}] \\
[k_{41}^{(4)}] & [0] & [k_{43}^{(3)}] & [k_{44}^{(3)}] + [k_{44}^{(4)}] & [k_{45}^{(3)}] + [k_{45}^{(4)}] \\
\hdashline
[k_{51}^{(1)}] + [k_{51}^{(4)}] & [k_{52}^{(1)}] + [k_{52}^{(2)}] & [k_{53}^{(2)}] + [k_{53}^{(3)}] & [k_{54}^{(3)}] + [k_{54}^{(4)}] & ([k_{55}^{(1)}] + [k_{55}^{(2)}]) + ([k_{55}^{(3)}] + [k_{55}^{(4)}])
\end{array}
\right]
\tag{8.5.9}
$$

where the orders of the degrees of freedom are shown above the columns of the stiffness matrix and the partitioning scheme used in static condensation is indicated by the dotted lines. Before static condensation is applied, the stiffness matrix is of order 10×10.

Example 8.1

Consider the quadrilateral with internal node 5 and dimensions as shown in Figure 8–15 to illustrate the application of static condensation.

Recall that the original stiffness matrix of the quadrilateral is 10×10, but static condensation will result in an 8×8 stiffness matrix after removal of the degrees of freedom (u_5, v_5) at node 5.

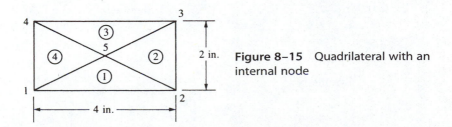

Figure 8–15 Quadrilateral with an internal node

Using the CST stiffness matrix of Eq. (7.4.3) for plane strain, we have

$$
[k^{(1)}] = [k^{(3)}] = \frac{E}{4.16}
\begin{array}{ccc}
\begin{array}{cc} 3 & 4 \\ 1 & 2 \end{array} & \begin{array}{c} 5 \\ 5 \end{array} &
\end{array}
\begin{bmatrix}
1.5 & 1.0 & 0.1 & 0.2 & -1.6 & -1.2 \\
 & 3.0 & -0.2 & 2.6 & -0.8 & -5.6 \\
 & & 1.5 & -1.0 & -1.6 & 1.2 \\
 & & & 3.0 & 0.8 & -5.6 \\
 & & & & 3.2 & 0.0 \\
\text{Symmetry} & & & & & 11.2
\end{bmatrix}
\tag{8.5.10}
$$

Similarly, from Figure 8–15, we can show that

$$
[k^{(2)}] = [k^{(4)}] = \frac{E}{4.16}
\begin{bmatrix}
1.5 & -1.0 & -0.1 & 0.2 & -1.4 & 0.8 \\
 & 3.0 & -0.2 & -2.6 & 1.2 & -0.4 \\
 & & 1.5 & 1.0 & -1.4 & -0.8 \\
 & & & 3.0 & -1.2 & -0.4 \\
 & & & & 2.8 & 0.0 \\
\text{Symmetry} & & & & & 0.8
\end{bmatrix}
\tag{8.5.11}
$$

with column headers $\begin{array}{cc} 2 & 3 \\ 4 & 1 \end{array}$ and $\begin{array}{c} 5 \\ 5 \end{array}$.

where the numbers above the columns in Eqs. (8.5.10) and (8.5.11) indicate the orders of the degrees of freedom associated with each stiffness matrix. Here the quantity in the denominator of Eq. (7.4.3), $4A(1 + v)(1 - 2v)$, is equal to 4.16 in Eqs. (8.5.10) and (8.5.11) because $A = 2$ in^2 and v is taken to be 0.3. Also, the thickness t of the element has been taken as 1 in. Now we can superimpose the stiffness terms as indicated by Eq. (8.5.9) to obtain the general expression for a four-CST element. The resulting assembled total stiffness matrix before static condensation is applied is given by

$$
[k] = \frac{E}{4.16}
\begin{bmatrix}
3.0 & 2.0 & 0.1 & 0.2 & 0.0 & 0.0 & -0.1 & -0.2 & -3.0 & -2.0 \\
 & 6.0 & -0.2 & 2.6 & 0.0 & 0.0 & 0.2 & -2.6 & -2.0 & -6.0 \\
 & & 3.0 & -2.0 & -0.1 & 0.2 & 0.0 & 0.0 & -3.0 & 2.0 \\
 & & & 6.0 & -0.2 & -2.6 & 0.0 & 0.0 & 2.0 & -6.0 \\
 & & & & 3.0 & 2.0 & 0.1 & 0.2 & -3.0 & -2.0 \\
 & & & & & 6.0 & -0.2 & 2.6 & -2.0 & -6.0 \\
 & & & & & & 3.0 & -2.0 & -3.0 & 2.0 \\
 & & & & & & & 6.0 & 2.0 & -6.0 \\
 & & & & & & & & 12.0 & 0.0 \\
\text{Symmetry} & & & & & & & & & 24.0
\end{bmatrix}
$$

$$\tag{8.5.12}$$

After we partition Eq. (8.5.12) and use Eq. (8.5.6), the condensed stiffness matrix is given by

$$
[k_c] = \frac{E}{4.16}
\begin{array}{cccccccc}
u_1 & v_1 & u_2 & v_2 & u_3 & v_3 & u_4 & v_4
\end{array}
\begin{bmatrix}
2.08 & 1.00 & -0.48 & 0.20 & -0.92 & -1.00 & -0.68 & -0.20 \\
 & 4.17 & -0.20 & 1.43 & -1.00 & -1.83 & 0.20 & -3.77 \\
 & & 2.08 & -1.00 & -0.68 & 0.20 & -0.92 & 1.00 \\
 & & & 4.17 & -0.20 & -3.77 & 1.00 & -1.83 \\
 & & & & 2.08 & 1.00 & -0.48 & 0.20 \\
 & & & & & 4.17 & -0.20 & 1.43 \\
 & & & & & & 2.08 & -1.00 \\
\text{Symmetry} & & & & & & & 4.17
\end{bmatrix}
$$

(8.5.13)

■

▲ 8.6 Flowcharts for the Solution of Plane Stress/Strain Problems and Typical Steps Using Algor

In Figure 8–16(a), we present a flowchart of a typical finite element process used for the analysis of plane stress and plane strain problems on the basis of the theory presented in Chapter 7. Figure 8–16(b) is a flowchart of the programs used in the Algor system to create the model, decode it, review it, process or analyze it, and finally examine the results.

Based on the flowchart of Figure 8–16(b), the following typical steps are used to create a two-dimensional model and analyze it with the Algor program.

Step 1 Define the Boundary of the Region Using Superdraw III

Each boundary can be made up of any combination of lines and arcs or a whole circle. Each boundary must be a simple closed curve, and the boundaries must not intersect or touch each other. There must be a single outer boundary, while there can be any number of inner boundaries. Figure 8–17 shows examples of valid and invalid models.

Step 2 Generate the Meshed Model

Select the "FEA Mesh:Two-Dimensional Mesh Generation:Generate Mesh" commands in sequence to generate the meshed model.

Step 3 Add Boundary Conditions and Nodal Forces

Using the Superdraw III "FEA Add:Stress and Vibration Analysis:Boundary Conditions" commands in sequence, you can add boundary conditions to the model. Text-

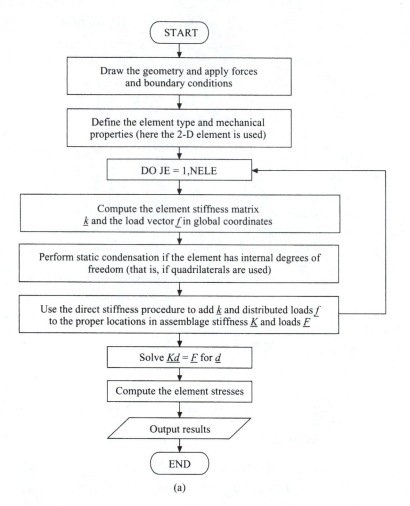

Figure 8–16 (a) Flowchart of plane stress/strain finite element process

only finite element items such as boundary conditions are treated in special ways by automatic meshers [12, 15] and allow you to add boundary conditions before automatic meshing in step 2 if so desired. The "Box Apply" command is often useful in facilitating the application of boundary conditions along an edge. Examples in Section 8.7 illustrate the "Box Apply" command applied to define nodal boundary conditions. Using the "Nodal Forces" command instead of the "Boundary Conditions" command within "FEA Add" allows nodal forces to be applied.

Step 4   Add Surface Numbers for Pressure Loading

Using the Superdraw III "Modify:Update Object Parameters:Surface Number" commands in sequence, you can change the surface number for sections of the outer and

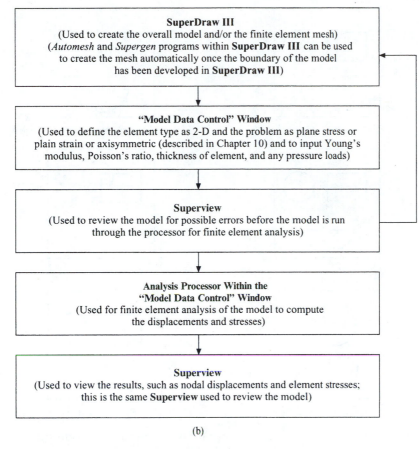

SuperDraw III
(Used to create the overall model and/or the finite element mesh)
(*Automesh* and *Supergen* programs within **SuperDraw III** can be used
to create the mesh automatically once the boundary of the model
has been developed in **SuperDraw III**)

"Model Data Control" Window
(Used to define the element type as 2-D and the problem as plane stress or
plain strain or axisymmetric (described in Chapter 10) and to input Young's
modulus, Poisson's ratio, thickness of element, and any pressure loads)

Superview
(Used to review the model for possible errors before the model is run
through the processor for finite element analysis)

**Analysis Processor Within the
"Model Data Control" Window**
(Used for finite element analysis of the model to compute
the displacements and stresses)

Superview
(Used to view the results, such as nodal displacements and element stresses;
this is the same **Superview** used to review the model)

(b)

Figure 8–16 (b) Flowchart of Algor process to perform plane stress/strain analysis

inner boundaries. Because we are using the "type four" element for plane stress/strain analysis, the "Surface Properties" screen can be used to apply pressure to the boundaries based on the surface number. A common way to indicate the boundary where surface loading will be applied is to use "Select:Box" and box (highlight) the side(s) that you want pressure on. Then use the "Modify:Update Object Parameters:Surface Number" sequence and enter a 2 to change the edge to surface 2. The side with the highest surface number will have the pressure applied to it within the "Surface Definition" screen inside the "Model Data Control" window.

Step 5 Enter Material Properties and Surface Pressure

When you select the "Model Data" button within Superdraw III, you can enter the kind of "Geometric Type" for the analysis you want, as plane stress, plane strain, or axisymmetric, the material properties, and the surface pressure.

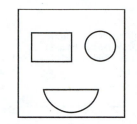

(a) Valid region—with three holes

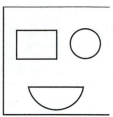

(b) Invalid region—break in outer boundary

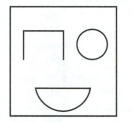

(c) Invalid region—break in inner boundary

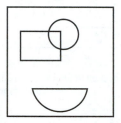

(d) Invalid region—intersecting boundaries

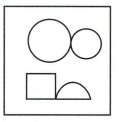

(e) Invalid region—touching boundaries

(f) Invalid region—touching boundaries

Figure 8–17   Valid and invalid regions for two-dimensional modeling

Step 6   Analyze the Model

The analyze the model, select "Analysis" and then "Analyze" in the "Model Data Control" window.

Step 7   Review the Results

You can review the results within Superview, which is accessed by selecting the "Results" button in the "Model Data Control" window. You can then decide if the model needs refinement. Use the "Precision" option to determine where the model should be refined. Usually you will want to refine the model in regions where the precision is bad or the stresses are highest. Precision as used in Algor is defined explicitly in Section 8.7. Examples in Section 8.7 illustrate use of the "Precision" option.

Step 8   Refine the Mesh Using the Two-Dimensional Mesh
Generator

You must first load the premeshed model into Superdraw. First select the "FEA Mesh:2-D Mesh:Generate Mesh" command sequence. This takes you to the "Two-

Dimensional Mesh Generator" screen. The premeshed model can now be loaded by selecting the "Premeshed" button on the "Two-Dimensional Mesh Generator" screen. If you want to refine the whole model, choose a greater mesh density from the screen. If you want to selectively refine the mesh at critical locations only, go back to the "FEA Mesh:Two-Dimensional Mesh Generation:Refinement Points" command sequence. You are now at the "Refine" window where you click "Factor" and then enter a factor, say, 2 or 4. Then click on the model and add the divide refinement points at those critical regions where you want to refine or divide the mesh further. If you choose a factor of 4, then "#d-4" appears at the location where you click on the model.

Typically, a divide factor from 2 to 4 should be sufficient for the first refinement. A divide factor of 2 means that the new element sides where the refinement is defined will be one-half the size of the previous element sides of the original model. You will probably want to use a divide factor of 4 around areas of large precision values (those larger than about 0.1), or where high stresses occur, and a divide factor of 2 where marginally bad precision values occur (those of about 0.05–0.1).

After refining the mesh, you then have to repeat steps 3–7 to solve and analyze the refined model. If you want to increase the density of the location of an existing refine point, use the "FEA:Two-Dimensional Mesh Generation:Refinement Points: Update" option to change the divide factor.

▲ 8.7 Algor Example Solutions for Plane Stress/Strain Analysis ▲

The Algor computer program can be used to solve both small and large plane stress/strain problems. The program is based on the flowchart of Section 8.6. We now solve a number of example problems. These examples illustrate numerous concepts: use of the "2-D Element"; use of "the Surface Number" command to enter the thickness of the element; use of arc, circle, and fillet commands in modeling; how to change "the surface number" of an edge of an element to apply pressure loading along an edge; modeling using symmetry; use of the "Automatic Mesh" and "Two-Dimensional Mesh Generation" menus from the "FEA Mesh" pull-down menu for automatic meshing; and use of the *Group* command to enter materials with different mechanical properties (see Example 8.7). Detailed steps will be listed, including specific keystrokes and commands and descriptions of the purpose of these keystrokes and commands.

Note that when you are using two-dimensional elasticity elements, the model should be formulated and analyzed in the *y-z* plane. Therefore, the transformation from *x-y* to *y-z* can be performed initially using the "Modify:Transform XY to YZ" commands in sequence; or you can select "View:Predefined Views:YZ Right" and draw the model in the *y-z* plane; or finally, you can draw the model in the *x-y* plane and within the "Two-Dimensional Mesh Generation" window change to the *y-z* plane before doing the analysis.

Also, any pressure loading is positive when pushing on the surface edge and is negative when pulling away on the edge. Examples 8.4 and 8.5 illustrate the commands used to create pressure loads along an edge. Also see Reference [12] for more on pressure loads.

Example 8.2

We now solve the thin plate previously solved longhand in Section 7.5 to illustrate the steps for an Algor solution of an elementary plane stress problem (see Figure 8–18). Here $E = 30 \times 10^6$ psi, $v = 0.30$, and thickness $t = 1$ in. The steps to obtain the solution follow. Note that for this simple example, the 1000-psi surface traction will be replaced by 5000-lb nodal loads, one at the top node and one at the bottom node of the free end. In other examples, we often let Algor handle surface traction internally.

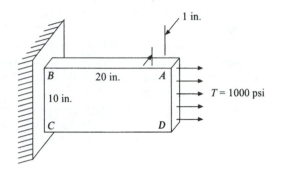

Figure 8–18 Thin plate subjected to a tensile load

Step 1   Start Superdraw III

Double click on the "Algor FEA" icon on your desktop computer screen.

Step 2   Create the Model

Click "Modify".
Click "Transform XY to YZ".
Click "Add" on the main menu bar.
Click "Rectangle". Then enter diagonal coordinates at C and A.
(0, 0, 0)
<Enter>
(0, 20, 10)
<Enter>
Select "View".
Select "Enclose".

Step 3   Mesh the Model

Click "FEA Mesh".
Click "3 Point Triangular".
Click "Division Values".
1
<Tab>

1

<Enter>

Right-click on point *A*, then on *B*, and finally on *C*.

Click "Done".

A two-triangle mesh is created.

Step 4   Add in Boundary Conditions

Select "FEA Add".

Select "Stress and Vibration Analysis".

Click "Boundary Conditions".

Click "Change Values".

Make sure all the boxes have check marks in them.

Click "Done".

Right-click on *B* and *C*. Nodes *B* and *C* are now fixed.

Click "Done".

Step 5   Add Nodal Forces and Moments

Select "FEA Add".

Select "Stress and Vibration Analysis".

Select "Nodal Forces".

Click "Vector".

Click "Y Direction".

Click "Done".

Click "Magnitude".

5000

<Enter>

Click "Done".

Right-click on *A* and *D*. Nodes *A* and *D* now have 5000 lb to them.

Click "Done".

Step 6   Go to the "Model Data Control" Window

Click on the "Model Data" button.

Click "Element".

Click "2-D".

Click "OK".

Select the box below "Data".

You are asked to save the model. Save the model under an appropriate name and continue.

ex82   The name of the model is *ex82*.

Click "Save".

Click "OK".

The "Units" window then pops up. Click "OK" to use the default units.

The "Elements Definition" window pops up.

Set "Thickness" to 1.

Click "OK".

Click "Material".

Select "Customer Defined".

Click "Edit Properties".

Set "Modulus of Elasticity" to 30e6.

Set "Poisson's Ratio" to 0.30.

Click "OK".

Click "OK".

Select "Global".

Set "Load Case Multipliers" to 1.

Click on the "Output" tab.

Click the boxes next to "Displacement Data" and "Stress Data".

Click "OK".

Step 7 Check the Model

Click "Check".

Your model should look like the accompanying figure.

Click "Done".

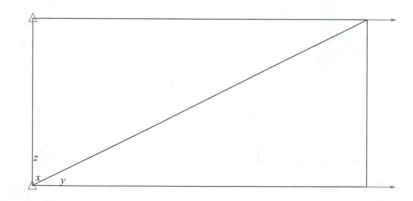

Step 8 Run the Analysis

Click on the "Analysis" button in the "Model Data Control" window.

Click "Analyze". Wait for a message telling you the analysis is complete.

Click "OK".
Click "Done".

Step 9   Review the Results

Click on the "Results" button in the "Model Data Control" window.
Click "OK".
Click "Displaced".
Click "Displ on".
Click "With undi".
Click "Calc scale".
Click "Nodes inq".
Click "Get".
Left-click on "Center Node".
Displacements of point D should be: "DX=0 DY=6.637E−004 DZ=1.041E−004"
`<Esc>`
`<Esc>`
Click "Stress-di".
Click "Post".
Click "Von-Mises".
Click "max-Prin".
`<Esc>`
Click "Done" to exit the Superview program.   ■

Table 8–3 lists the nodal displacements from the Algor solution. The answers compare closely with those from the longhand solution of Example 7.2; see Eq. (7.5.27).

Table 8–4 lists the principal stresses from the Algor solution for the elements of Example 8.2. These stresses compare closely with Example 7.2, Eqs. (7.5.31) and (7.5.34).

Table 8–3   Nodal displacements from Example 8.2

Displacements/Rotations(degrees) of nodes

NODE number	X− translation	Y− translation	Z− translation	X− rotation	Y− rotation	Z− rotation
1	0.0000E+00	0.0000E+00	0.0000E+00	0.0000E+00	0.0000E+00	0.0000E+00
2	0.0000E+00	6.6370E−04	1.0408E−04	0.0000E+00	0.0000E+00	0.0000E+00
3	0.0000E+00	0.0000E+00	0.0000E+00	0.0000E+00	0.0000E+00	0.0000E+00
4	0.0000E+00	6.0958E−04	4.1630E−06	0.0000E+00	0.0000E+00	0.0000E+00

Table 8–4 Element stresses from Example 8.2

**** Nodal stresses for 2-D elasticity elements:

El.	#	LC	ND	Sigma-11 Sigma-Int	Sigma-22	Sigma-33	Tau-12	Sigma-Max	Sigma-Min
1	1	I	9.952E+02 -5.684E-14	-1.201E+00	0.000E+00	-2.402E+00	9.952E+02	-1.207E+00
1	1	J	9.952E+02 5.684E-14	-1.201E+00	0.000E+00	-2.402E+00	9.952E+02	-1.207E+00
1	1	K	9.952E+02 5.684E-14	-1.201E+00	0.000E+00	-2.402E+00	9.952E+02	-1.207E+00
2	1	I	3.014E+02 3.014E+02	1.005E+03	0.000E+00	-2.402E+00	1.005E+03	0.000E+00
2	1	J	3.014E+02 3.014E+02	1.005E+03	0.000E+00	-2.402E+00	1.005E+03	0.000E+00
2	1	K	3.014E+02 3.014E+02	1.005E+03	0.000E+00	-2.402E+00	1.005E+03	0.000E+00

■

Example 8.3

We now solve the problem of a cantilever member subjected to an end load (Figure 8–19) to illustrate how to discretize the beam member into the desired 40 elements. The steps to model and analyze the member using the Algor program follow. Let $E = 30 \times 10^6$ psi, $v = 0.3$, and $t = 1$ in.

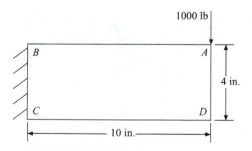

Figure 8–19 Example of cantilever beam modeled using plane stress elements

Step 1 Start Superdraw III

Double-click on the "Algor FEA" icon on your desktop computer screen.

Step 2 Create the Model

Select the "Add" menu from the top main menu bar.
Click "Rectangle".
Enter the nodal data points for the opposite corners of the rectangle.

0<Tab>0<Tab>0   These are the lower-left corner coordinates of the rectangle.

<Enter>

10<Tab>4<Tab>0   These are the upper-right corner coordinates of the rectangle.

<Enter>

Select "View:Enclose". You see the rectangle in the center of the screen.

Step 3 Go to Automesh to Mesh the Model

We will let the Algor program generate the mesh automatically.

Select "FEA Mesh" from the main menu bar.

Select "Two-Dimensional Mesh Generation".

Select "Generate Mesh".

Enter the model name in the "File_name" area that appears.

ex83   The name of the model is *ex83*.

Click "Save". By default, a "Model information does not contain a specified unit system" window appears.

Click "OK".

The "Units Definition" menu appears.

Click "OK" as you want to use the default units "English (in)".

You see a message: "2-D element requires the model in the Y-Z plane".

Click "Yes".

You are now sent to the "Two-Dimensional Mesh Generation" window.

Use the defaults "Quadrilateral", "Mesh Density 400", and so on.

Click "Generate" in the lower left corner of the window.

Click "OK".

Click "Done" to close the "Two-Dimensional Mesh Generation" window.

You now see the meshed model with 374 elements and 417 nodes.

Step 4 Eliminate Duplicate Lines

Select "Modify:Clean:Duplicate".

Click "Perform Cleaning" on the "Duplicate" menu.

You see the following message:

"790 Kept 0 Deleted. done"

Click "Done".

Step 5 Add in Boundary Conditions and Applied Force

Select the "FEA Add:Stress and Vibration Analysis" menu.

Click "Boundary Conditions".

Click "Change Values".

Click "Done". The default setting with all values constrained is needed.

Click "Box Apply". Now apply a box around the left edge nodes. These nodes are now fully constrained.

Click "Done".

Select "FEA:Stress and Vibration Analysis".

Select "Nodal Forces". You will now add the nodal force to the upper-right corner of the beam.

Select "Vector". You need to create the direction of the force.

Click "Z Direction". You see "DX=0 DY=0 DZ=0.97" at the bottom of the screen.

Click "Done".

Click "Magnitude". A prompt "Magnitude" appears at the bottom of the screen asking you to enter the magnitude of the force.

−1000   Enter the magnitude of the force.

<Enter>   A message appears: "Click on node to apply current values".

Move the mouse cursor to the upper-right corner node and click on it. The −1000-lb force appears.

Click "Done".

Select "View" and "Enclose" to see the beam with the applied load.

Step 6   Go to the "Model Data Control" Window to Enter Material Properties

Click on the "Model Data" button.

Click on the box below "Element".

Click "2D". You now have a two-dimensional element selected.

Click "OK".

Click on the box below "Data". You are now at the "Element Data" screen.

Enter a 1 in the "Thickness" box. This sets the thickness of the element to 1 in.

Click "OK". This takes you back to the "Model Data Control" window.

Click on the box below "Material" and scroll down the list of materials.

Click "Steel (4130)".

Click "OK". This takes you back to the "Model Data Control" window.

Click "Global". You are now at the "Global Data" menu screen.

1   Under "Load Case Multipliers", type a 1 under the "Pressure" column.

Click on the "Output" tab.

Click on the boxes next to "Displacement Data" and "Stress Data".

Click "OK".

Step 7   Check Your Model

Click "Check" on the "Global Data" screen.

You now see your model loaded with the boundary conditions and nodal force.

You can now check the model in the usual manner.

Click "Done".

Step 8   Analyze the Model

Click "Analysis" in the "Model Data Control" window.

Click "Analyze" in the lower left corner of the "Linear Stress" menu. Wait a short time for the analysis to be performed.

Click "OK" when the "Algor Analysis" window appears telling you the analysis is finished.

Click "Done" in the lower right of the "Linear Stress and Vibration-Static Stress" window.

Step 9   Review the Results

Click "Results" in the "Model Data Control" window. This takes you to the Superview program, where you can view the results.

Select "Displaced" from the main menu.

Click "Displ on" so that "*Displ on" appears. The asterisk means the displaced mode is active.

Click "With undi" so that "*With undi" appears. This allows the displaced and undisplaced models to be superimposed.

Click "Calc scal" if you want the program to enlarge or scale the displaced model.

Click "Nodes inq" on the "Displaced" menu.

Click "Get" to get a node.

Click on the upper-right corner node. The y and z displacements appear at the bottom of the screen. You see "DY=7.265e−004 DZ=−2.510e−003".

<Esc>

<Esc>

Click "Stress-di". You are now going to post the beam stress.

Click "Post". The Von Mises stress for the beam will appear. The maximum Von Mises stress is shown in red and is 9172.3 psi.

Click "max prin". The maximum principal stress for the beam appears. The maximum principal stress for the beam is 4536.9 psi.

Click "Precision" A plot of the precision is displayed. The maximum precision is 0.19462. A precision of 0.1 or less is desired. Precision higher than 0.1 is due to the point load and can be neglected.

<Esc>

<Esc>

Click "Done". This returns you to Superdraw III.

Close Algor.

Table 8–5 Free-end displacement of cantilever beam from *ex83.L* file

NODE number	X-Displacement	Y-Displacement	Z-Displacement	
55	0.0000E+00	7.265E-04	-2.510E-03	■

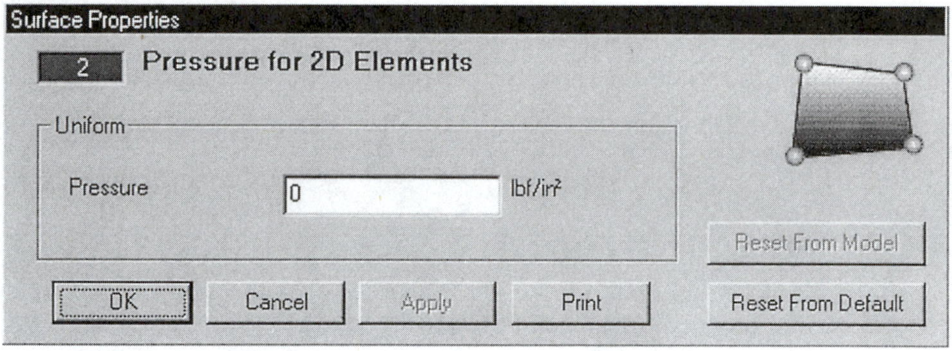

Figure 8–20 "2-D Surface Properties" screen

The free displacement from the Algor solution compares closely with a solution obtained using classical beam theory. Table 8–5 shows the free-end displacement from the *ex83.L* file.

Pressure can be applied to only one edge of an element. This edge must be the edge with the highest surface number. If more than one edge of an element has the highest surface number, then the pressure is applied to the *i-j* side of the element. A load case multiplier must also be entered in the "Global Data" screen under the pressure column as has been for concentrated nodal loadings in previous examples. The "2-D Surface Properties" screen is shown in Figure 8–20.

Example 8.4 now illustrates how we change the surface number of an edge and then apply surface pressure to this edge in the "Model Data Control" window. It should be noted that positive pressure points inward toward the edge of an element, while negative pressure points away from the edge. Figure 8–21 of Example 8.4 shows a negative pressure of 100 psi along the right edge.

Example 8.4

We now solve the problem of a plate clamped on the left edge with a center hole subjected to a tensile stress as shown in Figure 8–21. The model illustrates how to use the "Circle" and "Automesh" features of Algor for automatic meshing and how to change the surface numbers of a side of an element to then apply a pressure load to the new surface number side. Let $E = 30 \times 10^6$ psi, $v = 0.25$, and thickness $= 1$ in.

Step 1 Start Superdraw III

Click on the "Algor FEA" icon to start Superdraw III.

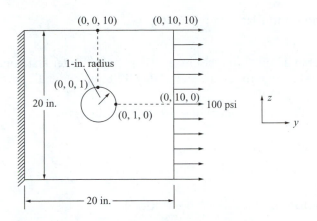

Figure 8–21 Plate with hole subjected to tensile loading

Step 2   Create the Model

Click "Modify" on the main menu bar.

Click "Transform XY to YZ". This is used to transform the drawing to the *y-z* plane at the outset. You could draw the model in the *x-y* plane and then change it to the *y-z* plane by using the "View:Predefined Views:YZ Right" menu.

Click "Add" on the main menu bar. We will create the square plate first.

Click "Line" and enter the coordinates.

(0, 0, 0)   First coordinate of model (lower left node).

<Enter>

(0, 20, 0)   Second coordinate of model (lower right node).

<Enter>

(0, 20, 20)   Third coordinate of model (upper right node).

<Enter>

(0, 0, 20)   Upper left node.

<Enter>

(0, 0, 0)   Enter the initial coordinates again. You are now finished.

<Enter>

Click "Done" on the "Line" menu.

Click "View:Enclose" to view the model. You see a square plate.

Step 3   Eliminate Duplicate Lines

Select "Modify:Clean:Duplicate".

Click "Perform Cleaning" on the "Duplicate" menu.

You see the following message:

                    "4 Kept 0 Deleted. done"

Click "Done".

Step 4   Create the Center Hole in the Plate

Select "Add".

Click "Circle:Center and Point" (on the circle). Then at the "Enter first point" prompt enter the center coordinate of the circle, and at the "Enter next point" prompt enter a point on the circle.

(0, 10, 10)   This is the center coordinate of the circle.

<Enter>

(0, 11, 10)   This is a point on the circle.

<Enter>   You see the plate with a center hole.

Step 5   Go to "Automesh" to Mesh the Model

We will let the Algor program generate the mesh automatically.

Click "FEA Mesh" on the main menu bar.

Select "Two-Dimensional Mesh Generation".

Click "Generate Mesh".

ex84   This is the name of the file.

Click "Save". A message appears: "Model does not contain a specified unit system. Please choose a unit system first."

Click "OK".

You are then asked for a unit system and taken to the "Units" menu.

Click "OK" as we want the current English units. This takes you to the "2-D Mesh Generation" menu. By default, the element shape is a quadrilateral. (You can also choose all triangles or a mix of triangles and quadrilaterals.) The mesh density is 400.

Click "Generate" located in the lower left of this menu. A message appears telling you the number of elements is 796 and the number of nodes is 844.

Click "OK".

Click "Done" on the "2-D Mesh Generation" menu. You see the discretized plate.

Select "Modify:Clean:Duplicate".

Click "Perform Cleaning".

You see "1640 Kept 0 Deleted. done" at bottom of screen.

Click "Done".

Step 6   Change the Surface Number of the Right Side to Apply a
             Pressure Load

Click "Select".

Click "Box" and then carefully box the right edge of the plate. Each element edge has a red highlighted box.

Click "Done".

Click "Modify".

Select "Update Object Parameters".

Click "Surface Number".

2   Enter a 2 for the red surface. The whole right edge has the surface number 2. The pressure load is applied to the edge with the highest surface number so that the right edge has the pressure acting on it.

\<Enter\>

Step 7   Apply Boundary Conditions

Select "FEA Add" from the main menu bar.

Select "Stress and Vibration Analysis".

Click "Boundary Conditions".

Click "Box Apply" and box the left edge of the plate. The "@" sign appears next to each node. This indicates complete fixity of these nodes, which is the default boundary condition.

Click "Done".

Step 8   Add the Material Properties and Pressure Load in the "Model Data Control" Window

Click on the "Model Data" button at the lower part of the screen.

In the "Model Data Control" window you see "Analysis Type-Linear Static Stress".

Click on the box below "Element".

Click "2D".

Click "OK".

Click on the box below "Data".

Select "Plane Stress" (the default type of analysis) and "Isotropic" (the default material).

1.0   Make the thickness 1 in.

Click "OK". You are now back in the "Model Data Control" window.

Click on the box below "Material". You are now in the "Element Material Selection" window.

Scroll down to "Steel ASTM-A36".

Click "Steel ASTM-A36".

Click "OK". You are back in the "Model Data Control" window.

Click on the box below "Surface". You are now at the "Surface" menu.

Click on the red box, which takes you to the "Surface Properties" window.

−100   Type in a pressure of −100 psi. The negative sign means the pressure acts away from the edge of the element.

Click "OK".

Click "Close" in the "Surface Definition" window.

Click on the "Global" box.

1   On the "Load Case Multipliers" menu, highlight "Pressure" and type a 1.

Click on the "Output" tab and click on the "Displacement data" and "Stress data" boxes.

Click "OK". This takes you back to the "Model Data Control" window.

Step 9 Check the Model

Click on the "Check" button in the "Model Data Control" window. This takes you to Superview, and you can check your model. You now see the model called "*ex84.asd*" as seen in upper part of the screen. The model appears with triangles at the left edge nodes and orange arrows along the right edge indicating pressure load.

Click "Done" when you want to exit this program.

Step 10 Run the Analysis

Click on the "Analysis" button under the "FEA Model" column in the "Model Data Control" window.

Click "Analyze" in the lower left corner of the window. Wait about a minute for the analysis to be performed (more or less time depending on the speed of your computer. The elapsed time was 1:22 min on a P5-133 desktop computer and 0:416 min on a G6-200 desktop computer.)

Click "OK".

Click "Done". This takes you back to the "Model Data Control" window.

Step 11 Review the Results

Click on the "Results" button in the "Model Data Control" window.

You now see the model with fixed left edge triangles and arrows on the right edge.

If you select "Nodes Inq" and "Get" and click on the center right edge, you see "DY=7.051e−005."

`<Esc>`

`<Esc>`

To move the legend box off the fringe carpet contour plot:

Click "Stress-di:Post:max prin:⟨Esc⟩:lgnd box:wind adj" and then ⟨Esc⟩ to post the principal stress with the legend window adjusted off the fringe carpet contour. The largest principal stress is 301 psi at the top and bottom of the hole.

To get contour lines instead of a fringe carpet contour plot:

Click "Stress-di:Post:max prin".

`<Esc>`

Click "Aux post:L contour:Display".

To get node numbers or element numbers use ⟨Esc⟩ three times to get back to the main menu:

Click "Options:General:Node num" (An asterisk appears in front of "Node num") to get node numbers or select "Ele num" to get element numbers.

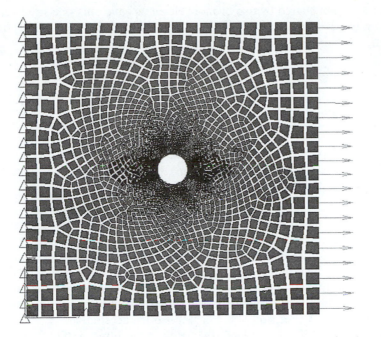

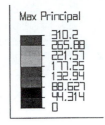

Figure 8–22   Shrink plot of maximum principal stresses for the plate of Example 8.4

To get a shrink-fit plot go back to the main menu:

Click "Options:General:Shrink" on the "Stress-di" menu to get a stress plot with elements taken apart as shown in Figure 8–22.

To get a plot of the precision go back to the "Stress-di" menu:

Click "Post:Precision". A fringe plot of the model precision appears. The largest precision of 0.1511 occurs near the hole. Use step 8 of the general instructions to refine the model near the hole.   ■

The *precision* (Pr) of a model can also be checked as indicated in Example 8.4 and plotted in Examples 8.5 and 8.6. Elements with shared nodes have their stresses calculated independently as we have seen in the longhand Example 7.2, for instance. In general, each element has nodal stresses calculated for it by the "local-least-squares" method in which the stresses are extrapolated from the Gauss points (see Chapter 11 for details of Gauss points) to associated nodes. These elements with shared nodes provide a precision estimate for the model based on the Von Mises stress. Precision is calculated at the nodes in the Algor program as follows:

$$\text{Pr} = \frac{0.5(\text{max value} - \text{min value})}{\text{global Von Mises max}}$$

Nodes that are not shared by two or more elements have only one stress estimate and a precision of zero. For example, assume the following nodal Von Mises stresses from elements.

Node Number	Von Mises Stress			Precision	Calculation of Precision
	Value 1	Value 2	Value 3		
1	40	—	—	0.0	Single-element node
2	40	80	50	0.05	$0.5(80 - 40)/400$
3	300	240	400	0.20	$0.5(400 - 240)/400$

The precision should be used with some caution, as when we have elements with different thicknesses or large stiffness changes. The stresses are then different, and a large local precision might be expected. The precision then depends on how well the actual part is modeled, how well the model is meshed, and other factors. Differences in stress smaller than the indicated precision are normally not significant. For instance, if the precision is 0.1 and the maximum stress is 400 psi, then differences of less than 40 psi are not significant. A precision of 0.1 or less is often acceptable.

For an infinitely long plate, the stress concentration factor approaches 3. Therefore, the maximum stress near the top or bottom of the hole approaches 300 psi. The finite element result for a finite long plate yields a maximum principal stress of 301.72 psi (Figure 8–22), which is very close to a theoretical value for an infinitely long plate.

Example 8.5

We now solve the problem of a plate with a fillet subjected to a tensile load as shown in Figure 8–23. In this model, symmetry will be used. Therefore, only one-fourth of the actual plate will be modeled, as shown in the cross-hatched part. Example 8.6 will show how to remesh areas of Example 8.5 to concentrate meshing on critical areas such as near the fillet. The maximum principal stresses will then be compared for Examples 8.5 and 8.6.

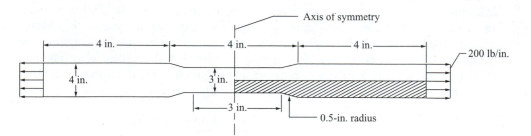

Figure 8–23 Plate with fillet

Step 1 Start Superdraw III

Double-click on the "Algor FEA" icon on your desktop computer screen.

Step 2 Create the Model

Click "Add".

Click "Line". Because of symmetry we will create only the lower right corner of the plate.

0<Tab>0<Tab>0

<Enter>

6<Tab>0<Tab>0

<Enter>

6<Tab>−2<Tab>0

<Enter>

2<Tab>−2<Tab>0

<Enter>

2<Tab>−1.5<Tab>0

<Enter>

0<Tab>−1.5<Tab>0

<Enter>

0<Tab>0<Tab>0

Click "Done" on the menu in the upper right corner.

Click "View:Enclose" You now see the model in the center of the screen.

Click "Construct".

Click "Fillet". You will now create the fillet.

Click "Radius" in the "Fillet" window.

0.5　　　　　This is the radius of the fillet.

<Enter>

To create the fillet you must select the lines near where the fillet is to be created. These lines are the shortest horizontal line near its left end and the shortest vertical line near its bottom.

Left-click on the first line (short horizontal) near its left end where the fillet begins.

Left-click on the short vertical line near its bottom end where the fillet ends.

The fillet should appear as desired. If not, keep trying by clicking in slightly different places on each line.

Click "Done".

Step 3 Mesh the Model

We will let the Algor program generate the mesh automatically.

Select "FEA Mesh" from the main menu bar.

Select "Two-Dimensional Mesh Generation".

Select "Generate Mesh".

Enter the model name in the "File_name" area that appears.

ex85 The name of the model is *ex85*.

Click "Save". By default, the "Model information does not contain a specified unit system" window appears.

Click "OK".

The "Units Definition" menu appears.

Click "OK" as you want to use the default units "English (in)".

You will see a message: "2-D element requires the model in the Y-Z plane".

Click "Yes".

You are now sent to the "Two-Dimensional Mesh Generation" window.

Use the defaults "Quadrilateral", "Mesh Density 400", and so on.

Click "Generate" in the lower left corner of the window.

Click "OK".

Click "Done" to close the "Two-Dimensional Mesh Generation" window.

You now see the meshed model with 346 elements and 391 nodes.

Step 4 Eliminate Duplicate Lines

Select "Modify:Clean:Duplicate".

Click "Perform Cleaning" on the "Duplicate" menu.

You should see the following message:

"736 Kept 0 Deleted. done"

Click "Done".

Step 5 Add in Boundary Conditions

Select the "FEA Add:Stress and Vibration Analysis" menu.

Click "Boundary Conditions".

Click "Change Values".

Click on "Tz Constrained" to release the check mark in front of "Tz Constrained".

This releases the *z*-direction translation.

Click "Done".

Click "Box Apply". Then draw a box around the left edge nodes. These nodes are free to move in the *z* direction.

Click "Change Values".

Click "Tz Constrained" to activate the constraint in the *z* direction.

Click "Ty Constrained" to release the *y*-direction constraint.

Click "Done".

Click "Box Apply". Now draw a box around the top nodes. These nodes are free to move in the *y* direction.

Click "Done".

Click "Select:Box". Now apply a box around the right edge nodes. This allows you to change the surface number of this side to add a pressure later in the "Global Data" window.

Click "Done".

Click "Modify".

Click "Update Object Parameters".

Click "Surface Number".

2   Enter a 2 to change the surface number of the right edge to 2.

`<Enter>`

Step 6   Go to the "Model Data Control" Window to Enter Material Properties

Click on the "Model Data" button.

Click on the box below "Element".

Click "2D". You now have a two-dimensional element selected.

Click "OK".

Click on box below "Data". You are now at the "Element Data" screen.

Enter 1 in the "Thickness" box. This sets the thickness of the element to 1 in.

Click "OK". This takes you back to the "Model Data Control" window.

Click the box below "Material" and scroll down the list of materials.

Click "Steel (4130)".

Click "OK". This takes you back to the "Model Data Control" window.

Click on the box below "Surface". The "Surface Definition" window appears.

Click on the box below "Data" for surface 2.

-200   Enter a pressure of negative 200 in the "Pressure for 2D Elements" window for surface 2. The negative number indicates that the pressure is away from the surface.

Click "OK".

Click "Close". You are back in the "Model Data Control" window.

Click "Global". You are now at the "Global Data" menu screen.

1   Under "Load Case Multipliers" type 1 under the "Pressure" column.

Click on the "Output" tab.

Click on the boxes next to "Displacement Data" and "Stress Data".

Click "OK".

Step 7   Check Your Model

Click "Check" on the "Global Data" screen.

You now see your model loaded with the boundary conditions and nodal forces.

You can now check the model in the usual manner.

Click "Done".

Step 8 Analyze the Model

Click "Analysis" in the "Model Data Control" window.

Click "Analyze" in the lower left corner of the "Linear Stress" menu. Wait a short time for the analysis to be performed.

Click "OK" when the "Algor Analysis" window appears telling you the analysis is finished.

Click "Done" in the lower right of the "Linear Stress and Vibration-Static Stress" window.

Step 9 Review the Results

Click "Results" in the "Model Data Control" window. This takes you to Superview, where you can view the results.

Select "Displaced" on the main menu.

Click "Displ on" so that "*Displ on" appears. The asterisk means the displaced mode is active.

Click "With undi" so that "*With undi" appears. This allows the displaced and undisplaced models to be superimposed.

Click "Calc scal" if you want the program to enlarge or scale the displaced model.

Click "Nodes inq" on the "Displaced" menu.

Click "Get" to get a node.

Click on the upper-right corner node. The y and z displacements appear at the bottom of the screen. You will see "DX=0 DY=4.535e−005 DZ=0".

`<Esc>`

`<Esc>`

Click "Stress di". You are now going to post the beam stress.

Click "Post". The Von Mises stress is automatically displayed with the maximum Von Mises stress equal to 452.82 psi (Figure 8–24).

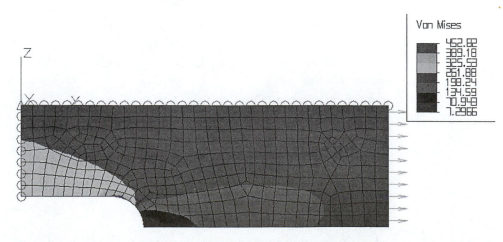

Figure 8–24 Plot of Von Mises stress for Example 8.5

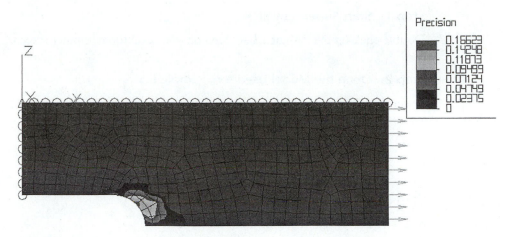

Figure 8–25 Precision plot for Example 8.5 (greatest precision near the fillet indicates a refined mesh is needed near the fillet)

Click "max Prin". The maximum principal stress is displayed with the maximum stress equal to 465.01 psi.

Click "Precision". A plot of the precision (Figure 8–25) is displayed. The maximum precision is 0.16623. A precision of 0.1 or less is desired. Example 8.6 shows how to refine the mesh in order to decrease the precision.

<Esc>

<Esc>

Click "Done". This returns you to Superdraw III.

Close Algor.   ■

Example 8.6

Resolve Example 8.5 by refining the mesh near the fillet in Figure 8–26. Compare these results with those for Example 8.5. Refer to Reference [12] for more details about refining the mesh.

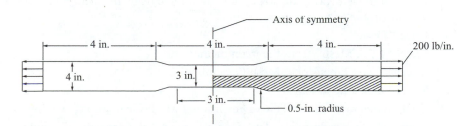

Figure 8–26 Plate with fillet (refined mesh to be used)

Step 1 Start Superdraw III

Double-click on the "Algor FEA" icon on your desktop computer screen.

Step 2 Open the Model Used for Example 8.5

Select "File" from the main menu.
Select "Open".
ex85.esd   Enter the filename of the model from Example 8.5.
`<Enter>`

Step 3 Refine the Mesh

Select "FEA Mesh" from the main menu bar.
Select "Two-Dimensional Mesh Generation".
Select "Generate Mesh".
Select "Premeshed". This clears the old mesh from the model.
Click "Done".
Select "FEA Mesh" from the main menu bar.
Select "Two-Dimensional Mesh Generation".
Select "Refinement Points".
A divide factor of 2 is used. This divides the mesh into 2 times as many elements in the region selected.
Click "Factor" in the "Refine" window.
2   Enter a value of 2 for "Size Factor".
`<Enter>`
Click on the right end of the bottom left horizontal line near the fillet. You then see "#d=2" appear on the edge.
Click on the arc of the fillet. "#d=2" appears where you click.
A divide factor of 4 is now used. This divides the mesh into 4 times as many elements in the region selected.
Click "Factor" in the "Refine" window.
4   Enter a value of 4 for "Size Factor".
`<Enter>`
Click on the left end of the lowest horizontal line near the fillet. You then see "#d=4" appear on the edge.
Click "Done".
Select "FEA Mesh" from the main menu bar.
Select "Two-Dimensional Mesh Generation".
Select "Generate Mesh".
Click "Generate" in the "Two-Dimensional Mesh Generation" window.
Click "Done" to close the "Two-Dimensional Mesh Generation" window.

You now see the meshed model with approximately 587 elements and 643 nodes. The number of elements and nodes can vary slightly depending on where you click to apply the refinement.

Step 4 Eliminate Duplicate Lines

Select "Modify:Clean:Duplicate".
Click "Perform Cleaning" on the "Duplicate" menu.
You should see the following message:

> "736 Kept 0 Deleted. done"

Click "Done".

Step 5 Add in Boundary Conditions

Select the "FEA Add:Stress and Vibration Analysis" menu.
Click "Boundary Conditions".
Click "Change Values".
Click on "Tz Constrained" to release the check mark in front of "Tz Constrained".
This releases the z-direction translation.
Click "Done".
Click "Box Apply". Then draw a box around the left edge nodes. These nodes are free to move in the z direction.
Click "Change Values".
Click "Tz Constrained" to activate the constraint in the z direction.
Click "Ty Constrained" to release the y-direction constraint.
Click "Done".
Click "Box Apply". Now draw a box around the top nodes. These nodes are free to move in the y direction.
Click "Done".
Click "Select:Box". Now apply a box around the right edge nodes. This allows you to change the surface number of this side to add a pressure later in the "Global Data" window.
Click "Done".
Click "Modify".
Click "Update Object Parameters".
Click "Surface Number".
2   Enter a 2 to change the surface number of the right edge to 2.
<Enter>

Step 6 Go to the "Model Data Control" Window to Enter the Material Properties

Click on the "Model Data" button
Click on the box below "Element".

Click "2D". You now have a two-dimensional element selected.

Click "OK".

Click on the box below "Data". You are now at the "Element Data" screen.

Enter 1 in the "Thickness" box. This sets the thickness of the element to 1 in.

Click "OK". This takes you back to the "Model Data Control" window.

Click on the box below "Material" and scroll down the list of materials.

Click "Steel (4130)".

Click "OK". This takes you back to the "Model Data Control" window.

Click on the box below "Surface". The "Surface Definition" window appears.

Click on the box below "Data" for surface "2".

−200   Enter a pressure of negative 200 in the "Pressure for 2D Elements" window for surface 2. The negative number indicates that the pressure is away from the surface.

Click "OK".

Click "Close". You are now in the "Model Data Control" window.

Click "Global". You are now at the "Global Data" menu screen.

1   Under "Load Case Multipliers", type 1 under the "Pressure" column.

Click on the "Output" tab.

Click on the boxes next to "Displacement Data" and "Stress Data".

Click "OK".

Step 7   Check Your Model

Click "Check" on the "Global Data" screen.

You now see your model loaded with the boundary conditions and nodal force.

You can now check the model in the usual manner.

Click "Done".

Step 8   Analyze the Model

Click "Analysis" in the "Model Data Control" window.

Click "Analyze" in the lower left corner of the "Linear Stress" menu. Wait a short time for the analysis to be performed.

Click "OK" when the "Algor Analysis" window appears telling you the analysis is finished.

Click "Done" in the lower right of the "Linear Stress and Vibration-Static Stress" window.

Step 9   Review the Results

Click "Results" in the "Model Data Control" window. This takes you to Superview, where you can view the results.

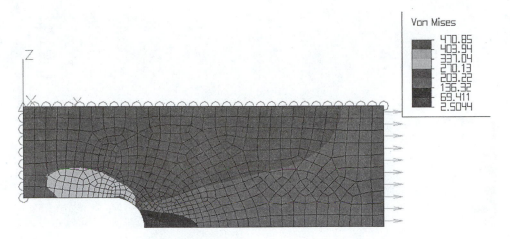

Figure 8–27   Von Mises stress plot for a plate with a fillet (refined mesh)

Select "Displaced" on the main menu.

Click "Displ on" so that "*Displ on" appears. The asterisk means the displaced mode is active.

Click "With undi" so that "*With undi" appears. This allows the displaced and undisplaced models to be superimposed.

Click "Calc scal" if you want the program to enlarge or scale the displaced model.

Click "Nodes inq" on the "Displaced" menu.

Click "Get" to get a node.

Click on the upper right corner node. The y and z displacements appear at the bottom of the screen. You will see "DX=0 DY=4.535e−005 DZ=0".

<Esc>

<Esc>

Click "Stress di". You are now going to post the beam stress.

Click "Post". The Von Mises stress is automatically displayed with the maximum Von Mises stress equal to 470.85 psi (Figure 8–27).

Click "max Prin". The maximum principal stress is displayed with the maximum stress equal to 478.57 psi.

Click "Precision". A plot of the precision is displayed (Figure 8–28). The maximum precision is 0.08275. Refining the mesh has lowered the precision into the acceptable level below 0.1. This also results in more accurate stresses being calculated for the part.

<Esc>

<Esc>

Click "Done". This returns you to Superdraw III.

Close Algor.   ■

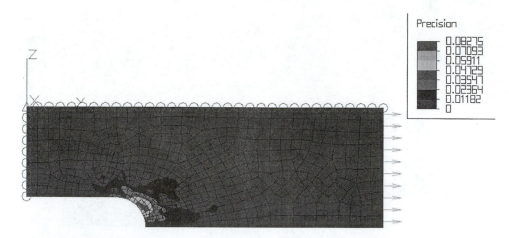

Figure 8–28 Precision plot for a plate with a fillet (refined mesh)

Comparing the results from Examples 8.5 and 8.6, we observe that the Von Mises stress for the refined mesh model is now 470.85 psi versus the 452.8 psi for the less refined model.

Example 8.7

To illustrate the use of two different groups in a plane stress example, a model showing an implant as group 1 and a bony material as group 2 is now solved (Figure 8–29).

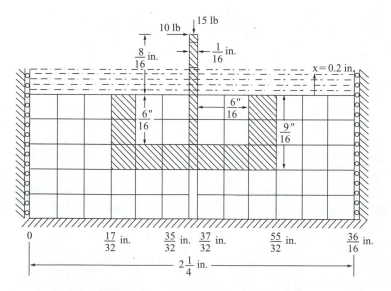

Figure 8–29 Model of a tooth implant within a bony material

Step 1   Start Superdraw III

Click on the "Algor FEA" icon to start Superdraw III.

Step 2   Create the Model

The model looks like Figure P8–13 with $x = 0.2$ in.

Start group 1.

Click "Add" on the main menu bar.

Click "Line".

(17/32, 15/16, 0)

<Enter>

(23/32, 15/16, 0)

<Enter>

(23/32, 9/16, 0)

<Enter>

(35/32, 9/16, 0)

<Enter>

(35/32, 23/16, 0)

<Enter>

(37/32, 23/16, 0)

<Enter>

(37/32, 9/16, 0)

<Enter>

(49/32, 9/16, 0)

<Enter>

(49/32, 15/16, 0)

<Enter>

(55/32, 15/16, 0)

<Enter>

(55/32, 3/8, 0)

<Enter>

(17/32, 3/8, 0)

<Enter>

(17/32, 15/16, 0)

<Enter>

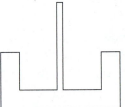

Group 1 model

Click "Done".

Start group 2.

Click "Model Data" button at the lower left edge of the screen.

Click "Group Options".

Click "Add".

Click "2".

Click "Close" on the "Model Data Control" screen.

Click "Snap" on the lower right edge of the screen.

Click the box next to "Point."

Click "Add".

Click "Line".

(0, 0, 0)

<Enter>

(0, 91/80, 0)

<Enter>

(35/32, 91/80, 0)

<Enter>

Group 1 and group 2

Left-click on each corner until you reach the lower left corner of the left side. The coordinates of this point are (37/32, 9/16, 0).

(37/32, 91/80, 0)

<Enter>

(9/4, 91/80, 0)

<Enter>

(9/4, 0, 0)

<Enter>

(0, 0, 0)

<Enter>

Click "Done". Model should look like accompanying figure.

Step 3   Mesh the Model

Select "FEA Mesh".

Select "Two-Dimensional Mesh Generation".

Click "Generate Mesh".

The program asks you to save the model. Call it *P8_13*.

You see the message "Model does not contain a specified unit system. Please choose a unit system first."

Click "OK". This takes you to the "Units Definition" window.

Click "OK" as you want to use the default units "English (in)".

A message appears: "Do you want Superdraw III to convert your model?"

Click "Yes".

Once you are in the "Two-Dimensional Mesh Generation" window,

Click "Generate".

Click "OK".

Click "Apply".

Click "Done".

Step 4   Add in the Boundary Conditions

Select "FEA Add:Stress and Vibration Analysis".

Click "Boundary Conditions".

Click "Change Values".

Click the box next to "Tz". This releases the constraint in the z direction.

Click "Done".

Click "Box Apply".

Now box all the nodes on the left edge of the model.

Click "Box Apply".

Now box all the nodes on the right edge of the model.

Click "Change Values".

Click on the box next to "Tz".

Click "Done".

Click "Box Apply".

Now box all the elements on the bottom edge of the model.

Click "Done".

Step 5   Add Nodal Forces

Select "FEA Add".

Select "Nodal Forces".

Click "Vector".

Click "Y Direction"

Click "Done"

Click "Magnitude"

10

<Enter>

Right-click on the upper right corner of the implant.

Click "Vector".

Click "Z Direction".

Click "Done".

Click "Magnitude".

-15

<Enter>

Right-click on the top of the implant on the central node.

Step 6   Go to the "Model Data Control" Window to Enter Material
　　　　　Properties

Click on the "Model Data" button.

In the group 1 row,

Click on the box below "Element".

Click "2D".
Click "OK".
Click "Data".
Set "Thickness" to 0.25.
Click "OK".
Click "Material".
Select "Customer Defined".
Click "Edit Properties".
Set $E = 1.6e6$.
Set $v = 0.3$.
Click "OK".
Click "OK".
In the group 2 row
Click on the box below "Element".
Click "2D".
Click "OK".
Click "Data".
Set "Thickness" to 0.25.
Click "OK".
Click "Material".
Select "Customer Defined".
Click "Edit Properties".
Set $E = 1e6$.
Set $v = 0.35$.
Click "OK".
Click "OK".
Select "Global".
Under "Load Case Multipliers", type 1 under "Pressure" column.
Click on the "Output" tab.
Click on the boxes next to "Displacement Data" and "Stress Data".
Click "OK".

Step 7   Run the Analysis

Click on the "Analysis" button in the "Model Data Control" window.
Click "Analyze". Wait for a message telling you the analysis is complete.
Click "OK".
Click "Done".

Step 8   Review the Results

Click on the "Results" button in the "Model Data Control" window.
Click "Displaced".

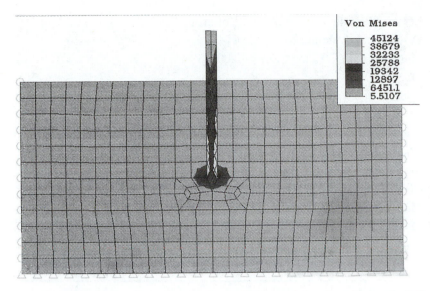

Figure 8–30 Von Mises stress (psi) plot for a tooth implant (Example 8.7)

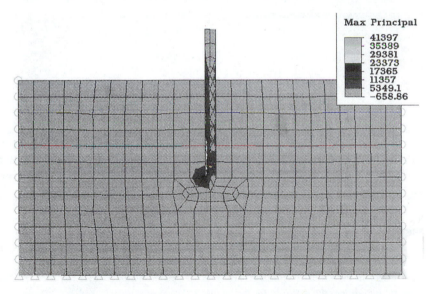

Figure 8–31 Maximum principal stress (psi) plot for a tooth implant (Example 8.7)

Click "Displ on".
Click "With undi".
Click "Calc scale".
Click "Nodes inq".
Click "Get".

Left-click on the upper right node of group 1

Displacements should be:

"DX=0" "DY=0.225" "DZ=−0.00123"

`<Esc>`

`<Esc>`

Click "Stress-di".

Click "Post".

Click "max prin".

Max should be 41,397 (Figure 8–31).

Click "Von Mises".

Max should be 45,124 (Figure 8–30).

`<Esc>`   ■

▲ References

[1] Desai, C. S., and Abel, J. F., *Introduction to the Finite Element Method*, Van Nostrand Reinhold, New York, 1972.

[2] Timoshenko, S., and Goodier, J., *Theory of Elasticity*, 3rd ed., McGraw-Hill, New York, 1970.

[3] Glockner, P. G., "Symmetry in Structural Mechanics," *Journal of the Structural Division*, American Society of Civil Engineers, Vol. 99, No. ST1, pp. 71–89, 1973.

[4] Yamada, Y., "Dynamic Analysis of Civil Engineering Structures," *Recent Advances in Matrix Methods of Structural Analysis and Design*, R. H. Gallagher, Y. Yamada, and J. T. Oden, eds., University of Alabama Press, AL, pp. 487–512, 1970.

[5] Koswara, H., *A Finite Element Analysis of Underground Shelter Subjected to Ground Shock Load*, M. S. Thesis, Rose-Hulman Institute of Technology, Terre Haute, IN, 1983.

[6] Dunlop, P., Duncan, J. M., and Seed, H. B., "Finite Element Analyses of Slopes in Soil," *Journal of the Soil Mechanics and Foundations Division*, Proceedings of the American Society of Civil Engineers, Vol. 96, No. SM2, March 1970.

[7] Cook, R. D., Malkus, D. S., and Plesha, M. E., *Concepts and Applications of Finite Element Analysis*, 3rd ed., Wiley, New York, 1989.

[8] Taylor, R. L., Beresford, P. J., and Wilson, E. L., "A Nonconforming Element for Stress Analysis," *International Journal for Numerical Methods in Engineering*, Vol. 10, No. 6, pp. 1211–1219, 1976.

[9] Melosh, R. J., "Basis for Derivation of Matrices for the Direct Stiffness Method," *Journal of the American Institute of Aeronautics and Astronautics*, Vol. 1, No. 7, pp. 1631–1637. July 1963.

[10] Fraeijes de Veubeke, B., "Upper and Lower Bounds in Matrix Structural Analysis," *Matrix Methods of Structural Analysis*, AGARDograph 72, B. Fraeijes de Veubeke, ed., Macmillan, New York, 1964.

[11] Dunder, V., and Ridlon, S., "Practical Applications of Finite Element Method," *Journal of the Structural Division*, American Society of Civil Engineers, No. ST1, pp. 9–21, 1978.

[12] *Linear Stress and Dynamics Reference Division*, Docutech On-line Documentation, Algor, Inc., Pittsburgh, PA 15238.

[13] Bettess, P., "More on Infinite Elements," *International Journal for Numerical Methods in Engineering*, Vol. 15, pp. 1613–1626, 1980.

[14] Gere, J. M., *Mechanics of Materials*, 5th ed., Brooks/Cole Publishers, Pacific Grove, CA, 2001.

[15] *Superdraw Reference Division*, Docutech On-line Documentation, Algor, Inc., Pittsburgh, PA 15238.

[16] Cook, R. D., and Young, W. C., *Advanced Mechanics of Materials*, Macmillan, New York, 1985.

[17] Cook, R. D., *Finite Element Modeling for Stress Analysis*, Wiley, New York, 1995.

▲ Problems

8.1 For the finite element mesh shown in Figure P8–1, comment on the goodness of the mesh. Indicate the mistakes in the model. Explain and show how to correct them.

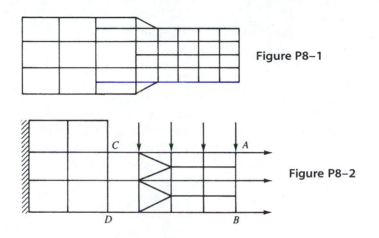

Figure P8–1

Figure P8–2

8.2 Comment on the mesh sizing in Figure P8–2. Is it reasonable? If not, explain why not.

8.3 What happens if the material property $v = 0.5$ in the plane strain case? Is this possible? Explain.

8.4 Under what conditions is the structure in Figure P8–4 a plane strain problem? Under what conditions is the structure a plane stress problem?

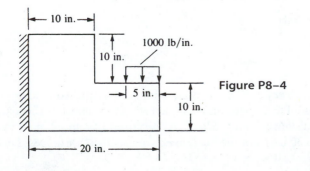

Figure P8–4

8.5 Show that Eq. (8.5.13) is obtained by static condensation of Eq. (8.5.12).

Solve the following problems using the Algor computer program. In some of these problems, we suggest that students be assigned separate parts (or models) to facilitate parametric studies.

8.6 Determine the free-end displacements and the element stresses for the plate discretized into four triangular elements and subjected to the tensile forces shown in Figure P8–6. Compare your results to the solution given in Section 7.5 Why are these results different? Let $E = 30 \times 10^6$ psi, $v = 0.30$, and $t = 1$ in.

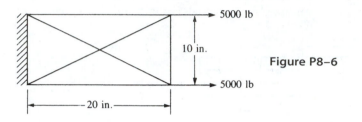

5000 lb

10 in.

Figure P8–6

5000 lb

—20 in.—

8.7 Determine the stresses in the plate with the hole subjected to the tensile stress shown in Figure P8–7. Graph the stress variation σ_x versus the distance y from the hole. Let $E = 30 \times 10^6$ psi, $v = 0.25$, and $t = 1$ in. (Use approximately 25, 50, 75, 100, and then 120 nodes in your finite element model.) Use symmetry as appropriate.

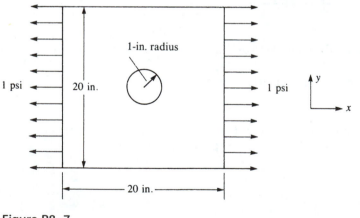

1-in. radius

1 psi   20 in.   1 psi

y

x

20 in.

Figure P8–7

8.8 Solve the following problem of a tensile plate with a concentrated load applied at the top, as shown in Figure P8–8. Determine at what depth the effect of the load dies out. Plot stress σ_y versus distance from the load. At distances of 1 in., 2 in., 4 in., 6 in., 10 in., 15 in., 20 in., and 30 in. from the load, list σ_y versus these distances. Let the width

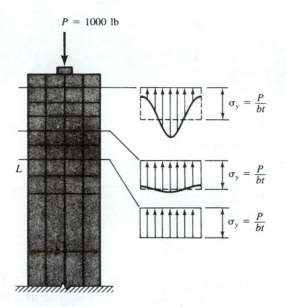

$$\sigma_y = \frac{P}{bt}$$

$$\sigma_y = \frac{P}{bt}$$

$$\sigma_y = \frac{P}{bt}$$

Figure P8–8

of the plate be $b = 4$ in., thickness of the plate be $t = 0.25$ in., and length be $L = 40$ in. Look up the concept of St. Venant's principle to see how it explains the stress behavior in this problem.

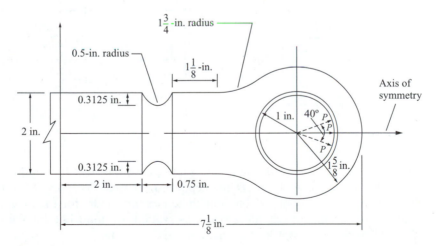

Figure P8–9

8.9 For the connecting rod shown in Figure P8–9, determine the maximum principal stresses and their location. Let $E = 30 \times 10^6$ psi, $v = 0.25$, $t = 1$ in., and $P = 1000$ lb.

8.10 Determine the maximum principal stresses and their locations for the member with fillet subjected to tensile forces shown in Figure P8–10. Let $E = 30 \times 10^6$ psi and $v = 0.25$. Then let $E = 10 \times 10^6$ psi and $v = 0.30$. Let $t = 1$ in. for both cases. Compare your answers for the two cases.

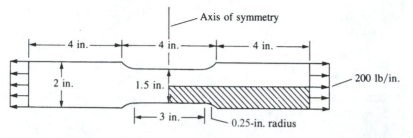

Figure P8–10

8.11 Determine the stresses in the member with a re-entrant corner as shown in Figure P8–11. At what location are the principal stresses largest? Let $E = 30 \times 10^6$ psi and $v = 0.25$. Use plane strain conditions.

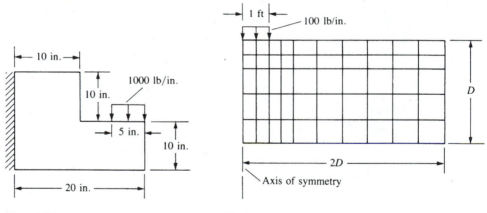

Figure P8–11

Figure P8–12

8.12 Determine the stresses in the soil mass subjected to the strip footing load shown in Figure P8–12. Use a width of $2D$ and depth of D, where D is 3, 4, 6, 8, and 10 ft. Plot the maximum stress contours on your finite element model for each case. Compare your results. Comment regarding your observations on modeling infinite media. Let $E = 30,000$ psi and $v = 0.30$. Use plane strain conditions.

8.13 For the tooth implant subjected to loads shown in Figure P8–13, determine the maximum principal stresses. Let $E = 1.6 \times 10^6$ psi and $v = 0.3$ for the dental restorative implant material (cross-hatched), and let $E = 1 \times 10^6$ psi and $v = 0.35$ for the bony material. Let $X = 0.05$ in., 0.1 in., 0.2 in., 0.3 in., and 0.5 in., where X represents the

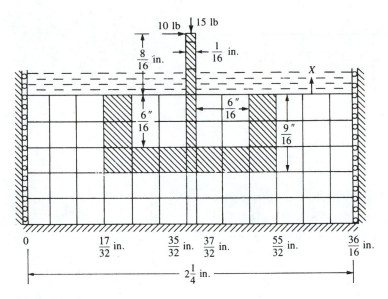

Figure P8–13

various depths of the implant beneath the bony surface. Rectangular elements are used in the finite element model shown in Figure P8–13. Assume the thickness of each element to be $t = 0.25$ in.

8.14 Determine the middepth deflection at the free end and the maximum principal stresses and their location for the beam subjected to the shear load variation shown in Figure P8–14. Do this using 64 rectangular elements all of size 12 in. $\times \frac{1}{2}$ in.; then all of size 6 in. $\times$ 1 in.; then all of size 3 in. $\times$ 2 in. Then use 60 rectangular elements all of size 2.4 in. $\times$ $2\frac{2}{3}$ in.; then all of size 4.8 in. $\times$ $1\frac{1}{3}$ in. Compare the free-end deflections and the maximum principal stresses in each case to the exact solution. Let $E = 30 \times 10^6$ psi, $v = 0.3$, and $t = 1$ in. Comment on the accuracy of both displacements and stresses.

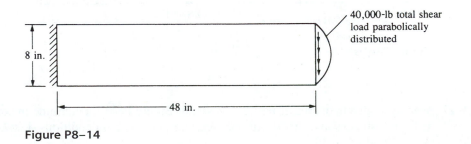

Figure P8–14

8.15 Determine the stresses in the shear wall shown in Figure P8–15. At what location are the principal stresses largest? Let $E = 21$ GPa, $v = 0.25$, $t_{\text{wall}} = 0.10$ m, and $t_{\text{beam}} = 0.20$ m.

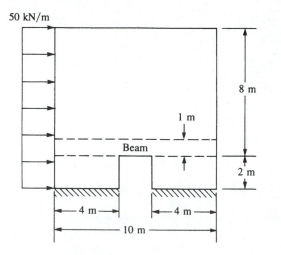

50 kN/m

8 m

1 m

Beam

2 m

4 m

4 m

10 m

Figure P8–15

 8.16 Determine the stresses in the plates with the round and square holes subjected to the tensile stresses shown in Figure P8–16. Compare the largest principal stresses for each plate. Let $E = 210$ GPa, $v = 0.25$, and $t = 5$ mm.

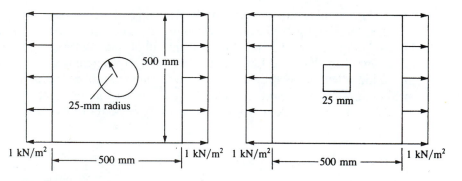

500 mm

25-mm radius

25 mm

1 kN/m²

500 mm

1 kN/m²   1 kN/m²

500 mm

1 kN/m²

Figure P8–16

 8.17 For the concrete overpass structure shown in Figure P8–17, determine the maximum principal stresses and their locations. Assume plane strain conditions. Let $E = 3.0 \times 10^6$ psi and $v = 0.30$.

8.18 For the steel culvert shown in Figure P8–18, determine the maximum principal stresses and their locations and the largest displacement and its location. Let $E_{\text{steel}} = 210$ GPa and let $v = 0.30$.

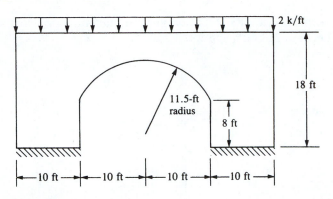

Figure P8–17

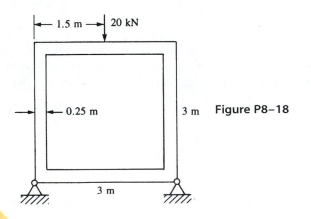

Figure P8–18

8.19 For the tensile member shown in Figure P8–19 with two holes, determine the maximum principal stresses and their locations. Let $E = 210$ GPa, $v = 0.25$, and $t = 10$ mm. Then let $E = 70$ GPa and $v = 0.30$. Compare your results.

Due
4/5/04

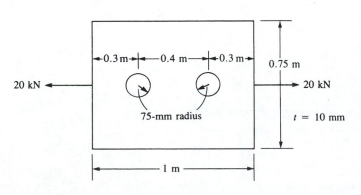

Figure P8–19

8.20 For the plate shown in Figure P8–20, determine the maximum principal stresses and their locations. Let $E = 210$ GPa and $v = 0.25$.

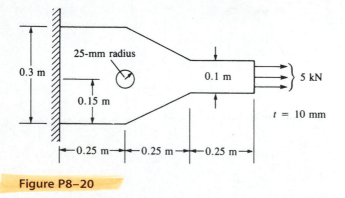

Figure P8–20

8.21 For the concrete dam shown subjected to water pressure in Figure P8–21, determine the principal stresses. Let $E = 3.5 \times 10^6$ psi and $v = 0.30$. Assume plane strain conditions. Perform the analysis for self-weight and then for hydrostatic (water) pressure against the dam vertical face as shown.

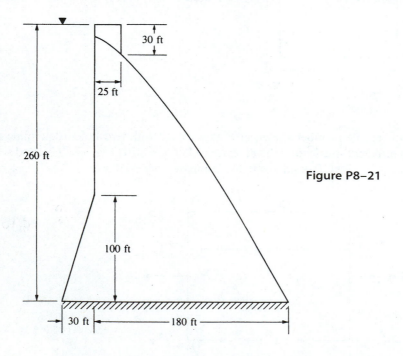

Figure P8–21

8.22 Determine the stresses in the wrench shown in Figure P8–22. Let $E = 200$ GPa and $v = 0.25$, and assume uniform thickness $t = 10$ mm.

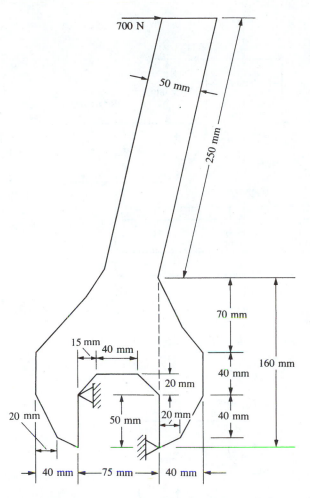

Figure P8–22

8.23 Determine the principal stresses in the blade implant and the bony material shown in Figure P8–23. Let $E_{blade} = 20$ GPa, $v_{blade} = 0.30$, $E_{bone} = 12$ GPa, and $v_{bone} = 0.35$. Assume plane stress conditions with $t = 5$ mm.

8.24 Determine the stresses in the plate shown in Figure P8–24. Let $E = 210$ GPa and $v = 0.25$. The element thickness is 10 mm.

Use the Algor computer program to help solve the design-type problems, 8.25–8.28.

8.25 The machine shown in Fig. P8–25 is an overload protection device that releases the load when the shear pin S fails. Determine the maximum Von Mises stress in the upper part ABE if the pin shears when its shear stress is 40 MPa. Assume the upper

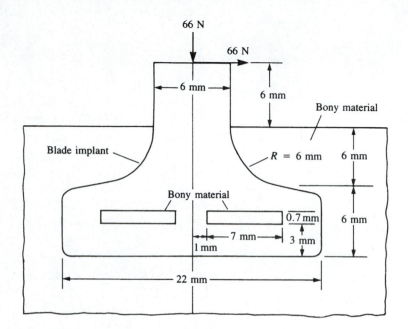

Figure P8–23

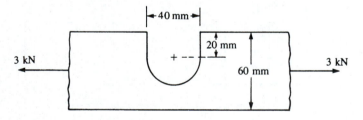

Figure P8–24

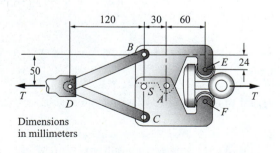
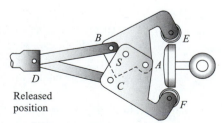

Figure P8–25 Overload protection device

part to have a uniform thickness of 6 mm. Assume plane stress conditions for the upper part. The part is made of 6061 aluminum alloy. Is the thickness sufficient to prevent failure based on the maximum distortion energy theory? If not, suggest a better thickness.

8.26 The steel triangular plate shown in Fig. P8–26 is bolted to a steel column with $\frac{3}{4}$-in.-diameter bolts in the pattern shown. Assuming the column and bolts are very rigid relative to the plate and neglecting friction forces between the column and plate, determine the highest load exerted on any bolt. The bolts should not be included in the model. Just fix the nodes around the bolt circles and consider the reactions at these nodes as the bolt loads. If $\frac{3}{4}$-in.-diameter bolts are not sufficient, recommend another standard diameter. Assume a standard material for the bolts. Compare the reactions from the finite element results to those found by classical methods.

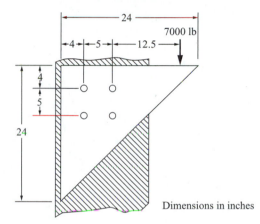

Figure P8–26 Steel triangular plate connection

Dimensions in inches

8.27 A machine part supports an end load of 1000 lb as shown in Fig. P8–27. Determine the stress concentration factors for the two changes in geometry located at the radii shown on the lower side of the part. Compare the stresses you get to classical beam theory results with and without the change in geometry, that is, with a uniform depth of 1 in. instead of the additional material depth of 1.5 in. Assume standard mild steel is used for the part. Recommend any changes you might make in the geometry.

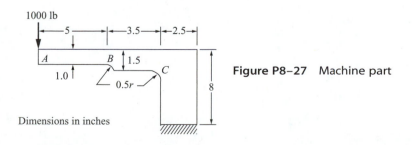

Figure P8–27 Machine part

Dimensions in inches

8.28 A plate with a hole off-centered is shown in Fig. P8–28. Determine how close to the top edge the hole can be placed before yielding of the A36 steel occurs (based on the maximum distortion energy theory). The applied tensile stress is 10,000 psi, and the plate thickness is $\frac{1}{4}$ in. Now if the plate is made of 6061-T6 aluminum alloy with a yield strength of 37 ksi, does this change your answer? If the plate thickness is changed to $\frac{1}{2}$ in., how does this change the results?

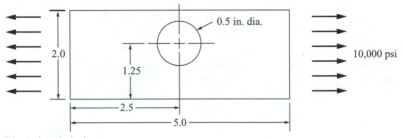

Dimensions in inches

Figure P8–28 Plate with off-centered hole

Development of the Linear-Strain Triangle Equations

Introduction

In this chapter, we consider the development of the stiffness matrix and equations for a higher-order triangular element, called the *linear-strain triangle* (LST). This element is available in many commercial computer programs and has some advantages over the constant-strain triangle described in Chapter 7.

The LST element has six nodes and twelve unknown displacement degrees of freedom. The displacement functions for the element are quadratic instead of linear (as in the CST).

The procedures for development of the equations for the LST element follow the same steps as those used in Chapter 7 for the CST element. However, the number of equations now becomes twelve instead of six, making a longhand solution extremely cumbersome. Hence, we will use a computer to perform many of the mathematical operations.

After deriving the element equations, we will compare results from problems solved using the LST element with those solved using the CST element. The introduction of the higher-order LST element will illustrate the possible advantages of higher-order elements and should enhance your general understanding of the concepts involved with finite element procedures.

▲ 9.1 Derivation of the Linear-Strain Triangular Element Stiffness Matrix and Equations ▲

We will now derive the LST stiffness matrix and element equations. The steps used here are identical to those used for the CST element, and much of the notation is the same.

Step 1 Select Element Type

Consider the triangular element shown in Figure 9–1 with the usual end nodes and three additional nodes conveniently located at the midpoints of the sides. Thus, a

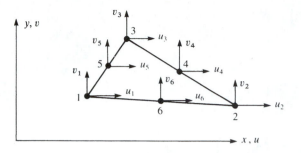

Figure 9–1 Basic six-node triangular element showing degrees of freedom

computer program can automatically compute the midpoint coordinates once the coordinates of the corner nodes are given as input.

The unknown nodal displacements are now given by

$$\{d\} = \begin{Bmatrix} \underline{d}_1 \\ \underline{d}_2 \\ \underline{d}_3 \\ \underline{d}_4 \\ \underline{d}_5 \\ \underline{d}_6 \end{Bmatrix} = \begin{Bmatrix} u_1 \\ v_1 \\ u_2 \\ v_2 \\ u_3 \\ v_3 \\ u_4 \\ v_4 \\ u_5 \\ v_5 \\ u_6 \\ v_6 \end{Bmatrix}$$

(9.1.1)

Step 2 Select a Displacement Function

We now select a quadratic displacement function in each element as

$$u(x, y) = a_1 + a_2 x + a_3 y + a_4 x^2 + a_5 xy + a_6 y^2$$

(9.1.2)

$$v(x, y) = a_7 + a_8 x + a_9 y + a_{10} x^2 + a_{11} xy + a_{12} y^2$$

Again, the number of coefficients $a_i(12)$ equals the total number of degrees of freedom for the element. The displacement compatibility among adjoining elements is satisfied because three nodes are located along each side and a parabola is defined by three points on its path. Since adjacent elements are connected at common nodes, their displacement compatibility across the boundaries will be maintained.

In general, when considering triangular elements, we can use a complete polynomial in Cartesian coordinates to describe the displacement field within an element. Using internal nodes as necessary for the higher-order cubic and quartic elements, we use all terms of a truncated Pascal triangle in the displacement field or, equivalently, the shape functions, as shown by Figure 9–2; that is, a complete linear function is used

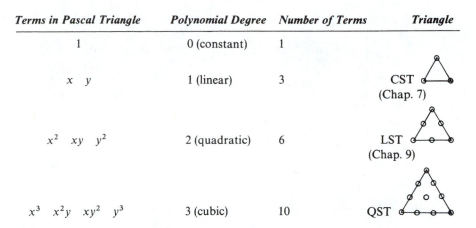

Terms in Pascal Triangle	Polynomial Degree	Number of Terms	Triangle
1	0 (constant)	1	
x y	1 (linear)	3	CST (Chap. 7)
x^2 xy y^2	2 (quadratic)	6	LST (Chap. 9)
x^3 x^2y xy^2 y^3	3 (cubic)	10	QST

Figure 9–2 Relation between type of plane triangular element and polynomial coefficients based on a Pascal triangle

for the CST element considered previously in Chapter 7. The complete quadratic function is used for the LST of this chapter. The complete cubic function is used for the quadratic-strain triangle (QST), with an internal node necessary as the tenth node.

The general displacement functions, Eqs. (9.1.2), expressed in matrix form are now

$$\{\psi\} = \left\{ \begin{array}{c} u \\ v \end{array} \right\} = \begin{bmatrix} 1 & x & y & x^2 & xy & y^2 & 0 & 0 & 0 & 0 & 0 & 0 \\ 0 & 0 & 0 & 0 & 0 & 0 & 1 & x & y & x^2 & xy & y^2 \end{bmatrix} \left\{ \begin{array}{c} a_1 \\ a_2 \\ \vdots \\ a_{12} \end{array} \right\} \tag{9.1.3}$$

Alternatively, we can express Eq. (9.1.3) as

$$\{\psi\} = [M^*]\{a\} \tag{9.1.4}$$

where $[M^*]$ is defined to be the first matrix on the right side of Eq. (9.1.3). The coefficients a_1 through a_{12} can be obtained by substituting the coordinates into u and v as follows:

$$\left\{ \begin{array}{c} u_1 \\ u_2 \\ \vdots \\ u_6 \\ v_1 \\ \vdots \\ v_5 \\ v_6 \end{array} \right\} = \begin{bmatrix} 1 & x_1 & y_1 & x_1^2 & x_1y_1 & y_1^2 & 0 & 0 & 0 & 0 & 0 & 0 \\ 1 & x_2 & y_2 & x_2^2 & x_2y_2 & y_2^2 & 0 & 0 & 0 & 0 & 0 & 0 \\ \vdots & \vdots & \vdots & \vdots & \vdots & \vdots & \vdots & \vdots & \vdots & \vdots & \vdots & \vdots \\ 1 & x_6 & y_6 & x_6^2 & x_6y_6 & y_6^2 & 0 & 0 & 0 & 0 & 0 & 0 \\ 0 & 0 & 0 & 0 & 0 & 0 & 1 & x_1 & y_1 & x_1^2 & x_1y_1 & y_1^2 \\ \vdots & \vdots & \vdots & \vdots & \vdots & \vdots & \vdots & \vdots & \vdots & \vdots & \vdots & \vdots \\ 0 & 0 & 0 & 0 & 0 & 0 & 1 & x_5 & y_5 & x_5^2 & x_5y_5 & y_5^2 \\ 0 & 0 & 0 & 0 & 0 & 0 & 1 & x_6 & y_6 & x_6^2 & x_6y_6 & y_6^2 \end{bmatrix} \left\{ \begin{array}{c} a_1 \\ a_2 \\ \vdots \\ a_6 \\ a_7 \\ \vdots \\ a_{11} \\ a_{12} \end{array} \right\}$$

$$\tag{9.1.5}$$

Solving for the a_i's, we have

$$
\begin{Bmatrix} a_1 \\ \vdots \\ a_6 \\ a_7 \\ \vdots \\ a_{12} \end{Bmatrix} = \begin{bmatrix} 1 & x_1 & y_1 & x_1^2 & x_1y_1 & y_1^2 & 0 & 0 & 0 & 0 & 0 & 0 \\ \vdots & \vdots & \vdots & \vdots & \vdots & \vdots & \vdots & \vdots & \vdots & \vdots & \vdots & \vdots \\ 1 & x_6 & y_6 & x_6^2 & x_6y_6 & y_6^2 & 0 & 0 & 0 & 0 & 0 & 0 \\ 0 & 0 & 0 & 0 & 0 & 0 & 1 & x_1 & y_1 & x_1^2 & x_1y_1 & y_1^2 \\ \vdots & \vdots & \vdots & \vdots & \vdots & \vdots & \vdots & \vdots & \vdots & \vdots & \vdots & \vdots \\ 0 & 0 & 0 & 0 & 0 & 0 & 1 & x_6 & y_6 & x_6^2 & x_6y_6 & y_6^2 \end{bmatrix}^{-1} \begin{Bmatrix} u_1 \\ \vdots \\ u_6 \\ v_1 \\ \vdots \\ v_6 \end{Bmatrix}
$$

$$(9.1.6)$$

or, alternatively, we can express Eq. (9.1.6) as

$$\{a\} = [X]^{-1}\{d\} \tag{9.1.7}$$

where $[X]$ is the 12×12 matrix on the right side of Eq. (9.1.6). It is best to invert the $[X]$ matrix by using a digital computer. Then the a_i's, in terms of nodal displacements, are substituted into Eq. (9.1.4). Note that only the 6×6 part of $[X]$ in Eq. (9.1.6) really must be inverted. Finally, using Eq. (9.1.7) in Eq. (9.1.4), we can obtain the general displacement expressions in terms of the shape functions and the nodal degrees of freedom as

$$\{\psi\} = [N]\{d\} \tag{9.1.8}$$

where

$$[N] = [M^*][X]^{-1} \tag{9.1.9}$$

Step 3 Define the Strain/Displacement and Stress/Strain Relationships

The element strains are again given by

$$\{\varepsilon\} = \begin{Bmatrix} \varepsilon_x \\ \varepsilon_y \\ \gamma_{xy} \end{Bmatrix} = \begin{Bmatrix} \dfrac{\partial u}{\partial x} \\[6pt] \dfrac{\partial v}{\partial y} \\[6pt] \dfrac{\partial v}{\partial x} + \dfrac{\partial u}{\partial y} \end{Bmatrix} \tag{9.1.10}$$

or, using Eq. (9.1.3) for u and v in Eq. (9.1.10), we obtain

$$\{\varepsilon\} = \begin{bmatrix} 0 & 1 & 0 & 2x & y & 0 & 0 & 0 & 0 & 0 & 0 & 0 \\ 0 & 0 & 0 & 0 & 0 & 0 & 0 & 0 & 1 & 0 & x & 2y \\ 0 & 0 & 1 & 0 & x & 2y & 0 & 1 & 0 & 2x & y & 0 \end{bmatrix} \begin{Bmatrix} a_1 \\ a_2 \\ \vdots \\ a_{12} \end{Bmatrix} \tag{9.1.11}$$

We observe that Eq. (9.1.11) yields a linear strain variation in the element. Therefore, the element is called a *linear-strain triangle* (LST). Rewriting Eq. (9.1.11), we have

$$\{\varepsilon\} = [M']\{a\} \tag{9.1.12}$$

where $[M']$ is the first matrix on the right side of Eq. (9.1.11). Substituting Eq. (9.1.6) for the a_i's into Eq. (9.1.12), we have $\{\varepsilon\}$ in terms of the nodal displacements as

$$\{\varepsilon\} = [B]\{d\} \tag{9.1.13}$$

where $[B]$ is a function of the variables x and y and the coordinates (x_1, y_1) through (x_6, y_6) given by

$$[B] = [M'][X]^{-1} \tag{9.1.14}$$

where Eq. (9.1.7) has been used in expressing Eq. (9.1.14). Note that $[B]$ is now a matrix of order 3×12.

The stresses are again given by

$$\left\{ \begin{array}{c} \sigma_x \\ \sigma_y \\ \tau_{xy} \end{array} \right\} = [D] \left\{ \begin{array}{c} \varepsilon_x \\ \varepsilon_y \\ \gamma_{xy} \end{array} \right\} = [D][B]\{d\} \tag{9.1.15}$$

where $[D]$ is given by Eq. (7.1.8) for plane stress or by Eq. (7.1.10) for plane strain. These stresses are now linear functions of x and y coordinates.

Step 4 Derive the Element Stiffness Matrix and Equations

We determine the stiffness matrix in a manner similar to that used in Section 7.2 by using Eq. (7.2.50) repeated here as

$$[k] = \iiint_V [B]^T [D][B]\, dV \tag{9.1.16}$$

However, the $[B]$ matrix is now a function of x and y as given by Eq. (9.1.14). Therefore, we must perform the integration in Eq. (9.1.16). Finally, the $[B]$ matrix is of the form

$$[B] = \frac{1}{2A} \begin{bmatrix} \beta_1 & 0 & \beta_2 & 0 & \beta_3 & 0 & \beta_4 & 0 & \beta_5 & 0 & \beta_6 & 0 \\ 0 & \gamma_1 & 0 & \gamma_2 & 0 & \gamma_3 & 0 & \gamma_4 & 0 & \gamma_5 & 0 & \gamma_6 \\ \gamma_1 & \beta_1 & \gamma_2 & \beta_2 & \gamma_3 & \beta_3 & \gamma_4 & \beta_4 & \gamma_5 & \beta_5 & \gamma_6 & \beta_6 \end{bmatrix} \tag{9.1.17}$$

where the β's and γ's are now functions of x and y as well as of the nodal coordinates, as is illustrated for a specific linear-strain triangle in Section 9.2 by Eq. (9.2.8). The stiffness matrix is then seen to be a 12×12 matrix on multiplying the matrices in Eq. (9.1.16). The stiffness matrix, Eq. (9.1.16), is very cumbersome to obtain in explicit form, so it will not be given here. However, if the origin of the coordinates is considered to be at the centroid of the element, the integrations become amenable [9]. Alternatively, area coordinates [3, 8, 9] can be used to obtain an explicit form of the stiffness matrix. However, even the use of area coordinates usually involves tedious calculations. Therefore, the integration is best carried out numerically. (Numerical integration is described in Section 11.4.)

The element body forces and surface forces should not be automatically lumped at the nodes, but for a consistent formulation (one that is formulated from the same shape functions used to formulate the stiffness matrix), Eqs. (7.3.1) and (7.3.7), respectively, should be used. (Problems 9.3 and 9.4 illustrate this concept.) These forces can be added to any concentrated nodal forces to obtain the element force matrix. Here the element force matrix is of order 12×1 because, in general, there could be an x and a y component of force at each of the six nodes associated with the element. The element equations are then given by

$$
\left\{
\begin{array}{c}
f_{1x} \\
f_{1y} \\
\vdots \\
f_{6y}
\end{array}
\right\}
\underset{(12 \times 1)}{} =
\underset{(12 \times 12)}{
\left[
\begin{array}{ccc}
k_{11} & \cdots & k_{1,12} \\
k_{21} & & k_{2,12} \\
\vdots & & \vdots \\
k_{12,1} & \cdots & k_{12,12}
\end{array}
\right]}
\left\{
\begin{array}{c}
u_1 \\
v_1 \\
\vdots \\
v_6
\end{array}
\right\}
\underset{(12 \times 1)}{}
\tag{9.1.18}
$$

Steps 5–7

Steps 5–7, which involve assembling the global stiffness matrix and equations, determining the unknown global nodal displacements, and calculating the stresses, are identical to those in Section 7.2 for the CST. However, instead of constant stresses in each element, we now have a linear variation of the stresses in each element. Common practice was to use the centroidal element stresses. Current practice is to use the average of the nodal element stresses.

▲ 9.2 Example LST Stiffness Determination

To illustrate some of the procedures outlined in Section 9.1 for deriving an LST stiffness matrix, consider the following example. Figure 9–3 shows a specific LST and its coordinates. The triangle is of base dimension b and height h, with midside nodes.

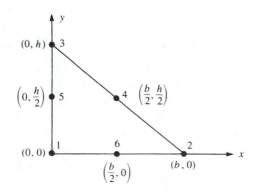

Figure 9–3 LST triangle for evaluation of a stiffness matrix

Using the first six equations of Eq. (9.1.5), we calculate the coefficients a_1 through a_6 by evaluating the displacement u at each of the six known coordinates of each node as follows:

$$u_1 = u(0,0) = a_1 \tag{9.2.1}$$

$$u_2 = u(b,0) = a_1 + a_2 b + a_4 b^2$$

$$u_3 = u(0,h) = a_1 + a_3 h + a_6 h^2$$

$$u_4 = u\left(\frac{b}{2}, \frac{h}{2}\right) = a_1 + a_2 \frac{b}{2} + a_3 \frac{h}{2} + a_4 \left(\frac{b}{2}\right)^2 + a_5 \frac{bh}{4} + a_6 \left(\frac{h}{2}\right)^2$$

$$u_5 = u\left(0, \frac{h}{2}\right) = a_1 + a_3 \frac{h}{2} + a_6 \left(\frac{h}{2}\right)^2$$

$$u_6 = u\left(\frac{b}{2}, 0\right) = a_1 + a_2 \frac{b}{2} + a_4 \left(\frac{b}{2}\right)^2$$

Solving Eqs. (9.2.1) simultaneously for the a_i's, we obtain

$$a_1 = u_1 \qquad a_2 = \frac{4u_6 - 3u_1 - u_2}{b} \qquad a_3 = \frac{4u_5 - 3u_1 - u_3}{h} \tag{9.2.2}$$

$$a_4 = \frac{2(u_2 - 2u_6 + u_1)}{b^2} \qquad a_5 = \frac{4(u_1 + u_4 - u_5 - u_6)}{bh}$$

$$a_6 = \frac{2(u_3 - 2u_5 + u_1)}{h^2}$$

Substituting Eqs. (9.2.2) into the displacement expression for u from Eqs. (9.1.2), we have

$$u = u_1 + \left[\frac{4u_6 - 3u_1 - u_2}{b}\right]x + \left[\frac{4u_5 - 3u_1 - u_3}{h}\right]y + \left[\frac{2(u_2 - 2u_6 + u_1)}{b^2}\right]x^2$$

$$+ \left[\frac{4(u_1 + u_4 - u_5 - u_6)}{bh}\right]xy + \left[\frac{2(u_3 - 2u_5 + u_1)}{h^2}\right]y^2 \tag{9.2.3}$$

Similarly, solving for a_7 through a_{12} by evaluating the displacement v at each of the six nodes and then substituting the results into the expression for v from Eqs. (9.1.2), we obtain

$$v = v_1 + \left[\frac{4v_6 - 3v_1 - v_2}{b}\right]x + \left[\frac{4v_5 - 3v_1 - v_3}{h}\right]y + \left[\frac{2(v_2 - 2v_6 + v_1)}{b^2}\right]x^2$$

$$+ \left[\frac{4(v_1 + v_4 - v_5 - v_6)}{bh}\right]xy + \left[\frac{2(v_3 - 2v_5 + v_1)}{h^2}\right]y^2 \tag{9.2.4}$$

Using Eqs. (9.2.3) and (9.2.4), we can express the general displacement expressions in terms of the shape functions as

$$\left\{ \begin{matrix} u \\ v \end{matrix} \right\} = \begin{bmatrix} N_1 & 0 & N_2 & 0 & N_3 & 0 & N_4 & 0 & N_5 & 0 & N_6 & 0 \\ 0 & N_1 & 0 & N_2 & 0 & N_3 & 0 & N_4 & 0 & N_5 & 0 & N_6 \end{bmatrix} \left\{ \begin{matrix} u_1 \\ v_1 \\ \vdots \\ v_6 \end{matrix} \right\}$$

(9.2.5)

where the shape functions are obtained by collecting coefficients that multiply each u_i term in Eq. (9.2.3). For instance, collecting all terms that multiply by u_1 in Eq. (9.2.3), we obtain N_1. These shape functions are then given by

$$N_1 = 1 - \frac{3x}{b} - \frac{3y}{h} + \frac{2x^2}{b^2} + \frac{4xy}{bh} + \frac{2y^2}{h^2} \qquad N_2 = \frac{-x}{b} + \frac{2x^2}{b^2}$$

(9.2.6)

$$N_3 = \frac{-y}{h} + \frac{2y^2}{h^2} \qquad N_4 = \frac{4xy}{bh} \qquad N_5 = \frac{4y}{h} - \frac{4xy}{bh} - \frac{4y^2}{h^2}$$

$$N_6 = \frac{4x}{b} - \frac{4x^2}{b^2} - \frac{4xy}{bh}$$

Using Eq. (9.2.5) in Eq. (9.1.10), and performing the differentiations indicated on u and v, we obtain

$$\underline{\varepsilon} = \underline{B}\underline{d} \qquad (9.2.7)$$

where $\underline{B}$ is of the form of Eq. (9.1.17), with the resulting β's and γ's in Eq. (9.1.17) given by

$$\beta_1 = -3h + \frac{4hx}{b} + 4y \qquad \beta_2 = -h + \frac{4hx}{b} \qquad \beta_3 = 0 \qquad (9.2.8)$$

$$\beta_4 = 4y \qquad \beta_5 = -4y \qquad \beta_6 = 4h - \frac{8hx}{b} - 4y$$

$$\gamma_1 = -3b + 4x + \frac{4by}{h} \qquad \gamma_2 = 0 \qquad \gamma_3 = -b + \frac{4by}{h}$$

$$\gamma_4 = 4x \qquad \gamma_5 = 4b - 4x - \frac{8by}{h} \qquad \gamma_6 = -4x$$

These β's and γ's are specific to the element in Figure 9–3. Specifically, using Eqs. (9.1.1) and (9.1.17) in Eq. (9.2.7), we obtain

$$\varepsilon_x = \frac{1}{2A}[\beta_1 u_1 + \beta_2 u_2 + \beta_3 u_3 + \beta_4 u_4 + \beta_5 u_5 + \beta_6 u_6]$$

$$\varepsilon_y = \frac{1}{2A}[\gamma_1 v_1 + \gamma_2 v_2 + \gamma_3 v_3 + \gamma_4 v_4 + \gamma_5 v_5 + \gamma_6 v_6]$$

$$\gamma_{xy} = \frac{1}{2A}[\gamma_1 u_1 + \beta_1 v_1 + \cdots + \beta_6 v_6]$$

The stiffness matrix for a constant-thickness element can now be obtained on substituting Eqs. (9.2.8) into Eq. (9.1.17) to obtain $\underline{B}$, then substituting $\underline{B}$ into Eq. (9.1.16) and using calculus to set up the appropriate integration. The explicit expression for the 12×12 stiffness matrix, being extremely cumbersome to obtain, is not given here. Stiffness matrix expressions for higher-order elements are found in References [1] and [2].

▲ 9.3 Comparison of Elements ▲

For a given number of nodes, a better representation of true stress and displacement is generally obtained using the LST element than is obtained with the same number of nodes using a much finer subdivision into simple CST elements. For example, using one LST yields better results than using four CST elements with the same number of nodes (Figure 9–4) and hence the same number of degrees of freedom (except for the case when constant stress exists).

We now present results to compare the CST of Chapter 7 with the LST of this chapter. Consider the cantilever beam subjected to a parabolic load variation acting as shown in Figure 9–5. Let $E = 30 \times 10^6$ psi, $v = 0.25$, and $t = 1.0$ in.

Table 9–1 lists the series of tests run to compare results using the CST and LST elements. Table 9–2 shows comparisons of free-end (tip) deflection and stress σ_x for each element type used to model the cantilever beam. From Table 9–2, we can observe that the larger the number of degrees of freedom for a given type of triangular

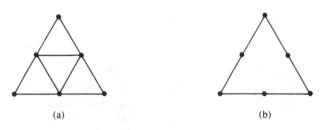

(a)   (b)

Figure 9–4   Basic triangular element: (a) four-CST and (b) one-LST

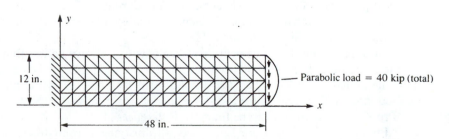

Figure 9–5   Cantilever beam used to compare the CST and LST elements with a 4 × 16 mesh

Table 9–1 Models used to compare CST and LST results for the cantilever beam of Figure 9–5

Series of Tests Run	Number of Nodes	Number of Degrees of Freedom, n_d	Number of Triangular Elements
A-1 4 × 16 mesh	85	160	128 CST
A-2 8 × 32	297	576	512 CST
B-1 2 × 8	85	160	32 LST
B-2 4 × 16	297	576	128 LST

Table 9–2 Comparison of CST and LST results for the cantilever beam of Figure 9–5

Run	n_d	Bandwidth[1] n_b	Tip Deflection (in.)	σ_x (ksi)	Location (in.), x, y
A-1	160	14	−0.29555	67.236	2.250, 11.250
A-2	576	22	−0.33850	81.302	1.125, 11.630
B-1	160	18	−0.33470	58.885	4.500, 10.500
B-2	576	22	−0.35159	69.956	2.250, 11.250
Exact solution			−0.36133	80.000	0, 12

[1] Bandwidth is described in Appendix B.4.

element, the closer the solution converges to the exact one (compare run A-1 to run A-2, and B-1 to B-2). For a given number of nodes, the LST analysis yields somewhat better results than the CST analysis (compare run A-1 to run B-1). Although the CST element is rather poor in modeling bending, we observe from Table 9–2 that the element can be used to model a beam in bending if a sufficient number of elements are used through the depth of the beam. In general, both LST and CST analyses yield results good enough for most plane stress/strain problems, provided a sufficient number of elements are used. In fact, most commercial programs incorporate the use of CST and/or LST elements for plane stress/strain problems, although these elements are used primarily as transition elements (usually during mesh generation). The four-sided isoparametric plane stress/strain element is most frequently used in commercial programs and is described in Chapter 11.

Also, recall that finite element displacements will always be less than (or equal to) the exact ones, because finite element models are normally predicted to be stiffer than the actual structures when the displacement formulation of the finite element method is used. (The reason for the stiffer model was discussed in Sections 3.10 and 8.3. Proof of this assertion can be found in References [4–7].

Finally, Figure 9–6 (from Reference [8]) illustrates a comparison of CST and LST models of a plate subjected to parabolically distributed edge loads. Figure 9–6 shows that the LST model converges to the exact solution for horizontal displacement at point *A* faster than does the CST model. However, the CST model is quite accept-

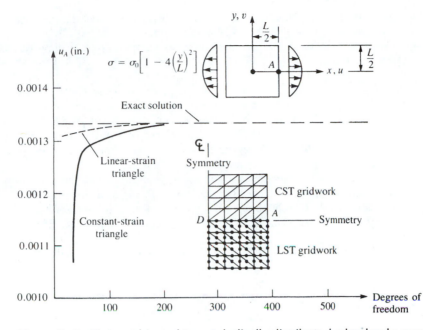

Figure 9–6 Plates subjected to parabolically distributed edge loads; comparison of results for triangular elements. (Gallagher, R. H. *Finite Element Analysis: Fundamentals*, © 1975, pp. 269, 270. Reprinted by permission of Prentice Hall, Inc., Englewood Cliffs, NJ)

able even for modest numbers of degrees of freedom. For example, a CST model with 100 nodes (200 degrees of freedom) often yields nearly as accurate a solution as does an LST model with the same number of degrees of freedom.

In conclusion, the results of Table 9–2 and Figure 9–6 indicate that the LST model might be preferred over the CST model for plane stress applications when relatively small numbers of nodes are used. However, the use of triangular elements of higher order, such as the LST, is not visibly advantageous when large numbers of nodes are used, particularly when the cost of formation of the element stiffnesses, equation bandwidth, and overall complexities involved in the computer modeling are considered.

▲ References

[1] Pederson, P., "Some Properties of Linear Strain Triangles and Optimal Finite Element Models," *International Journal for Numerical Methods in Engineering*, Vol. 7, pp. 415–430, 1973.

[2] Tocher, J. L., and Hartz, B. J., "Higher-Order Finite Element for Plane Stress," *Journal of the Engineering Mechanics Division*, Proceedings of the American Society of Civil Engineers, Vol. 93, No. EM4, pp. 149–174, Aug. 1967.

[3] Bowes, W. H., and Russell, L. T., *Stress Analysis by the Finite Element Method for Practicing Engineers*, Lexington Books, Toronto, 1975.

[4] Fraeijes de Veubeke, B., "Upper and Lower Bounds in Matrix Structural Analysis," *Matrix Methods of Structural Analysis*, AGAR-Dograph 72, B. Fraeijes de Veubeke, ed., Macmillan, New York, 1964.

[5] McLay, R. W., *Completeness and Convergence Properties of Finite Element Displacement Functions—A General Treatment*, American Institute of Aeronautics and Astronautics Paper No. 67–143, AIAA 5th Aerospace Meeting, New York, 1967.

[6] Tong, P., and Pian, T. H. H., "The Convergence of Finite Element Method in Solving Linear Elastic Problems," *International Journal of Solids and Structures*, Vol. 3, pp. 865–879, 1967.

[7] Cowper, G. R., "Variational Procedures and Convergence of Finite-Element Methods," *Numerical and Computer Methods in Structural Mechanics*, S. J. Fenves, N. Perrone, A. R. Robinson, and W. C. Schnobrich, eds., Academic Press, New York, 1973.

[8] Gallagher, R., *Finite Element Analysis Fundamentals*, Prentice Hall, Englewood Cliffs, NJ, 1975.

[9] Zienkiewicz, O. C., *The Finite Element Method*, 3rd ed., McGraw-Hill, New York, 1977.

▲ **Problems**

9.1 Evaluate the shape functions given by Eq. (9.2.6). Sketch the variation of each function over the surface of the triangular element shown in Figure 9–3.

9.2 Express the strains $\varepsilon_x, \varepsilon_y$, and γ_{xy} for the element of Figure 9–3 by using the results given in Section 9.2. Evaluate these strains at the centroid of the element; then evaluate the stresses at the centroid in terms of E and v. Assume plane stress conditions apply.

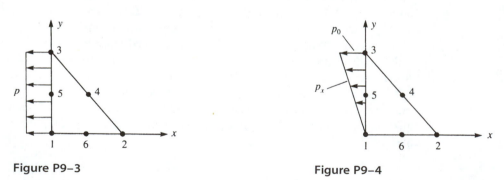

Figure P9–3   Figure P9–4

9.3 For the element of Figure 9–3 (shown again as Figure P9–3) subjected to the uniform pressure shown acting over the vertical side, determine the nodal force replacement system using Eq. (7.3.7). Assume an element thickness of t.

9.4 For the element of Figure 9–3 (shown as Figure P9–4) subjected to the linearly varying line load shown acting over the vertical side, determine the nodal force replacement system using Eq. (7.3.7). Compare this result to that of Problem 7.9. Are these results expected? Explain.

9.5 For the linear-strain elements shown in Figure P9–5, determine the strains $\varepsilon_x, \varepsilon_y$, and γ_{xy}. Evaluate the stresses σ_x, σ_y, and τ_{xy} at the centroids. The coordinates of the nodes are shown in units of inches. Let $E = 30 \times 10^6$ psi, $v = 0.25$, and $t = 0.25$ in. for both elements. Assume plane stress conditions apply. The nodal displacements are given as

$$u_1 = 0.0 \qquad\qquad v_1 = 0.0$$
$$u_2 = 0.001 \text{ in.} \qquad v_2 = 0.002 \text{ in.}$$
$$u_3 = 0.0005 \text{ in.} \qquad v_3 = 0.0002 \text{ in.}$$
$$u_4 = 0.0002 \text{ in.} \qquad v_4 = 0.0001 \text{ in.}$$
$$u_5 = 0.0 \qquad\qquad v_5 = 0.0001 \text{ in.}$$
$$u_6 = 0.0005 \text{ in.} \qquad v_6 = 0.001 \text{ in.}$$

(*Hint*: Use the results of Section 9.2.)

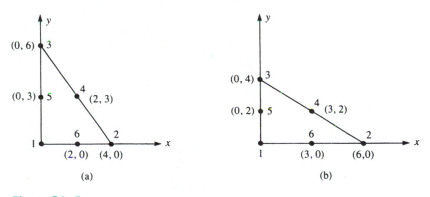

(a)   (b)

Figure P9–5

9.6 For the linear-strain element shown in Figure P9–6, determine the strains $\varepsilon_x, \varepsilon_y$, and γ_{xy}. Evaluate these strains at the centroid of the element; then evaluate the stresses σ_x, σ_y, and τ_{xy} at the centroid. The coordinates of the nodes are shown in units of millimeters. Let $E = 210$ GPa, $v = 0.25$, and $t = 10$ mm. Assume plane stress con-

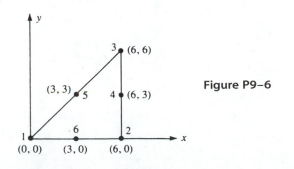

Figure P9–6

ditions apply. Use the nodal displacements given in Problem 9.5 (converted to milli-meters). Note that the β's and γ's from the example in Section 9.2 cannot be used here as the element in Figure P9–6 is oriented differently than the one in Figure 9–3.

9.7 Evaluate the shape functions for the linear-strain triangle shown in Figure P9–7. Then evaluate the $\underline{B}$ matrix. Units are millimeters.

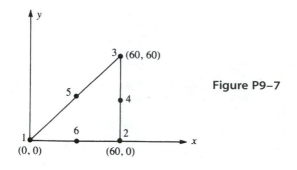

Figure P9–7

10

Axisymmetric Elements

Introduction

In previous chapters, we have been concerned with line or one-dimensional elements (Chapters 2–6) and two-dimensional elements (Chapters 7–9). In this chapter, we consider a special two-dimensional element called the *axisymmetric element*. This element is quite useful when symmetry with respect to geometry and loading exists about an axis of the body being analyzed. Problems that involve soil masses subjected to circular footing loads or thick-walled pressure vessels can often be analyzed using the element developed in this chapter.

We begin with the development of the stiffness matrix for the simplest axisymmetric element, the triangular torus, whose vertical cross section is a plane triangle.

We then present the longhand solution of a thick-walled pressure vessel to illustrate the use of the axisymmetric element equations. This is followed by a description of some typical large-scale problems that have been modeled using the axisymmetric element.

Finally, we describe how to solve axisymmetric problems using the Algor computer program and present example solutions.

▲ 10.1 Derivation of the Stiffness Matrix ▲

In this section, we will derive the stiffness matrix and the body and surface force matrices for the axisymmetric element. However, before the development, we will first present some fundamental concepts prerequisite to the understanding of the derivation. Axisymmetric elements are triangular tori such that each element is symmetric with respect to geometry and loading about an axis such as the z axis in Figure 10–1. Hence, the z axis is called the *axis of symmetry* or the *axis of revolution*. Each vertical cross section of the element is a plane triangle. The nodal points of an axisymmetric triangular element describe circumferential lines, as indicated in Figure 10–1.

In plane stress problems, stresses exist only in the x-y plane. In axisymmetric problems, the radial displacements develop circumferential strains that induce stresses σ_r, σ_θ, σ_z, and τ_{rz}, where r, θ, and z indicate the radial, circumferential, and longitudinal directions, respectively. Triangular torus elements are often used to idealize the axisymmetric system because they can be used to simulate complex surfaces and are

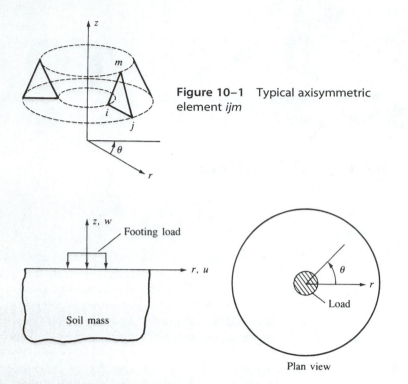

Figure 10–1 Typical axisymmetric element *ijm*

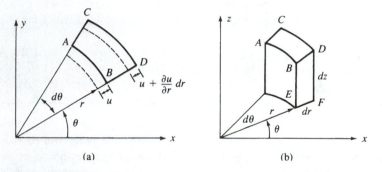

Plan view

Figure 10–2 Semi-infinite half-space modeled by axisymmetric elements

Figure 10–3 (a) Plane cross section of (b) axisymmetric element

simple to work with. For instance, the axisymmetric problem of a semi-infinite half-space loaded by a circular area (circular footing) shown in Figure 10–2 can be solved using the axisymmetric element developed in this chapter.

Because of symmetry about the z axis, the stresses are independent of the θ coordinate. Therefore, all derivatives with respect to θ vanish, and the displacement component v (tangent to the θ direction), the shear strains $\gamma_{r\theta}$ and $\gamma_{\theta z}$, and the shear stresses $\tau_{r\theta}$ and $\tau_{\theta z}$ are all zero.

Figure 10–3 shows an axisymmetric ring element and its cross section to represent the general state of strain for an axisymmetric problem. It is most convenient to express the displacements of an element *ABCD* in the plane of a cross section in cylindrical coordinates. We then let u and w denote the displacements in the radial and longitudinal directions, respectively. The side *AB* of the element is displaced an amount u, and side *CD* is then displaced an amount $u + (\partial u/\partial r)\, dr$ in the radial direction. The normal strain in the radial direction is then given by

$$\varepsilon_r = \frac{\partial u}{\partial r} \tag{10.1.1a}$$

In general, the strain in the tangential direction depends on the tangential displacement v and on the radial displacement u. However, for axisymmetric deformation behavior, recall that the tangential displacement v is equal to zero. Hence, the tangential strain is due only to the radial displacement. Having only radial displacement u, the new length of the arc $\widehat{AB}$ is $(r + u)\, d\theta$, and the tangential strain is then given by

$$\varepsilon_\theta = \frac{(r + u)\, d\theta - r\, d\theta}{r\, d\theta} = \frac{u}{r} \tag{10.1.1b}$$

Next, we consider the longitudinal element *BDEF* to obtain the longitudinal strain and the shear strain. In Figure 10–4, the element is shown to displace by amounts u and w in the radial and longitudinal directions at point *E*, and to displace additional amounts $(\partial w/\partial z)\, dz$ along line *BE* and $(\partial u/\partial r)\, dr$ along line *EF*. Furthermore, observing lines *EF* and *BE*, we see that point *F* moves upward an amount $(\partial w/\partial r)\, dr$ with respect to point *E* and point *B* moves to the right an amount $(\partial u/\partial z)\, dz$ with respect to point *E*. Again, from the basic definitions of normal and shear strain, we have the longitudinal normal strain given by

$$\varepsilon_z = \frac{\partial w}{\partial z} \tag{10.1.1c}$$

and the shear strain in the *r-z* plane given by

$$\gamma_{rz} = \frac{\partial u}{\partial z} + \frac{\partial w}{\partial r} \tag{10.1.1d}$$

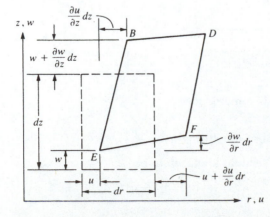

Figure 10–4 Displacement and rotations of lines of element in the *r-z* plane

Summarizing the strain/displacement relationships of Eqs. (10.1.1a–d) in one equation for easier reference, we have

$$\varepsilon_r = \frac{\partial u}{\partial r} \qquad \varepsilon_\theta = \frac{u}{r} \qquad \varepsilon_z = \frac{\partial w}{\partial z} \qquad \gamma_{rz} = \frac{\partial u}{\partial z} + \frac{\partial w}{\partial r} \qquad (10.1.1e)$$

The isotropic stress/strain relationship, obtained by simplifying the general stress/strain relationships given in Appendix C, is

$$\left\{ \begin{array}{c} \sigma_r \\ \sigma_z \\ \sigma_\theta \\ \tau_{rz} \end{array} \right\} = \frac{E}{(1+v)(1-2v)} \begin{bmatrix} 1-v & v & v & 0 \\ v & 1-v & v & 0 \\ v & v & 1-v & 0 \\ 0 & 0 & 0 & \dfrac{1-2v}{2} \end{bmatrix} \left\{ \begin{array}{c} \varepsilon_r \\ \varepsilon_z \\ \varepsilon_\theta \\ \gamma_{rz} \end{array} \right\} \qquad (10.1.2)$$

The theoretical development follows that of the plane stress/strain problem given in Chapter 7.

Step 1   Select Element Type

An axisymmetric solid is shown discretized in Figure 10–5(a), along with a typical triangular element. The element has three nodes with two degrees of freedom per node (that is, u_i, w_i at node i). The stresses in the axisymmetric problem are shown in Figure 10–5(b).

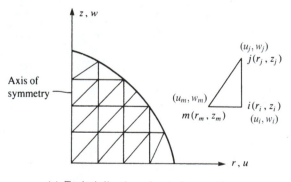

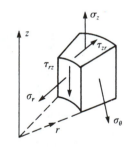

(a) Typical slice through an axisymmetric solid discretized into triangular elements

(b) Stresses in the axisymmetric problem

Figure 10–5   Discretized axisymmetric solid

Step 2   Select Displacement Functions

The element displacement functions are taken to be

$$u(r, z) = a_1 + a_2 r + a_3 z$$
$$w(r, z) = a_4 + a_5 r + a_6 z \qquad (10.1.3)$$

so that we have the same linear displacement functions as used in the plane stress, constant-strain triangle. Again, the total number of a_i's (six) introduced in the displacement functions is the same as the total number of degrees of freedom for the element. The nodal displacements are

$$\{d\} = \begin{Bmatrix} \underline{d}_i \\ \underline{d}_j \\ \underline{d}_m \end{Bmatrix} = \begin{Bmatrix} u_i \\ w_i \\ u_j \\ w_j \\ u_m \\ w_m \end{Bmatrix} \tag{10.1.4}$$

and u evaluated at node i is

$$u(r_i, z_i) = u_i = a_1 + a_2 r_i + a_3 z_i \tag{10.1.5}$$

Using Eq. (10.1.3), the general displacement function is then expressed in matrix form as

$$\{\psi\} = \begin{Bmatrix} u \\ w \end{Bmatrix} = \begin{Bmatrix} a_1 + a_2 r + a_3 z \\ a_4 + a_5 r + a_6 z \end{Bmatrix} = \begin{bmatrix} 1 & r & z & 0 & 0 & 0 \\ 0 & 0 & 0 & 1 & r & z \end{bmatrix} \begin{Bmatrix} a_1 \\ a_2 \\ a_3 \\ a_4 \\ a_5 \\ a_6 \end{Bmatrix} \tag{10.1.6}$$

Substituting the coordinates of the nodal points shown in Figure 10–5(a) into Eq. (10.1.6), we can solve for the a_i's in a manner similar to that in Section 7.2. The resulting expressions are

$$\begin{Bmatrix} a_1 \\ a_2 \\ a_3 \end{Bmatrix} = \begin{bmatrix} 1 & r_i & z_i \\ 1 & r_j & z_j \\ 1 & r_m & z_m \end{bmatrix}^{-1} \begin{Bmatrix} u_i \\ u_j \\ u_m \end{Bmatrix} \tag{10.1.7}$$

and

$$\begin{Bmatrix} a_4 \\ a_5 \\ a_6 \end{Bmatrix} = \begin{bmatrix} 1 & r_i & z_i \\ 1 & r_j & z_j \\ 1 & r_m & z_m \end{bmatrix}^{-1} \begin{Bmatrix} w_i \\ w_j \\ w_m \end{Bmatrix} \tag{10.1.8}$$

Performing the inversion operations in Eqs. (10.1.7) and (10.1.8), we have

$$\begin{Bmatrix} a_1 \\ a_2 \\ a_3 \end{Bmatrix} = \frac{1}{2A} \begin{bmatrix} \alpha_i & \alpha_j & \alpha_m \\ \beta_i & \beta_j & \beta_m \\ \gamma_i & \gamma_j & \gamma_m \end{bmatrix} \begin{Bmatrix} u_i \\ u_j \\ u_m \end{Bmatrix} \tag{10.1.9}$$

and

$$\begin{Bmatrix} a_4 \\ a_5 \\ a_6 \end{Bmatrix} = \frac{1}{2A} \begin{bmatrix} \alpha_i & \alpha_j & \alpha_m \\ \beta_i & \beta_j & \beta_m \\ \gamma_i & \gamma_j & \gamma_m \end{bmatrix} \begin{Bmatrix} w_i \\ w_j \\ w_m \end{Bmatrix} \tag{10.1.10}$$

where

$$\alpha_i = r_j z_m - z_j r_m \qquad \alpha_j = r_m z_i - z_m r_i \qquad \alpha_m = r_i z_j - z_i r_j$$

$$\beta_i = z_j - z_m \qquad \beta_j = z_m - z_i \qquad \beta_m = z_i - z_j \qquad (10.1.11)$$

$$\gamma_i = r_m - r_j \qquad \gamma_j = r_i - r_m \qquad \gamma_m = r_j - r_i$$

We define the shape functions, similar to Eqs. (7.2.18), as

$$N_i = \frac{1}{2A}(\alpha_i + \beta_i r + \gamma_i z)$$

$$N_j = \frac{1}{2A}(\alpha_j + \beta_j r + \gamma_j z) \qquad (10.1.12)$$

$$N_m = \frac{1}{2A}(\alpha_m + \beta_m r + \gamma_m z)$$

Substituting Eqs. (10.1.7) and (10.1.8) into Eq. (10.1.6), along with the shape function Eqs. (10.1.12), we find that the general displacement function is

$$\{\psi\} = \left\{ \begin{array}{c} u(r,z) \\ w(r,z) \end{array} \right\} = \begin{bmatrix} N_i & 0 & N_j & 0 & N_m & 0 \\ 0 & N_i & 0 & N_j & 0 & N_m \end{bmatrix} \left\{ \begin{array}{c} u_i \\ w_i \\ u_j \\ w_j \\ u_m \\ w_m \end{array} \right\} \qquad (10.1.13)$$

or

$$\{\psi\} = [N]\{d\} \qquad (10.1.14)$$

Step 3 Define the Strain/Displacement and Stress/Strain Relationships

When we use Eqs. (10.1.1) and (10.1.3), the strains become

$$\{\varepsilon\} = \left\{ \begin{array}{c} a_2 \\ a_6 \\ \dfrac{a_1}{r} + a_2 + \dfrac{a_3 z}{r} \\ a_3 + a_5 \end{array} \right\} \qquad (10.1.15)$$

Rewriting Eq. (10.1.15) with the a_i's as a separate column matrix, we have

$$\left\{ \begin{array}{c} \varepsilon_r \\ \varepsilon_z \\ \varepsilon_\theta \\ \gamma_{rz} \end{array} \right\} = \begin{bmatrix} 0 & 1 & 0 & 0 & 0 & 0 \\ 0 & 0 & 0 & 0 & 0 & 1 \\ \dfrac{1}{r} & 1 & \dfrac{z}{r} & 0 & 0 & 0 \\ 0 & 0 & 1 & 0 & 1 & 0 \end{bmatrix} \left\{ \begin{array}{c} a_1 \\ a_2 \\ a_3 \\ a_4 \\ a_5 \\ a_6 \end{array} \right\} \qquad (10.1.16)$$

Substituting Eqs. (10.1.7) and (10.1.8) into Eq. (10.1.16) and making use of Eq. (10.1.11), we obtain

$$\{\varepsilon\} = \frac{1}{2A} \begin{bmatrix} \beta_i & 0 & \beta_j & 0 & \beta_m & 0 \\ 0 & \gamma_i & 0 & \gamma_j & 0 & \gamma_m \\ \dfrac{\alpha_i}{r}+\beta_i+\dfrac{\gamma_i z}{r} & 0 & \dfrac{\alpha_j}{r}+\beta_j+\dfrac{\gamma_j z}{r} & 0 & \dfrac{\alpha_m}{r}+\beta_m+\dfrac{\gamma_m z}{r} & 0 \\ \gamma_i & \beta_i & \gamma_j & \beta_j & \gamma_m & \beta_m \end{bmatrix} \begin{Bmatrix} u_i \\ w_i \\ u_j \\ w_j \\ u_m \\ w_m \end{Bmatrix}$$

(10.1.17)

or, rewriting Eq. (10.1.17) in simplified matrix form,

$$\{\varepsilon\} = [\underline{B}_i \quad \underline{B}_j \quad \underline{B}_m] \begin{Bmatrix} u_i \\ w_i \\ u_j \\ w_j \\ u_m \\ w_m \end{Bmatrix}$$

(10.1.18)

where

$$[B_i] = \frac{1}{2A} \begin{bmatrix} \beta_i & 0 \\ 0 & \gamma_i \\ \dfrac{\alpha_i}{r}+\beta_i+\dfrac{\gamma_i z}{r} & 0 \\ \gamma_i & \beta_i \end{bmatrix}$$

(10.1.19)

Similarly, we obtain submatrices $\underline{B}_j$ and $\underline{B}_m$ by replacing the subscript i with j and then with m in Eq. (10.1.19). Rewriting Eq. (10.1.18) in compact matrix form, we have

$$\{\varepsilon\} = [B]\{d\}$$

(10.1.20)

where

$$[B] = [\underline{B}_i \quad \underline{B}_j \quad \underline{B}_m]$$

(10.1.21)

Note that $[B]$ is a function of the r and z coordinates. Therefore, in general, the strain ε_θ will not be constant.

The stresses are given by

$$\{\sigma\} = [D][B]\{d\}$$

(10.1.22)

where $[D]$ is given by the first matrix on the right side of Eq. (10.1.2). (As mentioned in Chapter 7, for $v = 0.5$, a special formula must be used; see Reference [9].)

Step 4 Derive the Element Stiffness Matrix and Equations

The stiffness matrix is

$$[k] = \iiint_V [B]^T[D][B]\, dV$$

(10.1.23)

or

$$[k] = 2\pi \iint\limits_{A} [B]^T [D][B] r \, dr \, dz \qquad (10.1.24)$$

after integrating along the circumferential boundary. The $[B]$ matrix, Eq. (10.1.21), is a function of r and z. Therefore, $[k]$ is a function of r and z and is of order 6×6.

We can evaluate Eq. (10.1.24) for $[k]$ by one of three methods:

1. Numerical integration (Gaussian quadrature) as discussed in Chapter 11.
2. Explicit multiplication and term-by-term integration [1].
3. Evaluate $[B]$ for a centroidal point $(\bar{r}, \bar{z})$ of the element

$$r = \bar{r} = \frac{r_i + r_j + r_m}{3} \qquad z = \bar{z} = \frac{z_i + z_j + z_m}{3} \qquad (10.1.25)$$

and define $[B(\bar{r}, \bar{z})] = [\bar{B}]$. Therefore, as a first approximation,

$$[k] = 2\pi \bar{r} A [\bar{B}]^T [D][\bar{B}] \qquad (10.1.26)$$

If the triangular subdivisions are consistent with the final stress distribution (that is, small elements in regions of high stress gradients), then acceptable results can be obtained by method 3.

Distributed Body Forces

Loads such as gravity (in the direction of the z axis) or centrifugal forces in rotating machine parts (in the direction of the r axis) are considered to be body forces (as shown in Figure 10–6). The body forces can be found by

$$\{f_b\} = 2\pi \iint\limits_{A} [N]^T \begin{Bmatrix} R_b \\ Z_b \end{Bmatrix} r \, dr \, dz \qquad (10.1.27)$$

where $R_b = \omega^2 \rho r$ for a machine part moving with a constant angular velocity ω about the z axis, with material mass density ρ and radial coordinate r, and where Z_b is the body force per unit volume due to the force of gravity.

Considering the body force at node i, we have

$$\{f_{bi}\} = 2\pi \iint\limits_{A} [N_i]^T \begin{Bmatrix} R_b \\ Z_b \end{Bmatrix} r \, dr \, dz \qquad (10.1.28)$$

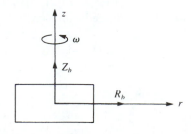

Figure 10–6 Axisymmetric element with body forces per unit volume

where
$$[N_i]^T = \begin{bmatrix} N_i & 0 \\ 0 & N_i \end{bmatrix} \tag{10.1.29}$$

Multiplying and integrating in Eq. (10.1.28), we obtain

$$\{f_{bi}\} = \frac{2\pi}{3} \begin{Bmatrix} \bar{R}_b \\ Z_b \end{Bmatrix} A\bar{r} \tag{10.1.30}$$

where the origin of the coordinates has been taken as the centroid of the element, and $\bar{R}_b$ is the radially directed body force per unit volume evaluated at the centroid of the element. The body forces at nodes j and m are identical to those given by Eq. (10.1.30) for node i. Hence, for an element, we have

$$\{f_b\} = \frac{2\pi\bar{r}A}{3} \begin{Bmatrix} \bar{R}_b \\ Z_b \\ \bar{R}_b \\ Z_b \\ \bar{R}_b \\ Z_b \end{Bmatrix} \tag{10.1.31}$$

where
$$\bar{R}_b = \omega^2 \rho\bar{r} \tag{10.1.32}$$

Equation (10.1.31) is a first approximation to the radially directed body force distribution.

Surface Forces

Surface forces can be found by

$$\{f_s\} = \iint_S [N_s]^T \{T\}\, dS \tag{10.1.33}$$

where again $[N_s]$ denotes the shape function matrix evaluated along the surface where the surface traction acts.

For radial and axial pressures p_r and p_z, respectively, we have

$$\{f_s\} = \iint_S [N_s]^T \begin{Bmatrix} p_r \\ p_z \end{Bmatrix}\, dS \tag{10.1.34}$$

For example, along the vertical face jm of an element, let uniform loads p_r and p_z be applied, as shown in Figure 10–7 along surface $r = r_j$. We can use Eq. (10.1.34) written for each node separately. For instance, for node j, substituting N_j from Eqs. (10.1.12) into Eq. (10.1.34), we have

$$\{f_{sj}\} = \int_{z_j}^{z_m} \frac{1}{2A} \begin{bmatrix} \alpha_j + \beta_j r + \gamma_j z & 0 \\ 0 & \alpha_j + \beta_j r + \gamma_j z \end{bmatrix} \begin{Bmatrix} p_r \\ p_z \end{Bmatrix} 2\pi r_j\, dz \tag{10.1.35}$$

evaluated at $r = r_j,\ z = z$

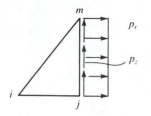

Figure 10–7 Axisymmetric element with surface forces

Performing the integration of Eq. (10.1.35) explicitly, along with similar evaluations for f_{si} and f_{sm}, we obtain the total distribution of surface force to nodes i, j, and m as

$$\{f_s\} = \frac{2\pi r_j(z_m - z_j)}{2} \begin{Bmatrix} 0 \\ 0 \\ p_r \\ p_z \\ p_r \\ p_z \end{Bmatrix} \tag{10.1.36}$$

Steps 5–7

Steps 5–7, which involve assembling the total stiffness matrix, total force matrix, and total set of equations; solving for the nodal degrees of freedom; and calculating the element stresses, are analogous to those of Chapter 7 for the CST element, except the stresses are not constant in each element. They are usually determined by one of two methods that we use to determine the LST element stresses. Either we determine the centroidal element stresses, or we determine the nodal stresses for the element and then average them. The latter method has been shown to be more accurate in some cases [2].

Example 10.1

For the element of an axisymmetric body rotating with a constant angular velocity $\omega = 100$ rev/min as shown in Figure 10–8, evaluate the approximate body force matrix. Include the weight of the material, where the weight density ρ_w is 0.283 lb/in³. The coordinates of the element (in inches) are shown in the figure.

We need to evaluate Eq. (10.1.31) to obtain the approximate body force matrix. Therefore, the body forces per unit volume evaluated at the centroid of the element are

$$Z_b = 0.283 \text{ lb/in}^3$$

and by Eq. (10.1.32), we have

$$\bar{R}_b = \omega^2 \rho \bar{r} = \left[\left(100 \frac{\text{rev}}{\text{min}} \right) \left(2\pi \frac{\text{rad}}{\text{rev}} \right) \left(\frac{1 \text{ min}}{60 \text{ s}} \right) \right]^2 \frac{(0.283 \text{ lb/in}^3)}{(32.2 \times 12) \text{ in./s}^2} (2.333 \text{ in.})$$

$$\bar{R}_b = 0.187 \text{ lb/in}^3$$

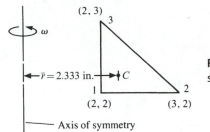

Figure 10–8 Axisymmetric element subjected to angular velocity

$$\frac{2\pi\bar{r}A}{3} = \frac{2\pi(2.333)(0.5)}{3} = 2.44 \text{ in}^3$$

$$f_{b1r} = (2.44)(0.187) = 0.457 \text{ lb}$$

$$f_{b1z} = -(2.44)(0.283) = -0.691 \text{ lb} \qquad \text{(downward)}$$

Because we are using the first approximation Eq. (10.1.31), all r-directed nodal body forces are equal, and all z-directed body forces are equal. Therefore,

$$f_{b2r} = 0.457 \text{ lb} \qquad f_{b2z} = -0.691 \text{ lb}$$

$$f_{b3r} = 0.457 \text{ lb} \qquad f_{b3z} = -0.691 \text{ lb} \qquad \blacksquare$$

▲ 10.2 Solution of an Axisymmetric Pressure Vessel ▲

To illustrate the use of the equations developed in Section 10.1, we will now solve an axisymmetric stress problem.

Example 10.2

For the long, thick-walled cylinder under internal pressure p equal to 1 psi shown in Figure 10–9, determine the displacements and stresses.

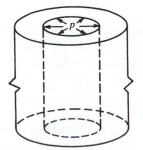

Figure 10–9 Thick-walled cylinder subjected to internal pressure

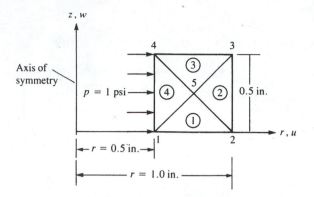

Figure 10–10 Discretized cylinder slice

Discretization

To illustrate the finite element solution for the cylinder, we first discretize the cylinder into four triangular elements, as shown in Figure 10–10. A horizontal slice of the cylinder represents the total cylinder behavior. Because we are performing a longhand solution, a coarse mesh of elements is used for simplicity's sake (but without loss of generality of the method). The governing global matrix equation is

$$
\begin{Bmatrix} F_{1r} \\ F_{1z} \\ F_{2r} \\ F_{2z} \\ F_{3r} \\ F_{3z} \\ F_{4r} \\ F_{4z} \\ F_{5r} \\ F_{5z} \end{Bmatrix} = [K] \begin{Bmatrix} u_1 \\ w_1 \\ u_2 \\ w_2 \\ u_3 \\ w_3 \\ u_4 \\ w_4 \\ u_5 \\ w_5 \end{Bmatrix}
\tag{10.2.1}
$$

where the $[K]$ matrix is of order 10×10.

Assemblage of the Stiffness Matrix

We assemble the $[K]$ matrix in the usual manner by superposition of the individual element stiffness matrices. For simplicity's sake, we will use the first approximation method given by Eq. (10.1.26) to evaluate the element matrices. Therefore,

$$
[k] = 2\pi \bar{r} A [\bar{B}]^T [D] [\bar{B}]
\tag{10.2.2}
$$

For element 1 (Figure 10–11), the coordinates are $r_i = 0.5$, $z_i = 0$, $r_j = 1.0$, $z_j = 0$, $r_m = 0.75$, and $z_m = 0.25$ ($i = 1, j = 2$, and $m = 5$ for element 1) for the global-coordinate axes as set up in Figure 10–10.

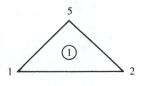

Figure 10–11 Element 1 of the discretized cylinder

We now evaluate $[\bar{B}]$, where $[\bar{B}]$ is given by Eq. (10.1.19) evaluated at the centroid of the element $r = \bar{r}$, $z = \bar{z}$, and expanded here as

$$[\bar{B}] = \frac{1}{2A} \begin{bmatrix} \beta_i & 0 & \beta_j & 0 & \beta_m & 0 \\ 0 & \gamma_i & 0 & \gamma_j & 0 & \gamma_m \\ \frac{\alpha_i}{\bar{r}} + \beta_i + \frac{\gamma_i \bar{z}}{\bar{r}} & 0 & \frac{\alpha_j}{\bar{r}} + \beta_j + \frac{\gamma_j \bar{z}}{\bar{r}} & 0 & \frac{\alpha_m}{\bar{r}} + \beta_m + \frac{\gamma_m \bar{z}}{\bar{r}} & 0 \\ \gamma_i & \beta_i & \gamma_j & \beta_j & \gamma_m & \beta_m \end{bmatrix} \quad (10.2.3)$$

where, using element coordinates in Eqs. (10.1.11), we have

$$\alpha_i = r_j z_m - z_j r_m = (1.0)(0.25) - (0.0)(0.75) = 0.25 \text{ in}^2$$

$$\alpha_j = r_m z_i - z_m r_i = (0.75)(0) - (0.25)(0.5) = -0.125 \text{ in}^2$$

$$\alpha_m = r_i z_j - z_i r_j = (0.5)(0.0) - (0)(1.0) = 0.0 \text{ in}^2$$

$$\beta_i = z_j - z_m = 0.0 - 0.25 = -0.25 \text{ in.}$$

$$\beta_j = z_m - z_i = 0.25 - 0 = 0.25 \text{ in.}$$

$$\beta_m = z_i - z_j = 0.0 - 0.0 = 0.0 \text{ in.} \quad (10.2.4)$$

$$\gamma_i = r_m - r_j = 0.75 - 1.0 = -0.25 \text{ in.}$$

$$\gamma_j = r_i - r_m = 0.5 - 0.75 = -0.25 \text{ in.}$$

$$\gamma_m = r_j - r_i = 1.0 - 0.5 = 0.5 \text{ in.}$$

and
$$\bar{r} = 0.5 + \tfrac{1}{2}(0.5) = 0.75 \text{ in.} \qquad \bar{z} = \tfrac{1}{3}(0.25) = 0.0833 \text{ in.}$$
$$A = \tfrac{1}{2}(0.5)(0.25) = 0.0625 \text{ in}^2$$

Substituting the results from Eqs. (10.2.4) into Eq. (10.2.3), we obtain

$$[\bar{B}] = \frac{1}{0.125} \begin{bmatrix} -0.25 & 0 & 0.25 & 0 & 0 & 0 \\ 0 & -0.25 & 0 & -0.25 & 0 & 0.5 \\ 0.0556 & 0 & 0.0556 & 0 & 0.0556 & 0 \\ -0.25 & -0.25 & -0.25 & 0.25 & 0.5 & 0 \end{bmatrix} \frac{1}{\text{in.}} \quad (10.2.5)$$

For the axisymmetric stress case, the matrix $[D]$ is given in Eq. (10.1.2) as

$$[D] = \frac{E}{(1+v)(1-2v)} \begin{bmatrix} 1-v & v & v & 0 \\ v & 1-v & v & 0 \\ v & v & 1-v & 0 \\ 0 & 0 & 0 & \dfrac{1-2v}{2} \end{bmatrix} \tag{10.2.6}$$

With $v = 0.3$ and $E = 30 \times 10^6$ psi, we obtain

$$[D] = \frac{30(10^6)}{(1+0.3)[1-2(0.3)]} \begin{bmatrix} 1-0.3 & 0.3 & 0.3 & 0 \\ 0.3 & 1-0.3 & 0.3 & 0 \\ 0.3 & 0.3 & 1-0.3 & 0 \\ 0 & 0 & 0 & \dfrac{1-2(0.3)}{2} \end{bmatrix} \tag{10.2.7}$$

or, simplifying Eq. (10.2.7),

$$[D] = 57.7(10^6) \begin{bmatrix} 0.7 & 0.3 & 0.3 & 0 \\ 0.3 & 0.7 & 0.3 & 0 \\ 0.3 & 0.3 & 0.7 & 0 \\ 0 & 0 & 0 & 0.2 \end{bmatrix} \text{psi} \tag{10.2.8}$$

Using Eqs. (10.2.5) and (10.2.8), we obtain

$$[\bar{B}]^T[D] = \frac{57.7(10^6)}{0.125} \begin{bmatrix} -0.158 & -0.0583 & -0.0361 & -0.05 \\ -0.075 & -0.175 & -0.075 & -0.05 \\ 0.192 & 0.0917 & 0.114 & -0.05 \\ -0.075 & -0.175 & -0.075 & 0.05 \\ 0.0167 & 0.0166 & 0.0388 & 0.1 \\ 0.15 & 0.35 & 0.15 & 0 \end{bmatrix} \tag{10.2.9}$$

Substituting Eqs. (10.2.5) and (10.2.9) into Eq. (10.2.2), we obtain the stiffness matrix for element 1 as

$$[k^{(1)}] = (10^6) \begin{matrix} i=1 \qquad\qquad j=2 \qquad\qquad m=5 \\ \begin{bmatrix} 54.46 & 29.45 & -31.63 & 2.26 & -29.37 & -31.71 \\ 29.45 & 61.17 & -11.33 & 33.98 & -31.72 & -95.15 \\ -31.63 & -11.33 & 72.59 & -38.52 & -20.31 & 49.84 \\ 2.26 & 33.98 & -38.52 & 61.17 & 22.66 & -95.15 \\ -29.37 & -31.72 & -20.31 & 22.66 & 56.72 & 9.06 \\ -31.71 & -95.15 & 49.84 & -95.15 & 9.06 & 190.31 \end{bmatrix} \end{matrix} \frac{\text{lb}}{\text{in.}} \tag{10.2.10}$$

where the numbers above the columns indicate the nodal orders of degrees of freedom in the element 1 stiffness matrix.

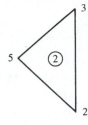

Figure 10–12 Element 2 of the discretized cylinder

For element 2 (Figure 10–12), the coordinates are $r_i = 1.0$, $z_i = 0.0$, $r_j = 1.0$, $z_j = 0.5$, $r_m = 0.75$, and $z_m = 0.25$ ($i = 2$, $j = 3$, and $m = 5$ for element 2). Therefore,

$$\alpha_i = (1.0)(0.25) - (0.5)(0.75) = -0.125 \text{ in}^2 \qquad (10.2.11)$$

$$\alpha_j = (0.75)(0.0) - (0.25)(1.0) = -0.25 \text{ in}^2$$

$$\alpha_m = (1.0)(0.5) - (0.0)(1.0) = 0.5 \text{ in}^2$$

$$\beta_i = 0.5 - 0.25 = 0.25 \text{ in.} \qquad \beta_j = 0.25 - 0.0 = 0.25 \text{ in.}$$

$$\beta_m = 0.0 - 0.5 = -0.5 \text{ in.} \qquad \gamma_i = 0.75 - 1.0 = -0.25 \text{ in.}$$

$$\gamma_j = 1.0 - 0.75 = 0.25 \text{ in.} \qquad \gamma_m = 1.0 - 1.0 = 0.0 \text{ in.}$$

and
$$\bar{r} = 0.9167 \text{ in.} \qquad \bar{z} = 0.25 \text{ in.} \qquad A = 0.0625 \text{ in}^2$$

Using Eqs. (10.2.11) in Eq. (10.2.2) and proceeding as for element 1, we obtain the stiffness matrix for element 2 as

$$[k^{(2)}] = (10^6) \begin{bmatrix} 85.75 & -46.07 & 52.52 & 12.84 & -118.92 & 33.23 \\ -46.07 & 74.77 & -12.84 & -41.54 & 45.32 & -33.23 \\ 52.52 & -12.84 & 85.74 & 46.07 & -118.92 & -33.23 \\ 12.84 & -41.54 & 46.07 & 74.77 & -45.32 & -33.23 \\ -118.92 & 45.32 & -118.92 & -45.32 & 216.41 & 0 \\ 33.23 & -33.23 & -33.23 & -33.23 & 0 & 66.46 \end{bmatrix} \frac{\text{lb}}{\text{in.}}$$

with column headers $i = 2$, $j = 3$, $m = 5$.

$$(10.2.12)$$

We obtain the stiffness matrices for elements 3 and 4 in a manner similar to that used to obtain the stiffness matrices for elements 1 and 2. Thus,

$$[k^{(3)}] = (10^6) \begin{bmatrix} 72.58 & 38.52 & -31.63 & 11.33 & -20.31 & -49.84 \\ 38.52 & 61.17 & -2.26 & 33.98 & -22.66 & -95.15 \\ -31.63 & -2.26 & 54.46 & -29.45 & -29.37 & 31.72 \\ 11.33 & 33.98 & -29.45 & 61.17 & 31.72 & -95.15 \\ -20.31 & -22.66 & -29.37 & 31.72 & 56.72 & -9.06 \\ -49.84 & -95.15 & 31.72 & -95.15 & -9.06 & 190.31 \end{bmatrix} \frac{\text{lb}}{\text{in.}}$$

with column headers $i = 3$, $j = 4$, $m = 5$.

$$(10.2.13)$$

and

$$i = 4 \qquad\qquad j = 1 \qquad\qquad m = 5$$

$$[k^{(4)}] = (10^6) \begin{bmatrix} 41.53 & -21.90 & 20.39 & 0.75 & -66.45 & 21.14 \\ -21.90 & 47.57 & -0.75 & -26.43 & 36.24 & -21.14 \\ 20.39 & -0.75 & 41.53 & 21.90 & -66.45 & -21.14 \\ 0.75 & -26.43 & 21.90 & 47.57 & -36.24 & -21.14 \\ -66.45 & 36.24 & -66.45 & -36.24 & 169.14 & 0 \\ 21.14 & -21.14 & -21.14 & -21.14 & 0 & 42.28 \end{bmatrix} \frac{\text{lb}}{\text{in.}}$$

$$(10.2.14)$$

Using superposition of the element stiffness matrices [Eqs. (10.2.10) and (10.2.12)–(10.2.14)], where we rearrange the elements of each stiffness matrix in order of increasing nodal degrees of freedom, we obtain the global stiffness matrix as

$$[K] = (10^6) \begin{bmatrix} 95.99 & 51.35 & -31.63 & 2.26 & 0 & 0 & 20.39 & -0.75 & -95.82 & -52.86 \\ 51.35 & 108.74 & -11.33 & 33.98 & 0 & 0 & 0.75 & -26.43 & -67.96 & -116.3 \\ -31.63 & -11.33 & 158.34 & -84.59 & 52.52 & 12.84 & 0 & 0 & -139.2 & 83.07 \\ 2.26 & 33.98 & -84.59 & 135.94 & -12.84 & -41.54 & 0 & 0 & 67.98 & -128.4 \\ 0 & 0 & 52.52 & -12.84 & 158.33 & 84.59 & -31.63 & 11.33 & -139.2 & -83.07 \\ 0 & 0 & 12.84 & -41.54 & 84.59 & 135.94 & -2.26 & 33.98 & -67.98 & -128.4 \\ 20.39 & 0.75 & 0 & 0 & -31.63 & -2.26 & 95.99 & -51.35 & -95.82 & 52.86 \\ -0.75 & -26.43 & 0 & 0 & 11.33 & 33.98 & -51.35 & 108.74 & 67.96 & -116.3 \\ -95.82 & -67.96 & -139.2 & 67.98 & -139.2 & -67.98 & -95.82 & 67.96 & 498.99 & 0 \\ -52.86 & -116.3 & 83.07 & -128.4 & -83.07 & -128.4 & 52.86 & -116.3 & 0 & 489.36 \end{bmatrix} \frac{\text{lb}}{\text{in.}}$$

$$(10.2.15)$$

The applied nodal forces are given by Eq.(10.1.36) as

$$F_{1r} = F_{4r} = \frac{2\pi(0.5)(0.5)}{2}(1) = 0.785 \text{ lb} \qquad (10.2.16)$$

All other nodal forces are zero. Using Eq. (10.2.15) for $[K]$ and Eq. (10.2.16) for the nodal forces in Eq. (10.2.1), and solving for the nodal displacements, we obtain

$$u_1 = 0.0322 \times 10^{-6} \text{ in.} \qquad w_1 = 0.00115 \times 10^{-6} \text{ in.}$$

$$u_2 = 0.0219 \times 10^{-6} \text{ in.} \qquad w_2 = 0.00206 \times 10^{-6} \text{ in.}$$

$$u_3 = 0.0219 \times 10^{-6} \text{ in.} \qquad w_3 = -0.00206 \times 10^{-6} \text{ in.} \qquad (10.2.17)$$

$$u_4 = 0.0322 \times 10^{-6} \text{ in.} \qquad w_4 = -0.00115 \times 10^{-6} \text{ in.}$$

$$u_5 = 0.0244 \times 10^{-6} \text{ in.} \qquad w_5 = 0$$

The results for nodal displacements are as expected because radial displacements at the inner edge are equal ($u_1 = u_4$) and those at the outer edge are equal ($u_2 = u_3$). In addition, the axial displacements at the outer nodes and inner nodes are equal but opposite in sign ($w_1 = -w_4$ and $w_2 = -w_3$) as a result of the Poisson effect and symmetry. Finally, the axial displacement at the center node is zero ($w_5 = 0$), as it should be because of symmetry.

By using Eq. (10.1.22), we now determine the stresses in each element as

$$\{\sigma\} = [D][\bar{B}]\{d\} \qquad (10.2.18)$$

For element 1, we use Eq. (10.2.5) for $[\bar{B}]$, Eq. (10.2.8) for $[D]$, and Eq. (10.2.17) for $\{d\}$ in Eq. (10.2.18) to obtain

$$\sigma_r = -0.338 \text{ psi} \qquad \sigma_z = -0.0126 \text{ psi}$$

$$\sigma_\theta = 0.942 \text{ psi} \qquad \tau_{rz} = -0.1037 \text{ psi}$$

Similarly, for element 2, we obtain

$$\sigma_r = -0.105 \text{ psi} \qquad \sigma_z = -0.0747 \text{ psi}$$

$$\sigma_\theta = 0.690 \text{ psi} \qquad \tau_{rz} = 0.000 \text{ psi}$$

For element 3, the stresses are

$$\sigma_r = -0.337 \text{ psi} \qquad \sigma_z = -0.0125 \text{ psi}$$

$$\sigma_\theta = 0.942 \text{ psi} \qquad \tau_{rz} = 0.1037 \text{ psi}$$

For element 4, the stresses are

$$\sigma_r = -0.470 \text{ psi} \qquad \sigma_z = 0.1493 \text{ psi}$$

$$\sigma_\theta = 1.426 \text{ psi} \qquad \tau_{rz} = 0.000 \text{ psi}$$

Figure 10–13 shows the exact solution [10] along with the results determined here and the results from Reference [5]. Observe that agreement with the exact solution is quite good except for the limited results due to the very coarse mesh used in the longhand example, and in case 1 of Reference [5]. In Reference [5], stresses have been plotted at the center of the quadrilaterals and were obtained by averaging the stresses in the four connecting triangles.   ■

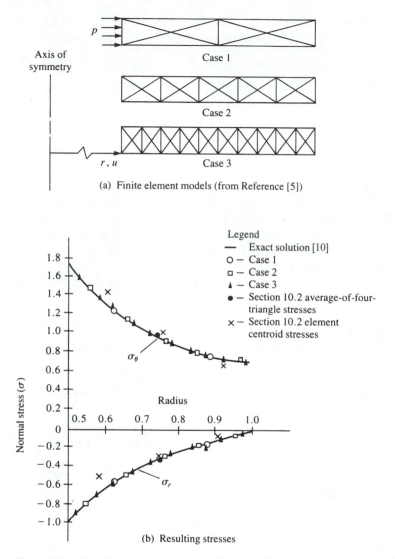

(a) Finite element models (from Reference [5])

(b) Resulting stresses

Figure 10–13 Finite element analysis of a thick-walled cylinder

▲ 10.3 Applications of Axisymmetric Elements ▲

Numerous structural (and nonstructural) systems can be classified as axisymmetric. Some typical structural systems whose behavior is modeled accurately using the axisymmetric element developed in this chapter are represented in Figures 10–14 and 10–15.

Figure 10–14 illustrates the finite element model of a steel-reinforced concrete pressure vessel. The vessel is a thick-walled cylinder with flat heads. An axis of symmetry (the z axis) exists such that only one-half of the r-z plane passing through the

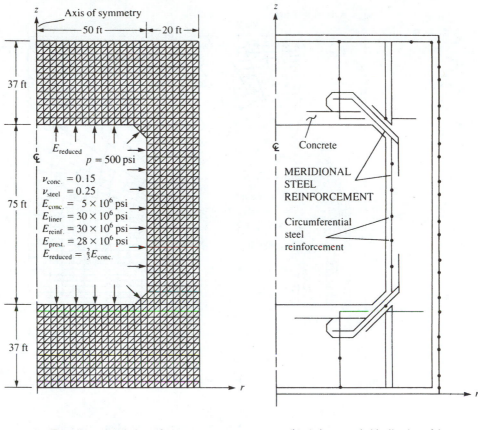

(a) Two-dimensional view of a
finite element idealization
for a prestressed concrete reactor
vessel (PCRV)

(b) Axisymmetric idealization of the
steel reinforcement

Figure 10–14 Model of steel-reinforced concrete pressure vessel (from Reference [4], North Holland Physics Publishing, Amsterdam)

middle of the structure need be modeled. The concrete was modeled by using the axisymmetric triangular element developed in this chapter. The steel elements were laid out along the boundaries of the concrete elements so as to maintain continuity (or perfect bond assumption) between the concrete and the steel. The vessel was then subjected to an internal pressure as shown in the figure. Note that the nodes along the axis of symmetry should be supported by rollers preventing motion perpendicular to the axis of symmetry.

Figure 10–15 shows a finite element model of a high-strength steel die used in a thin-plastic-film-making process [7]. The die is an irregularly shaped disk. An axis of symmetry with respect to geometry and loading exists as shown. The die was modeled by using simple quadrilateral axisymmetric elements. The locations of high stress were of primary concern. Figure 10–16 shows a plot of the Von Mises stress contours for

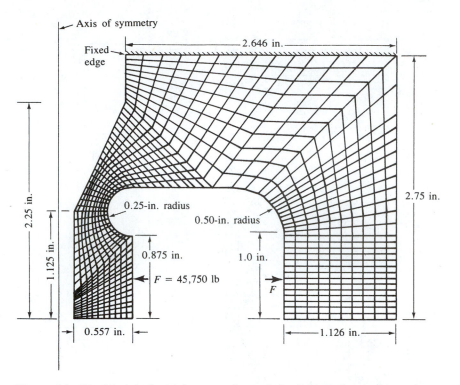

Figure 10–15 Model of a high-strength steel die (924 nodes and 830 elements)

the die of Figure 10–15. The Von Mises (or equivalent, or effective) stress [8] is often used as a failure criterion in design.

(Recall that the failure criterion based on the maximum distortion energy theory for ductile materials subjected to static loading predicts that a material will fail if the Von Mises stress reaches the yield strength of the material.) The Von Mises stress σ_{VM} may be related to the principal stresses by the expression

$$\sigma_{VM} = \frac{1}{\sqrt{2}} \sqrt{(\sigma_1 - \sigma_2)^2 + (\sigma_2 - \sigma_3)^2 + (\sigma_3 - \sigma_1)^2}$$

where the principal stresses are given by σ_1, σ_2, and σ_3. These results were obtained from the commercial computer code ANSYS.

Other dies with modifications in geometry were also studied to evaluate the most suitable die before the construction of an expensive prototype. Confidence in the acceptability of the prototype was enhanced by doing these comparison studies. Other examples of the use of the axisymmetric element can be found in References [2–6].

In this chapter, we have shown the finite element analysis of axisymmetric systems using a simple three-noded triangular element to be analogous to that of the two-dimensional plane stress problem using three-noded triangular elements as developed in Chapter 7. Therefore, the two-dimensional element in the Algor computer program

Axis of symmetry

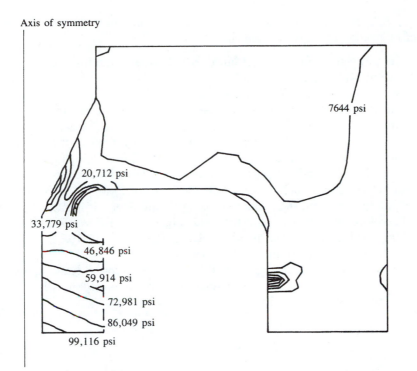

7644 psi

20,712 psi

33,779 psi

46,846 psi

59,914 psi

72,981 psi

86,049 psi

99,116 psi

Figure 10–16 Von Mises stress contour plot of axisymmetric model of Figure 10–15 (also producing a radial inward deflection of about 0.015 in.)

with the axisymmetric element selected will allow for the analysis of axisymmetric structures.

Finally, note that other axisymmetric elements, such as a simple quadrilateral (one with four corner nodes and two degrees of freedom per node, as used in the steel die analysis of Figure 10–15) or higher-order triangular elements, such as in Reference [6], in which a cubic polynomial involving ten terms (ten a's) for both u and w, could be used for axisymmetric analysis. The three-noded triangular element was described here because of its simplicity and ability to describe geometric boundaries rather easily.

▲ 10.4 Algor Example Solutions for Axisymmetric Problems ▲

To illustrate the use of the Algor computer program for axisymmetric problems, we now solve a number of examples. The program is based on the flowchart of Section 8.6. Longhand Example 10.2 is first solved using Algor. Next a four-element solution for Problem 10.10 is presented to describe how to handle centrifugal forces from a spinning disk. A stepped shaft is modeled to consider stress concentrations, and finally a steel-lined concrete pressure vessel is solved to illustrate modeling with two different materials by using the *Group* command.

Before describing the steps to create a solution for an axisymmetric problem, we must consider the following points:

1. The z direction must be constrained for all nodes at the top and bottom of the model for an infinitely long pressure vessel in order for us to obtain the desired symmetric deformation.

2. Radial forces must be in units of pounds per radian. For example, assume we have a uniform internal pressure of 1 psi, as in Example 10.3. For a vessel with a depth of 0.5 in., the radial force at the upper and lower interior nodes is $F = (1 \text{ psi})(0.5 \text{ in.})/2 = 0.25 \text{ lb/in.}$ per node. The total radial force around the circumferential direction for an inside radius of 0.5 in. is $F_c = 2(\pi)(0.50 \text{ in.})(0.25 \text{ lb/in.}) = 0.785 \text{ lb}$. However, the radial force is input into Algor as lb/rad. Therefore, the forces input at the inside nodes will be $F_i = 0.785/(2\pi) = 0.125 \text{ lb/rad}$. This is the number used in Example 10.3 for the radial force applied to the inside nodes.

3. The body must be transformed to the y-z plane (as for all plane stress/strain problems) with y corresponding to the radial direction. All y coordinates must be positive.

Example 10.3

Determine the circumferential stress in the thick-walled cylinder under an internal pressure of 1 psi shown in Figure 10–17. Compare this stress to the analytical value given in Reference [10].

Step 1   Start Superdraw III

Select the "Start" button of Windows NT/95/98 and then proceed to "Programs: Algor Software:Algor FEA".

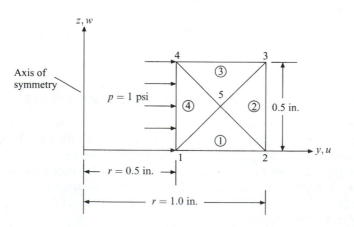

Figure 10–17   Discretized thick-walled cylinder slice

Click "Algor FEA" to start the program or double-click on the "Algor FEA" icon on your desktop computer screen. The Superdraw III program appears. This is the Algor main graphics user interface. All your work will be done from this program. You will construct, analyze, and review your results from Superdraw III. Each menu option and a short example are available in the Superdraw III Reference Division (accessed through "Docutech", the on-line "Technical User Documentation").

Step 2   Create the Axisymmetric Pressure Vessel Model

Click on the "Modify" menu on the top menu bar.

Click "Transform XY to YZ" to transform from the x-y plane to the y-z plane. You must perform the analysis in the y-z plane. For axisymmetric problems, the y axis represents the radial direction.

Click on the "Add" menu on the top main menu bar.

Click "Rectangle". A menu appears in the upper right part of the screen. Then enter the data points.

```
0<Tab>0.5<Tab>0          This is the first corner node.
<Enter>
0<Tab>1.0<Tab>0.5        This is the opposite corner node.
<Enter>
```

Click "Enclose" on the "View" menu.

Now the object needs to be discretized.

Click on the "Add" menu on the top main menu bar.

Click "Line". A menu appears in the upper right part of the screen.

Click "Single" on the menu in the upper right side of the screen. Then enter the data points.

```
0<Tab>0.5<Tab>0          Enter the first coordinate of the diagonal line.
<Enter>
0<Tab>0.75<Tab>0.25      This is the second coordinate of the diagonal line.
<Enter>
0<Tab>0.5<Tab>0.5        Enter the first coordinate of the diagonal line.
<Enter>
0<Tab>0.75<Tab>0.25      This is the second coordinate of the diagonal line.
<Enter>
0<Tab>1.0<Tab>0          Enter the first coordinate of the diagonal line.
<Enter>
0<Tab>0.75<Tab>0.25      This is the second coordinate of the diagonal line.
<Enter>
0<Tab>1.0<Tab>0.5        Enter the first coordinate of the diagonal line.
<Enter>
0<Tab>0.75<Tab>0.25      This is the second coordinate of the diagonal line.
<Enter>
```

Step 3 Eliminate Duplicate Lines

Click "Modify" on the main menu.

Click "Clean:Duplicate".

Click "Perform Cleaning" on the "Duplicate" menu. You see the following message in the lower part of the screen:

"6 Kept 0 Deleted. done"

Click "Done" on the "Duplicate" menu located in the upper right corner of the screen.

Step 4 Add in Boundary Conditions and Forces

Click "FEA Add" on the main menu and highlight "Stress and Vibration Analysis".

Click "Boundary Conditions".

The "Boundary Condition" menu appears. The default boundary condition is "Use @ Symbol for Full". "Full" means full constraint or prevention of translations and rotations (when relevant).

Click "Change Values" on the "Boundary Condition" menu.

Select the box with the check mark next to "Ty Constrained". This allows model translation in the y direction.

Click "Done".

Move the cursor to nodes 1, 2, 3, and 4 one at a time and left-click on each node. "TxzRxyz" appears at each node.

Click "FEA Add" on the main menu and highlight "Stress and Vibration Analysis".

Click "Nodal Forces".

Click "Vector".

Click "Y Direction" This applies a force in the positive y direction.

Click "Done".

Click "Magnitude".

Type in ⟨0.125⟩. This is calculated by obtaining the total radial force around the circumference of the cylinder and then dividing by 2π, that is, $0.785/2\pi = 0.125$ lb/rad.

Left-click on nodes 1 and 4. The value of the force with large arrows indicating the direction of each force now appears at each node.

Click "Done".

Click "View" and "Enclose" to see the entire model.

Step 5 Add the Material Properties Using the "Model Data Control" Window

After the nodal boundary conditions and the forces are applied,

Select "Tools" and the "Model Data" button from Superdraw III to complete the data input for analysis. The "Model Data Control" window appears. Alternatively,

Click on the "Model Data Control" button located at the lower part of the screen and highlighted in red.

For the "Analysis Type", "Linear Static Stress" appears as the default. You can scroll down for other analysis types as appropriate.

Click on the box below "Element". A table of group 1 element types appears.

Click "2-D".

Click "OK".

Click on the box below "Data". It responds with "Please enter the model name first".

Click "OK" and enter the model name in the "File_name" area that appears.

ex103

Click "Save". By default, the "Units Definition" menu appears.

Click "OK" if you accept the default "English (in)" units system. Otherwise, scroll down to select another choice of units. The "Element Definition" menu appears next.

In the box next to "Geometry Type" pick the down arrow and select "Axisymmetric". A thickness of 1 rad automatically comes up. Do not attempt to change this.

Click "OK". This sends you back to the "Material Data Control" window.

Click on the box below "Material". This brings up the "Element Material Selection" window.

Highlight the "[Customer Defined]" selection and

Click on the "Edit Properties" button. This brings up the "Element Material Specification" window.

Next to "Modulus of Elasticity" enter 30e6.

Next to "Poisson's Ratio" enter 0.3.

Click "OK". This brings back the "Element Material Selection" window.

Click "OK". This sends you back to the "Material Data Control" window.

Step 6 Enter the Global Data

Click on "Global" in the "Material Data Control" window.

Under "Load Case Multipliers" and under "Pressure", type in 1. You have to type in a 1 or else the program will not run. This indicates that one load case is defined and that it is an applied force (denoted by the 1 under the "Pressure" column).

While still in the "Global" menu,

Click "Output" on the right side of the menu and then

Click on the boxes next to "Displacement data" and "Stress data". This creates nodal displacement output in the *.L* file and the element forces and stresses in the *.S* file.

Click "OK". You are now out of the "Global" menu.

Step 7 Check the Model

Click on the "Check" button under the "FEA Model" column in the "Model Data Control" window. This takes you to Superview's main menu, and you can check your model. The model appears automatically, showing the boundary conditions and load arrow.

Click "Done". This takes you out of the Superview menu and back to the "Model Data Control" menu.

Step 8 Run the Analysis

Click on the "Analysis" button under the "FEA Model" column in the "Model Data Control" window.

Click "Analyze" in the lower left corner of the "Linear Stress and Vibration-Static Stress" menu. (See the green box next to "Analyze".) Wait a short time for the analysis to be performed.

Click "Done" in the lower right part of the screen.

Step 9 Review the Results

Click on the "Results" button in the "Model Data Control" window. This takes you to Superview again, and you can view the results.

Select "Stress-di" on the main menu.

Click "Post" on the "Stress-di" menu. Now you can choose what type of output you would like to be displayed.

Click "max prin". This gives you a color display of the maximum principal stress plot in the model. Also see Figure 10–18.

To view displacements, select "Disp vec" from the "Post-di" menu. This gives you a color displacement plot in the model. Also see Figure 10–19.

To exit Superview, select ⟨ESC⟩ twice and click "Done".

Figures 10–18 and 10–19 show the graphical outputs for maximum principal stresses and displacements, respectively, for the thick-walled pressure vessel. The

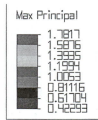

Figure 10–18 Maximum principal stresses (in psi) for the thick-walled cylinder of Example 10.3

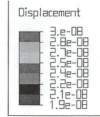

Figure 10–19 Displacements (in inches) for the thick-walled cylinder of Example 10.3

stresses compare closely with those from an analytical solution. The displacements from Figure 10–19 compare closely with the longhand solution given by Eq. (10.2.17).

■

Example 10.4

Determine the hoop stress in a uniformly thick disk rotating at a constant angular velocity of 50 rpm, as shown in Figure 10–20. Compare the hoop stress to the analytical value given in Reference [10].

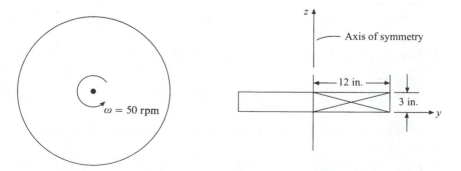

Figure 10–20 Discretized disk rotating with a constant angular velocity

Step 1   Start Superdraw III

Select the "Start" button of Windows NT/95/98 and then proceed to "Programs: Algor Software:Algor FEA".

Click "Algor FEA" to start the program or double-click on the "Algor FEA" icon on your desktop computer screen. The Superdraw III program appears. This is the Algor main graphics user interface. All your work will be done from this program. You will construct, analyze, and review your results from Superdraw III. Each menu option and a short example are available in the Superdraw III Reference Division (accessed through "Docutech", the on-line "Technical User Documentation").

Step 2   Create the Axisymmetric Model

Click on the "Modify" menu on the top menu bar.

Click "Transform XY to YZ" to transform from the x-y plane to the y-z plane. You must perform the analysis in the y-z plane. For axisymmetric problems, the y axis represents the radial direction.

Click on the "Add" menu on the top main menu bar.

Click "Rectangle". A menu appears in the upper right part of the screen. Then enter the data points.

0<Tab>0<Tab>0   This is the first corner node.

<Enter>

0<Tab>12.0<Tab>3.0   This is the opposite corner node.

<Enter>

Click "Enclose" on the "View" menu.

Now the object needs to be discretized.

Click on the "Add" menu on the top main menu bar.

Click "Line". A menu appears in the upper right part of the screen.

Click "Single" on the menu on the upper right side of the screen. Then enter the data points.

0<Tab>0<Tab>0   Enter the first coordinate of the diagonal line.

<Enter>

0<Tab>6.0<Tab>1.5   This is the second coordinate of the diagonal line.

<Enter>

0<Tab>12.0<Tab>0   Enter the first coordinate of the diagonal line.

<Enter>

0<Tab>6.0<Tab>1.5   This is the second coordinate of the diagonal line.

<Enter>

0<Tab>0<Tab>3.0   Enter the first coordinate of the diagonal line.

<Enter>

0<Tab>6.0<Tab>1.5   This is the second coordinate of the diagonal line.

<Enter>

`0<Tab>12.0<Tab>3.0`   Enter the first coordinate of the diagonal line.
`<Enter>`
`0<Tab>6.0<Tab>1.5`   This is the second coordinate of the diagonal line.
`<Enter>`

Step 3   Eliminate Duplicate Lines

Click "Modify" on the main menu.
Select "Clean:Duplicate".
Click "Perform Cleaning" on the "Duplicate" menu. You see the following message in the lower part of the screen:

"6 Kept 0 Deleted. done"

Click "Done" on the "Duplicate" menu located in the upper right corner of the screen.

Step 4   Add in Boundary Conditions

Click "FEA Add" on the main menu and highlight "Stress and Vibration Analysis".
Click "Boundary Conditions".
The "Boundary Condition" menu appears. The default boundary condition is "Use @ Symbol for Full". "Full" means full constraint or prevention of translations and rotations (when relevant).
Click "Change Values" on the "Boundary Condition" menu.
Select the box with the check mark in it next to "Ty Constrained". This allows the model translation in the y direction.
Click "Done".
Move the cursor to the four outside corner nodes one at a time and left-click on each node. "TxzRxyz" appears at each node.
Click "Done".
Click "View" and "Enclose" to see the entire model.

Step 5   Add the Material Properties Using the "Model Data Control" Window

After the nodal boundary conditions and the forces are applied, select "Tools" and "Model Data Control" from Superdraw III to complete the data input for analysis. The "Model Data Control" window appears. Alternatively,
Click on the "Model Data" button located in the lower part of the screen and highlighted in red.
For the "Analysis Type", "Linear Static Stress" appears as the default.
Click on the box below "Element". A table of group 1 element types appears.
Click "2-D".
Click "OK".
Click on the box below "Data". It responds with "Please enter the model name first".
Click "OK" and enter the model name in the "File_name" area that appears.

Ex104

Click "Save". By default, the "Units Definition" menu appears.

Click "OK" to accept the default "English (in)" units system.

The "Element Definition" menu appears next.

In the box next to "Geometry Type" pick the down arrow and select "Axisymmetric".

Click "OK". This sends you back to the "Material Data Control" window.

Click on the box below "Material". This brings up the "Element Material Selection" window.

Highlight the "[Customer Defined]" selection and

Click on the "Edit Properties" button. This brings up the "Element Material Specification" window.

Next to "Mass Density" enter 0.000732.

Next to "Modulus of Elasticity" enter 30e6.

Next to "Poisson's Ratio" enter 0.3.

Click "OK". This brings back the "Element Material Selection" window.

Click "OK". This sends you back to the "Material Data Control" window.

Step 6   Enter the Global Data

Click on "Global" in the "Material Data Control" window.

Under "Load Case Multipliers" and under "Pressure", type in 1. You have to type in a 1 or the program will not run. This indicates that one load case is defined and that it is an applied force (denoted by the 1 under the "Pressure" column).

Under "Centrifugal" the box next to "Include specified centrifugal load" must be checked.

Next to "Rotation Rate" under the "Centrifugal" tab enter ⟨50⟩ for 50 rev/min.

Next to "Axis Orientation" under the "Centrifugal" tab select "Z direction".

Next to "Z-coordinate of point on axis" under the "Centrifugal" tab enter ⟨1.5⟩.

While still under "Global", click "Output" on the right side of the menu and then click on the boxes next to "Displacement data" and "Stress data". This creates nodal displacement output in the *.L* file and the element forces and stresses in the *.S* file.

Click "OK". You are now out of the "Global" menu.

Step 7   Check the Model

Click on the "Check" button under the "FEA Model" column in the "Model Data Control" window. This takes you to Superview's main menu, and you can check your model. The model appears automatically, showing the boundary conditions and load arrow.

Click "Done". This takes you out of the Superview menu and back to the "Model Data Control" menu.

Step 8   Run the Analysis

Click on the "Analysis" button under the "FEA Model" column in the "Model Data Control" window.

Click "Analyze" in the lower left corner of the "Linear Stress and Vibration-Static Stress" menu. (See the green box next to "Analyze"). Wait a short time for the analysis to be performed.

Click "Done" on the lower right of the screen.

Step 9   Review the Results

Click on the "Results" button in the "Model Data Control" window. This takes you to the Superview program again, and you can view the results.

Select "Stress-di" on the main menu.

Click "Post" on the "Stress-di" menu. Now you can choose what type of output you would like to be displayed.

Click "max prin". This gives you a color plot of the maximum principal stress in the model.

To view displacements select "Disp vec" from the "Post-di" menu. This gives you the displacements in the model.

To exit Superview, select ⟨ESC⟩ twice and select "Done". ■

The circumferential or hoop stress, "Sigma-33," from the four-element Algor solution (from file *ex104.S*) is equal to 0.9388 psi, whereas the analytical solution [10] at $r = 2$ in. is 1.173 psi. The finite element solution will improve with increasing numbers of elements. You can verify this by working Problem 10.10.

Example 10.5

A stepped 4130 steel shaft with a fillet radius is subjected to an axial pressure of 1000 psi in tension (Figure 10–21). Fatigue analysis for reversed axial loading requires an accurate stress concentration factor to be applied to the average axial stress of 1000 psi. Determine the stress concentration factor for the geometry shown and compare your answer to published results.

Step 1   Start Superdraw III

Double-click on the "FEA Algor" icon on your desktop computer screen.

Step 2   Create the Boundary Perimeter of the Model

You can draw in either the *x-y* plane or the *y-z* plane. In the end the model must be viewed in the *y-z* plane.

Click "Modify".

Click "Transform XY to YZ" to transform to the *y-z* plane.

Now begin entering corner nodes around the perimeter of Figure 10–21 while disregarding the fillet.

Click "Add".

Click "Line".

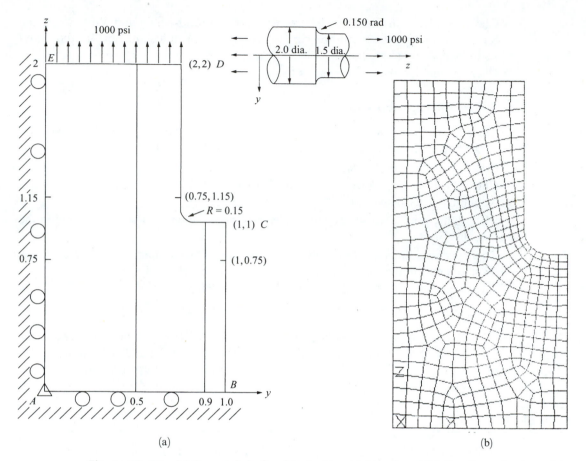

Figure 10–21 (a) Stepped shaft subjected to axial load and (b) the discretized model

```
(0, 0, 0)     Enter the first coordinate of the perimeter.
<Enter>
(0, 1, 0)
<Enter>
(0, 1, 1)
<Enter>
(0, 0.75, 1)
<Enter>
(0, 0.75, 2)
<Enter>
(0, 0, 2)
<Enter>
(0, 0, 0)
```

<Enter>   You have now created the perimeter of the model.

Click "Done".

Click "View:Enclose" to see the model enclosed on the screen without the fillet. Create the fillet.

Click "Construct".

Click "Fillet" to create the fillet.

Click "Radius".

0.15   This is the radius of the fillet.

<Enter>

To create the fillet, you must select the lines near where the fillet is to be created. These lines are the upper vertical line near its bottom and the short horizontal line near its left end.

Left-click on the first line (upper vertical) near its bottom where the fillet begins.

Left-click on the horizontal line near its left end where the fillet ends.

The fillet should appear as desired. If not, keep trying by clicking on slightly different places on each line.

Click "Close" on the "Fillet Menu".

Step 3   Mesh the Model

Select "FEA Mesh".

Select "Two-Dimensional Mesh Generation".

Click "Generate Mesh".

ex105   Enter the filename.

Click "Save".

You see a message: "Model information does not contain a specified unit system. Please choose a unit system first."

Click "OK". This takes you to the "Units Definition" window.

Click "OK" as you want to use the default units "English (in)".

You are now sent to the "Two-Dimensional Mesh Generation" window.

Use the defaults "Quadrilateral", "Mesh Density" 400, and so on.

Click "Generate" in the lower left corner of the window.

Click "OK".

Click "Done" to close the "Two-Dimensional Mesh Generation" window. You now see the meshed model with 404 elements and 449 nodes.

Step 4   Eliminate Duplicate Lines

Select "Modify:Clean:Duplicate:Perform Cleaning".

You now see the message

"Kept 852 Deleted 0. done"

Click "Done" in the "Duplicate" window.

Step 5 Add in Boundary Conditions

Select the "FEA Add:Stress and Vibration Analysis" menu.

Click "Boundary Conditions".

Click "Change Values".

Click "Tz Constrained" to release the check mark in front of the "Tz Constrained". This releases the z-direction translation.

Click "Done".

Click "Box Apply". Then box all the left edge nodes. These nodes are now free to move in the z direction.

Click "Change Values" again to add the check mark to "Tz Constrained" to activate the fixity.

Click "Ty Constrained" to release the y-direction constraint.

Click "Done".

Click "Box Apply". Now box all the bottom nodes. These nodes are now free to move in the y direction.

Click "Done".

Click "Select:Box". Now apply a box around the top edge nodes. You see ten sides selected on the top. This will allow you to change the surface number of this side to add pressure later in the "Global Data" window.

Click "Done". Change the surface number for pressure loading to be added later in the "Model Data Control" window.

Click "Modify".

Click "Update Object Parameters".

Click "Surface Number".

2   Enter a 2 to change the surface number of the top edge to 2.

<Enter>

Step 6 Go to the "Model Data Control" Window to Enter Material Properties

Click on the "Model Data" button.

Click on the box below "Element".

Click "2D". You now have a two-dimensional element selected.

Click "OK".

Click on the box below "Data". You are now at the "Element Data" screen.

You are at the "Geometric Type" command. Now scroll to "Axisymmetric" and

Click "Axisymmetric".

Click "OK". This takes you back to the "Model Data Control" window.

Click on the box below "Material".

Scroll down the list of materials and

Click "Steel (4130)".

Click "OK".

Click on the box below "Surface". (This is the box next to the "Material" box.) You are now at the "Surface Definition" screen.

Click the red box next to "Surface 2". The "Surface Properties" screen appears. Enter the surface pressure as

–1000   This is the surface pressure in psi along the top surface. The negative sign means the pressure is acting away from the top surface.

Click "OK".

Click "Close".

Click "Global". You are now at the "Global Data" menu screen.

Under the "Load Case Multipliers", type a 1 under the "Pressure" column.

1

Click on the "Output" tab.

Click on the boxes next to "Displacement data" and "Stress data".

Click "OK".

Step 7   Check Your Model

Click "Check" on the "Global Data" screen.

You now see your model loaded with the boundary conditions and orange lines at the top surface to represent the surface pressure. You can now check the model in the usual manner.

Step 8   Analyze the Model

Click "Analysis" in the "Model Data Control" window.

Click "Analyze". The total elapsed time is 0.854 min on a P5-133 computer.

Click "OK" to the message shown on the screen.

Click "Done" in the "Model Data Control" window.

Step 9   Review Your Results in Superview

Click "Results" in the "Model Data Control" window.

To obtain stresses:

Click "Stress-di".

Select "Post".

Select "max prin".

<Esc>

Select "Do Dither". A color plot appears, and you see the maximum principal stress as 1932.58. The red in the legend represents the largest principal stress of 1932.58 at the fillet [Figure 10–22(a)].

<Esc>

To obtain displacements:

Select "Displaced".

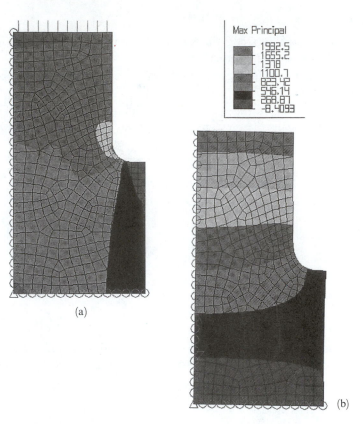

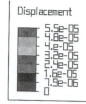

(a)

(b)

Figure 10–22 (a) Principal stresses and (b) displacement plots for shaft of Example 10.5

Select "Nodes inq".

Select "Get".

Click on the top center node and see "DZ=5.5494e-5" for the largest vertical displacement.

To get a color fringe plot of displacements:

Click "Post".

Click "Disp on".

Select "Disp vec".

Select "Do Dither". The plot should appear [Figure 10–22(b)].

Select "Quit" to exit the Superview program. ∎

For the model of the shaft shown subjected to a nominal axial stress of 1000 psi, the finite element analysis yields $\sigma_{\max} = 1932$ psi. The theoretical stress concentration factor is approximately $K = 1.8$ based on Reference [8] for $r/d = 0.1$ and $D/d = 1.33$. Therefore, using this stress concentration factor, we get $\sigma_{\max} = 1800$ psi, which is reasonably close to the 1932 psi obtained by the finite element solution. Again, the solution could be improved by a finer discretization of the model.

Example 10.6

Perform a stress analysis of a pressure vessel (Figure 10–23). Let $E = 5 \times 10^6$ psi and $v = 0.15$ of the concrete and let $E = 29 \times 10^6$ psi and $v = 0.25$ for the steel liner. The steel liner is 2 in. thick. Let the pressure p equal 500 psi.

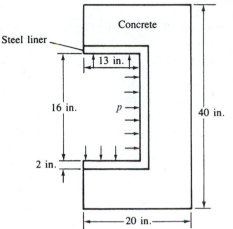

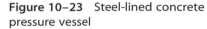

Figure 10–23 Steel-lined concrete pressure vessel

Step 1   Start Superdraw III

Double-click on the "FEA Algor" icon on your desktop computer screen.

Step 2   Create the Model

The model will be constructed using two groups. One group is for the concrete material, and the other group is for the steel material.

Click "Add" on the main menu bar.

Click "Line".

```
(0, 0, 0)    First coordinate of the concrete model.
<Enter>
(20, 0, 0)
<Enter>
(20, 40, 0)
<Enter>
(0, 40, 0)
<Enter>
(0, 30, 0)
<Enter>
```

(15, 30, 0)

\<Enter\>

(15, 10, 0)

\<Enter\>

(0, 10, 0)

\<Enter\>

(0, 0, 0)

\<Enter\>   The concrete model is complete.

Click "Done".

Click on the "G=1" button at the bottom of the screen. This allows you to change the group you are currently working with.

Click "2", which is the red box labeled "2".

Click "Add" on the main menu bar.

Click "Line".

(0, 10, 0) First coordinate of steel liner.

\<Enter\>

(0, 12, 0)

\<Enter\>

(13, 12, 0)

\<Enter\>

(13, 28, 0)

\<Enter\>

(0, 28, 0)

\<Enter\>

(0, 30, 0)

\<Enter\>

(15, 30, 0)

\<Enter\>

(15, 10, 0)

\<Enter\>

(0, 10, 0)

\<Enter\>   The steel liner model is complete. ·

Click "Done".

Step 3 Mesh the Model

We will let the Algor program generate the mesh automatically for both materials at the same time.

Click "Select" on the main menu bar.

Select "All".

Click "FEA Mesh" on the main menu bar.

Select "Two-Dimensional Mesh Generation".

Select "Generate Mesh".

Pressure Vessel   This is the name of your file.

Click "Save".

You are then asked for a unit system. Click "OK". You are then taken to the "Units" menu. Just click "OK" as we want the current English units. This takes you to the "2-D Mesh Generation" menu. By default, the element shape is a quadrilateral. (You can also choose all triangles or a mix of triangles and quadrilaterals.)

Click "Yes" to the attention menu that says "2-D element type requires the model in Y-Z plane. Do you want Superdraw III to convert your model?"

Click "Generate" in the lower left corner of this menu.

Click "OK" when the "Two-Dimensional Mesh Generation" window appears.

Click "Done". Now the whole model has been meshed in two different groups.

Step 4   Add the Material Properties and Pressure Load in the
"Model Data Control" Window

Click on the "Model Data" button at the bottom of the screen.

In the "Model Data Control" window you see "Analysis Type-Linear Static Stress", which is what you want.

Click on the box below "Element" in the same row as the green box labeled "1".

Click on the circle next to "2-D".

Click "OK".

Click on the box below "Data" in the same row as the green box labeled "1".

This brings you to the "Element Definition" window. Next to "Geometry Type" scroll to "Axisymmetric" and click on it.

Click "OK".

Click on the box below "Material" in the same row as the green box labeled "1".

This brings you to the "Element Material Selection" window. Select "[Customer Defined]" and then click "View Properties". Click "Unlock Properties".

Double-click on the box next to "Modulus of Elasticity".

5e6

Double-click on the box next to "Poisson's Ratio".

0.15

Click "OK".

Click "OK". Now the concrete properties have been assigned to group 1.

Click on the box below "Element" in the same row as the red box labeled "2".

Click on the circle next to "2-D".

Click "OK".

Click on the box below "Data" in the same row as the red box labeled "2".

This brings you to the "Element Definition" window. Next to "Geometry Type" scroll to "Axisymmetric" and click on it.

Click "OK".

Click on the box below "Material" in the same row as the red box labeled "2".

This brings you to the "Element Material Selection" window. Select "[Customer Defined]" and then click "View Properties". Click "Unlock Properties".

Double-click on the box next to "Modulus of Elasticity".

29e6

Double-click on the box next to "Poisson's Ratio".

0.25

Click "OK".

Click "OK".

Click "Close".

Click "Select" on the main menu bar.

Click "Box".

Create a box around the inside lines that make up the inside parameter of the pressure vessel. Make sure that only the inside lines that form the cutout are trapped inside the box.

Click "Done".

Click "Modify" on the main menu bar.

Select "Update Object Parameters".

Click "Surface Number".

This brings you to the "Surface" window. Click on the red box labeled "2". This allows you to add pressure to this surface.

Click on the "Model Data" button at the bottom of the screen.

Use the sliding bar (if necessary) to get to the "Surface" column.

Click on the box below "Surface" in the same row as the red box labeled "2".

This brings up the "Surface Definition" window for group 2.

Click on the box below "Data" in the same row as the red box labeled "2".

Double-click on the box next to "Pressure".

500

Click "OK". This creates the pressure inside the pressure vessel pushing outward. Now the steel properties have been assigned to group 1.

Click "Close".

Click "Global".

Locate and click on the "Multipliers" tab. Double-click on the box under "Pressure" and type a 1. Locate and click on the "Output" tab. Click on the boxes next to "Displacement Data" and "Stress Data".

Click "OK".

Click "Close" in the "Model Data Control" window.

Step 5   Apply Boundary Conditions

Click "FEA Add" on the main menu bar.

Select "Stress and Vibration Analysis".

Click "Boundary Conditions".

Click "Change Values".

Click on the box next to "Tz Constrained". This removes the check mark in the box, which allows motion to be carried out in the z direction.

Click "Done".

Click "Box Apply".

Create a box around the leftmost edge of model.

Click "Change Values".

Click on every box once. This removes each check mark and places one in the box labeled "Tz Constrained".

Click "Done".

Click "Box Apply".

Create a box around the lowermost right-hand corner of the model.

Click "Done".

Step 6   Check the Model

Click on the "Model Data" button at the bottom of the screen.

Click "Check". This takes you to Superview, and you can check your model. You now see the model called "Pressure Vessel.asd", in the upper part of the screen. The model appears with arrows pushing outward on the inner walls of the model, rollers on the left side, and a constraint located in the lower right corner that doesn't allow for motion in the z direction.

Click "Done" when you are ready to exit this program.

Step 7   Run the Analysis

Click on the "Analysis" button in the "Model Data Control" window.

Click "Analyze" in the lower left corner of the "Linear Stress and Vibration-Static Stress" window.

Click "OK" in the "Algor Analysis" window.

Click "Done".

Step 8   Review the Results

Click on the "Results" button in the "Model Data Control" window.

You now see the model back in Superview ready to be examined.

Click "Displaced".

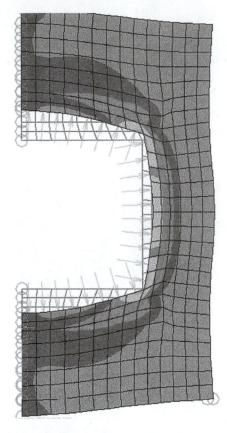

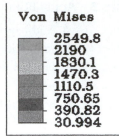

Von Mises

2549.8
2190
1830.1
1470.3
1110.5
750.65
390.82
30.994

Figure 10–24 Plot of Von Mises stress for the pressure vessel of Example 10.6

Click "Displ on". This shows the displacement the model goes through.

Click "With undi". This shows the original outline of the model without displacement.

Click "Calc scal". This gives you a better idea of what is going on with the model.

Click "Esc".

Click "Stress-di".

Click "Post". This gives you the colored plot of the Von Mises stress (in psi) shown in Figure 10–24. The largest stress occurs at the corners of the vessel.

Click "max prin". This gives you a colored plot of the maximum principal stresses shown in Figure 10–25. The largest principal stresses occur at the inner corners of the vessel.

Click "Disp vec". This gives you the displacement value for each colored section. The largest displacement occurs at the inner top and bottom nodes as shown in Figure 10–26.

Select "Quit" to exit the Superview program. ■

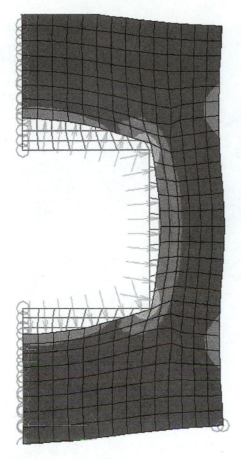

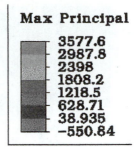

Max Principal

3577.6
2987.8
2398
1808.2
1218.5
628.71
38.935
−550.84

Figure 10–25 Plot of maximum principal stresses in the pressure vessel of Example 10.6

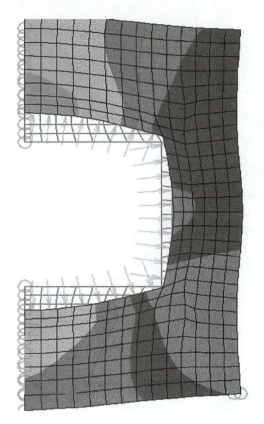

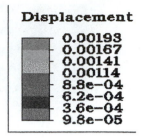

Figure 10–26 Displacement vector plot of Example 10.6

▲ References

[1] Utku, S., "Explicit Expressions for Triangular Torus Element Stiffness Matrix," *Journal of the American Institute of Aeronautics and Astronautics*, Vol. 6, No. 6, pp. 1174–1176, June 1968.

[2] Zienkiewicz, O. C., *The Finite Element Method*, 3rd ed., McGraw-Hill, London, 1977.

[3] Clough, R., and Rashid, Y., "Finite Element Analysis of Axisymmetric Solids," *Journal of the Engineering Mechanics Division*, American Society of Civil Engineers, Vol. 91, pp. 71–85, Feb. 1965.

[4] Rashid, Y., "Analysis of Axisymmetric Composite Structures by the Finite Element Method," *Nuclear Engineering and Design*, Vol. 3, pp. 163–182, 1966.

[5] Wilson, E., "Structural Analysis of Axisymmetric Solids," *Journal of the American Institute of Aeronautics and Astronautics*, Vol. 3, No. 12, pp. 2269–2274, Dec. 1965.

[6] Chacour, S., "A High Precision Axisymmetric Triangular Element Used in the Analysis of Hydraulic Turbine Components," Transactions of the American Society of Mechanical Engineers, *Journal of Basic Engineering*, Vol. 92, pp. 819–826, 1973.

[7] Greer, R. D., *The Analysis of a Film Tower Die Utilizing the ANSYS Finite Element Package*, M.S. Thesis, Rose-Hulman Institute of Technology, Terre Haute, IN, May 1989.

[8] Gere, J. M., and Timoshenko, S. P., *Mechanics of Materials*, 4th ed., Brooks/Cole Publishers, Pacific Grove, CA, 1997.

[9] Cook, R. D., Malkus, D. S., and Plesha, M. E., *Concepts and Applications of Finite Element Analysis*, 3rd ed., Wiley, New York, 1989.

[10] Cook, R. D., and Young, W. C., *Advanced Mechanics of Materials*, Macmillan, New York, 1985.

▲ Problems

10.1 For the elements shown in Figure P10–1, evaluate the stiffness matrices using Eq. (10.2.2). The coordinates are shown in the figures. Let $E = 30 \times 10^6$ psi and $v = 0.25$ for each element.

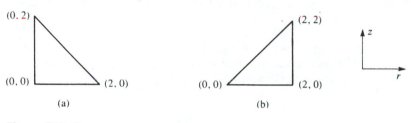

Figure P10–1

10.2 Evaluate the nodal forces used to replace the linearly varying surface traction shown in Figure P10–2.

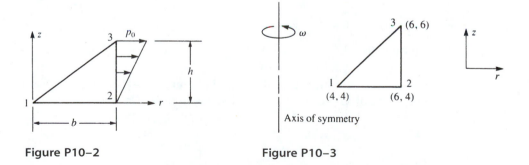

Figure P10–2 Figure P10–3

10.3 For an element of an axisymmetric body rotating with a constant angular velocity $\omega = 20$ rpm as shown in Figure P10–3, evaluate the body-force matrix. The coordinates of the element are shown in the figure. Let the weight density ρ_w be 0.283 lb/in^3.

10.4 For the axisymmetric elements shown in Figure P10–4, determine the element stresses. Let $E = 30 \times 10^6$ psi and $v = 0.25$. The coordinates (in inches) are shown in the fig-

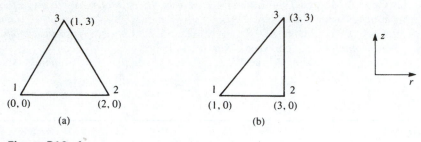

Figure P10–4

ures, and the nodal displacements for each element are $u_1 = 0.0001$ in., $w_1 = 0.0002$ in., $u_2 = 0.0005$ in., $w_2 = 0.0006$ in., $u_3 = 0$, and $w_3 = 0$.

10.5 Explicitly show that the integration of Eq. (10.1.35) yields the j surface forces given by Eq. (10.1.36).

10.6 For the elements shown in Figure P10–6, evaluate the stiffness matrices using Eq. (10.2.2). The coordinates (in millimeters) are shown in the figures. Let $E = 210$ GPa and $v = 0.25$ for each element.

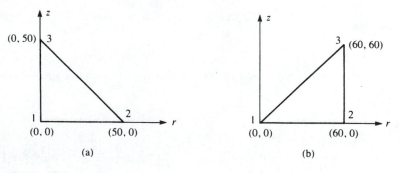

Figure P10–6

10.7 For the axisymmetric elements shown in Figure P10–7, determine the element stresses. Let $E = 210$ GPa and $v = 0.25$. The coordinates (in millimeters) are shown in the

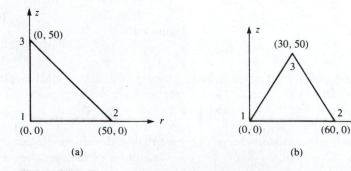

Figure P10–7

figures, and the nodal displacements for each element are

$$u_1 = 0.05 \text{ mm} \qquad w_1 = 0.03 \text{ mm}$$

$$u_2 = 0.02 \text{ mm} \qquad w_2 = 0.02 \text{ mm}$$

$$u_3 = 0.0 \text{ mm} \qquad w_3 = 0.0 \text{ mm}$$

Solve the following axisymmetric problems using the Algor computer program.

10.8 The soil mass in Figure P10–8 is loaded by a force transmitted through a circular footing as shown. Determine the stresses in the soil. Compare the values of σ_r using an axisymmetric model with the σ_y values using a plane stress model. Let $E = 3000$ psi and $v = 0.45$ for the soil mass.

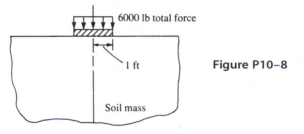

Figure P10–8

10.9 Perform a stress analysis of the pressure vessel shown in Figure P10–9. Let $E = 5 \times 10^6$ psi and $v = 0.15$ for the concrete, and let $E = 29 \times 10^6$ psi and $v = 0.25$ for the steel liner. The steel liner is 2 in. thick. Let the pressure p equal 500 psi.

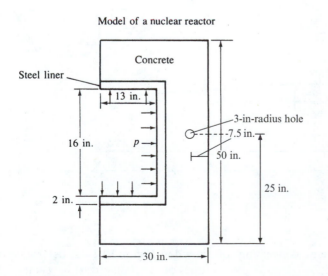

Figure P10–9

10.10 Perform a stress analysis of the disk shown in Figure P10–10 if it rotates with constant angular velocity of $\omega = 50$ rpm. Let $E = 30 \times 10^6$ psi, $v = 0.25$, and the weight density $\rho_w = 0.283$ lb/in^3. (Use 8 and then 16 elements symmetrically modeled similar to Example 10.4. Compare stress S_{33} from the Algor solution to the theoretical circumferential stress.)

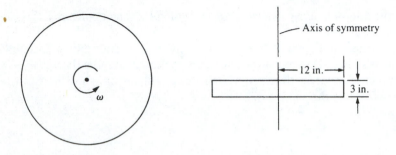

Figure P10–10

10.11 For the die casting shown in Figure P10–11, determine the maximum stresses and their locations. Let $E = 30 \times 10^6$ psi and $v = 0.25$. The dimensions are shown in the figure.

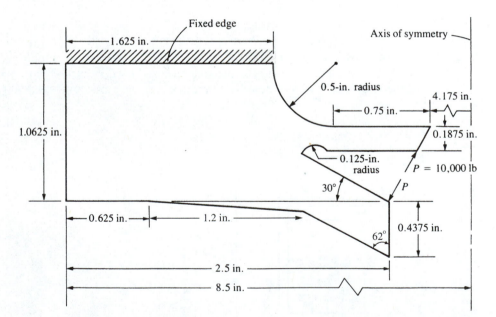

Figure P10–11

10.12 For the axisymmetric connecting rod shown in Figure P10–12, determine the stresses $\sigma_z, \sigma_r, \sigma_\theta$, and τ_{rz}. Plot stress contours (lines of constant stress) for each of the normal stresses. Let $E = 30 \times 10^6$ psi and $v = 0.25$. The applied loading and boundary con-

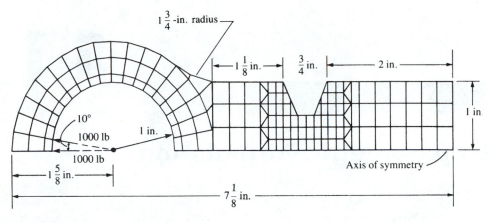

Figure P10–12

ditions are shown in the figure. A typical discretized rod is shown in the figure for illustrative purposes only.

10.13 For the thick-walled open-ended cylindrical pipe subjected to internal pressure shown in Figure P10–13, use five layers of elements to obtain the circumferential stress, σ_θ, and the principal stresses and maximum radial displacement. Compare these results to the exact solution. Let $E = 30 \times 10^6$ psi and $\nu = 0.3$.

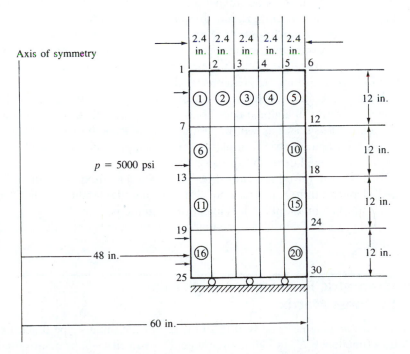

Figure P10–13

11

Isoparametric Formulation

Introduction

In this chapter, we introduce the isoparametric formulation of the element stiffness matrices. After considering the linear-strain triangular element in Chapter 9, we can see that the development of element matrices and equations expressed in terms of a global coordinate system becomes an enormously difficult task (if even possible) except for the simplest of elements such as the constant-strain triangle of Chapter 7. Hence, the isoparametric formulation was developed [1]. The isoparametric method may appear somewhat tedious (and confusing initially), but it will lead to a simple computer program formulation, and it is generally applicable for two- and three-dimensional stress analysis and for nonstructural problems. The isoparametric formulation allows elements to be created that are nonrectangular and have curved sides. Furthermore, numerous commercial computer programs (as described in Chapter 1) have adapted this formulation for their various libraries of elements.

We first illustrate the isoparametric formulation to develop the simple bar element stiffness matrix. Use of the bar element makes it relatively easy to understand the method because simple expressions result.

We then consider the development of the rectangular plane stress element stiffness matrix in terms of a global-coordinate system that will be convenient for use with the element. These concepts will be useful in understanding some of the procedures used with the isoparametric formulation of the simple quadrilateral element stiffness matrix, which we will develop subsequently.

Next, we will introduce a numerical integration method for evaluating the quadrilateral element stiffness matrix and illustrate the adaptability of the isoparametric formulation to common numerical integration methods.

Finally, we will consider some higher-order elements and their associated shape functions.

11.1 Isoparametric Formulation of the Bar Element Stiffness Matrix

The term *isoparametric* is derived from the use of the same shape functions (or interpolation functions) $[N]$ to define the element's geometric shape as are used to define

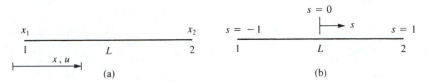

Figure 11–1 Linear bar element in (a) a global coordinate system x and (b) a natural coordinate system s

the displacements within the element. Thus, when the shape function is $u = a_1 + a_2 s$ for the displacement, we use $x = a_1 + a_2 s$ for the description of the nodal coordinate of a point on the bar element and, hence, the physical shape of the element.

Isoparametric element equations are formulated using a **natural** (or **intrinsic**) **coordinate system** s that is defined by element geometry and not by the element orientation in the global-coordinate system. In other words, axial coordinate s is attached to the bar and remains directed along the axial length of the bar, regardless of how the bar is oriented in space. There is a relationship (called a *transformation mapping*) between the natural coordinate system s and the global coordinate system x for each element of a specific structure, and this relationship must be used in the element equation formulations.

We will now develop the isoparametric formulation of the stiffness matrix of a simple linear bar element [with two nodes as shown in Figure 11–1(a)].

Step 1 Select Element Type

First, the natural coordinate s is attached to the element, with the origin located at the center of the element, as shown in Figure 11–1(b). The s axis need not be parallel to the x axis—this is only for convenience.

We consider the bar element to have two degrees of freedom—axial displacements u_1 and u_2 at each node associated with the global x axis.

For the special case when the s and x axes are parallel to each other, the s and x coordinates can be related by

$$x = x_c + \frac{L}{2}s \qquad (11.1.1)$$

where x_c is the global coordinate of the element centroid.

The shape functions used to define a position within the bar are found in a manner similar to that used in Chapter 3 to define displacement within a bar (Section 3.1). We begin by relating the natural coordinate to the global coordinate by

$$x = a_1 + a_2 s \qquad (11.1.2)$$

where we note that s is such that $-1 \leqslant s \leqslant 1$. Solving for the a_i's in terms of x_1 and x_2, we obtain

$$x = \tfrac{1}{2}[(1 - s)x_1 + (1 + s)x_2] \qquad (11.1.3)$$

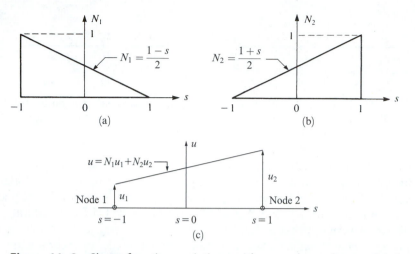

Figure 11–2 Shape function variations with natural coordinates: (a) shape function N_1, (b) shape function N_2, and (c) linear displacement field u plotted over element length

or, in matrix form, we can express Eq. (11.1.3) as

$$\{x\} = [N_1 \quad N_2] \begin{Bmatrix} x_1 \\ x_2 \end{Bmatrix} \tag{11.1.4}$$

where the shape functions in Eq. (11.1.4) are

$$N_1 = \frac{1-s}{2} \qquad N_2 = \frac{1+s}{2} \tag{11.1.5}$$

The linear shape functions in Eqs. (11.1.5) map the s coordinate of any point in the element to the x coordinate when used in Eq. (11.1.3). For instance, when we substitute $s = -1$ into Eq. (11.1.3), we obtain $x = x_1$. These shape functions are shown in Figure 11–2, where we can see that they have the same properties as defined for the interpolation functions of Section 3.1. Hence, N_1 represents the physical shape of the coordinate x when plotted over the length of the element for $x_1 = 1$ and $x_2 = 0$, and N_2 represents the coordinate x when plotted over the length of the element for $x_2 = 1$ and $x_1 = 0$. Again, we must have $N_1 + N_2 = 1$.

These shape functions must also be continuous throughout the element domain and have finite first derivatives within the element.

Step 2   Select a Displacement Function

The displacement function within the bar is now defined by the same shape functions, Eqs. (11.1.5), as are used to define the element shape; that is,

$$\{u\} = [N_1 \quad N_2] \begin{Bmatrix} u_1 \\ u_2 \end{Bmatrix} \tag{11.1.6}$$

When a particular coordinate s of the point of interest is substituted into $[N]$, Eq. (11.1.6) yields the displacement of a point on the bar in terms of the nodal degrees of freedom u_1 and u_2 as shown in Figure 11–2(c). Since u and x are defined by the same shape functions at the same nodes, comparing Eqs. (11.1.4) and (11.1.6), the element is called *isoparametric*.

Step 3 Define the Strain/Displacement and Stress/Strain Relationships

We now want to formulate element matrix $[B]$ to evaluate $[k]$. We use the isoparametric formulation to illustrate its manipulations. For a simple bar element, no real advantage may appear evident. However, for higher-order elements, the advantage will become clear because relatively simple computer program formulations will result.

To construct the element stiffness matrix, we must determine the strain, which is defined in terms of the derivative of the displacement with respect to x. The displacement u, however, is now a function of s as given by Eq. (11.1.6). Therefore, we must apply the chain rule of differentiation to the function u as follows:

$$\frac{du}{ds} = \frac{du}{dx}\frac{dx}{ds} \tag{11.1.7}$$

We can evaluate (du/ds) and (dx/ds) using Eqs. (11.1.6) and (11.1.3). We seek $(du/dx) = \varepsilon_x$. Therefore, we solve Eq. (11.1.7) for (du/dx) as

$$\frac{du}{dx} = \frac{\left(\dfrac{du}{ds}\right)}{\left(\dfrac{dx}{ds}\right)} \tag{11.1.8}$$

Using Eq. (11.1.6) for u, we obtain

$$\frac{du}{ds} = \frac{u_2 - u_1}{2} \tag{11.1.9a}$$

and using Eq. (11.1.3) for x, we have

$$\frac{dx}{ds} = \frac{x_2 - x_1}{2} = \frac{L}{2} \tag{11.1.9b}$$

because $x_2 - x_1 = L$.

Using Eqs. (11.1.9a) and (11.1.9b) in Eq. (11.1.8), we obtain

$$\{\varepsilon_x\} = \begin{bmatrix} -\dfrac{1}{L} & \dfrac{1}{L} \end{bmatrix} \begin{Bmatrix} u_1 \\ u_2 \end{Bmatrix} \tag{11.1.10}$$

Since $\{\varepsilon\} = [B]\{d\}$, the strain/displacement matrix $[B]$ is then given in Eq. (11.1.10) as

$$[B] = \begin{bmatrix} -\dfrac{1}{L} & \dfrac{1}{L} \end{bmatrix} \tag{11.1.11}$$

We recall that use of linear shape functions results in a constant $\underline{B}$ matrix, and hence, in a constant strain within the element. For higher-order elements, such as the quadratic bar with three nodes, $[B]$ becomes a function of natural coordinate s (see Problem 11.2).

The stress matrix is again given by Hooke's law as

$$\underline{\sigma} = E\underline{\varepsilon} = E\underline{B}\underline{d}$$

Step 4 Derive the Element Stiffness Matrix and Equations

The stiffness matrix is

$$[k] = \int_0^L [B]^T [D][B] A \, dx \tag{11.1.12}$$

However, in general, we must transform the coordinate x to s because $[B]$ is, in general, a function of s. This general type of transformation is given by References [4] and [5]

$$\int_0^L f(x) \, dx = \int_{-1}^1 f(s)|\underline{J}| \, ds \tag{11.1.13}$$

where $\underline{J}$ is called the *Jacobian*. In the one-dimensional case, we have $|\underline{J}| = \underline{J}$. For the simple bar element, from Eq. (11.1.9b), we have

$$|\underline{J}| = \frac{dx}{ds} = \frac{L}{2} \tag{11.1.14}$$

Observe that in Eq. (11.1.14), the Jacobian relates an element length in the global-coordinate system to an element length in the natural-coordinate system. In general, $|\underline{J}|$ is a function of s and depends on the numerical values of the nodal coordinates. This can be seen by working Problem 11.4 and by looking at Eq. (11.3.22) for the quadrilateral element. (Section 11.3 further discusses the Jacobian.) Using Eqs. (11.1.13) and (11.1.14) in Eq. (11.1.12), we obtain the stiffness matrix in natural coordinates as

$$[k] = \frac{L}{2} \int_{-1}^1 [B]^T E[B] A \, ds \tag{11.1.15}$$

where, for the one-dimensional case, we have used the modulus of elasticity $E = [D]$ in Eq. (11.1.15). Substituting Eq. (11.1.11) in Eq. (11.1.15) and performing the simple integration, we obtain

$$[k] = \frac{AE}{L} \begin{bmatrix} 1 & -1 \\ -1 & 1 \end{bmatrix} \tag{11.1.16}$$

which is the same as Eq. (3.1.14). For higher-order one-dimensional elements, the integration in closed form becomes difficult if not impossible (see Problem 11.4). Even the simple rectangular element stiffness matrix is difficult to evaluate in closed form (Section 11.3). However, the use of numerical integration, as described in Section 11.4, illustrates the distinct advantage of the isoparametric formulation of the equations.

Body Forces

We will now determine the body-force matrix using the natural coordinate system s. Using Eq. (3.10.20b), the body-force matrix is

$$\{\hat{f}_b\} = \iiint\limits_{V} [N]^T \{\hat{X}_b\}\, dV \tag{11.1.17}$$

Letting $dV = A\,dx$, we have

$$\{\hat{f}_b\} = A \int_0^L [N]^T \{\hat{X}_b\}\, dx \tag{11.1.18}$$

Substituting Eqs. (11.1.5) for N_1 and N_2 into $[N]$ and noting that by Eq. (11.1.9b), $dx = (L/2)\,ds$, we obtain

$$\{\hat{f}_b\} = A \int_{-1}^{1} \left\{ \begin{array}{c} \dfrac{1-s}{2} \\[2mm] \dfrac{1+s}{2} \end{array} \right\} \{\hat{X}_b\} \frac{L}{2}\, ds \tag{11.1.19}$$

On integrating Eq. (11.1.19), we obtain

$$\{\hat{f}_b\} = \frac{AL\hat{X}_b}{2} \left\{ \begin{array}{c} 1 \\ 1 \end{array} \right\} \tag{11.1.20}$$

The physical interpretation of the results for $\{\hat{f}_b\}$ is that since AL represents the volume of the element and $\hat{X}_b$ the body force per unit volume, then $AL\hat{X}_b$ is the total body force acting on the element. The factor $\frac{1}{2}$ indicates that this body force is equally distributed to the two nodes of the element.

Surface Forces

Surface forces can be found using Eq. (3.10.20a) as

$$\{\hat{f}_s\} = \iint\limits_{S} [N_s]^T \{\hat{T}_x\}\, dS \tag{11.1.21}$$

Assuming the cross section is constant and the traction is uniform over the perimeter and along the length of the element, we obtain

$$\{\hat{f}_s\} = \int_0^L [N_s]^T \{\hat{T}_x\}\, dx \tag{11.1.22}$$

where we now assume $\hat{T}_x$ is in units of force per unit length. Using the shape functions N_1 and N_2 from Eq. (11.1.5) in Eq. (11.1.22), we obtain

$$\{\hat{f}_s\} = \int_{-1}^{1} \left\{ \begin{array}{c} \dfrac{1-s}{2} \\[2mm] \dfrac{1+s}{2} \end{array} \right\} \{\hat{T}_x\} \frac{L}{2}\, ds \tag{11.1.23}$$

On integrating Eq. (11.1.23), we obtain

$$\{\hat{f}_s\} = \hat{T}_x \frac{L}{2} \begin{Bmatrix} 1 \\ 1 \end{Bmatrix} \tag{11.1.24}$$

The physical interpretation of Eq. (11.1.24) is that since $\hat{T}_x$ is in force-per-unit-length units, $\hat{T}_x L$ is now the total force. The $\frac{1}{2}$ indicates that the uniform surface traction is equally distributed to the two nodes of the element. Note that if $\hat{T}_x$ were a function of x (or s), then the amounts of force allocated to each node would generally not be equal and would be found through integration as in Example 3.10.

▲ 11.2 Rectangular Plane Stress Element ▲

We will now develop the rectangular plane stress element stiffness matrix. We will later refer to this element in the isoparametric formulation of a general quadrilateral element.

Two advantages of the rectangular element over the triangular element are ease of data input and simpler interpretation of output stresses. A disadvantage of the rectangular element is that the simple linear-displacement rectangle with its associated straight sides poorly approximates the real boundary condition edges.

The usual steps outlined in Chapter 1 will be followed to obtain the element stiffness matrix and related equations.

Step 1 Select Element Type

Consider the rectangular element shown in Figure 11–3 (all interior angles are 90°) with corner nodes 1–4 (again labeled counterclockwise) and base and height dimensions $2b$ and $2h$, respectively.

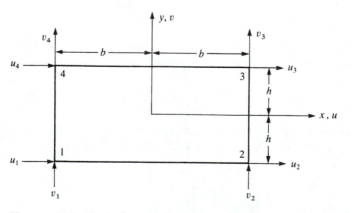

Figure 11–3 Basic four-node rectangular element with nodal degrees of freedom

The unknown nodal displacements are now given by

$$\{d\} = \begin{Bmatrix} u_1 \\ v_1 \\ u_2 \\ v_2 \\ u_3 \\ v_3 \\ u_4 \\ v_4 \end{Bmatrix} \tag{11.2.1}$$

Step 2   Select Displacement Functions

For a compatible displacement field, the element displacement functions u and v must be linear along each edge because only two points (the corner nodes) exist along each edge. We then select the linear displacement functions as

$$u(x, y) = a_1 + a_2 x + a_3 y + a_4 xy$$
$$v(x, y) = a_5 + a_6 x + a_7 y + a_8 xy \tag{11.2.2}$$

We can proceed in the usual manner to eliminate the a_i's from Eqs. (11.2.2) to obtain

$$u(x, y) = \frac{1}{4bh}[(b - x)(h - y)u_1 + (b + x)(h - y)u_2$$
$$+ (b + x)(h + y)u_3 + (b - x)(h + y)u_4]$$
$$v(x, y) = \frac{1}{4bh}[(b - x)(h - y)v_1 + (b + x)(h - y)v_2$$
$$+ (b + x)(h + y)v_3 + (b - x)(h + y)v_4] \tag{11.2.3}$$

These displacement expressions, Eqs. (11.2.3), can be expressed equivalently in terms of the shape functions and unknown nodal displacements as

$$\{\psi\} = [N]\{d\} \tag{11.2.4}$$

where the shape functions are given by

$$N_1 = \frac{(b - x)(h - y)}{4bh} \qquad N_2 = \frac{(b + x)(h - y)}{4bh}$$
$$N_3 = \frac{(b + x)(h + y)}{4bh} \qquad N_4 = \frac{(b - x)(h + y)}{4bh} \tag{11.2.5}$$

and the N_i's are again such that $N_1 = 1$ at node 1 and $N_1 = 0$ at all the other nodes, with similar requirements for the other shape functions. In expanded form, Eq. (11.2.4) becomes

$$
\begin{Bmatrix} u \\ v \end{Bmatrix} = \begin{bmatrix} N_1 & 0 & N_2 & 0 & N_3 & 0 & N_4 & 0 \\ 0 & N_1 & 0 & N_2 & 0 & N_3 & 0 & N_4 \end{bmatrix} \begin{Bmatrix} u_1 \\ v_1 \\ u_2 \\ v_2 \\ u_3 \\ v_3 \\ u_4 \\ v_4 \end{Bmatrix}
\tag{11.2.6}
$$

Step 3 Define the Strain/Displacement and Stress/Strain Relationships

Again the element strains for the two-dimensional stress state are given by

$$
\begin{Bmatrix} \varepsilon_x \\ \varepsilon_y \\ \gamma_{xy} \end{Bmatrix} = \begin{Bmatrix} \dfrac{\partial u}{\partial x} \\[2mm] \dfrac{\partial v}{\partial y} \\[2mm] \dfrac{\partial u}{\partial y} + \dfrac{\partial v}{\partial x} \end{Bmatrix}
\tag{11.2.7}
$$

Using Eq. (11.2.6) in Eq. (11.2.7) and taking the derivatives of u and v as indicated, we can express the strains in terms of the unknown nodal displacements as

$$
\{\varepsilon\} = [B]\{d\}
\tag{11.2.8}
$$

where
$$
[B] = \frac{1}{4bh} \begin{bmatrix} -(h-y) & 0 & (h-y) & 0 \\ 0 & -(b-x) & 0 & -(b+x) \\ -(b-x) & -(h-y) & -(b+x) & (h-y) \end{bmatrix}
$$

$$
\begin{bmatrix} (h+y) & 0 & -(h+y) & 0 \\ 0 & (b+x) & 0 & (b-x) \\ (b+x) & (h+y) & (b-x) & -(h+y) \end{bmatrix}
\tag{11.2.9}
$$

From Eqs. (11.2.8) and (11.2.9), we observe that ε_x is a function of y, ε_y is a function of x, and γ_{xy} is a function of both x and y. The stresses are again given by the formulas in Eq. (9.1.15), where $[B]$ is now that of Eq. (11.2.9) and $\{d\}$ is that of Eq. (11.2.1).

Step 4 Derive the Element Stiffness Matrix and Equations

The stiffness matrix is determined by

$$
[k] = \int_{-h}^{h} \int_{-b}^{b} [B]^T [D][B] t \, dx \, dy
\tag{11.2.10}
$$

with $[D]$ again given by the usual plane stress or plane strain conditions, Eq. (7.1.8) or (7.1.10). Because the $[B]$ matrix is a function of x and y, integration of Eq. (11.2.10) must be performed. The $[k]$ matrix for the rectangular element is now of order 8×8.

The element force matrix is determined by Eq. (7.2.46) as

$$\{f\} = \iiint_V [N]^T \{X\} \, dV + \{P\} + \iint_S [N_s]^T \{T\} \, dS \qquad (11.2.11)$$

where $[N]$ is the rectangular matrix in Eq. (11.2.6), and N_1 through N_4 are given by Eqs. (11.2.5). The element equations are then given by

$$\{f\} = [k]\{d\} \qquad (11.2.12)$$

Steps 5–7

Steps 5–7, which involve assembling the global stiffness matrix and equations, determining the unknown nodal displacements, and calculating the stress, are identical to those in Section 7.2 for the CST. However, the stresses within each element now vary in both the x and y directions.

▲ ## 11.3 Isoparametric Formulation of the Plane Element Stiffness Matrix   ▲

Recall that the term *isoparametric* is derived from the use of the same shape functions to define the element shape as are used to define the displacements within the element. Thus, when the shape function is $u = a_1 + a_2 s + a_3 t + a_4 st$ for the displacement, we use $x = a_1 + a_2 s + a_3 t + a_4 st$ for the description of a coordinate point in the plane element.

The natural-coordinate system s-t is defined by element geometry and not by the element orientation in the global-coordinate system x-y. Much as in the bar element example, there is a transformation mapping between the two coordinate systems for each element of a specific structure, and this relationship must be used in the element formulation.

We will now discuss the isoparametric formulation of the simple linear plane element stiffness matrix. This formulation is general enough to be applied to more complicated (higher-order) elements such as a quadratic plane element with three nodes along an edge, which can have straight or quadratic curved sides. Higher-order elements have additional nodes and use different shape functions as compared to the linear element, but the steps in the development of the stiffness matrices are the same. We will briefly discuss these elements after examining the linear plane element formulation.

Step 1   Select Element Type

First, the natural s-t coordinates are attached to the element, with the origin at the center of the element, as shown in Figure 11–4(a). The s and t axes need not be orthogonal, and neither has to be parallel to the x or y axis. The orientation of s-t

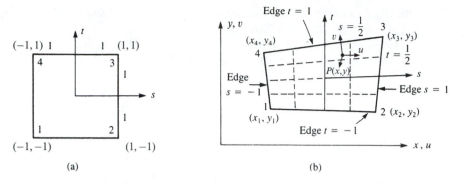

Figure 11–4 (a) Linear square element in *s-t* coordinates and (b) square element mapped into quadrilateral in *x-y* coordinates whose size and shape are determined by the eight nodal coordinates $x_1, y_1, \ldots, y_4$

coordinates is such that the four corner nodes and the edges of the quadrilateral are bounded by +1 or −1. This orientation will later allow us to take advantage more fully of common numerical integration schemes.

We consider the quadrilateral to have eight degrees of freedom, $u_1, v_1, \ldots, u_4$, and v_4 associated with the global x and y directions. The element then has straight sides but is otherwise of arbitrary shape, as shown in Figure 11–4(b).

For the special case when the distorted element becomes a rectangular element with sides parallel to the global x-y coordinates (see Figure 11–3), the s-t coordinates can be related to the global element coordinates x and y by

$$x = x_c + bs \qquad y = y_c + ht \tag{11.3.1}$$

where x_c and y_c are the global coordinates of the element centroid.

As we have shown for a rectangular element, the shape functions that define the displacements within the element are given by Eqs. (11.2.5). These same shape functions will now be used to map the square of Figure 11–4(a) in isoparametric coordinates s and t to the quadrilateral of Figure 11–4(b) in x and y coordinates whose size and shape are determined by the eight nodal coordinates $x_1, y_1, \ldots, x_4$, and y_4. That is, letting

$$x = a_1 + a_2 s + a_3 t + a_4 st$$
$$y = a_5 + a_6 s + a_7 t + a_8 st \tag{11.3.2}$$

and solving for the a_i's in terms of $x_1, x_2, x_3, x_4, y_1, y_2, y_3$, and y_4, we establish a form similar to Eqs. (11.2.3) such that

$$
\begin{aligned}
x = \tfrac{1}{4}[(1 - s)(1 - t)x_1 &+ (1 + s)(1 - t)x_2 \\
&+ (1 + s)(1 + t)x_3 + (1 - s)(1 + t)x_4] \\
y = \tfrac{1}{4}[(1 - s)(1 - t)y_1 &+ (1 + s)(1 - t)y_2 \\
&+ (1 + s)(1 + t)y_3 + (1 - s)(1 + t)y_4]
\end{aligned}
\tag{11.3.3}
$$

Or, in matrix form, we can express Eqs. (11.3.3) as

$$
\left\{ \begin{array}{c} x \\ y \end{array} \right\} = \left[\begin{array}{cccccccc} N_1 & 0 & N_2 & 0 & N_3 & 0 & N_4 & 0 \\ 0 & N_1 & 0 & N_2 & 0 & N_3 & 0 & N_4 \end{array} \right] \left\{ \begin{array}{c} x_1 \\ y_1 \\ x_2 \\ y_2 \\ x_3 \\ y_3 \\ x_4 \\ y_4 \end{array} \right\} \tag{11.3.4}
$$

where the shape functions of Eq. (11.3.4) are now

$$
N_1 = \frac{(1-s)(1-t)}{4} \qquad N_2 = \frac{(1+s)(1-t)}{4}
$$

$$
N_3 = \frac{(1+s)(1+t)}{4} \qquad N_4 = \frac{(1-s)(1+t)}{4} \tag{11.3.5}
$$

The shape functions of Eqs. (11.3.5) are linear. These shape functions are seen to map the s and t coordinates of any point in the square element of Figure 11–4(a) to those x and y coordinates in the quadrilateral element of Figure 11–4(b). For instance, consider square element node 1 coordinates, where $s = -1$ and $t = -1$. Using Eqs. (11.3.4) and (11.3.5), the left side of Eq. (11.3.4) becomes

$$
x = x_1 \qquad y = y_1 \tag{11.3.6}
$$

Similarly, we can map the other local nodal coordinates at nodes 2, 3, and 4 such that the square element in s-t isoparametric coordinates is mapped into a quadrilateral element in global coordinates. Also observe the property that $N_1 + N_2 + N_3 + N_4 = 1$ for all values of s and t.

We further observe that the shape functions in Eq. (11.3.5) are again such that N_1 through N_4 have the properties that N_i ($i = 1, 2, 3, 4$) is equal to one at node i and equal to zero at all other nodes. The physical shapes of N_i as they vary over the element with natural coordinates are shown in Figure 11–5. For instance, N_1 represents the geometric shape for $x_1 = 1$, $y_1 = 1$, and x_2, y_2, x_3, y_3, x_4, and y_4 all equal to zero.

Until this point in the discussion, we have always developed the element shape functions either by assuming some relationship between the natural and global coordinates in terms of the generalized coordinates (a_i's) as in Eqs. (11.3.2) or, similarly, by assuming a displacement function in terms of the a_i's. However, physical intuition can often guide us in directly expressing shape functions based on the following two criteria set forth in Section 3.2 and used on numerous occasions:

$$
\sum_{i=1}^{n} N_i = 1 \qquad (i = 1, 2, \dots, n)
$$

where $n =$ the number of shape functions corresponding to displacement shape functions N_i, and $N_i = 1$ at node i and $N_i = 0$ at all nodes other than i. In addition, a third

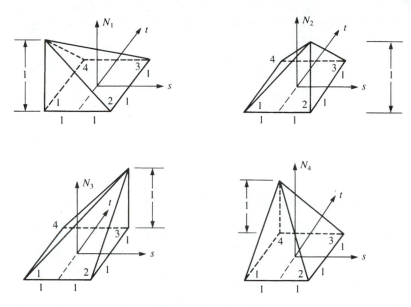

Figure 11–5 Variations of the shape functions over a linear square element

criterion is based on Lagrangian interpolation when displacement continuity is to be satisfied, or on Hermitian interpolation when additional slope continuity needs to be satisfied, as in the beam element of Chapter 5. (For a description of the use of Lagrangian and Hermitian interpolation to develop shape functions, consult References [4] and [6].)

Step 2 Select Displacement Functions

The displacement functions within an element are now similarly defined by the same shape functions as are used to define the element shape; that is,

$$
\begin{Bmatrix} u \\ v \end{Bmatrix} = \begin{bmatrix} N_1 & 0 & N_2 & 0 & N_3 & 0 & N_4 & 0 \\ 0 & N_1 & 0 & N_2 & 0 & N_3 & 0 & N_4 \end{bmatrix} \begin{Bmatrix} u_1 \\ v_1 \\ u_2 \\ v_2 \\ u_3 \\ v_3 \\ u_4 \\ v_4 \end{Bmatrix} \tag{11.3.7}
$$

where u and v are displacements parallel to the global x and y coordinates, and the shape functions are given by Eqs. (11.3.5). The displacement of an interior point p located at (x, y) in the element of Figure 11–4(b) is described by u and v in Eq. (11.3.7).

Comparing Eqs. (11.2.6) and (11.3.7), we see similarities between the rectangular element with sides of lengths $2b$ and $2h$ (Figure 11–3) and the square element with sides of length 2. If we let $b = 1$ and $h = 1$, the two sets of shape functions, Eqs. (11.2.5) and (11.3.5), are identical.

Step 3　Define the Strain/Displacement and Stress/Strain Relationships

We now want to formulate element matrix $\underline{B}$ to evaluate $\underline{k}$. However, because it becomes tedious and difficult (if not impossible) to write the shape functions in terms of the x and y coordinates, as seen in Chapter 9, we will carry out the formulation in terms of the isoparametric coordinates s and t. This may appear tedious, but it is easier to use the s- and t-coordinate expressions than to attempt to use the x- and y-coordinate expressions. This approach also leads to a simple computer program formulation.

To construct an element stiffness matrix, we must determine the strains, which are defined in terms of the derivatives of the displacements with respect to the x and y coordinates. The displacements, however, are now functions of the s and t coordinates, as given by Eq. (11.3.7), with the shape functions given by Eqs. (11.3.5). Before, we could determine $(\partial f/\partial x)$ and $(\partial f/\partial y)$, where, in general, f is a function representing the displacement functions u or v. However, u and v are now expressed in terms of s and t. Therefore, we need to apply the chain rule of differentiation because it will not be possible to express s and t as functions of x and y directly. For f as a function of x and y, the chain rule yields

$$
\begin{aligned}
\frac{\partial f}{\partial s} &= \frac{\partial f}{\partial x}\frac{\partial x}{\partial s} + \frac{\partial f}{\partial y}\frac{\partial y}{\partial s} \\
\frac{\partial f}{\partial t} &= \frac{\partial f}{\partial x}\frac{\partial x}{\partial t} + \frac{\partial f}{\partial y}\frac{\partial y}{\partial t}
\end{aligned}
\tag{11.3.8}
$$

In Eq. (11.3.8), $(\partial f/\partial s)$, $(\partial f/\partial t)$, $(\partial x/\partial s)$, $(\partial y/\partial s)$, $(\partial x/\partial t)$, and $(\partial y/\partial t)$ are all known using Eqs. (11.3.7) and (11.3.4). We still seek $(\partial f/\partial x)$ and $(\partial f/\partial y)$. The strains can then be found; for example, $\varepsilon_x = (\partial u/\partial x)$. Therefore, we solve Eqs. (11.3.8) for $(\partial f/\partial x)$ and $(\partial f/\partial y)$ using Cramer's rule, which involves evaluation of determinants (Appendix B), as

$$
\frac{\partial f}{\partial x} = \frac{\begin{vmatrix} \dfrac{\partial f}{\partial s} & \dfrac{\partial y}{\partial s} \\[2ex] \dfrac{\partial f}{\partial t} & \dfrac{\partial y}{\partial t} \end{vmatrix}}{\begin{vmatrix} \dfrac{\partial x}{\partial s} & \dfrac{\partial y}{\partial s} \\[2ex] \dfrac{\partial x}{\partial t} & \dfrac{\partial y}{\partial t} \end{vmatrix}}
\qquad
\frac{\partial f}{\partial y} = \frac{\begin{vmatrix} \dfrac{\partial x}{\partial s} & \dfrac{\partial f}{\partial s} \\[2ex] \dfrac{\partial x}{\partial t} & \dfrac{\partial f}{\partial t} \end{vmatrix}}{\begin{vmatrix} \dfrac{\partial x}{\partial s} & \dfrac{\partial y}{\partial s} \\[2ex] \dfrac{\partial x}{\partial t} & \dfrac{\partial y}{\partial t} \end{vmatrix}}
\tag{11.3.9}
$$

where the determinant in the denominator is the determinant of the *Jacobian* matrix $\underline{J}$. Hence, the Jacobian matrix is given by

$$[J] = \begin{bmatrix} \dfrac{\partial x}{\partial s} & \dfrac{\partial y}{\partial s} \\[2mm] \dfrac{\partial x}{\partial t} & \dfrac{\partial y}{\partial t} \end{bmatrix} \tag{11.3.10}$$

We now want to express the element strains as

$$\varepsilon = \underline{B}\underline{d} \tag{11.3.11}$$

where $\underline{B}$ must now be expressed as a function of s and t. We start with the usual relationship between strains and displacements given in matrix form as

$$\left\{ \begin{array}{c} \varepsilon_x \\ \varepsilon_y \\ \gamma_{xy} \end{array} \right\} = \begin{bmatrix} \dfrac{\partial(\ )}{\partial x} & 0 \\[2mm] 0 & \dfrac{\partial(\ )}{\partial y} \\[2mm] \dfrac{\partial(\ )}{\partial y} & \dfrac{\partial(\ )}{\partial x} \end{bmatrix} \left\{ \begin{array}{c} u \\ v \end{array} \right\} \tag{11.3.12}$$

where the rectangular matrix on the right side of Eq. (11.3.12) is an *operator matrix*; that is, $\partial(\ )/\partial x$ and $\partial(\ )/\partial y$ represent the partial derivatives of any variable we put inside the parentheses.

Using Eqs. (11.3.9) and evaluating the determinant in the numerators, we have

$$\begin{aligned} \frac{\partial(\ )}{\partial x} &= \frac{1}{|J|} \left[\frac{\partial y}{\partial t} \frac{\partial(\ )}{\partial s} - \frac{\partial y}{\partial s} \frac{\partial(\ )}{\partial t} \right] \\[2mm] \frac{\partial(\ )}{\partial y} &= \frac{1}{|J|} \left[\frac{\partial x}{\partial s} \frac{\partial(\ )}{\partial t} - \frac{\partial x}{\partial t} \frac{\partial(\ )}{\partial s} \right] \end{aligned} \tag{11.3.13}$$

where $|\underline{J}|$ is the determinant of $\underline{J}$ given by Eq. (11.3.10). Using Eq. (11.3.13) in Eq. (11.3.12) we obtain the strains expressed in terms of the natural coordinates $(s\text{-}t)$ as

$$\left\{ \begin{array}{c} \varepsilon_x \\ \varepsilon_y \\ \gamma_{xy} \end{array} \right\} = \frac{1}{|J|} \begin{bmatrix} \dfrac{\partial y}{\partial t} \dfrac{\partial(\ )}{\partial s} - \dfrac{\partial y}{\partial s} \dfrac{\partial(\ )}{\partial t} & 0 \\[3mm] 0 & \dfrac{\partial x}{\partial s} \dfrac{\partial(\ )}{\partial t} - \dfrac{\partial x}{\partial t} \dfrac{\partial(\ )}{\partial s} \\[3mm] \dfrac{\partial x}{\partial s} \dfrac{\partial(\ )}{\partial t} - \dfrac{\partial x}{\partial t} \dfrac{\partial(\ )}{\partial s} & \dfrac{\partial y}{\partial t} \dfrac{\partial(\ )}{\partial s} - \dfrac{\partial y}{\partial s} \dfrac{\partial(\ )}{\partial t} \end{bmatrix} \left\{ \begin{array}{c} u \\ v \end{array} \right\} \tag{11.3.14}$$

Using Eq. (11.3.7), we can express Eq. (11.3.14) in terms of the shape functions and global coordinates in compact matrix form as

$$\varepsilon = \underline{D}'\underline{N}\underline{d} \tag{11.3.15}$$

where $\underline{D}'$ is an operator matrix given by

$$\underline{D}' = \frac{1}{|J|}\begin{bmatrix} \dfrac{\partial y}{\partial t}\dfrac{\partial(\ )}{\partial s} - \dfrac{\partial y}{\partial s}\dfrac{\partial(\ )}{\partial t} & 0 \\[2ex] 0 & \dfrac{\partial x}{\partial s}\dfrac{\partial(\ )}{\partial t} - \dfrac{\partial x}{\partial t}\dfrac{\partial(\ )}{\partial s} \\[2ex] \dfrac{\partial x}{\partial s}\dfrac{\partial(\ )}{\partial t} - \dfrac{\partial x}{\partial t}\dfrac{\partial(\ )}{\partial s} & \dfrac{\partial y}{\partial t}\dfrac{\partial(\ )}{\partial s} - \dfrac{\partial y}{\partial s}\dfrac{\partial(\ )}{\partial t} \end{bmatrix} \quad (11.3.16)$$

and $\underline{N}$ is the 2×8 shape function matrix given as the first matrix on the right side of Eq. (11.3.7) and $\underline{d}$ is the column matrix on the right side of Eq. (11.3.7).

Defining $\underline{B}$ as

$$\begin{array}{cccc} \underline{B} & = & \underline{D}' & \underline{N} \\ (3 \times 8) & & (3 \times 2) & (2 \times 8) \end{array} \quad (11.3.17)$$

we have $\underline{B}$ expressed as a function of s and t and thus have the strains in terms of s and t. Here $\underline{B}$ is of order 3×8, as indicated in Eq. (11.3.17).

The explicit form of $\underline{B}$ can be obtained by substituting Eq. (11.3.16) for $\underline{D}'$ and Eqs. (11.3.5) for the shape functions into Eq. (11.3.17). The matrix multiplications yield

$$\underline{B}(s, t) = \frac{1}{|J|}[\underline{B}_1 \quad \underline{B}_2 \quad \underline{B}_3 \quad \underline{B}_4] \quad (11.3.18)$$

where the submatrices of $\underline{B}$ are given by

$$\underline{B}_i = \begin{bmatrix} a(N_{i,s}) - b(N_{i,t}) & 0 \\ 0 & c(N_{i,t}) - d(N_{i,s}) \\ c(N_{i,t}) - d(N_{i,s}) & a(N_{i,s}) - b(N_{i,t}) \end{bmatrix} \quad (11.3.19)$$

Here i is a dummy variable equal to 1, 2, 3, and 4, and

$$\begin{aligned} a &= \tfrac{1}{4}[y_1(s-1) + y_2(-1-s) + y_3(1+s) + y_4(1-s)] \\ b &= \tfrac{1}{4}[y_1(t-1) + y_2(1-t) + y_3(1+t) + y_4(-1-t)] \\ c &= \tfrac{1}{4}[x_1(t-1) + x_2(1-t) + x_3(1+t) + x_4(-1-t)] \\ d &= \tfrac{1}{4}[x_1(s-1) + x_2(-1-s) + x_3(1+s) + x_4(1-s)] \end{aligned} \quad (11.3.20)$$

Using the shape functions defined by Eqs. (11.3.5), we have

$$N_{1,s} = \tfrac{1}{4}(t-1) \qquad N_{1,t} = \tfrac{1}{4}(s-1) \qquad \text{(and so on)} \quad (11.3.21)$$

where the comma followed by the variable s or t indicates differentiation with respect to that variable; that is, $N_{1,s} \equiv \partial N_1/\partial s$, and so on. The determinant $|\underline{J}|$ is a polynomial in s and t and is tedious to evaluate even for the simplest case of the linear plane element. However, using Eq. (11.3.10) for $[J]$ and Eqs. (11.3.3) for x and y, we can evaluate $|\underline{J}|$ as

$$|\underline{J}| = \tfrac{1}{8}\{X_c\}^T \begin{bmatrix} 0 & 1-t & t-s & s-1 \\ t-1 & 0 & s+1 & -s-t \\ s-t & -s-1 & 0 & t+1 \\ 1-s & s+t & -t-1 & 0 \end{bmatrix} \{Y_c\} \qquad (11.3.22)$$

where

$$\{X_c\}^T = [x_1 \quad x_2 \quad x_3 \quad x_4] \qquad (11.3.23)$$

and

$$\{Y_c\} = \begin{Bmatrix} y_1 \\ y_2 \\ y_3 \\ y_4 \end{Bmatrix} \qquad (11.3.24)$$

We observe that $|\underline{J}|$ is a function of s and t and the known global coordinates $x_1, x_2, \ldots, y_4$. Hence, $\underline{B}$ is a function of s and t in both the numerator and the denominator [because of $|\underline{J}|$ given by Eq. (11.3.22)] and of the known global coordinates x_1 through y_4.

The stress/strain relationship is again $\underline{\sigma} = \underline{DB}\underline{d}$, where because the $\underline{B}$ matrix is a function of s and t, so also is the stress matrix $\underline{\sigma}$.

Step 4 Derive the Element Stiffness Matrix and Equations

We now want to express the stiffness matrix in terms of s-t coordinates. For an element with a constant thickness h, we have

$$[k] = \iint_A [B]^T [D][B] h \, dx \, dy \qquad (11.3.25)$$

However, $\underline{B}$ is now a function of s and t, as seen by Eqs. (11.3.18)–(11.3.20), and so we must integrate with respect to s and t. Once again, to transform the variables and the region from x and y to s and t, we must have a standard procedure that involves the determinant of $\underline{J}$. This general type of transformation [4, 5] is given by

$$\iint_A f(x, y) \, dx \, dy = \iint_A f(s, t)|\underline{J}| \, ds \, dt \qquad (11.3.26)$$

where the inclusion of $|\underline{J}|$ in the integrand on the right side of Eq. (11.3.26) results from a theorem of integral calculus (see Reference [5] for the complete proof of this theorem). Using Eq. (11.3.26) in Eq. (11.3.25), we obtain

$$[k] = \int_{-1}^{1} \int_{-1}^{1} [B]^T [D][B] h|\underline{J}| \, ds \, dt \qquad (11.3.27)$$

The $|\underline{J}|$ and $\underline{B}$ are such as to result in complicated expressions within the integral of Eq. (11.3.27), and so the integration to determine the element stiffness matrix is usually done numerically. A method for numerically integrating Eq. (11.3.27) is given in Section 11.4. The stiffness matrix in Eq. (11.3.27) is of the order 8×8.

Body Forces

The element body-force matrix will now be determined from

$$
\{f_b\} = \int_{-1}^{1} \int_{-1}^{1} [N]^T \{X\} \, h|J| \, ds \, dt
$$

$$
(8 \times 1) \qquad \qquad (8 \times 2)(2 \times 1)
$$

(11.3.28)

Like the stiffness matrix, the body-force matrix in Eq. (11.3.28) has to be evaluated by numerical integration.

Surface Forces

The surface-force matrix, say, along edge $t = 1$ (Figure 11–6) with overall length L, is

$$
\{f_s\} = \int_{-1}^{1} [N_s]^T \{T\} \, h\frac{L}{2} \, ds
$$

$$
(4 \times 1) \qquad \quad (4 \times 2)(2 \times 1)
$$

(11.3.29)

or

$$
\begin{Bmatrix} f_{s3s} \\ f_{s3t} \\ f_{s4s} \\ f_{s4t} \end{Bmatrix} = \int_{-1}^{1} \begin{bmatrix} N_3 & 0 & N_4 & 0 \\ 0 & N_3 & 0 & N_4 \end{bmatrix}^T \Bigg|_{\substack{\text{evaluated} \\ \text{along } t=1}} \begin{Bmatrix} p_s \\ p_t \end{Bmatrix} h\frac{L}{2} \, ds
$$

(11.3.30)

because $N_1 = 0$ and $N_2 = 0$ along edge $t = 1$, and hence, no nodal forces exist at nodes 1 and 2. For the case of uniform (constant) p_s and p_t along edge $t = 1$, the total surface-force matrix is

$$
\{f_s\} = h\frac{L}{2}[0 \quad 0 \quad 0 \quad 0 \quad p_s \quad p_t \quad p_s \quad p_t]^T
$$

(11.3.31)

Surface forces along other edges can be obtained similar to Eq. (11.3.30) by merely using the proper shape functions associated with the edge where the tractions are applied.

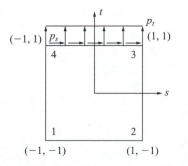

Figure 11–6   Surface traction: p_s and p_t acting at edge $t = 1$

▲ 11.4 Gaussian Quadrature (Numerical Integration) ▲

In this section, we will describe Gauss's method, one of the many schemes for numerical evaluation of definite integrals, because it has proved most useful for finite element work.

To evaluate the integral

$$I = \int_{-1}^{1} y \, dx \tag{11.4.1a}$$

where $y = y(x)$, we might choose (sample or evaluate) y at the midpoint $y(0) = y_1$ and multiply by the length of the interval, as shown in Figure 11–7, to arrive at $I = 2y_1$, a result that is exact if the curve happens to be a straight line. This is an example of what is called **one-point Gaussian quadrature** because only one sampling point was used. Therefore,

$$I = \int_{-1}^{1} y(x) \, dx \cong 2y(0) \tag{11.4.1b}$$

which is the familiar midpoint rule. Generalization of the formula [Eq. (11.4.1b)] leads to

$$I = \int_{-1}^{1} y \, dx = \sum_{i=1}^{n} W_i y_i \tag{11.4.2}$$

That is, to approximate the integral, we evaluate the function at several sampling points n, multiply each value y_i by the appropriate weight W_i, and add the terms. Gauss's method chooses the sampling points so that for a given number of points, the best possible accuracy is obtained. Sampling points are located symmetrically with respect to the center of the interval. Symmetrically paired points are given the same weight W_i. Table 11–1 gives appropriate sampling points and weighting coefficients for the first three orders—that is, one, two, or three sampling points (see Reference [2] for more complete tables). For example, using two points (Figure 11–8), we simply have $I = y_1 + y_2$ because $W_1 = W_2 = 1.000$. This is the exact result if $y = f(x)$ is a polynomial containing terms up to and including x^3. In general, Gaussian quadrature using n points (Gauss points) is exact if the integrand is a polynomial of degree $2n - 1$ or less. In using n points, we effectively replace the given function $y = f(x)$ by a polynomial of degree $2n - 1$. The accuracy of the numerical integration depends on how well the polynomial fits the given curve.

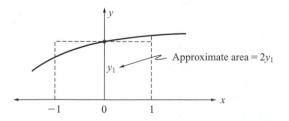

Figure 11–7 Gaussian quadrature using one sampling point

Table 11–1 Table for Gauss points for integration from minus one to one, $\int_{-1}^{1} y(x)\, dx = \sum_{i=1}^{n} W_i y_i$

Number of Points	Locations, x_i	Associated Weights, W_i
1	$x_1 = 0.000\ldots$	2.000
2	$x_1, x_2 = \pm 0.57735026918962$	1.000
3	$x_1, x_3 = \pm 0.77459666924148$	$\frac{5}{9} = 0.555\ldots$
	$x_2 = 0.000\ldots$	$\frac{8}{9} = 0.888\ldots$
4	$x_1, x_4 = \pm 0.8611363116$	0.3478548451
	$x_2, x_3 = \pm 0.3399810436$	0.6521451549

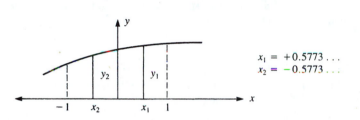

$$x_1 = +0.5773\ldots$$
$$x_2 = -0.5773\ldots$$

Figure 11–8 Gaussian quadrature using two sampling points

If the function $f(x)$ is not a polynomial, Gaussian quadrature is inexact, but it becomes more accurate as more Gauss points are used. Also, it is important to understand that the ratio of two polynomials is, in general, not a polynomial; therefore, Gaussian quadrature will not yield exact integration of the ratio.

Two-Point Formula

To illustrate the derivation of a two-point $(n = 2)$ Gauss formula based on Eq. (11.4.2), we have

$$I = \int_{-1}^{1} y\, dx = W_1 y_1 + W_2 y_2 = W_1 y(x_1) + W_2 y(x_2) \qquad (11.4.3)$$

There are four unknown parameters to determine: $W_1, W_2, x_1,$ and x_2. Therefore, we assume a cubic function for y as follows:

$$y = C_0 + C_1 x + C_2 x^2 + C_3 x^3 \qquad (11.4.4)$$

In general, with four parameters in the two-point formula, we would expect the Gauss formula to exactly predict the area under the curve. That is,

$$A = \int_{-1}^{1} (C_0 + C_1 x + C_2 x^2 + C_3 x^3)\, dx = 2C_0 + \frac{2C_2}{3} \qquad (11.4.5)$$

However, we will assume, based on Gauss's method, that $W_1 = W_2$ and $x_1 = x_2$ as we use two symmetrically located Gauss points at $x = \pm a$ with equal weights. The area predicted by Gauss's formula is

$$A_G = Wy(-a) + Wy(a) = 2W(C_0 + C_2 a^2) \tag{11.4.6}$$

where $y(-a)$ and $y(a)$ are evaluated using Eq. (11.4.4). If the error, $e = A - A_G$, is to vanish for any C_0 and C_2, we must have, using Eqs. (11.4.5) and (11.4.6) in the error expression,

$$\frac{\partial e}{\partial C_0} = 0 = 2 - 2W \qquad \text{or} \qquad W = 1 \tag{11.4.7a}$$

and

$$\frac{\partial e}{\partial C_2} = 0 = \frac{2}{3} - 2a^2 W \qquad \text{or} \qquad a = \sqrt{\frac{1}{3}} = 0.5773\ldots \tag{11.4.7b}$$

Now $W = 1$ and $a = 0.5773\ldots$ are the W_i's and a_i's (x_i's) for the two-point Gaussian quadrature given in Table 11–1.

Example 11.1

Evaluate the integral $I = \int_{-1}^{1}[x^2 + \cos(x/2)]\,dx$ using three-point Gaussian quadrature.

Using Table 11–1 for the three Gauss points and weights, we have $x_1 = x_3 = \pm 0.77459\ldots$, $x_2 = 0.000\ldots$, $W_1 = W_3 = \frac{5}{9}$, and $W_2 = \frac{8}{9}$. The integral then becomes

$$I = \left[(-0.77459)^2 + \cos\left(-\frac{0.77459}{2}\text{ rad}\right)\right]\frac{5}{9} + \left[0^2 + \cos\frac{0}{2}\right]\frac{8}{9}$$

$$+ \left[(0.77459)^2 + \cos\left(\frac{0.77459}{2}\text{ rad}\right)\right]\frac{5}{9}$$

$$= 1.918 + 0.667 = 2.585$$

Compared to the exact solution, we have $I_{\text{exact}} = 2.585$.

In this example, three-point Gaussian quadrature yields the exact answer to four significant figures. ■

In two dimensions, we obtain the quadrature formula by integrating first with respect to one coordinate and then with respect to the other as

$$I = \int_{-1}^{1}\int_{-1}^{1} f(s, t)\,ds\,dt = \int_{-1}^{1}\left[\sum_i W_i f(s_i, t)\right]dt$$

$$= \sum_j W_j\left[\sum_i W_i f(s_i, t_j)\right] = \sum_i \sum_j W_i W_j f(s_i, t_j) \tag{11.4.8}$$

In Eq. (11.4.8), we need not use the same number of Gauss points in each direction (that is, i does not have to equal j), but this is usually done. Thus, for example, a four-point Gauss rule (often described as a 2×2 rule) is shown in Figure 11–9. Equation

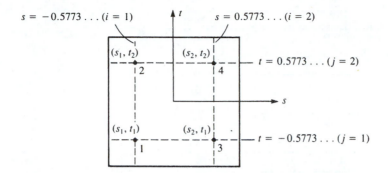

$s = -0.5773 \ldots (i = 1)$ $s = 0.5773 \ldots (i = 2)$

(s_1, t_2) (s_2, t_2) $t = 0.5773 \ldots (j = 2)$

(s_1, t_1) (s_2, t_1) $t = -0.5773 \ldots (j = 1)$

Figure 11–9   Four-point Gaussian quadrature in two dimensions

$(11.4.8)$ with $i = 1, 2$ and $j = 1, 2$ yields

$$I = W_1 W_1 f(s_1, t_1) + W_1 W_2 f(s_1, t_2) + W_2 W_1 f(s_2, t_1) + W_2 W_2 f(s_2, t_2) \quad (11.4.9)$$

where the four sampling points are at $s_i, t_i = \pm 0.5773 \ldots = \pm 1/\sqrt{3}$, and the weights are all 1.000. Hence, the double summation in Eq. (11.4.8) can really be interpreted as a single summation over the four points for the rectangle.

In general, in three dimensions, we have

$$I = \int_{-1}^{1} \int_{-1}^{1} \int_{-1}^{1} f(s, t, z) \, ds \, dt \, dz = \sum_i \sum_j \sum_k W_i W_j W_k f(s_i, t_j, z_k) \quad (11.4.10)$$

▲ 11.5 Evaluation of the Stiffness Matrix and Stress Matrix by Gaussian Quadrature

Evaluation of the Stiffness Matrix

For the two-dimensional element, we have shown in previous chapters that

$$\underline{k} = \iint_A \underline{B}^T(x, y) \underline{D} \underline{B}(x, y) h \, dx \, dy \quad (11.5.1)$$

where, in general, the integrand is a function of x and y and nodal coordinate values.

We have shown in Section 11.3 that $\underline{k}$ for a quadrilateral element can be evaluated in terms of a local set of coordinates s-t, with limits from minus one to one within the element, and in terms of global nodal coordinates as given by Eq. (11.3.27). We repeat Eq. (11.3.27) here for convenience as

$$\underline{k} = \int_{-1}^{1} \int_{-1}^{1} \underline{B}^T(s, t) \underline{D} \underline{B}(s, t) |J| h \, ds \, dt \quad (11.5.2)$$

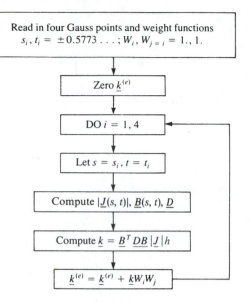

Figure 11–10 Flowchart to evaluate $\underline{k}^{(e)}$ by four-point Gaussian quadrature

where $|\underline{J}|$ is defined by Eq. (11.3.22) and $\underline{B}$ is defined by Eq. (11.3.18). In Eq. (11.5.2), each coefficient of the integrand $\underline{B}^T\underline{D}\underline{B}|\underline{J}|$ must be evaluated by numerical integration in the same manner as $f(s,t)$ was integrated in Eq. (11.4.9).

A flowchart to evaluate $\underline{k}$ of Eq. (11.5.2) for an element using four-point Gaussian quadrature is given in Figure 11–10. The four-point Gaussian quadrature rule is relatively easy to use. Also, it has been shown to yield good results [7]. In Figure 11–10, in explicit form for four-point Gaussian quadrature (now using the single summation notation with $i = 1, 2, 3, 4$), we have

$$\underline{k} = \underline{B}^T(s_1, t_1)\underline{D}\underline{B}(s_1, t_1)|\underline{J}(s_1, t_1)|hW_1W_1$$

$$+ \underline{B}^T(s_2, t_2)\underline{D}\underline{B}(s_2, t_2)|\underline{J}(s_2, t_2)|hW_2W_2$$

$$+ \underline{B}^T(s_3, t_3)\underline{D}\underline{B}(s_3, t_3)|\underline{J}(s_3, t_3)|hW_3W_3$$

$$+ \underline{B}^T(s_4, t_4)\underline{D}\underline{B}(s_4, t_4)|\underline{J}(s_4, t_4)|hW_4W_4 \qquad (11.5.3)$$

where $s_1 = t_1 = -0.5773$, $s_2 = -0.5773$, $t_2 = 0.5773$, $s_3 = 0.5773$, $t_3 = -0.5773$, and $s_4 = t_4 = 0.5773$ as shown in Figure 11–8, and $W_1 = W_2 = W_3 = W_4 = 1.000$.

Example 11.2

Evaluate the stiffness matrix for the quadrilateral element shown in Figure 11–11 using the four-point Gaussian quadrature rule. Let $E = 30 \times 10^6$ psi and $v = 0.25$. The global coordinates are shown in inches. Assume $h = 1$ in.

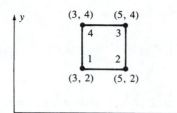

Figure 11–11 Quadrilateral element for stiffness evaluation

Using Eq. (11.5.3), we evaluate the $\underline{k}$ matrix. Using the four-point rule, the four points are

$$(s_1, t_1) = (-0.5773, -0.5773)$$
$$(s_2, t_2) = (-0.5773, 0.5773)$$
$$(s_3, t_3) = (0.5773, -0.5773)$$
$$(s_4, t_4) = (0.5773, 0.5773)$$

(11.5.4a)

with weights $W_1 = W_2 = W_3 = W_4 = 1.000$.

Therefore, by Eq. (11.5.3), we have

$$\underline{k} = \underline{B}^T(-0.5773, -0.5773)\underline{D}\underline{B}(-0.5773, -0.5773)$$
$$\times |\underline{J}(-0.5773, -0.5773)|(1)(1.000)(1.000)$$
$$+ \underline{B}^T(-0.5773, 0.5773)\underline{D}\underline{B}(-0.5773, 0.5773)$$
$$\times |\underline{J}(-0.5773, 0.5773)|(1)(1.000)(1.000)$$
$$+ \underline{B}^T(0.5773, -0.5773)\underline{D}\underline{B}(0.5773, -0.5773)$$
$$\times |\underline{J}(0.5773, -0.5773)|(1)(1.000)(1.000)$$
$$+ \underline{B}^T(0.5773, 0.5773)\underline{D}\underline{B}(0.5773, 0.5773)$$
$$\times |\underline{J}(0.5773, 0.5773)|(1)(1.000)(1.000)$$

(11.5.4b)

To evaluate $\underline{k}$, we first evaluate $|\underline{J}|$ at each Gauss point by using Eq. (11.3.22). For instance, one part of $|\underline{J}|$ is given by

$$|\underline{J}(-0.5773, -0.5773)| = \tfrac{1}{8}[3 \quad 5 \quad 5 \quad 3]$$

$$\times \begin{bmatrix} 0 & 1-(-0.5773) & -0.5773-(-0.5773) & -0.5773-1 \\ -0.5773-1 & 0 & -0.5773+1 & -0.5773-(-0.5773) \\ -0.5773-(-0.5773) & -0.5773-1 & 0 & -0.5773+1 \\ 1-(-0.5773) & -0.5773+(-0.5773) & -0.5773-1 & 0 \end{bmatrix}$$

$$\times \begin{Bmatrix} 2 \\ 2 \\ 4 \\ 4 \end{Bmatrix} = 1.000$$

(11.5.4c)

Similarly,
$$|J(-0.5773, 0.5773)| = 1.000$$

$$|J(0.5773, -0.5773)| = 1.000 \tag{11.5.4d}$$

$$|J(0.5773, 0.5773)| = 1.000$$

Even though $|J| = 1$ in this example, in general, $|J| \neq 1$ and varies in space.

Then, using Eqs. (11.3.18) and (11.3.19), we evaluate $\underline{B}$. For instance, one part of $\underline{B}$ is

$$\underline{B}(-0.5773, -0.5773) = \frac{1}{|J(-0.5773, -0.5773)|}[\underline{B_1} \quad \underline{B_2} \quad \underline{B_3} \quad \underline{B_4}]$$

where, by Eq. (11.3.19),

$$\underline{B_1} = \begin{bmatrix} aN_{1,s} - bN_{1,t} & 0 \\ 0 & cN_{1,t} - dN_{1,s} \\ cN_{1,t} - dN_{1,s} & aN_{1,s} - bN_{1,t} \end{bmatrix} \tag{11.5.4e}$$

and by Eqs. (11.3.20) and (11.3.21), $a, b, c, d, N_{1,s}$, and $N_{1,t}$ are evaluated. For instance,

$$a = \tfrac{1}{4}[y_1(s-1) + y_2(-1-s) + y_3(1+s) + y_4(1-s)]$$

$$= \tfrac{1}{4}\{2(-0.5773 - 1) + 2[-1 - (0.5773)]\}$$

$$+ 4[1 + (-0.5773)] + 4[1 - (-0.5773)]$$

$$= 1.00 \tag{11.5.4f}$$

with similar computations used to obtain b, c, and d. Also,

$$N_{1,s} = \tfrac{1}{4}(t-1) = \tfrac{1}{4}(-0.5773 - 1) = -0.3943$$

$$N_{1,t} = \tfrac{1}{4}(s-1) = \tfrac{1}{4}(-0.5773 - 1) = -0.3943 \tag{11.5.4g}$$

Similarly, $\underline{B_2}$, $\underline{B_3}$, and $\underline{B_4}$ must be evaluated like $\underline{B_1}$, at $(-0.5773, -0.5773)$. We then repeat the calculations to evaluate $\underline{B}$ at the other Gauss points [Eq. (11.5.4a)].

Using a computer program written specifically to evaluate $\underline{B}$ at each Gauss point and then $\underline{k}$, we obtain the final form of $\underline{B}(-0.5773, -0.5773)$ as

$$\underline{B}(-0.5773, -0.5773) =$$

$$\begin{bmatrix} -0.1057 & 0 & 0.1057 & 0 & 0 & -0.1057 & 0 & -0.3943 \\ -0.1057 & -0.1057 & -0.3943 & 0.1057 & 0.3943 & 0 & -0.3943 & 0 \\ 0 & 0.3943 & 0 & 0.1057 & 0.3943 & 0.3943 & 0.1057 & -0.3943 \end{bmatrix}$$

with similar expressions for $\underline{B}(-0.5773, 0.5773)$, and so on.

From Eq. (7.1.8), the matrix $\underline{D}$ is

$$\underline{D} = \frac{E}{1-v^2} \begin{bmatrix} 1 & v & 0 \\ v & 1 & 0 \\ 0 & 0 & \frac{1-v}{2} \end{bmatrix} = \begin{bmatrix} 32 & 8 & 0 \\ 8 & 32 & 0 \\ 0 & 0 & 12 \end{bmatrix} \times 10^6 \text{ psi} \qquad (11.5.4i)$$

Finally, using Eq. (11.5.4b), the matrix $\underline{k}$ becomes

$$\underline{k} = 10^4 \begin{bmatrix} 1466 & 500 & -866 & -99 & -733 & -500 & 133 & 99 \\ 500 & 1466 & 99 & 133 & -500 & -733 & -99 & -866 \\ -866 & 99 & 1466 & -500 & 133 & -99 & -733 & 500 \\ -99 & 133 & -500 & 1466 & 99 & -866 & 500 & -733 \\ -733 & -500 & 133 & 99 & 1466 & 500 & -866 & -99 \\ -500 & -733 & -99 & -866 & 500 & 1466 & 99 & 133 \\ 133 & -99 & -733 & 500 & -866 & 99 & 1466 & -500 \\ 99 & -866 & 500 & -733 & -99 & 133 & -500 & 1466 \end{bmatrix}$$

$$(11.5.4j) \quad \blacksquare$$

Evaluation of Element Stresses

The stresses $\underline{\sigma} = \underline{DBd}$ are not constant within the quadrilateral element. Because $\underline{B}$ is a function of s and t coordinates, $\underline{\sigma}$ is also a function of s and t. In practice, the stresses are evaluated at the same Gauss points used to evaluate the stiffness matrix $\underline{K}$. For a quadrilateral using 2×2 integration, we get four sets of stress data. To reduce the data, it is often practical to evaluate $\underline{\sigma}$ at $s = 0$, $t = 0$ instead. Another method mentioned in Section 8.4 is to evaluate the stresses in all elements at a shared (common) node and then use an average of these element nodal stresses to represent the stress at the node. Computer programs such as Algor use this method. Stress plots obtained in Algor are based on this average nodal stress method. Example 11.3 illustrates the use of Gaussian quadrature to evaluate the stress matrix at the $s = 0$, $t = 0$ location of the element.

Example 11.3

For the rectangular element shown in Figure 11–11, assume plane stress conditions with $E = 30 \times 10^6$ psi, $v = 0.3$, and displacements $u_1 = 0$, $v_1 = 0$, $u_2 = 0.001$ in., $v_2 = 0.0015$ in., $u_3 = 0.003$ in., $v_3 = 0.0016$ in., $u_4 = 0$, and $v_4 = 0$. Evaluate the stresses, σ_x, σ_y, and τ_{xy} at $s = 0$, $t = 0$.

Using Eqs. (11.3.18)–(11.3.20), we evaluate $\underline{B}$ at $s = 0$, $t = 0$.

$$\underline{B} = \frac{1}{|J|} [\underline{B}_1 \quad \underline{B}_2 \quad \underline{B}_3 \quad \underline{B}_4] \qquad (11.3.18)$$

$$\underline{B}(0,0) = \frac{1}{|\underline{J}(0,0)|} [\underline{B}_1(0,0) \quad \underline{B}_2(0,0) \quad \underline{B}_3(0,0) \quad \underline{B}_4(0,0)]$$

By Eq. (11.3.22), $|\underline{J}|$ is

$$|\underline{J}(0,0)| = \tfrac{1}{8} [0 \quad 2 \quad 2 \quad 0] \begin{bmatrix} 0 & 1 & 0 & -1 \\ -1 & 0 & 1 & 0 \\ 0 & -1 & 0 & 1 \\ 1 & 0 & -1 & 0 \end{bmatrix} \begin{Bmatrix} 0 \\ 0 \\ 1 \\ 1 \end{Bmatrix}$$

$$= \tfrac{1}{8} [-2 \quad -2 \quad 2 \quad 2] \begin{Bmatrix} 0 \\ 0 \\ 1 \\ 1 \end{Bmatrix}$$

$$|\underline{J}(0,0)| = \tfrac{1}{2} \tag{11.5.5a}$$

By Eq. (11.3.19), we have

$$\underline{B}_i = \begin{bmatrix} aN_{i,s} - bN_{i,t} & 0 \\ 0 & cN_{i,t} - dN_{i,s} \\ cN_{i,t} - dN_{i,s} & aN_{i,s} - bN_{i,t} \end{bmatrix} \tag{11.5.5b}$$

By Eq. (11.3.20), we obtain

$$a = \tfrac{1}{2} \qquad b = 0 \qquad c = 1 \qquad d = 0$$

Differentiating the shape functions in Eq. (11.3.5) with respect to s and t and then evaluating at $s = 0$, $t = 0$, we obtain

$$
\begin{aligned}
N_{1,s} &= -\tfrac{1}{4} & N_{1,t} &= -\tfrac{1}{4} & N_{2,s} &= \tfrac{1}{4} & N_{2,t} &= -\tfrac{1}{4} \\
N_{3,s} &= \tfrac{1}{4} & N_{3,t} &= \tfrac{1}{4} & N_{4,s} &= -\tfrac{1}{4} & N_{4,t} &= \tfrac{1}{4}
\end{aligned}
\tag{11.5.5c}
$$

Therefore, substituting Eqs. (11.5.5c) into Eq. (11.5.5b), we obtain

$$\underline{B}_1 = \begin{bmatrix} -\tfrac{1}{8} & 0 \\ 0 & -\tfrac{1}{4} \\ -\tfrac{1}{4} & -\tfrac{1}{8} \end{bmatrix} \quad \underline{B}_2 = \begin{bmatrix} \tfrac{1}{8} & 0 \\ 0 & -\tfrac{1}{4} \\ -\tfrac{1}{4} & \tfrac{1}{8} \end{bmatrix} \quad \underline{B}_3 = \begin{bmatrix} \tfrac{1}{8} & 0 \\ 0 & \tfrac{1}{4} \\ \tfrac{1}{4} & \tfrac{1}{8} \end{bmatrix} \quad \underline{B}_4 = \begin{bmatrix} -\tfrac{1}{8} & 0 \\ 0 & \tfrac{1}{4} \\ \tfrac{1}{4} & -\tfrac{1}{8} \end{bmatrix}$$

$$\tag{11.5.5d}$$

The element stress matrix $\underline{\sigma}$ is then obtained by substituting Eqs. (11.5.5a) for $|\underline{J}| = \tfrac{1}{2}$ and (11.5.5d) into Eq. (11.3.18) for $\underline{B}$ and the plane stress $\underline{D}$ matrix from Eq. (7.1.8) into the definition for $\underline{\sigma}$ as

$$\underline{\sigma} = \underline{DBd} = (30) \frac{10^6 \begin{bmatrix} 1 & 0.3 & 0 \\ 0.3 & 1 & 0 \\ 0 & 0 & 0.35 \end{bmatrix}}{1 - 0.09}$$

(*continued*)

$$\times \begin{bmatrix} 0.25 & 0 & 0.25 & 0 & 0.25 & 0 & -0.25 & 0 \\ 0 & -0.5 & 0 & -0.5 & 0 & 0.5 & 0 & 0.5 \\ -0.5 & -0.25 & -0.5 & 0.25 & 0.5 & 0.25 & 0.5 & -0.25 \end{bmatrix} \begin{Bmatrix} 0 \\ 0 \\ 0.001 \\ 0.0015 \\ 0.003 \\ 0.0016 \\ 0 \\ 0 \end{Bmatrix}$$

$$\underline{\sigma} = \begin{Bmatrix} 3.346 \cdot 10^4 \\ 1.154 \cdot 10^4 \\ 2.048 \cdot 10^4 \end{Bmatrix} \text{psi} \qquad ■$$

11.6 Higher-Order Shape Functions

In general, higher-order element shape functions can be developed by adding additional nodes to the sides of the linear element. These elements result in higher-order strain variations within each element, and convergence to the exact solution thus occurs at a faster rate using fewer elements. (However, a trade-off exists because a more complicated element takes up so much computation time that even with few elements in the model, the computation time can become larger than for the simple linear element.) Another advantage of the use of higher-order elements is that curved boundaries of irregularly shaped bodies can be approximated more closely than by the use of simple straight-sided linear elements.

To illustrate the concept of higher-order elements, we will consider the quadratic and cubic element shape functions as described in Reference [3]. Figure 11–12 shows a quadratic isoparametric element with four corner nodes and four additional midside

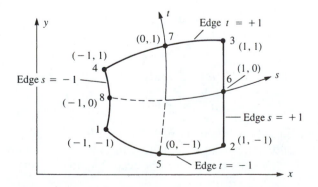

Figure 11–12 Quadratic isoparametric element

nodes. The shape functions of the quadratic element are based on the incomplete cubic polynomial such that coordinates x and y are

$$x = a_1 + a_2s + a_3t + a_4st + a_5s^2 + a_6t^2 + a_7s^2t + a_8st^2$$

$$y = a_9 + a_{10}s + a_{11}t + a_{12}st + a_{13}s^2 + a_{14}t^2 + a_{15}s^2t + a_{16}st^2$$

(11.6.1)

Again, we have the same number of degrees of freedom (2 per node times 8 nodes equals 16) as we have a_i's. To describe the shape functions, two forms are required—one for corner nodes and one for midside nodes, as given in Reference [3]. For the corner nodes ($i = 1, 2, 3, 4$),

$$N_1 = \tfrac{1}{4}(1 - s)(1 - t)(-s - t - 1)$$

$$N_2 = \tfrac{1}{4}(1 + s)(1 - t)(s - t - 1)$$

$$N_3 = \tfrac{1}{4}(1 + s)(1 + t)(s + t - 1)$$

$$N_4 = \tfrac{1}{4}(1 - s)(1 + t)(-s + t - 1)$$

(11.6.2)

or, in compact index notation, we express Eqs. (11.6.2) as

$$N_i = \tfrac{1}{4}(1 + ss_i)(1 + tt_i)(ss_i + tt_i - 1)$$

(11.6.3)

where i is the number of the shape function and

$$s_i = -1, 1, 1, -1 \quad (i = 1, 2, 3, 4)$$

$$t_i = -1, -1, 1, 1 \quad (i = 1, 2, 3, 4)$$

(11.6.4)

For the midside nodes ($i = 5, 6, 7, 8$),

$$N_5 = \tfrac{1}{2}(1 - t)(1 + s)(1 - s)$$

$$N_6 = \tfrac{1}{2}(1 + s)(1 + t)(1 - t)$$

$$N_7 = \tfrac{1}{2}(1 + t)(1 + s)(1 - s)$$

$$N_8 = \tfrac{1}{2}(1 - s)(1 + t)(1 - t)$$

(11.6.5)

or, in index notation,

$$N_i = \tfrac{1}{2}(1 - s^2)(1 + tt_i) \quad t_i = -1, 1 \quad (i = 5, 7)$$

$$N_i = \tfrac{1}{2}(1 + ss_i)(1 - t^2) \quad s_i = 1, -1 \quad (i = 6, 8)$$

(11.6.6)

We can observe from Eqs. (11.6.2) and (11.6.5) that an edge (and displacement) can vary with s^2 (along t constant) or with t^2 (along s constant). Furthermore, $N_i = 1$ at node i and $N_i = 0$ at the other nodes, as it must be according to our usual definition of shape functions.

The displacement functions are given by

$$\begin{Bmatrix} u \\ v \end{Bmatrix} = \begin{bmatrix} N_1 & 0 & N_2 & 0 & N_3 & 0 & N_4 & 0 & N_5 & 0 & N_6 & 0 & N_7 & 0 & N_8 & 0 \\ 0 & N_1 & 0 & N_2 & 0 & N_3 & 0 & N_4 & 0 & N_5 & 0 & N_6 & 0 & N_7 & 0 & N_8 \end{bmatrix}$$

$$\times \begin{Bmatrix} u_1 \\ v_1 \\ u_2 \\ v_2 \\ \vdots \\ v_8 \end{Bmatrix} \tag{11.6.7}$$

and the strain matrix is now

$$\underline{\varepsilon} = \underline{D}'\underline{N}\underline{d}$$

with

$$\underline{B} = \underline{D}'\underline{N}$$

We can develop the matrix $\underline{B}$ using Eq. (11.3.17) with $\underline{D}'$ from Eq. (11.3.16) and with $\underline{N}$ now the 2×16 matrix given in Eq. (11.6.7), where the N's are defined in explicit form by Eq. (11.6.2) and (11.6.5).

To evaluate the matrix $\underline{B}$ and the matrix $\underline{k}$ for the eight-noded quadratic isoparametric element, we now use the nine-point Gauss rule (often described as a 3×3 rule). Results using 2×2 and 3×3 rules have shown significant differences, and the 3×3 rule is recommended by Bathe and Wilson [7]. Table 11–1 indicates the locations of points and the associated weights. The 3×3 rule is shown in Figure 11–13.

The cubic element in Figure 11–14 has four corner nodes and additional nodes taken to be at one-third and two-thirds of the length along each side. The shape functions of the cubic element (as derived in Reference [3]) are based on the incomplete quartic polynomial such that

$$x = a_1 + a_2 s + a_3 t + a_4 s^2 + a_5 st + a_6 t^2 + a_7 s^2 t + a_8 st^2$$

$$+ a_9 s^3 + a_{10} t^3 + a_{11} s^3 t + a_{12} st^3 \tag{11.6.8}$$

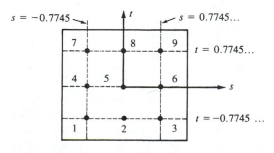

Figure 11–13 3×3 rule in two dimensions

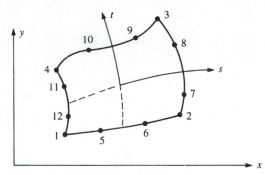

Figure 11–14 Cubic isoparametric element

with a similar polynomial for y. For the corner nodes ($i = 1, 2, 3, 4$),

$$N_i = \tfrac{1}{32}(1 + ss_i)(1 + tt_i)[9(s^2 + t^2) - 10] \tag{11.6.9}$$

with s_i and t_i given by Eqs. (11.6.4). For the nodes on sides $s = \pm 1$ ($i = 7, 8, 11, 12$),

$$N_i = \tfrac{9}{32}(1 + ss_i)(1 + 9tt_i)(1 - t^2) \tag{11.6.10}$$

with $s_i = \pm 1$ and $t_i = \pm \tfrac{1}{3}$. For the nodes on sides $t = \pm 1$ ($i = 5, 6, 9, 10$),

$$N_i = \tfrac{9}{32}(1 + tt_i)(1 + 9ss_i)(1 - s^2) \tag{11.6.11}$$

with $t_i = \pm 1$ and $s_i = \pm \tfrac{1}{3}$.

Having the shape functions for the quadratic element given by Eqs. (11.6.2) and (11.6.5) or for the cubic element given by Eqs. (11.6.9)–(11.6.11), we can again use Eq. (11.3.17) to obtain $\underline{B}$ and then Eq. (11.3.27) to set up $\underline{k}$ for numerical integration for the plane element. The cubic element requires a 3×3 rule (nine points) to evaluate the matrix $\underline{k}$ exactly. We then conclude that what is really desired is a library of shape functions that can be used in the general equations developed for stiffness matrices, distributed load, and body force and can be applied not only to stress analysis but to nonstructural problems as well.

Since in this discussion the element shape functions N_i relating x and y to nodal coordinates x_i and y_i are of the same form as the shape functions relating u and v to nodal displacements u_i and v_i, this is said to be an *isoparametric formulation*. For instance, for the linear element $x = \sum_{i=1}^{4} N_i x_i$ and the displacement function $u = \sum_{i=1}^{4} N_i u_i$, use the same shape functions N_i given by Eq. (11.3.5). If instead the shape functions for the coordinates are of lower order (say, linear for x) than the shape functions used for displacements (say, quadratic for u), this is called a *subparametric formulation*.

Finally, referring to Figure 11–14, note that an element can have a linear shape along, say, one edge (1-2), a quadratic along, say, two edges (2-3 and 1-4), and a cubic along the other edge (3-4). Hence, the simple linear element can be mixed with different higher-order elements in regions of a model where rapid stress variation is expected. The advantage of the use of higher-order elements is further illustrated in Reference [3].

▲ References

[1] Irons, B. M., "Engineering Applications of Numerical Integration in Stiffness Methods," *Journal of the American Institute of Aeronautics and Astronautics*, Vol. 4, No. 11, pp. 2035–2037, 1966.

[2] Stroud, A. H., and Secrest, D., *Gaussian Quadrature Formulas*, Prentice-Hall, Englewood Cliffs, NJ, 1966.

[3] Ergatoudis, I., Irons, B. M., and Zienkiewicz, O. C., "Curved Isoparametric, Quadrilateral Elements for Finite Element Analysis," *International Journal of Solids and Structures*, Vol. 4, pp. 31–42, 1968.

[4] Zienkiewicz, O. C., *The Finite Element Method*, 3rd ed., McGraw-Hill, London, 1977.

[5] Thomas, B. G., and Finney, R. L., *Calculus and Analytic Geometry*, Addison-Wesley, Reading, MA, 1984.

[6] Gallagher, R., *Finite Element Analysis Fundamentals*, Prentice-Hall, Englewood Cliffs, NJ, 1975.

[7] Bathe, K. J., and Wilson, E. L., *Numerical Methods in Finite Element Analysis*, Prentice-Hall, Englewood Cliffs, NJ, 1976.

▲ Problems

11.1 For the three-noded linear strain bar with three coordinates of nodes $x_1, x_2,$ and $x_3,$ shown in Figure P11–1 in the global-coordinate system and in Figure P11–2 in the natural-coordinate system, show that the Jacobian is $J = L/2$.

Figure P11–1   **Figure P11–2**

11.2 For the three-noded linear-strain one-dimensional isoparametric element shown in Figure P11–2, determine (a) the shape functions $N_1, N_2,$ and N_3 and (b) the strain/displacement matrix $[B]$. Assume $u = a_1 + a_2 s + a_3 s^2$.

11.3 For the three-noded bar element in Figures P11–1 and P11–2, evaluate the stiffness matrix analytically. Use the $\underline{B}$ matrix from Problem 11.2.

11.4 For the four-noded quadratic-strain one-dimensional isoparametric element shown in Figure P11–4, determine (a) the shape functions $N_1, N_2, N_3,$ and $N_4,$ and (b) the strain/displacement matrix $[B]$. Assume $u = a_1 + a_2 s + a_3 s^2 + a_4 s^3$.

Figure P11–4

11.5 Show that the sum $N_1 + N_2 + N_3 + N_4$ is equal to 1 anywhere on a rectangular element, where N_1 through N_4 are defined by Eqs. (11.2.5).

11.6 For the rectangular element of Figure 11–3, the nodal displacements are given by

$$u_1 = 0 \qquad\qquad v_1 = 0 \qquad\qquad u_2 = 0.005 \text{ in.}$$

$$v_2 = 0.0025 \text{ in.} \qquad u_3 = 0.0025 \text{ in.} \qquad v_3 = -0.0025 \text{ in.}$$

$$u_4 = 0 \qquad\qquad v_4 = 0$$

For $b = 2$ in., $h = 1$ in., $E = 30 \times 10^6$ psi, and $v = 0.3$, determine the element strains and stresses at the centroid of the element and at the corner nodes.

11.7 Derive $|J|$ given by Eq. (11.3.22) for a four-noded isoparametric quadrilateral element.

11.8 Show that for the quadrilateral element described in Section 11.3, $[J]$ can be expressed as

$$[J] = \begin{bmatrix} N_{1,s} & N_{2,s} & N_{3,s} & N_{4,s} \\ N_{1,t} & N_{2,t} & N_{3,t} & N_{4,t} \end{bmatrix} \begin{bmatrix} x_1 & y_1 \\ x_2 & y_2 \\ x_3 & y_3 \\ x_4 & y_4 \end{bmatrix}$$

11.9 Derive Eq. (11.3.18) with B_i given by Eq. (11.3.19) by substituting Eq. (11.3.16) for $\underline{D}'$ and Eqs. (11.3.5) for the shape functions into Eq. (11.3.17).

11.10 Use Eq. (11.3.30) with $p_s = 0$ and $p_t = p$ (a constant) alongside 3-4 of the element shown in Figure 11–4(b) to obtain the nodal forces.

11.11 For the element shown in Figure P11–11, replace the distributed load with the energy equivalent nodal forces by evaluating a force matrix similar to Eq. (11.3.29). Let $h = 0.1$ in thick.

11.12 Use Gaussian quadature with two and three Gauss points and Table 11–1 to evaluate the following integrals:

(a) $\displaystyle\int_{-1}^{1} \cos\frac{s}{2}\, ds$ (b) $\displaystyle\int_{-1}^{1} s^2\, ds$ (c) $\displaystyle\int_{-1}^{1} s^4\, ds$

(d) $\displaystyle\int_{-1}^{1} \frac{\cos s}{1 - s^2}\, ds$ (e) $\displaystyle\int_{-1}^{1} s^3\, ds$ (f) $\displaystyle\int_{-1}^{1} s\cos s\, ds$

11.13 For the quadrilateral elements shown in Figure P11–13, write a computer program to evaluate the stiffness matrices using four-point Gaussian quadrature as outlined in Section 11.5. Let $E = 30 \times 10^6$ psi and $v = 0.25$. The global coordinates (in inches) are shown in the figures.

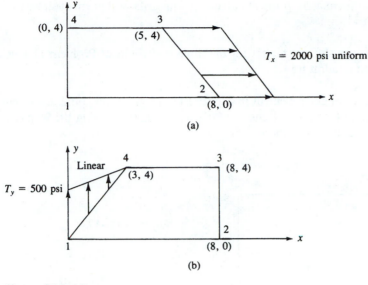

Figure P11–11

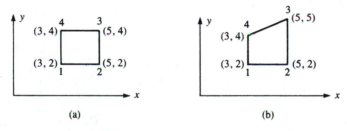

Figure P11–13

11.14 For the quadrilateral elements shown in Figure P11–14, evaluate the stiffness matrices using four-point Gaussian quadrature. Let $E = 210$ GPa and $v = 0.25$. The global coordinates (in millimeters) are shown in the figures.

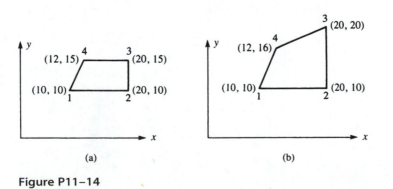

Figure P11–14

11.15 Evaluate the matrix $\underline{B}$ for the quadratic quadrilateral element shown in Figure 11.12 (Section 11.6).

11.16 Evaluate the stiffness matrix for the three-noded bar of Problem 11.1 using two-point Gaussian quadrature.

11.17 For the rectangular element in Figure P11–17, with the nodal displacements given in Problem 11.6, determine the $\underline{\sigma}$ matrix at $s = 0$, $t = 0$ using the isoparametric formulation described in Section 11.5.

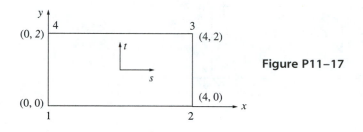

Figure P11–17

12

Three-Dimensional Stress Analysis

Introduction

In this chapter, we consider the three-dimensional, or solid, element. This element is useful for the stress analysis of general three-dimensional bodies that require more precise analysis than is possible through two-dimensional and/or axisymmetric analysis. Examples of three-dimensional problems are arch dams, thick-walled pressure vessels, and solid forging parts as used, for instance, in the heavy equipment and automotive industries.

The tetrahedron is the basic three-dimensional element, and it is used in the development of the shape functions, stiffness matrix, and force matrices in terms of a global coordinate system. We follow this development with the isoparametric formulation of the stiffness matrix for the hexahedron, or brick element. Finally, we will provide some typical three-dimensional applications.

In the last section of this chapter, we describe how to model and solve three-dimensional problems using the Algor computer program.

▲ 12.1 Three-Dimensional Stress and Strain ▲

We begin by considering the three-dimensional infinitesimal element in Cartesian coordinates with dimensions $dx, dy,$ and dz and normal and shear stresses as shown in Figure 12–1. This element conveniently represents the state of stress on three mutually perpendicular planes of a body in a state of three-dimensional stress. As usual, normal stresses are perpendicular to the faces of the element and are represented by $\sigma_x, \sigma_y,$ and σ_z. Shear stresses act in the faces (planes) of the element and are represented by $\tau_{xy}, \tau_{yz}, \tau_{zx},$ and so on.

From moment equilibrium of the element, we show in Appendix C that

$$\tau_{xy} = \tau_{yx} \qquad \tau_{yz} = \tau_{zy} \qquad \tau_{zx} = \tau_{xz}$$

Hence, there are only three independent shear stresses, along with the three normal stresses.

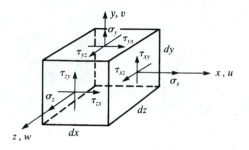

Figure 12–1 Three-dimensional stresses on an element

The element strain/displacement relationships are obtained in Appendix C. They are repeated here, for convenience, as

$$\varepsilon_x = \frac{\partial u}{\partial x} \qquad \varepsilon_y = \frac{\partial v}{\partial y} \qquad \varepsilon_z = \frac{\partial w}{\partial z} \tag{12.1.1}$$

where $u, v,$ and w are the displacements associated with the $x, y,$ and z directions. The shear strains γ are now given by

$$\gamma_{xy} = \frac{\partial u}{\partial y} + \frac{\partial v}{\partial x} = \gamma_{yx}$$

$$\gamma_{yz} = \frac{\partial v}{\partial z} + \frac{\partial w}{\partial y} = \gamma_{zy} \tag{12.1.2}$$

$$\gamma_{zx} = \frac{\partial w}{\partial x} + \frac{\partial u}{\partial z} = \gamma_{xz}$$

where, as for shear stresses, only three independent shear strains exist.

We again represent the stresses and strains by column matrices as

$$\{\sigma\} = \begin{Bmatrix} \sigma_x \\ \sigma_y \\ \sigma_z \\ \tau_{xy} \\ \tau_{yz} \\ \tau_{zx} \end{Bmatrix} \qquad \{\varepsilon\} = \begin{Bmatrix} \varepsilon_x \\ \varepsilon_y \\ \varepsilon_z \\ \gamma_{xy} \\ \gamma_{yz} \\ \gamma_{zx} \end{Bmatrix} \tag{12.1.3}$$

The stress/strain relationships for an isotropic material are again given by

$$\{\sigma\} = [D]\{\varepsilon\} \tag{12.1.4}$$

where $\{\sigma\}$ and $\{\varepsilon\}$ are defined by Eqs. (12.1.3), and the constitutive matrix $[D]$ (see also Appendix C) is now given by

$$[D] = \frac{E}{(1+v)(1-2v)} \begin{bmatrix} 1-v & v & v & 0 & 0 & 0 \\ & 1-v & v & 0 & 0 & 0 \\ & & 1-v & 0 & 0 & 0 \\ & & & \dfrac{1-2v}{2} & 0 & 0 \\ & & & & \dfrac{1-2v}{2} & 0 \\ \text{Symmetry} & & & & & \dfrac{1-2v}{2} \end{bmatrix} \quad (12.1.5)$$

12.2 Tetrahedral Element

We now develop the tetrahedral stress element stiffness matrix by again using the steps outlined in Chapter 1. The development is seen to be an extension of the plane element previously described in Chapter 7. This extension was suggested in References [1] and [2].

Step 1 Select Element Type

Consider the tetrahedral element shown in Figure 12–2 with corner nodes 1–4. This element is a four-noded solid. The nodes of the element must be numbered such that when viewed from the last node (say, node 4), the first three nodes are numbered in a counterclockwise manner, such as 1, 2, 3, 4 or 2, 3, 1, 4. This ordering of nodes avoids the calculation of negative volumes and is consistent with the counterclockwise node numbering associated with the CST element in Chapter 7. (Using an isoparametric formulation to evaluate the k matrix for the tetrahedral element enables us to use the element node numbering in any order. The isoparametric formulation of k is left to

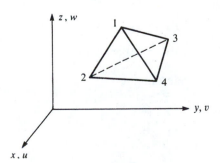

Figure 12–2 Tetrahedral solid element

Section 12.3.) The unknown nodal displacements are now given by

$$\{d\} = \begin{Bmatrix} u_1 \\ v_1 \\ w_1 \\ \vdots \\ u_4 \\ v_4 \\ w_4 \end{Bmatrix} \tag{12.2.1}$$

Hence, there are 3 degrees of freedom per node, or 12 total degrees of freedom per element.

Step 2   Select Displacement Functions

For a compatible displacement field, the element displacement functions u, v, and w must be linear along each edge because only two points (the corner nodes) exist along each edge, and the functions must be linear in each plane side of the tetrahedron. We then select the linear displacement functions as

$$u(x, y, z) = a_1 + a_2 x + a_3 y + a_4 z$$

$$v(x, y, z) = a_5 + a_6 x + a_7 y + a_8 z \tag{12.2.2}$$

$$w(x, y, z) = a_9 + a_{10} x + a_{11} y + a_{12} z$$

In the same manner as in Chapter 7, we can express the a_i's in terms of the known nodal coordinates $(x_1, y_1, z_1, \ldots, z_4)$ and the unknown nodal displacements $(u_1, v_1, w_1, \ldots, w_4)$ of the element. Skipping the straightforward but tedious details, we obtain

$$u(x, y, z) = \frac{1}{6V} \{ (\alpha_1 + \beta_1 x + \gamma_1 y + \delta_1 z) u_1$$

$$+ (\alpha_2 + \beta_2 x + \gamma_2 y + \delta_2 z) u_2$$

$$+ (\alpha_3 + \beta_3 x + \gamma_3 y + \delta_3 z) u_3$$

$$+ (\alpha_4 + \beta_4 x + \gamma_4 y + \delta_4 z) u_4 \} \tag{12.2.3}$$

where $6V$ is obtained by evaluating the determinant

$$6V = \begin{vmatrix} 1 & x_1 & y_1 & z_1 \\ 1 & x_2 & y_2 & z_2 \\ 1 & x_3 & y_3 & z_3 \\ 1 & x_4 & y_4 & z_4 \end{vmatrix} \tag{12.2.4}$$

and V represents the volume of the tetrahedron. The coefficients $\alpha_i, \beta_i, \gamma_i,$ and δ_i $(i = 1, 2, 3, 4)$ in Eq. (12.2.3) are given by

$$\alpha_1 = \begin{vmatrix} x_2 & y_2 & z_2 \\ x_3 & y_3 & z_3 \\ x_4 & y_4 & z_4 \end{vmatrix} \qquad \beta_1 = -\begin{vmatrix} 1 & y_2 & z_2 \\ 1 & y_3 & z_3 \\ 1 & y_4 & z_4 \end{vmatrix}$$

$$\gamma_1 = \begin{vmatrix} 1 & x_2 & z_2 \\ 1 & x_3 & z_3 \\ 1 & x_4 & z_4 \end{vmatrix} \qquad \delta_1 = -\begin{vmatrix} 1 & x_2 & y_2 \\ 1 & x_3 & y_3 \\ 1 & x_4 & y_4 \end{vmatrix}$$

$$(12.2.5)$$

$$\alpha_2 = -\begin{vmatrix} x_1 & y_1 & z_1 \\ x_3 & y_3 & z_3 \\ x_4 & y_4 & z_4 \end{vmatrix} \qquad \beta_2 = \begin{vmatrix} 1 & y_1 & z_1 \\ 1 & y_3 & z_3 \\ 1 & y_4 & z_4 \end{vmatrix}$$

and

$$\gamma_2 = -\begin{vmatrix} 1 & x_1 & z_1 \\ 1 & x_3 & z_3 \\ 1 & x_4 & z_4 \end{vmatrix} \qquad \delta_2 = \begin{vmatrix} 1 & x_1 & y_1 \\ 1 & x_3 & y_3 \\ 1 & x_4 & y_4 \end{vmatrix}$$

$$(12.2.6)$$

$$\alpha_3 = \begin{vmatrix} x_1 & y_1 & z_1 \\ x_2 & y_2 & z_2 \\ x_4 & y_4 & z_4 \end{vmatrix} \qquad \beta_3 = -\begin{vmatrix} 1 & y_1 & z_1 \\ 1 & y_2 & z_2 \\ 1 & y_4 & z_4 \end{vmatrix}$$

and

$$\gamma_3 = \begin{vmatrix} 1 & x_1 & z_1 \\ 1 & x_2 & z_2 \\ 1 & x_4 & z_4 \end{vmatrix} \qquad \delta_3 = -\begin{vmatrix} 1 & x_1 & y_1 \\ 1 & x_2 & y_2 \\ 1 & x_4 & y_4 \end{vmatrix}$$

$$(12.2.7)$$

$$\alpha_4 = -\begin{vmatrix} x_1 & y_1 & z_1 \\ x_2 & y_2 & z_2 \\ x_3 & y_3 & z_3 \end{vmatrix} \qquad \beta_4 = \begin{vmatrix} 1 & y_1 & z_1 \\ 1 & y_2 & z_2 \\ 1 & y_3 & z_3 \end{vmatrix}$$

and

$$\gamma_4 = -\begin{vmatrix} 1 & x_1 & z_1 \\ 1 & x_2 & z_2 \\ 1 & x_3 & z_3 \end{vmatrix} \qquad \delta_4 = \begin{vmatrix} 1 & x_1 & y_1 \\ 1 & x_2 & y_2 \\ 1 & x_3 & y_3 \end{vmatrix}$$

$$(12.2.8)$$

Expressions for v and w are obtained by simply substituting v_i's for all u_i's and then w_i's for all u_i's in Eq. (12.2.3).

The displacement expression for u given by Eq. (12.2.3), with similar expressions for v and w, can be written equivalently in expanded form in terms of the shape functions and unknown nodal displacements as

$$\left\{ \begin{array}{c} u \\ v \\ w \end{array} \right\} = \begin{bmatrix} N_1 & 0 & 0 & N_2 & 0 & 0 & N_3 & 0 & 0 & N_4 & 0 & 0 \\ 0 & N_1 & 0 & 0 & N_2 & 0 & 0 & N_3 & 0 & 0 & N_4 & 0 \\ 0 & 0 & N_1 & 0 & 0 & N_2 & 0 & 0 & N_3 & 0 & 0 & N_4 \end{bmatrix} \left\{ \begin{array}{c} u_1 \\ v_1 \\ w_1 \\ \vdots \\ u_4 \\ v_4 \\ w_4 \end{array} \right\}$$

(12.2.9)

where the shape functions are given by

$$N_1 = \frac{(\alpha_1 + \beta_1 x + \gamma_1 y + \delta_1 z)}{6V} \qquad N_2 = \frac{(\alpha_2 + \beta_2 x + \gamma_2 y + \delta_2 z)}{6V}$$

(12.2.10)

$$N_3 = \frac{(\alpha_3 + \beta_3 x + \gamma_3 y + \delta_3 z)}{6V} \qquad N_4 = \frac{(\alpha_4 + \beta_4 x + \gamma_4 y + \delta_4 z)}{6V}$$

and the rectangular matrix on the right side of Eq. (12.2.9) is the shape function matrix $[N]$.

Step 3 Define the Strain/Displacement and Stress/Strain Relationships

The element strains for the three-dimensional stress state are given by

$$\{\varepsilon\} = \left\{ \begin{array}{c} \varepsilon_x \\ \varepsilon_y \\ \varepsilon_z \\ \gamma_{xy} \\ \gamma_{yz} \\ \gamma_{zx} \end{array} \right\} = \left\{ \begin{array}{c} \dfrac{\partial u}{\partial x} \\[2mm] \dfrac{\partial v}{\partial y} \\[2mm] \dfrac{\partial w}{\partial z} \\[2mm] \dfrac{\partial u}{\partial y} + \dfrac{\partial v}{\partial x} \\[2mm] \dfrac{\partial v}{\partial z} + \dfrac{\partial w}{\partial y} \\[2mm] \dfrac{\partial w}{\partial x} + \dfrac{\partial u}{\partial z} \end{array} \right\}$$

(12.2.11)

Using Eq. (12.2.9) in Eq. (12.2.11), we obtain

$$\{\varepsilon\} = [B]\{d\}$$

(12.2.12)

where

$$[B] = [\underline{B}_1 \quad \underline{B}_2 \quad \underline{B}_3 \quad \underline{B}_4]$$

(12.2.13)

The submatrix $\underline{B}_1$ in Eq. (12.2.13) is defined by

$$\underline{B}_1 = \begin{bmatrix} N_{1,x} & 0 & 0 \\ 0 & N_{1,y} & 0 \\ 0 & 0 & N_{1,z} \\ N_{1,y} & N_{1,x} & 0 \\ 0 & N_{1,z} & N_{1,y} \\ N_{1,z} & 0 & N_{1,x} \end{bmatrix} \tag{12.2.14}$$

where, again, the comma after the subscript indicates differentation with respect to the variable that follows. Submatrices $\underline{B}_2, \underline{B}_3$, and $\underline{B}_4$ are defined by simply indexing the subscript in Eq. (12.2.14) from 1 to 2, 3, and then 4, respectively. Substituting the shape functions from Eqs. (12.2.10) into Eq. (12.2.14), $\underline{B}_1$ is expressed as

$$\underline{B}_1 = \frac{1}{6V} \begin{bmatrix} \beta_1 & 0 & 0 \\ 0 & \gamma_1 & 0 \\ 0 & 0 & \delta_1 \\ \gamma_1 & \beta_1 & 0 \\ 0 & \delta_1 & \gamma_1 \\ \delta_1 & 0 & \beta_1 \end{bmatrix} \tag{12.2.15}$$

with similar expressions for $\underline{B}_2, \underline{B}_3$, and $\underline{B}_4$.

The element stresses are related to the element strains by

$$\{\sigma\} = [D]\{\varepsilon\} \tag{12.2.16}$$

where the constitutive matrix for an elastic material is now given by Eq. (12.1.5).

Step 4 Derive the Element Stiffness Matrix and Equations

The element stiffness matrix is given by

$$[k] = \iiint_V [B]^T [D][B] \, dV \tag{12.2.17}$$

Because both matrices $[B]$ and $[D]$ are constant for the simple tetrahedral element, Eq. (12.2.17) can be simplified to

$$[k] = [B]^T [D][B] V \tag{12.2.18}$$

where, again, V is the volume of the element. The element stiffness matrix is now of order 12×12.

Body Forces

The element body force matrix is given by

$$\{f_b\} = \iiint_V [N]^T \{X\} \, dV \tag{12.2.19}$$

where $[N]$ is given by the 3×12 matrix in Eq. (12.2.9), and

$$\{X\} = \begin{Bmatrix} X_b \\ Y_b \\ Z_b \end{Bmatrix} \tag{12.2.20}$$

For constant body forces, the nodal components of the total resultant body forces can be shown to be distributed to the nodes in four equal parts. That is,

$$\{f_b\} = \tfrac{1}{4}[X_b \ Y_b \ Z_b \ X_b \ Y_b \ Z_b \ X_b \ Y_b \ Z_b \ X_b \ Y_b \ Z_b]^T$$

The element body force is then a 12×1 matrix.

Surface Forces

Again, the surface forces are given by

$$\{f_s\} = \iint_S [N_s]^T \{T\} \, dS \tag{12.2.21}$$

where $[N_s]$ is the shape function matrix evaluated on the surface where the surface traction occurs.

For example, consider the case of uniform pressure p acting on the face with nodes 1–3 of the element shown in Figure 12–2. The resulting nodal forces become

$$\{f_s\} = \iint_S [N]^T \Big|_{\substack{\text{evaluated on} \\ \text{surface } 1,2,3}} \begin{Bmatrix} p_x \\ p_y \\ p_z \end{Bmatrix} dS \tag{12.2.22}$$

where p_x, p_y, and p_z are the x, y, and z components, respectively, of p. Simplifying and integrating Eq. (12.2.22), we can show that

$$\{f_s\} = \frac{S_{123}}{3} \begin{Bmatrix} p_x \\ p_y \\ p_z \\ p_x \\ p_y \\ p_z \\ p_x \\ p_y \\ p_z \\ 0 \\ 0 \\ 0 \end{Bmatrix} \tag{12.2.23}$$

where S_{123} is the area of the surface associated with nodes 1–3. The use of volume coordinates, as explained in Reference [8], facilitates the integration of Eq. (12.2.22).

Example 12.1

Evaluate the matrices necessary to determine the stiffness matrix for the tetrahedral element shown in Figure 12–3. Let $E = 30 \times 10^6$ psi and $v = 0.30$. The coordinates are shown in the figure in units of inches.

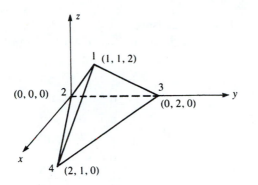

Figure 12–3 Tetrahedral element

To evaluate the element stiffness matrix, we first determine the element volume V and all α's, β's, γ's, and δ's from Eqs. (12.2.4)–(12.2.8). From Eq. (12.2.4), we have

$$6V = \begin{vmatrix} 1 & 1 & 1 & 2 \\ 1 & 0 & 0 & 0 \\ 1 & 0 & 2 & 0 \\ 1 & 2 & 1 & 0 \end{vmatrix} = 8 \text{ in}^3 \tag{12.2.24}$$

From Eqs. (12.2.5), we obtain

$$\alpha_1 = \begin{vmatrix} 0 & 0 & 0 \\ 0 & 2 & 0 \\ 2 & 1 & 0 \end{vmatrix} = 0 \qquad \beta_1 = - \begin{vmatrix} 1 & 0 & 0 \\ 1 & 2 & 0 \\ 1 & 1 & 0 \end{vmatrix} = 0 \tag{12.2.25}$$

and similarly,

$$\gamma_1 = 0 \qquad \delta_1 = 4$$

From Eqs. (12.2.6)–(12.2.8), we obtain

$$\begin{array}{cccc} \alpha_2 = 8 & \beta_2 = -2 & \gamma_2 = -4 & \delta_2 = -1 \\ \alpha_3 = 0 & \beta_3 = -2 & \gamma_3 = 4 & \delta_3 = -1 \\ \alpha_4 = 0 & \beta_4 = 4 & \gamma_4 = 0 & \delta_4 = -2 \end{array} \tag{12.2.26}$$

Note that α's typically have units of cubic inches or cubic meters, where β's, γ's, and δ's have units of square inches or square meters.

Next, the shape functions are determined using Eqs. (12.2.10) and the results from Eqs. (12.2.25) and (12.2.26) as

$$N_1 = \frac{4z}{8} \qquad\qquad N_2 = \frac{8 - 2x - 4y - z}{8} \tag{12.2.27}$$

$$N_3 = \frac{-2x + 4y - z}{8} \qquad N_4 = \frac{4x - 2z}{8}$$

Note that $N_1 + N_2 + N_3 + N_4 = 1$ is again satisfied.

The 6×3 submatrices of the matrix $\underline{B}$, Eq. (12.2.13), are now evaluated using Eqs. (12.2.14) and (12.2.27) as

$$\underline{B}_1 = \begin{bmatrix} 0 & 0 & 0 \\ 0 & 0 & 0 \\ 0 & 0 & \frac{1}{2} \\ 0 & 0 & 0 \\ 0 & \frac{1}{2} & 0 \\ \frac{1}{2} & 0 & 0 \end{bmatrix} \qquad \underline{B}_2 = \begin{bmatrix} -\frac{1}{4} & 0 & 0 \\ 0 & -\frac{1}{2} & 0 \\ 0 & 0 & -\frac{1}{8} \\ -\frac{1}{2} & -\frac{1}{4} & 0 \\ 0 & -\frac{1}{8} & -\frac{1}{2} \\ -\frac{1}{8} & 0 & -\frac{1}{4} \end{bmatrix} \tag{12.2.28}$$

$$\underline{B}_3 = \begin{bmatrix} -\frac{1}{4} & 0 & 0 \\ 0 & \frac{1}{2} & 0 \\ 0 & 0 & -\frac{1}{8} \\ \frac{1}{2} & -\frac{1}{4} & 0 \\ 0 & -\frac{1}{8} & \frac{1}{2} \\ -\frac{1}{8} & 0 & -\frac{1}{4} \end{bmatrix} \qquad \underline{B}_4 = \begin{bmatrix} \frac{1}{2} & 0 & 0 \\ 0 & 0 & 0 \\ 0 & 0 & -\frac{1}{4} \\ 0 & \frac{1}{2} & 0 \\ 0 & -\frac{1}{4} & 0 \\ -\frac{1}{4} & 0 & \frac{1}{2} \end{bmatrix}$$

Next, the matrix $\underline{D}$ is evaluated using Eq. (12.1.5) as

$$[D] = \frac{30 \times 10^6}{(1 + 0.3)(1 - 0.6)} \begin{bmatrix} 0.7 & 0.3 & 0.3 & 0 & 0 & 0 \\ & 0.7 & 0.3 & 0 & 0 & 0 \\ & & 0.7 & 0 & 0 & 0 \\ & & & 0.2 & 0 & 0 \\ & & & & 0.2 & 0 \\ \text{Symmetry} & & & & & 0.2 \end{bmatrix} \tag{12.2.29}$$

Finally, substituting the results from Eqs. (12.2.24), (12.2.28), and (12.2.29) into Eq. (12.2.18), we obtain the element stiffness matrix. The resulting 12×12 matrix, being cumbersome to obtain by longhand calculations, is best left for the computer to evaluate. ∎

▲ ## 12.3 Isoparametric Formulation ▲

We now describe the isoparametric formulation of the stiffness matrix for some three-dimensional hexahedral elements.

Linear Hexahedral Element

The basic (linear) hexahedral element [Figure 12–4(a)] now has eight corner nodes with isoparametric natural coordinates given by s, t, and z' as shown in Figure 12–4(b). The element faces are now defined by $s, t, z' = \pm 1$. (We use s, t, and z' for the coordinate axes because they are probably simpler to use than Greek letters ξ, η, and ζ).

The formulation of the stiffness matrix follows steps analogous to the isoparametric formulation of the stiffness matrix for the plane element in Chapter 11.

First, we expand Eq. (11.3.4) to include the z coordinate as follows:

$$\left\{ \begin{array}{c} x \\ y \\ z \end{array} \right\} = \sum_{i=1}^{8} \left(\begin{bmatrix} N_i & 0 & 0 \\ 0 & N_i & 0 \\ 0 & 0 & N_i \end{bmatrix} \left\{ \begin{array}{c} x_i \\ y_i \\ z_i \end{array} \right\} \right) \tag{12.3.1}$$

where the shape functions are now given by

$$N_i = \frac{(1 + ss_i)(1 + tt_i)(1 + z'z_i')}{8} \tag{12.3.2}$$

with $s_i, t_i, z_i' = \pm 1$ and $i = 1, 2, \ldots, 8$. For instance,

$$N_1 = \frac{(1 + ss_1)(1 + tt_1)(1 + z'z_1')}{8} \tag{12.3.3}$$

and when, from Figure 12–4, $s_1 = -1$, $t_1 = -1$, and $z_1' = +1$ are used in Eq. (12.3.3), we obtain

$$N_1 = \frac{(1 - s)(1 - t)(1 + z')}{8} \tag{12.3.4}$$

Explicit forms of the other shape functions follow similarly. The shape functions in Eq. (12.3.2) map the natural coordinates (s, t, z') of any point in the element to any point in the global coordinates (x, y, z) when used in Eq. (12.3.1). For instance, when we let $i = 8$ and substitute $s_8 = 1$, $t_8 = 1$, $z_8' = 1$ into Eq. (12.3.2) for N_8, we obtain

$$N_8 = \frac{(1 + s)(1 + t)(1 + z')}{8}$$

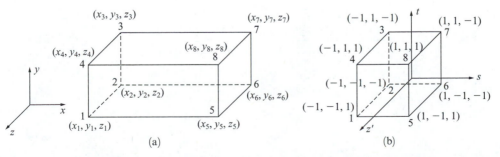

Figure 12–4 Linear hexahedral element (a) in a global-coordinate system and (b) element mapped into a cube of two unit sides placed symmetrically with natural or intrinsic coordinates s, t, and z'

Similar expressions are obtained for the other shape functions. Then evaluating all shape functions at node 8, we obtain $N_8 = 1$, and all other shape functions equal zero at node 8. [From Eq. (12.3.4), we see that $N_1 = 0$ when $s = 1$ or when $t = 1$.] Therefore, using Eq. (12.3.1), we obtain

$$x = x_8 \qquad y = y_8 \qquad z = z_8$$

We see that indeed Eq. (12.3.1) maps any point in the natural-coordinate system to one in the global-coordinate system.

The displacement functions now include w such that

$$\left\{ \begin{array}{c} u \\ v \\ w \end{array} \right\} = \sum_{i=1}^{8} \left(\begin{bmatrix} N_i & 0 & 0 \\ 0 & N_i & 0 \\ 0 & 0 & N_i \end{bmatrix} \left\{ \begin{array}{c} u_i \\ v_i \\ w_i \end{array} \right\} \right) \tag{12.3.5}$$

with the same shape functions as defined by Eq. (12.3.2) and the size of the shape function matrix now 3×24.

The Jacobian matrix [Eq. (11.3.10)] is now expanded to

$$[J] = \begin{bmatrix} \dfrac{\partial x}{\partial s} & \dfrac{\partial y}{\partial s} & \dfrac{\partial z}{\partial s} \\[2mm] \dfrac{\partial x}{\partial t} & \dfrac{\partial y}{\partial t} & \dfrac{\partial z}{\partial t} \\[2mm] \dfrac{\partial x}{\partial z'} & \dfrac{\partial y}{\partial z'} & \dfrac{\partial z}{\partial z'} \end{bmatrix} \tag{12.3.6}$$

Because the strain/displacement relationships, given by Eq. (12.2.11) in terms of global coordinates, include differentiation with respect to z, we expand Eq. (11.3.9) as follows:

$$\frac{\partial f}{\partial x} = \frac{\begin{vmatrix} \dfrac{\partial f}{\partial s} & \dfrac{\partial y}{\partial s} & \dfrac{\partial z}{\partial s} \\[2mm] \dfrac{\partial f}{\partial t} & \dfrac{\partial y}{\partial t} & \dfrac{\partial z}{\partial t} \\[2mm] \dfrac{\partial f}{\partial z'} & \dfrac{\partial y}{\partial z'} & \dfrac{\partial z}{\partial z'} \end{vmatrix}}{|J|} \qquad \frac{\partial f}{\partial y} = \frac{\begin{vmatrix} \dfrac{\partial x}{\partial s} & \dfrac{\partial f}{\partial s} & \dfrac{\partial z}{\partial s} \\[2mm] \dfrac{\partial x}{\partial t} & \dfrac{\partial f}{\partial t} & \dfrac{\partial z}{\partial t} \\[2mm] \dfrac{\partial x}{\partial z'} & \dfrac{\partial f}{\partial z'} & \dfrac{\partial z}{\partial z'} \end{vmatrix}}{|J|}$$

$$\frac{\partial f}{\partial z} = \frac{\begin{vmatrix} \dfrac{\partial x}{\partial s} & \dfrac{\partial y}{\partial s} & \dfrac{\partial f}{\partial s} \\[2mm] \dfrac{\partial x}{\partial t} & \dfrac{\partial y}{\partial t} & \dfrac{\partial f}{\partial t} \\[2mm] \dfrac{\partial x}{\partial z'} & \dfrac{\partial y}{\partial z'} & \dfrac{\partial f}{\partial z'} \end{vmatrix}}{|J|}$$

$$\tag{12.3.7}$$

Table 12-1 Table of Gauss points for linear hexahedral element with associated weights[a]

Points, i	s_i	t_i	z_i'	Weight, W_i
1	$-1/\sqrt{3}$	$-1/\sqrt{3}$	$1/\sqrt{3}$	1
2	$1/\sqrt{3}$	$-1/\sqrt{3}$	$1/\sqrt{3}$	1
3	$1/\sqrt{3}$	$1/\sqrt{3}$	$1/\sqrt{3}$	1
4	$-1/\sqrt{3}$	$1/\sqrt{3}$	$1/\sqrt{3}$	1
5	$-1/\sqrt{3}$	$-1/\sqrt{3}$	$-1/\sqrt{3}$	1
6	$1/\sqrt{3}$	$-1/\sqrt{3}$	$-1/\sqrt{3}$	1
7	$1/\sqrt{3}$	$1/\sqrt{3}$	$-1/\sqrt{3}$	1
8	$-1/\sqrt{3}$	$1/\sqrt{3}$	$-1/\sqrt{3}$	1

[a] $1/\sqrt{3} = 0.57735$.

Using Eqs. (12.3.7) by substituting u, v, and then w for f and using the definitions of the strains, we can express the strains in terms of natural coordinates (s, t, z') to obtain an equation similar to Eq. (11.3.14). In compact form, we can again express the strains in terms of the shape functions and global nodal coordinates similar to Eq. (11.3.15). The matrix $\underline{B}$, given by a form similar to Eq. (11.3.17), is now a function of $s, t,$ and z' and is of order 6×24.

The 24×24 stiffness matrix is now given by

$$[k] = \int_{-1}^{1} \int_{-1}^{1} \int_{-1}^{1} [B]^{T}[D][B]|\underline{J}|\, ds\, dt\, dz' \tag{12.3.8}$$

Again, it is best to evaluate $[k]$ by numerical integration (also see Section 11.4); that is, we evaluate (integrate) the eight-node hexahedral element stiffness matrix using a $2 \times 2 \times 2$ rule (or two-point rule). Actually, eight points defined in Table 12–1 are used to evaluate $\underline{k}$ as

$$\underline{k} = \sum_{i=1}^{8} \underline{B}^{T}(s_i, t_i, z_i')\underline{DB}(s_i, t_i, z_i')|\underline{J}(s_i, t_i, z_i')|\, W_i W_j W_k$$

where $W_i = W_j = W_k$ for the two-point rule.

Quadratic Hexahedral Element

For the quadratic hexahedral element shown in Figure 12–5, we have a total of 20 nodes with the inclusion of a total of 12 midside nodes. The development of the stiffness matrix follows the same steps we outlined before for the linear hexahedral element, where the shape functions now take on new forms. Again, letting $s_i, t_i, z_i' = \pm 1$, we have for the corner nodes ($i = 1, 2, \ldots, 8$),

$$N_i = \frac{(1 + ss_i)(1 + tt_i)(1 + z'z_i')}{8}(ss_i + tt_i + z'z_i' - 2) \tag{12.3.9}$$

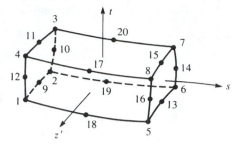

Figure 12–5 Quadratic hexahedral isoparametric element

For the midside nodes at $s_i = 0$, $t_i = \pm 1$, $z_i' = \pm 1$ $(i = 17, 18, 19, 20)$, we have

$$N_i = \frac{(1 - s^2)(1 + tt_i)(1 + z'z_i')}{4} \tag{12.3.10}$$

For the midside nodes at $s_i = \pm 1$, $t_i = 0$, $z_i' = \pm 1$ $(i = 10, 12, 14, 16)$, we have

$$N_i = \frac{(1 + ss_i)(1 - t^2)(1 + z'z_i')}{4} \tag{12.3.11}$$

Finally, for the midside nodes at $s_i = \pm 1$, $t_i = \pm 1$, $z_i' = 0$ $(i = 9, 11, 13, 15)$, we have

$$N_i = \frac{(1 + ss_i)(1 + tt_i)(1 - z'^2)}{4} \tag{12.3.12}$$

Note that the stiffness matrix for the quadratic solid element is of order 60×60 because it has 20 nodes and 3 degrees of freedom per node.

The stiffness matrix for this 20-node quadratic solid element can be evaluated using a $3 \times 3 \times 3$ rule (27 points). However, a special 14-point rule may be a better choice [9, 10].

Figures 1–7 and 12–6 show applications of the use of linear and quadratic (curved sides) solid elements to model three-dimensional solids.

It has been shown [3] that use of the simple eight-node hexahedral element yields better results than use of the tetrahedral element discussed in Section 12.1. In any case, the use of three-dimensional elements results in a large number of equations to be solved simultaneously. Large bandwidths are usually created, which cause large computer storage requirements. For instance, a model using a simple cube with, say, 20 by 20 by 20 nodes ($= 8000$ total nodes) for a region requires 8000 times 3 degrees of freedom per node ($= 24{,}000$) simultaneous equations. Furthermore, the bandwidth now involves an interconnection of some 20 by 20 ($= 400$) nodes or 400 times 3 degrees of freedom per node ($= 1200$) variables.

Finally, References [4–7] report on three-dimensional programs and analysis procedures using solid elements such as a family of subparametric curvilinear elements, linear tetrahedral elements, and 8-node linear and 20-node quadratic isoparametric elements.

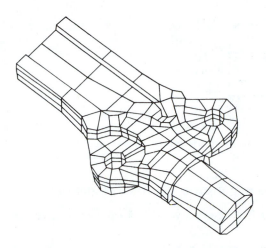

Figure 12–6 Finite element model of a forging using linear and quadratic solid elements

▲ 12.4 Algor Example Solutions of Three-Dimensional Stress Analysis ▲

To illustrate the use of the Algor program for three-dimensional stress analysis, we now solve four examples. To help the reader become familiar with the basic commands used for three-dimensional modeling, we begin by solving a simple cantilever beam problem in two ways. First we use basic commands such as "8 pt" and let the user control the discretization. Then the same problem is solved by using *Supersurf* [12] and *Hexagen* (Brick) with *Automeshing* doing the discretizing. We then describe how to model and solve two general problems: a hollow cast iron member with a side hole used in a machine tool structure and a foot pedal used in a machine. The basic tetrahedral element described in Section 12.2 and the eight-node hexahedral (brick) described in Section 12.3 are available in Algor.

Example 12.2

To become familiar with some basic commands used for three-dimensional modeling, along with the *Modify*, *Copy*, and *Join* commands [13], we now model and analyze the cantilever beam shown in Figure 12–7. The beam is 100 in. long, 4 in. high, and 1 in. thick. Let $E = 30 \times 10^6$ psi and $v = 0.30$. The load of 1000 lb is applied at the free end.

The beam will be divided into ten equal segments in the long direction, four equal segments through the depth, and one segment through the thickness. Commands used to discretize the beam are shown in the steps that follow. In other examples, the use of the *Automesh* command, along with the *Hexagen* command, will be illustrated to show how Algor automatically meshes a solid body.

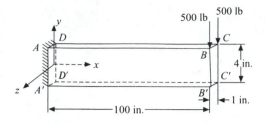

Figure 12–7 Cantilever beam to be modeled using 3-D elements

Step 1 Start Superdraw III to Construct the Basic Model

Double-click on the "Algor FEA" icon on your desktop computer screen.

Step 2 Create the Three-Dimensional Model

Click "Add".

Click "Rectangle". We will now create the back side rectangle $DD'C'C$.

(0, 2, -0.5)   Enter the first corner node of the rectangle. This is point D in Figure 12–7.

<Enter>

(100, -2, -0.5)   Enter the remaining corner node of the rectangle. This is point C'.

<Enter>

Click "View".

Click "Enclose". You now see the rectangle on the screen.

Select "View".

Select "Predefined Views".

Click "Isometric".

Click "View:Enclose".

Click "Modify". We will copy the rectangle 1 in. in front and then join the two rectangles $DD'C'C$ and $AA'B'B$.

Select "Copy" to copy the rectangle.

Activate the "Join all copies" command to join the two rectangles.

(0, 2, -0.5)   Enter the "Select start point for move operation". These are the coordinates of point D. This is done while in the "Copy Option" menu.

<Enter>

(0, 2, 0.5)   Enter the "Select end position of move". These are the coordinates of point A.

<Enter>   You see the three-dimensional model on the screen.

Click "FEA Mesh".

Select "Automatic Mesh".

Click on the lowest number under "Values" in the three-dimensional part of the menu and set "AB Divisions" at 10, "BC Divisions" at 1, and "AA′ Divisions" at 4.

<Enter>

Click "Done".

Click "FEA Mesh" again.

Select "Automatic Mesh" again.

Click on "8 Point 3-D Mesh". You see the prompt "Select Point A". This command allows you to select the first set of points (A, B, C, D) by right-clicking on each corner of the upper rectangle and going counterclockwise from A at the upper far left to $B, C,$ and D (Figure 12–7). You are tracing the upper top surface. Select the far-left outer corner of the lower rectangle as $A′$ and right-click; then proceed counterclockwise to $B′, C′,$ and $D′$. You are now tracing the lower bottom surface. (The ⟨F2⟩ function key can be used to undo any step that seems incorrect.)

Step 3   Eliminate Duplicate Lines

Click "Modify".

Select "Clean".

Click "Duplicate".

Click "Performing Cleaning".

You now see the message:

$$\text{"Kept 243 Deleted 0. done"}$$

Click "Done".

Step 4   Add in Boundary Conditions and Forces

Click "FEA Add".

Select "Stress And Vibration Analysis".

Click "Boundary Conditions". By default, all conditions are fixed.

Select "View".

Select "Zoom in". Then zoom in on the left edge to see the nodes clearly.

Click "Done" within the "Zoom in" menu.

Right-click on the ten nodes one at a time on the left edge to fix them from translation. An ampersand sign "(@)" appears next to each clicked node. You might try the "Box Apply" command or the "Polyline Apply" command to create a box around the left edge nodes to fix these nodes in one step.

Click "Done".

Click "View" and "Enclose". You should see the whole beam again.

Click "FEA Add" again.

Select "Stress and Vibration Analysis" again.

Click "Nodal Forces".

Select "Vector".

Select "Y dir". This is actually the default direction. You see "...$DY = 0.97$...".

Click "Done".

Click "Magnitude".

−500   Enter the force in pounds.

<Enter>

Right-click on the two outer right or top edge nodes at the free end. You may want to zoom in to get a better view of the right end first.

Click "Done".

Step 5   Add the Material Properties Using the "Model Data Control" Window

Click on the "Model Data" button. The "Linear Static Stress" screen appears.

Click on the box below "Element".

Click "Brick".

Click "OK".

Click on the box below "Data". A message appears: "Please enter the model name first".

Click "OK". Then enter the filename as

ex122   The filename is *ex122*.

Click "Save". This takes you to the "Units Definition" window.

Click "OK" as you want to use the default English system of units. This takes you to the "Element Definition:Linear Brick" window.

Click "OK" to use the defaults in the window.

Click on the box below "Material" and scroll to and select "ASTM A36" steel material.

Click "OK".

Step 6   Enter the Global Data

Click on the "Global" button in the "Model Data Control" window. You are now at the "Global" screen.

1   Enter a 1 under the "Pressure" column on the "Load Case Multipliers" menu.

Click on the "Output" tab.

Click on the "Displacement data" and "Stress data" boxes to have the displacements and stresses written to the *.L* and *.S* files, respectively.

Click "OK".

Step 7   Run the Analysis

Click on the "Analysis" button under the "FEA Model" column in the "Model Data Control" window.

Click "Analyze" in lower left corner of the menu.

Click "OK".

Click "Done".

Step 8   Review the Results

Click on the "Results" button in the "Model Data Control" window. This takes you to the Superview program.

To view the stress:

Click "Stress-di".

Click "Post". The "Von Mises" stress is plotted by default with the maximum Von Mises stress as 36,831 psi [Figure 12–8(a)].

Click "maX prin" for maximum principal stress.

<Esc>

Click "Do Dither" to see the color plot of maximum principal stress. The largest principal stress is 36,103 psi. The classical bending stress is 37,500 psi.

For line contours of stress go back to the "Stress-di" menu:

Click "Aux Post".

Click "L contour".

Select "Display", and the line contour of stresses will appear on the model.

The displaced model can be plotted in the usual manner.

Go back to the main menu of Superview. Then

Click "Displaced".

Click "Displ on".

Click "With undi"

Click "Calc scal". You should see the displaced beam along with the undisplaced beam.

Click "Get" to get a node.

Click on the free end node and see "DY=−2.111 in." This is the largest displacement. The analytical solution is −2.152 in. The finite element model is slightly stiffer than the actual beam. Figure 12–8(b) shows the deflected beam.

Click "Quit" to exit the program.   ■

Example 12.3

This example illustrates how to use the commands *Modify*, *Copy*, and *Join* [13] to extrude a rectangle into a three-dimensional beam. Then the **Supersurf** program [12] is used to create the discretized model of the beam from the surface model of the beam. Then either the brick element or the tetrahedral element is used in the finite element analysis. The actual beam is a cantilever that has a 1000-lb end load applied to it and has the same dimensions and properties as the beam in Example 12.2.

The Supersurf program [12] allows you to create and mesh three-dimensional surfaces. Using Supersurf you can model relatively complex surfaces using Non-Uniform Rational Basis Splines (NURBS). Supersurf supports the versatile general NURBS surface representation, which can exactly represent quadric surfaces including cylinders and generalized surfaces of revolution.

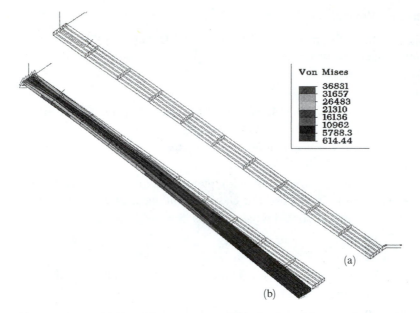

Von Mises

	36831
	31657
	26483
	21310
	16136
	10962
	5788.3
	614.44

(a)

(b)

Figure 12–8 (a) Von Mises stress and (b) displaced beam of Example 12.2

Supersurf complements Superdraw III in providing a complete basis for subsequent decoding. Many of the menus are similar to those in Superdraw. However, loads and boundary conditions cannot be applied in Supersurf. You must go back to Superdraw III to apply boundary conditions and forces.

Supersurf is basically a two-step process. First, create a NURBS surface using one of five basic surface creation commands, "2-obj/t," "3-patch," "Patch/t," "Multi," and "Gpatch." Here we will describe only "Gpatch," which is used in the following examples.

The "Gpatch" command creates surfaces between all of the three or four connected curve combinations of a wire frame model that it can find that can form surfaces. There are options available to control which surfaces are generated. In addition, there are built-in options that help prevent the creation of surfaces inside objects that are intended to be solid.

Once the surface has been generated in Supersurf, the generation of a surface mesh can be carried out. The commands are "Construct" and then "Quick msh." These commands take you to a menu where the number of divisions desired on the surface's edge is set using either "Edge" menu submenu commands or the "Number-u" (number of divisions for undefined edges) command. Setting "Number-u" to 5, for example, creates five divisions in the mesh. This is illustrated in Examples 12.3–12.5. Then, selecting the "Test" command will show you the mesh that has been generated. Then, selecting the "File msh" command will give the file the name previously created in Superdraw III. Step 5 of Example 12.3 illustrates the above commands.

Step 1   Start Superdraw III

Click on the "Algor FEA" icon on your desktop computer screen.

Step 2   Create the Three-Dimensional Model

Click "Add".
Click "Rectangle".
(0, 2, -0.5)   Enter the first corner node (point D in Figure 12–7).
<Enter>
(100, -2, -0.5)   Enter the other corner node (point C' in Figure 12–7).
<Enter>
Click "Done".
Click "View".
Click "Enclose".
Click "View".
Click "Predefined views".
Select "Isometric" view.
Click "View"
Click "Enclose".
Click "Done".

Step 3   Eliminate Duplicate Lines

Follow the usual steps in eliminating duplicate lines.

Step 4   Copy One Rectangle and Join It with the Other

Select "Modify".
Select "Copy".
Select "Number". This is used to set the number of copies.
1   Type 1 as one copy will be made. This is also the default value.
<Enter>
Select "Select" and select all four sides of the model.
Click "All" and four red-highlighted boxes, one on each line, appear.
Click the "Join All Copies" option so that a check mark appears in the box.
(0, 2, -0.5)   Enter the "Select starting point for move operation" point.
<Enter>
(0, 2, 0.5)   Enter the "Select end position of move" point.
<Enter>
Click "Done".

Step 5   Transfer to Supersurf

Select "FEA Mesh".
Click "Supersurf".
Ex123   Enter the name of the file to be meshed.

Click "Save". You will see a message "Model does not contain a unit system. Please choose a unit system first".

Click "OK".

Click "OK" to use the default English units.

Select "Construct" from the main menu of Supersurf.

Select "Gpatch" from the "Construct" menu.

Select "Select".

Select "All". This command puts IGES surfaces on the solid beam.

<Esc>   Escape back to the "Gpatch" menu.

Select "Gpatch" again.

<Esc>   Escape back to the "Construct" menu.

Click "Quick msh" to create a surface mesh from the IGES surface.

Click "Number-u" to set the number of elements for each edge of the surface.

5   Enter the new division number, 5, for the undefined edges. This puts a 5×5 surface mesh on each face of the beam.

<Enter>

Click "Test" within the "Quick msh" menu to see a temporary test view of the model.

Select "File msh" to create a Superdraw III.esd file of the mesh. You see the drawing divided into 5×5 parts as shown in the accompanying figure.

<Esc>

<Esc>   Return to the main menu of Supersurf.

Click "Done".

Step 6 Solid Mesh Generation

Click "FEA Mesh".

Click "Solid Mesh Generation".

Click "Generate" and use the defaults "Standard" for "Type" and "1" for "Mesh Group".

The mesh statistics appear with 198 nodes and 350 elements.

Click "OK".

Click "Done".

Step 7 Add Boundary Conditions and Forces

Select "FEA Add".

Click "Stress and Vibration Analysis".

Click "Boundary Conditions".

Select "Box Apply" and carefully box the left end of the model.

Depending on the model, 30 nodes are indicated to be fixed.

Click "Done".

Click "View".

Click "Enclose".

Click "View".

Click "Zoom:In" and zoom in on the right end of the model to clearly see the nodes to apply the forces to.

Click "FEA Add".

Click "Stress and Vibration Analysis".

Click "Nodal Forces".

Select "Vector".

Select "Y Direction" (the default direction is y).

Click "Done".

Select "Magnitude".

–500   Enter the force value of 500 lb.

<Enter>

Right-click on the two outer nodes at the free end of the beam. You should see the arrows and "−500;L1;lbf" next to the arrows.

Click "Done".

Click "View".

Click "Enclose".

Step 8 Add Material Properties in the "Model Data Control" Window

Click "Tools:Model Data Control".

Click on the box below "Elements".

Select "Brick".
Click "OK".
Click on the box below "Data".
Click "OK".
Click on the box below "Material".
Select "ASTM A36" for the material.
Click "OK".

Step 9 Enter the Global Data

Click "Global" in the "Model Data Control" window.

1   Enter a 1 under the "Pressure" column of the "Load Case Multipliers" menu.

Select the "Output" tab.
Click on the "Displacement data" and "Stress data" boxes to get the displacements in the .L file and the stresses in the .S file.
Click "OK".

Step 10 Run the Analysis

Click "Analysis". Make sure the "Linear Stress Analysis" processor is active.
Click "Analyze".
Click "OK".
Click "Done".

Step 11 Review the Results

Click "Results".
To view the stresses:
Select "Stress-di".
Click "Post". The Von Mises stress is posted by default.
Click "Max prin". The maximum principal stress appears as 36,155 psi.
Click "Disp vec" to see a plot of the displacement vector.
Click "⟨Esc⟩" three times to exit.
Click "Done".
Select "File:Exit" to exit the program.   ■

You should view a maximum principal stress of 36,155 psi and a maximum Von Mises stress of 33,474 psi. The maximum displacement is $DY = -1.981$ in. at the free end node 216. Recall that the classical solution yields $DY = -2.152$ in. Further refinement of the mesh (keeping in mind the concept of aspect ratio) in the Supersurf section (step 5) would increase the accuracy.

Example 12.4

We now solve the problem of a foot pedal subjected to a point load (Figure 12–9). The problem is a three-dimensional structure. The keystrokes needed to create the model and analyze it follow. Let $E = 30 \times 10^6$ psi and $v = 0.30$.

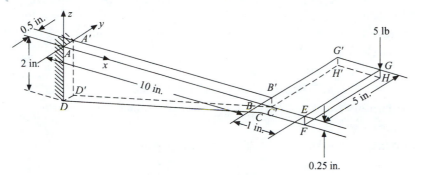

Figure 12–9   Three-dimensional foot pedal

Step 1   Start Superdraw III

Select the "Start" button of Windows NT/95/98 and then proceed to "Programs: Algor Software:Algor FEA".

Then click on "Algor FEA" to start the program.

Step 2   Create the Three-Dimensional Model Side *ABCD*

Select "Add:Line".

0<Tab>0<Tab>0	Enter the coordinates of point *A* in Figure 12–9. (Begin to create the 10-in.-long section clamped to the wall.) The following points refer to Figure 12–9.
<Enter>	
10<Tab>0<Tab>0	Enter the coordinates of point *B*.
<Enter>	
10<Tab>0<Tab>-0.25	Enter the coordinates of point *C*.
<Enter>	
0<Tab>0<Tab>-2	Enter the coordinates of point *D*.
<Enter>	
0<Tab>0<Tab>0	Enter the coordinates of point *A* again to create the front side of the section clamped to the wall.
<Enter>	You have finished creating side *ABCD*.

Select "Done".
Select "View:Enclose".
Select "View:Predefined Views".
Select "Isometric". You see an isometric view of the quadrilateral *ABCD*.

Step 3 Copy Side *ABCD* and Join It with the Back Side and Create
Rest of Model

Select "Modify:Copy".

Activate the "Join all copies" command.

0<Tab>0<Tab>0   Enter the start point (point *A*).

<Enter>

0<Tab>0.5<Tab>0   Enter the endpoint (point *A'*)

<Enter>

Select "View:Enclose".

Select "Add:Line". Begin to create the side *BEFC*.

10<Tab>0<Tab>0   Enter the coordinates of point *B*.

<Enter>

11<Tab>0<Tab>0   Enter the coordinates of point *E*.

<Enter>

11<Tab>0<Tab>-0.25   Enter the coordinates of point *F*.

<Enter>

10<Tab>0<Tab>-0.25   Enter the coordinates of point *C*.

<Enter>   You have finished creating side *BEFC*.

Select "Done".

Select "Add:Line". Create line *B'E*.

10<Tab>0.5<Tab>0   Enter the coordinates of point *B'*

<Enter>

11<Tab>0<Tab>0   Enter the coordinates of point *E*.

<Enter>

Select "Done".

Select "Add:Line". Create line *C'F*.

10<Tab>0.5<Tab>-0.25   Enter the coordinates of point *C'*.

<Enter>

11<Tab>0<Tab>-0.25   Enter the coordinates of point *F*.

<Enter>

Select "Done".

Select "View:Enclose".

Select "Add:Line". The next set of steps finishes the wire frame for the model by
adding an extension off the triangular section of the model *EFGHB'C'G'H'*.

10<Tab>0.5<Tab>0   Enter the coordinates of point *B'*.

<Enter>

10<Tab>5<Tab>0   Enter the coordinates of point *G'*.

<Enter>

11<Tab>5<Tab>0   Enter the coordinates of point *G*.

<Enter>

`11<Tab>0<Tab>0`	Enter the coordinates of point E.
`<Enter>`	

Select "Done".

Select "Add:Line". Create the lines from C' to H' to H to F.

`10<Tab>0.5<Tab>-0.25`	Enter the coordinates of point C'.
`<Enter>`	
`10<Tab>5<Tab>-0.25`	Enter the coordinates of point H'.
`<Enter>`	
`11<Tab>5<Tab>-0.25`	Enter the coordinates of point H.
`<Enter>`	
`11<Tab>0<Tab>-0.25`	Enter the coordinates of point F.
`<Enter>`	

Select "Done".

Select "Add:Line". Create line $G'H'$.

`10<Tab>5<Tab>0`	Enter the coordinates of point G'.
`<Enter>`	
`10<Tab>5<Tab>-0.25`	Enter the coordinates of point H'.
`<Enter>`	

Select "Done".

Select "Add:Line". Create line GH.

`11<Tab>5<Tab>0`	Enter the coordinates of point G.
`<Enter>`	
`11<Tab>5<Tab>-0.25`	Enter the coordinates of point H.

Select "Done".

Select "View:Enclose".

Step 4   Eliminate Duplicate Lines

Select "Modify:Clean".

Select "Duplicate". You see the following message appear in the lower part of the screen: "3 Kept 0 Deleted. done."

Select "Done".

Step 5   Transfer to Supersurf

Before you transfer to the Supersurf program, be sure to save the model with a short file name. This must be done for the Supersurf program to be able to transfer the file.

Select "FEA Mesh:Supersurf".

Select "Construct:Gpatch:Select".

Select "All".

`<Esc>`

Select "Gpatch". You see the three sections divided into two parts each.

`<Esc>`

Select "Quick msh".

Select "Number-u".

6   Enter the number 6 for a 6 × 6 mesh.

`<Enter>`

Select "Test".

Select "File msh".

`<Esc>`

`<Esc>`

Select "Done". You now return to Superdraw.

Step 6   Solid Mesh Generation

Select "FEA Mesh:Solid Mesh Generation".

Select "Generate". Leave default settings for "Type" and "Mesh Group".

Select "Done". A "Mesh Statistics" box appears showing 889 nodes and 648 total elements.

Select "Done". Exit Supersurf.

Step 7   Add Boundary Conditions and Force

Select "FEA Add:Stress and Vibration Analysis:Boundary Conditions".

Select "Box Apply". Carefully box the left edge nodes of the model.

Select "Done".

Select "FEA Add:Stress and Vibration Analysis:Nodal Forces".

Select "Vector:Z Direction" The direction of the vector is along the z axis.

Select "Done".

Select "View:Zoom In". Zoom in on the far right end of the foot pedal near point G to see node G clearly to apply the force properly.

Select "Done".

Select "Magnitude".

−5   Enter the force in pounds.

`<Enter>`

Right-click on node G. The force is added at the far edge front corner of the pedal.

Select "Done".

Select "View:Enclose". The model is seen enclosed on the screen.

Step 8   Add Material Properties Using the "Model Data Control" Window

Select the "Model Data" button. The "Linear Static Stress" screen appears.

Select the box below "Element".

Select "Brick".

Select "OK".

Select the box below "Data".

Select "OK". You want to use the defaults.

Select the box under "Material".

Select "[Customer Defined]"

Select "Edit Properties".

30e6   Enter the modulus of elasticity.

0.3   Enter Poisson's ratio.

Select "OK:OK".

Step 9 Enter the Global Data

Select the "Global" button in the "Model Data Control" window. You are now at the "Global" screen.

1   Enter a 1 in "Load Case Multipliers" under "Pressure".

Select the "Output" tab.

Click on the "Displacement data" and "Stress data" boxes to have the displacements and stresses written to the .L and .S files, respectively.

Select "OK".

Step 10 Run the Analysis

Select the "Analysis" button under the "FEA Model" column in the "Model Data Control" window.

Select "Analyze" in the lower left corner of the menu.

Select "OK". A box appears saying "The analysis of E:\FEA*Ex12-4* finished!" Also enter the model name.

Ex12-4

Select "Done".

Step 11 Review the Results

Select the "Results" button from the "Model Data Control" window. This takes you to Superview.

To view stress:

Select "Stress-di:Post". The "Von Mises" stress is plotted by default with the maximum Von Mises stress as 4023.4 psi at the interior corner of the elbow.

Select "maX prin". The maximum principal stress is 4111 psi at the inside corner of the elbow.

Select "Precision". This shows the validity of your model. The precision has a maximum value of 0.28203 at the interior corner. This relatively high value of the precision indicates that the mesh should probably be refined near the elbow.

Select "Disp vec". This shows the maximum displacement of 0.01239 in. at the free endpoint H in Figure 12–9. (Click on a node at point F, and a vertical displacement of −0.001 in. appears.

```
<Esc>
<Esc>
<Esc>
```
Select "Done".
Select "File:Exit" to exit the program. ∎

Example 12.5

As a final example, we solve the problem of a cast iron hollow member subjected to end loading (Figure 12–10). Let $E = 165$ GPa and $v = 0.25$.

Step 1 Start the Algor Program

Select the "Start" button of Windows NT/95/98.
Select "Programs:Algor Software:Algor FEA".
Click "Algor FEA" to start the program.
Or double-click on the "Algor FEA" icon on your desktop computer screen.

The Superdraw III program appears. This is the main graphics user interface. All your work will be done from this program. You will construct, analyze, and review the results from Superdraw III. Each menu option and a short example are available in the Superdraw III Reference Division. This can be accessed through "Docutech", the on-line "Technical User Documentation".

Step 2 Create the Model

Click "Modify:Transform:XY-YZ". This changes the drawing plane from x-y to y-z.
Click "Add:Rectangle".

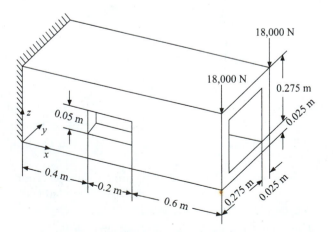

Figure 12–10 Cast iron hollow member

Enter the diagonal corners of the outer rectangle.

<0, 0, 0>

<0, 0.3, 0.3>

Click "Add:Rectangle".

Enter the diagonal corners of the inner rectangle.

<0, 0.025, 0.025>

<0, 0.0275, 0.0275>

Click "View".

Click "Enclose".

Select "View:XY rig".

The base section of the model has now been created.

Click "Add:Line".

Enter the lower left and right corner points.

<0.4, 0, 0.125>

<0.4, 0, 0.175>

Enter the upper right and left corner points.

<0.6, 0, 0.175>

<0.6, 0, 0.125>

Enter the start point again.

<0.4, 0, 0.125>

Click "View:Enclose:Predefined Views:Isometric".

Click "View:Enclose".

Click "Modify:Clean:Duplicate:Perform Cleaning".

Click "Done".

Click "Select:Point:Toggle Mode". Click on all eight lines that make up the base.

Click "Done".

Click "Modify:Copy". Make sure "Join all copies" is selected.

Enter the "Select move start point" and "Select move endpoint".

<0, 0, 0>

<1.2, 0, 0>

Click "View:Enclose".

Click "Select" and make sure the toggle option is still highlighted.

Left-click on all four lines that make up the base of the window. These lines should be the only ones with red-highlighted selection boxes on them.

Enter the "Select move start point" and "Select move endpoint".

<0, 0, 0>

<0, 0.25, 0>

Click "View:Enclose". The model should look like the figure on page 626.

In order for the program to create surfaces, each model surface must be bounded by only three or four entities, such as arcs or lines. Inspection of the model shows that the

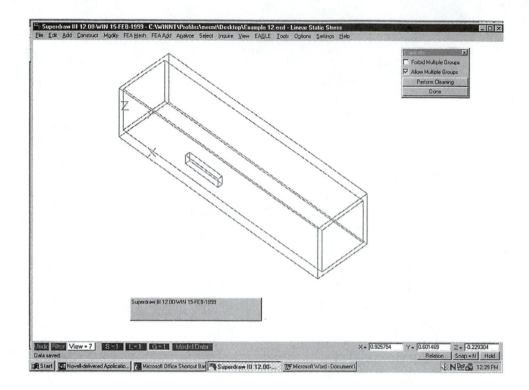

front surface and all the end surfaces are bounded by more than four lines. These surfaces must be manually divided into acceptable sections. To do this:

Click "Add:Line:Single".

Right-click on each of the corners until your figure looks like the one shown on page 627.

Step 3   Mesh the Model

Click "FEA Mesh:Automatic Mesh".

Click on the lowest number under "Values" in the three-dimensional part of the menu and use the default divisions.

Click "FEA Mesh:Automatic Mesh:8 Point 3-D Mesh".

Click on the eight corners of each of the seven sections that make up the model. This is accomplished by selecting the four corners of a planar section and then selecting the accompanying section that defines the thickness. The "Zoom" and "Predefined view" commands found on the "View" menu will aid in this process.

Click "Done".

Step 4   Eliminate Duplicate Lines

Click "Modify:Clean:Duplicate".

Click "Perform Cleaning".

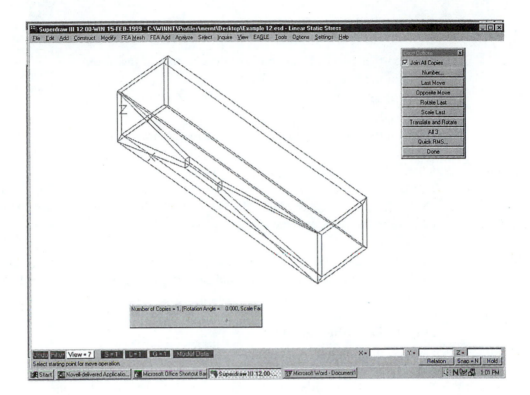

You should receive the message: "1840 Kept 0 deleted. done"
Click "Done".

Step 5　Add in Boundary Conditions and Forces

Click "FEA Add:Stress and Vibration Analysis:Boundary Conditions".

By default, these boundary conditions are fixed in all directions.

Highlight the nodes of the left end of the model by right-clicking on each of them. This procedure may be accomplished by switching views and using the "Box Apply" command to fix all the nodes in one step.

Click "Done".

Click "View:Enclose", and the entire beam should be visible again.

Click "FEA Add:Stress and Vibration Analysis:Nodal Forces" to add the forces to the right end.

Click "Vector:Z dir". This orientates the force along the *z* axis of the model.

Click "Done".

Click "Magnitude" and enter a value of −18,000.

−18000

<Enter>

Right-click on the outer two upper right-hand corners of the model to apply the force to them. You may want to zoom in to make this easier to do.

Click "Done".

Step 6 Add the Material Properties Using the "Model Data Control" Window

Click on the "Model Data" button at the bottom of the screen.

Click on the box below "Element".

Click "Brick".

Click "OK".

Click on the box below "Data". A message appears instructing you to enter a model name.

Click "OK" and enter an appropriate filename.

Ex125

Click "Save". This takes you to the "Units Definition" window. Select "Metric mks (SI)" from the pull-down menu.

Click "OK".

Click on the box below "Material" and scroll to and select "Customer Defined".

Click on the "Edit Properties" box and enter the values of $E = 165e9$ and $v = 0.25$.

Click "OK".

Step 7 Enter the Global Data

Click on the "Global" button in the "Model Data Control" window. You are now at the "Global" screen.

Enter a 1 under the "Pressure" column on the "Load Case Multipliers" menu.

1

Click on the "Output" tab and click on the "Displacement data" and "Stress data" boxes.

Click "OK".

Step 8 Run the Analysis

Click on the "Analysis" button in the "Model Data Control" window.

Click "Analyze" in the lower left corner of the menu.

Click "OK".

Click "Done".

Step 9 Review the Results

Click on the "Results" button in the "Model Data Control" window. This takes you to Superview.

To view stresses:

Click "Stess-di:Post:Max prin". This plots the maximum principle stress.

Click "⟨Esc⟩".

Click "Do Dither" to see a color plot of the maximum principle stress.

To view the displacements:

Click "〈Esc〉" to get back to the main menu.

Click "Displaced:Displ on:With undi:Calc scal" and a displaced and an undisplaced model appear on the screen.

To determine the displacements:

Click "Get" and right-click near the desired nodes. The displacement is displayed at the bottom of the screen.

When you are finished exploring the model,

Click "Quit" to exit the program. ∎

The maximum principal stress is 19 MPa and the maximum Von Mises stress is 23.2 MPa. Both occur at the top near the fixed wall support. Using the beam bending stress formula, the classical bending stress is 18.54 MPa. The free-end vertical displacement from the finite element result is −3.000e-4 m, while the classical beam deflection equation yields −3.595e-4 m. If a finer mesh were used, the results would approach the classical values.

▲ References

[1] Martin, H. C., "Plane Elasticity Problems and the Direct Stiffness Method." *The Trend in Engineering*, Vol. 13, pp. 5–19, Jan. 1961.

[2] Gallagher, R. H., Padlog, J., and Bijlaard, P. P., "Stress Analysis of Heated Complex Shapes," *Journal of the American Rocket Society*, pp. 700–707, May 1962.

[3] Melosh, R. J., "Structural Analysis of Solids," *Journal of the Structural Division*, American Society of Civil Engineers, pp. 205–223, Aug. 1963.

[4] Chacour, S., "DANUTA, a Three-Dimensional Finite Element Program Used in the Analysis of Turbo-Machinery," Transactions of the American Society of Mechanical Engineers, *Journal of Basic Engineering*, March 1972.

[5] Rashid, Y. R., "Three-Dimensional Analysis of Elastic Solids-I: Analysis Procedure," *International Journal of Solids and Structures*, Vol. 5, pp. 1311–1331, 1969.

[6] Rashid, Y. R., "Three-Dimensional Analysis of Elastic Solids-II: The Computational Problem," *International Journal of Solids and Structures*, Vol. 6, pp. 195–207, 1970.

[7] *Three-Dimensional Continuum Computer Programs for Structural Analysis*, Cruse, T. A., and Griffin, D. S., eds., American Society of Mechanical Engineers, 1972.

[8] Zienkiewicz, O. C., *The Finite Element Method*, 3rd ed., McGraw-Hill, London, 1977.

[9] Irons, B. M., "Quadrature Rules for Brick Based Finite Elements," *International Journal for Numerical Methods in Engineering*, Vol. 3, No. 2, pp. 293–294, 1971.

[10] Hellen, T. K., "Effective Quadrature Rules for Quadratic Solid Isoparametric Finite Elements," *International Journal for Numerical Methods in Engineering*, Vol. 4, No. 4, pp. 597–599, 1972.

[11] Linear Stress and Dynamics Reference Division, Docutech On-line Documentation, Algor, Inc., Pittsburgh, PA.

[12] Supersurf Reference Division, Docutech On-line Documentation, Algor, Inc., Pittsburgh, PA.

[13] Superdraw Reference Division, Docutech On-line Documentation, Algor, Inc., Pittsburgh, PA.

▲ Problems

12.1 Evaluate the matrix $\underline{B}$ for the tetrahedral solid element shown in Figure P12–1.

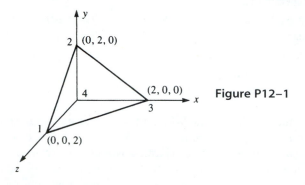

Figure P12–1

12.2 For the element shown in Figure P12–1, assume the nodal displacements have been determined to be

$$u_1 = 0.005 \text{ in.} \qquad v_1 = 0.0 \qquad w_1 = 0.0$$

$$u_2 = 0.001 \text{ in.} \qquad v_2 = 0.0 \qquad w_2 = 0.001 \text{ in.}$$

$$u_3 = 0.005 \text{ in.} \qquad v_3 = 0.0 \qquad w_3 = 0.0$$

$$u_4 = -0.001 \text{ in.} \qquad v_4 = 0.0 \qquad w_4 = 0.005 \text{ in.}$$

Determine the strains and then the stresses in the element. Let $E = 30 \times 10^6$ psi and $v = 0.3$.

12.3 Show that for constant body force Z_b acting on an element ($X_b = 0$ and $Y_b = 0$),

$$\{f_{bi}\} = \frac{V}{4} \left\{ \begin{array}{c} 0 \\ 0 \\ Z_b \end{array} \right\}$$

where $\{f_{bi}\}$ represents the body forces at node i of the element with volume V.

12.4 Evaluate the $\underline{B}$ matrix for the tetrahedral solid element shown in Figure P12–4. The coordinates are in units of millimeters.

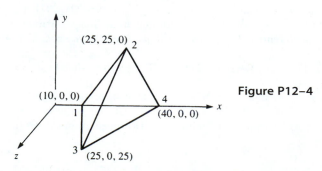

Figure P12–4

12.5 For the element shown in Figure P12–4, assume the nodal displacements have been determined to be

$$u_1 = 0.0 \qquad v_1 = 0.0 \qquad w_1 = 0.0$$

$$u_2 = 0.01 \text{ mm} \qquad v_2 = 0.02 \text{ mm} \qquad w_2 = 0.01 \text{ mm}$$

$$u_3 = 0.02 \text{ mm} \qquad v_3 = 0.01 \text{ mm} \qquad w_3 = 0.005 \text{ mm}$$

$$u_4 = 0.0 \qquad v_4 = 0.01 \text{ mm} \qquad w_4 = 0.01 \text{ mm}$$

Determine the strains and then the stresses in the element. Let $E = 210$ GPa and $v = 0.3$.

12.6 Express the explicit shape functions N_2 through N_8, similar to N_1 given by Eq. (12.3.4), for the linear hexahedral element shown in Figure 12–4.

12.7 Express the explicit shape functions for the corner nodes of the quadratic hexahedral element shown in Figure 12–5.

12.8 Write a computer program to evaluate $\underline{k}$ of Eq. (12.3.8) using a $2 \times 2 \times 2$ Gaussian quadrature rule.

Solve the following problems using the Algor computer program.

 12.9 Determine the deflections at the four corners of the free end of the structural steel cantilever beam shown in Figure P12–9. Also determine the maximum principal stress.

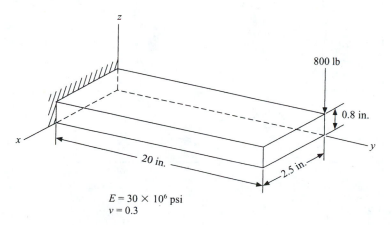

800 lb

0.8 in.

20 in.

2.5 in.

$E = 30 \times 10^6$ psi
$v = 0.3$

Figure P12–9

12.10 A portion of a structural steel brake pedal in a vehicle is modeled as shown in Figure P12–10. Determine the maximum deflection at the pedal under a line load of 5 lb/in. as shown.

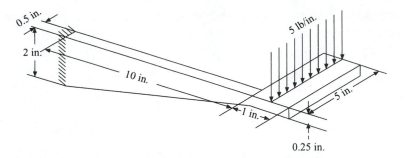

Figure P12–10

12.11 For the compressor flap valve shown in Figure P12–11, determine the maximum operating pressure such that the material yield stress is not exceeded with a factor of safety of two. The value is made of hardened 1020 steel with a modulus of elasticity of 30 million psi and a yield strength of 62,000 psi. The valve thickness is a uniform 0.018 in. The value clip ears support the valve at opposite diameters. The pressure load is applied uniformly around the annular region.

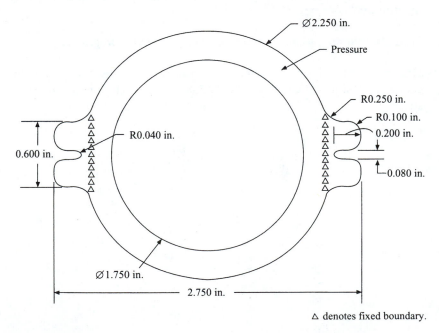

△ denotes fixed boundary.

Figure P12–11

12.12 An S-shaped block used in force measurement as shown in Figure P12.12 is to be designed for a pressure of 1000 psi applied uniformly to the top surface. Determine the uniform thickness of the block needed such that the sensor is compressed no more than 0.05 in. Also make sure that the maximum stress from the maximum distortion

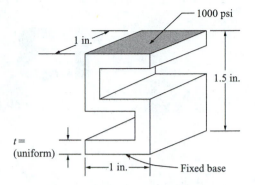

Figure P12–12   S-shaped block

energy failure theory is less than the yield strength of the material. Use a factor of safety of 1.5 on the stress only. The overall size of the block must fit in a 1.5-in.-high, 1-in.-wide, 1-in.-deep volume. The block should be made of steel.

12.13   A device is to be hydraulically loaded to resist an upward force $P = 6000$ lb as shown in Figure P12–13. Determine the thickness of the device such that the maximum deflection is 0.1 in. vertically and the maximum stress is less than the yield strength using a factor of safety of 2 (only on the stress). The device must fit in a space 7 in. high, 3 in. wide, and 2.3 in. deep. The top flange is bent vertically as shown, and the device is clamped to the floor. Use steel for the material.

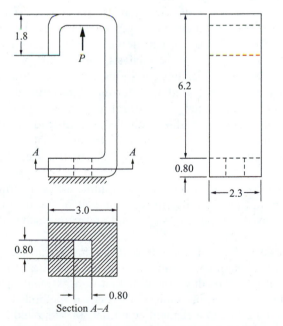

Figure P12–13   Hydraulically loaded device

13

Heat Transfer and Mass Transport

Introduction

In this chapter, we present the first use in this text of the finite element method for solution of nonstructural problems. We first consider the heat-transfer problem, although many similar problems, such as seepage through porous media, torsion of shafts, and magnetostatics [3], can also be treated by the same form of equations (but with different physical characteristics) as that for heat transfer.

Familiarity with the heat-transfer problem makes possible determination of the temperature distribution within a body. We can then determine the amount of heat moving into or out of the body and the thermal stresses.

We begin with a derivation of the basic differential equation for heat conduction in one dimension and then extend this derivation to the two-dimensional case. We will then review the units used for the physical quantities involved in heat transfer.

In preceding chapters dealing with stress analysis, we used the principle of minimum potential energy to derive the element equations, where an assumed displacement function within each element was used as a starting point in the derivation. We will now use a similar procedure for the nonstructural heat-transfer problem. We define an assumed temperature function within each element. Instead of minimizing a potential energy functional, we minimize a similar functional to obtain the element equations. Matrices analogous to the stiffness and force matrices of the structural problem result.

We will consider both one- and two-dimensional finite element formulations of the heat-transfer problem and provide illustrative examples of the determination of the temperature distribution along the length of a rod and within a two-dimensional body.

Next, we will consider the contribution of fluid mass transport. The one-dimensional mass-transport phenomenon is included in the basic heat-transfer differential equation. Because it is not readily apparent that a variational formulation is possible for this problem, we will apply Galerkin's residual method directly to the differential equation to obtain the finite element equations. (You should note that the mass transport stiffness matrix is asymmetric.) We will compare an analytical solution to the finite element solution for a heat exchanger design/analysis problem to show the excellent agreement.

Finally, we will describe how to use the Algor computer program for two-dimensional heat transfer and will present examples of its use.

▲ 13.1 Derivation of the Basic Differential Equation ▲

One-Dimensional Heat Conduction (Without Convection)

We now consider the derivation of the basic differential equation for the one-dimensional problem of heat conduction without convection. The purpose of this derivation is to present a physical insight into the heat-transfer phenomena, which must be understood so that the finite element formulation of the problem can be fully understood. (For additional information on heat transfer, consult texts such as References [1] and [2].) We begin with the control volume shown in Figure 13–1. By conservation of energy, we have

$$E_{in} + E_{generated} = \Delta U + E_{out} \tag{13.1.1}$$

or

$$q_x A \, dt + Q A \, dx \, dt = \Delta U + q_{x+dx} A \, dt \tag{13.1.2}$$

where

E_{in} is the energy entering the control volume, in units of joules (J) or kW · h or Btu.

ΔU is the change in stored energy, in units of kW · h (kWh) or Btu.

q_x is the heat conducted (heat flux) into the control volume at surface edge x, in units of kW/m^2 or Btu/(h-ft^2).

q_{x+dx} is the heat conducted out of the control volume at the surface edge $x + dx$.

t is time, in h or s (in U.S. customary units) or s (in SI units).

Q is the internal heat source (heat generated per unit time per unit volume is positive), in kW/m^3 or Btu/(h-ft^3) (a heat sink, heat drawn out of the volume, is negative).

A is the cross-sectional area perpendicular to heat flow q, in m^2 or ft^2.

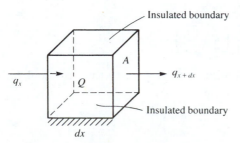

Figure 13–1 Control volume for one-dimensional heat conduction

By Fourier's law of heat conduction,

$$q_x = -K_{xx}\frac{dT}{dx} \tag{13.1.3}$$

where

K_{xx} is the thermal conductivity in the x direction, in kW/(m · °C) or Btu/(h-ft-°F).

T is the temperature, in °C or °F.

dT/dx is the temperature gradient, in °C/m or °F/ft.

Equation (13.1.3) states that the heat flux in the x direction is proportional to the gradient of temperature in the x direction. The minus sign in Eq. (13.1.3) implies that, by convention, heat flow is positive in the direction opposite the direction of temperature increase. Equation (13.1.3) is analogous to the one-dimensional stress/strain law for the stress analysis problem—that is, to $\sigma_x = E(du/dx)$. Similarly,

$$q_{x+dx} = -K_{xx}\frac{dT}{dx}\bigg|_{x+dx} \tag{13.1.4}$$

where the gradient in Eq. (13.1.4) is evaluated at $x + dx$. By Taylor series expansion, for any general function $f(x)$, we have

$$f_{x+dx} = f_x + \frac{df}{dx}dx + \frac{d^2f}{dx^2}\frac{dx^2}{2} + \cdots$$

Therefore, using a two-term Taylor series, Eq. (13.1.4) becomes

$$q_{x+dx} = -\left[K_{xx}\frac{dT}{dx} + \frac{d}{dx}\left(K_{xx}\frac{dT}{dx}\right)dx\right] \tag{13.1.5}$$

The change in stored energy can be expressed by

$$\Delta U = \text{specific heat} \times \text{mass} \times \text{change in temperature}$$

$$= c(\rho A\,dx)\,dT \tag{13.1.6}$$

where c is the specific heat in kW · h/(kg · °C) or Btu/(slug-°F), and ρ is the mass density in kg/m^3 or slug/ft^3. On substituting Eqs. (13.1.3), (13.1.5), and (13.1.6) into Eq. (13.1.2), dividing Eq. (13.1.2) by $A\,dx\,dt$, and simplifying, we have the one-dimensional heat conduction equation as

$$\frac{\partial}{\partial x}\left(K_{xx}\frac{\partial T}{\partial x}\right) + Q = \rho c\frac{\partial T}{\partial t} \tag{13.1.7}$$

For steady state, any differentiation with respect to time is equal to zero, so Eq. (13.1.7) becomes

$$\frac{d}{dx}\left(K_{xx}\frac{dT}{dx}\right) + Q = 0 \tag{13.1.8}$$

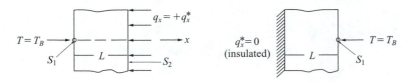

Figure 13-2 Examples of boundary conditions in one-dimensional heat conduction

For constant thermal conductivity and steady state, Eq. (13.1.7) becomes

$$K_{xx}\frac{d^2 T}{dx^2} + Q = 0 \tag{13.1.9}$$

The boundary conditions are of the form

$$T = T_B \qquad \text{on } S_1 \tag{13.1.10}$$

where T_B represents a known boundary temperature and S_1 is a surface where the temperature is known, and

$$q_x^* = -K_{xx}\frac{dT}{dx} = \text{constant} \qquad \text{on } S_2 \tag{13.1.11}$$

where S_2 is a surface where the prescribed heat flux q_x^* or temperature gradient is known. On an insulated boundary, $q_x^* = 0$. These different boundary conditions are shown in Figure 13-2, where by sign convention, positive q_x^* occurs when heat is flowing into the body, and negative q_x^* when heat is flowing out of the body.

Two-Dimensional Heat Conduction (Without Convection)

Consider the two-dimensional heat conduction problem in Figure 13-3. In a manner similar to the one-dimensional case, for steady-state conditions, we can show that for material properties coinciding with the global x and y directions,

$$\frac{\partial}{\partial x}\left(K_{xx}\frac{\partial T}{\partial x}\right) + \frac{\partial}{\partial y}\left(K_{yy}\frac{\partial T}{\partial y}\right) + Q = 0 \tag{13.1.12}$$

with boundary conditions

$$T = T_B \qquad \text{on } S_1 \tag{13.1.13}$$

$$q_n = q_n^* = K_{xx}\frac{\partial T}{\partial x}C_x + K_{yy}\frac{\partial T}{\partial y}C_y = \text{constant} \qquad \text{on } S_2 \tag{13.1.14}$$

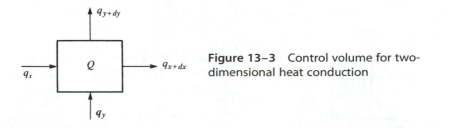

Figure 13-3 Control volume for two-dimensional heat conduction

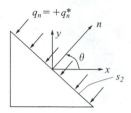

Figure 13–4 Unit vector normal to surface S_2

where C_x and C_y are the direction cosines of the unit vector $\boldsymbol{n}$ normal to the surface S_2 shown in Figure 13–4. Again, q_n^* is by sign convention, positive if heat is flowing into the edge of the body.

▲ 13.2 Heat Transfer with Convection

For a conducting solid in contact with a fluid, there will be a heat transfer taking place between the fluid and solid surface when a temperature difference occurs.

The fluid will be in motion either through external pumping action (**forced convection**) or through the buoyancy forces created within the fluid by the temperature differences within it (**natural** or **free convection**).

We will now consider the derivation of the basic differential equation for one-dimensional heat conduction with convection. Figure 13–5 shows the control volume used in the derivation. Again, by Eq. (13.1.1) for conservation of energy, we have

$$q_x A\, dt + QA\, dx\, dt = c(\rho A\, dx)\, dT + q_{x+dx} A\, dt + q_h P\, dx\, dt \qquad (13.2.1)$$

In Eq. (13.2.1), all terms have the same meaning as in Section 13.1, except the heat flow by convective heat transfer is given by [1, 2]

$$q_h = h(T - T_\infty) \qquad (13.2.2)$$

where

h is the heat-transfer or convection coefficient, in $kW/(m^2 \cdot {}^\circ C)$ or $Btu/(h\text{-}ft^2\text{-}{}^\circ F)$.

T is the temperature of the solid surface at the solid/fluid interface.

T_∞ is the temperature of the fluid (here the free-stream fluid temperature).

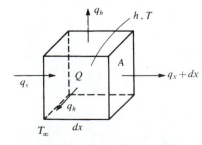

Figure 13–5 Control volume for one-dimensional heat conduction with convection

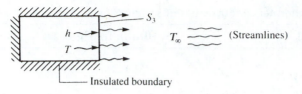

Figure 13–6 Model illustrating convective heat transfer (arrows on surface S_3 indicate heat transfer by convection)

P in Eq. (13.2.1) denotes the perimeter around the constant cross-sectional area A.

Again, using Eqs. (13.1.3)–(13.1.6) and (13.2.2) in Eq. (13.2.1), dividing by $A\,dx\,dt$, and simplifying, we obtain the equation for one-dimensional heat conduction with convection as

$$\frac{\partial}{\partial x}\left(K_{xx}\frac{\partial T}{\partial x}\right) + Q = \rho c\frac{\partial T}{\partial t} + \frac{hP}{A}(T - T_\infty) \qquad (13.2.3)$$

with possible boundary conditions on (1) temperature, given by Eq. (13.1.10), and/or (2) temperature gradient, given by Eq. (13.1.11), and/or (3) loss of heat by convection from the ends of the one-dimensional body, as shown in Figure 13–6. Equating the heat flow in the solid wall to the heat flow in the fluid at the solid/fluid interface, we have

$$-K_{xx}\frac{dT}{dx} = h(T - T_\infty) \qquad \text{on } S_3 \qquad (13.2.4)$$

as a boundary condition for the problem of heat conduction with convection.

13.3 Typical Units; Thermal Conductivities, K; and Heat-Transfer Coefficients, h

Table 13–1 lists some typical units used for the heat-transfer problem.

Table 13–1 Typical units for heat transfer

Variable	SI	U.S. Customary
Thermal conductivity, K	kW/(m · °C)	Btu/(h-ft-°F)
Temperature, T	°C or K	°F or °R
Internal heat source, Q	kW/m^3	Btu/(h-ft^3)
Heat flux, q	kW/m^2	Btu/(h-ft^2)
Convection coefficient, h	kW/(m^2 · °C)	Btu/(h-ft^2-°F)
Energy, E	kW · h	Btu
Specific heat, c	(kW · h)/(kg · °C)	Btu/(slug-°F)
Mass density, ρ	kg/m^3	slug/ft^3

Table 13–2 Typical thermal conductivities of some solids and liquids

Material	K [Btu/(h-ft-°F)]	K [W/(m · °C)]
Solids		
Aluminum, $0\,°C$ ($32\,°F$)	117	202
Steel (1% carbon), $0\,°C$	20	35
Fiberglass, $20\,°C$ ($68\,°F$)	0.020	0.035
Concrete, $0\,°C$	0.468–0.81	0.81–1.40
Earth, coarse gravelly, $20\,°C$	0.300	0.520
Wood, oak, radial direction, $20\,°C$	0.098	0.17
Fluids		
Engine oil, $20\,°C$	0.084	0.145
Dry air, atmospheric pressure, $20\,°C$	0.014	0.0243

Table 13–3 Approximate values of convection heat-transfer coefficients (from Reference [1])

Mode	h [Btu/(h-ft^2-°F)]	h [W/(m^2 · °C)]
Free convection, air	1–5	5–25
Forced convection, air	2–100	10–500
Forced convection, water	20–3,000	100–15,000
Boiling water	500–5,000	2,500–25,000
Condensation of water vapor	1,000–20,000	5,000–100,000

Table 13–2 lists some typical thermal conductivities of various solids and liquids. The thermal conductivity K, in Btu/(h-ft-°F) or W/(m · °C), measures the amount of heat energy (Btu or W · h) that will flow through a unit length (ft or m) of a given substance in a unit time (h) to raise the temperature one degree (°F or °C).

Table 13–3 lists approximate ranges of values of convection coefficients for various conditions of convection. The heat transfer coefficient h, in Btu/(h-ft^2-°F) or W/(m^2 · °C), measures the amount of heat energy (Btu or W · h) that will flow across a unit area (ft^2 or m^2) of a given substance in a unit time (h) to raise the temperature one degree (°F or °C).

Natural or **free convection** occurs when, for instance, a heated plate is exposed to ambient room air without an external source of motion. This movement of the air, experienced as a result of the density gradients near the plate, is called *natural* or *free convection*. **Forced convection** is experienced, for instance, in the case of a fan blowing air over a plate.

▲ **13.4 One-Dimensional Finite Element Formulation Using a Variational Method** ▲

The temperature distribution influences the amount of heat moving into or out of a body and also influences the stresses in a body. Thermal stresses occur in all bodies that experience a temperature gradient from some equilibrium state but are not free to expand in all directions. To evaluate thermal stresses, we need to know the temperature distribution in the body. The finite element method is a realistic method for predicting quantities such as temperature distribution and thermal stresses in a body. In this section, we formulate the one-dimensional heat-transfer equations using a variational method. Examples are included to illustrate the solution of this type of problem.

Step 1   Select Element Type

The basic element with nodes 1 and 2 is shown in Figure 13–7(a).

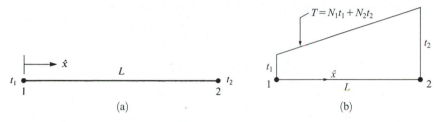

(a)   (b)

Figure 13–7   (a) Basic one-dimensional temperature element and (b) temperature variation along length of element

Step 2   Choose a Temperature Function

We choose the temperature function T [Figure 13–7(b)] within each element similar to the displacement function of Chapter 3, as

$$T(x) = N_1 t_1 + N_2 t_2 \tag{13.4.1}$$

where t_1 and t_2 are the nodal temperatures to be determined, and

$$N_1 = 1 - \frac{\hat{x}}{L} \qquad N_2 = \frac{\hat{x}}{L} \tag{13.4.2}$$

are again the same shape functions as used for the bar element. The $[N]$ matrix is then given by

$$[N] = \left[1 - \frac{\hat{x}}{L} \quad \frac{\hat{x}}{L} \right] \tag{13.4.3}$$

and the nodal temperature matrix is

$$\{t\} = \begin{Bmatrix} t_1 \\ t_2 \end{Bmatrix} \tag{13.4.4}$$

In matrix form, we express Eq. (13.4.1) as

$$\{T\} = [N]\{t\} \tag{13.4.5}$$

Step 3 Define the Temperature Gradient/Temperature and Heat Flux/Temperature Gradient Relationships

The temperature gradient matrix $\{g\}$, analogous to the strain matrix $\{\varepsilon\}$, is given by

$$\{g\} = \left\{\frac{dT}{d\hat{x}}\right\} = [B]\{t\} \tag{13.4.6}$$

where $[B]$ is obtained by substituting Eq. (13.4.1) for $T(\hat{x})$ into Eq. (13.4.6) and differentiating with respect to $\hat{x}$, that is,

$$[B] = \left[\frac{dN_1}{d\hat{x}} \quad \frac{dN_2}{d\hat{x}}\right]$$

Using Eqs. (13.4.2) in the definition for $[B]$, we have

$$[B] = \left[-\frac{1}{L} \quad \frac{1}{L}\right] \tag{13.4.7}$$

The heat flux/temperature gradient relationship is given by

$$q_x = -[D]\{g\} \tag{13.4.8}$$

where the material property matrix is now given by

$$[D] = [K_{xx}] \tag{13.4.9}$$

Step 4 Derive the Element Conduction Matrix and Equations

Equations (13.1.9)–(13.1.11) and (13.2.3) can be shown to be derivable (as shown, for instance, in References [4–6]) by the minimization of the following functional (analogous to the potential energy functional π_p):

$$\pi_h = U + \Omega_Q + \Omega_q + \Omega_h \tag{13.4.10}$$

where

$$U = \frac{1}{2}\iiint_V \left[K_{xx}\left(\frac{dT}{dx}\right)^2\right] dV$$

$$\Omega_Q = -\iiint_V QT\, dV \qquad \Omega_q = -\iint_{S_2} q^*T\, dS \qquad \Omega_h = \frac{1}{2}\iint_{S_3} h(T - T_\infty)^2\, dS \tag{13.4.11}$$

and where S_2 and S_3 are separate surface areas over which heat flow (flux) q^* (q^* is positive into the surface) and convection loss $h(T - T_\infty)$ are specified. We cannot specify q^* and h on the same surface because they cannot occur simultaneously on the same surface, as indicated by Eqs. (13.4.11).

Using Eqs. (13.4.5), (13.4.6), and (13.4.9) in Eq. (13.4.11) and then using Eq. (13.4.10), we can write π_h in matrix form as

$$\pi_h = \frac{1}{2}\iiint_V [\{g\}^T[D]\{g\}]\, dV - \iiint_V \{t\}^T[N]^T Q\, dV$$

$$- \iint_{S_2} \{t\}^T[N]^T q^*\, dS + \frac{1}{2}\iint_{S_3} h[(\{t\}^T[N]^T - T_\infty)^2]\, dS \tag{13.4.12}$$

On substituting Eq. (13.4.6) into Eq. (13.4.12) and using the fact that the nodal temperatures $\{t\}$ are independent of the general coordinates x and y and can therefore be taken outside the integrals, we have

$$\pi_h = \frac{1}{2}\{t\}^T \iiint_V [B]^T[D][B]\,dV\{t\} - \{t\}^T \iiint_V [N]^T Q\,dV$$

$$- \{t\}^T \iint_{S_2} [N]^T q^* \, dS + \frac{1}{2}\iint_{S_3} h[\{t\}^T[N]^T[N]\{t\}$$

$$- (\{t\}^T[N]^T + [N]\{t\})T_\infty + T_\infty^2]\,dS \tag{13.4.13}$$

In Eq. (13.4.13), the minimization is most easily accomplished by explicitly writing the surface integral S_3 with $\{t\}$ left inside the integral as shown. On minimizing Eq. (13.4.13) with respect to $\{t\}$, we obtain

$$\frac{\partial \pi_h}{\partial \{t\}} = \iiint_V [B]^T[D][B]\,dV\{t\} - \iiint_V [N]^T Q\,dV$$

$$- \iint_{S_2} [N]^T q^* \, dS + \iint_{S_3} h[N]^T[N]\,dS\{t\}$$

$$- \iint_{S_3} [N]^T h T_\infty \, dS = 0 \tag{13.4.14}$$

where the last term hT_∞^2 in Eq. (13.4.13) is a constant that drops out while minimizing π_h. Simplifying Eq. (13.4.14), we obtain

$$\left[\iiint_V [B]^T[D][B]\,dV + \iint_{S_3} h[N]^T[N]\,ds \right]\{t\} = \{f_Q\} + \{f_q\} + \{f_h\} \tag{13.4.15}$$

where the force matrices have been defined by

$$\{f_Q\} = \iiint_V [N]^T Q\,dV \qquad \{f_q\} = \iint_{S_2} [N]^T q^* \, dS$$

$$\tag{13.4.16}$$

$$\{f_h\} = \iint_{S_3} [N]^T h T_\infty \, dS$$

In Eq. (13.4.16), the first term $\{f_Q\}$ (heat source positive, sink negative) is of the same form as the body-force term, and the second term $\{f_q\}$ (heat flux, positive into the surface) and third term $\{f_h\}$ (heat transfer or convection) are similar to surface tractions (distributed loading) in the stress analysis problem. You can observe this fact by comparing Eq. (13.4.16) with Eq. (7.2.46). Because we are formulating element equations of the form $f = kt$, we have the element conduction matrix* for the heat-transfer problem given in Eq. (13.4.15) by

$$[k] = \iiint_V [B]^T [D][B] \, dV + \iint_{S_3} h[N]^T [N] \, dS \tag{13.4.17}$$

where the first and second integrals in Eq. (13.4.17) are the contributions of conduction and convection, respectively. Using Eq. (13.4.17) in Eq. (13.4.15), for each element, we have

$$\{f\} = [k]\{t\} \tag{13.4.18}$$

Using the first term of Eq. (13.4.17), along with Eqs. (13.4.7) and (13.4.9), the conduction part of the $[k]$ matrix for the one-dimensional element becomes

$$[k_c] = \iiint_V [B]^T [D][B] \, dV = \int_0^L \left\{ \begin{array}{c} -\dfrac{1}{L} \\ \dfrac{1}{L} \end{array} \right\} [K_{xx}] \left[-\dfrac{1}{L} \quad \dfrac{1}{L} \right] A \, dx$$

$$= \frac{AK_{xx}}{L^2} \int_0^L \begin{bmatrix} 1 & -1 \\ -1 & 1 \end{bmatrix} dx \tag{13.4.19}$$

or, finally,

$$[k_c] = \frac{AK_{xx}}{L} \begin{bmatrix} 1 & -1 \\ -1 & 1 \end{bmatrix} \tag{13.4.20}$$

The convection part of the $[k]$ matrix becomes

$$[k_h] = \iint_{S_3} h[N]^T [N] \, dS = hP \int_0^L \left\{ \begin{array}{c} 1 - \dfrac{\hat{x}}{L} \\ \dfrac{\hat{x}}{L} \end{array} \right\} \left[1 - \dfrac{\hat{x}}{L} \quad \dfrac{\hat{x}}{L} \right] d\hat{x}$$

or, on integrating,

$$[k_h] = \frac{hPL}{6} \begin{bmatrix} 2 & 1 \\ 1 & 2 \end{bmatrix} \tag{13.4.21}$$

where

$$dS = P \, d\hat{x}$$

---

* The element conduction matrix is often called the *stiffness matrix* because *stiffness matrix* is becoming a generally accepted term used to describe the matrix of known coefficients multiplied by the unknown degrees of freedom, such as temperatures, displacements, and so on.

and P is the perimeter of the element (assumed to be constant). Therefore, adding Eqs. (13.4.20) and (13.4.21), we find that the $[k]$ matrix is

$$[k] = \frac{AK_{xx}}{L}\begin{bmatrix} 1 & -1 \\ -1 & 1 \end{bmatrix} + \frac{hPL}{6}\begin{bmatrix} 2 & 1 \\ 1 & 2 \end{bmatrix} \qquad (13.4.22)$$

When h is zero on the boundary of an element, the second term on the right side of Eq. (13.4.22) (convection portion of $[k]$) is zero. This corresponds, for instance, to an insulated boundary.

The force matrix terms, on simplifying Eq. (13.4.16), are

$$\{f_Q\} = \iiint_V [N]^T Q \, dV = QA \int_0^L \begin{Bmatrix} 1 - \dfrac{\hat{x}}{L} \\ \dfrac{\hat{x}}{L} \end{Bmatrix} d\hat{x} = \frac{QAL}{2}\begin{Bmatrix} 1 \\ 1 \end{Bmatrix} \qquad (13.4.23)$$

and

$$\{f_q\} = \iint_{S_2} q^*[N]^T \, dS = q^*P \int_0^L \begin{Bmatrix} 1 - \dfrac{\hat{x}}{L} \\ \dfrac{\hat{x}}{L} \end{Bmatrix} d\hat{x} = \frac{q^*PL}{2}\begin{Bmatrix} 1 \\ 1 \end{Bmatrix} \qquad (13.4.24)$$

and

$$\{f_h\} = \iint_{S_3} hT_\infty[N]^T \, dS = \frac{hT_\infty PL}{2}\begin{Bmatrix} 1 \\ 1 \end{Bmatrix} \qquad (13.4.25)$$

Therefore, adding Eqs. (13.4.23)–(13.4.25), we obtain

$$\{f\} = \frac{QAL + q^*PL + hT_\infty PL}{2}\begin{Bmatrix} 1 \\ 1 \end{Bmatrix} \qquad (13.4.26)$$

Equation (13.4.26) indicates that one-half of the assumed uniform heat source Q goes to each node, one-half of the prescribed uniform heat flux q^* (positive q^* enters the body) goes to each node, and one-half of the convection from the perimeter surface hT_∞ goes to each node of an element.

Finally, we must consider the convection from the free end of an element. For simplicity's sake, we will assume convection occurs only from the right end of the element, as shown in Figure 13–8. The additional convection term contribution to the stiffness matrix is given by

$$[k_h]_{\text{end}} = \iint_{S_{\text{end}}} h[N]^T [N] \, dS \qquad (13.4.27)$$

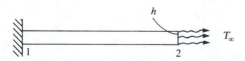

Figure 13–8 Convection force from the end of an element

Now $N_1 = 0$ and $N_2 = 1$ at the right end of the element. Substituting the N's into Eq. (13.4.27), we obtain

$$[k_h]_{\text{end}} = \iint_{S_{\text{end}}} h \begin{Bmatrix} 0 \\ 1 \end{Bmatrix} [0 \quad 1] \, dS = hA \begin{bmatrix} 0 & 0 \\ 0 & 1 \end{bmatrix} \tag{13.4.28}$$

The convection force from the free end of the element is obtained from the application of Eq. (13.4.25) with the shape functions now evaluated at the right end (where convection occurs) and with S_3 (the surface over which convection occurs) now equal to the cross-sectional area A of the rod. Hence,

$$\{f_h\}_{\text{end}} = hT_\infty A \begin{Bmatrix} N_1(\hat{x} = L) \\ N_2(\hat{x} = L) \end{Bmatrix} = hT_\infty A \begin{Bmatrix} 0 \\ 1 \end{Bmatrix} \tag{13.4.29}$$

represents the convective force from the right end of an element where $N_1(\hat{x} = L)$ represents N_1 *evaluated at* $\hat{x} = L$, and so on.

Step 5 Assemble the Element Equations to Obtain the Global Equations and Introduce Boundary Conditions

We obtain the global or total structure conduction matrix using the same procedure as for the structural problem (called the *direct stiffness method* as described in Section 2.4); that is,

$$[K] = \sum_{e=1}^{N} [k^{(e)}] \tag{13.4.30}$$

typically in units of kW/°C or Btu/(h-°F). The global force matrix is the sum of all element heat sources and is given by

$$\{F\} = \sum_{e=1}^{N} \{f^{(e)}\} \tag{13.4.31}$$

typically in units of kW or Btu/h. The global equations are then

$$\{F\} = [K]\{t\} \tag{13.4.32}$$

with the prescribed nodal temperature boundary conditions given by Eq. (13.1.13). Note that the boundary conditions on heat flux, Eq. (13.1.11), and convection, Eq. (13.2.4), are actually accounted for in the same manner as distributed loading was accounted for in the stress analysis problem; that is, they are included in the column of force matrices through a consistent approach (using the same shape functions used to derive $[k]$), as given by Eqs. (13.4.2).

The heat-transfer problem is now amenable to solution by the finite element method. The procedure used for solution is similar to that for the stress analysis problem. In Section 13.5, we will derive the specific equations used to solve the two-dimensional heat-transfer problem.

Step 6 Solve for the Nodal Temperatures

We now solve for the global nodal temperature, $\{t\}$, where the appropriate nodal temperature boundary conditions, Eq. (13.1.13), are specified.

Step 7 Solve for the Element Temperature Gradients and Heat Fluxes

Finally, we calculate the element temperature gradients from Eq. (13.4.6), and the heat fluxes, typically from Eq. (13.4.8).

To illustrate the use of the equations developed in this section, we will now solve some one-dimensional heat-transfer problems.

Example 13.1

Determine the temperature distribution along the length of the rod shown in Figure 13–9 with an insulated perimeter. The temperature at the left end is 100°F and the free-stream temperature is 10°F. Let $h = 10$ Btu/(h-ft²-°F) and $K_{xx} = 20$ Btu/(h-ft-°F). The value of h is typical for forced air convection and the value of K_{xx} is a typical conductivity for carbon steel (Tables 13–2 and 13–3).

The finite element discretization is shown in Figure 13–10. For simplicity's sake, we will use four elements, each 10 in. long. There will be convective heat loss only over the right end of the rod because we consider the left end to have a known temperature and the perimeter to be insulated. We calculate the stiffness matrices for each element as follows:

$$\frac{AK_{xx}}{L} = \frac{\pi(1 \text{ in.})^2[20 \text{ Btu}/(\text{h-ft-°F})](1 \text{ ft}^2)}{\left(\dfrac{10 \text{ in.}}{12 \text{ in./ft}}\right)(144 \text{ in}^2)}$$

$$= 0.5236 \text{ Btu}/(\text{h-°F}) \tag{13.4.33}$$

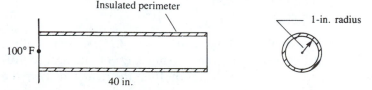

Figure 13–9 One-dimensional rod subjected to temperature variation

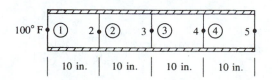

Figure 13–10 Finite element discretized rod

$$\frac{hPL}{6} = \frac{[10 \text{ Btu}/(\text{h-ft}^2\text{-}°\text{F})](2\pi)}{6}\left(\frac{1 \text{ in.}}{12 \text{ in.}/\text{ft}}\right)\left(\frac{10 \text{ in.}}{12 \text{ in.}/\text{ft}}\right)$$

$$= 0.7272 \text{ Btu}/(\text{h-}°\text{F})$$

$$hT_\infty PL = [10 \text{ Btu}/(\text{h-ft}^2\text{-}°\text{F})](10\,°\text{F})(2\pi)\left(\frac{1 \text{ in.}}{12 \text{ in.}/\text{ft}}\right)\left(\frac{10 \text{ in.}}{12 \text{ in.}/\text{ft}}\right)$$

$$= 43.63 \text{ Btu}/\text{h}$$

In general, from Eqs. (13.4.22) and (13.4.27), we have

$$[k] = \frac{AK_{xx}}{L}\begin{bmatrix} 1 & -1 \\ -1 & 1 \end{bmatrix} + \frac{hPL}{6}\begin{bmatrix} 2 & 1 \\ 1 & 2 \end{bmatrix} + \iint_{S_{\text{end}}} h[N]^T[N]\,dS \tag{13.4.34}$$

Substituting Eqs. (13.4.33) into Eq. (13.4.34) for element 1, we have

$$[k^{(1)}] = 0.5236\begin{bmatrix} 1 & -1 \\ -1 & 1 \end{bmatrix} \text{ Btu}/(\text{h-}°\text{F}) \tag{13.4.35}$$

where the second and third terms on the right side of Eq. (13.4.34) are zero because there are no convection terms associated with element 1. Similarly, for elements 2 and 3, we have

$$[k^{(2)}] = [k^{(3)}] = [k^{(1)}] \tag{13.4.36}$$

However, element 4 has an additional (convection) term owing to heat loss from the flat surface at its right end. Hence, using Eq. (13.4.28), we have

$$[k^{(4)}] = [k^{(1)}] + hA\begin{bmatrix} 0 & 0 \\ 0 & 1 \end{bmatrix}$$

$$= 0.5236\begin{bmatrix} 1 & -1 \\ -1 & 1 \end{bmatrix} + [10 \text{ Btu}/(\text{h-ft}^2\text{-}°\text{F})]\pi\left(\frac{1 \text{ in.}}{12 \text{ in.}/\text{ft}}\right)^2\begin{bmatrix} 0 & 0 \\ 0 & 1 \end{bmatrix}$$

$$= \begin{bmatrix} 0.5236 & -0.5236 \\ -0.5236 & 0.7418 \end{bmatrix} \text{ Btu}/(\text{h-}°\text{F}) \tag{13.4.37}$$

In general, we would use Eqs. (13.4.23)–(13.4.25), and (13.4.29) to obtain the element force matrices. However, in this example, $Q = 0$ (no heat source), $q^* = 0$ (no heat flux), and there is no convection except from the right end. Therefore,

$$\{f^{(1)}\} = \{f^{(2)}\} = \{f^{(3)}\} = 0 \tag{13.4.38}$$

and

$$\{f^{(4)}\} = hT_\infty A\begin{Bmatrix} 0 \\ 1 \end{Bmatrix}$$

$$= [10 \text{ Btu}/(\text{h-ft}^2\text{-}°\text{F})](10\,°\text{F})\pi\left(\frac{1 \text{ in.}}{12 \text{ in.}/\text{ft}}\right)^2\begin{Bmatrix} 0 \\ 1 \end{Bmatrix}$$

$$= 2.182\begin{Bmatrix} 0 \\ 1 \end{Bmatrix} \text{ Btu}/\text{h} \tag{13.4.39}$$

The assembly of the element stiffness matrices [Eqs. (13.4.35)–(13.4.37)] and the element force matrices [Eqs. (13.4.38) and (13.4.39)], using the direct stiffness method, produces the following system of equations:

$$
\begin{bmatrix}
0.5236 & -0.5236 & 0 & 0 & 0 \\
-0.5236 & 1.0472 & -0.5236 & 0 & 0 \\
0 & -0.5236 & 1.0472 & -0.5236 & 0 \\
0 & 0 & -0.5236 & 1.0472 & -0.5236 \\
0 & 0 & 0 & -0.5236 & 0.7418
\end{bmatrix}
\begin{Bmatrix}
t_1 \\ t_2 \\ t_3 \\ t_4 \\ t_5
\end{Bmatrix}
=
\begin{Bmatrix}
F_1 \\ 0 \\ 0 \\ 0 \\ 2.182
\end{Bmatrix}
$$

(13.4.40)

where F_1 corresponds to an unknown rate of heat flow at node 1 (analogous to an unknown support force in the stress analysis problem). We have a known nodal temperature boundary condition of $t_1 = 100\,°F$. This nonhomogeneous boundary condition must be treated in the same manner as was described for the stress analysis problem (see Section 2.5 and Appendix B.4). We modify the stiffness (conduction) matrix and force matrix as follows:

$$
\begin{bmatrix}
1 & 0 & 0 & 0 & 0 \\
0 & 1.0472 & -0.5236 & 0 & 0 \\
0 & -0.5236 & 1.0472 & -0.5236 & 0 \\
0 & 0 & -0.5236 & 1.0472 & -0.5236 \\
0 & 0 & 0 & -0.5236 & 0.7418
\end{bmatrix}
\begin{Bmatrix}
t_1 \\ t_2 \\ t_3 \\ t_4 \\ t_5
\end{Bmatrix}
=
\begin{Bmatrix}
100 \\ 52.36 \\ 0 \\ 0 \\ 2.182
\end{Bmatrix}
$$

(13.4.41)

where the terms in the first row and column of the stiffness matrix corresponding to the known temperature condition, $t_1 = 100\,°F$, have been set equal to 0 except for the main diagonal, which has been set equal to 1, and the first row of the force matrix has been set equal to the known nodal temperature at node 1. Also, the term $(-0.5236) \times (100\,°F) = -52.36$ on the left side of the second equation of Eq. (13.4.40) has been transposed to the right side in the second row (as $+52.36$) of Eq. (13.4.41). The second through fifth equations of Eq. (13.4.41) corresponding to the rows of unknown nodal temperatures can now be solved (typically by Gaussian elimination). The resulting solution is given by

$$t_2 = 85.93\,°F \qquad t_3 = 71.87\,°F \qquad t_4 = 57.81\,°F \qquad t_5 = 43.75\,°F \qquad (13.4.42)$$

For this elementary problem, the closed-form solution of the differential equation for conduction, Eq. (13.1.9), with the left-end boundary condition given by Eq. (13.1.10) and the right-end boundary condition given by Eq. (13.2.4) yields a linear temperature distribution through the length of the rod. The evaluation of this linear temperature function at 10-in. intervals (corresponding to the nodal points used in the finite element model) yields the same temperatures as obtained in this example by the finite element method. Because the temperature function was assumed to be linear in each finite element, this comparison is as expected. Note that F_1 could be determined by the first of Eqs. (13.4.40). ■

Example 13.2

To illustrate more fully the use of the equations developed in Section 13.4, we will now solve the heat-transfer problem shown in Figure 13–11. For the one-dimensional rod, determine the temperatures at 3-in. increments along the length of the rod and the rate of heat flow through element 1. Let $K_{xx} = 3$ Btu/(h-in.-°F), $h = 1.0$ Btu/(h-in.²-°F), and $T_\infty = 0$ °F. The temperature at the left end of the rod is 200 °F.

The finite element discretization is shown in Figure 13–12. Three elements are sufficient to enable us to determine temperatures at the four points along the rod, although more elements would yield answers more closely approximating the analytical solution obtained by solving the differential equation such as Eq. (13.2.3) with the partial derivative with respect to time equal to zero. There will be convective heat loss over the perimeter and the right end of the rod. The left end will not have convective heat loss. Using Eqs. (13.4.22) and (13.4.28), we calculate the stiffness matrices for the elements as follows:

$$\frac{AK_{xx}}{L} = \frac{(4\pi)(3)}{3} = 4\pi \text{ Btu/(h-°F)}$$

$$\frac{hPL}{6} = \frac{(1)(4\pi)(3)}{6} = 2\pi \text{ Btu/(h-°F)} \qquad (13.4.43)$$

$$hA = (1)(4\pi) = 4\pi \text{ Btu/(h-°F)}$$

Substituting the results of Eqs. (13.4.43) into Eq. (13.4.22), we obtain the stiffness matrix for element 1 as

$$[k^{(1)}] = 4\pi \begin{bmatrix} 1 & -1 \\ -1 & 1 \end{bmatrix} + 2\pi \begin{bmatrix} 2 & 1 \\ 1 & 2 \end{bmatrix}$$

$$= 4\pi \begin{bmatrix} 2 & -\frac{1}{2} \\ -\frac{1}{2} & 2 \end{bmatrix} \text{ Btu/(h-°F)} \qquad (13.4.44)$$

Because there is no convection across the ends of element 1 (its left end has a known temperature and its right end is inside the whole rod and thus not exposed to fluid motion), the contribution to the stiffness matrix owing to convection from an end of

Figure 13–11   One-dimensional rod subjected to temperature variation

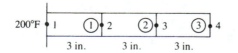

Figure 13–12   Finite element discretized rod of Figure 13–11

the element, such as given by Eq. (13.4.28), is zero. Similarly,

$$[k^{(2)}] = [k^{(1)}] = 4\pi \begin{bmatrix} 2 & -\frac{1}{2} \\ -\frac{1}{2} & 2 \end{bmatrix} \text{ Btu/(h-°F)} \tag{13.4.45}$$

However, element 3 has an additional (convection) term owing to heat loss from the exposed surface at its right end. Therefore, Eq. (13.4.28) yields a contribution to the element 3 stiffness matrix, which is then given by

$$[k^{(3)}] = [k^{(1)}] + hA \begin{bmatrix} 0 & 0 \\ 0 & 1 \end{bmatrix}$$

$$= 4\pi \begin{bmatrix} 2 & -\frac{1}{2} \\ -\frac{1}{2} & 2 \end{bmatrix} + 4\pi \begin{bmatrix} 0 & 0 \\ 0 & 1 \end{bmatrix}$$

$$= 4\pi \begin{bmatrix} 2 & -\frac{1}{2} \\ -\frac{1}{2} & 3 \end{bmatrix} \text{ Btu/(h-°F)} \tag{13.4.46}$$

In general, we calculate the force matrices by using Eqs. (13.4.26) and (13.4.29). Because $Q = 0$, $q^* = 0$, and $T_\infty = 0\,°F$, all force terms are equal to zero.

The assembly of the element matrices, Eqs. (13.4.44)–(13.4.46), using the direct stiffness method, produces the following system of equations:

$$4\pi \begin{bmatrix} 2 & -\frac{1}{2} & 0 & 0 \\ -\frac{1}{2} & 4 & -\frac{1}{2} & 0 \\ 0 & -\frac{1}{2} & 4 & -\frac{1}{2} \\ 0 & 0 & -\frac{1}{2} & 3 \end{bmatrix} \begin{Bmatrix} t_1 \\ t_2 \\ t_3 \\ t_4 \end{Bmatrix} = \begin{Bmatrix} F_1 \\ 0 \\ 0 \\ 0 \end{Bmatrix} \tag{13.4.47}$$

We have a known nodal temperature boundary condition of $t_1 = 200\,°F$. As in Example 13.1, we modify the conduction matrix and force matrix as follows:

$$4\pi \begin{bmatrix} 1 & 0 & 0 & 0 \\ 0 & 4 & -\frac{1}{2} & 0 \\ 0 & -\frac{1}{2} & 4 & -\frac{1}{2} \\ 0 & 0 & -\frac{1}{2} & 3 \end{bmatrix} \begin{Bmatrix} t_1 \\ t_2 \\ t_3 \\ t_4 \end{Bmatrix} = \begin{Bmatrix} 800\pi \\ 400\pi \\ 0 \\ 0 \end{Bmatrix} \tag{13.4.48}$$

where the terms in the first row and column of the conduction matrix corresponding to the known temperature condition, $t_1 = 200\,°F$, have been set equal to zero except for the main diagonal, which has been set to equal one, and the row of the force matrix has been set equal to the known nodal temperature at node 1. That is, the first row force is $(200)(4\pi) = 800\pi$, as we have left the 4π term as a multiplier of the elements inside the stiffness matrix. Also, the term $(-1/2)(200)(4\pi) = -400\pi$ on the left side of the second equation of Eq. (13.4.47) has been transposed to the right side in the second row (as $+400\pi$) of Eq. (13.4.48). The second through fourth equations of Eq. (13.4.48), corresponding to the rows of unknown nodal temperatures, can now be solved. The resulting solution is given by

$$t_2 = 25.4\,°F \qquad t_3 = 3.24\,°F \qquad t_4 = 0.54\,°F \tag{13.4.49}$$

Next, we determine the heat flux for element 1 by using Eqs. (13.4.6) in (13.4.8) as

$$q^{(1)} = -K_{xx}[B]\{t\} \qquad (13.4.50)$$

Using Eq. (13.4.7) in Eq. (13.4.50), we have

$$q^{(1)} = -K_{xx}\left[-\frac{1}{L} \quad \frac{1}{L}\right]\begin{Bmatrix} t_1 \\ t_2 \end{Bmatrix} \qquad (13.4.51)$$

Substituting the numerical values into Eq. (13.4.51), we obtain

$$q^{(1)} = -3\left[-\frac{1}{3} \quad \frac{1}{3}\right]\begin{Bmatrix} 200 \\ 25.4 \end{Bmatrix}$$

or

$$q^{(1)} = 174.6 \text{ Btu/(h-in}^2) \qquad (13.4.52)$$

We then determine the rate of heat flow $\bar{q}$ by multiplying Eq. (13.4.52) by the cross-sectional area over which q acts. Therefore,

$$\bar{q}^{(1)} = 174.6(4\pi) = 2194 \text{ Btu/h} \qquad (13.4.53) \quad \blacksquare$$

Example 13.3

The plane wall shown in Figure 13–13 is 1 m thick. The left surface of the wall ($x = 0$) is maintained at a constant temperature of 200 °C, and the right surface ($x = L = 1$ m) is insulated. The thermal conductivity is $K_{xx} = 25$ W/(m · °C) and there is a uniform generation of heat inside the wall of $Q = 400$ W/m³. Determine the temperature distribution through the wall thickness.

This problem is assumed to be approximated as a one-dimensional heat-transfer problem. The discretized model of the wall is shown in Figure 13–14. For simplicity, we use four equal-length elements all with unit cross-sectional area ($A = 1$ m²). The unit area represents a typical cross section of the wall. The perimeter of the wall model is then insulated to obtain the correct conditions.

Using Eqs. (13.4.22) and (13.4.28), we calculate the element stiffness matrices as follows:

$$\frac{AK_{xx}}{L} = \frac{(1 \text{ m}^2)[25 \text{ W/(m · °C)}]}{0.25 \text{ m}} = 100 \text{ W/°C}$$

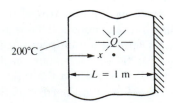

Figure 13–13 Conduction in a plane wall subjected to uniform heat generation

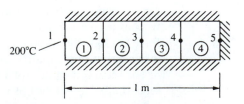

Figure 13–14 Discretized model of Figure 13–13

For each identical element, we have

$$[k] = 100 \begin{bmatrix} 1 & -1 \\ -1 & 1 \end{bmatrix} \text{W/°C} \tag{13.4.54}$$

Because no convection occurs, h is equal to zero; therefore, there is no convection contribution to $\underline{k}$.

The element force matrices are given by Eq. (13.4.26). With $Q = 400 \text{ W/m}^3$, $q = 0$, and $h = 0$, Eq. (13.4.26) becomes

$$\{f\} = \frac{QAL}{2} \begin{Bmatrix} 1 \\ 1 \end{Bmatrix} \tag{13.4.55}$$

Evaluating Eq. (13.4.55) for a typical element, such as element 1, we obtain

$$\begin{Bmatrix} f_{1x} \\ f_{2x} \end{Bmatrix} = \frac{(400 \text{ W/m}^3)(1 \text{ m}^2)(0.25 \text{ m})}{2} \begin{Bmatrix} 1 \\ 1 \end{Bmatrix} = \begin{Bmatrix} 50 \\ 50 \end{Bmatrix} \text{W} \tag{13.4.56}$$

The force matrices for all other elements are equal to Eq. (13.4.56).

The assemblage of the element matrices, Eqs. (13.4.54) and (13.4.56) and the other force matrices similar to Eq. (13.4.56), yields

$$100 \begin{bmatrix} 1 & -1 & 0 & 0 & 0 \\ -1 & 2 & -1 & 0 & 0 \\ 0 & -1 & 2 & -1 & 0 \\ 0 & 0 & -1 & 2 & -1 \\ 0 & 0 & 0 & -1 & 1 \end{bmatrix} \begin{Bmatrix} t_1 \\ t_2 \\ t_3 \\ t_4 \\ t_5 \end{Bmatrix} = \begin{Bmatrix} F_1 + 50 \\ 100 \\ 100 \\ 100 \\ 50 \end{Bmatrix} \tag{13.4.57}$$

Substituting the known temperature $t_1 = 200\,°C$ into Eq. (13.4.57), dividing both sides of Eq. (13.4.57) by 100, and transposing known terms to the right side, we have

$$\begin{bmatrix} 1 & 0 & 0 & 0 & 0 \\ 0 & 2 & -1 & 0 & 0 \\ 0 & -1 & 2 & -1 & 0 \\ 0 & 0 & -1 & 2 & -1 \\ 0 & 0 & 0 & -1 & 1 \end{bmatrix} \begin{Bmatrix} t_1 \\ t_2 \\ t_3 \\ t_4 \\ t_5 \end{Bmatrix} = \begin{Bmatrix} 200\,°C \\ 201 \\ 1 \\ 1 \\ 0.5 \end{Bmatrix} \tag{13.4.58}$$

The second through fifth equations of Eq. (13.4.58) can now be solved simultaneously to yield

$$t_2 = 203.5\,°C \qquad t_3 = 206\,°C \qquad t_4 = 207.5\,°C \qquad t_5 = 208\,°C \tag{13.4.59}$$

Using the first of Eqs. (13.4.57) yields the rate of heat flow out the left end:

$$F_1 = 100(t_1 - t_2) - 50$$
$$F_1 = 100(200 - 203.5) - 50$$
$$F_1 = -400 \text{ W}$$

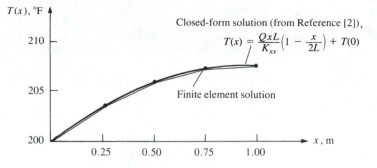

Figure 13–15 Comparison of the finite element and closed-form solutions for Example 13.3

The closed-form solution of the differential equation for conduction, Eq. (13.1.9), with the left-end boundary condition given by Eq. (13.1.10) and the right-end boundary condition given by Eq. (13.1.11), and with $q_x^* = 0$, is shown in Reference [2] to yield a parabolic temperature distribution through the wall. Evaluating the expression for the temperature function given in Reference [2] for values of x corresponding to the node points of the finite element model, we obtain

$$t_2 = 203.5\,°C \qquad t_3 = 206\,°C \qquad t_4 = 207.5\,°C \qquad t_5 = 208\,°C \qquad (13.4.60)$$

Figure 13–15 is a plot of the closed-form solution and the finite element solution for the temperature variation through the wall. The finite element nodal values and the closed-form values are equal, because the consistent equivalent force matrix has been used. (This was also discussed in Sections 3.10 and 3.11 for the axial bar subjected to distributed loading, and in Section 5.5 for the beam subjected to distributed loading.) However, recall that the finite element model predicts a linear temperature distribution within each element as indicated by the straight lines connecting the nodal temperature values in Figure 13–15. ∎

Finally, remember that the most important advantage of the finite element method is that it enables us to approximate, with high confidence, more complicated problems, such as those with more then one thermal conductivity, for which closed-form solutions are difficult (if not impossible) to obtain. The automation of the finite element method through general computer programs makes the method extremely powerful.

▲ 13.5 Two-Dimensional Finite Element Formulation ▲

Because many bodies can be modeled as two-dimensional heat-transfer problems, we now develop the equations for an element appropriate for these problems. Examples using this element then follow.

Step 1   Select Element Type

The three-noded triangular element with nodal temperatures shown in Figure 13–16 is the basic element for solution of the two-dimensional heat-transfer problem.

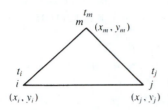

Figure 13–16   Basic triangular element with nodal temperatures

Step 2   Select a Temperature Function

The temperature function is given by

$$\{T\} = [N_i \quad N_j \quad N_m] \begin{Bmatrix} t_i \\ t_j \\ t_m \end{Bmatrix} \tag{13.5.1}$$

where t_i, t_j, and t_m are the nodal temperatures, and the shape functions are again given by Eqs. (7.2.18); that is,

$$N_i = \frac{1}{2A}(\alpha_i + \beta_i x + \gamma_i y) \tag{13.5.2}$$

with similar expressions for N_j and N_m. Here the α's, β's, and γ's are defined by Eqs. (7.2.10).

Step 3   Define the Temperature Gradient/Temperature and Heat Flux/Temperature Gradient Relationships

We define the gradient matrix analogous to the strain matrix used in the stress analysis problem as

$$\{g\} = \begin{Bmatrix} \dfrac{\partial T}{\partial x} \\[2mm] \dfrac{\partial T}{\partial y} \end{Bmatrix} \tag{13.5.3}$$

Using Eq. (13.5.1) in Eq. (13.5.3), we have

$$\{g\} = \begin{bmatrix} \dfrac{\partial N_i}{\partial x} & \dfrac{\partial N_j}{\partial x} & \dfrac{\partial N_m}{\partial x} \\[3mm] \dfrac{\partial N_i}{\partial y} & \dfrac{\partial N_j}{\partial y} & \dfrac{\partial N_m}{\partial y} \end{bmatrix} \begin{Bmatrix} t_i \\ t_j \\ t_m \end{Bmatrix} \tag{13.5.4}$$

The gradient matrix $\{g\}$, written in compact matrix form analogously to the strain matrix $\{\varepsilon\}$ of the stress analysis problem, is given by

$$\{g\} = [B]\{t\} \tag{13.5.5}$$

where the $[B]$ matrix is obtained by substituting the three equations suggested by Eq. (13.5.2) in the rectangular matrix on the right side of Eq. (13.5.4) as

$$[B] = \frac{1}{2A}\begin{bmatrix} \beta_i & \beta_j & \beta_m \\ \gamma_i & \gamma_j & \gamma_m \end{bmatrix} \tag{13.5.6}$$

The heat flux/temperature gradient relationship is now

$$\left\{ \begin{array}{c} q_x \\ q_y \end{array} \right\} = -[D]\{g\} \tag{13.5.7}$$

where the material property matrix is

$$[D] = \begin{bmatrix} K_{xx} & 0 \\ 0 & K_{yy} \end{bmatrix} \tag{13.5.8}$$

Step 4   Derive the Element Conduction Matrix and Equations

The element stiffness matrix from Eq. (13.4.17) is

$$[k] = \iiint_V [B]^T [D][B]\, dV + \iint_{S_3} h[N]^T [N]\, dS \tag{13.5.9}$$

where
$$[k_c] = \iiint_V [B]^T [D][B]\, dV$$

$$= \iiint_V \frac{1}{4A^2}\begin{bmatrix} \beta_i & \gamma_i \\ \beta_j & \gamma_j \\ \beta_m & \gamma_m \end{bmatrix}\begin{bmatrix} K_{xx} & 0 \\ 0 & K_{yy} \end{bmatrix}\begin{bmatrix} \beta_i & \beta_j & \beta_m \\ \gamma_i & \gamma_j & \gamma_m \end{bmatrix} dV \tag{13.5.10}$$

Assuming constant thickness in the element and noting that all terms of the integrand of Eq. (13.5.10) are constant, we have

$$[k_c] = \iiint_V [B]^T [D][B]\, dV = tA[B]^T [D][B] \tag{13.5.11}$$

Equation (13.5.11) is the true conduction portion of the total stiffness matrix Eq. (13.5.9). The second integral of Eq. (13.5.9) (the convection portion of the total stiffness matrix) is defined by

$$[k_h] = \iint_{S_3} h[N]^T [N]\, dS \tag{13.5.12}$$

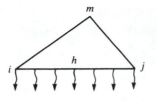

Figure 13–17 Heat loss by convection from side *i-j*

We can explicitly multiply the matrices in Eq. (13.5.12) to obtain

$$[k_h] = h \iint_{S_3} \begin{bmatrix} N_iN_i & N_iN_j & N_iN_m \\ N_jN_i & N_jN_j & N_jN_m \\ N_mN_i & N_mN_j & N_mN_m \end{bmatrix} dS \qquad (13.5.13)$$

To illustrate the use of Eq. (13.5.13), consider the side between nodes i and j of the triangular element to be subjected to convection (Figure 13–17). Then $N_m = 0$ along side *i-j*, and we obtain

$$[k_h] = \frac{hL_{i\text{-}j}t}{6} \begin{bmatrix} 2 & 1 & 0 \\ 1 & 2 & 0 \\ 0 & 0 & 0 \end{bmatrix} \qquad (13.5.14)$$

where $L_{i\text{-}j}$ is the length of side *i-j*.

The evaluation of the force matrix integrals in Eq. (13.4.16) is as follows:

$$\{f_Q\} = \iiint_V Q[N]^T \, dV = Q \iiint_V [N]^T \, dV \qquad (13.5.15)$$

for constant heat source Q. Thus it can be shown (left to your discretion) that this integral is equal to

$$\{f_Q\} = \frac{QV}{3} \begin{Bmatrix} 1 \\ 1 \\ 1 \end{Bmatrix} \qquad (13.5.16)$$

where $V = At$ is the volume of the element. Equation (13.5.16) indicates that heat is generated by the body in three equal parts to the nodes (like body forces in the elasticity problem). The second force matrix in Eq. (13.4.16) is

$$\{f_q\} = \iint_{S_2} q^*[N]^T \, dS = \iint_{S_2} q^* \begin{Bmatrix} N_i \\ N_j \\ N_m \end{Bmatrix} dS \qquad (13.5.17)$$

This reduces to

$$\frac{q^*L_{i\text{-}j}t}{2} \begin{Bmatrix} 1 \\ 1 \\ 0 \end{Bmatrix} \qquad \text{on side } i\text{-}j \qquad (13.5.18)$$

$$\frac{q^* L_{j\text{-}m} t}{2} \begin{Bmatrix} 0 \\ 1 \\ 1 \end{Bmatrix} \qquad \text{on side } j\text{-}m \qquad (13.5.19)$$

$$\frac{q^* L_{m\text{-}i} t}{2} \begin{Bmatrix} 1 \\ 0 \\ 1 \end{Bmatrix} \qquad \text{on side } m\text{-}i \qquad (13.5.20)$$

where $L_{i\text{-}j}$, $L_{j\text{-}m}$, and $L_{m\text{-}i}$ are the lengths of the sides of the element, and q^* is assumed constant over each edge. The integral $\iint_{S_3} hT_\infty [N]^T dS$ can be found in a manner similar to Eq. (13.5.17) by simply replacing q^* with hT_∞ in Eqs. (13.5.18)–(13.5.20).

Steps 5–7

Steps 5–7 are identical to those described in Section 13.4.

To illustrate the use of the equations presented in Section 13.5, we will now solve some two-dimensional heat-transfer problems.

Example 13.4

For the two-dimensional body shown in Figure 13–18, determine the temperature distribution. The temperature at the left side of the body is maintained at $100\,°F$. The edges on the top and bottom of the body are insulated. There is heat convection from the right side with convection coefficient $h = 20$ Btu/(h-ft²-°F). The free-stream temperature is $T_\infty = 50\,°F$. The coefficients of thermal conductivity are $K_{xx} = K_{yy} = 25$ Btu/(h-ft-°F). The dimensions are shown in the figure. Assume the thickness to be 1 ft.

The finite element discretization is shown in Figure 13–19. We will use four triangular elements of equal size for simplicity of the longhand solution. There will be convective heat loss only over the right side of the body because the other faces are insulated. We now calculate the element stiffness matrices using Eq. (13.5.11) applied

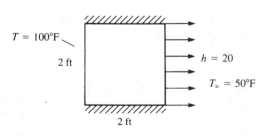

Figure 13–18 Two-dimensional body subjected to temperature variation and convection

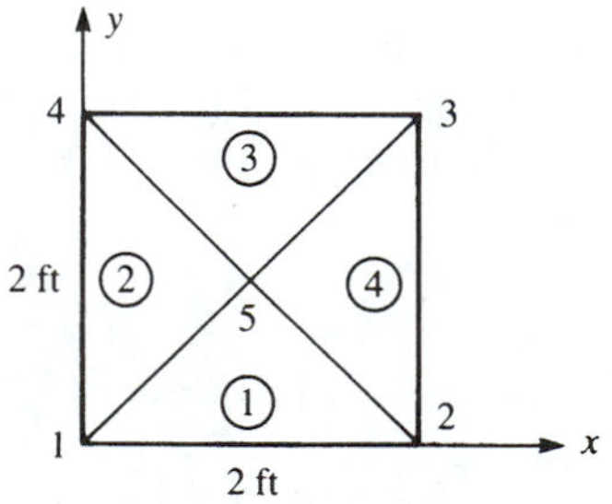

Figure 13–19 Discretized two-dimensional body of Figure 13–18

for all elements and using Eq. (13.5.14) applied for element 4 only, because convection is occurring only across one edge of element 4.

Element 1

The coordinates of the element 1 nodes are $x_1 = 0$, $y_1 = 0$, $x_2 = 2$, $y_2 = 0$, $x_5 = 1$, and $y_5 = 1$. Using these coordinates and Eqs. (7.2.10), we obtain

$$\beta_1 = 0 - 1 = -1 \qquad \beta_2 = 1 - 0 = 1 \qquad \beta_5 = 0 - 0 = 0$$
$$\gamma_1 = 1 - 2 = -1 \qquad \gamma_2 = 0 - 1 = -1 \qquad \gamma_5 = 2 - 0 = 2 \tag{13.5.21}$$

Using Eqs. (13.5.21) in Eq. (13.5.11), we have

$$[k_c^{(1)}] = \frac{1(1)}{2(2)} \begin{bmatrix} -1 & -1 \\ 1 & -1 \\ 0 & 2 \end{bmatrix} \begin{bmatrix} 25 & 0 \\ 0 & 25 \end{bmatrix} \begin{bmatrix} -1 & 1 & 0 \\ -1 & -1 & 2 \end{bmatrix} \tag{13.5.22}$$

Simplifying Eq. (13.5.22), we obtain

$$[k_c^{(1)}] = \begin{matrix} 1 & 2 & 5 \\ \begin{bmatrix} 12.5 & 0 & -12.5 \\ 0 & 12.5 & -12.5 \\ -12.5 & -12.5 & 25 \end{bmatrix} \end{matrix} \text{Btu}/(\text{h-}^\circ\text{F}) \tag{13.5.23}$$

where the numbers above the columns indicate the node numbers associated with the matrix.

Element 2

The coordinates of the element 2 nodes are $x_1 = 0$, $y_1 = 0$, $x_5 = 1$, $y_5 = 1$, $x_4 = 0$, and $y_4 = 2$. Using these coordinates, we obtain

$$\beta_1 = 1 - 2 = -1 \qquad \beta_5 = 2 - 0 = 2 \qquad \beta_4 = 0 - 1 = -1$$
$$\gamma_1 = 0 - 1 = -1 \qquad \gamma_5 = 0 - 0 = 0 \qquad \gamma_4 = 1 - 0 = 1 \tag{13.5.24}$$

Using Eqs. (13.5.24) in Eq. (13.5.11), we have

$$[k_c^{(2)}] = \frac{1}{4} \begin{bmatrix} -1 & -1 \\ 2 & 0 \\ -1 & 1 \end{bmatrix} \begin{bmatrix} 25 & 0 \\ 0 & 25 \end{bmatrix} \begin{bmatrix} -1 & 2 & -1 \\ -1 & 0 & 1 \end{bmatrix} \tag{13.5.25}$$

Simplifying Eq. (13.5.25), we obtain

$$[k_c^{(2)}] = \begin{matrix} 1 & 5 & 4 \\ \begin{bmatrix} 12.5 & -12.5 & 0 \\ -12.5 & 25 & -12.5 \\ 0 & -12.5 & 12.5 \end{bmatrix} \end{matrix} \text{Btu}/(\text{h-}^\circ\text{F}) \tag{13.5.26}$$

Element 3

The coordinates of the element 3 nodes are $x_4 = 0$, $y_4 = 2$, $x_5 = 1$, $y_5 = 1$, $x_3 = 2$, and $y_3 = 2$. Using these coordinates, we obtain

$$\beta_4 = 1 - 2 = -1 \qquad \beta_5 = 2 - 2 = 0 \qquad \beta_3 = 2 - 1 = 1$$

$$\gamma_4 = 2 - 1 = 1 \qquad \gamma_5 = 0 - 2 = -2 \qquad \gamma_3 = 1 - 0 = 1 \tag{13.5.27}$$

Using Eqs. (13.5.27) in Eq. (13.5.11), we obtain

$$[k_c^{(3)}] = \begin{matrix} & 4 & 5 & 3 \\ & \begin{bmatrix} 12.5 & -12.5 & 0 \\ -12.5 & 25 & -12.5 \\ 0 & -12.5 & 12.5 \end{bmatrix} \end{matrix} \text{Btu/(h-°F)} \tag{13.5.28}$$

Element 4

The coordinates of the element 4 nodes are $x_2 = 2$, $y_2 = 0$, $x_3 = 2$, $y_3 = 2$, $x_5 = 1$, and $y_5 = 1$. Using these coordinates, we obtain

$$\beta_2 = 2 - 1 = 1 \qquad \beta_3 = 1 - 0 = 1 \qquad \beta_5 = 0 - 2 = -2$$

$$\gamma_2 = 1 - 2 = -1 \qquad \gamma_3 = 2 - 1 = 1 \qquad \gamma_5 = 2 - 2 = 0 \tag{13.5.29}$$

Using Eqs. (13.5.29) in Eq. (13.5.11), we obtain

$$[k_c^{(4)}] = \begin{matrix} & 2 & 3 & 5 \\ & \begin{bmatrix} 12.5 & 0 & -12.5 \\ 0 & 12.5 & -12.5 \\ -12.5 & -12.5 & 25 \end{bmatrix} \end{matrix} \text{Btu/(h-°F)} \tag{13.5.30}$$

For element 4, we have a convection contribution to the total stiffness matrix because side 2–3 is exposed to the free-stream temperature. Using Eq. (13.5.14) with $i = 2$ and $j = 3$, we obtain

$$[k_h^{(4)}] = \frac{(20)(2)(1)}{6} \begin{bmatrix} 2 & 1 & 0 \\ 1 & 2 & 0 \\ 0 & 0 & 0 \end{bmatrix} \tag{13.5.31}$$

Simplifying Eq. (13.5.31) yields

$$[k_h^{(4)}] = \begin{matrix} & 2 & 3 & 5 \\ & \begin{bmatrix} 13.3 & 6.67 & 0 \\ 6.67 & 13.3 & 0 \\ 0 & 0 & 0 \end{bmatrix} \end{matrix} \text{Btu/(h-°F)} \tag{13.5.32}$$

Adding Eqs. (13.5.30) and (13.5.32), we obtain the element 4 total stiffness matrix as

$$[k^{(4)}] = \begin{matrix} 2 & 3 & 5 \\ \begin{bmatrix} 25.83 & 6.67 & -12.5 \\ 6.67 & 25.83 & -12.5 \\ -12.5 & -12.5 & 25 \end{bmatrix} \end{matrix} \text{Btu/(h-°F)} \qquad (13.5.33)$$

Superimposing the stiffness matrices given by Eqs. (13.5.23), (13.5.26), (13.5.28), and (13.5.33), we obtain the total stiffness matrix for the body as

$$\underline{K} = \begin{bmatrix} 25 & 0 & 0 & 0 & -25 \\ 0 & 38.33 & 6.67 & 0 & -25 \\ 0 & 6.67 & 38.33 & 0 & -25 \\ 0 & 0 & 0 & 25 & -25 \\ -25 & -25 & -25 & -25 & 100 \end{bmatrix} \text{Btu/(h-°F)} \qquad (13.5.34)$$

Next, we determine the element force matrices by using Eqs. (13.5.18)–(13.5.20) with q^* replaced by hT_∞. Because $Q = 0$, $q^* = 0$, and we have convective heat transfer only from side 2–3, element 4 is the only one that contributes nodal forces. Hence,

$$\{f^{(4)}\} = \begin{Bmatrix} f_2 \\ f_3 \\ f_5 \end{Bmatrix} = \frac{hT_\infty L_{2-3}t}{2} \begin{Bmatrix} 1 \\ 1 \\ 0 \end{Bmatrix} \qquad (13.5.35)$$

Substituting the appropriate numerical values into Eq. (13.5.35) yields

$$\{f^{(4)}\} = \frac{(20)(50)(2)(1)}{2} \begin{Bmatrix} 1 \\ 1 \\ 0 \end{Bmatrix} = \begin{Bmatrix} 1000 \\ 1000 \\ 0 \end{Bmatrix} \frac{\text{Btu}}{\text{h}} \qquad (13.5.36)$$

Using Eqs. (13.5.34) and (13.5.36), we find that the total assembled system of equations is

$$\begin{bmatrix} 25 & 0 & 0 & 0 & -25 \\ 0 & 38.33 & 6.67 & 0 & -25 \\ 0 & 6.67 & 38.33 & 0 & -25 \\ 0 & 0 & 0 & 25 & -25 \\ -25 & -25 & -25 & -25 & 100 \end{bmatrix} \begin{Bmatrix} t_1 \\ t_2 \\ t_3 \\ t_4 \\ t_5 \end{Bmatrix} = \begin{Bmatrix} F_1 \\ 1000 \\ 1000 \\ F_4 \\ 0 \end{Bmatrix} \qquad (13.5.37)$$

We have known nodal temperature boundary conditions of $t_1 = 100\,°F$ and $t_4 = 100\,°F$. We again modify the stiffness and force matrices as follows:

$$\begin{bmatrix} 1 & 0 & 0 & 0 & 0 \\ 0 & 38.33 & 6.67 & 0 & -25 \\ 0 & 6.67 & 38.33 & 0 & -25 \\ 0 & 0 & 0 & 1 & 0 \\ 0 & -25 & -25 & 0 & 100 \end{bmatrix} \begin{Bmatrix} t_1 \\ t_2 \\ t_3 \\ t_4 \\ t_5 \end{Bmatrix} = \begin{Bmatrix} 100 \\ 1000 \\ 1000 \\ 100 \\ 5000 \end{Bmatrix} \qquad (13.5.38)$$

The terms in the first and fourth rows and columns corresponding to the known temperature conditions $t_1 = 100\,°F$ and $t_4 = 100\,°F$ have been set equal to zero except for the main diagonal, which has been set equal to one, and the first and fourth rows of the force matrix have been set equal to the known nodal temperatures. Also, the term $(-25)(100\,°F) + (-25) \times (100\,°F) = -5000$ on the left side of the fifth equation of Eq. (13.5.37) has been transposed to the right side in the fifth row (as $+5000$) of Eq. (13.5.38). The second, third and fifth equations of Eq. (13.5.38), corresponding to the rows of unknown nodal temperatures, can now be solved in the usual manner. The resulting solution is given by

$$t_2 = 69.33\,°F \qquad t_3 = 69.33\,°F \qquad t_5 = 84.62\,°F \qquad (13.5.39) \quad \blacksquare$$

Example 13.5

For the two-dimensional body shown in Figure 13–20, determine the temperature distribution. The temperature of the top side of the body is maintained at 100 °C. The body is insulated on the other edges. A uniform heat source of $Q = 1000$ W/m^3 acts over the whole plate, as shown in the figure. Assume a constant thickness of 1 m. Let $K_{xx} = K_{yy} = 25$ W/(m · °C).

We need consider only the left half of the body, because we have a vertical plane of symmetry passing through the body 2 m from both the left and right edges. This vertical plane can be considered to be an insulated boundary. The finite element model is shown in Figure 13–21.

We will now calculate the element stiffness matrices. Because the magnitudes of the coordinates are the same as in Example 13.4, the element stiffness matrices are the same as Eqs. (13.5.23), (13.5.26), (13.5.28), and (13.5.30). Remember that there is no convection from any side of an element, so the convection contribution $[k_h]$ to the stiffness matrix is zero. Superimposing the element stiffness matrices, we obtain the total stiffness matrix as

$$
\underline{K} =
\begin{bmatrix}
25 & 0 & 0 & 0 & -25 \\
0 & 25 & 0 & 0 & -25 \\
0 & 0 & 25 & 0 & -25 \\
0 & 0 & 0 & 25 & -25 \\
-25 & -25 & -25 & -25 & 100
\end{bmatrix}
\text{W/°C}
\qquad (13.5.40)
$$

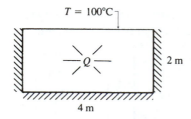

Figure 13–20 Two-dimensional body subjected to a heat source

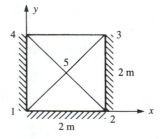

Figure 13–21 Discretized body of Figure 13–20

Because the heat source Q is acting uniformly over each element, we use Eq. (13.5.16) to evaluate the nodal forces for each element as

$$\{f^{(e)}\} = \frac{QV}{3}\begin{Bmatrix} 1 \\ 1 \\ 1 \end{Bmatrix} = \frac{1000(1\text{ m}^3)}{3}\begin{Bmatrix} 1 \\ 1 \\ 1 \end{Bmatrix} = \begin{Bmatrix} 333 \\ 333 \\ 333 \end{Bmatrix}\text{ W} \qquad (13.5.41)$$

We then use Eqs. (13.5.40) and (13.5.41) applied to each element, to assemble the total system of equations as

$$\begin{bmatrix} 25 & 0 & 0 & 0 & -25 \\ 0 & 25 & 0 & 0 & -25 \\ 0 & 0 & 25 & 0 & -25 \\ 0 & 0 & 0 & 25 & -25 \\ -25 & -25 & -25 & -25 & 100 \end{bmatrix}\begin{Bmatrix} t_1 \\ t_2 \\ t_3 \\ t_4 \\ t_5 \end{Bmatrix} = \begin{Bmatrix} 666 \\ 666 \\ 666 + F_3 \\ 666 + F_4 \\ 1333 \end{Bmatrix} \qquad (13.5.42)$$

We have known nodal temperature boundary conditions of $t_3 = 100\,^\circ\text{C}$ and $t_4 = 100\,^\circ\text{C}$. In the usual manner, as was shown in Example 13.4, we modify the stiffness and force matrices of Eq. (13.5.42) to obtain

$$\begin{bmatrix} 25 & 0 & 0 & 0 & -25 \\ 0 & 25 & 0 & 0 & -25 \\ 0 & 0 & 1 & 0 & 0 \\ 0 & 0 & 0 & 1 & 0 \\ -25 & -25 & 0 & 0 & 100 \end{bmatrix}\begin{Bmatrix} t_1 \\ t_2 \\ t_3 \\ t_4 \\ t_5 \end{Bmatrix} = \begin{Bmatrix} 666 \\ 666 \\ 100 \\ 100 \\ 6333 \end{Bmatrix} \qquad (13.5.43)$$

Equation (13.5.43) satisfies the boundary temperature conditions and is equivalent to Eq. (13.5.42); that is, the first, second, and fifth equations of Eq. (13.5.43) are the same as the first, second, and fifth equations of Eq. (13.5.42), and the third and fourth equations of Eq. (13.5.43) identically satisfy the boundary temperature conditions at nodes 3 and 4. The first, second, and fifth equations of Eq. (13.5.43) corresponding to the rows of unknown nodal temperatures, can now be solved simultaneously. The resulting solution is given by

$$t_1 = 180\,^\circ\text{C} \qquad t_2 = 180\,^\circ\text{C} \qquad t_5 = 153\,^\circ\text{C} \qquad (13.5.44) \quad \blacksquare$$

We then use the results from Eq. (13.5.44) in Eq. (13.5.42) to obtain the rates of heat flow at nodes 3 and 4 (that is, F_3 and F_4).

▲ 13.6 Line or Point Sources ▲

A common practical heat-transfer problem is that of a source of heat generation present within a very small volume or area of some larger medium. When such heat sources exist within small volumes or areas, they may be idealized as **line** or **point**

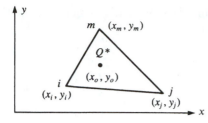

Figure 13–22 Line source located within a typical triangular element

sources. Practical examples that can be modeled as line sources include hot-water pipes embedded within a medium such as concrete or earth, and conducting electrical wires embedded within a material.

A line or point source can be considered by simply including a node at the location of the source when the discretized finite element model is created. The value of the line source can then be added to the row of the global force matrix corresponding to the global degree of freedom assigned to the node. However, another procedure can be used to treat the line source when it is more convenient to leave the source within an element.

We now consider the line source of magnitude Q^*, with typical units of Btu/(h-ft), located at (x_o, y_o) within the two-dimensional element shown in Figure 13–22. The heat source Q is no longer constant over the element volume.

Using Eq. (13.4.16), we can express the heat source matrix as

$$\{f_Q\} = \iiint_V \begin{Bmatrix} N_i \\ N_j \\ N_m \end{Bmatrix} \Bigg|_{x=x_o, y=y_o} \frac{Q^*}{A^*} dV \qquad (13.6.1)$$

where A^* is the cross-sectional area over which Q^* acts, and the N's are evaluated at $x = x_o$ and $y = y_o$. Equation (13.6.1) can be rewritten as

$$\{f_Q\} = \iint_{A^*} \int_0^t \begin{Bmatrix} N_i \\ N_j \\ N_m \end{Bmatrix} \Bigg|_{x=x_o, y=y_o} \frac{Q^*}{A^*} dA \, dz \qquad (13.6.2)$$

Because the N's are evaluated at $x = x_o$ and $y = y_o$, they are no longer functions of x and y. Thus, we can simplify Eq. (13.6.2) to

$$\{f_Q\} = \begin{Bmatrix} N_i \\ N_j \\ N_m \end{Bmatrix} \Bigg|_{x=x_o, y=y_o} Q^* t \text{ Btu/h} \qquad (13.6.3)$$

From Eq. (13.6.3), we can see that the portion of the line source Q^* distributed to each node is based on the values of $N_i, N_j,$ and N_m, which are evaluated using the coordinates (x_o, y_o) of the line source. Recalling that the sum of the N's at any point within an element is equal to one [that is, $N_i(x_o, y_o) + N_j(x_o, y_o) + N_m(x_o, y_o) = 1$], we see that no more than the total amount of Q^* is distributed and that

$$Q_i^* + Q_j^* + Q_m^* = Q^* \qquad (13.6.4)$$

Example 13.6

A line source $Q^* = 65$ Btu/(h-in.) is located at coordinates $(5, 2)$ in the element shown in Figure 13–23. Determine the amount of Q^* allocated to each node. All nodal coordinates are in units of inches. Assume an element thickness of $t = 1$ in.

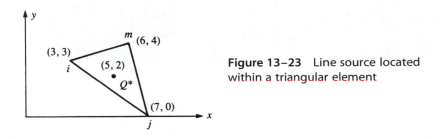

Figure 13–23   Line source located within a triangular element

We first evaluate the α's, β's, and γ's, defined by Eqs. (7.2.10), associated with each shape function as follows:

$$\alpha_i = x_j y_m - x_m y_j = 7(4) - 6(0) = 28$$
$$\alpha_j = x_m y_i - x_i y_m = 6(3) - 3(4) = 6$$
$$\alpha_m = x_i y_j - x_j y_i = 3(0) - 7(3) = -21$$
$$\beta_i = y_j - y_m = 0 - 4 = -4$$
$$\beta_j = y_m - y_i = 4 - 3 = 1 \qquad (13.6.5)$$
$$\beta_m = y_i - y_j = 3 - 0 = 3$$
$$\gamma_i = x_m - x_j = 6 - 7 = -1$$
$$\gamma_j = x_i - x_m = 3 - 6 = -3$$
$$\gamma_m = x_j - x_i = 7 - 3 = 4$$

Also,
$$2A = \begin{vmatrix} 1 & x_i & y_i \\ 1 & x_j & y_j \\ 1 & x_m & y_m \end{vmatrix} = \begin{vmatrix} 1 & 3 & 3 \\ 1 & 7 & 0 \\ 1 & 6 & 4 \end{vmatrix} = 13 \qquad (13.6.6)$$

Substituting the results of Eqs. (13.6.5) and (13.6.6) into Eq. (13.5.2) yields

$$N_i = \tfrac{1}{13}[28 - 4x - 1y]$$
$$N_j = \tfrac{1}{13}[6 + x - 3y] \qquad (13.6.7)$$
$$N_m = \tfrac{1}{13}[-21 + 3x + 4y]$$

Equations (13.6.7) for $N_i, N_j,$ and N_m evaluated at $x = 5$ and $y = 2$ are

$$N_i = \tfrac{1}{13}[28 - 4(5) - 1(2)] = \tfrac{6}{13}$$

$$N_j = \tfrac{1}{13}[6 + 5 - 3(2)] = \tfrac{5}{13} \qquad (13.6.8)$$

$$N_m = \tfrac{1}{13}[-21 + 3(5) + 4(2)] = \tfrac{2}{13}$$

Therefore, using Eq. (13.6.3), we obtain

$$\left\{ \begin{array}{c} f_{Qi} \\ f_{Qj} \\ f_{Qm} \end{array} \right\} = Q^* t \left\{ \begin{array}{c} N_i \\ N_j \\ N_m \end{array} \right\}_{\substack{x=x_0=5 \\ y=y_0=2}} = \frac{65(1)}{13} \left\{ \begin{array}{c} 6 \\ 5 \\ 2 \end{array} \right\} = \left\{ \begin{array}{c} 30 \\ 25 \\ 10 \end{array} \right\} \text{Btu/h} \qquad (13.6.9)$$

■

▲ 13.7 One-Dimensional Heat Transfer with Mass Transport ▲

We now consider the derivation of the basic differential equation for one-dimensional heat flow where the flow is due to conduction, convection, and **mass transport** (or **transfer**) of the fluid. The purpose of this derivation including mass transport is to show how Galerkin's residual method can be directly applied to a problem for which the variational method is difficult to apply.

The control volume used in the derivation is shown in Figure 13–24. Again, from Eq. (13.1.1) for conservation of energy, we obtain

$$q_x A \, dt + QA \, dx \, dt = cpA \, dx \, dT + q_{x+dx} A \, dt + q_h P \, dx \, dt + q_m \, dt \qquad (13.7.1)$$

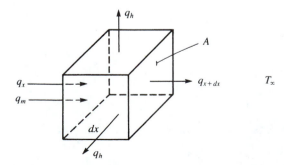

Figure 13–24 Control volume for one-dimensional heat conduction with convection and mass transport

All of the terms in Eq. (13.7.1) have the same meaning as in Sections 13.1 and 13.2, except the additional mass-transport term is given by [1]

$$q_m = \dot{m}cT \tag{13.7.2}$$

where the additional variable $\dot{m}$ is the *mass flow rate* in typical units of kg/h or slug/h.

Again, using Eqs. (13.1.3)–(13.1.6), (13.2.2), and (13.7.2) in Eq. (13.7.1) and differentiating with respect to x and t, we obtain

$$\frac{\partial}{\partial x}\left(K_{xx}\frac{\partial T}{\partial x}\right) + Q = \frac{\dot{m}c}{A}\frac{\partial T}{\partial x} + \frac{hP}{A}(T - T_\infty) + \rho c\frac{\partial T}{\partial t} \tag{13.7.3}$$

Equation (13.7.3) is the basic one-dimensional differential equation for heat transfer with mass transport.

13.8 Finite Element Formulation of Heat Transfer with Mass Transport by Galerkin's Method

Having obtained the differential equation for heat transfer with mass transport, Eq. (13.7.3), we now derive the finite element equations by applying Galerkin's residual method, as outlined in Section 3.12, directly to the differential equation. We assume here that $Q = 0$ and that we have steady-state conditions so that differentiation with respect to time is zero.

The residual R is now given by

$$R(T) = -\frac{d}{dx}\left(K_{xx}\frac{dT}{dx}\right) + \frac{\dot{m}c}{A}\frac{dT}{dx} + \frac{hP}{A}(T - T_\infty) \tag{13.8.1}$$

Applying Galerkin's criterion, Eq. (3.12.3), to Eq. (13.8.1), we have

$$\int_0^L\left[-\frac{d}{dx}\left(K_{xx}\frac{dT}{dx}\right) + \frac{\dot{m}c}{A}\frac{dT}{dx} + \frac{hP}{A}(T - T_\infty)\right]N_i\,dx = 0 \qquad (i = 1, 2) \tag{13.8.2}$$

where the shape functions are given by Eqs. (13.4.2). Applying integration by parts to the first term of Eq. (13.8.2), we obtain

$$u = N_i \qquad du = \frac{dN_i}{dx}dx$$

$$dv = -\frac{d}{dx}\left(K_{xx}\frac{dT}{dx}\right)dx \qquad v = -K_{xx}\frac{dT}{dx} \tag{13.8.3}$$

Using Eqs. (13.8.3) in the general formula for integration by parts [see Eq. (3.12.6)], we obtain

$$\int_0^L\left[-\frac{d}{dx}\left(K_{xx}\frac{dT}{dx}\right)\right]N_i\,dx = -K_{xx}\frac{dT}{dx}N_i\Big|_0^L + \int_0^L K_{xx}\frac{dT}{dx}\frac{dN_i}{dx}dx \tag{13.8.4}$$

Substituting Eq. (13.8.4) into Eq. (13.8.2), we obtain

$$\int_0^L \left(K_{xx} \frac{dT}{dx} \frac{dN_i}{dx} \right) dx + \int_0^L \left[\frac{\dot{m}c}{A} \frac{dT}{dx} + \frac{hP}{A} (T - T_\infty) \right] N_i \, dx = K_{xx} \frac{dT}{dx} N_i \bigg|_0^L \qquad (13.8.5)$$

Using Eq. (13.4.2) in (13.4.1) for T, we obtain

$$\frac{dT}{dx} = -\frac{t_1}{L} + \frac{t_2}{L} \qquad (13.8.6)$$

From Eq. (13.4.2), we obtain

$$\frac{dN_1}{dx} = -\frac{1}{L} \qquad \frac{dN_2}{dx} = \frac{1}{L} \qquad (13.8.7)$$

By letting $N_i = N_1 = 1 - (x/L)$ and substituting Eqs. (13.8.6) and (13.8.7) into Eq. (13.8.5), along with Eq. (13.4.1) for T, we obtain the first finite element equation

$$\int_0^L K_{xx} \left(-\frac{t_1}{L} + \frac{t_2}{L} \right) \left(-\frac{1}{L} \right) dx + \int_0^L \frac{\dot{m}c}{A} \left(-\frac{t_1}{L} + \frac{t_2}{L} \right) \left(1 - \frac{x}{L} \right) dx$$

$$+ \int_0^L \frac{hP}{A} \left[\left(1 - \frac{x}{L} \right) t_1 + \left(\frac{x}{L} \right) t_2 - T_\infty \right] \left(\frac{1 - x}{L} \right) dx = q_{x1}^* \qquad (13.8.8)$$

where the definition for q_x given by Eq. (13.1.3) has been used in Eq. (13.8.8). Equation (13.8.8) has a boundary condition q_{x1}^* at $x = 0$ only because $N_1 = 1$ at $x = 0$ and $N_1 = 0$ at $x = L$. Integrating Eq. (13.8.8), we obtain

$$\left(\frac{K_{xx}A}{L} - \frac{\dot{m}c}{2} + \frac{hPL}{3} \right) t_1 + \left(-\frac{K_{xx}A}{L} + \frac{\dot{m}c}{2} + \frac{hPL}{6} \right) t_2 = q_{x1}^* + \frac{hPL}{2} T_\infty \qquad (13.8.9)$$

where q_{x1}^* is defined to be q_x evaluated at node 1.

To obtain the second finite element equation, we let $N_i = N_2 = x/L$ in Eq. (13.8.5) and again use Eqs. (13.8.6), (13.8.7), and (13.4.1) in Eq. (13.8.5) to obtain

$$\left(-\frac{K_{xx}A}{L} - \frac{\dot{m}c}{2} + \frac{hPL}{6} \right) t_1 + \left(\frac{K_{xx}A}{L} + \frac{\dot{m}c}{2} + \frac{hPL}{3} \right) t_2 = q_{x2}^* + \frac{hPL}{2} T_\infty \qquad (13.8.10)$$

where q_{x2}^* is defined to be q_x evaluated at node 2. Rewriting Eqs. (13.8.9) and (13.8.10) in matrix form yields

$$\left[\frac{K_{xx}A}{L} \begin{bmatrix} 1 & -1 \\ -1 & 1 \end{bmatrix} + \frac{\dot{m}c}{2} \begin{bmatrix} -1 & 1 \\ -1 & 1 \end{bmatrix} + \frac{hPL}{6} \begin{bmatrix} 2 & 1 \\ 1 & 2 \end{bmatrix} \right] \begin{Bmatrix} t_1 \\ t_2 \end{Bmatrix}$$

$$= \frac{hPLT_\infty}{2} \begin{Bmatrix} 1 \\ 1 \end{Bmatrix} + \begin{Bmatrix} q_{x1}^* \\ q_{x2}^* \end{Bmatrix} \qquad (13.8.11)$$

Applying the element equation $\{f\} = [k]\{t\}$ to Eq. (13.8.11), we see that the element stiffness (conduction) matrix is now composed of three parts:

$$[k] = [k_c] + [k_h] + [k_m] \tag{13.8.12}$$

where

$$[k_c] = \frac{K_{xx}A}{L}\begin{bmatrix} 1 & -1 \\ -1 & 1 \end{bmatrix} \qquad [k_h] = \frac{hPL}{6}\begin{bmatrix} 2 & 1 \\ 1 & 2 \end{bmatrix} \qquad [k_m] = \frac{\dot{m}c}{2}\begin{bmatrix} -1 & 1 \\ -1 & 1 \end{bmatrix} \tag{13.8.13}$$

and the element nodal force and unknown nodal temperature matrices are

$$\{f\} = \frac{hPLT_\infty}{2}\begin{Bmatrix} 1 \\ 1 \end{Bmatrix} + \begin{Bmatrix} q^*_{x1} \\ q^*_{x2} \end{Bmatrix} \qquad \{t\} = \begin{Bmatrix} t_1 \\ t_2 \end{Bmatrix} \tag{13.8.14}$$

We observe from Eq. (13.8.13) that the mass transport stiffness matrix $[k_m]$ is asymmetric and, hence, $[k]$ is asymmetric. Also, if heat flux exists, it usually occurs across the free ends of a system. Therefore, q_{x1} and q_{x2} usually occur only at the free ends of a system modeled by this element. When the elements are assembled, the heat fluxes q_{x1} and q_{x2} are usually equal but opposite at the node common to two elements, unless there is an internal concentrated heat flux in the system. Furthermore, for insulated ends, the q^*_x's also go to zero.

To illustrate the use of the finite element equations developed in this section for heat transfer with mass transport, we will now solve the following problem.

Example 13.7

Air is flowing at a rate of 4.72 lb/h inside a round tube with a diameter of 1 in. and length of 5 in., as shown in Figure 13–25. The initial temperature of the air entering the tube is 100 °F. The wall of the tube has a uniform constant temperature of 200 °F. The specific heat of the air is 0.24 Btu/(lb-°F), the convection coefficient between the air and the inner wall of the tube is 2.7 Btu/(h-ft²-°F), and the thermal conductivity is

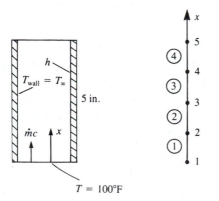

Figure 13–25 Air flowing through a tube, and the finite element model

0.017 Btu/(h-ft-°F). Determine the temperature of the air along the length of the tube and the heat flow at the inlet and outlet of the tube. Here the flow rate and specific heat are given in force units (pounds) instead of mass units (slugs). This is not a problem because the units cancel in the $\dot{m}c$ product in the formulation of the equations.

We first determine the element stiffness and force matrices using Eqs. (13.8.13) and (13.8.14). To do this, we evaluate the following factors:

$$\frac{K_{xx}A}{L} = \frac{(0.017)\left[\dfrac{\pi(1)}{4(144)}\right]}{1.25/12} = 0.891 \times 10^{-3}\ \text{Btu}/(\text{h-°F})$$

$$\dot{m}c = (4.72)(0.24) = 1.133\ \text{Btu}/(\text{h-°F})$$

$$\frac{hPL}{6} = \frac{(2.7)(0.262)(0.104)}{6} = 0.0123\ \text{Btu}/(\text{h-°F})$$

$$hPLT_\infty = (2.7)(0.262)(0.104)(200) = 14.71\ \text{Btu/h}$$

(13.8.15)

We can see from Eqs. (13.8.15) that the conduction portion of the stiffness matrix is negligible. Therefore, we neglect this contribution to the total stiffness matrix and obtain

$$\underline{k}^{(1)} = \frac{1.133}{2}\begin{bmatrix} -1 & 1 \\ -1 & 1 \end{bmatrix} + 0.0123\begin{bmatrix} 2 & 1 \\ 1 & 2 \end{bmatrix} = \begin{bmatrix} -0.542 & 0.579 \\ -0.554 & 0.591 \end{bmatrix} \quad (13.8.16)$$

Similarly, because all elements have the same properties,

$$\underline{k}^{(2)} = \underline{k}^{(3)} = \underline{k}^{(4)} = \underline{k}^{(1)} \quad (13.8.17)$$

Using Eqs. (13.8.14) and (13.8.15), we obtain the element force matrices as

$$\underline{f}^{(1)} = \underline{f}^{(2)} = \underline{f}^{(3)} = \underline{f}^{(4)} = \begin{Bmatrix} 7.35 \\ 7.35 \end{Bmatrix} \quad (13.8.18)$$

Assembling the global stiffness matrix using Eqs. (13.8.16) and (13.8.17) and the global force matrix using Eq. (13.8.18), we obtain the global equations as

$$\begin{bmatrix} -0.542 & 0.579 & 0 & 0 & 0 \\ -0.554 & 0.591 - 0.542 & 0.579 & 0 & 0 \\ 0 & -0.554 & 0.591 - 0.542 & 0.579 & 0 \\ 0 & 0 & -0.554 & 0.591 - 0.542 & 0.579 \\ 0 & 0 & 0 & -0.554 & 0.591 \end{bmatrix} \begin{Bmatrix} t_1 \\ t_2 \\ t_3 \\ t_4 \\ t_5 \end{Bmatrix}$$

$$= \begin{Bmatrix} F_1 + 7.35 \\ 14.7 \\ 14.7 \\ 14.7 \\ 7.35 \end{Bmatrix} \quad (13.8.19)$$

Applying the boundary condition $t_1 = 100\,°F$, we rewrite Eq. (13.8.19) as

$$\begin{bmatrix} 1 & 0 & 0 & 0 & 0 \\ 0 & 0.049 & 0.579 & 0 & 0 \\ 0 & -0.554 & 0.049 & 0.579 & 0 \\ 0 & 0 & -0.554 & 0.049 & 0.579 \\ 0 & 0 & 0 & -0.554 & 0.591 \end{bmatrix} \begin{Bmatrix} t_1 \\ t_2 \\ t_3 \\ t_4 \\ t_5 \end{Bmatrix} = \begin{Bmatrix} 100 \\ 14.7 + 55.4 \\ 14.7 \\ 14.7 \\ 7.35 \end{Bmatrix} \qquad (13.8.20)$$

Solving the second through fifth equations of Eq. (13.8.20) for the unknown temperatures, we obtain

$$t_2 = 106.1\,°F \qquad t_3 = 112.1\,°F \qquad t_4 = 117.6\,°F \qquad t_5 = 122.6\,°F \qquad (13.8.21)$$

Using Eq. (13.7.2), we obtain the heat flow into and out of the tube as

$$q_{in} = \dot{m}ct_1 = (4.72)(0.24)(100) = 113.28 \text{ Btu/h}$$

$$q_{out} = \dot{m}ct_5 = (4.72)(0.24)(122.6) = 138.9 \text{ Btu/h} \qquad (13.8.22)$$

where, again, the conduction contribution to q is negligible; that is, $-kA\Delta T$ is negligible. The analytical solution in Reference [7] yields

$$t_5 = 123.0\,°F \qquad q_{out} = 139.33 \text{ Btu/h} \qquad (13.8.23)$$

The finite element solution is then seen to compare quite favorably with the analytical solution. ■

The element with the stiffness matrix given by Eq. (13.8.13) has been used in Reference [8] to analyze heat exchangers. Both double-pipe and shell-and-tube heat exchangers were modeled to predict the length of tube needed to perform the task of proper heat exchange between two counterflowing fluids. Excellent agreement was found between the finite element solution and the analytical solutions described in Reference [9].

Finally, remember that when the variational formulation of a problem is difficult to obtain but the differential equation describing the problem is available, a residual method such as Galerkin's method can be used to solve the problem.

▲ 13.9 Flowchart of a Heat-Transfer Program ▲

Figure 13–26 is a flowchart of the finite element process used for the analysis of two-dimensional heat-transfer problems.

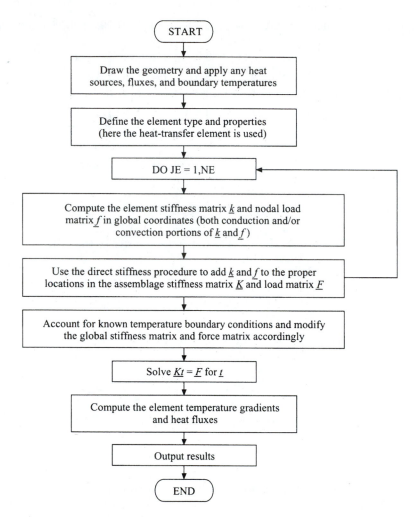

Figure 13–26 Flowchart of two-dimensional heat-transfer process

---

▲ ## 13.10 Algor Example Solutions for Heat-Transfer Problems ▲

To illustrate how to use the Algor computer program for steady-state heat-transfer analysis [10], we now solve four example problems.

The model geometry is drawn in Superdraw III. The *Surface number* command is used to select the lines for two-dimensional elements that have convection boundary conditions. The *Group* command is used to denote different material properties (that is, different thermal conductivities k) associated with each element. The solver processor is now called **SSAP10** instead of SSAP0. Just select "Steady-State Heat Transfer" analysis instead of "Linear Static Stress" analysis within the "Model Data

Control" window. This solver solves for nodal temperatures, whereas SSAP0 solved the static stress analysis problem for nodal displacements.

The available element types used with the thermal decoder are the two-dimensional thermal element with 3 or 4 nodes formulated in the y-z plane, the three-dimensional thermal brick element with 6 and 8 nodes formulated in three-dimensional space, the three-dimensional thermal tetrahedral element with 4 and 10 nodes formulated in three-dimensional space, and the thermal boundary element with 1 node formulated in space and used to apply a time-dependent temperature boundary condition to a node. The boundary element is defined using Superdraw III's "FEA Add:Heat Transfer Analysis:Applied Temperature..." commands. In the "Model Data Control" window, within the "Global" menu you set "Load case:A(Bdry)" to 1. Setting "A(Bdry)" to 0 does not disable the thermal boundary element; it just changes the applied temperature to 0.

The following examples are solved using the Algor program: (1) a plate with known boundary temperatures, convection from one surface, and two insulated sides, (2) a plate with symmetric temperature boundary conditions (Problem 13.14), (3) a square duct with an insulated perimeter, and (4) Problem 13.5 with an element added to the right end with constant heat generation acting on it to simulate the heat flux.

Example 13.8

The plate shown in Figure 13–27 has a constant temperature of $100\,°F$ on the left side and is exposed to convection heat transfer on the right side. The ambient temperature on the right side is $70\,°F$, and the heat transfer coefficient on the right side is $h = 20$ Btu/(h-ft^2-°F) $= 0.3603$ in.-lb/(s-in.-°F). The thermal conductivity of the plate is 25 Btu/(h-ft-°F) $= 5.404$ in.-lb/(s-in^2-°F). Determine the temperature variation throughout the plate.

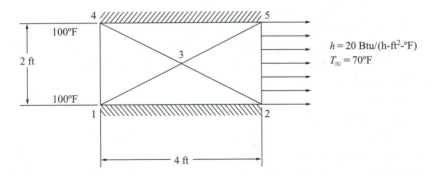

Figure 13–27 Two-dimensional heat-transfer problem for Algor analysis

Step 1 Start Superdraw III

Select the "Start" button of Windows NT/95/98 and then proceed to "Programs: Algor Software:Algor FEA".

Then click "Algor FEA" to start the program.

Or double-click on the "Algor FEA" icon on your desktop computer screen. The Superdraw III program appears. This is the Algor main graphics user interface. All your work will be done from this program. You will construct, analyze, and review your results from Superdraw III. Each menu option and a short example are available in the Superdraw III Reference Division (accessed through "Docutech", the on-line "Technical User Documentation").

Step 2   Create the Model

Select the "Modify" menu from the top main menu bar.

Click "Transform XY to YZ". This is used to transform the drawing to the *y-z* plane at the outset.

Select the "Add" menu from the top main menu bar.

Click "Line". A "Line" menu appears in the upper right part of the screen. Enter the data points for the outside rectangle in succession.

```
0<Tab>0<Tab>0
<Enter>
0<Tab>48<Tab>0
<Enter>
0<Tab>48<Tab>24
<Enter>
0<Tab>0<Tab>24
<Enter>
0<Tab>0<Tab>0
<Enter>                        You are finished with the outside rectangle.
```

Click "Done" on the menu in the upper right corner.

Note: You still must create the diagonal lines of the model.

Select the "Add" menu from the main menu bar.

Click "Line". A "Line" menu appears in the upper right part of the screen. Enter the data points for the first diagonal line of the model in succession.

```
0<Tab>0<Tab>0
<Enter>
0<Tab>24<Tab>12
<Enter>
0<Tab>48<Tab>24
<Enter>
```

Click "Done" on the menu in the upper right corner.

Note: You still must create the second diagonal line of the model.

Select the "Add" menu from the main menu bar.

Click "Line". A "Line" menu appears in the upper right part of the screen. Enter the data points for the second diagonal line of the model in succession.

```
0<Tab>0<Tab>24
```

```
<Enter>
0<Tab>24<Tab>12
<Enter>
0<Tab>48<Tab>0
<Enter>
```
                    All data points have been entered.

Click "Done" on the menu in the upper right corner.

Select "View". This takes you to the view options.

Click on "Enclose". This encloses the model. You now see the line drawing of the model on the screen. It should look like Figure 13–27 shown in this example.

Select "Select". This takes you to the select options.

Click on "None". This ensures that none of the previously drawn elements is selected.

Select "Select" again. This takes you back to the select options.

Click "Box" and then left-click near the upper location of the right side of the model, enclose the right side of the model, and again click to end the box. All other lines should not be selected.

Click "Done".

Click "Modify".

Select "Update Object Parameters".

Click "Surface Number".

2   Enter a 2 to make the right-side element surface 2.

```
<Enter>
```

Step 3 Eliminate Duplicate Lines

Click "Modify" on the main menu.

Select "Clean:Duplicate". This notation means to select "Clean" and then click "Duplicate" in succession. This command brings up the on-screen "Duplicate" menu.

Click "Perform Cleaning" on the "Duplicate" menu. You see the following message on the lower part of the screen:

"8 Kept 0 Deleted. done"

Click "Done" on the "Duplicate" menu in the upper right corner of the screen.

Step 4 Apply Temperature Boundary Conditions

Click "FEA Add".

Select "Heat Transfer Analysis".

Click "Applied Temperature".

Click "Temperature Value".

100   The left boundary has a temperature of $100\,°F$.

```
<Enter>
```

Click "Stiffness".

1e6   Make the stiffness 1e6 for good results.

```
<Enter>
```

Select "Box Apply" and create a box around the left edge of the model. You see "TB;100;L1:1:1e+006".

Click "Done".

Step 5 Go to the "Model Data Control" Window to Enter Material Properties

Click on the "Model Data" button at the bottom of the screen.

Select "Steady-State Heat Transfer" by scrolling down the "Analysis Type" menu.

Select the box below "Element".

Click "2D".

Click "OK".

Click on the box below "Data". A prompt appears: "Please enter model name first".

Click "OK".

Prob138

Click "Save".

Now put in the units you want.

Accept the default "English (in)" units system.

Click "OK". This takes you to the "Element Definitions" window.

1   Enter a 1 for the element thickness (1 in. thick).

Click "OK".

Click on the box below "Material".

Select "[Customer Defined]".

Select "View Properties".

Click "Unlock Properties".

5.404   Enter the thermal conductivity K in in.-lb/(s-in.-°F).

Click "OK".

Click "OK".

Click on the box below "Surface".

Click on the box next to "Surface 2".

0.3603   The convection coefficient h for surface 2 in in.-lb/(s-°F-in^2).

70   Ambient temperature T_∞ for surface 2 in °F.

Click "Apply".

Click "OK".

Click "Close" in the "Surface Definition" window.

Click on the "Global" box on the right side of the "Model Data Control" window.

Under "Multipliers" type a 1 next to "Boundary temperature multiplier" and next to "Convection multiplier".

1

1

Click on the "Output" tab and click on the "Nodal temperature data" box. The temperatures will be output into a *Prob138.L10* file. This file includes the echo of the input data and the output of the nodal temperatures.
Click "OK".

Step 6 Analyze the Model Using the SSAP10 Processor

The model is now analyzed using the SSAP10 processor for steady-state heat-transfer analysis.
Click "Analysis" on the right of the "Model Data Control" window.
Click "Analyze". Wait a short time for the analysis to be completed.
Click "OK".
Click "Done".

Step 7 Review the Results

Click "Results" on the right side of the "Model Data Control" window. This takes you to Superview.
Click "Show T-ht".
Click "Temp-ht". The temperature contours appear on the model as a color plot (Figure 13–28). The temperatures match the hand solution (particularly if you look at the *Prob138.L10* file). That is, $t_1 = 100\,°F$, $t_2 = 77.14\,°F$, $t_3 = 88.57\,°F$, $t_4 = 100\,°F$, $t_5 = 77.14\,°F$. (Also see Table 13–4.)
<Esc>

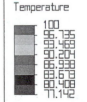

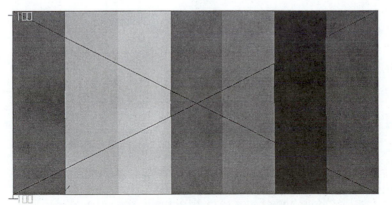

Figure 13–28 Temperature contours of Example 13.8

Table 13–4 Temperatures of nodes

Node Number	Temperature (°F)
1	100.00
2	77.14
3	88.57
4	100.00
5	77.14

`<Esc>`

Click "Done" to exit Superview. ■

Example 13.9

For the square two-dimensional body shown in Figure 13–29, determine the temperature distribution. Let $K = 10$ Btu/(h-ft-°F). The plate's top surface is maintained at 500°F, and the other three sides are maintained at 100°F.

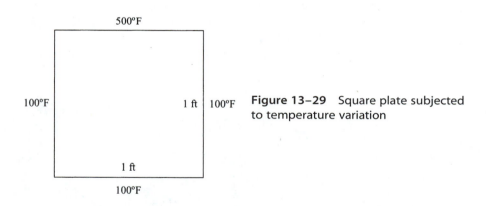

Figure 13–29 Square plate subjected to temperature variation

Step 1 Start Superdraw III

Double-click on the "FEA Algor" icon on your desktop computer screen.

Step 2 Create the Model

Click on the "Add" menu on the top main menu bar.

Click "Rectangle". Then enter the data points.

`0<Tab>0`   This is point 1 of the rectangle.

`<Enter>`   Press the ⟨Enter⟩ key to actually enter the data.

`1<Tab>1`   This is the opposite point of the rectangle.

`<Enter>`

Click "View". This takes you to the draw options menu.

Click on "Enclose". This encloses the model. You now see the body enclosed.

Step 3 Mesh the Model

Click "FEA Mesh".

Select "Two-Dimensional Mesh Generation".

Click "Generate Mesh". This takes you to the "Save As" menu. Enter the filename.

Ex139　　　　This is the filename.

Click "Save".

You will see a message: "Model information does not contain a specified unit system. Please choose a unit system first."

Click "OK". This takes you to the "Units Definition" window.

Select "English (ft)" as these are the units in which you drew the model.

Click "OK". You are now given the warning "Attention: 2-D element type requires the model in Y-Z plane. Do you want Superdraw III to convert your model?"

Click "YES". You are now sent to the "Two-Dimensional Mesh Generation" window.

Use the defaults "Quadrilateral", "Mesh Density 400", and so on.

Click "Generate" in the lower left corner of the window.

Click "OK".

Click "Done" to close the "Two-Dimensional Mesh Generation" window.

You now see the meshed model with 342 elements and 379 nodes.

Step 4 Eliminate Duplicate Lines

Click "Modify" on the main menu.

Select "Clean:Duplicate".

Click on "Perform Cleaning" on the "Duplicate" menu. You will see the following message at the lower part of the screen:

"720 Kept 0 Deleted. done"

Click "Done" on the "Duplicate" menu located in the upper right corner of the screen.

Step 5 Set Temperatures

Select the "FEA Add:Heat Transfer Analysis" menu.

Click "Applied Temperature".

Click "Temperature Values". A prompt "Temperature" appears at the bottom of the screen asking you to enter the temperature you wish to add.

`100`　　　　Enter the temperature 100.

`<Enter>`　　　You might have to move the screen to see the message "Click on node to apply the current values."

Click "Stiffness". A prompt "Stiffness" appears at the bottom of the screen asking you for the stiffness. Enter a very large number to make the boundary temperature remain constant.

`1e6` This is the recommended stiffness.

`<Enter>`

Click "Box Apply". Now box all the nodes on the left side of the model.

Click "Box Apply". This time box all the nodes on the bottom, repeat one more time, and do all the nodes on the right side.

Click "Temperature Values". This time change the value to 500.

`500`

`<Enter>` This is the temperature you will use for the top of the plate.

Click "Box Apply". Now box all the nodes on the top of the model.

Click "Done".

Step 6 Add the Material Properties Using the "Model Data Control" Window

After the mesh has been created and the temperatures have been applied, you can select "Tools" and "Model Data Control" from Superdraw III to complete the data input for analysis. The "Model Data Control" window appears.

Alternatively click on the "Model Data" button located in the lower part of the screen and highlighted in red.

For the "Analysis Type", "Linear Static Stress" appears highlighted as the default. You have to scroll down to

Select "Steady State Heat Transfer" and click on it.

Click on the box below "Element". A table of group 1 element types appears.

Click on "2-D". You have now created a two-dimensional mesh of a square plate.

Click "OK".

Click on the box below "Data". You are now at the "Element Definition" screen. Leave all the variables the same except make the thickness 1 ft.

Click on the box to the right of "Thickness".

`1`

`<Enter>`

Click "OK". This takes you back to the "Model Data Control" window.

Click on the box below "Material". This takes you to the "Element Material Selection" window.

Click "[Customer Defined]".

Click "Edit Properties" and enter the thermal conductivity of the material in the box next to "Thermal Conductivity" as

`10`

`<Enter>`

Click "OK". This takes you back to the "Element Material Selection" window.
Click "OK". This takes you back to the "Model Data Control" window.

Step 7 Enter the Global Data

Click "Global". You are now at the "Global Data" menu.
Under "Multipliers" type a 1 next to "Boundary temperature".
1
Click on the "Output" tab and then select "Nodal temperature data".
Click "OK". This takes you back to the "Model Data Control" window.

Step 8 Check Your Model

Click "Check" under "FEA Model".
You now see your model loaded with the temperatures listed along the edges in orange. You can now check the model in the usual manner.

Step 9 Analyze the Model

Click "Analysis" in the "Model Data Control" window.
Click "Analyze".
Click "OK" to the message shown on the screen.
Click "Done".

Step 10 Review Your Results in Superview

Click "Results" in the "Model Data Control" window.
Click "Show T-ht".
Click "Temp-ht". A color plot appears, and you see the temperature distribution of the block as shown in Figure 13–30.
Click "flux Mag". The flux magnitude contour plot appears with a maximum value of 50911 at the upper corners of the model.
<Esc>
<Esc>
Click "Done".
Select "Quit" to exit Superview.   ■

Example 13.10

The square duct shown in Figure 13–31 carries hot gases such that its surface temperature is 570°F. The duct is wrapped by a layer of circular fiberglass that has a thermal conductivity of $K = 0.020$ Btu/(h-ft-°F). The outer surface is exposed to convective heat transfer with a temperature of 110°F and a convection coefficient of 5 Btu/(h-ft^2-°F). Determine the temperature distribution within the fiberglass.

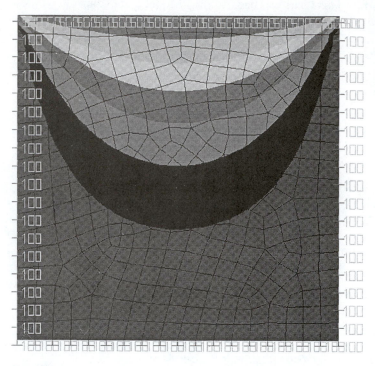

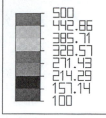

Figure 13–30 Temperature variation in plate of Example 13.9

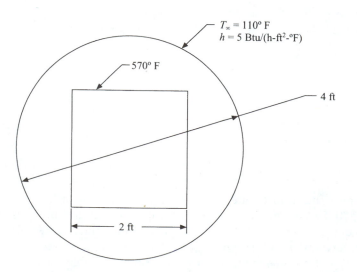

Figure 13–31 Square duct wrapped in insulation

Step 1 Start Superdraw III

Double-click on the "FEA Algor" icon on your desktop computer screen.

Step 2 Create the Model

Click on the "Add" menu on the top main menu bar.

Click "Rectangle". Then enter the data points (coordinates of opposite points) in succession.

0<Tab>0   This is point 1 of the rectangle.

<Enter>   Press the ⟨Enter⟩ key to actually enter the data.

2<Tab>2   This is the opposite point of the rectangle.

<Enter>

Select the "Add" menu from the top main menu bar.

Click "Circle" and "Center and Point". Then enter the data points (coordinates of center point and a point on the circle) in succession.

1<Tab>1   This is the center of the circle.

<Enter>

1<Tab>3   This is a point on the circle.

<Enter>

Select "View". This takes you to the draw options menu.

Click on "Enclose". This encloses the model. You now see the body enclosed.

Step 3 Mesh the Model

Select "FEA Mesh".

Select "Two-Dimensional Mesh Generation".

Click "Generate Mesh". This takes you to the "Save As" menu. Enter the filename.

Ex1310   This is the filename.

Click "Save".

You will see a message: "Model information does not contain a specified unit system. Please choose a unit system first."

Click "OK". This takes you to the "Units Definition" window.

Select "English (ft)" as these are the units in which you drew the model.

Click "OK". You are now given the warning "Attention: 2-D element type requires the model in Y-Z plane. Do you want Superdraw III to convert your model?"

Click "Yes". You are now sent to the "Two-Dimensional Mesh Generation" window.

Use the defaults "Quadrilateral", "Mesh Density 400", and so on.

Click "Generate" in the lower left corner of the window.

Click "OK".

Click "Done" to close the "Two-Dimensional Mesh Generation" window.

You now see the meshed model with 238 elements and 288 nodes.

Step 4   Eliminate Duplicate Lines

Click "Modify" on the main menu.

Select "Clean:Duplicate".

Click on "Perform Cleaning" on the "Duplicate" menu. You will see the following message at the lower part of the screen:

"526 Kept 0 Deleted. Done"

Click "Done" on the "Duplicate" menu located in the upper right corner of the screen.

Step 5   Change Surfaces of Sides

Click "Select" from the main menu, then "Box select", and place a box around the center square.

Select "Modify", "Update Object Parameters", and "Surface Number".

Click on the red "2" box. This sets the inner square to a separate surface number.

Click "Select", then click "Point". Click on the box next to "Toggle", and now click on each of the elements around the outside edge of the circle.

Click "Done".

Select "Modify", "Update Object Parameters", and "Surface Number".

Click on the yellow "3" box. This sets the outer circle to a separate surface number.

Step 6   Set Temperatures

Click on the "FEA Add:Heat Transfer Analysis" menu.

Click "Applied Temperature".

Click "Temperature Values". A prompt "Temperature" appears at the bottom of the screen asking you to enter the temperature you wish to add.

570   Enter the temperature.

<Enter>   You might have to move the screen to see the message "Click on node to apply the current values."

Click "Stiffness". A prompt "Stiffness" appears at the bottom of the screen asking you for the stiffness. Enter a very large number to make the boundary temperature remain at 570.

1e6

<Enter>

Click "Box Apply".

Then box around the inside of the square duct in the usual manner.

Nodal temperatures of 570 appear at each node on the inside edge of the duct.

Click "Done".

Step 7   Add the Material Properties Using the "Model Data
              Control" Window

After the mesh has been created and the temperatures have been applied, you can select "Tools" and "Model Data Control" from Superdraw III to complete the data input for analysis. The "Model Data Control" window appears. Alternatively, click on the "Model Data" button located in the lower part of the screen and highlighted in red.

For "Analysis Type", "Linear Static Stress" appears highlighted as the default. You have to scroll down to "Steady State Heat Transfer".

Click "Steady State Heat Transfer".

Click on the box below "Element". A table of group 1 element types appears.

Click on "2-D". You have now created a two-dimensional mesh.

Click "OK".

Click on the box below "Data". You are now at the "Element Data" screen.

You are now in the "Element Data" window. Leave all the other variables the same, but change the thickness to 1 ft.

Click on the box to the right of "Thickness".

1

<Enter>

Click "OK". This takes you back to the "Model Data Control" window.

Repeat this step for the two remaining colors.

Click on the box below "Material". This takes you to the "Element Material Selection" window.

Click "[Customer Defined]".

Then Select "Edit Properties" and enter the thermal conductivity of the material in the box next to "Thermal Conductivity".

0.020

<Enter>

Click "OK". This takes you back to the "Element Material Selection" window.

Click "OK". This takes you back to the "Model Data Control" window.

Click on the box below "Surface". This takes you to the "Surface Definition" window.

Click on the box beside the yellow "3". This takes you to the "Heat Surface 2D Element" window. Enter the convection coefficient next to its box, and the ambient temperature next to its box.

5

<Tab>

110

Click "OK". This takes you back to the "Surface Definition" window.

Click "Close". This takes you back to the "Model Data Control" window.

Step 8   Enter the Global Data

Click "Global". You are now at the "Global Data" menu screen.

Under "Multipliers", type a 1 next to "Boundary temperature" and "Convection" multipliers.

```
1
<Tab>
1
<Tab>
```

Click "OK". This takes you back to the "Model Data Control" window.

Step 9   Check Your Model

Click "Check" under "FEA Model".

You now see your model loaded with the temperatures along the edges of the internal square. You can now check the model in the usual manner.

Step 10   Analyze the Model

Click "Analysis" in the "Model Data Control" window.
Click "Analyze".
Click "OK" to the message shown on the screen.
Click "Done".

Step 10   Review Your Results in Superview

Click "Results" in the "Model Data Control" window.
Click "Show T-ht".

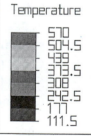

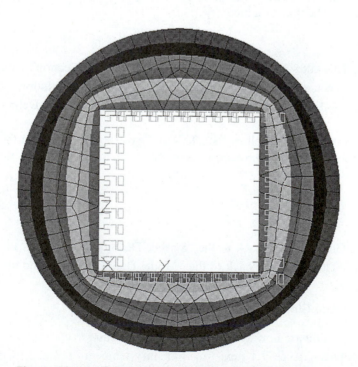

Figure 13–32   Temperature variation through insulation

Click "Temp-ht". A color plot appears, and you see the temperature distribution of the block with temperatures ranging from 570 °F at the duct to 111.5 °F at the outside edge of the insulation as shown in Figure 13–32.

Click "flux Mag". The flux magnitude contour plot appears with a maximum value of 20.234 Btu/(h-ft^2) at the inner corners of the duct.

`<Esc>`

`<Esc>`

Select "Quit" to exit Superview. ■

Example 13.11

The fin shown in Figure 13–33 is insulated on the perimeter. The left end has a constant temperature of 100 °C. A positive heat flux of 5000 W/m^2 acts on the right end. This flux will be modeled in Algor by using a heat generation within an element added to the right end of the body. Let $K_{xx} = 6$ W/(m · °C) and cross-sectional area $A = 0.1$ m^2. Determine the temperatures at $L/4, L/2, 3L/4$, and L, where $L = 0.4$ m. We will need to change the right-end added element to another group to include the heat generation. Also, remember that the nodal temperatures will be found in the *.L10* file now instead of the *.L* file where nodal displacements are stored.

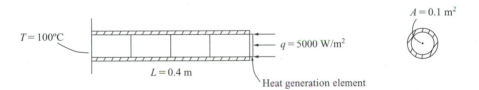

Figure 13–33 Fin insulated around the perimeter and subjected to flux

Step 1 Start Superdraw III

Click on the "Algor FEA" icon to start Superdraw III.

Step 2 Create the Model

The model will look like Figure P13.5 with a 0.01-m-long element added to the right end to apply a uniform heat generation to. The thickness of the model will be 1 m.

Click "Add" on the main menu bar.

Click "Line".

(0, 0, 0)   First coordinate.

`<Enter>`

(0.1, 0, 0)

`<Enter>`

(0.2, 0, 0)

`<Enter>`

(0.3, 0, 0)

```
<Enter>
(0.4, 0, 0)
<Enter>
(0.40, 0.1, 0)
<Enter>
(0.30, 0.1, 0)
<Enter>
(0.20, 0.1, 0)
<Enter>
(0.10, 0.10, 0)
<Enter>
(0, 0.10, 0)
<Enter>
(0, 0, 0)
<Enter>
```

Click "Done". Now add the vertical lines.

Click "Add".

Click "Line" and check "Single".

```
(0.1, 0, 0)
<Enter>
(0.1, 0.1, 0)
<Enter>
```

Click "Done".

Click "Add".

Click "Line" and check "Single".

```
(0.2, 0, 0)
<Enter>
(0.2, 0.1, 0)
<Enter>
```

Click "Done".

Click "Add".

Click "Line" and check "Single".

```
(0.3, 0, 0)
<Enter>
(0.3, 0.1, 0)
<Enter>
```

Click "Done". The main model is now complete. Note that the vertical lines could have been created by clicking "Add" and then "Line" and then right-clicking near the bottom and top nodes at the quarter-points along the horizontal lines.

Click "Add". Add the heat generation element to the right end of the model. This will be group 2.

Click "Line".

(0.4, 0, 0)

<Enter>

(0.41, 0, 0)

<Enter>

(0.41, 0.1, 0)

<Enter>

(0.40, 0.1, 0)

<Enter>

(0.40, 0, 0)   Last coordinate of the heat generation element.

<Enter>

Click "Done".

Click "View".

Click "Enclose". You now see the model in Figure 13–13.

Make sure to highlight only the outer line of the model. All other lines should not be selected.

Click "Modify".

Select "Update Object Parameters".

Click "Group Number".

2   Enter a 2 to make the outer last element group 2.

<Enter>   The last element turns red. You now have the original model as group 1 and the thin right-end model as group 2.

Step 3   Eliminate Duplicate Lines

Click "Modify".

Click "Clean".

Click "Duplicate".

Click "Perform Cleaning".

A message appears:

                    "Kept 17 Deleted 0. Done"

Note that the overlapping lines for groups 1 and 2 are maintained, as they should be.

Click "Done".

Step 4   Apply Temperature Boundary Conditions

Click "FEA Add".

Select "Heat Transfer Analysis".

Click "Applied Temperature".

Click "Temperature Value".

100   The left boundary has a temperature of 100 °C.

<Enter>

Click "Stiffness".

1e6   Make the stiffness 1e6 for good results.

<Enter>

Select "Box Apply" and create a box around the left edge of the model. You see "TB;100;L1:1:1e+006".

Click "Done".

Step 5   Go to the "Model Data Control" Window to Enter Material
           Properties

Click on the "Model Data" button at the bottom of the screen. You now see two groups, group 1 in green and group 2 in red.

Select "Steady-State Heat Transfer" by scrolling down the "Analysis Type" menu.

Select the first box below "Element".

Click "2D" for group 1.

Click "OK". A message appears: "2-D element type requires the model in y-z plane. Do you want Superdraw III to convert your model?"

Click "Yes". The model is now in the y-z plane.

Click the second box below "Element".

Click "2D".

Click "OK".

Click the box below "Data". A prompt appears: "Please enter model name first".

Click "OK".

Ex1311

Click "Save".

Now put in the units you want.

Change to metric units by clicking "Metric mks (SI)".

Click "OK". This takes you to "Element Definition" window.

1   Enter a 1 for the element thickness (1 m thick).

Click "OK".

Click on the first box below "Material".

Select "[Customer Defined]".

Select "View Properties".

Click "Unlock Properties".

6   Enter the thermal conductivity K in W/(m · °C) units.

Click "OK".

Click "OK".

Click the second box below "Data".

The "Element Definition" window appears.

1   This is the thickness of the group 2 element.

Click on the "Loading" tab. This takes you to the "Heat Generation" window.

500000   For internal heat generation we want 500 W at the end. To get this value, we multiply the element volume by the heat generation per unit volume: $(500,000 \ W/m^3)(0.01 \ m \ long \times 1 \ m \ thick \times 0.1 \ m \ high) = 500 \ W.$

Click "OK".

Click on the box below "Material".

Select "[Customer Defined]".

6   Enter the thermal conductivity of the second group material.

Click "OK".

Click "OK".

Click "Global" box.

Under "Multipliers" type a 1 next to "Boundary temperature multiplier" and next to "Heat generation multiplier".

1

1

Click on the "Output" tab and click on the "Nodal temperature data" box. The temperatures are output into a *prob135.L10* file. This file includes the echo of the input data and the output of the nodal temperatures.

Click "OK".

Step 6   Analyze the Model Using the SSAP10 Processor

The model is now analyzed using the SSAP10 processor for steady-state heat-transfer analysis.

Click "Analysis".

Click "Analyze". Wait a short time for the analysis to be completed.

Click "OK".

Click "Done".

Step 7   Review the Results

Click on the "Results" button in the "Model Data Control" window. This takes you to Superview.

Click "Show T-ht".

Click "Temp-ht". The temperature contours appear on the model as a color plot. The temperatures match the hand solution (particularly if you look at the *prob135.L10* file). That is, $t_2 = 183 \,°C$, $t_3 = 266.7 \,°C$, $t_4 = 350 \,°C$, and $t_5 = 433.3 \,°C$.

Click "flux Mag" for the magnitude of the heat flux throughout the model.

`<Esc>`

`<Esc>`

Click "Done" to exit Superview. ■

▲ References

[1] Holman, J. P., *Heat Transfer*, 8th ed., McGraw-Hill, New York, 1997.

[2] Kreith, F., and Black, W. Z., *Basic Heat Transfer*, Harper & Row, New York, 1980.

[3] Lyness, J. F., Owen, D. R. J., and Zienkiewicz, O. C., "The Finite Element Analysis of Engineering Systems Governed by a Non-Linear Quasi-Harmonic Equation," *Computers and Structures*, Vol. 5, pp. 65–79, 1975.

[4] Zienkiewicz, O. C., and Cheung, Y. K., "Finite Elements in the Solution of Field Problems," *The Engineer*, pp. 507–510, Sept. 24, 1965.

[5] Wilson, E. L., and Nickell, R. E., "Application of the Finite Element Method to Heat Conduction Analysis," *Nuclear Engineering and Design*, Vol. 4, pp. 276–286, 1966.

[6] Emery, A. F., and Carson, W. W., "An Evaluation of the Use of the Finite Element Method in the Computation of Temperature," *Journal of Heat Transfer*, American Society of Mechanical Engineers, pp. 136–145, May 1971.

[7] Rohsenow, W. M., and Choi, H. Y., *Heat, Mass, and Momentum Transfer*, Prentice-Hall, Englewood Cliffs, NJ, 1963.

[8] Goncalves, L., *Finite Element Analysis of Heat Exchangers*, M.S. Thesis, Rose-Hulman Institute of Technology, Terre Haute, IN, 1984.

[9] Kern, D. Q., and Kraus, A. D., *Extended & Surface Heat Transfer*, McGraw-Hill, New York, 1972.

[10] *Heat Transfer Reference Division*, Docutech On-line Documentation, Algor, Inc., Pittsburgh, PA.

▲ Problems

13.1 For the one-dimensional composite bar shown in Figure P13–1, determine the interface temperatures. For element 1, let $K_{xx} = 200$ W/(m · °C); for element 2, let $K_{xx} = 100$ W/(m · °C); and for element 3, let $K_{xx} = 50$ W/(m · °C). Let $A = 0.1$ m². The left end has a constant temperature of 100 °C and the right end has a constant temperature of 300 °C.

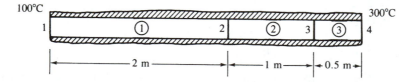

Figure P13–1

13.2 For the one-dimensional rod shown in Figure P13–2 (insulated except at the ends), determine the temperatures at $L/3, 2L/3$, and L. Let $K_{xx} = 3$ Btu/(h.-in.-°F), $h = 1.0$ Btu/(h-in²-°F), and $T_\infty = 0$ °F. The temperature at the left end is 200 °F.

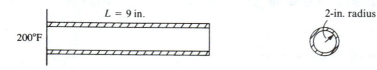

Figure P13–2

13.3 A rod with uniform cross-sectional area of 2 in² and thermal conductivity of 3 Btu/(h-in.-°F) has heat flow in the x direction only (Figure P13–3). The right end is insulated. The left end is maintained at 50°F, and the system has the linearly distributed heat flux shown.

Use a two-element model and estimate the temperature at the node points and the heat flow at the left boundary.

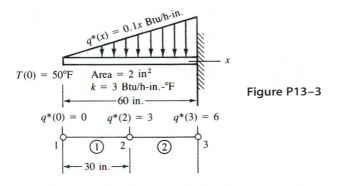

Figure P13–3

13.4 The rod of 1-in. radius shown in Figure P13–4 generates heat internally at the rate of uniform $Q = 10,000$ Btu/(h-ft³) throughout the rod. The left edge and perimeter of the rod are insulated, and the right edge is exposed to an environment of $T_\infty = 100$°F. The convection heat-transfer coefficient between the wall and the environment is $h = 100$ Btu/(h-ft²-°F). The thermal conductivity of the rod is $K_{xx} = 12$ Btu/(h-ft-°F). The length of the rod is 3 in. Calculate the temperature distribution in the rod. Use at least three elements in your finite element model.

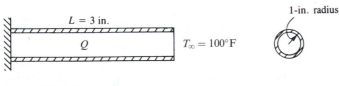

Figure P13–4

13.5 The fin shown in Figure P13–5 is insulated on the perimeter. The left end has a constant temperature of 100°C. A positive heat flux of $q^* = 5000$ W/m² acts on the right end. Let $K_{xx} = 6$ W/(m · °C) and cross-sectional area $A = 0.1$ m². Determine the temperatures at $L/4, L/2, 3L/4,$ and L, where $L = 0.4$ m.

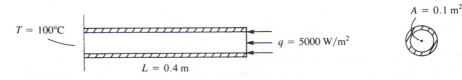

Figure P13–5

13.6 For the composite wall shown in Figure P13–6, determine the interface temperatures. What is the heat flux through the 8-cm portion? Use the finite element method. Use three elements with the nodes shown. 1 cm = 0.01 m.

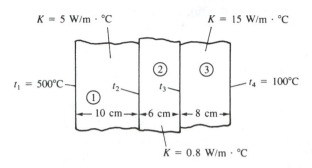

Figure P13–6

13.7 For the composite wall idealized by the one-dimensional model shown in Figure P13–7, determine the interface temperatures. For element 1, let $K_{xx} = 5$ W/(m · °C); for element 2, $K_{xx} = 10$ W/(m · °C); and for element 3, $K_{xx} = 15$ W/(m · °C). The left end has a constant temperature of $200\,°C$ and the right end has a constant temperature of $600\,°C$.

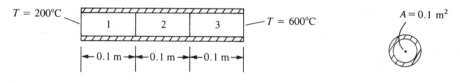

Figure P13–7

13.8 Use the direct method to derive the element equations for the one-dimensional steady-state conduction heat-transfer problem shown in Figure P13–8. The bar is insulated all around and has cross-sectional area A, length L, and thermal conductivity K_{xx}. Determine the relationship between nodal temperatures t_1 and t_2 (°F) and the thermal inputs F_1 and F_2 (in Btu). Use Fourier's law of heat conduction for this case.

Figure P13–8

13.9 For the element shown in Figure P13–9, determine the $\underline{k}$ and $\underline{f}$ matrices. The conductivities are $K_{xx} = K_{yy} = 15$ Btu/(h-ft-°F) and the convection coefficient is $h = 20$

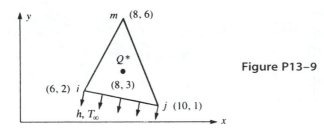

Figure P13-9

Btu/(h-ft²-°F). Convection occurs across the *i-j* surface. The free-stream temperature is $T_\infty = 70°F$. The coordinates are expressed in units of feet. Let the line source be $Q^* = 150$ Btu/(h-ft) as located in the figure. Take the thickness of the element to be 1 ft.

13.10 Calculate the $\underline{k}$ and $\underline{f}$ matrices for the element shown in Figure P13–10. The conductivities are $K_{xx} = K_{yy} = 15$ W/(m · °C) and the convection coefficient is $h = 20$ W/(m² · °C). Convection occurs across the *i-m* surface. The free-stream temperature is $T_\infty = 15°C$. The coordinates are shown expressed in units of meters. Let the line source be $Q^* = 100$ W/m as located in the figure. Take the thickness of the element to be 1 m.

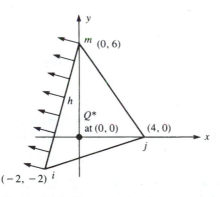

Figure P13-10

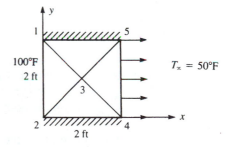

Figure P13-11

13.11 For the square two-dimensional body shown in Figure P13–11, determine the temperature distribution. Let $K_{xx} = K_{yy} = 25$ Btu/(h-ft-°F) and $h = 10$ Btu/(h-ft²-°F). Convection occurs across side 4–5. The free-stream temperature is $T_\infty = 50°F$. The temperatures at nodes 1 and 2 are $100°F$. The dimensions of the body are shown in the figure. Take the thickness of the body to be 1 ft.

13.12 For the square plate shown in Figure P13–12, determine the temperature distribution. Let $K_{xx} = K_{yy} = 10$ W/(m · °C) and $h = 20$ W/(m² · °C). The temperature along the left side is maintained at $100°C$ and that along the top side is maintained at $200°C$.

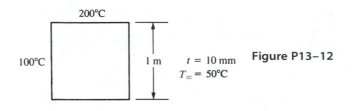

Figure P13–12

Use the computer program Algor to calculate the temperature distribution in the following two-dimensional bodies.

13.13 For the body shown in Figure P13–13, determine the temperature distribution. Surface temperatures are shown in the figure. The body is insulated along the top and bottom edges, and $K_{xx} = K_{yy} = 1.0$ Btu/(h-in.-°F). No internal heat generation is present.

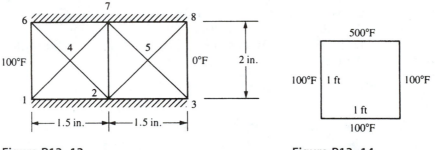

Figure P13–13   Figure P13–14

13.14 For the square two-dimensional body shown in Figure P13–14, determine the temperature distribution. Let $K_{xx} = K_{yy} = 10$ Btu/(h-ft-°F). The top surface is maintained at 500°F and the other three sides are maintained at 100°F. Also, plot the temperature contours on the body.

13.15 For the square two-dimensional body shown in Figure P13–15, determine the temperature distribution. Let $K_{xx} = K_{yy} = 10$ Btu/(h-ft-°F) and $h = 10$ Btu/(h-ft²-°F). The top face is maintained at 500°F, the left face is maintained at 100°F, and the other two faces are exposed to an environmental (free-stream) temperature of 100°F. Also, plot the temperature contours on the body.

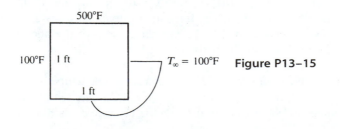

Figure P13–15

13.16 Hot water pipes are located on 2.0-ft centers in a concrete slab with $K_{xx} = K_{yy} = 0.80$ Btu/(h-ft-°F), as shown in Figure P13–16. If the outside surfaces of the concrete are at 85 °F and the water has an average temperature of 200 °F, determine the temperature distribution in the concrete slab. Plot the temperature contours through the concrete. Use symmetry in your finite element model.

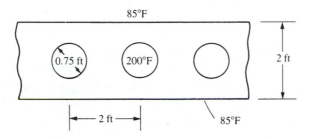

Figure P13–16

13.17 The cross section of a tall chimney shown in Figure P13–17 has an inside surface temperature of 330 °F and an exterior temperature of 130 °F. The thermal conductivity is $K = 0.5$ Btu/(h-ft-°F). Determine the temperature distribution within the chimney per unit length. Remember that the two circular openings cannot touch when modeling in Superdraw III.

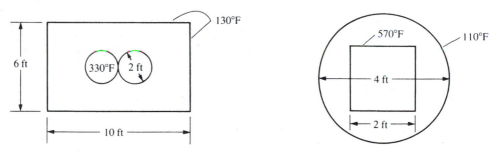

Figure P13–17   **Figure P13–18**

13.18 The square duct shown in Figure P13–18 carries hot gases such that its surface temperature is 570 °F. The duct is insulated by a layer of circular fiberglass that has a thermal conductivity of $K = 0.020$ Btu/(h-ft-°F). The outside surface temperature of the fiberglass is maintained at 110 °F. Determine the temperature distribution within the fiberglass.

13.19 The buried pipeline in Figure P13–19 transports oil with an average temperature of 60 °F. The pipe is located 15 ft below the surface of the earth. The thermal conductivity of the earth is 0.6 Btu/(h-ft-°F). The surface of the earth is 50 °F. Determine the temperature distribution in the earth.

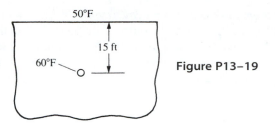

Figure P13–19

13.20 A 10-in.-thick concrete bridge deck is embedded with heating cables, as shown in Figure P13–20. If the lower surface is at 0 °F, the rate of heat generation (assumed to be the same in each cable) is 100 Btu/(h-in.) and the top surface of the concrete is at 35 °F. The thermal conductivity of the concrete is 0.500 Btu/(h-ft-°F). What is the temperature distribution in the slab? Use symmetry in your model. To model the heat source in Algor, use a small element of uniform heat source acting on it.

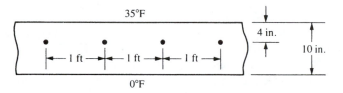

Figure P13–20

13.21 For the circular body with holes shown in Figure P13–21, determine the temperature distribution. The inside surfaces of the holes have temperatures of 150 °C. The outside of the circular body has a temperature of 30 °C. Let $K_{xx} = K_{yy} = 10$ W/(m · °C).

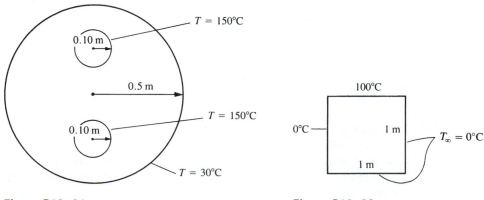

Figure P13–21

Figure P13–22

13.22 For the square two-dimensional body shown in Figure P13–22, determine the temperature distribution. Let $K_{xx} = K_{yy} = 10$ W/(m · °C) and $h = 10$ W/(m² · °C). The top face is maintained at 100 °C, the left face is maintained at 0 °C, and the other two faces are exposed to a free-stream temperature of $0\,°C$. Also, plot the temperature contours on the body.

13.23 A 200-mm-thick concrete bridge deck is embedded with heating cables as shown in Figure P13–23. If the lower surface is at $-10\,°C$ and the upper surface is at $5\,°C$, what is the temperature distribution in the slab? The heating cables are line sources generating heat of $Q^* = 50$ W/m. The thermal conductivity of the concrete is 1.2 W/(m · °C). Use symmetry in your model. To model the heat source in Algor, use a small element of uniform heat source acting on it.

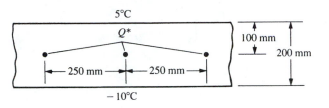

Figure P13–23

13.24 For the two-dimensional body shown in Figure P13–24, determine the temperature distribution. Let the left and right ends have constant temperatures of $200\,°C$ and $100\,°C$, respectively. Let $K_{xx} = K_{yy} = 5$ W/(m · °C). The body is insulated along the top and bottom.

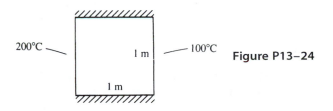

Figure P13–24

13.25 For the two-dimensional body shown in Figure P13–25, determine the temperature distribution. The top and bottom sides are insulated. The right side is subjected to heat transfer by convection. Let $K_{xx} = K_{yy} = 10$ W/(m · °C).

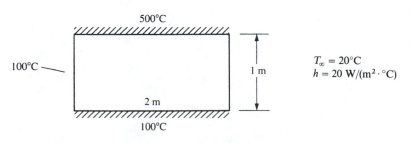

Figure P13–25

13.26 For the two-dimensional body shown in Figure P13–26, determine the temperature distribution. The left and right sides are insulated. The top surface is subjected to heat transfer by convection. The bottom and internal portion surfaces are maintained at 300 °C.

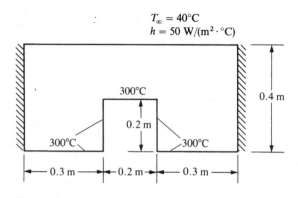

Figure P13–26

Fluid Flow

Introduction

In this chapter, we consider the flow of fluid through porous media, such as the flow of water through an earthen dam, through pipes, and around solid bodies. We will observe that the form of the equations is the same as that for heat transfer described in Chapter 13.

We begin with a derivation of the basic differential equation in one dimension for an ideal fluid in a steady state, not rotating (that is, the fluid particles are translating only), incompressible (constant mass density), and inviscid (having no viscosity). We then extend this derivation to the two-dimensional case. We also consider the units used for the physical quantities involved in fluid flow. For more advanced topics, such as viscous flow, compressible flow, and three-dimensional problems, consult Reference [1].

We will use the same procedure to develop the element equations as in the heat-transfer problem; that is, we define an assumed fluid head for the flow through porous media (seepage) problem or velocity potential for flow of fluid through pipes and around solid bodies within each element. Then, to obtain the element equations, we use both a direct approach similar to that used in Chapters 2, 3, and 5 to develop the element equations and the minimization of a functional as used in Chapter 13. These equations result in matrices analogous to the stiffness and force matrices of the stress analysis problem or the conduction and associated force matrices of the heat-transfer problem.

Next, we consider both one- and two-dimensional finite element formulations of the fluid-flow problem and provide examples of one-dimensional fluid flow through porous media and through pipes and of flow within a two-dimensional region.

Finally, we describe how the Algor computer program is applied to solve two-dimensional fluid-flow problems.

▲ 14.1 Derivation of the Basic Differential Equations ▲

Fluid Flow Through a Porous Medium

Let us first consider the derivation of the basic differential equation for the one-dimensional problem of fluid flow through a porous medium. The purpose of this

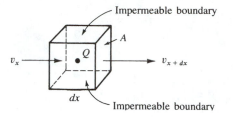

Figure 14–1　Control volume for one-dimensional fluid flow

derivation is to present a physical insight into the fluid-flow phenomena, which must be understood so that the finite element formulation of the problem can be fully comprehended. (For additional information on fluid flow, consult References [2] and [3]). We begin by considering the control volume shown in Figure 14–1. By conservation of mass, we have

$$M_{\text{in}} + M_{\text{generated}} = M_{\text{out}} \tag{14.1.1}$$

or

$$\rho v_x A \, dt + \rho Q \, dt = \rho v_{x+dx} A \, dt \tag{14.1.2}$$

where

M_{in} is the mass entering the control volume, in units of kilograms or slugs.

$M_{\text{generated}}$ is the mass generated within the body.

M_{out} is the mass leaving the control volume.

v_x is the velocity of the fluid flow at surface edge x, in units of m/s or in./s.

v_{x+dx} is the velocity of the fluid leaving the control volume at surface edge $x + dx$.

t is time, in s.

Q is an internal fluid source (an internal volumetric flow rate), in m^3/s or in^3/s.

ρ is the mass density of the fluid, in kg/m^3 or $slugs/in^3$.

A is the cross-sectional area perpendicular to the fluid flow, in m^2 or in^2.

By Darcy's law, we relate the velocity of fluid flow to the hydraulic gradient (the change in fluid head with respect to x) as

$$v_x = -K_{xx}\frac{d\phi}{dx} = -K_{xx}g_x \tag{14.1.3}$$

where

K_{xx} is the permeability coefficient of the porous medium in the x direction, in m/s or in./s.

ϕ is the fluid head, in m or in.

$d\phi/dx = g_x$ is the fluid head gradient or hydraulic gradient, which is a unitless quantity in the seepage problem.

Equation (14.1.3) states that the velocity in the x direction is proportional to the gradient of the fluid head in the x direction. The minus sign in Eq. (14.1.3) implies that fluid flow is positive in the direction opposite the direction of fluid head increase, or that the fluid flows in the direction of lower fluid head. Equation (14.1.3) is analogous to Fourier's law of heat conduction, Eq. (13.1.3).

Similarly,

$$v_{x+dx} = -K_{xx} \frac{d\phi}{dx}\bigg|_{x+dx} \tag{14.1.4}$$

where the gradient is now evaluated at $x + dx$. By Taylor series expansion, similar to that used in obtaining Eq. (13.1.5), we have

$$v_{x+dx} = -\left[K_{xx} \frac{d\phi}{dx} + \frac{d}{dx}\left(K_{xx} \frac{d\phi}{dx} \right) dx \right] \tag{14.1.5}$$

where a two-term Taylor series has been used in Eq. (14.1.5). On substituting Eqs. (14.1.3) and (14.1.5) into Eq. (14.1.2), dividing Eq. (14.1.2) by $\rho A\, dx\, dt$, and simplifying, we have the equation for one-dimensional fluid flow through a porous medium as

$$\frac{d}{dx}\left(K_{xx} \frac{d\phi}{dx} \right) + \bar{Q} = 0 \tag{14.1.6}$$

where $\bar{Q} = Q/A\, dx$ is the volume flow rate per unit volume in units 1/s. For a constant permeability coefficient, Eq. (14.1.6) becomes

$$K_{xx} \frac{d^2\phi}{dx^2} + \bar{Q} = 0 \tag{14.1.7}$$

The boundary conditions are of the form

$$\phi = \phi_B \qquad \text{on } S_1 \tag{14.1.8}$$

where ϕ_B represents a known boundary fluid head and S_1 is a surface where this head is known and

$$v_x^* = -K_{xx} \frac{d\phi}{dx} = \text{constant} \qquad \text{on } S_2 \tag{14.1.9}$$

where S_2 is a surface where the prescribed velocity v_x^* or gradient is known. On an impermeable boundary, $v_x^* = 0$.

Comparing this derivation to that for the one-dimensional heat conduction problem in Section 13.1, we observe numerous analogies among the variables; that is, ϕ is analogous to the temperature function T, v_x is analogous to heat flux, and K_{xx} is analogous to thermal conductivity.

Now consider the two-dimensional fluid flow through a porous medium, as shown in Figure 14–2. As in the one-dimensional case, we can show that for material

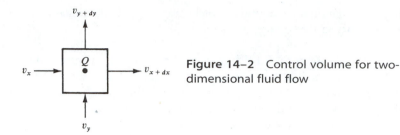

Figure 14–2 Control volume for two-dimensional fluid flow

properties coinciding with the global x and y directions,

$$\frac{\partial}{\partial x}\left(K_{xx}\frac{\partial \phi}{\partial x}\right) + \frac{\partial}{\partial y}\left(K_{yy}\frac{\partial \phi}{\partial y}\right) + \bar{Q} = 0 \tag{14.1.10}$$

with boundary conditions

$$\phi = \phi_B \qquad \text{on } S_1 \tag{14.1.11}$$

and

$$K_{xx}\frac{\partial \phi}{\partial x}C_x + K_{yy}\frac{\partial \phi}{\partial y}C_y = \text{constant} \qquad \text{on } S_2 \tag{14.1.12}$$

where C_x and C_y are direction cosines of the unit vector normal to the surface S_2, as previously shown in Figure 13–4.

Fluid Flow in Pipes and Around Solid Bodies

We now consider the steady-state irrotational flow of an incompressible and inviscid fluid. For the ideal fluid, the fluid particles do not rotate; they only translate, and the friction between the fluid and the surfaces is ignored. Also, the fluid does not penetrate into the surrounding body or separate from the surface of the body, which could create voids.

The equations for this fluid motion can be expressed in terms of the stream function or the velocity potential function. We will use the velocity potential analogous to the fluid head that was used for the derivation of the differential equation for flow through a porous medium in the preceding subsection.

The velocity v of the fluid is related to the velocity potential function ϕ by

$$v_x = -\frac{\partial \phi}{\partial x} \qquad v_y = -\frac{\partial \phi}{\partial y} \tag{14.1.13}$$

where v_x and v_y are the velocities in the x and y directions, respectively. In the absence of sources or sinks Q, conservation of mass in two dimensions yields the two-dimensional differential equation as

$$\frac{\partial^2 \phi}{\partial x^2} + \frac{\partial^2 \phi}{\partial y^2} = 0 \tag{14.1.14}$$

Equation (14.1.14) is analogous to Eq. (14.1.10) when we set $K_{xx} = K_{yy} = 1$ and $Q = 0$. Hence, Eq. (14.1.14) is just a special form of Eq. (14.1.10). The boundary

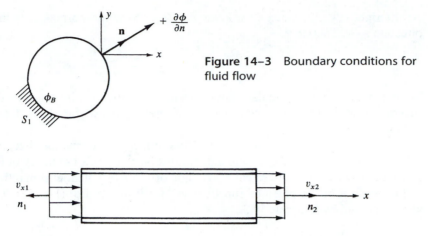

Figure 14-3 Boundary conditions for fluid flow

Figure 14-4 Known velocities at left and right edges of a pipe

conditions are

$$\phi = \phi_B \qquad \text{on } S_1 \qquad (14.1.15)$$

and

$$\frac{\partial \phi}{\partial x} C_x + \frac{\partial \phi}{\partial y} C_y = \text{constant} \qquad \text{on } S_2 \qquad (14.1.16)$$

where C_x and C_y are again direction cosines of unit vector $\mathbf{n}$ normal to surface S_2. Also see Figure 14–3. That is, Eq. (14.1.15) states that the velocity potential ϕ_B is known on a boundary surface S_1, whereas Eq. (14.1.16) states that the potential gradient or velocity is known normal to a surface S_2, as indicated for flow out of the pipe shown in Figure 14–3.

To clarify the sign convention on the S_2 boundary condition, consider the case of fluid flowing through a pipe in the positive x direction, as shown in Figure 14–4. Assume we know the velocities at the left edge (1) and the right edge (2). By Eq. (14.1.13) the velocity of the fluid is related to the velocity potential by

$$v_x = -\frac{\partial \phi}{\partial x}$$

At the left edge (1) assume we know $v_x = v_{x1}$. Then

$$v_{x1} = -\frac{\partial \phi}{\partial x}$$

But the normal is always positive away, or outward, from the surface. Therefore, positive n_1 is directed to the left, whereas positive x is to the right, resulting in

$$\frac{\partial \phi}{\partial n_1} = -\frac{\partial \phi}{\partial x} = v_{x1} = v_{n1}$$

At the right edge (2) assume we know $v_x = v_{x2}$. Now the normal n_2 is in the same direction as x. Therefore,

$$\frac{\partial \phi}{\partial n_2} = \frac{\partial \phi}{\partial x} = -v_{x2} = -v_{n2}$$

We conclude that the boundary flow velocity is positive if directed into the surface (region), as at the left edge, and is negative if directed away from the surface, as at the right edge.

At an impermeable boundary, the flow velocity and thus the derivative of the velocity potential normal to the boundary must be zero. At a boundary of uniform or constant velocity, any convenient magnitude of velocity potential ϕ may be specified as the gradient of the potential function; see, for instance, Eq. (14.1.13). This idea is also illustrated by Example 14.3.

▲ 14.2 One-Dimensional Finite Element Formulation ▲

We can proceed directly to the one-dimensional finite element formulation of the fluid-flow problem by now realizing that the fluid-flow problem is analogous to the heat-conduction problem of Chapter 13. We merely substitute the fluid velocity potential function ϕ for the temperature function T, the vector of nodal potentials denoted by $\{p\}$ for the nodal temperature vector $\{t\}$, fluid velocity v for heat flux q, and permeability coefficient K for flow through a porous medium instead of the conduction coefficient K. If fluid flow through a pipe or around a solid body is considered, then K is taken as unity. The steps are as follows.

Step 1   Select Element Type

The basic two-node element is again used, as shown in Figure 14–5, with nodal fluid heads, or potentials, denoted by p_1 and p_2.

Step 2   Choose a Potential Function

We choose the potential function ϕ similarly to the way we chose the temperature function of Section 13.4, as

$$\phi = N_1 p_1 + N_2 p_2 \tag{14.2.1}$$

where p_1 and p_2 are the nodal potentials (or fluid heads in the case of the seepage problem) to be determined, and

$$N_1 = 1 - \frac{\hat{x}}{L} \qquad N_2 = \frac{\hat{x}}{L} \tag{14.2.2}$$

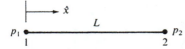

Figure 14–5   Basic one-dimensional fluid-flow element

Table 14–1 Permeabilities of granular materials

Material	K (cm/s)
Clay	1×10^{-8}
Sandy clay	1×10^{-3}
Ottawa sand	$2-3 \times 10^{-2}$
Coarse gravel	1

are again the same shape functions used for the temperature element. The matrix $[N]$ is then

$$[N] = \left[1 - \frac{\hat{x}}{L} \quad \frac{\hat{x}}{L} \right] \tag{14.2.3}$$

Step 3 Define the Gradient/Potential and Velocity/Gradient Relationships

The hydraulic gradient matrix $\{g\}$ is given by

$$\{g\} = \left\{ \frac{d\phi}{d\hat{x}} \right\} = [B]\{p\} \tag{14.2.4}$$

where $[B]$ is identical to Eq. (13.4.7), given by

$$[B] = \left[-\frac{1}{L} \quad \frac{1}{L} \right] \tag{14.2.5}$$

and

$$\{p\} = \left\{ \begin{array}{c} p_1 \\ p_2 \end{array} \right\} \tag{14.2.6}$$

The velocity/gradient relationship based on Darcy's law is given by

$$v_x = -[D]\{g\} \tag{14.2.7}$$

where the material property matrix is now given by

$$[D] = [K_{xx}] \tag{14.2.8}$$

with K_{xx} the permeability of the porous medium in the x direction. Typical permeabilities of some granular materials are listed in Table 14–1. High permeabilities occur when $K > 10^{-1}$ cm/s, and when $K < 10^{-7}$ the material is considered to be nearly impermeable. For ideal flow through a pipe or over a solid body, we arbitrarily—but conveniently—let $K = 1$.

Step 4 Derive the Element Stiffness Matrix and Equations

The fluid-flow problem has a stiffness matrix that can be found using the first term on the right side of Eq. (13.4.17). That is, the fluid-flow stiffness matrix is analogous to the conduction part of the stiffness matrix in the heat-transfer problem. There is no

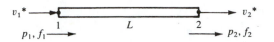

Figure 14–6 Fluid element subjected to nodal velocities

comparable convection matrix to be added to the stiffness matrix. However, we will choose to use a direct approach similar to that used initially to develop the stiffness matrix for the bar element in Chapter 3.

Consider the fluid element shown in Figure 14–6 with length L and uniform cross-sectional area A. Recall that the stiffness matrix is defined in the structure problem to relate nodal forces to nodal displacements or in the temperature problem to relate nodal rates of heat flow to nodal temperatures. In the fluid-flow problem, we define the stiffness matrix to relate nodal volumetric fluid-flow rates to nodal potentials or fluid heads as $\underline{f} = \underline{k}\underline{p}$. Therefore,

$$f = v^* A \tag{14.2.9}$$

defines the volumetric flow rate f in units of cubic meters or cubic inches per second. Now, using Eqs. (14.2.7) and (14.2.8) in Eq. (14.2.9), we obtain

$$f = -K_{xx}Ag \text{ m}^3/\text{s or in}^3/\text{s} \tag{14.2.10}$$

in scalar form; based on Eqs. (14.2.4) and (14.2.5), g is given in explicit form by

$$g = \frac{p_2 - p_1}{L} \tag{14.2.11}$$

Applying Eqs. (14.2.10) and (14.2.11) at nodes 1 and 2, we obtain

$$f_1 = -K_{xx}A\frac{p_2 - p_1}{L} \tag{14.2.12}$$

and

$$f_2 = K_{xx}A\frac{p_2 - p_1}{L} \tag{14.2.13}$$

where f_1 is directed into the element, indicating fluid flowing into the element (p_1 must be greater than p_2 to push the fluid through the element, actually resulting in positive f_1), whereas f_2 is directed away from the element, indicating fluid flowing out of the element; hence the negative sign changes to a positive one in Eq. (14.2.13). Expressing Eqs. (14.2.12) and (14.2.13) together in matrix form, we have

$$\begin{Bmatrix} f_1 \\ f_2 \end{Bmatrix} = \frac{AK_{xx}}{L} \begin{bmatrix} 1 & -1 \\ -1 & 1 \end{bmatrix} \begin{Bmatrix} p_1 \\ p_2 \end{Bmatrix} \tag{14.2.14}$$

The stiffness matrix is then

$$\underline{k} = \frac{AK_{xx}}{L} \begin{bmatrix} 1 & -1 \\ -1 & 1 \end{bmatrix} \text{ m}^2/\text{s or in}^2/\text{s} \tag{14.2.15}$$

for flow through a porous medium.

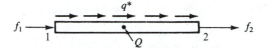

Figure 14–7 Additional sources of volumetric fluid-flow rates

Equation (14.2.15) is analogous to Eq. (13.4.20) for the heat-conduction element or to Eq. (3.1.14) for the one-dimensional (axial stress) bar element. The permeability or stiffness matrix will have units of square meters or square inches per second.

In general, the basic element may be subjected to internal sources or sinks, such as from a pump, or to surface-edge flow rates, such as from a river or stream. To include these or similar effects, consider the element of Figure 14–6 now to include a uniform internal source Q acting over the whole element and a uniform surface flow-rate source q^* acting over the surface, as shown in Figure 14–7. The force matrix terms are

$$\{f_Q\} = \iiint_V [N]^T Q \, dV = \frac{QAL}{2} \begin{Bmatrix} 1 \\ 1 \end{Bmatrix} \text{ m}^3/\text{s or in}^3/\text{s} \qquad (14.2.16)$$

where Q will have units of $\text{m}^3/(\text{m}^3 \cdot \text{s})$, or $1/\text{s}$, and

$$\{f_q\} = \iint_{S_2} q^* [N]^T \, dS = \frac{q^* Lt}{2} \begin{Bmatrix} 1 \\ 1 \end{Bmatrix} \text{ m}^3/\text{s or in}^3/\text{s} \qquad (14.2.17)$$

where q^* will have units of m/s or in./s. Equations (14.2.16) and (14.2.17) indicate that one-half of the uniform volumetric flow rate per unit volume Q (a source being positive and a sink being negative) is allocated to each node and one-half the surface flow rate (again a source is positive) is allocated to each node.

Step 5 Assemble the Element Equations to Obtain the Global Equations and Introduce Boundary Conditions

We assemble the total stiffness matrix $[K]$, total force matrix $\{F\}$, and total set of equations as

$$[K] = \sum [k^{(e)}] \qquad \{F\} = \sum \{f^{(e)}\} \qquad (14.2.18)$$

and $$\{F\} = [K]\{p\} \qquad (14.2.19)$$

The assemblage procedure is similar to the direct stiffness approach, but it is now based on the requirement that the potentials at a common node between two elements be equal. The boundary conditions on nodal potentials are given by Eq. (14.1.15).

Step 6 Solve for the Nodal Potentials

We now solve for the global nodal potentials, $\{p\}$, where the appropriate nodal potential boundary conditions, Eq. (14.1.15), are specified.

Step 7 Solve for the Element Velocities and Volumetric Flow
Rates

Finally, we calculate the element velocities from Eq. (14.2.7) and the volumetric flow
rate Q_f as

$$Q_f = (v)(A) \text{ m}^3/\text{s or in}^3/\text{s} \qquad (14.2.20)$$

Example 14.1

Determine (a) the fluid head distribution along the length of the coarse gravelly medium
shown in Figure 14–8, (b) the velocity in the upper part, and (c) the volumetric flow
rate in the upper part. The fluid head at the top is 10 in. and that at the bottom is 1 in.
Let the permeability coefficient be $K_{xx} = 0.5$ in./s. Assume a cross-sectional area of
$A = 1$ in^2.

The finite element discretization is shown in Figure 14–9. For simplicity, we will
use three elements, each 10 in. long.

We calculate the stiffness matrices for each element as follows:

$$\frac{AK_{xx}}{L} = \frac{(1 \text{ in}^2)(0.5 \text{ in./s})}{10 \text{ in.}} = 0.05 \text{ in}^2/\text{s}$$

Using Eq. (14.2.15) for elements 1, 2, and 3, we have

$$[k^{(1)}] = [k^{(2)}] = [k^{(3)}] = 0.05 \begin{bmatrix} 1 & -1 \\ -1 & 1 \end{bmatrix} \text{ in}^2/\text{s} \qquad (14.2.21)$$

In general, we would use Eqs. (14.2.16) and (14.2.17) to obtain element forces. However, in this example $Q = 0$ (no sources or sinks) and $q^* = 0$ (no applied surface flow
rates). Therefore,

$$\{f^{(1)}\} = \{f^{(2)}\} = \{f^{(3)}\} = 0 \qquad (14.2.22)$$

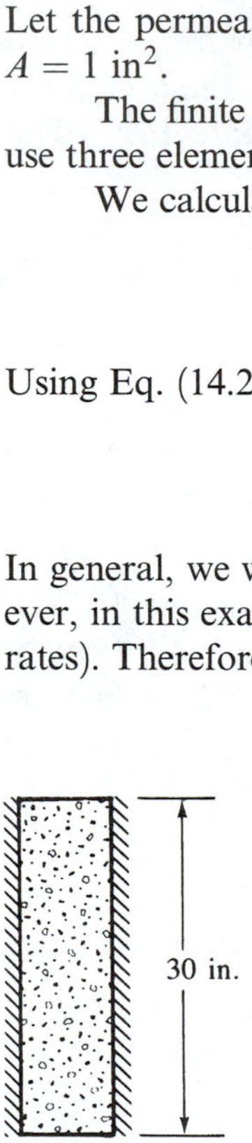

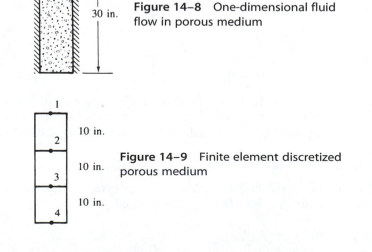

30 in.

Figure 14–8 One-dimensional fluid
flow in porous medium

1

2   10 in.

Figure 14–9 Finite element discretized
3   10 in.   porous medium

4   10 in.

The assembly of the element stiffness matrices from Eq. (14.2.21), via the direct stiffness method, produces the following system of equations:

$$0.05 \begin{bmatrix} 1 & -1 & 0 & 0 \\ -1 & 2 & -1 & 0 \\ 0 & -1 & 2 & -1 \\ 0 & 0 & -1 & 1 \end{bmatrix} \begin{Bmatrix} p_1 \\ p_2 \\ p_3 \\ p_4 \end{Bmatrix} = \begin{Bmatrix} 0 \\ 0 \\ 0 \\ 0 \end{Bmatrix} \tag{14.2.23}$$

Known nodal fluid head boundary conditions are $p_1 = 10$ in. and $p_4 = 1$ in. These nonhomogeneous boundary conditions are treated as described for the stress analysis and heat-transfer problems. We modify the stiffness (permeability) matrix and force matrix as follows:

$$\begin{bmatrix} 1 & 0 & 0 & 0 \\ 0 & 0.1 & -0.05 & 0 \\ 0 & -0.05 & 0.1 & 0 \\ 0 & 0 & 0 & 1 \end{bmatrix} \begin{Bmatrix} p_1 \\ p_2 \\ p_3 \\ p_4 \end{Bmatrix} = \begin{Bmatrix} 10 \\ 0.5 \\ 0.05 \\ 1 \end{Bmatrix} \tag{14.2.24}$$

where the terms in the first and fourth rows and columns of the stiffness matrix corresponding to the known fluid heads $p_1 = 10$ in. and $p_4 = 1$ in. have been set equal to 0 except for the main diagonal, which has been set equal to 1, and the first and fourth rows of the force matrix have been set equal to the known nodal fluid heads at nodes 1 and 4. Also the terms $(-0.05) \times (10 \text{ in.}) = -0.5$ in. on the left side of the second equation of Eq. (14.2.24) and $(-0.05) \times (1 \text{ in.}) = -0.05$ in. on the left side of the third equation of Eq. (14.2.24) have been transposed to the right side in the second and third rows (as $+0.5$ and $+0.05$). The second and third equations of Eq. (14.2.24) can now be solved. The resulting solution is given by

$$p_2 = 7 \text{ in.} \qquad p_3 = 4 \text{ in.} \tag{14.2.25}$$

Next we use Eq. (14.2.7) to determine the fluid velocity in element 1 as

$$v_x^{(1)} = -K_{xx}[B]\{p^{(1)}\} \tag{14.2.26}$$

$$= -K_{xx}\begin{bmatrix} -\dfrac{1}{L} & \dfrac{1}{L} \end{bmatrix} \begin{Bmatrix} p_1 \\ p_2 \end{Bmatrix} \tag{14.2.27}$$

or

$$v_x^{(1)} = 0.15 \text{ in./s} \tag{14.2.28}$$

You can verify that the velocities in the other elements are also 0.15 in./s because the cross section is constant and the material properties are uniform. We then determine the volumetric flow rate Q_f in element 1 using Eq. (14.2.20) as

$$Q_f = (0.15 \text{ in./s})(1 \text{ in}^2) = 0.15 \text{ in}^3/\text{s} \tag{14.2.29}$$

This volumetric flow rate is constant throughout the length of the medium. ■

Example 14.2

For the smooth pipe of variable cross section shown in Figure 14–10, determine the potential at the junctions, the velocities in each section of pipe and the volumetric flow rate. The potential at the left end is $p_1 = 10$ m²/s and that at the right end is $p_4 = 1$ m²/s.

For the fluid flow through a smooth pipe, $K_{xx} = 1$. The pipe has been discretized into three elements and four nodes, as shown in Figure 14–11. Using Eq. (14.2.15), we find that the element stiffness matrices are

$$\underline{k}^{(1)} = \frac{3}{1}\begin{bmatrix} 1 & -1 \\ -1 & 1 \end{bmatrix} \text{m} \qquad \underline{k}^{(2)} = \frac{2}{1}\begin{bmatrix} 1 & -1 \\ -1 & 1 \end{bmatrix} \text{m} \qquad \underline{k}^{(3)} = \frac{1}{1}\begin{bmatrix} 1 & -1 \\ -1 & 1 \end{bmatrix} \text{m}$$

(14.2.30)

where the units on $\underline{k}$ are now meters for fluid flow through a pipe.

There are no applied fluid sources. Therefore, $\underline{f}^{(1)} = \underline{f}^{(2)} = \underline{f}^{(3)} = 0$. The assembly of the element stiffness matrices produces the following system of equations:

$$\begin{bmatrix} 3 & -3 & 0 & 0 \\ -3 & 5 & -2 & 0 \\ 0 & -2 & 3 & -1 \\ 0 & 0 & -1 & 1 \end{bmatrix} \begin{Bmatrix} 10 \\ p_2 \\ p_3 \\ 1 \end{Bmatrix} = \begin{Bmatrix} 0 \\ 0 \\ 0 \\ 0 \end{Bmatrix} \frac{\text{m}^3}{\text{s}}$$

(14.2.31)

Solving the second and third of Eqs. (14.2.31) for p_2 and p_3 in the usual manner, we obtain

$$p_2 = 8.365 \text{ m}^2/\text{s} \qquad p_3 = 5.91 \text{ m}^2/\text{s}$$ (14.2.32)

Using Eqs. (14.2.7) and (14.2.20), the velocities and volumetric flow rates in each element are

$$v_x^{(1)} = -[B]\{p^{(1)}\}$$

$$= -\begin{bmatrix} -\dfrac{1}{L} & \dfrac{1}{L} \end{bmatrix} \begin{Bmatrix} 10 \\ 8.365 \end{Bmatrix}$$

$$= 1.635 \text{ m/s}$$

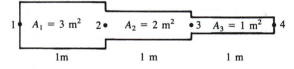

1 ◆ $A_1 = 3$ m² 2 ● $A_2 = 2$ m² ● 3 $A_3 = 1$ m² ◆ 4

1m   1 m   1 m

Figure 14–10 Variable-cross-section pipe subjected to fluid flow

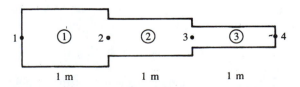

1 ◆ ① 2 ● ② 3 ● ③ ◆ 4

1 m   1 m   1 m

Figure 14–11 Discretized pipe

$$Q_f^{(1)} = Av_x^{(1)} = 3(1.635) = 4.91 \text{ m}^3/\text{s}$$

$$v_x^{(2)} = -(-8.365 + 5.91) = 2.455 \text{ m/s}$$

$$Q_f^{(2)} = 2.455(2) = 4.91 \text{ m}^3/\text{s}$$

$$v_x^{(3)} = -(-5.91 + 1) = 4.91 \text{ m/s}$$

$$Q_f^{(3)} = 4.91(1) = 4.91 \text{ m}^3/\text{s}$$

The potential, being higher at the left and decreasing to the right, indicates that the velocities are to the right. The volumetric flow rate is constant throughout the pipe, as conservation of mass would indicate. ∎

We now illustrate how you can solve a fluid-flow problem where the boundary condition is a known fluid velocity, but none of the p's are initially known.

Example 14.3

For the smooth pipe shown discretized in Figure 14–12 with uniform cross section of 1 in^2, determine the flow velocities at the center and right end, knowing the velocity at the left end is $v_x = 2$ in./s.

Using Eq. (14.2.15), the element stiffness matrices are

$$\underline{k}^{(1)} = \frac{1}{10} \begin{bmatrix} 1 & -1 \\ -1 & 1 \end{bmatrix} \text{ in.} \qquad \underline{k}^{(2)} = \frac{1}{10} \begin{bmatrix} 1 & -1 \\ -1 & 1 \end{bmatrix} \text{ in.} \tag{14.2.33}$$

where now the units on $\underline{k}$ are inches for fluid flow through a pipe.

Assembling the element stiffness matrices produces the following equations:

$$\frac{1}{10} \begin{bmatrix} 1 & -1 & 0 \\ -1 & 2 & -1 \\ 0 & -1 & 1 \end{bmatrix} \begin{Bmatrix} p_1 \\ p_2 \\ p_3 \end{Bmatrix} = \begin{Bmatrix} f_1 \\ f_2 \\ f_3 \end{Bmatrix} \tag{14.2.34}$$

The specified boundary condition is $v_x = 2$ in./s, so that by Eq. (14.2.9), we have

$$f_1 = v_1 A = (2 \text{ in./s})(1 \text{ in}^2) = 2 \text{ in}^3/\text{s} \tag{14.2.35}$$

Because $p_1, p_2,$ and p_3 in Eq. (14.2.34) are not known, we cannot determine these potentials directly. The problem is similar to that occurring if we try to solve the structural problem without prescribing displacements sufficient to prevent rigid body motion of the structure. This was discussed in Chapter 2. Because the p's correspond

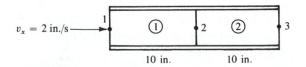

Figure 14–12 Discretized pipe for fluid-flow problem

to displacements in the structural problem, it appears that we must specify at least one value of p in order to obtain a solution. We then proceed as follows. Select a convenient value for p_3 (for instance set $p_3 = 0$). (The velocities are functions of the derivatives or differences in p's, so a value of $p_3 = 0$ is acceptable.) Then p_1 and p_2 are the unknowns. The solution will yield p_1 and p_2 relative to $p_3 = 0$. Therefore, from the first two of Eqs. (14.2.34), we have

$$\frac{1}{10}\begin{bmatrix} 1 & -1 \\ -1 & 2 \end{bmatrix}\begin{Bmatrix} p_1 \\ p_2 \end{Bmatrix} = \begin{Bmatrix} 2 \\ 0 \end{Bmatrix} \qquad (14.2.36)$$

where $f_1 = 2$ in^3/s from Eq. (14.2.35) and $f_2 = 0$, because there is no applied fluid force at node 2.

Solving Eq. (14.2.36), we obtain

$$p_1 = 40 \qquad p_2 = 20 \qquad (14.2.37)$$

These are not absolute values for p_1 and p_2; rather, they are relative to p_3. The fluid velocities in each element are absolute values, because velocities depend on the differences in p's. These differences are the same no matter what value for p_3 was chosen. You can verify this by choosing $p_3 = 10$, for instance, and re-solving for the velocities. [You would find $p_1 = 50$ and $p_2 = 30$ and the same v's as in Eq. (14.2.38).]

$$v_x^{(1)} = -\begin{bmatrix} -\dfrac{1}{L} & \dfrac{1}{L} \end{bmatrix}\begin{Bmatrix} 40 \\ 20 \end{Bmatrix} = 2 \text{ in./s}$$

$$(14.2.38)$$

and

$$v_x^{(2)} = -\begin{bmatrix} -\dfrac{1}{L} & \dfrac{1}{L} \end{bmatrix}\begin{Bmatrix} 20 \\ 0 \end{Bmatrix} = 2 \text{ in./s} \qquad ■$$

14.3 Two-Dimensional Finite Element Formulation

Because many fluid-flow problems can be modeled as two-dimensional problems, we now develop the equations for an element appropriate for these problems. Examples using this element then follow.

Step 1

The three-node triangular element in Figure 14–13 is the basic element for the solution of the two-dimensional fluid-flow problem.

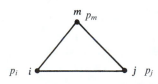

Figure 14–13 Basic triangular element with nodal potentials

Step 2

The potential function is

$$[\phi] = [N_i \quad N_j \quad N_m] \begin{Bmatrix} p_i \\ p_j \\ p_m \end{Bmatrix} \tag{14.3.1}$$

where $p_i, p_j,$ and p_m are the nodal potentials (for groundwater flow, ϕ is the piezo-metric fluid head function, and the p's are the nodal heads), and the shape functions are again given by Eq. (7.2.18) or (13.5.2) as

$$N_i = \frac{1}{2A}(\alpha_i + \beta_i x + \gamma_i y) \tag{14.3.2}$$

with similar expressions for N_j and N_m. The α's, β's, and γ's are defined by Eqs. (7.2.10).

Step 3

The gradient matrix $\{g\}$ is given by

$$\{g\} = [B]\{p\} \tag{14.3.3}$$

where the matrix $[B]$ is again given by

$$[B] = \frac{1}{2A} \begin{bmatrix} \beta_i & \beta_j & \beta_m \\ \gamma_i & \gamma_j & \gamma_m \end{bmatrix} \tag{14.3.4}$$

and

$$\{g\} = \begin{Bmatrix} g_x \\ g_y \end{Bmatrix} \tag{14.3.5}$$

with

$$g_x = \frac{\partial \phi}{\partial x} \qquad g_y = \frac{\partial \phi}{\partial y} \tag{14.3.6}$$

The velocity/gradient matrix relationship is now

$$\begin{Bmatrix} v_x \\ v_y \end{Bmatrix} = -[D]\{g\} \tag{14.3.7}$$

where the material property matrix is

$$[D] = \begin{bmatrix} K_{xx} & 0 \\ 0 & K_{yy} \end{bmatrix} \tag{14.3.8}$$

and the K's are permeabilities (for the seepage problem) of the porous medium in the x and y directions. For fluid flow around a solid object or through a smooth pipe, $K_{xx} = K_{yy} = 1$.

Step 4

The element stiffness matrix is given by

$$[k] = \iiint\limits_{V} [B]^T [D][B] \, dV \tag{14.3.9}$$

Assuming constant-thickness (t) triangular elements and noting that the integrand terms are constant, we have

$$[k] = tA[B]^T [D][B] \ \text{m}^2/\text{s or in}^2/\text{s} \tag{14.3.10}$$

which can be simplified to

$$[k] = \frac{tK_{xx}}{4A}
\begin{bmatrix}
\beta_i^2 & \beta_i\beta_j & \beta_i\beta_m \\
\beta_i\beta_j & \beta_j^2 & \beta_j\beta_m \\
\beta_i\beta_m & \beta_j\beta_m & \beta_m^2
\end{bmatrix}
+ \frac{tK_{yy}}{4A}
\begin{bmatrix}
\gamma_i^2 & \gamma_i\gamma_j & \gamma_i\gamma_m \\
\gamma_i\gamma_j & \gamma_j^2 & \gamma_j\gamma_m \\
\gamma_i\gamma_m & \gamma_j\gamma_m & \gamma_m^2
\end{bmatrix} \tag{14.3.11}$$

The force matrices are

$$\{f_Q\} = \iiint\limits_{V} Q[N]^T \, dV = Q \iiint\limits_{V} [N]^T \, dV \tag{14.3.12}$$

for constant volumetric flow rate per unit volume over the whole element. On evaluating Eq. (14.3.12), we obtain

$$\{f_Q\} = \frac{QV}{3}
\begin{Bmatrix}
1 \\ 1 \\ 1
\end{Bmatrix}
\frac{\text{m}^3}{\text{s}} \ \text{or} \ \frac{\text{in}^3}{\text{s}} \tag{14.3.13}$$

We find that the second force matrix is

$$\{f_q\} = \iint\limits_{S_2} q^*[N]^T \, dS = \iint\limits_{S_2} q^*
\begin{Bmatrix}
N_i \\ N_j \\ N_m
\end{Bmatrix}
dS \tag{14.3.14}$$

This reduces to

$$\{f_q\} = \frac{q^* L_{i-j} t}{2}
\begin{Bmatrix}
1 \\ 1 \\ 0
\end{Bmatrix}
\frac{\text{m}^3}{\text{s}} \ \text{or} \ \frac{\text{in}^3}{\text{s}} \ \text{on side } i\text{-}j \tag{14.3.15}$$

with similar terms on sides j-m and m-i [see Eqs. (13.5.19) and (13.5.20)]. Here L_{i-j} is the length of side i-j of the element and q^* is the assumed constant surface flow rate. Both Q and q^* are positive quantities if fluid is being added to the element. The units on Q and q^* are $\text{m}^3/(\text{m}^3 \cdot \text{s})$ and m/s. The total force matrix is then the sum of $\{f_Q\}$ and $\{f_q\}$.

Example 14.4

For the two-dimensional sandy soil region shown in Figure 14–14, determine the potential distribution. The potential (fluid head) on the left side is a constant 10.0 m

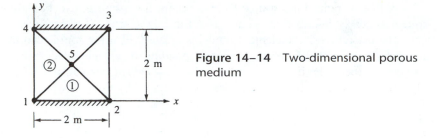

Figure 14–14 Two-dimensional porous medium

and that on the right side is 0.0. The upper and lower edges are impermeable. The permeabilities are $K_{xx} = K_{yy} = 25 \times 10^{-5}$ m/s. Assume unit thickness.

The finite element model is shown in Figure 14–14. We use only the four triangular elements of equal size for simplicity of the longhand solution. For increased accuracy in results, we would need to refine the mesh. This body has the same magnitude of coordinates as Figure 13–19. Therefore, the total stiffness matrix is given by Eq. (13.5.40) as

$$
\underline{K} =
\begin{bmatrix}
25 & 0 & 0 & 0 & -25 \\
0 & 25 & 0 & 0 & -25 \\
0 & 0 & 25 & 0 & -25 \\
0 & 0 & 0 & 25 & -25 \\
-25 & -25 & -25 & -25 & 100
\end{bmatrix}
\times 10^{-5} \; \frac{\text{m}^2}{\text{s}}
\tag{14.3.16}
$$

The force matrices are zero, because $Q = 0$ and $q^* = 0$. Applying the boundary conditions, we have

$$
p_1 = p_4 = 10.0 \text{ m} \qquad p_2 = p_3 = 0
$$

The assembled total system of equations is then

$$
10^{-5}
\begin{bmatrix}
25 & 0 & 0 & 0 & -25 \\
0 & 25 & 0 & 0 & -25 \\
0 & 0 & 25 & 0 & -25 \\
0 & 0 & 0 & 25 & -25 \\
-25 & -25 & -25 & -25 & 100
\end{bmatrix}
\begin{Bmatrix}
10 \\ 0 \\ 0 \\ 10 \\ p_5
\end{Bmatrix}
=
\begin{Bmatrix}
0 \\ 0 \\ 0 \\ 0 \\ 0
\end{Bmatrix}
\tag{14.3.17}
$$

Solving the fifth of Eqs. (14.3.17) for p_5, we obtain

$$
p_5 = 5 \text{ m}
$$

Using Eqs. (14.3.7) and (14.3.3) we obtain the velocity in element 2 as

$$
\begin{Bmatrix}
v_x^{(2)} \\
v_y^{(2)}
\end{Bmatrix}
=
\begin{bmatrix}
+25 & 0 \\
0 & +25
\end{bmatrix}
\times 10^{-5} \frac{1}{2A}
\begin{bmatrix}
-1 & 2 & -1 \\
-1 & 0 & 1
\end{bmatrix}
\begin{Bmatrix}
p_1 \\ p_5 \\ p_4
\end{Bmatrix}
\tag{14.3.18}
$$

where $\beta_1 = -1$, $\beta_5 = 2$, $\beta_4 = -1$, $\gamma_1 = -1$, $\gamma_5 = 0$, and $\gamma_4 = 1$ were obtained from Eq. (13.5.24). Simplifying Eq. (14.3.18), we obtain

$$
v_x^{(2)} = 125 \times 10^{-5} \text{ m/s} \qquad v_y^{(2)} = 0 \qquad \blacksquare
$$

A line or point fluid source from a pump, for instance, can be handled in the same manner as described in Section 13.6 for heat sources. If the source is at a node when the discretized finite element model is created, then the source can be added to the row of the global force matrix corresponding to the global degree of freedom assigned to the node. If the source is within an element, we can use Section 13.6 to allocate the source to the proper nodes, as illustrated by the following example.

Example 14.5

A pump, pumping fluid at $Q^* = 6500$ m^2/h, is located at coordinates (5, 2) in the element shown in Figure 14–15. Determine the amount of Q^* allocated to each node. All nodal coordinates are in units of meters. Assume unit thickness of $t = 1$ mm.

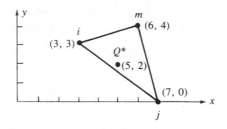

Figure 14–15 Triangular element with pump located within element

The magnitudes of the numbers are the same as in Example 13.6. Therefore, the shape functions are identical to Eq. (13.6.7); when evaluated at the source $x = 5$ m, $y = 2$ m, they are equal to Eq. (13.6.8). Using Eq. (13.6.3), we obtain the amount of Q^* allocated to each node or equivalently the force matrix as

$$\begin{Bmatrix} f_{Qi} \\ f_{Qj} \\ f_{Qm} \end{Bmatrix} = Q^* t \begin{Bmatrix} N_i \\ N_j \\ N_m \end{Bmatrix} \Bigg|_{\substack{x=x_0=5 \text{ m} \\ y=y_0=2 \text{ m}}}$$

$$= \frac{(6500 \text{ m}^2/\text{h})(1 \text{ mm})}{(13)\left(\dfrac{1000 \text{ mm}}{1 \text{ m}}\right)} \begin{Bmatrix} 6 \\ 5 \\ 2 \end{Bmatrix} = \begin{Bmatrix} 3.0 \\ 2.5 \\ 1.0 \end{Bmatrix} \frac{\text{m}^3}{\text{h}} \qquad ■$$

▲ 14.4 Flowchart of a Fluid-Flow Program ▲

Figure 14–16 is a flowchart of a finite element process used for the analysis of two-dimensional steady-state fluid flow through a porous medium or through a pipe. Because flow through a porous medium is analogous to heat transfer by conduction, we can use the heat-transfer processor **SSAP10** to analyze the problem. For more complicated fluid flows, other processors are available [6].

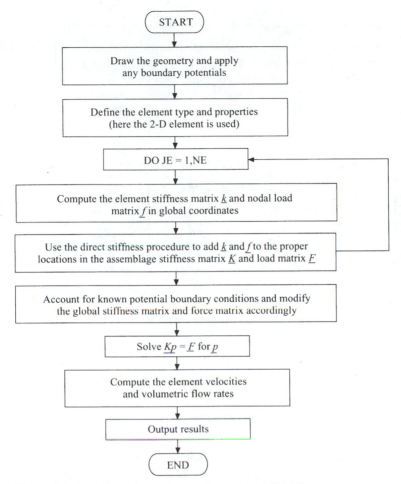

Figure 14–16 Flowchart of two-dimensional fluid-flow process

▲ 14.5 Algor Example Solutions for Two-Dimensional Steady-State Fluid Flow ▲

To use the Algor computer program for two-dimensional steady-state, incompressible fluid flow, we now solve a number of examples. The program is based on the flow-chart of Figure 14–16.

For flow through a porous medium, we recall the analogies between conductive heat transfer and flow through a porous medium and use the heat-transfer processor SSAP10 to solve the problem. That is, we input any known nodal potentials or fluid heads using the "FEA Add", "Heat Transfer Analysis", "Applied Temperature", and "Temperature Value" menus (see step 4 of Examples 14.6 and 14.7). The permeability coefficient is input in the "Model Data Control" window on the "Material" screen.

Example 14.6

We now consider the fluid-flow problem shown descretized in Figure 14–17 to illustrate how to input data into the Algor program. The top and bottom sides are impervious, whereas the right side has a constant head of 3 cm and the left side has a constant head of 4 cm. We will use the analogies between heat transfer and fluid flow during this analysis. The step-by-step commands follow.

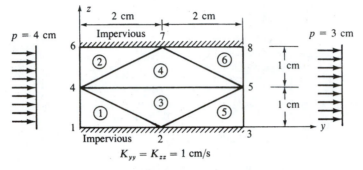

$$K_{yy} = K_{zz} = 1 \text{ cm/s}$$

Figure 14–17 Two-dimensional fluid-flow problem

Step 1   Start the Algor Program

Click on the "Algor FEA" icon on your desktop computer screen.

Step 2   Create the Fluid-Flow Model

Click "Modify".

Click "Transform XY to YZ". You could put the coordinates in the x-y plane and convert them later as we have sometimes done. However, we will transform to the y-z plane initially instead.

Click "Add". Draw the outer lines of the model.

Click "Line". Begin entering the coordinates of the lines in the usual manner.

(0, 0, 0)   Enter the first coordinate (node 1 in Figure 14–17).

<Enter>

(0, 2, 0)   Enter the second coordinate (node 2 in Figure 14–17).

<Enter>

(0, 4, 0)   Enter node 3 coordinates.

<Enter>

(0, 4, 1)   Enter node 5 coordinates.

<Enter>

(0, 4, 2)   Enter node 8 coordinates.

<Enter>

(0, 2, 2)   Enter node 7 coordinates.

```
<Enter>
(0, 0, 2)    Enter node 6 coordinates.
<Enter>
(0, 0, 1)    Enter node 4 coordinates.
<Enter>
(0, 0, 0)    Enter the first coordinate again to finish creating the perimeter.
<Enter>
```
Click "Done".
Click "View:Enclose" to enlarge the model.
Click "Add". Now add the interior lines.
Click "Line".
```
(0, 0, 1)    Enter node 4 coordinate.
<Enter>
(0, 2, 0)    Enter node 2 coordinates.
<Enter>
(0, 4, 1)    Enter node 5 coordinates.
<Enter>
(0, 2, 2)    Enter node 7 coordinates.
<Enter>
(0, 0, 1)    Enter node 4 coordinates.
<Enter>
(0, 4, 1)    Enter node 5 coordinates.
<Enter>
```
Click "Done". The full model appears on the screen.

Step 3 Eliminate Duplicate Lines

Click "Modify:Clean:Duplicate".
Click "Perform Cleaning". You see the following message:
                    "13 kept 0 deleted. done"
Click "Done".

Step 4 Set the Potential Boundary Conditions

Click "FEA Add'.
Click "Heat Transfer Analysis". Remember we are using the heat-transfer program and making use of the analogies between heat transfer and fluid flow as indicated in this chapter.
Click "Applied Temperature". You are now using the boundary element.
Click "Temperature Value".
```
4                 Type a 4 for a boundary head of 4 cm.
<Enter>
```

Click "Stiffness".

1e6   Enter a large stiffness as recommended by the Algor program.

<Enter>

Click on each left-side node (nodes 1, 4, and 6). A potential appears at each node. Remember that potential is analogous to temperature.

Click "Temperature Value" again.

3   Type a 3 for a boundary head of 3 cm.

<Enter>

Click "Stiffness".

1e6   Make the stiffness large.

<Enter>

Click on each right-side node (nodes 3, 5, and 8). The potential appears at each node.

Click "Done".

Click "View:Enclose" to view the model.

Step 5   Add the Material Properties Using the "Model Data Control" Window

Click on the "Model Data" button.

Select "Steady State Heat Transfer" for "Analysis Type". Scroll down in this window until you see "Steady State Heat Transfer" and click on it.

Click on the box under "Element".

Click on the circle next to "2D".

Click "OK".

Click on the box under "Data". You see a prompt "Please enter the model name first".

Click "OK". Then enter the model name.

Ex146   This is the model name.

Click "Save". The "Units Definition" window appears next.

Click on "Metric cgs". The units will be centimeters/grams/seconds.

Click "OK". This takes you to the "Element Definition" window. Use the defaults for "Geometry Type" ("Planar") and "Material Model" ("Isotropic") and make the thickness 1 cm.

1   This is the material thickness.

Click "OK". You are now back to the "Model Data Control" window.

Click the box under "Material".

Click "[Customer Defined]".

Click "Edit Properties".

1   Enter a 1 for the permeability. This is analogous to the thermal conductivity in heat transfer.

Click "OK".

Click "OK" again. You are now back in the "Model Data Control" window.

Table 14–2   Nodal potentials for
Example 14.6

Node Number	Potential
1	4.0000D+00
2	3.5000D+00
3	3.0000D+00
4	4.0000D+00
5	3.0000D+00
6	4.0000D+00
7	3.5000D+00
8	3.0000D+00

Step 6   Enter the Global Data

Click "Global" in the "Model Data Control" window.

1   Type a 1 on the "Load Case Multiplier" screen in the "Boundary temperature multiplier" column. This indicates that you have a boundary condition load case and that the multiplier on this boundary condition is 1.

Click "OK". You are now back in the "Model Data Control" window.

Step 7   Analyze the Model

Click "Analysis" in the "Model Data Control" window.
Click "Analyze". The analysis will now be performed.
Click "OK".
Click "Done".

Step 8   Review the Results

Click "Results" in the "Model Data Control" window. This takes you to Superview.
Click "Show T-ht".
Click "Temp-ht" to obtain a color plot of the potential variation within the model.
<Esc>
<Esc>
Click "Done" to exit Superview.

Results for the nodal potentials are shown in Table 14–2. They compare exactly with solutions obtained using another computer program [5]. ■

Example 14.7

For the one-dimensional flow through the porous medium shown in Figure 14–18, determine the potentials at one-third and two-thirds of the length. Also determine the

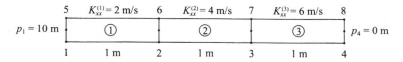

Figure 14–18 One-dimensional flow through a porous medium modeled with two-dimensional elements

volumetric flow rate through the medium. Note the different permeabilities in each element. Let $A = 0.2 \text{ m}^2$.

Step 1   Start the Algor Program

Click on the "Algor FEA" icon on your desktop computer screen.

Step 2   Create the Fluid-Flow Model

Click "Modify".

Click "Transform XY to YZ". You could put the coordinates in the x-y plane and convert them later as we have sometimes done. However, we will transform to the y-z plane initially instead.

Click "Add".

Click "Rectangle". Create the first of three rectangles.

(0, 0, 0)	Enter the first coordinate (node 1 in Figure 14–18).
\<Enter\>	
(0, 1, 0.2)	Enter the second coordinate (node 6 in Figure 14–18).
\<Enter\>	You have created a rectangle.

Click "Add".

Click "Rectangle". Create the second rectangle.

(0, 1, 0)	Enter node 2 coordinates.
\<Enter\>	
(0, 2, 0.2)	Enter node 7 coordinates.
\<Enter\>	You see the second rectangle. Make sure it is the only one highlighted.

Click "Modify:Update Object Parameters:Group".

2	Type a 2. This gives rectangle 2 group 2 status, and it turns red.

Click "Rectangle". Create the third rectangle.

(0, 2, 0)	Enter node 3 coordinates.
\<Enter\>	
(0, 3, 0.2)	Enter node 8 coordinates.
\<Enter\>	You see the third rectangle. Make sure it is now the only rectangle highlighted.

Click "Modify:Update Object Parameters:Group".

3   Type a 3. This gives rectangle 3 group 3 status, and it turns tan.
Click "View:Enclose" to enlarge the model.
Click "Done". The full model appears on the screen.

Step 3   Eliminate Duplicate Lines

Click "Modify:Clean:Duplicate".
Click "Perform Cleaning". You see the following message:

> "12 kept 0 deleted. done"

Click "Done".

Step 4   Set the Potential Boundary Conditions

Click "FEA Add".
Click "Heat Transfer Analysis". Remember we are using the heat-transfer program and making use of the analogies between heat transfer and fluid flow as indicated in this chapter.
Click "Applied Temperature". You are now using the boundary element.
Click "Temperature Value".

10   Type a 10 for a boundary potential of 10 m.

<Enter>
Click "Stiffness".

1e6   Enter a large stiffness as recommended by the Algor program.

<Enter>
Click on each left-side node (nodes 1 and 5). A potential appears at each node. Remember potential is analogous to temperature.
Click "Temperature Value" again.

0   Type a 0 for a boundary potential of 0 m.

<Enter>
Click "Stiffness".

1e6   Make the stiffness large. This is now the default value from the previous setting.

<Enter>
Click on each right-side nodes (nodes 4 and 8). The potential appears at each node.
Click "Done".
Click "View:Enclose" to view the model.

Step 5   Add the Material Properties Using the "Model Data Control" Window

Click on the "Model Data" button. Three group lines appear.
Select "Steady State Heat Transfer" for "Analysis Type". Scroll down in this window until you see "Steady State Heat Transfer" and click on it.

Click on the first box under "Element".

Click on the circle next to "2D".

Click "OK".

Click on the second box under "Element".

Click on the circle next to "2D".

Click "OK".

Click on the third box under "Element".

Click "OK".

Click on the first box under "Data". You see a prompt "Please enter the model name first".

Click "OK". Then enter the model name.

Ex147   This is the model name.

Click "Save". The "Units Definition" window appears next.

Click "Metric mks SI". The units are meters/kilograms/seconds.

Click "OK". This takes you to the "Element Definition" window. Use the defaults for "Geometry Type" ("Planar") and "Material Model" ("Isotropic") and make the thickness 1 m.

1   This is the material thickness.

Click "OK". Repeat this process for the other two groups; that is, give these rectangles a thickness of 1 m. You are now back to the "Model Data Control" window.

Click on the box under "Material".

Click "[Customer Defined]".

Click "Edit Properties".

2   Enter a permeability of 2 m/s. This is analogous to the thermal conductivity in heat transfer.

Click "OK". Repeat this step for the other rectangle groups.

Click "[Customer Defined]".

Click "Edit properties".

4   Enter a permeability of 4 m/s for the second-group material.

Click "OK".

Click "[Customer Defined]".

Click "Edit Properties".

6   Enter a permeability of 6 m/s for the third-group material.

Click "OK".

Click "OK" again. You are now back to the "Model Data Control" window.

Step 6   Enter the Global Data

Click "Global" in the "Model Data Control" window.

1   Type a 1 on the "Load Case Multiplier" screen under the "Boundary temperature multiplier" column. This indicates that you have a

> boundary condition load case and that the multiplier on this boundary condition is 1.

Click "OK". You are now back in the "Model Data Control" window.

Step 7 Analyze the Model

Click "Analysis" in the "Model Data Control" window.
Click "Analyze". The analysis will now be performed.
Click "OK".
Click "Done".

Step 8 Review the Results

Click "Results" in the "Model Data Control" window. This takes you to Superview.
Click "Show T-ht".
Click "Temp-ht" to obtain a color plot of the potential variation within the model.

`<Esc>`

`<Esc>`

Click "Done" to exit Superview.

Table 14–3 lists the potentials of the nodes from the *ex147.L10* file. The answers compare exactly with the longhand solution for Problem 14.1.

Table 14–3 Nodal potentials for Example 14.7

Node Number	Potential
1	1.0000D+01
2	4.5455D+00
3	1.8182D+00
4	1.0909D−06
5	1.0000D+01
6	4.5455D+00
7	1.8182D+00
8	1.0909D−06

Remember that the actual headings from file *ex147.L10* read "NODE NUMBER" and "TEMPERATURE" because the heat-transfer processor and decoder were used in this analysis although the input data were for a flow through a porous medium. ■

▲ References

[1] Chung, T. J., *Finite Element Analysis in Fluid Dynamics*, McGraw-Hill, New York, 1978.
[2] John, J. E. A., and Haberman, W. L., *Introduction to Fluid Mechanics*, Prentice-Hall, Englewood Cliffs, NJ, 1988.

[3] Harr, M. E., *Ground Water and Seepage*, McGraw-Hill, New York, 1962.

[4] *Heat Transfer Reference Division*, Algor, Inc., Pittsburgh, PA, 1999.

[5] Logan, D. L., *A First Course in the Finite Element Method*, 2nd ed., PWS-Kent Publishers, Boston, MA, 1992.

[6] *Fluid Flow Reference Division*, Algor, Inc., Pittsburgh, PA, 1999.

▲ Problems

14.1 For the one-dimensional flow through the porous media shown in Figure P14–1, determine the potentials at one-third and two-thirds of the length. Also determine the velocities in each element. Let $A = 0.2$ m^2.

Figure P14–1

14.2 For the one-dimensional flow through the porous medium shown in Figure P14–2 with fluid flux at the right end, determine the potentials at the third points. Also determine the velocities in each element. Let $A = 2$ m^2.

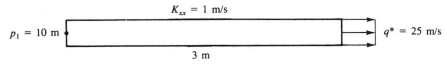

Figure P14–2

14.3 For the one-dimensional fluid flow through the stepped porous medium shown in Figure P14–3, determine the potentials at the junction of each area. Also determine the velocities in each element. Let $K_{xx} = 1$ in./s.

Figure P14–3

14.4 For the one-dimensional fluid-flow problem (Figure P14–4) with velocity known at the right end, determine the velocities and the volumetric flow rates at nodes 1 and 2. Let $K_{xx} = 2$ cm/s.

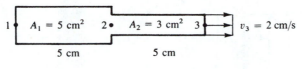

Figure P14–4

14.5 Derive the stiffness matrix, Eq. (14.2.15), using the first term on the right side of Eq. (13.5.17).

14.6 For the one-dimensional fluid-flow problem in Figure P14–6, determine the velocities and volumetric flow rates at nodes 2 and 3. Let $K_{xx} = 10^{-1}$ in./s.

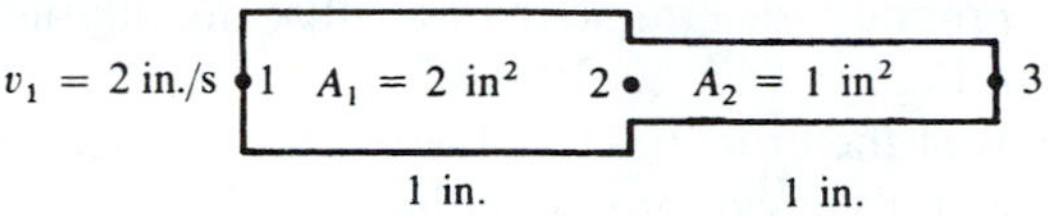

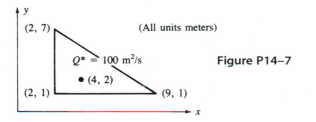

Figure P14–6

14.7 For the triangular element subjected to a fluid source shown in Figure P14–7, determine the amount of Q^* allocated to each node.

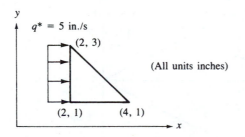

Figure P14–7

14.8 For the triangular element subjected to the surface fluid source shown in Figure P14–8, determine the amount of fluid force at each node.

Figure P14–8

14.9 For the two-dimensional fluid flow shown in Figure P14–9, determine the potentials at the center and right edge.

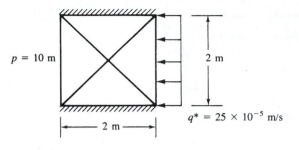

$p = 10$ m   2 m   2 m

$q* = 25 \times 10^{-5}$ m/s

Figure P14–9

14.10–
14.15   Using the Algor computer program, determine the potential distribution in the two-dimensional bodies shown in Figures P14–10–P14–15.

Use the heat transfer portion of the program with the proper analogies between conductive heat transfer and flow through a porous medium.

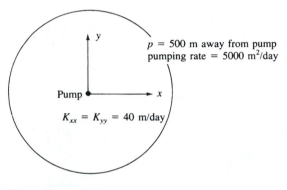

y

$p = 500$ m away from pump
pumping rate $= 5000$ m^2/day

Pump ● ——→ x

$K_{xx} = K_{yy} = 40$ m/day

Figure P14–10

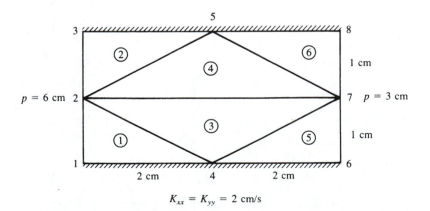

5

3   8

②
④   ⑥   1 cm

$p = 6$ cm   2 ——————————————————— 7 $p = 3$ cm

③
①   ⑤   1 cm

1   6
    2 cm   4   2 cm

$K_{xx} = K_{yy} = 2$ cm/s

Figure P14–11

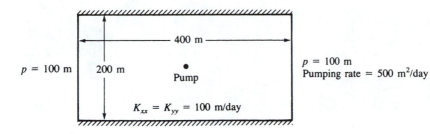

Figure P14–12

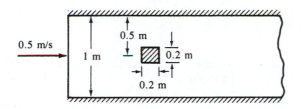

Figure P14–13

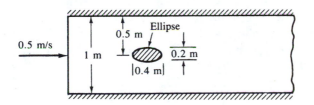

Figure P14–14

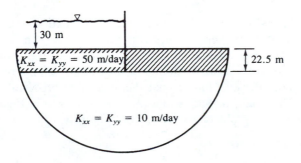

Figure P14–15

15

Thermal Stress

Introduction

In this chapter, we consider the problem of thermal stresses within a body. First, we will discuss the strain energy due to thermal stresses (stresses resulting from the constrained motion of a body or part of a body during a temperature change in the body).

The minimization of the thermal strain energy equation is shown to result in the thermal force matrix. We will then develop this thermal force matrix for the one-dimensional bar element and the two-dimensional plane stress and plane strain elements.

We will outline the procedures for solving both one- and two-dimensional problems and then provide solutions of specific problems, including illustrations of using the Algor computer program to solve thermal stress problems for trusses, frames, two-dimensional plane stress, and axisymmetric stress.

▲ 15.1 Formulation of the Thermal Stress Problem and Examples ▲

In addition to the strains associated with the displacement functions, there may be other strains within a body due to temperature variations, swelling (moisture differential), or other causes. We will concern ourselves only with the strains due to temperature variation, ε_T, and will consider both one- and two-dimensional problems.

We will first consider the one-dimensional thermal stress problem. The linear stress/strain diagram with initial (thermal) strain ($\varepsilon_0 = \varepsilon_T$) is shown in Figure 15–1.

For the one-dimensional problem, we have, from Figure 15–1,

$$\varepsilon_x = \frac{\sigma_x}{E} + \varepsilon_T \tag{15.1.1}$$

If, in general, we let $1/E = \underline{D}^{-1}$, then in general matrix form Eq. (15.1.1) can be written as

$$\underline{\varepsilon} = [D]^{-1}\underline{\sigma} + \underline{\varepsilon}_T \tag{15.1.2}$$

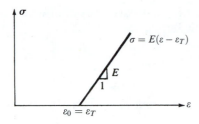

Figure 15–1 Linear stress/strain law with initial thermal strain

From Eq. (15.1.2), we solve for $\underline{\sigma}$ as

$$\underline{\sigma} = \underline{D}(\underline{\varepsilon} - \underline{\varepsilon}_T) \tag{15.1.3}$$

The strain energy per unit volume (called strain energy density) is the area under the $\sigma - \varepsilon$ diagram in Figure 15–1 and is given by

$$u_0 = \tfrac{1}{2}\underline{\sigma}(\underline{\varepsilon} - \underline{\varepsilon}_T) \tag{15.1.4}$$

Using Eq. (15.1.3) in Eq. (15.1.4), we have

$$u_0 = \tfrac{1}{2}(\underline{\varepsilon} - \underline{\varepsilon}_T)^T \underline{D}(\underline{\varepsilon} - \underline{\varepsilon}_T) \tag{15.1.5}$$

where, in general, the transpose is needed on the strain matrix to multiply the matrices properly.

The total strain energy is then

$$U = \int_V u_0 \, dV \tag{15.1.6}$$

Substituting Eq. (15.1.5) into Eq. (15.1.6), we obtain

$$U = \int_V \frac{1}{2}(\underline{\varepsilon} - \underline{\varepsilon}_T)^T \underline{D}(\underline{\varepsilon} - \underline{\varepsilon}_T) \, dV \tag{15.1.7}$$

Now, using $\underline{\varepsilon} = \underline{B}\underline{d}$ in Eq. (15.1.7), we obtain

$$U = \frac{1}{2} \int_V (\underline{B}\underline{d} - \underline{\varepsilon}_T)^T \underline{D}(\underline{B}\underline{d} - \underline{\varepsilon}_T) \, dV \tag{15.1.8}$$

Simplifying Eq. (15.1.8) yields

$$U = \frac{1}{2} \int_V (\underline{d}^T \underline{B}^T \underline{D}\underline{B}\underline{d} - \underline{d}^T \underline{B}^T \underline{D}\underline{\varepsilon}_T - \underline{\varepsilon}_T^T \underline{D}\underline{B}\underline{d} + \underline{\varepsilon}_T^T \underline{D}\underline{\varepsilon}_T) \, dV \tag{15.1.9}$$

The first term in Eq. (15.1.9) is the usual strain energy due to stress produced from mechanical loading—that is,

$$U_L = \frac{1}{2} \int_V \underline{d}^T \underline{B}^T \underline{D}\underline{B}\underline{d} \, dV \tag{15.1.10}$$

Terms 2 and 3 in Eq. (15.1.9) are identical and can be written together as

$$U_T = \int_V \underline{d}^T \underline{B}^T \underline{D}\underline{\varepsilon}_T \, dV \tag{15.1.11}$$

The last (fourth) term in Eq. (15.1.9) is a constant and drops out when we apply the principle of minimum potential energy by setting

$$\frac{\partial U}{\partial \underline{d}} = 0 \qquad (15.1.12)$$

Therefore, letting $U = U_L + U_T$ and substituting Eqs. (15.1.10) and (15.1.11) into Eq. (15.1.12), we obtain two contributions as

$$\frac{\partial U_L}{\partial \underline{d}} = \int_V \underline{B}^T \underline{D} \underline{B} \, dV \underline{d} \qquad (15.1.13)$$

and
$$\frac{\partial U_T}{\partial \underline{d}} = \int_V \underline{B}^T \underline{D} \varepsilon_T \, dV = \{f_T\} \qquad (15.1.14)$$

We recognize the integral term in Eq. (15.1.13) as the general form of the element stiffness matrix $\underline{k}$, whereas Eq. (15.1.14) is the load or force vector due to temperature change in the element.

We will now consider the one-dimensional thermal stress problem. We define the **thermal strain matrix** for the one-dimensional bar made of isotropic material with coefficient of thermal expansion α, and subjected to a uniform temperature rise T, as

$$\{\varepsilon_T\} = \{\varepsilon_{xT}\} = \{\alpha T\} \qquad (15.1.15)$$

where the units on α are typically (in./in.)/°F or (mm/mm)/°C.

For the simple one-dimensional bar (with a node at each end), we substitute Eq. (15.1.15) into Eq. (15.1.14) to obtain the thermal force matrix as

$$\{f_T\} = A \int_0^L [B]^T [D] \{\alpha T\} \, dx \qquad (15.1.16)$$

Recall that for the one-dimensional case, from Eqs. (3.10.15) and (3.10.13), we have

$$[D] = [E] \qquad [B] = \left[-\frac{1}{L} \quad \frac{1}{L} \right] \qquad (15.1.17)$$

Substituting Eqs. (15.1.17) into Eq. (15.1.16) and simplifying, we obtain the thermal force matrix as

$$\{f_T\} = \left\{ \begin{array}{c} f_{T1} \\ f_{T2} \end{array} \right\} = \left\{ \begin{array}{c} -E\alpha TA \\ E\alpha TA \end{array} \right\} \qquad (15.1.18)$$

For the two-dimensional thermal stress problem, the thermal strain matrix for an anisotropic material is

$$\{\varepsilon_T\} = \left\{ \begin{array}{c} \varepsilon_{xT} \\ \varepsilon_{yT} \\ \gamma_{xyT} \end{array} \right\} \qquad (15.1.19)$$

For the case of plane stress in an isotropic material with coefficient of thermal expansion α subjected to a temperature rise T, the thermal strain matrix is

$$\{\varepsilon_T\} = \begin{Bmatrix} \alpha T \\ \alpha T \\ 0 \end{Bmatrix} \tag{15.1.20}$$

No shear strains are caused by a change in temperature of isotropic materials, only expansion or contraction.

For the case of plane strain in an isotropic material, the thermal strain matrix is

$$\{\varepsilon_T\} = (1+v) \begin{Bmatrix} \alpha T \\ \alpha T \\ 0 \end{Bmatrix} \tag{15.1.21}$$

For a constant-thickness (t), constant-strain triangular element, Eq. (15.1.14) can be simplified to

$$\{f_T\} = [B]^T [D]\{\varepsilon_T\} t A \tag{15.1.22}$$

The forces in Eq. (15.1.22) are contributed to the nodes of an element in an unequal manner and require precise evaluation. It can be shown that substituting Eq. (7.1.8) for $[D]$, Eq. (7.2.34) for $[B]$, and Eq. (15.1.20) for $\{\varepsilon_T\}$ for a plane stress condition into Eq. (15.1.22) reveals the constant-strain triangular element thermal force matrix to be

$$\{f_T\} = \begin{Bmatrix} f_{Tix} \\ f_{Tiy} \\ \vdots \\ f_{Tmy} \end{Bmatrix} = \frac{\alpha E t T}{2(1-v)} \begin{Bmatrix} \beta_i \\ \gamma_i \\ \beta_j \\ \gamma_j \\ \beta_m \\ \gamma_m \end{Bmatrix} \tag{15.1.23}$$

where the β's and γ's are defined by Eqs. (7.2.10).

We will now describe the solution procedure for both one- and two-dimensional thermal stress problems.

Step 1

Evaluate the thermal force matrix, such as Eq. (15.1.18) or Eq. (15.1.23). Then treat this force matrix as an equivalent (or initial) force matrix $\underline{F}_0$ analogous to that obtained when we replace a distributed load acting on an element by equivalent nodal forces (Chapters 5 and 6 and Appendix D).

Step 2

Apply $\underline{F} = \underline{K}\underline{d} - \underline{F}_0$, where if only thermal loading is considered, we solve $\underline{F}_0 = \underline{K}\underline{d}$ for the nodal displacements. Recall that when we formulate the set of simultaneous equations, $\underline{F}$ represents the applied nodal forces, which here are assumed to be zero.

Step 3

Back-substitute the now known $\underline{d}$ into step 2 to obtain the actual nodal forces, $\underline{F}(=\underline{K}\underline{d}-\underline{F}_0)$.

Hence, the thermal stress problem is solved in a manner similar to the distributed load problem discussed for beams and frames in Chapters 5 and 6. We will now solve the following examples to illustrate the general procedure.

Example 15.1

For the one-dimensional bar fixed at both ends and subjected to a uniform temperature rise $T = 50\,°F$ as shown in Figure 15–2, determine the reactions at the fixed ends and the axial stress in the bar. Let $E = 30 \times 10^6$ psi, $A = 4$ in^2, $L = 4$ ft, and $\alpha = 7.0 \times 10^{-6}$ (in./in.)/°F.

Two elements will be sufficient to represent the bar because internal nodal displacements are not of importance here. To solve $\underline{F}_0 = \underline{K}\underline{d}$, we must determine the global stiffness matrix for the bar. Hence, for each element, we have

$$
\underset{1\quad\; 2}{k^{(1)} = \frac{AE}{L/2}\begin{bmatrix} 1 & -1 \\ -1 & 1 \end{bmatrix}\frac{\text{lb}}{\text{in.}}} \qquad \underset{2\quad\; 3}{k^{(2)} = \frac{AE}{L/2}\begin{bmatrix} 1 & -1 \\ -1 & 1 \end{bmatrix}\frac{\text{lb}}{\text{in.}}} \tag{15.1.24}
$$

where the numbers above the columns in the $\underline{k}$'s indicate the nodal displacements associated with each element.

Step 1

Using Eq. (15.1.18), the thermal force matrix for each element is given by

$$
\underline{f}^{(1)} = \begin{Bmatrix} -E\alpha TA \\ E\alpha TA \end{Bmatrix} \qquad \underline{f}^{(2)} = \begin{Bmatrix} -E\alpha TA \\ E\alpha TA \end{Bmatrix} \tag{15.1.25}
$$

where these forces are considered to be equivalent nodal forces.

Step 2

Applying the direct stiffness method to Eqs. (15.1.24) and (15.1.25), we assemble the global equations as

$$
\begin{Bmatrix} -E\alpha TA \\ 0 \\ E\alpha TA \end{Bmatrix} = \frac{AE}{L/2}\begin{bmatrix} 1 & -1 & 0 \\ -1 & 1+1 & -1 \\ 0 & -1 & 1 \end{bmatrix}\begin{Bmatrix} d_{1x} \\ d_{2x} \\ d_{3x} \end{Bmatrix} \tag{15.1.26}
$$

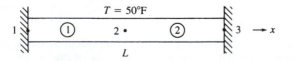

Figure 15–2 Bar subjected to a uniform temperature rise

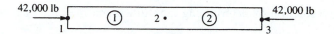

Figure 15–3 Free-body diagram of the bar of Figure 15–2

Applying the boundary conditions $d_{1x} = 0$ and $d_{3x} = 0$ and solving the second of Eq. (15.1.26), we obtain

$$d_{2x} = 0 \tag{15.1.27}$$

Step 3

Back-substituting Eq. (15.1.27) into the global equation (step 2) for the nodal forces, we obtain

$$\begin{Bmatrix} F_{1x} \\ F_{2x} \\ F_{3x} \end{Bmatrix} = \begin{Bmatrix} 0 \\ 0 \\ 0 \end{Bmatrix} - \begin{Bmatrix} -E\alpha TA \\ 0 \\ E\alpha TA \end{Bmatrix} = \begin{Bmatrix} E\alpha TA \\ 0 \\ -E\alpha TA \end{Bmatrix} \tag{15.1.28}$$

Using the numerical quantities for E, α, T, and A in Eq. (15.1.28), we obtain

$$F_{1x} = 42{,}000 \text{ lb} \qquad F_{2x} = 0 \qquad F_{3x} = -42{,}000 \text{ lb}$$

as shown in Figure 15–3. The stress in the bar is then

$$\sigma = \frac{42{,}000}{4} = 10{,}500 \text{ psi} \qquad \text{(compressive)} \tag{15.1.29} \quad \blacksquare$$

Example 15.2

For the bar assemblage shown in Figure 15–4, determine the reactions at the fixed ends and the axial stress in each bar. Bar 1 is subjected to a temperature drop of $10\,^\circ$C. Let bar 1 be aluminum with $E = 70$ GPa, $\alpha = 23 \times 10^{-6}$ (mm/mm)/$^\circ$C, $A = 12 \times 10^{-4}$ m^2, and $L = 2$ m. Let bars 2 and 3 be brass with $E = 100$ GPa, $\alpha = 20 \times 10^{-6}$ (mm/mm)/$^\circ$C, $A = 6 \times 10^{-4}$ m^2, and $L = 2$ m.

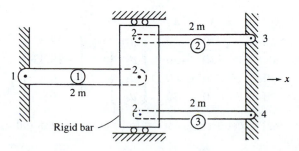

Figure 15–4 Bar assemblage for thermal stress analysis

We begin the solution by determining the stiffness matrices for each element.

Element 1

$$k^{(1)} = \frac{(12 \times 10^{-4})(70 \times 10^6)}{2} \begin{bmatrix} 1 & -1 \\ -1 & 1 \end{bmatrix} = 42,000 \begin{matrix} & 1 & 2 \\ & \begin{bmatrix} 1 & -1 \\ -1 & 1 \end{bmatrix} \end{matrix} \frac{kN}{m} \qquad (15.1.30a)$$

Elements 2 and 3

$$k^{(2)} = k^{(3)} = \frac{(6 \times 10^{-4})(100 \times 10^6)}{2} \begin{bmatrix} 1 & -1 \\ -1 & 1 \end{bmatrix} = 30,000 \begin{matrix} \begin{matrix} 2 & 3 \\ 2 & 4 \end{matrix} \\ \begin{bmatrix} 1 & -1 \\ -1 & 1 \end{bmatrix} \end{matrix} \frac{kN}{m}$$

$$(15.1.30b)$$

Step 1

We obtain the element thermal force matrices by evaluating Eq. (15.1.18). First, evaluating $-E\alpha TA$ for element 1, we have

$$-E\alpha TA = -(70 \times 10^6)(23 \times 10^{-6})(-10)(12 \times 10^{-4}) = 19.32 \text{ kN} \qquad (15.1.31)$$

where the -10 term in Eq. (15.1.31) is due to the temperature drop in element 1. Using the result of Eq. (15.1.31) in Eq. (15.1.18), we obtain

$$f^{(1)} = \begin{Bmatrix} f_{1x} \\ f_{2x} \end{Bmatrix} = \begin{Bmatrix} 19.32 \\ -19.32 \end{Bmatrix} \text{ kN} \qquad (15.1.32)$$

There is no temperature change in elements 2 and 3, and so

$$f^{(2)} = f^{(3)} = \begin{Bmatrix} 0 \\ 0 \end{Bmatrix} \qquad (15.1.33)$$

Step 2

Assembling the global equations using Eqs. (15.1.30), (15.1.32), and (15.1.33), we obtain

$$1000 \begin{matrix} & 1 & 2 & 3 & 4 \\ & \begin{bmatrix} 42 & -42 & 0 & 0 \\ -42 & 42+30+30 & -30 & -30 \\ 0 & -30 & 30 & 0 \\ 0 & -30 & 0 & 30 \end{bmatrix} \end{matrix} \begin{Bmatrix} d_{1x} \\ d_{2x} \\ d_{3x} \\ d_{4x} \end{Bmatrix} = \begin{Bmatrix} 19.32 \\ -19.32 \\ 0 \\ 0 \end{Bmatrix} \qquad (15.1.34)$$

where the right-side thermal forces are considered to be equivalent nodal forces. Using the boundary conditions

$$d_{1x} = 0 \qquad d_{3x} = 0 \qquad d_{4x} = 0 \qquad (15.1.35)$$

we obtain, from the second equation of Eq. (15.1.34),

$$1000(102)d_{2x} = -19.32$$

Solving for d_{2x}, we obtain

$$d_{2x} = -1.89 \times 10^{-4} \text{ m} \tag{15.1.36}$$

Step 3

Back-substituting Eq. (15.1.36) into the global equation for the nodal forces, $\underline{F} = \underline{K}\underline{d} - \underline{F}_0$, we have

$$\begin{Bmatrix} F_{1x} \\ F_{2x} \\ F_{3x} \\ F_{4x} \end{Bmatrix} = 1000 \begin{bmatrix} 42 & -42 & 0 & 0 \\ -42 & 102 & -30 & -30 \\ 0 & -30 & 30 & 0 \\ 0 & -30 & 0 & 30 \end{bmatrix} \begin{Bmatrix} 0 \\ -1.89 \times 10^{-4} \\ 0 \\ 0 \end{Bmatrix} - \begin{Bmatrix} 19.32 \\ -19.32 \\ 0 \\ 0 \end{Bmatrix} \tag{15.1.37}$$

Simplifying Eq. (15.1.37), we obtain

$$F_{1x} = -11.38 \text{ kN}$$

$$F_{2x} = 0.0 \text{ kN}$$

$$F_{3x} = 5.69 \ \text{ kN} \tag{15.1.38}$$

$$F_{4x} = 5.69 \text{ kN}$$

A free-body diagram of the bar assemblage is shown in Figure 15–5. The stresses in each bar are then

$$\sigma^{(1)} = \frac{11.38}{12 \times 10^{-4}} = 9.48 \times 10^3 \text{ kN/m}^2 \quad (9.48 \text{ MPa})$$

$$\tag{15.1.39}$$

$$\sigma^{(2)} = \sigma^{(3)} = \frac{5.69}{6 \times 10^{-4}} = 9.48 \times 10^3 \text{ kN/m}^2 \quad (9.48 \text{ MPa})$$

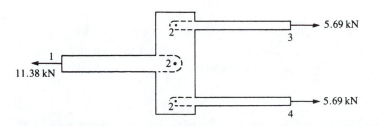

Figure 15–5 Free-body diagram of the bar assemblage of Figure 15–4 ■

Example 15.3

For the plane truss shown in Figure 15–6, determine the displacements at node 1 and the axial stresses in each bar. Bar 1 is subjected to a temperature rise of 75°F. Let $E = 30 \times 10^6$ psi, $\alpha = 7 \times 10^{-6}$ (in./in.)/°F, and $A = 2$ in^2 for both bar elements.

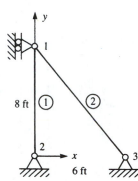

Figure 15–6 Plane truss for thermal stress analysis

First, using Eq. (3.4.23), we determine the stiffness matrices for each element.

Element 1

Choosing $\hat{x}$ from node 2 to node 1, $\theta = 90°$, and so $\cos\theta = 0$, $\sin\theta = 1$, and

$$\underline{k}^{(1)} = \frac{(2)(30 \times 10^6)}{(8 \times 12)} \begin{array}{c} \\ \begin{bmatrix} 0 & 0 & 0 & 0 \\ & 1 & 0 & -1 \\ & & 0 & 0 \\ \text{Symmetry} & & & 1 \end{bmatrix} \end{array} \frac{\text{lb}}{\text{in.}} \qquad (15.1.40)$$

Element 2

Choosing $\hat{x}$ from node 3 to node 1, $\theta = 180° - 53.13° = 126.87°$, and so $\cos\theta = -0.6$, $\sin\theta = 0.8$, and

$$\underline{k}^{(2)} = \frac{(2)(30 \times 10^6)}{(10 \times 12)} \begin{bmatrix} 0.36 & -0.48 & -0.36 & 0.48 \\ & 0.64 & 0.48 & -0.64 \\ & & 0.36 & -0.48 \\ \text{Symmetry} & & & 0.64 \end{bmatrix} \frac{\text{lb}}{\text{in.}} \qquad (15.1.41)$$

Step 1

We obtain the element thermal force matrices by evaluating Eq. (15.1.18) as follows:

$$-E\alpha TA = -(30 \times 10^6)(7 \times 10^{-6})(75)(2) = -31{,}500 \text{ lb} \qquad (15.1.42)$$

Using the result of Eq. (15.1.42) for element 1, we then have the local thermal force matrix as

$$\hat{\underline{f}}^{(1)} = \begin{Bmatrix} \hat{f}_{2x} \\ \hat{f}_{1x} \end{Bmatrix} = \begin{Bmatrix} -31{,}500 \\ 31{,}500 \end{Bmatrix} \text{ lb} \qquad (15.1.43)$$

There is no temperature change in element 2, so

$$\hat{\underline{f}}^{(2)} = \begin{Bmatrix} \hat{f}_{3x} \\ \hat{f}_{1x} \end{Bmatrix} = \begin{Bmatrix} 0 \\ 0 \end{Bmatrix} \qquad (15.1.44)$$

Recall that by Eq. (3.4.16), $\hat{\underline{f}} = \underline{T}\underline{f}$. Since we have shown that $\underline{T}^{-1} = \underline{T}^T$, we can obtain the global forces by premultiplying Eq. (3.4.16) by $\underline{T}^T$ to obtain the element nodal forces in the global reference frame as

$$\underline{f} = \underline{T}^T \hat{\underline{f}} \qquad (15.1.45)$$

Using Eq. (15.1.45), the element 1 global nodal forces are then

$$\begin{Bmatrix} f_{2x} \\ f_{2y} \\ f_{1x} \\ f_{1y} \end{Bmatrix} = \begin{bmatrix} C & -S & 0 & 0 \\ S & C & 0 & 0 \\ 0 & 0 & C & -S \\ 0 & 0 & S & C \end{bmatrix} \begin{Bmatrix} \hat{f}_{2x} \\ \hat{f}_{2y} \\ \hat{f}_{1x} \\ \hat{f}_{1y} \end{Bmatrix} \qquad (15.1.46)$$

where the order of terms in Eq. (15.1.46) is due to the choice of the $\hat{x}$ axis from node 2 to node 1 and where $\underline{T}$, given by Eq. (3.4.15), has been used.

Substituting the numerical quantities $C = 0$ and $S = 1$ (consistent with $\hat{x}$ for element 1), and $\hat{f}_{1x} = 31{,}500$, $\hat{f}_{1y} = 0$, $\hat{f}_{2x} = -31{,}500$, and $\hat{f}_{2y} = 0$ into Eq. (15.1.46), we obtain

$$f_{2x} = 0 \qquad f_{2y} = -31{,}500 \text{ lb} \qquad f_{1x} = 0 \qquad f_{1y} = 31{,}500 \text{ lb} \qquad (15.1.47)$$

These element forces are now the only equivalent global nodal forces, because element 2 is not subjected to a change in temperature.

Step 2

Assembling the global equations using Eqs. (15.1.40), (15.1.41), and (15.1.47), we obtain

$$0.50 \times 10^6 \begin{bmatrix} 0.36 & -0.48 & 0 & 0 & 0 & 0 \\ & 1.89 & 0 & -1.25 & 0 & 0 \\ & & 0 & 0 & 0 & 0 \\ & & & 1.25 & 0 & 0 \\ & & & & 0.36 & -0.48 \\ \text{Symmetry} & & & & & 0.64 \end{bmatrix} \begin{Bmatrix} d_{1x} \\ d_{1y} \\ d_{2x} \\ d_{2y} \\ d_{3x} \\ d_{3y} \end{Bmatrix} = \begin{Bmatrix} F_{1x} + 0 \\ 31{,}500 \\ F_{2x} + 0 \\ -31{,}500 + F_{2y} \\ F_{3x} + 0 \\ F_{3y} + 0 \end{Bmatrix}$$

$$(15.1.48)$$

The boundary conditions are given by

$$d_{1x} = 0 \qquad d_{2x} = 0 \qquad d_{2y} = 0 \qquad d_{3x} = 0 \qquad d_{3y} = 0 \qquad (15.1.49)$$

Using the boundary condition Eqs. (15.1.49) and the second equation of Eq. (15.1.48), we obtain

$$(0.945 \times 10^6)d_{1y} = 31,500$$

or
$$d_{1y} = 0.0333 \text{ in.} \qquad (15.1.50)$$

Step 3

We now illustrate the procedure used to obtain the local element forces in local coordinates; that is, the local element forces are

$$\hat{f} = \hat{k}\hat{d} - \hat{f}_0 \qquad (15.1.51)$$

We determine the actual local element nodal forces by using the relationship $\hat{d} = \underline{T}^* d$, the usual bar element $\hat{k}$ matrix [Eq. (3.1.14)], the transformation matrix $\underline{T}^*$ [Eq. (3.4.8)], and the calculated displacements and initial thermal forces applicable for the element under consideration. Substituting the numerical quantities for element 1, from Eq. (15.1.51), we have

$$\begin{Bmatrix} \hat{f}_{2x} \\ \hat{f}_{1x} \end{Bmatrix} = \frac{2(30 \times 10^6)}{8 \times 12} \begin{bmatrix} 1 & -1 \\ -1 & 1 \end{bmatrix} \begin{bmatrix} 0 & 1 & 0 & 0 \\ 0 & 0 & 0 & 1 \end{bmatrix} \begin{Bmatrix} d_{2x} = 0 \\ d_{2y} = 0 \\ d_{1x} = 0 \\ d_{1y} = 0.0333 \end{Bmatrix} - \begin{Bmatrix} -31,500 \\ 31,500 \end{Bmatrix}$$
$$(15.1.52)$$

Simplifying Eq. (15.1.52), we obtain

$$\hat{f}_{2x} = 10,700 \text{ lb} \qquad \hat{f}_{1x} = -10,700 \text{ lb} \qquad (15.1.53)$$

Dividing the local element force $\hat{f}_{1x}$ (which is the far-end force consistent with the convention used in Section 3.5) by the cross-sectional area, we obtain the stress as

$$\sigma^{(1)} = \frac{-10,700}{2} = -5350 \text{ psi} \qquad (15.1.54)$$

Similarly, for element 2, we have

$$\begin{Bmatrix} \hat{f}_{3x} \\ \hat{f}_{1x} \end{Bmatrix} = \frac{2(30 \times 10^6)}{10 \times 12} \begin{bmatrix} 1 & -1 \\ -1 & 1 \end{bmatrix} \begin{bmatrix} -0.6 & 0.8 & 0 & 0 \\ 0 & 0 & -0.6 & 0.8 \end{bmatrix} \begin{Bmatrix} 0 \\ 0 \\ 0 \\ 0.0333 \end{Bmatrix} \qquad (15.1.55)$$

Simplifying, Eq. (15.1.55), we obtain

$$\hat{f}_{3x} = -13,310 \text{ lb} \qquad \hat{f}_{1x} = 13,310 \text{ lb} \qquad (15.1.56)$$

where no initial thermal forces were present for element 2 because the element was not subjected to a temperature change. Dividing the far-end force $\hat{f}_{1x}$ by the cross-sectional area results in

$$\sigma^{(2)} = 6660 \text{ psi} \tag{15.1.57}$$

For two- and three-dimensional stress problems, this direct division of force by cross-sectional area is not permissible. Hence, the total stress due to both applied loading and temperature change must be determined by

$$\underline{\sigma} = \underline{\sigma}_L - \underline{\sigma}_T \tag{15.1.58}$$

We now illustrate Eq. (15.1.58) for bar element 1 of the truss of Example 15.3. For the bar, σ_L can be obtained using Eq. (3.5.6), and σ_T is obtained from

$$\sigma_T = \underline{D}\underline{\varepsilon}_T = E\alpha T \tag{15.1.59}$$

because $\underline{D} = E$ and $\varepsilon_T = \alpha T$ for the bar element. The stress in bar element 1 is then determined to be

$$\sigma^{(1)} = \frac{E}{L}[-C \quad -S \quad C \quad S]\begin{Bmatrix} d_{2x} \\ d_{2y} \\ d_{1x} \\ d_{1y} \end{Bmatrix} - E\alpha T \tag{15.1.60}$$

Substituting the numerical quantities for element 1 into Eq. (15.1.60), we obtain

$$\sigma^{(1)} = \frac{30 \times 10^6}{8 \times 12}[0 \quad -1 \quad 0 \quad 1]\begin{Bmatrix} 0 \\ 0 \\ 0 \\ 0.0333 \end{Bmatrix} - (30 \times 10^6)(7 \times 10^{-6})(75) \tag{15.1.61}$$

or

$$\sigma^{(1)} = -5350 \text{ psi} \tag{15.1.62} \quad ■$$

We will now illustrate the solutions of two plane thermal stress problems.

Example 15.4

For the plane stress element shown in Figure 15–7, determine the element equations. The element has a 2000-lb/in^2 pressure acting perpendicular to side j-m and is subjected to a 30°F temperature rise.

Recall that the stiffness matrix is given by [Eq. (7.2.52) or (7.4.1)]

$$[k] = [B]^T[D][B]tA \tag{15.1.63}$$

and

$$\begin{aligned} \beta_i = y_j - y_m = -3 \qquad & \gamma_i = x_m - x_j = -1 \\ \beta_j = y_m - y_i = 3 \qquad & \gamma_j = x_i - x_m = -1 \\ \beta_m = y_i - y_j = 0 \qquad & \gamma_m = x_j - x_i = 2 \end{aligned} \tag{15.1.64}$$

and

$$A = \frac{(3)(2)}{2} = 3 \text{ in}^2$$

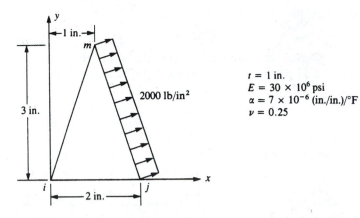

Figure 15–7 Plane stress element subjected to mechanical loading and a temperature change

Therefore, substituting the results of Eqs. (15.1.64) into Eq. (7.2.34) for $[B]$, we obtain

$$[B] = \frac{1}{6} \begin{bmatrix} -3 & 0 & 3 & 0 & 0 & 0 \\ 0 & -1 & 0 & -1 & 0 & 2 \\ -1 & -3 & -1 & 3 & 2 & 0 \end{bmatrix} \qquad (15.1.65)$$

Assuming plane stress conditions to be valid, we have

$$[D] = \frac{E}{1 - v^2} \begin{bmatrix} 1 & v & 0 \\ v & 1 & 0 \\ 0 & 0 & \dfrac{1-v}{2} \end{bmatrix} = \frac{30 \times 10^6}{1 - (0.25)^2} \begin{bmatrix} 1 & 0.25 & 0 \\ 0.25 & 1 & 0 \\ 0 & 0 & 0.375 \end{bmatrix}$$

$$= (4 \times 10^6) \begin{bmatrix} 8 & 2 & 0 \\ 2 & 8 & 0 \\ 0 & 0 & 3 \end{bmatrix} \text{ psi} \qquad (15.1.66)$$

Also,
$$[B]^T [D] = \frac{1}{6} \begin{bmatrix} -3 & 0 & -1 \\ 0 & -1 & -3 \\ 3 & 0 & -1 \\ 0 & -1 & 3 \\ 0 & 0 & 2 \\ 0 & 2 & 0 \end{bmatrix} (4 \times 10^6) \begin{bmatrix} 8 & 2 & 0 \\ 2 & 8 & 0 \\ 0 & 0 & 3 \end{bmatrix} \qquad (15.1.67)$$

Simplifying Eq. (15.1.67), we obtain

$$[B]^T[D] = \frac{4 \times 10^6}{6} \begin{bmatrix} -24 & -6 & -3 \\ -2 & -8 & -9 \\ 24 & 6 & -3 \\ -2 & -8 & 9 \\ 0 & 0 & 6 \\ 4 & 16 & 0 \end{bmatrix}$$

(15.1.68)

Therefore, substituting the results of Eqs. (15.1.65) and (15.1.68) into Eq. (15.1.63) yields the element stiffness matrix as

$$[k] = (1 \text{ in.}) \frac{(3 \text{ in}^2)}{6} \frac{4 \times 10^6}{6} \begin{bmatrix} -24 & -6 & -3 \\ -2 & -8 & -9 \\ 24 & 6 & -3 \\ -2 & -8 & 9 \\ 0 & 0 & 6 \\ 4 & 16 & 0 \end{bmatrix} \begin{bmatrix} -3 & 0 & 3 & 0 & 0 & 0 \\ 0 & -1 & 0 & -1 & 0 & 2 \\ -1 & -3 & -1 & 3 & 2 & 0 \end{bmatrix}$$

(15.1.69)

Simplifying Eq. (15.1.69), we have the element stiffness matrix as

$$[k] = \frac{1 \times 10^6}{3} \begin{bmatrix} 75 & 15 & -69 & -3 & -6 & -12 \\ 15 & 35 & 3 & -19 & -18 & -16 \\ -69 & 3 & 75 & -15 & -6 & 12 \\ -3 & -19 & -15 & 35 & 18 & -16 \\ -6 & -18 & -6 & 18 & 12 & 0 \\ -12 & -16 & 12 & -16 & 0 & 32 \end{bmatrix} \frac{\text{lb}}{\text{in.}}$$

(15.1.70)

Using Eq. (15.1.23), the thermal force matrix is given by

$$\{f_T\} = \frac{\alpha E t T}{2(1-v)} \begin{Bmatrix} \beta_i \\ \gamma_i \\ \beta_j \\ \gamma_j \\ \beta_m \\ \gamma_m \end{Bmatrix} = \frac{(7 \times 10^{-6})(30 \times 10^6)(1)(30)}{2(1-0.25)} \begin{Bmatrix} -3 \\ -1 \\ 3 \\ -1 \\ 0 \\ 2 \end{Bmatrix} = 4200 \begin{Bmatrix} -3 \\ -1 \\ 3 \\ -1 \\ 0 \\ 2 \end{Bmatrix}$$

or

$$\{f_T\} = \begin{Bmatrix} -12{,}600 \\ -4200 \\ 12{,}600 \\ -4200 \\ 0 \\ 8400 \end{Bmatrix} \text{lb}$$

(15.1.71)

The force matrix due to the pressure applied alongside *j-m* is determined as follows:

$$L_{j\text{-}m} = [(2-1)^2 + (3-0)^2]^{1/2} = 3.163 \text{ in.}$$

$$p_x = p\cos\theta = 2000\left(\frac{3}{3.163}\right) = 1896 \text{ lb/in}^2 \tag{15.1.72}$$

$$p_y = p\sin\theta = 2000\left(\frac{1}{3.163}\right) = 632 \text{ lb/in}^2$$

where θ is the angle measured from the x axis to the normal to surface *j-m*. Using Eq. (7.3.7) to evaluate the surface forces, we have

$$\{f_L\} = \iint\limits_{S_{j\text{-}m}} [N_s]^T \begin{Bmatrix} p_x \\ p_y \end{Bmatrix} dS$$

$$= \iint\limits_{S_{j\text{-}m}} \begin{bmatrix} N_i & 0 \\ 0 & N_i \\ N_j & 0 \\ 0 & N_j \\ N_m & 0 \\ 0 & N_m \end{bmatrix}_{\substack{\text{evaluated} \\ \text{alongside } j\text{-}m}} \begin{Bmatrix} p_x \\ p_y \end{Bmatrix} dS = \frac{tL_{j\text{-}m}}{2} \begin{bmatrix} 0 & 0 \\ 0 & 0 \\ 1 & 0 \\ 0 & 1 \\ 1 & 0 \\ 0 & 1 \end{bmatrix} \begin{Bmatrix} p_x \\ p_y \end{Bmatrix} \tag{15.1.73}$$

Evaluating Eq. (15.1.73), we obtain

$$\{f_L\} = \frac{(1 \text{ in.})(3.163 \text{ in.})}{2} \begin{bmatrix} 0 & 0 \\ 0 & 0 \\ 1 & 0 \\ 0 & 1 \\ 1 & 0 \\ 0 & 1 \end{bmatrix} \begin{Bmatrix} 1896 \\ 632 \end{Bmatrix} = \begin{Bmatrix} 0 \\ 0 \\ 3000 \\ 1000 \\ 3000 \\ 1000 \end{Bmatrix} \text{ lb} \tag{15.1.74}$$

Using Eqs. (15.1.70), (15.1.71), and (15.1.74), we find that the complete set of element equations is

$$\frac{1 \times 10^6}{3} \begin{bmatrix} 75 & 15 & -69 & -3 & -6 & -12 \\ & 35 & 3 & -19 & -18 & -16 \\ & & 75 & -15 & -6 & 12 \\ & & & 35 & 18 & -16 \\ & & & & 12 & 0 \\ \text{Symmetry} & & & & & 32 \end{bmatrix} \begin{Bmatrix} u_i \\ v_i \\ u_j \\ v_j \\ u_m \\ v_m \end{Bmatrix} = \begin{Bmatrix} -12{,}600 \\ -4200 \\ 15{,}600 \\ -3200 \\ 3000 \\ 9400 \end{Bmatrix} \tag{15.1.75}$$

where the force matrix is $\{f_T\} + \{f_L\}$, obtained by adding Eqs. (15.1.71) and (15.1.74). ∎

Example 15.5

For the plane stress plate fixed along one edge and subjected to a uniform temperature rise of 50 °C as shown in Figure 15–8, determine the nodal displacements and the stresses in each element. Let $E = 210$ GPa, $v = 0.30$, $t = 5$ mm, and $\alpha = 12 \times 10^{-6}$ (mm/mm)/°C.

The discretized plate is shown in Figure 15–8. We begin by evaluating the stiffness matrix of each element using Eq. (7.2.52).

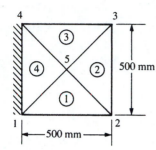

Figure 15–8 Discretized plate subjected to a temperature change

Element 1

Element 1 has coordinates $x_1 = 0$, $y_1 = 0$, $x_2 = 0.5$, $y_2 = 0$, $x_5 = 0.25$, and $y_5 = 0.25$. From Eqs. (7.2.10), we obtain

$$\beta_1 = y_2 - y_5 = -0.25 \text{ m} \qquad \beta_2 = y_5 - y_1 = 0.25 \text{ m} \qquad \beta_5 = y_1 - y_2 = 0$$

$$\gamma_1 = x_5 - x_2 = -0.25 \text{ m} \qquad \gamma_2 = x_1 - x_5 = -0.25 \text{ m} \qquad \gamma_5 = x_2 - x_1 = 0.5 \text{ m}$$

$$(15.1.76)$$

Using Eqs. (7.2.32) in Eq. (7.2.34), we have

$$[B] = \frac{1}{2A} \begin{bmatrix} \beta_1 & 0 & \beta_2 & 0 & \beta_5 & 0 \\ 0 & \gamma_1 & 0 & \gamma_2 & 0 & \gamma_5 \\ \gamma_1 & \beta_1 & \gamma_2 & \beta_2 & \gamma_5 & \beta_5 \end{bmatrix}$$

$$= \frac{1}{0.125} \begin{bmatrix} -0.25 & 0 & 0.25 & 0 & 0 & 0 \\ 0 & -0.25 & 0 & -0.25 & 0 & 0.5 \\ -0.25 & -0.25 & -0.25 & 0.25 & 0.5 & 0 \end{bmatrix} \frac{1}{m} \qquad (15.1.77)$$

For plane stress, $[D]$ is given by

$$D = \frac{E}{(1 - v^2)} \begin{bmatrix} 1 & v & 0 \\ v & 1 & 0 \\ 0 & 0 & \dfrac{1 - v}{2} \end{bmatrix} = \frac{210 \times 10^9}{0.91} \begin{bmatrix} 1 & 0.3 & 0 \\ 0.3 & 1 & 0 \\ 0 & 0 & 0.35 \end{bmatrix} \frac{N}{m^2} \qquad (15.1.78)$$

We obtain the element stiffness matrix using

$$[k] = tA[B]^T[D][B] \tag{15.1.79}$$

Substituting the results of Eqs. (15.1.77) and (15.1.78) into Eq. (15.1.79) and carrying out the multiplications, we have

$$
\underset{\sim}{k} = 4.615 \times 10^7
\begin{array}{c}
\begin{array}{cccccc}
d_{1x} & d_{1y} & d_{2x} & d_{2y} & d_{5x} & d_{5y}
\end{array} \\
\begin{bmatrix}
8.4375 & 4.0625 & -4.0625 & -0.3125 & -4.375 & -3.75 \\
4.0625 & 8.4375 & 0.3125 & 4.0625 & -4.375 & -12.5 \\
-4.0625 & 0.3125 & 8.4375 & -4.0625 & -4.375 & 3.75 \\
-0.3125 & 4.0625 & -4.0625 & 8.4375 & 4.375 & -12.5 \\
-4.375 & -4.375 & -4.375 & 4.375 & 8.75 & 0 \\
-3.75 & -12.5 & 3.75 & -12.5 & 0 & 25
\end{bmatrix}
\end{array}
\dfrac{N}{m}
$$

$$\tag{15.1.80}$$

Element 2

For element 2, the coordinates are $x_2 = 0.5$, $y_2 = 0$, $x_3 = 0.5$, $y_3 = 0.5$, $x_5 = 0.25$, and $y_5 = 0.25$. Proceeding as for element 1, we obtain

$$\beta_2 = 0.25 \text{ m} \qquad \beta_3 = 0.25 \text{ m} \qquad \beta_5 = -0.5 \text{ m}$$

$$\gamma_2 = -0.25 \text{ m} \qquad \gamma_3 = 0.25 \text{ m} \qquad \gamma_5 = 0$$

The element stiffness matrix then becomes

$$
\underset{\sim}{k} = 4.615 \times 10^7
\begin{array}{c}
\begin{array}{cccccc}
d_{2x} & d_{2y} & d_{3x} & d_{3y} & d_{5x} & d_{5y}
\end{array} \\
\begin{bmatrix}
8.4375 & -4.0625 & 4.0625 & -0.3125 & -12.5 & 4.375 \\
-4.0625 & 8.4375 & 0.3125 & -4.0625 & 3.75 & -4.375 \\
4.0625 & 0.3125 & 8.437 & 4.0625 & -12.5 & -4.375 \\
-0.3125 & -4.0625 & 4.0625 & 8.4375 & -3.75 & -4.375 \\
-12.5 & 3.75 & -12.5 & -3.75 & 25 & 0 \\
4.375 & -4.375 & -4.375 & -4.375 & 0 & 8.75
\end{bmatrix}
\end{array}
\dfrac{N}{m}
$$

$$\tag{15.1.81}$$

Element 3

For element 3, using the same steps as for element 1, we obtain the stiffness matrix as

$$
\underset{\sim}{k} = 4.615 \times 10^7
\begin{array}{c}
\begin{array}{cccccc}
d_{3x} & d_{3y} & d_{4x} & d_{4y} & d_{5x} & d_{5y}
\end{array} \\
\begin{bmatrix}
8.437 & 4.0625 & -4.0625 & -0.3125 & -4.375 & -3.75 \\
4.0625 & 8.437 & 0.3125 & 4.0625 & -4.375 & -12.5 \\
-4.0625 & 0.3125 & 8.437 & -4.0625 & -4.375 & 3.75 \\
-0.3125 & 4.0625 & -4.0625 & 8.4375 & 4.375 & -12.5 \\
-4.375 & -4.375 & -4.375 & 4.375 & 8.75 & 0 \\
-3.75 & -12.5 & 3.75 & -12.5 & 0 & 25
\end{bmatrix}
\end{array}
\dfrac{N}{m}
$$

$$\tag{15.1.82}$$

Element 4

Finally, for element 4, we obtain

$$
\underline{k} = 4.615 \times 10^7
\begin{array}{c}
\begin{array}{cccccc}
d_{4x} & d_{4y} & d_{1x} & d_{1y} & d_{5x} & d_{5y}
\end{array} \\
\left[
\begin{array}{cccccc}
8.437 & -4.0625 & 4.0625 & -0.3125 & -12.5 & 4.375 \\
-4.0625 & 8.4375 & 0.3125 & -4.0625 & 3.75 & -4.375 \\
4.0625 & 0.3125 & 8.437 & 4.0625 & -12.5 & -4.375 \\
-0.3125 & -4.0625 & 4.0625 & 8.431 & -3.75 & -4.375 \\
-12.5 & 3.75 & -12.5 & -3.75 & 25 & 0 \\
4.375 & -4.375 & -4.375 & -4.375 & 0 & 8.75
\end{array}
\right] \frac{N}{m}
\end{array}
$$

(15.1.83)

Using the direct stiffness method, we assemble the element stiffness matrices, Eqs. (15.1.80)–(15.1.83), to obtain the global stiffness matrix as

$$
\underline{K} = 4.615 \times 10^7
\begin{array}{c}
\begin{array}{cccc}
d_{1x} & d_{1y} & d_{2x} & d_{2y}
\end{array} \\
\left[
\begin{array}{cccc}
16.874 & 8.125 & -4.0625 & -0.3125 \\
8.125 & 16.874 & 0.3125 & 4.0625 \\
-4.0625 & 0.3125 & 16.874 & -8.125 \\
-0.3125 & 4.0625 & -8.125 & 16.875 \\
0 & 0 & 4.0625 & 0.3125 \\
0 & 0 & -0.3125 & -4.0625 \\
4.0625 & -0.3125 & 0 & 0 \\
0.3125 & -4.0625 & 0 & 0 \\
-16.875 & -8.125 & -16.875 & 8.125 \\
-8.125 & -16.875 & 8.125 & -16.875
\end{array}
\right.
\end{array}
$$

$$
\begin{array}{c}
\begin{array}{cccccc}
d_{3x} & d_{3y} & d_{4x} & d_{4y} & d_{5x} & d_{5y}
\end{array} \\
\left.
\begin{array}{cccccc}
0 & 0 & 4.0625 & 0.3125 & -16.875 & -8.125 \\
0 & 0 & -0.3125 & -4.0625 & -8.125 & -16.875 \\
4.0625 & -0.3125 & 0 & 0 & -16.875 & 8.125 \\
0.3125 & -4.0625 & 0 & 0 & 8.125 & -16.875 \\
16.875 & 8.125 & -4.0625 & -0.3125 & -16.875 & -8.125 \\
8.125 & 16.875 & 0.3125 & 4.0625 & -8.125 & -16.875 \\
-4.0625 & 0.3125 & 16.875 & -8.125 & -16.875 & 8.125 \\
-0.3125 & 4.0625 & -8.125 & 16.875 & 8.125 & -16.875 \\
-16.875 & -8.125 & -16.875 & 8.125 & 67.5 & 0 \\
-8.125 & -16.875 & 8.125 & -16.875 & 0 & 67.5
\end{array}
\right] \frac{N}{m}
\end{array}
$$

(15.1.84)

Next, we determine the thermal force matrices for each element by using Eq. (15.1.23) as follows:

Element 1

$$\{f_T\} = \frac{\alpha E t T}{2(1-v)} \begin{Bmatrix} \beta_1 \\ \gamma_1 \\ \beta_2 \\ \gamma_2 \\ \beta_5 \\ \gamma_5 \end{Bmatrix} = \frac{(12 \times 10^{-6})(210 \times 10^9)(0.005 \text{ m})(50)}{2(1-0.3)} \begin{Bmatrix} -0.25 \\ -0.25 \\ 0.25 \\ -0.25 \\ 0 \\ 0.5 \end{Bmatrix}$$

$$= 450{,}000 \begin{Bmatrix} -0.25 \\ -0.25 \\ 0.25 \\ -0.25 \\ 0 \\ 0.5 \end{Bmatrix} = \begin{Bmatrix} f_{T1x} \\ f_{T1y} \\ f_{T2x} \\ f_{T2y} \\ f_{T5x} \\ f_{T5y} \end{Bmatrix} = \begin{Bmatrix} -112{,}500 \\ -112{,}500 \\ 112{,}500 \\ -112{,}500 \\ 0 \\ 225{,}000 \end{Bmatrix} \text{ N} \qquad (15.1.85)$$

Element 2

$$\{f_T\} = 450{,}000 \begin{Bmatrix} 0.25 \\ -0.25 \\ 0.25 \\ 0.25 \\ -0.5 \\ 0 \end{Bmatrix} = \begin{Bmatrix} f_{T2x} \\ f_{T2y} \\ f_{T3x} \\ f_{T3y} \\ f_{T5x} \\ f_{T5y} \end{Bmatrix} = \begin{Bmatrix} 112{,}500 \\ -112{,}500 \\ 112{,}500 \\ 112{,}500 \\ -225{,}000 \\ 0 \end{Bmatrix} \text{ N} \qquad (15.1.86)$$

Element 3

$$\{f_T\} = 450{,}000 \begin{Bmatrix} 0.25 \\ 0.25 \\ -0.25 \\ 0.25 \\ 0 \\ -0.5 \end{Bmatrix} = \begin{Bmatrix} f_{T3x} \\ f_{T3y} \\ f_{T4x} \\ f_{T4y} \\ f_{T5x} \\ f_{T5y} \end{Bmatrix} = \begin{Bmatrix} 112{,}500 \\ 112{,}500 \\ -112{,}500 \\ 112{,}500 \\ 0 \\ -225{,}000 \end{Bmatrix} \text{ N} \qquad (15.1.87)$$

Element 4

$$\{f_T\} = 450{,}000 \begin{Bmatrix} -0.25 \\ 0.25 \\ -0.25 \\ -0.25 \\ 0.5 \\ 0 \end{Bmatrix} = \begin{Bmatrix} f_{T4x} \\ f_{T4y} \\ f_{T1x} \\ f_{T1y} \\ f_{T5x} \\ f_{T5y} \end{Bmatrix} = \begin{Bmatrix} -112{,}500 \\ 112{,}500 \\ -112{,}500 \\ -112{,}500 \\ 225{,}000 \\ 0 \end{Bmatrix} \text{ N} \qquad (15.1.88)$$

We then obtain the global thermal force matrix by direct assemblage of the element force matrices. The resulting matrix is

$$
\begin{Bmatrix}
f_{T1x} \\
f_{T1y} \\
f_{T2x} \\
f_{T2y} \\
f_{T3x} \\
f_{T3y} \\
f_{T4x} \\
f_{T4y} \\
f_{T5x} \\
f_{T5y}
\end{Bmatrix}
=
\begin{Bmatrix}
-225{,}000 \\
-225{,}000 \\
225{,}000 \\
-225{,}000 \\
225{,}000 \\
225{,}000 \\
-225{,}000 \\
225{,}000 \\
0 \\
0
\end{Bmatrix} \text{N}
\tag{15.1.89}
$$

Using Eqs. (15.1.84) and (15.1.89) and imposing the boundary conditions $d_{1x} = d_{1y} = d_{4x} = d_{4y} = 0$, we obtain the system of equations for solution as

$$
\begin{Bmatrix}
f_{T2x} = 225{,}000 \\
f_{T2y} = -225{,}000 \\
f_{T3x} = 225{,}000 \\
f_{T3y} = 225{,}000 \\
f_{T5x} = 0 \\
f_{T5y} = 0
\end{Bmatrix}
= 4.615 \times 10^7
$$

$$
\begin{bmatrix}
16.874 & -8.125 & 4.0625 & -0.3125 & -16.875 & 8.125 \\
-8.125 & 16.875 & 0.3125 & -4.0625 & 8.125 & -16.875 \\
4.0625 & 0.3125 & 16.875 & 8.125 & -16.875 & -8.125 \\
-0.3125 & -4.0625 & 8.125 & 16.875 & -8.125 & -16.875 \\
-16.875 & 8.125 & -16.875 & -8.125 & 67.5 & 0 \\
8.125 & -16.875 & -8.125 & -16.875 & 0 & 67.5
\end{bmatrix}
\begin{Bmatrix}
d_{2x} \\
d_{2y} \\
d_{3x} \\
d_{3y} \\
d_{5x} \\
d_{5y}
\end{Bmatrix}
\tag{15.1.90}
$$

Solving Eq. (15.1.90) for the nodal displacements, we have

$$
\begin{Bmatrix}
d_{2x} \\
d_{2y} \\
d_{3x} \\
d_{3y} \\
d_{5x} \\
d_{5y}
\end{Bmatrix}
=
\begin{Bmatrix}
3.327 \times 10^{-4} \\
-1.911 \times 10^{-4} \\
3.327 \times 10^{-4} \\
1.911 \times 10^{-4} \\
2.123 \times 10^{-4} \\
6.654 \times 10^{-9}
\end{Bmatrix} \text{m}
\tag{15.1.91}
$$

We now use Eq. (15.1.58) to obtain the stresses in each element. Using Eqs. (7.2.36) and (15.1.59), we write Eq. (15.1.58) as

$$
\{\sigma\} = [D][B]\{d\} - [D]\{\varepsilon_T\}
\tag{15.1.92}
$$

Element 1

$$
\left\{
\begin{array}{c}
\sigma_x \\
\sigma_y \\
\tau_{xy}
\end{array}
\right\}
=
\frac{E}{1-v^2}
\begin{bmatrix}
1 & v & 0 \\
v & 1 & 0 \\
0 & 0 & \dfrac{1-v}{2}
\end{bmatrix}
\frac{1}{2A}
\left\{
\begin{array}{cccccc}
\beta_1 & 0 & \beta_2 & 0 & \beta_5 & 0 \\
0 & \gamma_1 & 0 & \gamma_2 & 0 & \gamma_5 \\
\gamma_1 & \beta_1 & \gamma_2 & \beta_2 & \gamma_5 & \beta_5
\end{array}
\right\}
\left\{
\begin{array}{c}
d_{1x} \\
d_{1y} \\
d_{2x} \\
d_{2y} \\
d_{5x} \\
d_{5y}
\end{array}
\right\}
$$

$$
-\frac{E}{1-v^2}
\begin{bmatrix}
1 & v & 0 \\
v & 1 & 0 \\
0 & 0 & \dfrac{1-v}{2}
\end{bmatrix}
\left\{
\begin{array}{c}
\alpha T \\
\alpha T \\
0
\end{array}
\right\}
\tag{15.1.93}
$$

Using Eqs. (15.1.76) and (15.1.91) along with the mechanical properties E, v, and α in Eq. (15.1.93), we obtain

$$
\left\{
\begin{array}{c}
\sigma_x \\
\sigma_y \\
\tau_{xy}
\end{array}
\right\}
=
\frac{210 \times 10^9}{0.91}
\begin{bmatrix}
1 & 0.3 & 0 \\
0.3 & 1 & 0 \\
0 & 0 & 0.35
\end{bmatrix}
$$

$$
\times \frac{1}{0.125}
\begin{bmatrix}
-0.25 & 0 & 0.25 & 0 & 0 & 0 \\
0 & -0.25 & 0 & -0.25 & 0 & 0.5 \\
-0.25 & -0.25 & -0.25 & 0.25 & 0.5 & 0
\end{bmatrix}
\left\{
\begin{array}{c}
0 \\
0 \\
3.327 \times 10^{-4} \\
-1.911 \times 10^{-4} \\
2.123 \times 10^{-4} \\
6.654 \times 10^{-9}
\end{array}
\right\}
$$

$$
-\frac{210 \times 10^9}{0.91}
\begin{bmatrix}
1 & 0.3 & 0 \\
0.3 & 1 & 0 \\
0 & 0 & 0.35
\end{bmatrix}
\left\{
\begin{array}{c}
(12 \times 10^{-6})(50) \\
(12 \times 10^{-6})(50) \\
0
\end{array}
\right\}
\tag{15.1.94}
$$

Simplifying Eq. (15.1.94) yields

$$
\left\{
\begin{array}{c}
\sigma_x \\
\sigma_y \\
\tau_{xy}
\end{array}
\right\}
=
\left\{
\begin{array}{c}
1.800 \times 10^8 \\
1.342 \times 10^8 \\
-1.600 \times 10^7
\end{array}
\right\}
-
\left\{
\begin{array}{c}
1.8 \times 10^8 \\
1.8 \times 10^8 \\
0
\end{array}
\right\}
=
\left\{
\begin{array}{c}
0 \\
-4.57 \times 10^7 \\
-1.60 \times 10^7
\end{array}
\right\}
\text{ Pa}
\tag{15.1.95}
$$

Similarly, we obtain the stresses in the other elements as follows:

Element 2

$$
\left\{
\begin{array}{c}
\sigma_x \\
\sigma_y \\
\tau_{xy}
\end{array}
\right\}
=
\left\{
\begin{array}{c}
1.640 \times 10^8 \\
2.097 \times 10^8 \\
-2150
\end{array}
\right\}
-
\left\{
\begin{array}{c}
1.8 \times 10^8 \\
1.8 \times 10^8 \\
0
\end{array}
\right\}
=
\left\{
\begin{array}{c}
-1.6 \times 10^7 \\
2.973 \times 10^7 \\
-2150
\end{array}
\right\}
\text{ Pa}
\tag{15.1.96}
$$

Element 3

$$\begin{Bmatrix} \sigma_x \\ \sigma_y \\ \tau_{xy} \end{Bmatrix} = \begin{Bmatrix} 1.800 \times 10^8 \\ 1.342 \times 10^8 \\ 1.600 \times 10^7 \end{Bmatrix} - \begin{Bmatrix} 1.8 \times 10^8 \\ 1.8 \times 10^8 \\ 0 \end{Bmatrix} = \begin{Bmatrix} 0 \\ -4.57 \times 10^7 \\ -1.60 \times 10^7 \end{Bmatrix} \text{ Pa} \qquad (15.1.97)$$

Element 4

$$\begin{Bmatrix} \sigma_x \\ \sigma_y \\ \tau_{xy} \end{Bmatrix} = \begin{Bmatrix} 1.960 \times 10^8 \\ 5.880 \times 10^7 \\ 2150 \end{Bmatrix} - \begin{Bmatrix} 1.8 \times 10^8 \\ 1.8 \times 10^8 \\ 0 \end{Bmatrix} = \begin{Bmatrix} 1.6 \times 10^7 \\ -1.212 \times 10^8 \\ 2150 \end{Bmatrix} \text{ Pa} \qquad (15.1.98) \quad \blacksquare$$

15.2 Algor Example Solutions for Thermal Stress Problems

To illustrate how to solve thermal stress problems using the Algor program, we solve a number of example problems. These examples include a truss where one element is uniformly heated (see Example 15.4, to compare to the longhand solution); a plate with two elements at different temperatures (see Example 15.7); and a clamped plate subjected to uniform heating (see the longhand solution, Example 15.5). Even though the theory for axisymmetric thermal stress has not been presented, it is similar enough to the plane thermal stress problem that we include an axisymmetric tube fitted snugly into a rigid frame and then uniformly heated as in Example 15.9. Detailed steps are listed, including specific keystrokes and descriptions of the purposes of these keystrokes.

These examples illustrate that when element temperature changes are desired, proper nodal temperature changes associated with an element must be input so that when they are averaged, the desired element temperature changes result. This idea is illustrated in the examples to follow. Another procedure often used is to run the heat transfer problem to determine the nodal temperatures and then use the same mesh for the thermal stress analysis with the previously obtained nodal temperatures input for the thermal stress analysis.

Example 15.6

For the truss shown in Figure 15–9, bar element 1 is subjected to a uniform temperature rise of 30 °F. Let $E = 30 \times 10^6$ psi, $A = 2$ in^2, and $\alpha = 7.0 \times 10^{-6}$ (in./in.)/°F. The lengths of the bars are calculated from the geometry shown in the figure. Determine the horizontal and vertical displacements of the free node.

Step 1   Start Superdraw III

Select the "Start" button of Windows NT/95/98 and then proceed to "Programs: Algor Software:Algor FEA.

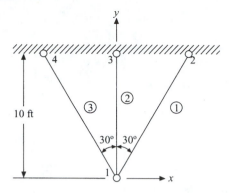

Figure 15–9 Plane truss for thermal stress analysis

Then click "Algor FEA" to start the program.

Or double-click the "Algor FEA" icon on your desktop computer screen. The Superdraw III program appears. This is the Algor main graphics user interface. All your work will be done from this program. You will construct, analyze, and review your results from Superdraw III. Each menu option and a short example are available in the Superdraw III Reference Division (accessed through "Docutech", the on-line "Technical User Documentation").

Step 2   Create the Truss Model

Select the "Add" menu from the top main menu bar.

Click "Line". A "Line" menu appears on the screen.

Click "Single" on the "Line" menu and begin entering the data points (coordinates of the elements) in succession.

0<Tab>0	This is the location of the free node.
<Enter>	Press the ⟨Enter⟩ key to actually enter the data.
–69.28<Tab>120	This is the endpoint of element 3.
<Enter>	
0<Tab>0	Repeat the process for elements 2 and 1.
<Enter>	
0<Tab>120	
<Enter>	
0<Tab>0	
<Enter>	
69.28<Tab>120	
<Enter>	All data points have been entered.

Click "Done" on the "Line" menu on the screen.

Click "View" on the main menu.

Click "Enclose" on the "View" menu to see the full view of the model. It should look like Figure 15–9.

Step 3 Eliminate Duplicate Lines

Click "Modify" on the main menu.

Click "Clean:Duplicate". This notation means to select "Clean" and then click "Duplicate" in succession. This command brings up the on-screen "Duplicate" menu.

Click "Perform Cleaning" on the "Duplicate" menu. You see the following message within the text bar near the bottom of the screen.

"3 Kept 0 Deleted. done"

Click "Done" on the "Duplicate" menu located on the screen.

Step 4 Apply the Boundary Conditions and Nodal Temperatures

Click on the "FEA Add" menu on the main menu.

Click "Stress and Vibration Analysis:Boundary Conditions" and the on-screen "Boundary Conditions" menu appears. The default boundary condition is "Use @ Symbol for Full". "Full" means full constraints on translation and rotation when this condition is applied to a particular node. The model needs full constraints at the three upper nodes of the elements.

Right-click on node 2. The "@" sign appears on screen next to the node indicating that node 2 is fully constrained.

Right-click on node 3.

Right-click on node 4. Nodes 2–4 have been fully constrained.

Click "Done" on the on-screen "Boundary Conditions" menu.

If there is difficulty seeing the "@" symbol on-screen, you may wish to zoom in closer to each node to view the boundary conditions.

Click "View:Zoom:In" and create a box on-screen around node 2 with the mouse and the left mouse button. You may have to repeat this process of zooming in one or more times to see the "@" sign.

Click "Done".

Note: If you have created one or more spurious or misplaced boundary condition "@" signs, click "Select:Box" on the main menu and draw a box around the object (in this case a boundary condition) that you wish to delete. After the object is highlighted, click "Modify:Delete". At this time the "@" symbol should have disappeared.

Click "View:Enclose" to view the entire model again.

Click "FEA Add:Stress and Vibration Analysis:Nodal Temperatures". This should bring up the on-screen menu "Nodal Temperature".

Click "Value". Enter the temperature value of the first element. Algor determines the temperature of an element by taking the average of the temperatures located at the nodes bounding the element. In this case the element temperature needs to be $30\,°F$. Placing two temperatures of $30\,°F$ at nodes 1 and 2 yields an average temperature of $30\,°F$ within element 1.

30

<Enter>

Right-click on node 1.

Right-click on node 2. Make sure to click exactly on node 2 or the nodal temperature will not be applied to the node. Use the zoom technique described earlier to see if the nodal temperatures have been placed in the correct locations. You should see "NT 30" at node 2 with the "N" and the "@" sign superimposed, one on the other.

Click "Value". We now give nodes 3 and 4 temperatures. The temperatures of elements 2 and 3 need to be 0°F. We must add −30°F to these nodes to average the element temperatures to 0°F.

−30

<Enter>

Right-click on node 3.

Right-click on node 4.

Use the zoom technique described earlier to see if the nodal temperatures have been placed in the correct locations.

Click "Done" on the on-screen "Nodal Temperature" menu.

Click "View:Enclose" to view the entire model again if necessary.

Step 5 Add the Material Properties Using the "Model Data Control" Window

Click on the "Model Data" button located in the lower part of the screen and highlighted in red. This brings up the on-screen menu "Model Data Control". Alternatively, you can access the "Model Data Control" menu by clicking "Tools:Model Data Control".

For "Analysis Type", "Linear Static Stress" appears as the default. You could select another analysis type by scrolling down the menu to an appropriate analysis algorithm. But in our case "Linear Static Stress" will do nicely.

Click on the box below the "Element" column. An on-screen menu appears called "Group 1" elements. These are the different types of elements we can choose from. For this analysis a truss element is needed.

Click "Truss".

Click "OK".

Click on the box below "Data". It responds with "Please enter the model name first".

Click "OK". Then enter the model name in the "File_name" area that appears.

ex15_6   The name of the file is *ex15_6*.

Click "Save". By default, the "Units Definition" menu appears on-screen.

Click "OK" to accept the default "English (in)" units system. Otherwise scroll down to select another choice of units.

The "Element Definition" menu appears next on-screen.

Enter the cross-sectional area of the truss elements. All elements have the same area in this example. Hold the left click of the mouse over the default value of "0" until a blue box appears on the "0". Now enter the cross-sectional area of the elements.

2

Click "OK". This should satisfy the "Data" box on the "Model Data Control" menu with a check mark.

Click on the box below the "Material" column. This brings up the on-screen menu "Element Material Selection". We can select one of the predefined materials or customer-defined materials. In this case we select customer defined materials.

Click "[Customer Defined]".

Click "View Properties:Unlock Properties" and enter the modulus of elasticity and the thermal coefficient of expansion.

30e6   Modulus of elasticity.

<Tab>

7e-6   Thermal coefficient of expansion.

Click "OK" to apply the properties to the "[Customer Defined]" material.

Click "OK" to close the "Element Material Selection" menu and return to the "Model Data Control" menu.

Step 6   Enter the Global Data

Click on the "Global" button on the right-side menu of the "Model Data Control" window. Under "Load Case Multipliers" and under "Pressure", type a 1, and under "Thermal", type a 1 as well.

1

<Tab>

<Tab>

<Tab>

1   This enters values of 1 under the "Pressure" and "Thermal" for "Load Case Multipliers".

Click on the "Output" tab to view the output options.

Click on the box next to "Displacement data".

Click on the box next to "Stress data".

This creates nodal displacement output in the .*L* file and element forces and stresses in the .*S* file. These files can be viewed later with a word processor.

Click "OK". You are out of the "Global" menu and back in the "Model Data Control" menu.

Step 7   Check the Model

Click "Check" under the "FEA Model" column of the "Model Data Control" window. This takes you to Superview's main menu, and you can check your model as explained in Section 4.1 and detailed by example in Appendix F. The truss model appears automatically, showing the boundary conditions in red and any forces that may be applied in yellow.

Click "Done". This takes you back to the "Model Data Control" menu.

Step 8 Run the Analysis

Click on the "Analysis" button under the "FEA Model" column in the "Model Data Control" window.

Click "Analyze" in the lower left corner of the "Linear Stress and Vibration-Static Stress" menu. A warning message appears after the analysis has been completed but has no bearing on the outcome of the analysis.

Click "OK" to close the warning message.

Click "Done" to return to the "Model Data Control" menu.

Step 9 Review the Results

Click on the "Results" button in the "Model Data Control" window. This takes you to Superview once again where you can view the results.

Click "Displaced" on the main menu of Superview.

Click "Displ on" so that "*With undi" appears. This allows the displaced and undisplaced models to be superimposed.

Click "Calc scal" to enlarge the scale of the displacement for easier viewing.

Click "Nodes inq" on the "Displaced" menu.

Click "Get" to get a node.

Click on node 1. Move the cursor to node 1 and click on it to see the x and y displacements at the bottom of the screen. You see "DX$=-2.91^*10^{-2}$ DY$=-9.493^*10^{-3}$" in inches.

<Esc>

<Esc>   You are now back in the main menu of Superview.

Click "Stress-di". You are now going to post a true stress.

Click "Post".

Click "Beam-trus". You must select "Beam-trus" as you have a truss model.

Click "1)P/A". The axial stress is posted. You now see the P/A stress plot for the model with colored boxes representing different stresses on each element. For instance, red on the vertical element indicates a stress of 2373.1 psi.

<Esc>

Click "Done". This takes you out of Superview.

Click "Close" to close the "Model Data Control" window.

You should be able to find the *ex15_6.L* file (storing the echo check of your data and the nodal displacements) and the *ex15_6.S* file (storing the element forces and stresses) and print a hard copy of each.

Table 15–1 shows the nodal displacements and element stresses and forces taken from the Algor program *ex15_6.L* and *ex15_6.S* files. The results are the same as those obtained from the longhand solution of Problem 15.4. ■

Table 15–1 Nodal displacements and element stresses for Example 15.6

Displacements/Rotations(degrees) of nodes

NODE number	X- translation	Y- translation	Z- translation	X- rotation	Y- rotation	Z- rotation
1	-2.9099E-02	-9.4925E-03	0.0000E+00	0.0000E+00	0.0000E+00	0.0000E+00
2	0.0000E+00	0.0000E+00	0.0000E+00	0.0000E+00	0.0000E+00	0.0000E+00
3	0.0000E+00	0.0000E+00	0.0000E+00	0.0000E+00	0.0000E+00	0.0000E+00
4	0.0000E+00	0.0000E+00	0.0000E+00	0.0000E+00	0.0000E+00	0.0000E+00

**** Nodal stresses for 3-D truss elements:

El. #	LC	ND	Stress	Force
1	1	I	1.370E+03	2.740E+03
1	1	J	-1.370E+03	-2.740E+03
2	1	I	-2.373E+03	-4.746E+03
2	1	J	2.373E+03	4.746E+03
3	1	I	1.370E+03	2.740E+03
3	1	J	-1.370E+03	-2.740E+03

■

Example 15.7

For the plate shown in Figure 15–10, the lower triangular element 1 is heated $80\,°F$ above the ambient temperature while the upper triangular element 2 is heated $50\,°F$ above the ambient temperature. Let $E = 30 \times 10^6$ psi, $v = 0.33$, and $\alpha = 10 \times 10^{-6}/°F$ for element 1 and let $E = 15 \times 10^6$ psi, $v = 0.25$, and $\alpha = 50 \times 10^{-6}/°F$ for element 2. The plate thickness is 0.1 in. Determine the nodal displacements.

This example illustrates use of the *Group* command to differentiate between the Young's moduli of the elements. Also, element temperatures are calculated as the

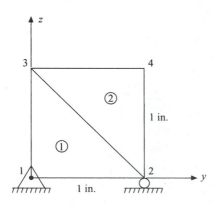

Figure 15–10 Plate subjected to nonuniform heating

average of the nodal temperatures associated with the element. This is illustrated where the nodal temperatures are defined within Algor.

Step 1   Start Superdraw III

Select the "Start" button of Windows NT/95/98 and then proceed to
                          "Programs:Algor Software:Algor FEA".
Click "Algor FEA" to start the program.
Or click on the "Algor FEA" icon on your desktop computer screen. The Superdraw III program appears. This is the Algor main graphics user interface. All your work will be done from this program. You will construct, analyze, and review your results from Superdraw III. Each menu option and a short example are available in the Superdraw III Reference Division (accessed through "Docutech", the on-line "Technical User Documentation").

Step 2   Create the Plate Model

Select "View:Predefined Views:YZ Right". This allows the *y-z* plane to be viewed.
Select the "Add" menu from the top main menu bar.
Click "Line". A "Line" menu appears in the upper right part of the screen.
0<Tab>0<Tab>1   The first point of group 1.
<Enter>
0<Tab>1<Tab>0   The endpoint of the line.
<Enter>
0<Tab>0<Tab>0
<Enter>
0<Tab>0<Tab>1   The endpoint of Group 1.
<Enter>
Click "Done" on the menu in the upper right corner.
Select "View". This takes you to the draw options menu.
Click on "Enclose". This encloses the model.
Select the "Add" menu from the top main menu bar.
Click "Line". A "Line" menu appears in the upper right part of the screen.
0<Tab>0<Tab>1   The first point of group 2.
<Enter>
0<Tab>1<Tab>1
<Enter>
0<Tab>1<Tab>0
<Enter>
0<Tab>0<Tab>1
<Enter>
Click "Done".

Make sure the last three lines created are highlighted.

Click "Modify".

Select "Update Object Parameters:Group Number".

2	Enter a 2 to give element 2 group 2 status.
<Enter>	The last group turns red.

Step 3 Eliminate Duplicate Lines

Click "Modify" on the main menu.

Select "Clean:Duplicate". This notation means to select "Clean" and then "Duplicate" in succession.

Click "Perform Cleaning" on the "Duplicate" menu. You see the following message in the lower part of the screen.

"5 Kept 0 Deleted. done"

Click "Done" on the "Duplicate" menu located in the upper right corner of the screen.

Step 4 Add the Boundary Conditions and Nodal Temperatures

Click on the "FEA Add" menu on the main menu bar and highlight "Stress and Vibration Analysis".

Click "Boundary Conditions". The "Boundary Conditions" menu appears.

The default boundary condition is "Use @ Symbol for Full". "Full" means full constraint or prevention of translations and rotations.

Right-click on node 1. The "@" sign appears next to the node, indicating node 1 is completely fixed.

Click "Change Values" on the "Boundary Conditions" menu.

Click on the check mark next to "Ty". This releases translation in the y direction.

Click "Done".

Right-click on node 2. "TxzRxyz" appears next to the node indicating node 2 is constrained.

Click "Done" on the menu in the upper right corner.

Click on the "FEA Add" menu on the main menu bar and highlight "Stress and Vibration Analysis".

Click "Nodal Temperatures". The "Nodal Temperature" menu appears.

Select "Value".

Enter a temperature of 80°.

<Enter>

Right-click on nodes 1–3. "NT 80" appears next to the nodes. This results in element 1 having a temperature increase of 80 °F.

Select "Value".

Enter a temperature of −10°.

<Enter>

Right-click on node 4. "NT −10" appears next to the node. This results in element 2 having a temperature of $(80 + 80 − 10)/3 = 50\,°F$.

Select "Done" on the "Nodal Temperatures" menu.

Step 5 Add the Material Properties Using the "Model Data Control" Window

Click on the "Model Data" button at the bottom of the screen. You see two groups, group 1 in green and group 2 in red.

Select "Linear Static Stress" by scrolling down the "Analysis Type" menu.

Click on the first box below "Element".

Click "2D" for group 1.

Click "OK".

Repeat the last three steps for the second group.

Click on the box below "Data". A prompt appears: "Enter model name first".

Click "OK".

Ex157

Click "Save".

Click "OK". Use the default unit system, English (in./lb).

0.1   Enter 0.1 for thickness (0.1 in. thick).

Click "OK".

Click on the box below "Material".

Select "[Customer Defined]".

Select "View Properties".

Click "Unlock Properties".

30e6   Enter 30e6 for "Modulus of Elasticity".

0.33   Enter 0.33 for "Poisson's Ratio".

10e-6   Enter 10e-6 for "Thermal Coefficient of Expansion".

Click "OK".

Click "OK"

Click on the second box under "Data".

0.1   Enter 0.1 for thickness (0.1 in. thick).

Click "OK".

Click on the second box under "Material".

Repeat the process of specifying material properties for group 2.

Step 6 Enter the Global Data

Click on the "Global" button in the "Model Data Control" window.

1   Enter a 1 under the "Thermal" column.

Select the "Output" tab.

Highlight the "Stress Data" and "Displacement Data" boxes.
Click "OK".

Step 7   Check the Model

Click "Check" under the "FEA Model" column on the right side of the "Model Data Control" window. This takes you to Superview.
Click "Done" to return to the "Model Data Control" window.

Step 8   Run the Analysis

Click on the "Analysis" button in the "Model Data Control" window.
Click "Analyze". Wait for a short time for the analysis to be completed.
Click "OK".
Click "Done".

Step 9   Review the Results

Click "Results" in the "Model Data Control" window. Review the results in the usual manner.

Open the *ex157.L* file using a word-processing program. All the nodal displacements shown in Table 15–2 can now be seen.

Table 15–2   Displacements of nodes for Example 15.7

```
Displacements/Rotations(degrees) of nodes
```

NODE number	X– translation	Y– translation	Z– translation	X– rotation	Y– rotation	Z– rotation
1	0.0000E+00	0.0000E+00	0.0000E+00	0.0000E+00	0.0000E+00	0.0000E+00
2	0.0000E+00	9.8983E-04	0.0000E+00	0.0000E+00	0.0000E+00	0.0000E+00
3	0.0000E+00	-7.5367E-04	9.8983E-04	0.0000E+00	0.0000E+00	0.0000E+00
4	0.0000E+00	1.3213E-03	2.0750E-03	0.0000E+00	0.0000E+00	0.0000E+00	■

Table 15–2 lists the displacements of the nodes as obtained from the Algor program. Answers compare closely with those from another program [1].   ■

Example 15.8

A plate clamped at the left end and subjected to a uniform temperature increase of $50\,°C$ is shown in Figure 15–11. Determine the displacement of the free end and the stress in the plate. Let $E = 210$ GPa, $v = 0.30$, $t = 5$ mm, and $\alpha = 12 \times 10^{-6}/°C$.

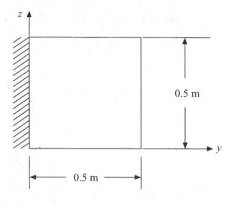

Figure 15–11 Clamped-free plate subjected to uniform heating

Step 1 Start Superdraw III

Select the "Start" button of Windows NT/95/98 and then proceed to "Programs: Algor Software:Algor FEA".

Then click "Algor FEA" to start the program.

Or double-click on the "Algor FEA" icon on your desktop computer screen. The Superdraw III program appears. This is the Algor main graphics user interface. All your work will be done from this program. You will construct, analyze, and review your results from Superdraw III. Each menu option and a short example are available in the Superdraw III Reference Division (accessed through "Docutech", the on-line "Technical User Documentation".)

Step 2 Create the Model

Select the "Modify" menu on the top menu bar.

Click "Transform XY to YZ" to convert the axes from x-y to y-z.

Select the "Add" menu from the top main menu bar.

Click "Line". A "Line" menu appears in the upper right part of the screen. Then enter the data points in succession.

0<Tab>0<Tab>0	Enter the first corner point of the model.
<Enter>	
0<Tab>0<Tab>0.5	Enter the next corner point of the model.
<Enter>	
0<Tab>0.5<Tab>0.5	Enter the next corner point of the model.
<Enter>	
0<Tab>0.5<Tab>0	Enter the next corner point of the model.
<Enter>	
0<Tab>0<Tab>0	Re-enter the original corner point to complete the model.
<Enter>	You are finished entering points.

Click "Done" on the menu in the upper right corner.

Select "View". This takes you to the draw options menu.

Click on "Enclose". This encloses the model.

Step 3   Mesh the Model

Select "FEA Mesh".

Select "Two-Dimensional Mesh Generation".

Click "Generate Mesh". It responds with "Please enter the model name first".

Click "OK" and then enter the model name in the "File_name" area that appears.

ex158   The name of the file is *ex158*.

Click "Save". By default, the "Units Definition" menu appears. In the "Units Definition" window, next to "Unit System",

Select "Metric mks (SI)". This changes the units to the SI unit system.

Click "OK". This takes you back to the "Two-Dimensional Mesh Generation" window.

Use the defaults "Quadrilateral", "Mesh Density 400", and so on.

Click "Generate" in the lower left corner of the window.

Click "OK".

Click "Done" to close the "Two-Dimensional Mesh Generation" window. You now see 342 elements and 379 nodes.

Step 4   Eliminate Duplicate Lines

Click "Modify" on main menu.

Select "Clean:Duplicate". This notation means to select "Clean" and then "Duplicate" in succession.

Click "Perform Cleaning" on the "Duplicate" menu. You see the following message in the lower part of the screen:

"720 Kept, 0 Deleted. done."

Click "Done" on the "Duplicate" menu in the upper right corner of the screen.

Step 5   Add the Boundary Conditions and Nodal Temperatures

Select the "FEA Add" menu from the main menu bar and highlight "Stress and Vibration Analysis".

Click "Boundary Conditions". The "Boundary Conditions" menu appears.

The default boundary condition is "Use @ Symbol for Full". "Full" means full constraint or prevention of translations and rotations.

Click "Text Attributes". This takes you to the "Text Setup" window.

Click "Height".

0.025   This changes the text height to 0.025 m.

<Enter>

Click "Done". This takes you back to the "Boundary Conditions" window.

Click "Box Apply". Then box all the left edge nodes. The "@" sign appears next to each node, indicating that each node is completely fixed.

Click "Done".

Select the "FEA Add" menu from the main menu bar and highlight "Stress and Vibration Analysis".

Click "Nodal Temperatures", and the "Nodal Temperature" window appears.

Select "Value" and then enter the new temperature in the space next to "Temperature Value" at the bottom of the screen.

50   This increases the temperature of the entire plate by 50 °C.

<Enter>

Click "Text Attributes". This takes you to the "Text Setup" window.

Click "Height" and enter the new value for the text height at the bottom of the screen next to "Character Height".

0.00625   This changes the text height to 0.00625.

Click "Done".

Click "Box Apply". Then box the entire plate, and "NT50" appears at each node of the mesh.

Click "Done".

Step 6   Add the Material Properties Using the "Model Data Control" Window

After the mesh has been created and the nodal boundary conditions and forces have been applied, you can select "Tools" and "Model Data Control" from Superdraw III to complete the data input for analysis. The "Model Data Control" window appears. Alternatively,

Click on the "Model Data" button located in the lower part of the screen and highlighted in red.

For "Analysis Type", "Linear Static Stress" appears highlighted as the default.

You could scroll down for other analysis types as appropriate.

Click on the box below the "Element" column. A table of group 1 element types appears.

Click on "2-D".

Click "OK".

Click on the box below the "Data" column. The "Element Definition" window for group 1 appears. In the box next to "Thickness" enter

0.005   This makes the plate 0.005 m thick.

Click "OK".

Click on the box under the "Material" column, and the "Element Material Selection" window pops up.

Click "[Customer Defined]" and then

Select "View Properties:Unlock Properties" and enter the modulus of elasticity, 210 GPa.

210e9   Next, enter the value for Poisson's ratio, 0.30.

0.30   Last, enter the coefficient of thermal expansion.

12e-6   Enter 12e-6 next to "Thermal Coefficient of Expansion". No other properties need to be entered.

Click "OK". This takes you back to the "Element Material Selection" window.

Click "OK".

Step 7   Enter the Global Data

Click "Global" in the "Model Data Control" window under the column "FEA Model". The "Global Data" window appears. Under "Load Case Multipliers" and under "Thermal", type a 1.

1   Type a 1 under the "Thermal" column as you have applied thermal loads to the plate. The program will not run without the load multiplier. The load case multiplier is what all the loads are multiplied by. Usually this value is 1. While still under "Global",

Click on the "Output" tab on the right side of the menu and then

Click on the boxes next to "Displacement data" and "Stress data". This creates nodal displacement output in the .L file and element forces and stresses in the .S file. You can also click on "Element data" to have element data written to the .L file as well. The element data consists of the nodes assigned to the elements, element loads, and material properties. You can click on "Nodal Data" to have nodal data written to the .L file. The nodal data consists of node numbers, boundary conditions, locations of nodes, and nodal forces.

Click "OK". You are now out of the "Global" menu.

Step 8   Check the Model

Click "Check" under the "FEA Model" column on the right side of the "Model Data Control" window. This takes you to Superview's main menu, and you can check your model as explained in Section 4.1 and detailed by example in Appendix F. The plate model appears automatically, showing the full constraint boundary condition along the left edge as red triangles.

Click "Done". This takes you out of the Superview menu and back to the "Model Data Control" window.

Step 9   Run the Analysis

Click on the "Analysis" button under the "FEA Model" column in the "Model Data Control" window.

Click "Analyze" in lower left corner of the "Linear Stress and Vibration-Static Stress" menu. (See green box next to "Analyze".) Wait a short time for the analysis to be performed.

Click "Done" on the lower right of the screen.

Step 10 Review the Results

Click on the "Results" button under the "FEA Model" column in the "Model Data Control" window.

Click "Displaced" on the main menu of Superview.

Click "Displ on" so that "*Displ on" appears. The asterisk means the displaced mode is active.

Click "With undi" so that "*With undi" appears.

Click "undis Top" so that "*undis Top" appears. This allows the displaced and undisplaced models to be superimposed as shown in Figure 15–12.

Click "Nodes inq:Get" and click on the top free edge node to see the following displacements: "DY=3.076e−04 and DZ=1.553e−04" in meters.

<Esc>

Click on "Stress-di". You are now going to post a stress.

Click on "Post". The Von Mises stresses are shown.

Click on "maX prin". You now see the maximum principal stresses, as plotted in Figure 15–12, superimposed on the displaced model with colored boxes representing different stresses throughout the element. For instance, the red shows a maximum of 31 MPa at the fixed edge.

<Esc>

<Esc>

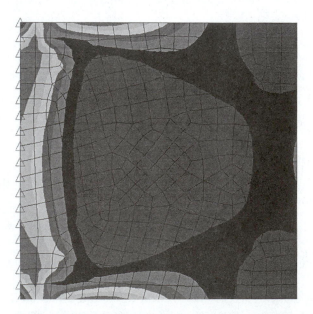

Figure 15–12 Displaced plate superimposed with maximum principal stress plot for Example 15.8

Click "Done". This takes you out of Superview.

Click "Close" to close the "Model Data Control" window.

You should be able to find the *ex158.L* file (storing the echo check of your data and the nodal displacements) and the *ex158.S* file (storing the element forces and stresses) and print a hard copy of each.

These results are similar to those that emerge from the longhand solution of Example 15.5 obtained with a very coarse mesh. The Algor solution with 342 elements is naturally more accurate than the longhand solution of Example 15.5 with 4 elements. ■

Example 15.9

The aluminum tube shown in Figure 15–13 fits snugly into a hole at room temperature. If the temperature is then increased by 40 °C, determine the deformed configuration and the stress distribution. Let $E = 70$ GPa, $v = 0.33$, and $\alpha = 23 \times 10^{-6}/°C$ for the tube.

Step 1   Start Superdraw III

Select the "Start" button of Windows NT/95/98 and then proceed to "Programs: Algor Software:Algor FEA".

Then click "Algor FEA" to start the program.

Or click on the "Algor FEA" icon on your desktop computer screen. The Superdraw III program appears. This is the Algor main graphics user interface. All your work will be done from this program. You will construct, analyze, and review your results from Superdraw III. Each menu option and a short example are available in the Superdraw III Reference Division (accessed through "Docutech", the on-line "Technical User Documentation").

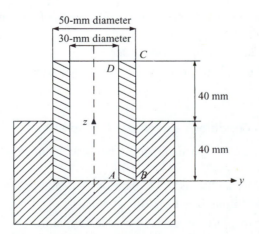

Figure 15–13   Tube fit into rigid hole

Step 2 Create the Model

Select "View:Predefined Views:YZ Right". This allows the *y-z* plane to be viewed.

Select the "Add" menu from the top main menu bar.

Click "Rectangle".

0<Tab>0.015<Tab>0   The first corner of the rectangle.

<Enter>

0<Tab>0.025<Tab>0.080   The opposite corner of the rectangle.

<Enter>

Select "View". This takes you to the draw options menu.

Click "Enclose". This encloses the model.

Select the "FEA Mesh" menu from the top main menu bar.

Click "Automatic Mesh:4 Point". A "Mesh" menu appears in the right part of the screen.

Click "Division Values" on the "Mesh" menu.

5   Enter 5 under "AB Divisions."

20   Enter 20 under "BC Divisions."

Right-click on point *A* in Figure 15–13.

Right-click on point *B* in Figure 15–13.

Right-click on point *C* in Figure 15–13.

Right-click on point *D* in Figure 15–13.

Click "Done" on the "Mesh" menu.

Step 3 Eliminate Duplicate Lines

Click "Modify" on the main menu.

Select "Clean:Duplicate". This notation means to select "Clean" and then "Duplicate" in succession.

Click "Perform Cleaning" on the "Duplicate" menu. You see the following message in the lower part of the screen:

                    "225 Kept 0 Deleted. done"

Click "Done" on the "Duplicate" menu located in the upper right corner of the screen.

Step 4 Add the Boundary Conditions and Nodal Temperatures

Click on the "FEA Add" menu on the main menu bar and highlight "Stress and Vibration Analysis".

Click "Boundary Conditions". The "Boundary Conditions" menu appears.

Click "Change Values" on the "Boundary Conditions" menu.

Remove the check mark next to "Tz Constrained" by clicking on it.

Click "Done".

Select "Box Apply" on the "Boundary Conditions" menu.

Create a box around the lowest 11 nodes on the outer edge.

Click "Change Values" on the "Boundary Conditions" menu.

Remove the check mark next to "Ty Constrained" by clicking on it.

Replace the check mark next to "Tz Constrained" by clicking on the empty box.

Click "Done".

Select "Box Apply" on the "Boundary Conditions" menu.

Create a box around the nodes on the lower edge.

Click "Done".

Click on the "FEA Add" menu on the main menu bar and highlight "Stress and Vibration Analysis".

Click "Nodal Temperatures". The "Nodal Temperatures" menu appears.

Click "Values" on the "Nodal Temperatures" menu.

40   Enter 40 for the nodal temperature.

Select "Box Apply" on the "Boundary Conditions" menu.

Create a box around the whole model.

Select "Done" on the "Nodal Temperatures" menu.

Step 5   Add the Material Properties Using the "Model Data Control" Window

Click on the "Model Data" button at the bottom of the screen.

Select "Linear Static Stress" by scrolling down the "Analysis Type" menu.

Click on the box below "Element".

Click "2D".

Click "OK".

Click on the box below "Data". A prompt appears: "Enter model name first".

Click "OK".

ex159   The model name is *ex159*.

Click "Save".

Select the SI units system.

Click "OK".

Click "Axisymmetric" next to "Geometry Type".

Click "OK".

Click on the box below "Material".

Select "Customer Defined".

Select "View Properties".

Click "Unlock Properties".

70e9   Enter 70e9 for "Modulus of Elasticity".

0.33   Enter 0.33 for "Poisson's Ratio".

23e-6   Enter 23e-6 for "Thermal Coefficient of Expansion".

Click "OK".
Click "OK".

Step 6 Enter the Global Data

Click on the "Global" button in the "Model Data Control" window.

1   Enter 1 under the "Pressure" column.

1   Enter 1 under the "Thermal" column.

Select the "Output" tab.
Highlight the "Stress Data" and "Displacement Data" boxes.
Click "OK".

Step 7 Check the Model

Click "Check" under the "FEA Model" column on the right side of the "Model Data Control" window. This takes you to Superview.

Click "Done" to return to the "Model Data Control" window.

Step 8 Run the Analysis

Click on the "Analysis" button in the "Model Data Control" window.

Click "Analyze". Wait for a short time for the analysis to be completed.

Click "OK".

Click "Done".

Step 9 Review the Results

Click "Results" in the "Model Data Control" window

Select "Displaced" from the main menu of Superview.

Click "Displ on". You see the displaced model on the screen.

Click "Nodes Inq:Get" and click on the upper inside node at point D to see the displacements "DY=1.47e−5" and "DZ=9.76e−5" and click on the upper outer node at point C to see "DY=2.83e−5" and "DZ=9.47e−5".

<Esc>

Click "Stress-di".

Click "Post".

Click "Von Mises". The Von Mises stress distribution is displayed. The largest Von Mises stress in 1.2e8 Pa. The largest principal stress is 4.2e7 Pa.

Click "Done" to exit Superview.   ■

▲ Reference

[1] Logan, D. L., *A First Course in the Finite Element Method*, 2nd ed., PWS-Kent Publishers, Boston, MA, 1993.

▲ Problems

15.1 For the one-dimensional bar fixed at the left end, free at the right end, and subjected to a uniform temperature rise $T = 50\,°F$ as shown in Figure P15–1, determine the free-end displacement, the displacement 60 in. from the fixed end, the reactions at the fixed end, and the axial stress. Let $E = 30 \times 10^6$ psi, $A = 4$ in^2, and $\alpha = 7.0 \times 10^{-6}$ (in./in.)/°F.

Figure P15–1   **Figure P15–2**

15.2 For the one-dimensional bar fixed at each end and subjected to a uniform temperature drop of $T = 20\,°C$ as shown in Figure P15–2, determine the reactions at the fixed ends and the stress in the bar. Let $E = 210$ GPa, $A = 1 \times 10^{-2}$ m^2, and $\alpha = 12 \times 10^{-6}$ (mm/mm)/°C.

15.3 For the plane truss shown in Figure P15–3, bar element 2 is subjected to a uniform temperature rise of $T = 50\,°F$. Let $E = 30 \times 10^6$ psi, $A = 2$ in^2, and $\alpha = 7.0 \times 10^{-6}$ (in./in.)/°F. The lengths of the truss elements are shown in the figure. Determine the stresses in each bar. [*Hint:* See Eqs. (3.6.4) and (3.6.6) in Example 3.5 for the global and reduced K matrices.]

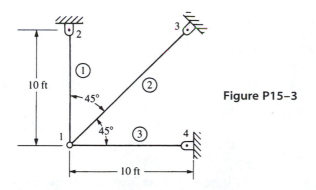

Figure P15–3

15.4 For the plane truss shown in Figure P15–4, bar element 1 is subjected to a uniform temperature rise of 30 °F. Let $E = 30 \times 10^6$ psi, $A = 2$ in^2, and $\alpha = 7.0 \times 10^{-6}$ (in./in.)/°F. The lengths of the truss elements are shown in the figure. Determine the stresses in each bar. (*Hint:* Use Problem 3.21 for K.)

15.5 For the structure shown in Figure P15–5, bar element 1 is subjected to a uniform temperature rise of $T = 20\,°C$. Let $E = 210$ GPa, $A = 2 \times 10^{-2}$ m^2, and $\alpha = 12 \times 10^{-6}$ (mm/mm)/°C. Determine the stresses in each bar.

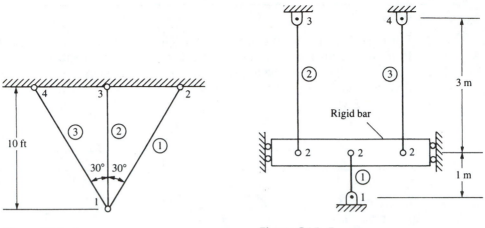

Figure P15–4   Figure P15–5

15.6 For the plane truss shown in Figure P15–6, bar element 2 is subjected to a uniform temperature drop of $T = 20\,°\text{C}$. Let $E = 70$ GPa, $A = 4 \times 10^{-2}$ m^2, and $\alpha = 23 \times 10^{-6}$ (mm/mm)/$°$C. Determine the stresses in each bar and the displacement of node 1.

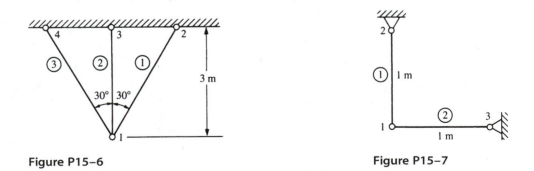

Figure P15–6   Figure P15–7

15.7 For the bar structure shown in Figure P15–7, element 1 is subjected to a uniform temperature rise of $T = 30\,°\text{C}$. Let $E = 210$ GPa, $A = 3 \times 10^{-2}$ m^2, and $\alpha = 12 \times 10^{-6}$ (mm/mm)/$°$C. Determine the displacement of node 1 and the stresses in each bar.

15.8 A bar assemblage consists of two outer steel bars and an inner brass bar. The three-bar assemblage is then heated to raise the temperature by an amount $T = 40\,°\text{F}$. Let all cross-sectional areas be $A = 2$ in^2 and $L = 60$ in., $E_{\text{steel}} = 30 \times 10^6$ psi, $E_{\text{brass}} = 15 \times 10^6$ psi, $\alpha_{\text{steel}} = 6.5 \times 10^{-6}/°\text{F}$, and $\alpha_{\text{brass}} = 10 \times 10^{-6}/°\text{F}$. Determine (a) the displacement of node 2 and (b) the stress in the steel and brass bars. See Figure P15–8.

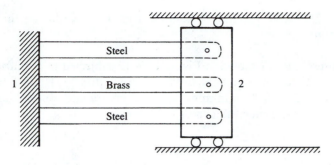

Figure P15–8

15.9 For the plane stress element shown in Figure P15–9 subjected to a uniform temperature rise of $T = 50\,°\text{F}$, determine the thermal force matrix $\{f_T\}$. Let $E = 10 \times 10^6$ psi, $\nu = 0.30$, and $\alpha = 12.5 \times 10^{-6}$ (in./in.)/°F. The coordinates (in inches) are shown in the figure. The element thickness is $t = 1$ in.

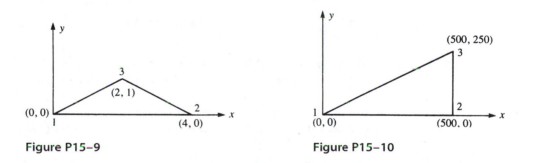

Figure P15–9   **Figure P15–10**

15.10 For the plane stress element shown in Figure P15–10 subjected to a uniform temperature rise of $T = 30\,°\text{C}$, determine the thermal force matrix $\{f_T\}$. Let $E = 70$ GPa, $\nu = 0.3$, $\alpha = 23 \times 10^{-6}$ (mm/mm)/°C, and $t = 5$ mm. The coordinates (in millimeters) are shown in the figure.

15.11 For the plane stress element shown in Figure P15–11 subjected to a uniform temperature rise of $T = 100\,°\text{F}$, determine the thermal force matrix $\{f_T\}$. Let $E = 30 \times 10^6$

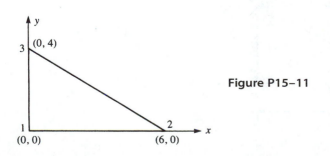

Figure P15–11

psi, $v = 0.3$, $\alpha = 7.0 \times 10^{-6}$ (in./in.)/°F, and $t = 1$ in. The coordinates (in inches) are shown in the figure.

15.12 For the plane stress element shown in Figure P15–12 subjected to a uniform temperature drop of $T = 20\,°C$, determine the thermal force matrix $\{f_T\}$. Let $E = 210$ GPa, $v = 0.25$, and $\alpha = 12 \times 10^{-6}$ (mm/mm)/°C. The coordinates (in millimeters) are shown in the figure. The element thickness is 10 mm.

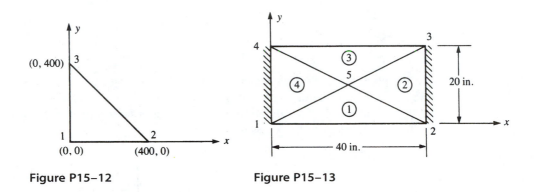

Figure P15–12   **Figure P15–13**

15.13 For the plane stress plate fixed along the left and right sides and subjected to a uniform temperature rise of 50 °F as shown in Figure P15–13, determine the stresses in each element. Let $E = 10 \times 10^6$ psi, $v = 0.30$, $\alpha = 12.5 \times 10^{-6}$ (in./in.)/°F, and $t = \frac{1}{4}$ in. The coordinates (in inches) are shown in the figure. (*Hint:* The nodal displacements are all equal to zero. Therefore, the stresses can be determined from $\{\sigma\} = -[D]\{\varepsilon_T\}$.)

15.14 For the plane stress plate fixed along all edges and subjected to a uniform temperature decrease of 20 °C as shown in Figure P15–14, determine the stresses in each element. Let $E = 210$ GPa, $v = 0.25$, and $\alpha = 12 \times 10^{-6}$ (mm/mm)/°C. The coordinates of the plate are shown in the figure. The plate thickness is 10 mm. (*Hint:* The nodal displacements are all equal to zero. Therefore, the stresses can be determined from $\{\sigma\} = -[D]\{\varepsilon_T\}$.)

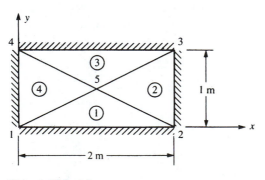

Figure P15–14

15.15 If the thermal expansion coefficient of a bar is given by $\alpha = \alpha_0(1 + x/L)$, determine the thermal force matrix. Let the bar have length L, modulus of elasticity E, and cross-sectional area A.

15.16 Assume the temperature function to vary linearly over the length of a bar as $T = t_1 + t_2 x$; that is, express the temperature function as $\{T\} = [N]\{t\}$, where $[N]$ is the shape function matrix for the two-node bar element. In other words, $[N] = [1 - x/L \quad x/L]$. Determine the force matrix in terms of E, A, α, L, t_1, and t_2. [*Hint:* Use Eq. (15.1.16).]

15.17 Derive the thermal force matrix for the axisymmetric element of Chapter 10.

Using the Algor computer program, solve the following problems.

15.18 The square plate in Figure P15–18 is subjected to uniform heating of $80\,°\text{F}$. Determine the nodal displacements and element stresses. Let the element thickness be $t = 0.1$ in., $E = 30 \times 10^6$ psi, $v = 0.33$, and $\alpha = 10 \times 10^{-6}/°\text{F}$.

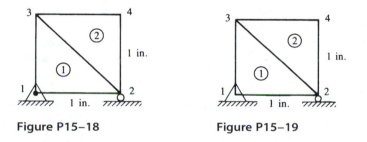

Figure P15–18 **Figure P15–19**

15.19 The square plate in Figure P15–19 has element 1 made of steel with $E = 30 \times 10^6$ psi, $v = 0.33$, and $\alpha = 10 \times 10^{-6}/°\text{F}$ and element 2 made of a material with $E = 15 \times 10^6$ psi, $v = 0.25$, and $\alpha = 50 \times 10^{-6}/°\text{F}$. Let the plate thickness be $t = 0.1$ in. Determine the nodal displacements and element stresses for element 1 subjected to an $80\,°\text{F}$ temperature increase and element 2 subjected to a $50\,°\text{F}$ temperature increase.

15.20 Solve Problem 15.3 using the Algor program.

15.21 Solve Problem 15.6 using the Algor program.

Structural Dynamics and Time-Dependent Heat Transfer

Introduction

This chapter provides an elementary introduction to time-dependent problems. We will introduce the basic concepts using the single-degree-of-freedom spring-mass system. We will include discussion of the stress analysis of the one-dimensional bar, beam, truss, and plane frame. This is followed by the analysis of one-dimensional heat transfer.

We will provide the basic equations necessary for structural dynamics analysis and develop both the lumped- and the consistent-mass matrices involved in the analyses of the bar, beam, truss, and plane frame. We will describe the assembly of the global mass matrix for truss and plane frame analysis and then present numerical integration methods for handling the time derivative. We also present the mass matrices for the constant strain triangle and quadrilateral plane elements, for the axisymmetric element, and for the tetrahedral solid element.

We will provide longhand solutions for the determination of the natural frequencies for bars and beams and then illustrate the time-step integration process involved with the stress analysis of a bar subjected to a time-dependent forcing function.

We will next derive the basic equations for the time-dependent one-dimensional heat-transfer problem and discuss their applications. This chapter provides the basic concepts necessary for the solution of time-dependent problems. We conclude with a section on the use of the Algor computer program to solve structural dynamics and time-dependent heat-transfer problems.

▲ 16.1 Dynamics of a Spring-Mass System ▲

In this section, we discuss the motion of a single-degree-of-freedom spring-mass system to introduce the important concepts necessary for the later study of continuous systems such as bars, beams, and plane frames. In Figure 16–1, we show the single-degree-of-freedom spring-mass system subjected to a time-dependent force $F(t)$. Here k represents the spring stiffness or constant, and m represents the mass of the system.

The free-body diagram of the mass is shown in Figure 16–2. The spring force $T = kx$ and the applied force $F(t)$ act on the mass, and the mass-times-acceleration term is shown separately.

Applying Newton's second law of motion, $f = ma$, to the mass, we obtain the equation of motion in the x direction as

$$F(t) - kx = m\ddot{x} \tag{16.1.1}$$

where a dot over a variable denotes differentiation with respect to time; that is, $(\dot{\ }) = d(\ )/dt$. Rewriting Eq. (16.1.1) in standard form, we have

$$m\ddot{x} + kx = F(t) \tag{16.1.2}$$

Equation (16.1.2) is a linear differential equation of the second order whose standard solution for the displacement x consists of a homogeneous solution and a particular solution. Standard analytical solutions for this forced vibration can be found in texts on dynamics or vibrations such as Reference [1]. The analytical solution will not be presented here as our intent is to introduce basic concepts in vibration behavior. However, we will solve the problem defined by Eq. (16.1.2) by an approximate numerical technique in Section 16.3 (see Examples 16.1 and 16.2).

The homogeneous solution to Eq. (16.1.2) is the solution obtained when the right side is set equal to zero. A number of useful concepts regarding vibrations are obtained by considering this free vibration of the mass—that is, when $F(t) = 0$. Hence, defining

$$\omega^2 = \frac{k}{m} \tag{16.1.3}$$

and setting the right side of Eq. (16.1.2) equal to zero, we have

$$\ddot{x} + \omega^2 x = 0 \tag{16.1.4}$$

where ω is called the **natural circular frequency** of the free vibration of the mass, expressed in units of radians per second or revolutions per minute (rpm). Hence, the

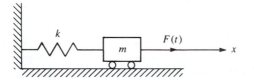

Figure 16–1 Spring-mass system subjected to a time-dependent force

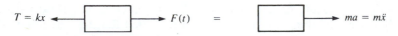

Figure 16–2 Free-body diagram of the mass of Figure 16–1

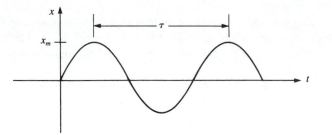

Figure 16–3 Displacement/time curve for simple harmonic motion

natural circular frequency defines the number of cycles per unit time of the mass vibration. We observe from Eq. (16.1.3) that ω depends only on the spring stiffness k and the mass m of the body.

The motion defined by Eq. (16.1.4) is called **simple harmonic motion**. The displacement and acceleration are seen to be proportional but of opposite direction. Again, a standard solution to Eq. (16.1.4) can be found in Reference [1]. A typical displacement/time curve is represented by the sine curve shown in Figure 16–3, where x_m denotes the maximum displacement (called the **amplitude** of the vibration). The time interval required for the mass to complete one full cycle of motion is called the **period** of the vibration τ and is given by

$$\tau = \frac{2\pi}{\omega} \qquad (16.1.5)$$

where τ is measured in seconds. Also the frequency in hertz (Hz = 1/s) is $f = 1/\tau = \omega/(2\pi)$.

Finally, note that all vibrations are damped to some degree by friction forces. These forces may be caused by dry or Coulomb friction between rigid bodies, by internal friction between molecules within a deformable body, or by fluid friction when a body moves in a fluid. Damping results in natural circular frequencies that are smaller than those for undamped systems; maximum displacements also are smaller when damping occurs. A basic treatment of damping can be found in Reference [1] and additional discussion is included in Example 16.12.

▲ 16.2 Direct Derivation of the Bar Element Equations

We will now derive the finite element equations for the time-dependent (dynamic) stress analysis of the one-dimensional bar. Recall that the time-independent (static) stress analysis of the bar was considered in Chapter 3. The steps used in deriving the dynamic equations are the same as those used for the derivation of the static equations.

Step 1 Select Element Type

Figure 16–4 shows the typical bar element of length L, cross-sectional area A, and mass density ρ (with typical units of lb-s^2/in^4), with nodes 1 and 2 subjected to external time-dependent loads $\hat{f}_x^e(t)$.

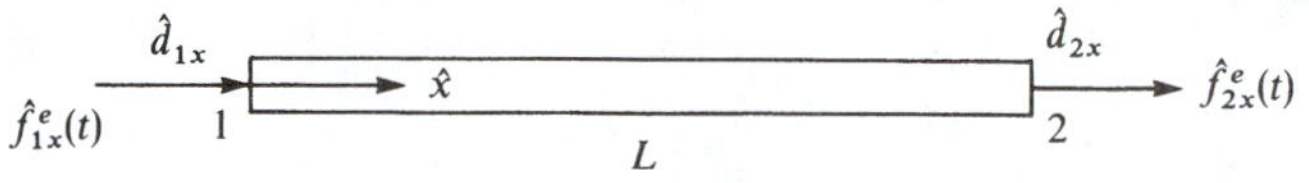

Figure 16–4 Bar element subjected to time-dependent loads

Step 2 Select a Displacement Function

Again, we assume a linear displacement function along the $\hat{x}$ axis of the bar [see Eq. (3.1.1)]; that is, we let

$$\hat{u} = a_1 + a_2\hat{x} \tag{16.2.1}$$

As was shown in Chapter 3, Eq. (16.2.1) can be expressed in terms of the shape functions as

$$\hat{u} = N_1\hat{d}_{1x} + N_2\hat{d}_{2x} \tag{16.2.2}$$

where

$$N_1 = 1 - \frac{\hat{x}}{L} \qquad N_2 = \frac{\hat{x}}{L} \tag{16.2.3}$$

Step 3 Define the Strain/Displacement and Stress/Strain Relationships

Again, the strain/displacement relationship is given by

$$\{\varepsilon_x\} = \frac{\partial\hat{u}}{\partial\hat{x}} = [B]\{\hat{d}\} \tag{16.2.4}$$

where

$$[B] = \begin{bmatrix} -\dfrac{1}{L} & \dfrac{1}{L} \end{bmatrix} \qquad \{\hat{d}\} = \begin{Bmatrix} \hat{d}_{1x} \\ \hat{d}_{2x} \end{Bmatrix} \tag{16.2.5}$$

and the stress/strain relationship is given by

$$\{\sigma_x\} = [D]\{\varepsilon_x\} = [D][B]\{\hat{d}\} \tag{16.2.6}$$

Step 4 Derive the Element Stiffness and Mass Matrices and Equations

The bar is generally not in equilibrium under a time-dependent force; hence, $f_{1x} \neq f_{2x}$. Therefore, we again apply Newton's second law of motion, $\boldsymbol{f} = \boldsymbol{ma}$, to each node. In general, the law can be written for each node as "the external (applied) force f_x^e minus the internal force is equal to the nodal mass times acceleration." Equivalently, adding the internal force to the **ma** term, we have

$$\hat{f}^e_{1x} = \hat{f}_{1x} + m_1 \frac{\partial^2 \hat{d}_{1x}}{\partial t^2} \qquad \hat{f}^e_{2x} = \hat{f}_{2x} + m_2 \frac{\partial^2 \hat{d}_{2x}}{\partial t^2} \qquad (16.2.7)$$

where the masses m_1 and m_2 are obtained by lumping the total mass of the bar equally at the two nodes such that

$$m_1 = \frac{\rho A L}{2} \qquad m_2 = \frac{\rho A L}{2} \qquad (16.2.8)$$

In matrix form, we express Eqs. (16.2.7) as

$$\left\{ \begin{matrix} \hat{f}^e_{1x} \\ \hat{f}^e_{2x} \end{matrix} \right\} = \left\{ \begin{matrix} \hat{f}_{1x} \\ \hat{f}_{2x} \end{matrix} \right\} + \begin{bmatrix} m_1 & 0 \\ 0 & m_2 \end{bmatrix} \left\{ \begin{matrix} \dfrac{\partial^2 \hat{d}_{1x}}{\partial t^2} \\ \dfrac{\partial^2 \hat{d}_{2x}}{\partial t^2} \end{matrix} \right\} \qquad (16.2.9)$$

Using Eqs. (3.1.13) and (3.1.14), we replace $\{\hat{f}\}$ with $[\hat{k}]\{\hat{d}\}$ in Eq. (16.2.9) to obtain the element equations

$$\{\hat{f}^e(t)\} = [\hat{k}]\{\hat{d}\} + [\hat{m}]\{\ddot{\hat{d}}\} \qquad (16.2.10)$$

where

$$[\hat{k}] = \frac{AE}{L} \begin{bmatrix} 1 & -1 \\ -1 & 1 \end{bmatrix} \qquad (16.2.11)$$

is the bar element stiffness matrix, and

$$[\hat{m}] = \frac{\rho A L}{2} \begin{bmatrix} 1 & 0 \\ 0 & 1 \end{bmatrix} \qquad (16.2.12)$$

is called the **lumped-mass matrix**. Also,

$$\{\ddot{\hat{d}}\} = \frac{\partial^2 \{\hat{d}\}}{\partial t^2} \qquad (16.2.13)$$

Observe that the lumped-mass matrix has diagonal terms only. This facilitates the computation of the global equations. However, solution accuracy is usually not as good as when a consistent-mass matrix is used [2].

We will now develop the **consistent-mass matrix** for the bar element. Numerous methods are available to obtain the consistent-mass matrix. The generally applicable virtual work principle (which is the basis of many energy principles, such as the principle of minimum potential energy for elastic bodies previously used in this text) provides a relatively simple method for derivation of the element equations and is included in Appendix E. However, an even simpler approach is to use D'Alembert's principle; thus, we introduce an effective body force X^e as

$$\{X^e\} = -\rho\{\ddot{u}\} \qquad (16.2.14)$$

where the minus sign is due to the fact that the acceleration produces D'Alembert's body forces in the direction opposite the acceleration. The nodal forces associated with $\{X^e\}$ are then found by using Eq. (7.3.1), repeated here as

$$\{f_b\} = \iiint_V [N]^T \{X\} \, dV \tag{16.2.15}$$

Substituting $\{X^e\}$ given by Eq. (16.2.14) into Eq. (16.2.15) for $\{X\}$, we obtain

$$\{f_b\} = -\iiint_V \rho [N]^T \{\ddot{u}\} \, dV \tag{16.2.16}$$

Recalling from Eq. (16.2.2) that $\{\hat{u}\} = [N]\{\hat{d}\}$, we find that the first and second derivatives with respect to time are

$$\{\dot{\hat{u}}\} = [N]\{\dot{\hat{d}}\} \qquad \{\ddot{\hat{u}}\} = [N]\{\ddot{\hat{d}}\} \tag{16.2.17}$$

where $\{\dot{\hat{d}}\}$ and $\{\ddot{\hat{d}}\}$ are the nodal velocities and accelerations, respectively. Substituting Eqs. (16.2.17) into Eq. (16.2.16), we obtain

$$\{f_b\} = -\iiint_V \rho [N]^T [N] \, dV \{\ddot{\hat{d}}\} = -[\hat{m}]\{\ddot{\hat{d}}\} \tag{16.2.18}$$

where the element mass matrix is defined as

$$[\hat{m}] = \iiint_V \rho [N]^T [N] \, dV \tag{16.2.19}$$

This mass matrix is called the *consistent-mass matrix* because it is derived from the same shape functions $[N]$ that are used to obtain the stiffness matrix $[\hat{k}]$. In general, $[\hat{m}]$ given by Eq. (16.2.19) will be a full but symmetric matrix. Equation (16.2.19) is a general form of the consistent-mass matrix; that is, substituting the appropriate shape functions, we can generate the mass matrix for such elements as the bar, beam, and plane stress.

We will now develop the consistent-mass matrix for the bar element of Figure 16–4 by substituting the shape function Eqs. (16.2.3) into Eq. (16.2.19) as follows:

$$[\hat{m}] = \iiint_V \rho \begin{Bmatrix} 1 - \dfrac{\hat{x}}{L} \\ \dfrac{\hat{x}}{L} \end{Bmatrix} \begin{bmatrix} 1 - \dfrac{\hat{x}}{L} & \dfrac{\hat{x}}{L} \end{bmatrix} dV \tag{16.2.20}$$

Simplifying Eq. (16.2.20), we obtain

$$[\hat{m}] = \rho A \int_0^L \begin{Bmatrix} 1 - \dfrac{\hat{x}}{L} \\ \dfrac{\hat{x}}{L} \end{Bmatrix} \begin{bmatrix} 1 - \dfrac{\hat{x}}{L} & \dfrac{\hat{x}}{L} \end{bmatrix} d\hat{x} \tag{16.2.21}$$

or, on multiplying the matrices of Eq. (16.2.21),

$$[\hat{m}] = \rho A \int_0^L \begin{bmatrix} \left(1 - \dfrac{\hat{x}}{L}\right)^2 & \left(1 - \dfrac{\hat{x}}{L}\right)\dfrac{\hat{x}}{L} \\[2ex] \left(1 - \dfrac{\hat{x}}{L}\right)\dfrac{\hat{x}}{L} & \left(\dfrac{\hat{x}}{L}\right)^2 \end{bmatrix} d\hat{x} \tag{16.2.22}$$

On integrating Eq. (16.2.22) term by term, we obtain the consistent-mass matrix for a bar element as

$$[\hat{m}] = \frac{\rho A L}{6} \begin{bmatrix} 2 & 1 \\ 1 & 2 \end{bmatrix} \tag{16.2.23}$$

Step 5   Assemble the Element Equations to Obtain the Global Equations and Introduce Boundary Conditions

We assemble the element equations using the direct stiffness method such that inter-element continuity of displacements is again satisfied at common nodes and, in addition, interelement continuity of accelerations is also satisfied; that is, we obtain the global equations

$$\{F(t)\} = [K]\{d\} + [M]\{\ddot{d}\} \tag{16.2.24}$$

where

$$[K] = \sum_{e=1}^{N}[k^{(e)}] \qquad [M] = \sum_{e=1}^{N}[m^{(e)}] \qquad \{F\} = \sum_{e=1}^{N}\{f^{(e)}\} \tag{16.2.25}$$

are the global stiffness, mass, and force matrices, respectively. Note that the global mass matrix is assembled in the same manner as the global stiffness matrix. Equation (16.2.24) represents a set of matrix equations discretized with respect to space. To obtain the solution of the equations, discretization in time is also necessary. We will describe this process in Section 16.3 and will later present representative solutions illustrating these equations.

▲ 16.3 Numerical Integration in Time   ▲

We now introduce procedures for the discretization of Eq. (16.2.24) with respect to time. These procedures will enable us to determine the nodal displacements at different time increments for a given dynamic system. The general method used is called *direct integration*. There are two classifications of direct integration: explicit and implicit. We will formulate the equations for three direct integration methods. The first, and simplest, is an explicit method known as the *central difference method* [3, 4]. The second and third, more complicated but more versatile than the central difference method, are implicit methods known as the Newmark-Beta (or Newmark's) method [5] and the Wilson-Theta (or Wilson's) method [7, 8]. The versatility of both Newmark's and Wilson's methods is evidenced by their adaptation in many commercially available computer programs. Wilson's method is used in the Algor computer pro-

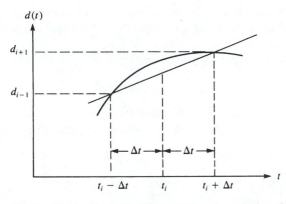

Figure 16–5 Numerical integration (approximation of derivative at t_i)

gram. Numerous other integration methods are available in the literature. Among these are Houboldt's method [8] and the alpha method [13].

Central Difference Method

The central difference method is based on finite difference expressions in time for velocity and acceleration at time t given by

$$\underline{\dot{d}}_i = \frac{\underline{d}_{i+1} - \underline{d}_{i-1}}{2(\Delta t)} \qquad (16.3.1)$$

$$\underline{\ddot{d}}_i = \frac{\underline{\dot{d}}_{i+1} - \underline{\dot{d}}_{i-1}}{2(\Delta t)} \qquad (16.3.2)$$

where the subscripts indicate the time step; that is, for a time increment of Δt, $\underline{d}_i = \underline{d}(t)$ and $\underline{d}_{i+1} = \underline{d}(t + \Delta t)$. The procedure used in deriving Eq. (16.3.1) is illustrated by use of the displacement/time curve shown in Figure 16–5. Graphically, Eq. (16.3.1) represents the slope of the line shown in Figure 16–5; that is, given two points at increments $i - 1$ and $i + 1$ on the curve, two Δt increments apart, an approximation of the first derivative at the midpoint i of the increment is given by Eq. (16.3.1). Similarly, using a velocity/time curve, we could obtain Eq. (16.3.2), or we can see that Eq. (16.3.2) is obtained simply by differentiating Eq. (16.3.1) with respect to time.

It has been shown using, for instance, Taylor series expansions [3] that the acceleration can also be expressed in terms of the displacements by

$$\underline{\ddot{d}}_i = \frac{\underline{d}_{i+1} - 2\underline{d}_i + \underline{d}_{i-1}}{(\Delta t)^2} \qquad (16.3.3)$$

Because we want to evaluate the nodal displacements, it is most suitable to use Eq. (16.3.3) in the form

$$\underline{d}_{i+1} = 2\underline{d}_i - \underline{d}_{i-1} + \underline{\ddot{d}}_i(\Delta t)^2 \qquad (16.3.4)$$

Equation (16.3.4) will be used to determine the nodal displacements in the next time step $i + 1$ knowing the displacements at time steps i and $i - 1$ and the acceleration at time i.

From Eq. (16.2.24), we express the acceleration as

$$\underline{\ddot{d}}_i = \underline{M}^{-1}(\underline{F}_i - \underline{K}\underline{d}_i) \tag{16.3.5}$$

To obtain an expression for $\underline{d}_{i+1}$, we first multiply Eq. (16.3.4) by the mass matrix $\underline{M}$ and then substitute Eq. (16.3.5) for $\underline{\ddot{d}}_i$ into this equation to obtain

$$\underline{M}\underline{d}_{i+1} = 2\underline{M}\underline{d}_i - \underline{M}\underline{d}_{i-1} + (\underline{F}_i - \underline{K}\underline{d}_i)(\Delta t)^2 \tag{16.3.6}$$

Combining like terms of Eq. (16.3.6), we obtain

$$\underline{M}\underline{d}_{i+1} = (\Delta t)^2 \underline{F}_i + [2\underline{M} - (\Delta t)^2 \underline{K}]\underline{d}_i - \underline{M}\underline{d}_{i-1} \tag{16.3.7}$$

To start the computations to determine $\underline{d}_{i+1}, \underline{\dot{d}}_{i+1}$, and $\underline{\ddot{d}}_{i+1}$, we need the displacement $\underline{d}_{i-1}$ initially, as indicated by Eq. (16.3.7). Using Eqs. (16.3.1) and (16.3.4), we solve for $\underline{d}_{i-1}$ as

$$\underline{d}_{i-1} = \underline{d}_i - (\Delta t)\underline{\dot{d}}_i + \frac{(\Delta t)^2}{2}\underline{\ddot{d}}_i \tag{16.3.8}$$

The procedure for solution is then as follows:

1. Given: $\underline{d}_0, \underline{\dot{d}}_0$, and $\underline{F}_i(t)$.
2. If $\underline{\ddot{d}}_0$ is not initially given, solve Eq. (16.3.5) at $t = 0$ for $\underline{\ddot{d}}_0$; that is,

$$\underline{\ddot{d}}_0 = \underline{M}^{-1}(\underline{F}_0 - \underline{K}\underline{d}_0)$$

3. Solve Eq. (16.3.8) at $t = -\Delta t$ for $\underline{d}_{-1}$; that is,

$$\underline{d}_{-1} = \underline{d}_0 - (\Delta t)\underline{\dot{d}}_0 + \frac{(\Delta t)^2}{2}\underline{\ddot{d}}_0$$

4. Having solved for $\underline{d}_{-1}$ in step 3, now solve for $\underline{d}_1$ using Eq. (16.3.7) as

$$\underline{d}_1 = \underline{M}^{-1}\{(\Delta t)^2 \underline{F}_0 + [2\underline{M} - (\Delta t)^2 \underline{K}]\underline{d}_0 - \underline{M}\underline{d}_{-1}\}$$

5. With $\underline{d}_0$ initially given, and $\underline{d}_1$ determined from step 4, use Eq. (16.3.7) to obtain

$$\underline{d}_2 = \underline{M}^{-1}\{(\Delta t)^2 \underline{F}_1 + [2\underline{M} - (\Delta t)^2 \underline{K}]\underline{d}_1 - \underline{M}\underline{d}_0\}$$

6. Using Eq. (16.3.5), solve for $\underline{\ddot{d}}_1$ as

$$\underline{\ddot{d}}_1 = \underline{M}^{-1}(\underline{F}_1 - \underline{K}\underline{d}_1)$$

7. Using the result of step 5 and the boundary condition for $\underline{d}_0$ given in step 1, determine the velocity at the first time step by Eq. (16.3.1) as

$$\underline{\dot{d}}_1 = \frac{\underline{d}_2 - \underline{d}_0}{2(\Delta t)}$$

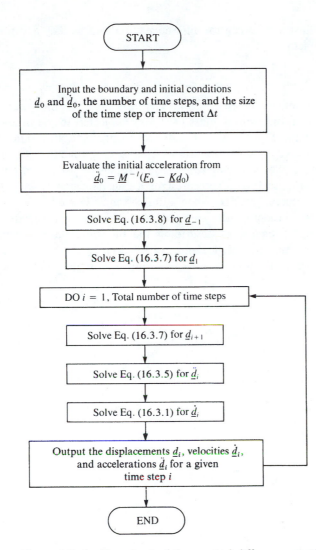

Figure 16–6 Flowchart of the central difference method

8. Use steps 5–7 repeatedly to obtain the displacement, acceleration, and velocity for all other time steps.

Figure 16–6 is a flowchart of the solution procedure using the central difference equations. Note that the recurrence formulas given by equations such as Eqs. (16.3.1) and (16.3.2) are approximate but yield sufficiently accurate results provided the time step Δt is taken small in relation to the variations in acceleration. Methods for determining proper time steps for the numerical integration process are described in Section 16.5.

We will now illustrate the central difference equations as they apply to the following example problem.

Example 16.1

Determine the displacement, velocity, and acceleration at 0.05-s time intervals up to 0.2 s for the one-dimensional spring-mass oscillator subjected to the time-dependent forcing function shown in Figure 16–7. [Guidelines regarding appropriate time intervals (or time steps) are given in Section 16.5.] This forcing function is a typical one assumed for blast loads. The restoring spring force versus displacement curve is also provided. [Note that Figure 16–7 also represents a one-element bar with its left end fixed and right node subjected to $F(t)$ when a lumped mass is used.]

Because we are considering the single degree of freedom associated with the mass, the general matrix equations describing the motion reduce to single scalar equations. We will represent this single degree of freedom by d.

The solution procedure follows the steps outlined in this section and in the flowchart of Figure 16–6.

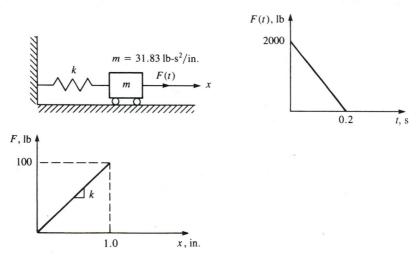

Figure 16–7 Spring-mass oscillator subjected to a time-dependent force

Step 1

At time $t = 0$, the initial displacement and velocity are zero; therefore,

$$d_0 = 0 \qquad \dot{d}_0 = 0$$

Step 2

The initial acceleration at $t = 0$ is obtained as

$$\ddot{d}_0 = \frac{2000 - 100(0)}{31.83} = 62.83 \text{ in./s}^2$$

where we have used $\underline{F}(0) = 2000$ lb and $\underline{K} = 100$ lb/in.

Step 3

The displacement d_{-1} is obtained as

$$d_{-1} = 0 - 0 + \frac{(0.05)^2}{2}(62.83) = 0.0785 \text{ in.}$$

Step 4

The displacement at time $t = 0.05$ s is

$$d_1 = \frac{1}{31.83}\{(0.05)^2(2000) + [2(31.83) - (0.05)^2(100)]0 - (31.83)(0.0785)\}$$

$$= 0.0785 \text{ in.}$$

Step 5

Having obtained d_1, we now determine the displacement at time $t = 0.10$ s as

$$d_2 = \frac{1}{31.83}\{(0.05)^2(1500) + [2(31.83) - (0.05)^2(100)](0.0785) - (31.83)(0)\}$$

$$= 0.274 \text{ in.}$$

Step 6

The acceleration at time $t = 0.05$ s is

$$\ddot{d}_1 = \frac{1}{31.83}[1500 - 100(0.0785)] = 46.88 \text{ in.}/\text{s}^2$$

Step 7

The velocity at time $t = 0.05$ s is

$$\dot{d}_1 = \frac{0.274 - 0}{2(0.05)} = 2.74 \text{ in.}/\text{s}$$

Step 8

Repeated use of steps 5–7 will result in the displacement, acceleration, and velocity for additional time steps as desired. We will now perform one more time-step iteration of the procedure.

Repeating step 5 for the next time step, we have

$$d_3 = \frac{1}{31.83}\{(0.05)^2(1000) + [2(31.83) - (0.05)^2(100)](0.274)$$

$$- (31.83)(0.0785)\} = 0.546 \text{ in.}$$

Repeating step 6 for the next time step, we have

$$\ddot{d}_2 = \frac{1}{31.83}[1000 - 100(0.274)] = 30.56 \text{ in.}/\text{s}^2$$

Table 16–1 Results of the analysis of Example 16.1

t (s)	$F(t)$ (lb)	d_i (in.)	Q (lb)	$\ddot{d}_i$ (in./s^2)	$\dot{d}_i$ (in./s)	d_i (exact)
0	2000	0	0	62.83	0	0
0.05	1500	0.0785	7.85	46.88	2.74	0.0718
0.10	1000	0.274	27.40	30.56	4.68	0.2603
0.15	500	0.546	54.64	13.99	5.79	0.5252
0.20	0	0.854	85.35	−2.68	6.07	0.8250
0.25	0	1.154	115.4	−3.63	5.91	1.132

Finally, repeating step 7 for the next time step, we obtain

$$\dot{d}_2 = \frac{0.546 - 0.0785}{2(0.05)} = 4.68 \text{ in./s}$$

Table 16–1 summarizes the results obtained through time $t = 0.25$ s. In Table 16–1, $Q = kd_i$ is the restoring spring force. Also, the exact analytical solution for displacement based on the equation in Reference [14] is given by

$$y = \frac{F_0}{k}(1 - \cos \omega t) + \frac{F_0}{kt_d}\left(\frac{\sin \omega t}{\omega} - t\right)$$

where $F_0 = 2000$ lb, $k = 100$ lb/in., $t_d = 0.2$ s, and

$$\omega = \sqrt{\frac{k}{m}} = \sqrt{\frac{100}{31.83}} = 1.77 \text{ rad/s} \quad ■$$

Newmark's Method

We will now outline Newmark's numerical method, which, because of its general versatility, has been adopted into numerous commercially available computer programs for purposes of structural dynamics analysis. (Complete development of the equations can be found in Reference [5].) Newmark's equations are given by

$$\dot{\underline{d}}_{i+1} = \dot{\underline{d}}_i + (\Delta t)[(1 - \gamma)\ddot{\underline{d}}_i + \gamma \ddot{\underline{d}}_{i+1}] \tag{16.3.9}$$

$$\underline{d}_{i+1} = \underline{d}_i + (\Delta t)\dot{\underline{d}}_i + (\Delta t)^2[(\tfrac{1}{2} - \beta)\ddot{\underline{d}}_i + \beta \ddot{\underline{d}}_{i+1}] \tag{16.3.10}$$

where β and γ are parameters chosen by the user. The parameter β is generally chosen between 0 and $\frac{1}{4}$, and γ is often taken to be $\frac{1}{2}$. For instance, choosing $\gamma = \frac{1}{2}$ and $\beta = 0$, it can be shown that Eqs. (16.3.9) and (16.3.10) reduce to the central difference Eqs. (16.3.1) and (16.3.2). If $\gamma = \frac{1}{2}$ and $\beta = \frac{1}{6}$ are chosen, Eqs. (16.3.9) and (16.3.10) correspond to those for which a linear acceleration assumption is valid within each time interval. For $\gamma = \frac{1}{2}$ and $\beta = \frac{1}{4}$, it has been shown that the numerical analysis is stable; that is, computed quantities such as displacement and velocities do not become unbounded regardless of the time step chosen. Furthermore, it has been found [5] that a time step of approximately $\frac{1}{10}$ of the shortest natural frequency of the structure being analyzed usually yields the best results.

To find $\underline{d}_{i+1}$, we first multiply Eq. (16.3.10) by the mass matrix $\underline{M}$ and then substitute Eq. (16.3.5) for $\underline{d}_{i+1}$ into this equation to obtain

$$\underline{M}\underline{d}_{i+1} = \underline{M}\underline{d}_i + (\Delta t)\underline{M}\underline{\dot{d}}_i + (\Delta t)^2\underline{M}(\tfrac{1}{2} - \beta)\underline{\ddot{d}}_i] + \beta(\Delta t)^2[\underline{F}_{i+1} - \underline{K}\underline{d}_{i+1}] \quad (16.3.11)$$

Combining like terms of Eq. (16.3.11), we obtain

$$(\underline{M} + \beta(\Delta t)^2\underline{K})\underline{d}_{i+1} = \beta(\Delta t)^2\underline{F}_{i+1} + \underline{M}\underline{d}_i + (\Delta t)\underline{M}\underline{\dot{d}}_i + (\Delta t)^2\underline{M}(\tfrac{1}{2} - \beta)\underline{\ddot{d}}_i$$

$$(16.3.12)$$

Finally, dividing Eq. (16.3.12) by $\beta(\Delta t)^2$, we obtain

$$\underline{K}'\underline{d}_{i+1} = \underline{F}'_{i+1} \quad (16.3.13)$$

where

$$\underline{K}' = \underline{K} + \frac{1}{\beta(\Delta t)^2}\underline{M} \quad (16.3.14)$$

$$\underline{F}'_{i+1} = \underline{F}_{i+1} + \frac{\underline{M}}{\beta(\Delta t)^2}\left[\underline{d}_i + (\Delta t)\underline{\dot{d}}_i + \left(\frac{1}{2} - \beta\right)(\Delta t)^2\underline{\ddot{d}}_i\right]$$

The solution procedure using Newmark's equations is as follows:

1. Starting at time $t = 0$, $\underline{d}_0$ is known from the given boundary conditions on displacement, and $\underline{\dot{d}}_0$ is known from the initial velocity conditions.
2. Solve Eq. (16.3.5) at $t = 0$ for $\underline{\ddot{d}}_0$ (unless $\underline{\ddot{d}}_0$ is known from an initial acceleration condition); that is,

$$\underline{\ddot{d}}_0 = \underline{M}^{-1}(\underline{F}_0 - \underline{K}\underline{d}_0)$$

3. Solve Eq. (16.3.13) for $\underline{d}_1$, because $\underline{F}_{i+1}$ is known for all time steps and $\underline{d}_0, \underline{\dot{d}}_0$, and $\underline{\ddot{d}}_0$ are now known from steps 1 and 2.
4. Use Eq. (16.3.10) to solve for $\underline{\dot{d}}_1$ as

$$\underline{\ddot{d}}_1 = \frac{1}{\beta(\Delta t)^2}\left[\underline{d}_1 - \underline{d}_0 - (\Delta t)\underline{\dot{d}}_0 - (\Delta t)^2\left(\frac{1}{2} - \beta\right)\underline{\ddot{d}}_0\right]$$

5. Solve Eq. (16.3.9) directly for $\underline{\dot{d}}_1$.
6. Using the results of steps 4 and 5, go back to step 3 to solve for $\underline{d}_2$ and then to steps 4 and 5 to solve for $\underline{\dot{d}}_2$ and $\underline{\ddot{d}}_2$. Use steps 3–5 repeatedly to solve for $\underline{d}_{i+1}, \underline{\dot{d}}_{i+1}$, and $\underline{\ddot{d}}_{i+1}$.

Figure 16–8 is a flowchart of the solution procedure using Newmark's equations. The advantages of Newmark's method over the central difference method are that Newmark's method can be made unconditionally stable (for instance, if $\beta = \frac{1}{4}$ and $\gamma = \frac{1}{2}$) and that larger time steps can be used with better results because, in general, the difference expressions more closely approximate the true acceleration and displacement time behavior [8] to [11]. Other difference formulas, such as Wilson's and Houboldt's, also yield unconditionally stable algorithms.

We will now illustrate the use of Newmark's equations as they apply to the following example problem.

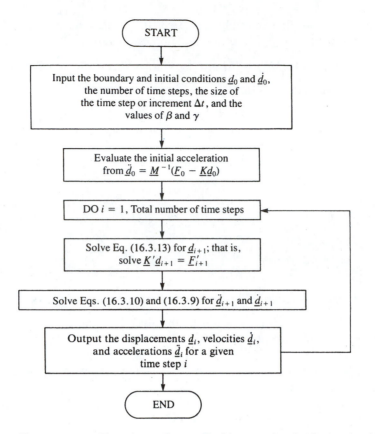

Figure 16–8 Flowchart of numerical integration in time using Newmark's equations

Example 16.2

Determine the displacement, velocity, and acceleration at 0.1-s time increments up to a time of 0.5 s for the one-dimensional spring-mass oscillator subjected to the time-dependent forcing function shown in Figure 16–9, along with the restoring spring force versus displacement curve. Assume the oscillator is initially at rest. Let $\beta = \frac{1}{6}$ and $\gamma = \frac{1}{2}$, which corresponds to an assumption of linear acceleration within each time step.

Because we are again considering the single degree of freedom associated with the mass, the general matrix equations describing the motion reduce to single scalar equations. Again, we represent this single degree of freedom by d.

The solution procedure follows the steps outlined in this section and in the flowchart of Figure 16–8.

Step 1

At time $t = 0$, the initial displacement and velocity are zero; therefore,

$$d_0 = 0 \qquad \dot{d}_0 = 0$$

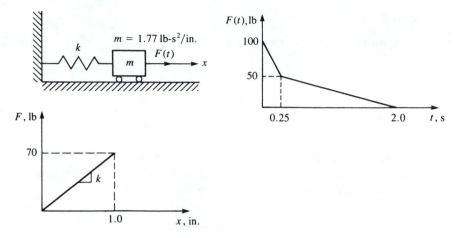

Figure 16–9 Spring-mass oscillator subjected to a time-dependent force

Step 2

The initial acceleration at $t = 0$ is obtained as

$$\ddot{d}_0 = \frac{100 - 70(0)}{1.77} = 56.5 \text{ in./s}^2$$

where we have used $\underline{F}_0 = 100$ lb and $\underline{K} = 70$ lb/in.

Step 3

We now solve for the displacement at time $t = 0.1$ s as

$$K' = 70 + \frac{1}{(\frac{1}{6})(0.1)^2}(1.77) = 1132 \text{ lb/in.}$$

$$F_1' = 80 + \frac{1.77}{(\frac{1}{6})(0.1)^2}\left[0 + (0.1)(0) + \left(\frac{1}{2} - \frac{1}{6}\right)(0.1)^2(56.5)\right] = 280 \text{ lb}$$

$$d_1 = \frac{280}{1132} = 0.248 \text{ in.}$$

Step 4

Solve for the acceleration at time $t = 0.1$ s as

$$\ddot{d}_1 = \frac{1}{(\frac{1}{6})(0.1)^2}\left[0.248 - 0 - (0.1)(0) - (0.1)^2\left(\frac{1}{2} - \frac{1}{6}\right)(56.5)\right]$$

$$\ddot{d}_1 = 35.4 \text{ in./s}^2$$

Step 5

Solve for the velocity at time $t = 0.1$ s as

$$\dot{d}_1 = 0 + (0.1)[(1 - \tfrac{1}{2})(56.5) + (\tfrac{1}{2})(35.4)]$$

$$\dot{d}_1 = 4.59 \text{ in./s}$$

Step 6

Repeated use of steps 3–5 will result in the displacement, acceleration, and velocity for additional time steps as desired. We will now perform one more time-step iteration.

Repeating step 3 for the next time step ($t = 0.2$ s), we have

$$F_2' = 60 + \frac{1.77}{(\tfrac{1}{6})(0.1)^2}\left[0.248 + (0.1)(4.59) + \left(\frac{1}{2} - \frac{1}{6}\right)(0.1)^2(35.4)\right]$$

$$F_2' = 934 \text{ lb}$$

$$d_2 = \frac{934}{1132} = 0.825 \text{ in.}$$

Repeating step 4 for time step $t = 0.2$ s, we obtain

$$\ddot{d}_2 = \frac{1}{(\tfrac{1}{6})(0.1)^2}\left[0.825 - 0.248 - (0.1)(4.59) - (0.1)^2\left(\frac{1}{2} - \frac{1}{6}\right)(35.4)\right]$$

$$\ddot{d}_2 = 1.27 \text{ in./s}^2$$

Finally, repeating step 5 for time step $t = 0.2$ s, we have

$$\dot{d}_2 = 4.59 + (0.1)[(1 - \tfrac{1}{2})(35.4) + \tfrac{1}{2}(1.27)]$$

$$\dot{d}_2 = 6.42 \text{ in./s}$$

Table 16–2 summarizes the results obtained through time $t = 0.5$ s.

Table 16–2 Results of the analysis of Example 16.2

t (s)	$F(t)$ (lb)	d_i (in.)	Q (lb)	$\ddot{d}_i$ (in./s^2)	$\dot{d}_i$ (in./s)
0.	100	0	0	56.5	0
0.1	80	0.248	17.3	35.4	4.59
0.2	60	0.825	57.8	1.27	6.42
0.3	48.6	1.36	95.2	−26.2	5.17
0.4	45.7	1.72	120.4	−42.2	1.75
0.5	42.9	1.68	117.6	−42.2	−2.45

■

Wilson's Method

We will now outline Wilson's method (also called the Wilson-Theta method). Because of its general versatility, it has been adopted into the Algor computer program for purposes of structural dynamics analysis. Wilson's method is an extension of the linear acceleration method wherein the acceleration is assumed to vary linearly within each time interval now taken from t to $t + \Theta \Delta t$, where $\Theta \geqslant 1.0$. For $\Theta = 1.0$, the method reduces to the linear acceleration scheme. However, for unconditional stability in the numerical analysis, we must use $\Theta \geqslant 1.37$ [7, 8]. In practice, $\Theta = 1.40$ is often selected. The Wilson equations are given in a form similar to the previous Newmark's equations, Eqs. (16.3.9) and (16.3.10), as

$$\dot{d}_{i+1} = \dot{d}_i + \frac{\Theta \Delta t}{2}(\ddot{d}_{i+1} + \ddot{d}_i) \tag{16.3.15}$$

$$d_{i+1} = d_i + \Theta \Delta t \dot{d}_i + \frac{\Theta^2 (\Delta t)^2}{6}(\ddot{d}_{i+1} + 2\ddot{d}_i) \tag{16.3.16}$$

where $\ddot{d}_{i+1}, \dot{d}_{i+1}$, and d_{i+1} represent the acceleration, velocity, and displacement, respectively, at time $t + \Theta \Delta t$.

We seek a matrix equation of the form of Eq. (16.3.13) that can be solved for displacement $\underline{d}_{i+1}$. To obtain this equation, first solve Eqs. (16.3.15) and (16.3.16) for $\ddot{d}_{i+1}$ and $\dot{d}_{i+1}$ in terms of d_{i+1} as follows:

Solve Eq. (16.3.16) for $\ddot{d}_{i+1}$ to obtain

$$\ddot{\underline{d}}_{i+1} = \frac{6}{\Theta^2 (\Delta t)^2}(\underline{d}_{i+1} - \underline{d}_i) - \frac{6}{\Theta \Delta t}\dot{\underline{d}}_i - 2\ddot{\underline{d}}_i \tag{16.3.17}$$

Now use Eq. (16.3.17) in Eq. (16.3.15) and solve for $\dot{\underline{d}}_{i+1}$ to obtain

$$\dot{\underline{d}}_{i+1} = \frac{3}{\Theta \Delta t}(\underline{d}_{i+1} - \underline{d}_i) - 2\dot{\underline{d}}_i - \frac{\Theta \Delta t}{2}\ddot{\underline{d}}_i \tag{16.3.18}$$

To obtain the displacement $\underline{d}_{i+1}$ (at time $t + \Theta \Delta t$), we use the equation of motion Eq. (16.2.24) rewritten as

$$\underline{F}_{i+1} = \underline{M}\ddot{d}_{i+1} + \underline{K}d_{i+1} \tag{16.3.19}$$

Now, substituting Eq. (16.3.17) for $\ddot{d}_{i+1}$ into Eq. (16.3.19), we obtain

$$\underline{M}\left[\frac{6}{\Theta^2 (\Delta t)^2}(\underline{d}_{i+1} - \underline{d}_i) - \frac{6}{\Theta \Delta t}\dot{\underline{d}}_i - 2\ddot{\underline{d}}_i\right] + \underline{K}\underline{d}_{i+1} = \underline{F}_{i+1} \tag{16.3.20}$$

Combining like terms and rewriting in a form similar to Eq. (16.3.13), we obtain

$$\underline{K}'\underline{d}_{i+1} = \underline{F}'_{i+1} \tag{16.3.21}$$

where
$$\underline{K}' = \underline{K} + \frac{6}{(\Theta \Delta t)^2}\underline{M}$$

$$\tag{16.3.22}$$

$$\underline{F}'_{i+1} = \underline{F}_{i+1} + \frac{\underline{M}}{(\Theta \Delta t)^2}[6\underline{d}_i + 6\Theta \Delta t \dot{\underline{d}}_i + 2(\Theta \Delta t)^2 \ddot{\underline{d}}_i]$$

You will note the similarities between Wilson's Eqs. (16.3.22) and Newmark's Eqs. (16.3.14). Because the acceleration is assumed to vary linearly, the load vector is expressed as

$$\bar{\underline{F}}_{i+1} = \underline{F}_i + \Theta(\underline{F}_{i+1} - \underline{F}_i) \tag{16.3.23}$$

where $\bar{\underline{F}}_{i+1}$ replaces $\underline{F}_{i+1}$ in Eq. (16.3.22). Note that if $\Theta = 1$, $\bar{\underline{F}}_{i+1} = \underline{F}_{i+1}$.

Also, Wilson's method (like Newmark's) is an implicit integration method, because the displacements show up as multiplied by the stiffness matrix and we implicitly solve for the displacements at time $t + \Theta \Delta t$.

The solution procedure using Wilson's equations is as follows:

1. Starting at time $t = 0$, d_0 is known from the given boundary conditions on displacement, and $\dot{d}_0$ is known from the initial velocity conditions.
2. Solve Eq. (16.3.5) for $\ddot{d}_0$ (unless $\ddot{d}_0$ is known from an initial acceleration condition).
3. Solve Eq. (16.3.21) for d_1, because $\underline{F}'_{i+1}$ is known for all time steps, and $d_0, \dot{d}_0, \ddot{d}_0$ are now known from steps 1 and 2.
4. Solve Eq. (16.3.17) for $\ddot{d}_1$.
5. Solve Eq. (16.3.18) for $\dot{d}_1$.
6. Using the results of steps 4 and 5, go back to step 3 to solve for d_2, and then return to steps 4 and 5 to solve for $\ddot{d}_2$ and $\dot{d}_2$. Use steps 3–5 repeatedly to solve for $d_{i+1}, \dot{d}_{i+1}$, and $\ddot{d}_{i+1}$.

A flowchart similar to Figure 16–8, based on Newmark's equation, is left to your discretion. Again, note that the advantage of Wilson's method is that it can be made unconditionally stable by setting $\Theta \geqslant 1.37$. Finally, the time step, Δt, recommended is approximately $\frac{1}{10}$ to $\frac{1}{20}$ of the shortest natural period τ_n of the finite element assemblage with n degrees of freedom; that is, $\Delta t \doteq \tau_n/10$. In comparing the Newmark and Wilson methods, we observe little difference in the computational effort, because they both require about the same time step. Wilson's method is very similar to Newmark's, so hand solutions will not be presented. However, we suggest that you rework Example 16.1 by Wilson's method and compare your displacement results with the exact solution listed in Table 16–1.

▲ 16.4 Natural Frequencies of a One-Dimensional Bar ▲

Before solving the structural stress dynamics analysis problem, we will first describe how to determine the natural frequencies of continuous elements (specifically the bar element). The natural frequencies are necessary in a vibration analysis and also are important when choosing a proper time step for a structural dynamics analysis (as will be discussed in Section 16.5).

Natural frequencies are determined by solving Eq. (16.2.24) in the absence of a forcing function $F(t)$. Therefore, we solve the matrix equation

$$\underline{M}\underline{\ddot{d}} + \underline{K}\underline{d} = 0 \qquad (16.4.1)$$

The standard solution for $\underline{d}(t)$ is given by the harmonic equation in time

$$\underline{d}(t) = \underline{d}'e^{i\omega t} \qquad (16.4.2)$$

where $\underline{d}'$ is the part of the nodal displacement matrix called *natural modes* that is assumed to be independent of time, i is the standard imaginary number given by $i = \sqrt{-1}$, and ω is a natural frequency.

Differentiating Eq. (16.4.2) twice with respect to time, we obtain

$$\underline{\ddot{d}}(t) = \underline{d}'(-\omega^2)e^{i\omega t} \qquad (16.4.3)$$

Substitution of Eqs. (16.4.2) and (16.4.3) into Eq. (16.4.1) yields

$$-\underline{M}\omega^2\underline{d}'e^{i\omega t} + \underline{K}\underline{d}'e^{i\omega t} = 0 \qquad (16.4.4)$$

Combining terms in Eq. (16.4.4), we obtain

$$e^{i\omega t}(\underline{K} - \omega^2\underline{M})\underline{d}' = 0 \qquad (16.4.5)$$

Because $e^{i\omega t}$ is not zero, from Eq. (16.4.5) we obtain

$$(\underline{K} - \omega^2\underline{M})\underline{d}' = 0 \qquad (16.4.6)$$

Equation (16.4.6) is a set of linear homogeneous equations in terms of displacement mode $\underline{d}'$. Hence, Eq. (16.4.6) has a nontrivial solution if and only if the determinant of the coefficient matrix of $\underline{d}'$ is zero; that is, we must have

$$|\underline{K} - \omega^2\underline{M}| = 0 \qquad (16.4.7)$$

In general, Eq. (16.4.7) is a set of n algebraic equations, where n is the number of degrees of freedom associated with the problem.

To illustrate the procedure for determining the natural frequencies, we will solve the following example problem.

Example 16.3

For the bar shown in Figure 16–10 with length $2L$, modulus of elasticity E, mass density ρ, and cross-sectional area A, determine the first two natural frequencies.

For simplicity, the bar is discretized into two elements each of length L as shown in Figure 16–11. To solve Eq. (16.4.7), we must develop the total stiffness matrix for the bar by using Eq. (16.2.11). Either the lumped-mass matrix Eq. (16.2.12) or the

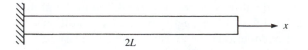

Figure 16–10 One-dimensional bar used for natural frequency determination

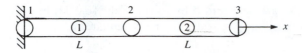

Figure 16–11 Discretized bar of Figure 16–10

consistent-mass matrix Eq. (16.2.23) can be used. In general, using the consistent-mass matrix has resulted in solutions that compare more closely to available analytical and experimental results than those found using the lumped-mass matrix. However, the longhand calculations are more tedious using the consistent-mass matrix than using the lumped-mass matrix because the consistent-mass matrix is a full symmetric matrix, whereas the lumped-mass matrix has nonzero terms only along the main diagonal. Hence, the lumped-mass matrix will be used in this analysis.

Using Eq. (16.2.11), the stiffness matrices for each element are given by

$$
[\hat{k}^{(1)}] = \frac{AE}{L}
\begin{array}{cc} 1 & 2 \end{array}
\begin{bmatrix} 1 & -1 \\ -1 & 1 \end{bmatrix}
\qquad
[\hat{k}^{(2)}] = \frac{AE}{L}
\begin{array}{cc} 2 & 3 \end{array}
\begin{bmatrix} 1 & -1 \\ -1 & 1 \end{bmatrix}
\qquad (16.4.8)
$$

The usual direct stiffness method for assembling the element matrices, Eqs. (16.4.8), yields the global stiffness matrix for the whole bar as

$$
[K] = \frac{AE}{L}
\begin{bmatrix} 1 & -1 & 0 \\ -1 & 2 & -1 \\ 0 & -1 & 1 \end{bmatrix}
\qquad (16.4.9)
$$

Using Eq. (16.2.12), the mass matrices for each element are given by

$$
[\hat{m}^{(1)}] = \frac{\rho AL}{2}
\begin{array}{cc} 1 & 2 \end{array}
\begin{bmatrix} 1 & 0 \\ 0 & 1 \end{bmatrix}
\qquad
[\hat{m}^{(2)}] = \frac{\rho AL}{2}
\begin{array}{cc} 2 & 3 \end{array}
\begin{bmatrix} 1 & 0 \\ 0 & 1 \end{bmatrix}
\qquad (16.4.10)
$$

The mass matrices for each element are assembled in the same manner as for the stiffness matrices. Therefore, by assembling Eqs. (16.4.10), we obtain the global mass matrix as

$$
[M] = \frac{\rho AL}{2}
\begin{bmatrix} 1 & 0 & 0 \\ 0 & 2 & 0 \\ 0 & 0 & 1 \end{bmatrix}
\qquad (16.4.11)
$$

We observe from the resulting global mass matrix that there are two mass contributions at node 2 because node 2 is common to both elements.

Substituting the global stiffness matrix Eq. (16.4.9) and the global mass matrix Eq. (16.4.11) into Eq. (16.4.6), and using the boundary condition $\hat{d}_{1x} = 0$ (or now $d_1' = 0$) to reduce the set of equations in the usual manner, we obtain

$$
\left(\frac{AE}{L} \begin{bmatrix} 2 & -1 \\ -1 & 1 \end{bmatrix} - \omega^2 \frac{\rho AL}{2} \begin{bmatrix} 2 & 0 \\ 0 & 1 \end{bmatrix} \right) \begin{Bmatrix} d_2' \\ d_3' \end{Bmatrix} = \begin{Bmatrix} 0 \\ 0 \end{Bmatrix}
\qquad (16.4.12)
$$

To obtain a solution to the set of homogeneous equations in Eq. (16.4.12), we set the determinant of the coefficient matrix equal to zero as indicated by Eq. (16.4.7). We then have

$$\left| \frac{AE}{L} \begin{bmatrix} 2 & -1 \\ -1 & 1 \end{bmatrix} - \lambda \frac{\rho AL}{2} \begin{bmatrix} 2 & 0 \\ 0 & 1 \end{bmatrix} \right| = 0 \tag{16.4.13}$$

where $\lambda = \omega^2$ has been used in Eq. (16.4.13). Dividing Eq. (16.4.13) by ρAL and letting $\mu = E/(\rho L^2)$, we obtain

$$\begin{vmatrix} 2\mu - \lambda & -\mu \\ -\mu & \mu - \dfrac{\lambda}{2} \end{vmatrix} = 0 \tag{16.4.14}$$

Evaluating the determinant in Eq. (16.4.14), we obtain

$$\lambda = 2\mu \pm \mu\sqrt{2}$$

or $\qquad\qquad \lambda_1 = 0.60\mu \qquad \lambda_2 = 3.41\mu \tag{16.4.15}$

For comparison, the exact solution is given by $\lambda = 0.616\mu$, whereas the consistent-mass approach yields $\lambda = 0.648\mu$. Therefore, for bar elements, the lumped-mass approach can yield results as good as, or even better than, the results for the consistent-mass approach. However, the consistent-mass approach can be mathematically proved to yield an upper bound on the frequencies, whereas the lumped-mass approach yields results that can be below or above the exact frequencies with no mathematical proof of boundedness. From Eqs. (16.4.15), the first and second natural frequencies are given by

$$\omega_1 = \sqrt{\lambda_1} = 0.77\sqrt{\mu} \qquad \omega_2 = \sqrt{\lambda_2} = 1.85\sqrt{\mu}$$

Letting $E = 30 \times 10^6$ psi, $\rho = 0.00073$ lb-s^2/in^4, and $L = 100$ in., we obtain

$$\mu = E/(\rho L^2) = (30 \times 10^6)/[(0.00073)(100)^2] = 4.12 \times 10^6 \text{ s}^{-2}$$

Therefore, we obtain the natural circular frequencies as

$$\omega_1 = 1.56 \times 10^3 \text{ rad/s} \qquad \omega_2 = 3.76 \times 10^3 \text{ rad/s} \tag{16.4.16}$$

or in Hertz (1/s) units

$$f_1 = \omega_1/2\pi = 248 \text{ Hz}, \qquad \text{and so on}$$

In conclusion, note that for a bar discretized such that two nodes are free to displace, there are two natural modes and two frequencies. When a system vibrates with a given natural frequency ω_i, that unique shape with arbitrary amplitude corresponding to ω_i is called the *mode*. In general, for an n-degrees-of-freedom discrete system, there are n natural modes and frequencies. A continuous system actually has an infinite number of natural modes and frequencies. When the system is discretized, only n degrees of freedom are created. The lowest modes and frequencies are approximated most often; the higher frequencies are damped out more rapidly and are usually of less importance.

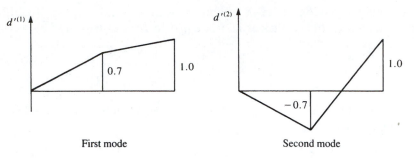

Figure 16–12 First and second modes of longitudinal vibration for the cantilever bar of Figure 16–10

Substituting λ_1 from Eqs. (16.4.15) into Eq. (16.4.12) and simplifying, the first modal equations are given by

$$1.4\mu d_2''^{(1)} - \mu d_3''^{(1)} = 0$$

$$-\mu d_2''^{(1)} + 0.7\mu d_3''^{(1)} = 0$$

$$(16.4.17)$$

It is customary to specify the value of one of the natural modes $\underline{d}'$ for a given ω_i or λ_i. Letting $d_3''^{(1)} = 1$ and solving Eq. (16.4.17), we find $d_2''^{(1)} = 0.7$. Similarly, substituting λ_2 from Eqs. (16.4.15) into Eq. (16.4.12), we obtain the second modal equations. For brevity's sake, these equations are not presented here. Now letting $d_3''^{(2)} = 1$ results in $d_2''^{(2)} = -0.7$. The modal response for the first and second natural frequencies of longitudinal vibration are plotted in Figure 16–12. The first mode means that the bar is completely in tension or compression, depending on the excitation direction. The second mode means the bar is in compression and tension or in tension and compression. ■

▲ 16.5 Time-Dependent One-Dimensional Bar Analysis ▲

Example 16.4

To illustrate the finite element solution of a time-dependent problem, we will solve the problem of the one-dimensional bar shown in Figure 16–13(a) subjected to the force shown in Figure 16–13(b). We will assume the boundary condition $d_{1x} = 0$ and the initial conditions $\underline{d}_0 = 0$ and $\underline{\dot{d}}_0 = 0$. For later numerical computation purposes, we let parameters $\rho = 0.00073$ lb-s^2/in^4, $A = 1$ in^2, $E = 30 \times 10^6$ psi, and $L = 100$ in. These parameters are the same values as used in Section 16.4.

Because the bar is discretized into two elements of equal length, the global stiffness and mass matrices determined in Section 16.4 and given by Eqs. (16.4.9) and (16.4.11) are applicable. We will again use the lumped-mass matrix because of its

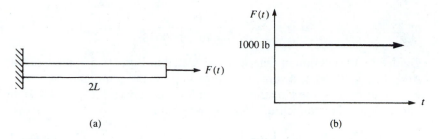

Figure 16–13   (a) Bar subjected to a time-dependent force and (b) the forcing function applied to the end of the bar

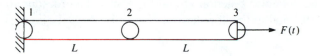

Figure 16–14   Discretized bar with lumped masses

resulting computational simplicity. Figure 16–14 shows the discretized bar and the associated lumped masses.

For illustration of the numerical time integration scheme, we will use the central difference method because it is easier to apply for longhand computations (and without loss of generality).

We next select the time step to be used in the integration process. It has been mathematically shown that the time step must be less than or equal to 2 divided by the highest natural frequency when the central difference method is used [7]; that is, $\Delta t \leqslant 2/\omega_{max}$. However, for practical results, we must use a time step of less than or equal to three-fourths of this value; that is,

$$\Delta t \leqslant \frac{3}{4}\left(\frac{2}{\omega_{max}}\right) \qquad (16.5.1)$$

This time step ensures stability of the integration method. This criterion for selecting a time step demonstrates the usefulness of determining the natural frequencies of vibration, as previously described in Section 16.4, before performing the dynamic stress analysis. An alternative guide (used only for a bar) for choosing the approximate time step is

$$\Delta t = \frac{L}{c_x} \qquad (16.5.2)$$

where L is the element length, and $c_x = \sqrt{E_x/\rho}$ is called the *longitudinal wave velocity*. Evaluating the time step by using both criteria, Eqs. (16.5.1) and (16.5.2), from Eqs. (16.4.16) for ω, we obtain

$$\Delta t = \frac{3}{4}\left(\frac{2}{\omega_{max}}\right) = \frac{1.5}{3.76 \times 10^3} = 0.40 \times 10^{-3} \text{ s} \qquad (16.5.3)$$

or
$$\Delta t = \frac{L}{c_x} = \frac{100}{\sqrt{30 \times 10^6 / 0.00073}} = 0.48 \times 10^{-3} \text{ s} \qquad (16.5.4)$$

Guided by the maximum time steps calculated in Eqs. (16.5.3) and (16.5.4), we choose $\Delta t = 0.25 \times 10^{-3}$ s as a convenient time step for the computations.

Substituting the global stiffness and mass matrices, Eqs. (16.4.9) and (16.4.11), into the global dynamic Eq. (16.2.24), we obtain

$$\frac{AE}{L} \begin{bmatrix} 1 & -1 & 0 \\ -1 & 2 & -1 \\ 0 & -1 & 1 \end{bmatrix} \begin{Bmatrix} d_{1x} \\ d_{2x} \\ d_{3x} \end{Bmatrix} + \frac{\rho AL}{2} \begin{bmatrix} 1 & 0 & 0 \\ 0 & 2 & 0 \\ 0 & 0 & 1 \end{bmatrix} \begin{Bmatrix} \ddot{d}_{1x} \\ \ddot{d}_{2x} \\ \ddot{d}_{3x} \end{Bmatrix} = \begin{Bmatrix} R_1 \\ 0 \\ F_3(t) \end{Bmatrix} \qquad (16.5.5)$$

where R_1 denotes the unknown reaction at node 1. Using the procedure for solution outlined in Section 16.3 and in the flowchart of Figure 16–6, we begin as follows:

Step 1

Given: $d_{1x} = 0$ because of the fixed support at node 1, and all nodal displacements and velocities are zero at time $t = 0$; that is, $\underline{\dot{d}}_0 = 0$ and $\underline{d}_0 = 0$. Also, assume $\ddot{d}_{1x} = 0$ at all times.

Step 2

Solve for $\underline{\ddot{d}}_0$ using Eq. (16.3.5) as

$$\underline{\ddot{d}}_0 = \begin{Bmatrix} \ddot{d}_{2x} \\ \ddot{d}_{3x} \end{Bmatrix}_{t=0} = \frac{2}{\rho AL} \begin{bmatrix} \frac{1}{2} & 0 \\ 0 & 1 \end{bmatrix} \left[\begin{Bmatrix} 0 \\ 1000 \end{Bmatrix} - \frac{AE}{L} \begin{bmatrix} 2 & -1 \\ -1 & 1 \end{bmatrix} \begin{Bmatrix} 0 \\ 0 \end{Bmatrix} \right] \qquad (16.5.6)$$

where Eq. (16.5.6) accounts for the conditions $d_{1x} = 0$ and $\ddot{d}_{1x} = 0$. Simplifying Eq. (16.5.6), we obtain

$$\underline{\ddot{d}}_0 = \frac{2000}{\rho AL} \begin{Bmatrix} 0 \\ 1 \end{Bmatrix} = \begin{Bmatrix} 0 \\ 27,400 \end{Bmatrix} \text{ in./s}^2 \qquad (16.5.7)$$

where the numerical values for ρ, A, and L have been substituted into the final numerical result in Eq. (16.5.7), and

$$\underline{M}^{-1} = \frac{2}{\rho AL} \begin{bmatrix} \frac{1}{2} & 0 \\ 0 & 1 \end{bmatrix} \qquad (16.5.8)$$

has been used in Eq. (16.5.6). The computational advantage of using the lumped-mass matrix for longhand calculations is now evident. The inverse of a diagonal matrix, such as the lumped-mass matrix, is obtained simply by inverting the diagonal elements of the matrix.

Step 3

Using Eq. (16.3.8), we solve for $\underline{d}_{-1}$ as

$$\underline{d}_{-1} = \underline{d}_0 - (\Delta t)\underline{\dot{d}}_0 + \frac{(\Delta t)^2}{2} \underline{\ddot{d}}_0 \qquad (16.5.9)$$

Substituting the initial conditions on $\underline{\dot{d}}_0$ and $\underline{d}_0$ from step 1 and Eq. (16.5.7) for the initial acceleration $\underline{\ddot{d}}_0$ from step 2 into Eq. (16.5.9), we obtain

$$\underline{d}_{-1} = 0 - (0.25 \times 10^{-3})(0) + \frac{(0.25 \times 10^{-3})^2}{2}(27,400)\begin{Bmatrix} 0 \\ 1 \end{Bmatrix}$$

or, on simplification,

$$\begin{Bmatrix} d_{2x} \\ d_{3x} \end{Bmatrix}_{-1} = \begin{Bmatrix} 0 \\ 0.856 \times 10^{-3} \end{Bmatrix} \text{ in.} \tag{16.5.10}$$

Step 4

On premultiplying Eq. (16.3.7) by $\underline{M}^{-1}$, we now solve for $\underline{d}_1$ by

$$\underline{d}_1 = \underline{M}^{-1}\{(\Delta t)^2 \underline{F}_0 + [2\underline{M} - (\Delta t)^2 \underline{K}]\underline{d}_0 - \underline{M}\underline{d}_{-1}\} \tag{16.5.11}$$

Substituting the numerical values for $\rho, A, L,$ and E and the results of Eq. (16.5.10) into Eq. (16.5.11), we obtain

$$\begin{Bmatrix} d_{2x} \\ d_{3x} \end{Bmatrix}_1 = \frac{2}{0.073}\begin{bmatrix} \frac{1}{2} & 0 \\ 0 & 1 \end{bmatrix}\left\{(0.25 \times 10^{-3})^2\begin{Bmatrix} 0 \\ 1000 \end{Bmatrix} + \begin{bmatrix} \frac{2(0.073)}{2}\begin{bmatrix} 2 & 0 \\ 0 & 1 \end{bmatrix} \end{bmatrix}\right.$$

$$- (0.25 \times 10^{-3})^2(30 \times 10^4)\begin{bmatrix} 2 & -1 \\ -1 & 1 \end{bmatrix}\right]\begin{Bmatrix} 0 \\ 0 \end{Bmatrix}$$

$$\left. - \frac{0.073}{2}\begin{bmatrix} 2 & 0 \\ 0 & 1 \end{bmatrix}\begin{Bmatrix} 0 \\ 0.856 \times 10^{-3} \end{Bmatrix}\right\}$$

Simplifying, we obtain

$$\begin{Bmatrix} d_{2x} \\ d_{3x} \end{Bmatrix}_1 = \frac{2}{0.073}\begin{bmatrix} \frac{1}{2} & 0 \\ 0 & 1 \end{bmatrix}\left[\begin{Bmatrix} 0 \\ 0.0625 \times 10^{-3} \end{Bmatrix} - \begin{Bmatrix} 0 \\ 0.0312 \times 10^{-3} \end{Bmatrix}\right]$$

Finally, the nodal displacements at time $t = 0.25 \times 10^{-3}$ s become

$$\begin{Bmatrix} d_{2x} \\ d_{3x} \end{Bmatrix}_1 = \begin{Bmatrix} 0 \\ 0.858 \times 10^{-3} \end{Bmatrix} \text{ in.} \quad (\text{at } t = 0.25 \times 10^{-3} \text{ s}) \tag{16.5.12}$$

Step 5

With $\underline{d}_0$ initially given and $\underline{d}_1$ determined from step 4, we use Eq. (16.3.7) to obtain

$$\underline{d}_2 = \underline{M}^{-1}\{(\Delta t)^2 \underline{F}_1 + [2\underline{M} - (\Delta t)^2 \underline{K}]\underline{d}_1 - \underline{M}\underline{d}_0\}$$

$$= \frac{2}{0.073}\begin{bmatrix} \frac{1}{2} & 0 \\ 0 & 1 \end{bmatrix}\left\{(0.25 \times 10^{-3})^2\begin{Bmatrix} 0 \\ 1000 \end{Bmatrix} + \begin{bmatrix} \frac{2(0.073)}{2}\begin{bmatrix} 2 & 0 \\ 0 & 1 \end{bmatrix} \end{bmatrix}\right.$$

$$- (0.25 \times 10^{-3})^2(30 \times 10^4)\begin{bmatrix} 2 & -1 \\ -1 & 1 \end{bmatrix}\right]$$

$$\times \left\{ \begin{array}{c} 0 \\ 0.858 \times 10^{-3} \end{array} \right\} - \frac{0.073}{2} \begin{bmatrix} 2 & 0 \\ 0 & 1 \end{bmatrix} \left\{ \begin{array}{c} 0 \\ 0 \end{array} \right\} \right\}$$

$$= \frac{2}{0.073} \begin{bmatrix} \frac{1}{2} & 0 \\ 0 & 1 \end{bmatrix} \left[\left\{ \begin{array}{c} 0 \\ 0.0625 \times 10^{-3} \end{array} \right\} + \left\{ \begin{array}{c} 0.0161 \times 10^{-3} \\ 0.0466 \times 10^{-3} \end{array} \right\} \right]$$

Simplifying, we obtain the nodal displacements at time $t = 0.50 \times 10^{-3}$ s as

$$\left\{ \begin{array}{c} d_{2x} \\ d_{3x} \end{array} \right\}_2 = \left\{ \begin{array}{c} 0.221 \times 10^{-3} \\ 2.99 \times 10^{-3} \end{array} \right\} \text{ in.} \qquad (\text{at } t = 0.50 \times 10^{-3} \text{ s}) \qquad (16.5.13)$$

Step 6

Solve for the nodal accelerations $\ddot{\underline{d}}_1$ again using Eq. (16.3.5) as

$$\ddot{\underline{d}}_1 = \frac{2}{0.073} \begin{bmatrix} \frac{1}{2} & 0 \\ 0 & 1 \end{bmatrix} \left[\left\{ \begin{array}{c} 0 \\ 1000 \end{array} \right\} - (30 \times 10^4) \begin{bmatrix} 2 & -1 \\ -1 & 1 \end{bmatrix} \left\{ \begin{array}{c} 0 \\ 0.858 \times 10^{-3} \end{array} \right\} \right]$$

Simplifying, we then obtain the nodal accelerations at time $t = 0.25 \times 10^{-3}$ s as

$$\left\{ \begin{array}{c} \ddot{d}_{2x} \\ \ddot{d}_{3x} \end{array} \right\}_1 = \left\{ \begin{array}{c} 3526 \\ 20{,}345 \end{array} \right\} \text{ in./s}^2 \qquad (\text{at } t = 0.25 \times 10^{-3} \text{ s}) \qquad (16.5.14)$$

The reaction R_1 could be found by using the results of Eqs. (16.5.12) and (16.5.14) in Eq. (16.5.5).

Step 7

Using Eq. (16.5.13) from step 5 and the boundary condition for $\underline{d}_0$ given in step 1, we obtain $\dot{\underline{d}}_1$ as

$$\dot{\underline{d}}_1 = \frac{\left[\left\{ \begin{array}{c} 0.221 \times 10^{-3} \\ 2.99 \times 10^{-3} \end{array} \right\} - \left\{ \begin{array}{c} 0 \\ 0 \end{array} \right\} \right]}{2(0.25 \times 10^{-3})}$$

Simplifying, we obtain

$$\left\{ \begin{array}{c} \dot{d}_{2x} \\ \dot{d}_{3x} \end{array} \right\} = \left\{ \begin{array}{c} 0.442 \\ 5.98 \end{array} \right\} \text{ in./s} \qquad (\text{at } t = 0.25 \times 10^{-3} \text{ s})$$

Step 8

We now use steps 5–7 repeatedly to obtain the displacement, acceleration, and velocity for all other time steps. For simplicity, we calculate the acceleration only.

Repeating step 6 with $t = 0.50 \times 10^{-3}$ s, we obtain the nodal accelerations as

$$\ddot{\underline{d}}_2 = \frac{2}{0.073} \begin{bmatrix} \frac{1}{2} & 0 \\ 0 & 1 \end{bmatrix} \left[\left\{ \begin{array}{c} 0 \\ 1000 \end{array} \right\} - 30 \times 10^4 \begin{bmatrix} 2 & -1 \\ -1 & 1 \end{bmatrix} \left\{ \begin{array}{c} 0.221 \times 10^{-3} \\ 2.99 \times 10^{-3} \end{array} \right\} \right]$$

On simplifying, the nodal accelerations at $t = 0.50 \times 10^{-3}$ s are

$$\left\{ \begin{array}{c} \ddot{d}_{2x} \\ \ddot{d}_{3x} \end{array} \right\}_2 = \left\{ \begin{array}{c} 0 \\ 27,400 \end{array} \right\} + \left\{ \begin{array}{c} 10,500 \\ -22,800 \end{array} \right\}$$

$$= \left\{ \begin{array}{c} 10,500 \\ 4600 \end{array} \right\} \text{ in./s}^2 \qquad (\text{at } t = 0.5 \times 10^{-3} \text{ s}) \qquad (16.5.15) \quad \blacksquare$$

16.6 Beam Element Mass Matrices and Natural Frequencies

We now consider the lumped- and consistent-mass matrices appropriate for time-dependent beam analysis. The development of the element equations follows the same general steps as used in Section 16.2 for the bar element.

The beam element with the associated nodal degrees of freedom (transverse displacement and rotation) is shown in Figure 16–15.

The basic element equations are given by the general form, Eq. (16.2.10), with the appropriate nodal force, stiffness, and mass matrices for a beam element. The stiffness matrix for the beam element is that given by Eq. (5.1.14). A lumped-mass matrix is obtained as

$$
[\hat{m}] = \frac{\rho A L}{2}
\begin{array}{c}
\begin{array}{cccc} \hat{d}_{1y} & \hat{\phi}_1 & \hat{d}_{2y} & \hat{\phi}_2 \end{array} \\
\begin{bmatrix}
1 & 0 & 0 & 0 \\
0 & 0 & 0 & 0 \\
0 & 0 & 1 & 0 \\
0 & 0 & 0 & 0
\end{bmatrix}
\end{array}
\qquad (16.6.1)
$$

where one-half of the total beam mass has been lumped at each node, corresponding to the translational degrees of freedom. In the lumped mass approach, the inertial effect associated with possible rotational degrees of freedom has been assumed to be zero in obtaining Eq. (16.6.1), although a value may be assigned to these rotational degrees of freedom by calculating the mass moment of inertia of a fraction of the beam segment about the nodal points. For a uniform beam we could then calculate the mass moment of inertia of half of the beam segment about each end node using

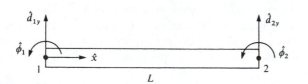

Figure 16–15 Beam element with nodal degrees of freedom

basic dynamics as

$$I = \tfrac{1}{3}(\rho AL/2)(L/2)^2$$

Again, the lumped-mass matrix given by Eq. (16.6.1) is a diagonal matrix, making matrix numerical calculations easier to perform than when using the consistent-mass matrix. The consistent-mass matrix can be obtained by applying the general Eq. (16.2.19) for the beam element, where the shape functions are now given by Eqs. (5.1.7). Therefore,

$$[\hat{m}] = \iiint_V \rho [N]^T [N] \, dV \tag{16.6.2}$$

$$[\hat{m}] = \int_0^L \iint_A \rho \begin{Bmatrix} N_1 \\ N_2 \\ N_3 \\ N_4 \end{Bmatrix} [N_1 \quad N_2 \quad N_3 \quad N_4] \, dA \, d\hat{x} \tag{16.6.3}$$

with

$$N_1 = \frac{1}{L^3}(2\hat{x}^3 - 3\hat{x}^2 L + L^3)$$

$$N_2 = \frac{1}{L^3}(\hat{x}^3 L - 2\hat{x}^2 L^2 + \hat{x}L^3)$$

$$N_3 = \frac{1}{L^3}(-2\hat{x}^3 + 3\hat{x}^2 L) \tag{16.6.4}$$

$$N_4 = \frac{1}{L^3}(\hat{x}^3 L - \hat{x}^2 L^2)$$

On substituting the shape function Eqs. (16.6.4) into Eq. (16.6.3) and performing the integration, the consistent-mass matrix becomes

$$[\hat{m}] = \frac{\rho AL}{420} \begin{bmatrix} 156 & 22L & 54 & -13L \\ 22L & 4L^2 & 13L & -3L^2 \\ 54 & 13L & 156 & -22L \\ -13L & -3L^2 & -22L & 4L^2 \end{bmatrix} \tag{16.6.5}$$

Having obtained the mass matrix for the beam element, we could proceed to formulate the global stiffness and mass matrices and equations of the form given by Eq. (16.2.24) to solve the problem of a beam subjected to a time-dependent load. We will not illustrate the procedure for solution here because it is tedious and similar to that used to solve the one-dimensional bar problem in Section 16.5. However, the Algor computer program can be used for the analysis of beams and frames subjected to time-dependent forces. Section 16.7 provides descriptions of plane frame analysis, and Section 16.9 describes the use of the computer program Algor for dynamics analysis of bars, beams, and frames.

To clarify the procedure for beam analysis, we will now determine the natural frequencies of a beam.

Example 16.5

We now consider the determination of the natural frequencies of vibration for a beam fixed at both ends as shown in Figure 16–16. The beam has mass density ρ, modulus of elasticity E, cross-sectional area A, area moment of inertia I, and length $2L$. For simplicity of the longhand calculations, the beam is discretized into two elements of length L.

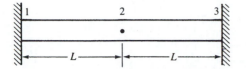

Figure 16–16   Beam for determination of natural frequencies

We can obtain the natural frequencies by using the general Eq. (16.4.7). First, we assemble the global stiffness and mass matrices (using the boundary conditions $d_{1x} = 0$, $\phi_1 = 0$, $d_{3x} = 0$, and $\phi_3 = 0$ as

$$\underline{K} = \frac{EI}{L^3} \begin{matrix} d_{2y} & \phi_2 \\ \begin{bmatrix} 24 & 0 \\ 0 & 8L^2 \end{bmatrix} \end{matrix} \qquad \underline{M} = \frac{\rho AL}{2} \begin{bmatrix} 2 & 0 \\ 0 & 0 \end{bmatrix} \qquad (16.6.6)$$

where Eq. (16.6.1) has been used to calculate the lumped-mass matrix. On substituting Eqs. (16.6.6) into Eq. (16.4.7), we obtain

$$\left| \frac{EI}{L^3} \begin{bmatrix} 24 & 0 \\ 0 & 8L^2 \end{bmatrix} - \omega^2 \rho AL \begin{bmatrix} 1 & 0 \\ 0 & 0 \end{bmatrix} \right| = 0 \qquad (16.6.7)$$

Dividing Eq. (16.6.7) by ρAL and simplifying, we obtain

$$\omega^2 = \frac{24EI}{\rho AL^4}$$

or

$$\omega = \frac{4.90}{L^2} \left(\frac{EI}{A\rho} \right)^{1/2}$$

The exact solution for the first natural frequency, from simple beam theory, is given by Reference [6]. It is

$$\omega = \frac{5.59}{L^2} \left(\frac{EI}{A\rho} \right)^{1/2}$$

The large discrepancy between the exact solution and the finite element solution is assumed to be accounted for by the coarseness of the finite element model. In Example 16.6 we show for a clamped-free beam that as the number of degrees of freedom increases, convergence to the exact solution results. Furthermore, if we had used the consistent-mass matrix for the beam [Eq. (16.6.5)], the results would have been more accurate than with the lumped-mass matrix as consistent-mass matrices yield more accurate results for flexural elements such as beams. ∎

Example 16.6

Determine the first natural frequency of vibration of the cantilever beam shown in Figure 16–17 with the following data:

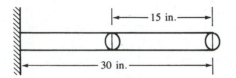

Figure 16–17 Fixed-free beam (two-element model, lumped-mass matrix)

Length of the beam: $L = 30$ in.
Modulus of elasticity: $E = 3 \times 10^7$ psi
Moment of inertia: $I = 0.0833$ in^4
Cross-sectional area: $A = 1$ in^2
Mass density: $\rho = 0.00073$ lb-s^2/in^4
Poisson's ratio: $v = 0.3$

The finite element longhand solution result for the first natural frequency is obtained similarly to that of Example 16.5 as

$$\omega = \frac{3.148}{L^2} \left(\frac{EI}{A\rho} \right)^{1/2}$$

The exact solution according to beam theory [1] is

$$\omega = \frac{3.516}{L^2} \left(\frac{EI}{\rho A} \right)^{1/2}$$

According to vibration theory for a clamped-free beam [1], we relate the second and third natural frequencies to the first natural frequency by

$$\frac{\omega_2}{\omega_1} = 6.2669 \qquad \frac{\omega_3}{\omega_1} = 17.5475$$

Figure 16–18 shows the first, second, and third mode shapes corresponding to the first three natural frequencies for the cantilever beam of Example 16.6 as obtained from the Algor program. Note that each mode shape has one fewer node where a node is a point of zero displacement. That is, the first mode has all the elements of the beam of

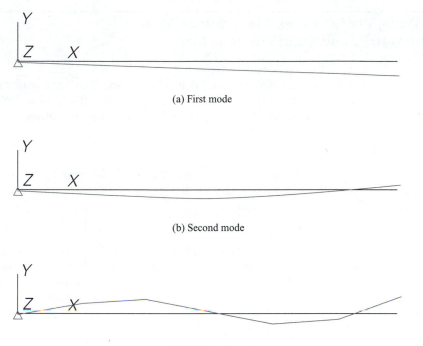

(a) First mode

(b) Second mode

(c) Third mode

Figure 16–18 First, second, and third mode shapes of flexural vibration for a cantilever beam

Table 16–3 Finite element computer solution compared to exact solution for Example 16.6

	ω_1 (rad/s)	ω_2 (rad/s)
Exact solution from beam theory	228	1434
Finite element solution		
Using 2 elements	205	1286
Using 6 elements	226	1372
Using 10 elements	227.5	1410
Using 30 elements	228.5	1430
Using 60 elements	228.5	1432

the same sign [Figure 16–18(a)], the second mode has one sign change and at some point along the beam the displacement is zero [Figure 16–18(b)], and the third mode has two sign changes and at two points along the beam the displacement is zero [Figure 16–18(c)].

Table 16–3 shows the computer solution compared with the exact solution. ■

▲ 16.7 Truss, Plane Frame, Plane Stress/Strain, Axisymmetric, and Solid Element Mass Matrices ▲

The dynamic analysis of the truss and that of the plane frame are performed by extending the concepts presented in Sections 16.2 and 16.6 to the truss and plane frame, as has previously been done for the static analysis of trusses and frames.

Truss Element

The truss analysis requires the same transformation of the mass matrix from local to global coordinates as in Eq. (3.4.22) for the stiffness matrix; that is, the global mass matrix for a truss element is given by

$$\underline{m} = \underline{T}^T \underline{\hat{m}} \underline{T} \tag{16.7.1}$$

We are now dealing with motion in two or three dimensions. Therefore, we must reformulate a bar element mass matrix with both axial and transverse inertial properties because mass is included in both the global x and y directions in plane truss analysis (Figure 16–19). Considering two-dimensional motion, we express both local axial displacement $\hat{u}$ and transverse displacement $\hat{v}$ for the element in terms of the local axial and transverse nodal displacements as

$$\begin{Bmatrix} \hat{u} \\ \hat{v} \end{Bmatrix} = \frac{1}{L} \begin{bmatrix} L - \hat{x} & 0 & \hat{x} & 0 \\ 0 & L - \hat{x} & 0 & \hat{x} \end{bmatrix} \begin{Bmatrix} \hat{d}_{1x} \\ \hat{d}_{1y} \\ \hat{d}_{2x} \\ \hat{d}_{2y} \end{Bmatrix} \tag{16.7.2}$$

In general, $\underline{\hat{\psi}} = \underline{N}\underline{\hat{d}}$; therefore, the shape function matrix from Eq. (16.7.2) is

$$[N] = \frac{1}{L} \begin{bmatrix} L - \hat{x} & 0 & \hat{x} & 0 \\ 0 & L - \hat{x} & 0 & \hat{x} \end{bmatrix} \tag{16.7.3}$$

We can then substitute Eq. (16.7.3) into the general expression given by Eq. (16.2.19) to evaluate the local *truss element consistent-mass matrix* as

$$[\hat{m}] = \frac{\rho A L}{6} \begin{bmatrix} 2 & 0 & 1 & 0 \\ 0 & 2 & 0 & 1 \\ 1 & 0 & 2 & 0 \\ 0 & 1 & 0 & 2 \end{bmatrix} \tag{16.7.4}$$

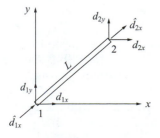

Figure 16–19 Truss element arbitrarily oriented in *x-y* plane showing nodal degrees of freedom

The truss element lumped-mass matrix for two-dimensional motion is obtained by simply lumping mass at each node and remembering that mass is the same in both the $\hat{x}$ and $\hat{y}$ directions. The local *truss element lumped-mass matrix* is then

$$[\hat{m}] = \frac{\rho A L}{2} \begin{bmatrix} 1 & 0 & 0 & 0 \\ 0 & 1 & 0 & 0 \\ 0 & 0 & 1 & 0 \\ 0 & 0 & 0 & 1 \end{bmatrix} \tag{16.7.5}$$

Plane Frame Element

The plane frame analysis requires first expanding and then combining the bar and beam mass matrices to obtain the local mass matrix. Because we recall there are six total degrees of freedom associated with a plane frame element (Figure 16–20), the bar and beam mass matrices are expanded to order 6×6 and superimposed. On combining the local axes consistent-mass matrices for the bar and beam from Eqs. (16.2.23) and (16.6.5), we obtain

$$\underline{\hat{m}} = \rho A L \begin{bmatrix} 2/6 & 0 & 0 & 1/6 & 0 & 0 \\ & 156/420 & 22L/420 & 0 & 54/420 & -13L/420 \\ & & 4L^2/420 & 0 & 13L/420 & -3L^2/420 \\ & & & 2/6 & 0 & 0 \\ & & & & 156/420 & -22L/420 \\ \text{Symmetry} & & & & & 4L^2/420 \end{bmatrix} \tag{16.7.6}$$

On combining the lumped-mass matrices Eqs. (16.2.12) and (16.6.1) for the bar and beam, respectively, the resulting local axes plane frame lumped-mass matrix is

$$\underline{\hat{m}} = \frac{\rho A L}{2} \begin{matrix} \hat{d}_{1x} & \hat{d}_{1y} & \hat{\phi}_1 & \hat{d}_{2x} & \hat{d}_{2y} & \hat{\phi}_2 \\ \begin{bmatrix} 1 & 0 & 0 & 0 & 0 & 0 \\ 0 & 1 & 0 & 0 & 0 & 0 \\ 0 & 0 & 0 & 0 & 0 & 0 \\ 0 & 0 & 0 & 1 & 0 & 0 \\ 0 & 0 & 0 & 0 & 1 & 0 \\ 0 & 0 & 0 & 0 & 0 & 0 \end{bmatrix} \end{matrix} \tag{16.7.7}$$

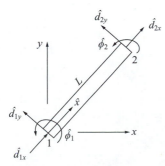

Figure 16–20 Frame element arbitrarily oriented in local coordinate system showing nodal degrees of freedom

The global mass matrix $\underline{m}$ for a plane frame element arbitrarily oriented in x-y coordinates is transformed according to Eq. (16.7.1), where the transformation matrix $\underline{T}$ is now given by Eq. (6.1.10) and either Eq. (16.7.6) for consistent-mass or (16.7.7) for lumped-mass matrices.

Because a longhand solution of the time-dependent plane frame problem is quite lengthy, only an Algor computer program solution will be presented in Section 16.9.

Plane Stress/Strain Element

The plane stress, plane strain, constant-strain triangle element (Figure 16–21) consistent-mass matrix is obtained by using the shape functions from Eq. (7.2.18) and the shape function matrix given by substituting

$$\underline{N} = \begin{bmatrix} N_1 & 0 & N_2 & 0 & N_3 & 0 \\ 0 & N_1 & 0 & N_2 & 0 & N_3 \end{bmatrix}$$

into Eq. (16.2.19) to obtain

$$[m] = \rho \int_V \underline{N}^T \underline{N} \, dV \tag{16.7.8}$$

Letting $dV = t \, dA$ and noting that $\int_A N_1^2 \, dA = \frac{1}{6} A$, $\int_A N_1 N_2 \, dA = \frac{1}{12} A$, and so on, we obtain the *CST* global *consistent-mass matrix* as

$$\underline{m} = \frac{\rho t A}{12} \begin{bmatrix} 2 & 0 & 1 & 0 & 1 & 0 \\ & 2 & 0 & 1 & 0 & 1 \\ & & 2 & 0 & 1 & 0 \\ & & & 2 & 0 & 1 \\ & & & & 2 & 0 \\ \text{Symmetry} & & & & & 2 \end{bmatrix} \tag{16.7.9}$$

For the isoparametric quadrilateral element for plane stress and plane strain considered in Chapter 11, we use the shape functions given by Eq. (11.3.5) with the shape function matrix given in Eq. (11.3.4) substituted into Eq. (16.7.10). This yields the *quadrilateral element consistent-mass matrix* as

$$\underline{m} = \rho t \int_{-1}^{1} \int_{-1}^{1} \underline{N}^T \underline{N} \det \underline{J} \, ds \, dt \tag{16.7.10}$$

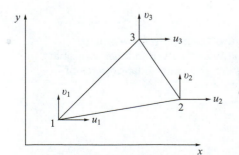

Figure 16–21 CST element with nodal degrees of freedom

The integral in Eq. (16.7.10) is evaluated best by numerical integration as described in Section 11.5.

Axisymmetric Element

The *axisymmetric triangular element* (considered in Chapter 10 and shown in Figure 16–22) *consistent-mass matrix* is given by

$$\underline{m} = \int_V \rho \underline{N}^T \underline{N} \, dV = \int_A \rho \underline{N}^T \underline{N} 2\pi r \, dA \tag{16.7.11}$$

Since $r = N_1 r_1 + N_2 r_2 + N_3 r_3$, we have

$$\underline{m} = 2\pi\rho \int_A (N_1 r_1 + N_2 r_2 + N_3 r_3) \underline{N}^T \underline{N} \, dA \tag{16.7.12}$$

Noting that

$$\int_A N_1^3 \, dA = \frac{2A}{20} \qquad \int_A N_1^2 N_2 \, dA = \frac{2A}{60}$$

$$\int_A N_1 N_2 N_3 \, dA = \frac{2A}{120} \qquad \text{and so on} \tag{16.7.13}$$

we obtain

$$\underline{m} = \frac{\pi \rho A}{10} \begin{bmatrix} \frac{4}{3}r_1 + 2\bar{r} & 0 & 2\bar{r} - \frac{r_3}{3} & 0 & 2\bar{r} - \frac{r_2}{3} & 0 \\ & \frac{4}{3}r_1 + 2\bar{r} & 0 & 2\bar{r} - \frac{r_3}{3} & 0 & 2\bar{r} - \frac{r_2}{3} \\ & & \frac{4}{3}r_2 + 2\bar{r} & 0 & 2\bar{r} - \frac{r_1}{3} & 0 \\ & & & \frac{4}{3}r_2 + 2\bar{r} & 0 & 2\bar{r} - \frac{r_1}{3} \\ & & & & \frac{4}{3}r_3 + 2\bar{r} & 0 \\ \text{Symmetry} & & & & & \frac{4}{3}r_3 + 2\bar{r} \end{bmatrix} \tag{16.7.14}$$

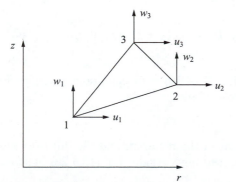

Figure 16–22 Axisymmetric triangular element showing nodal degrees of freedom

where
$$\bar{r} = \frac{r_1 + r_2 + r_3}{3}$$

Tetrahedral Solid Element

Finally, the *tetrahedral solid element* (considered in Chapter 12) *consistent-mass matrix* is obtained by substituting the shape function matrix Eq. (12.2.9) with shape functions defined in Eq. (12.2.10) into Eq. (16.2.19) and performing the integration to obtain

$$\underline{m} = \frac{\rho V}{20}
\begin{bmatrix}
2 & 0 & 0 & 1 & 0 & 0 & 1 & 0 & 0 & 1 & 0 & 0 \\
  & 2 & 0 & 0 & 1 & 0 & 0 & 1 & 0 & 0 & 1 & 0 \\
  &   & 2 & 0 & 0 & 1 & 0 & 0 & 1 & 0 & 0 & 1 \\
  &   &   & 2 & 0 & 0 & 1 & 0 & 0 & 1 & 0 & 0 \\
  &   &   &   & 2 & 0 & 0 & 1 & 0 & 0 & 1 & 0 \\
  &   &   &   &   & 2 & 0 & 0 & 1 & 0 & 0 & 1 \\
  &   &   &   &   &   & 2 & 0 & 0 & 1 & 0 & 0 \\
  &   &   &   &   &   &   & 2 & 0 & 0 & 1 & 0 \\
  &   &   &   &   &   &   &   & 2 & 0 & 0 & 1 \\
  &   &   &   &   &   &   &   &   & 2 & 0 & 0 \\
  &   &   &   &   &   &   &   &   &   & 2 & 0 \\
\text{Symmetry} & & & & & & & & & & & 2
\end{bmatrix} \tag{16.7.15}$$

▲ 16.8 Time-Dependent Heat Transfer ▲

In this section, we consider the time-dependent heat transfer problem in one dimension only. The basic differential equation for time-dependent heat transfer in one dimension was given previously by Eq. (13.1.7) with the boundary conditions given by Eqs. (13.1.10) and (13.1.11).

    The finite element formulation of the equations can be obtained by minimization of the following functional:

$$\pi_h = \frac{1}{2} \iiint\limits_{V} \left[K_{xx} \left(\frac{\partial T}{\partial x} \right)^2 - 2(Q - c\rho \dot{T}) T \right] dV$$

$$- \iint\limits_{S_2} q^* T \, dS + \frac{1}{2} \iint\limits_{S_3} h(T - T_\infty)^2 \, ds \tag{16.8.1}$$

Equation (16.8.1) is similar to Eq. (13.4.10) with definitions given by Eq. (13.4.11) except that the Q term is now replaced by

$$Q - c\rho \dot{T} \tag{16.8.2}$$

where, again, c is the specific heat of the material, and the dot over the variable T denotes differentiation with respect to time. Again, Eq. (13.4.22) obtained in Section

13.4 for the conductivity or stiffness matrix and Eqs. (13.4.23)–(13.4.25) for the force matrix terms are applicable here.

The term given by Eq. (16.8.2) yields an additional contribution to the basic element equations previously obtained for the time-independent problem as follows:

$$\Omega_Q = -\iiint_V T(Q - c\rho\dot{T})\,dV \tag{16.8.3}$$

Again, the temperature function is given by

$$\{T\} = [N]\{\hat{t}\} \tag{16.8.4}$$

where $[N]$ is the shape function matrix given by Eq. (13.4.3) or Eqs. (16.2.3) for the simple one-dimensional element, and $\{\hat{t}\}$ is the nodal temperature matrix. Substituting Eq. (16.8.4) into Eq. (16.8.3) and differentiating with respect to time where indicated yield

$$\Omega_Q = -\iiint_V ([N]\{\hat{t}\}Q - c\rho[N]\{\hat{t}\}[N]\{\dot{\hat{t}}\})\,dV \tag{16.8.5}$$

where the fact that $[N]$ is a function only of the coordinate system has been taken into account. Equation (16.8.5) must be minimized with respect to the nodal temperatures as follows:

$$\frac{\partial \Omega_Q}{\partial \{\hat{t}\}} = -\iiint_V [N]^T Q\,dV + \iiint_V c\rho[N]^T[N]\,dV\{\dot{\hat{t}}\} \tag{16.8.6}$$

where we have assumed that $\{\hat{t}\}$ remains constant during the differentiation with respect to $\{\hat{t}\}$. Equation (16.8.6) results in the additional time-dependent term added to Eq. (13.4.18). Hence, using previous definitions for the stiffness and force matrices, we obtain the element equations as

$$\{\hat{f}\} = [\hat{k}]\{\hat{t}\} + [\hat{m}]\{\dot{\hat{t}}\} \tag{16.8.7}$$

where now

$$[\hat{m}] = \iiint_V c\rho[N]^T[N]\,dV \tag{16.8.8}$$

For an element with constant cross-sectional area A, the differential volume is $dV = A\,d\hat{x}$. Substituting the one-dimensional shape function matrix Eq. (13.4.3) into Eq. (16.8.8) yields

$$[\hat{m}] = c\rho A \int_0^L \left\{ \begin{array}{c} 1 - \dfrac{\hat{x}}{L} \\[2mm] \dfrac{\hat{x}}{L} \end{array} \right\} \left[1 - \dfrac{\hat{x}}{L} \quad \dfrac{\hat{x}}{L} \right] d\hat{x}$$

or

$$[\hat{m}] = \frac{c\rho AL}{6} \begin{bmatrix} 2 & 1 \\ 1 & 2 \end{bmatrix} \tag{16.8.9}$$

Equation (16.8.9) is analogous to the consistent-mass matrix Eq. (16.2.23). The lumped-mass matrix for the heat conduction problem is then

$$[\hat{m}] = \frac{c\rho A L}{2}\begin{bmatrix} 1 & 0 \\ 0 & 1 \end{bmatrix} \tag{16.8.10}$$

which is analogous to Eq. (16.2.12) for the one-dimensional stress element.

The time-dependent heat-transfer problem can now be solved in a manner analogous to that for the stress analysis problem. We present the numerical time integration scheme.

Numerical Time Integration

The numerical time integration method described here is similar to Newmark's method used for structural dynamics analysis and can be used to solve time-dependent or transient heat-transfer problems.

We begin by assuming that two temperature states $\underline{T}_i$ at time t_i and $\underline{T}_{i+1}$ at time t_{i+1} are related by

$$\underline{T}_{i+1} = \underline{T}_i + [(1-\beta)\underline{\dot{T}}_i + \beta\underline{\dot{T}}_{i+1}](\Delta t) \tag{16.8.11}$$

Equation (16.8.11) is known as the *generalized trapezoid rule*. Much like Newmark's method for numerical time integration of the second-order equations of structural dynamics, Eq. (16.8.11) includes a parameter β that is chosen by the user.

Next we express Eq. (16.8.7) in global form as

$$\{F\} = [K]\{T\} + [M]\{\dot{T}\} \tag{16.8.12}$$

We now write Eq. (16.8.12) for time t_i and then for time t_{i+1}. We then multiply the first of these two equations by $1-\beta$ and the second by β to obtain

$$(1-\beta)(\underline{K}\underline{T}_i + \underline{M}\underline{\dot{T}}_i) = (1-\beta)\underline{F}_i \tag{16.8.13a}$$

$$\beta(\underline{K}\underline{T}_{i+1} + \underline{M}\underline{\dot{T}}_{i+1}) = \beta\underline{F}_{i+1} \tag{16.8.13b}$$

Next we add Eqs. (16.8.13a and b) together to obtain

$$\underline{M}[(1-\beta)\underline{\dot{T}}_i + \beta\underline{\dot{T}}_{i+1}] + \underline{K}[(1-\beta)\underline{T}_i + \beta\underline{T}_{i+1}] = (1-\beta)\underline{F}_i + \beta\underline{F}_{i+1} \tag{16.8.14}$$

Now, using Eq. (16.8.11), we can eliminate the time derivative terms from Eq. (16.8.14) to write

$$\frac{\underline{M}(\underline{T}_{i+1} - \underline{T}_i)}{\Delta t} + \underline{K}[(1-\beta)\underline{T}_i + \beta\underline{T}_{i+1}] = (1-\beta)\underline{F}_i + \beta\underline{F}_{i+1} \tag{16.8.15}$$

Rewriting Eq. (16.8.15) by grouping the $\underline{T}_{i+1}$ terms on the left side, we have

$$\left(\frac{1}{\Delta t}\underline{M} + \beta\underline{K}\right)\underline{T}_{i+1} = \left[\frac{1}{\Delta t}\underline{M} - (1-\beta)\underline{K}\right]\underline{T}_i + (1-\beta)\underline{F}_i + \beta\underline{F}_{i+1} \tag{16.8.16}$$

The time integration to solve for $\underline{T}$ begins as follows. Given a known initial temperature $\underline{T}_0$ at time $t = 0$ and a time step Δt, we solve Eq. (16.8.16) for $\underline{T}_1$ at $t = \Delta t$. Then, using $\underline{T}_1$, we determine $\underline{T}_2$ at $t = 2(\Delta t)$, and so on. For a constant Δt, the left-side coefficient of T_{i+1} need be evaluated only one time (assuming $\underline{M}$ and $\underline{K}$ do not vary with time). The matrix Eq. (16.8.16) can then be solved in the usual manner, such as by Gauss elimination. For a one-dimensional heat-transfer analysis, element $\underline{k}$ is given by Eqs. (13.4.22) and (13.4.28), whereas $\underline{f}$ is given by Eqs. (13.4.26) and (13.4.29).

It has been shown that depending on the value of β, the time step Δt may have an upper limit for the numerical analysis to be stable. If $\beta < \frac{1}{2}$, the largest Δt for stability as shown in Reference [12] is

$$\Delta t = \frac{2}{(1 - 2\beta)\lambda_{\max}} \tag{16.8.17}$$

where $\lambda_{\max}$ is the largest eigenvalue of

$$(\underline{K} - \lambda \underline{M})\underline{T}' = 0 \tag{16.8.18}$$

in which, as in Eq. (16.4.2), we have

$$\underline{T}(t) = \underline{T}'e^{i\lambda t} \tag{16.8.19}$$

with $\underline{T}'$ representing the natural modes. If $\beta \geqslant \frac{1}{2}$, the numerical analysis is unconditionally stable; that is, stability of solution (but not accuracy) is guaranteed for Δt greater than that given by Eq. (16.8.17), or as Δt becomes indefinitely large. Various numerical integration methods result, depending on specific values of β:

$\beta = 0$: Forward difference, or Euler [3], which is said to be conditionally stable (that is, Δt must be no greater than that given by Eq. (16.8.17) to obtain a stable solution).

$\beta = \frac{1}{2}$: Crank-Nicolson, or trapezoid, rule, which is unconditionally stable.

$\beta = \frac{2}{3}$: Galerkin, which is unconditionally stable.

$\beta = 1$: Backward difference, which is unconditionally stable.

If $\beta = 0$, the numerical integration method is called *explicit*; that is, we can solve for $\underline{T}_{i+1}$ directly at time Δt knowing only previous information at $t = \underline{T}_i$. If $\beta > 0$, the method is called implicit. If a diagonal mass-type matrix $\underline{M}$ exists and $\beta = 0$, the computational effort for each time step is small (see Example 16.4, where a lumped-mass matrix was used), but so must be Δt. The choice of $\beta > \frac{1}{2}$ is often used. However, if $\beta = \frac{1}{2}$ and sharp transients exist, the method generates spurious oscillations in the solution. Using $\beta > \frac{1}{2}$, along with smaller Δt [12], is probably better. Example 16.7 illustrates the solution of a one-dimensional time-dependent heat-transfer problem using the numerical time integration scheme [Eq. (16.8.16)].

Example 16.7

A circular fin (Figure 16–23) is made of pure copper with a thermal conductivity of $K_{xx} = 400$ W/(m · °C), $h = 150$ W/(m² · °C), mass density $\rho = 8900$ kg/m³, and specific heat $c = 375$ J/(kg · °C) (1 J = 1 W · s). The initial temperature of the fin is 25 °C.

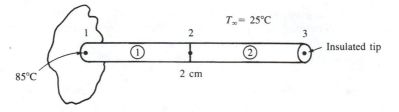

Figure 16–23 Rod subjected to time-dependent temperature

The fin length is 2 cm, and the diameter is 0.4 cm. The right tip of the fin is insulated. The base of the fin is then suddenly increased to a temperature of 85 °C and maintained at this temperature. Use the consistent form of the capacitance matrix, a time step of 0.1 s, and $\beta = \frac{2}{3}$. Use two elements of equal length. Determine the temperature distribution up to 3 s.

Using Eq. (13.4.10), the stiffness matrix is

$$
\underline{k}^{(1)} = \underline{k}^{(2)} = \frac{AK_{xx}}{L} \begin{matrix} 1 & 2 \\ 2 & 3 \\ \begin{bmatrix} 1 & -1 \\ -1 & 1 \end{bmatrix} \end{matrix} + \frac{hPL}{6} \begin{matrix} 1 & 2 \\ 2 & 3 \\ \begin{bmatrix} 2 & 1 \\ 1 & 2 \end{bmatrix} \end{matrix}
$$

$$
\underline{k}^{(1)} = \underline{k}^{(2)} = \frac{\pi(0.004)^2(400)}{4(0.01)} \begin{bmatrix} 1 & -1 \\ -1 & 1 \end{bmatrix} + \frac{150(2\pi)(0.002)(0.01)}{6} \begin{bmatrix} 2 & 1 \\ 1 & 2 \end{bmatrix}
$$

$$
\text{(16.8.20)}
$$

Assembling the element stiffness matrices, Eq. (16.8.20), we obtain the global stiffness matrix as

$$
\underline{K} = \begin{matrix} 1 & 2 & 3 \\ \begin{bmatrix} 0.50894 & -0.49951 & 0 \\ -0.49951 & 1.01788 & -0.49951 \\ 0 & -0.49951 & 0.50894 \end{bmatrix} \end{matrix} \frac{W}{{}^\circ C} \qquad \text{(16.8.21)}
$$

Using Eq. (16.8.12), we obtain each element force matrix as

$$
\{f_h^{(1)}\} = \{f_h^{(2)}\} = \frac{hT_\infty PL}{2} \begin{Bmatrix} 1 \\ 1 \end{Bmatrix} = \frac{(150)(25\,{}^\circ C)(2\pi)(0.002)(0.01)}{2} \begin{Bmatrix} 1 \\ 1 \end{Bmatrix}
$$

$$
\underline{f}_h^{(1)} = \underline{f}_h^{(2)} = \begin{Bmatrix} 0.23561 \\ 0.23561 \end{Bmatrix} \qquad \text{(16.8.22)}
$$

Using Eq. (16.8.22), we find that the assembled global force matrix is

$$
\{F\} = \begin{Bmatrix} 0.23561 \\ 0.47122 \\ 0.23561 \end{Bmatrix} W \qquad \text{(16.8.23)}
$$

Next using Eq. (16.8.9), we obtain each element mass (capacitance) matrix as

$$[m] = \frac{c\rho AL}{6}\begin{bmatrix} 2 & 1 \\ 1 & 2 \end{bmatrix}$$

$$\underline{m}^{(1)} = \underline{m}^{(2)} = \frac{(375)(8900)\frac{\pi(0.004)^2}{4}(0.01)}{6}\begin{bmatrix} 2 & 1 \\ 1 & 2 \end{bmatrix}$$

$$= 0.06990\begin{bmatrix} 2 & 1 \\ 1 & 2 \end{bmatrix} \text{ W} \cdot \text{s}/°\text{C} \tag{16.8.24}$$

Using Eq. (16.8.24), the assembled capacitance matrix is

$$\underline{M} = \begin{bmatrix} 0.13980 & 0.06990 & 0 \\ 0.06990 & 0.27960 & 0.06990 \\ 0 & 0.06990 & 0.13980 \end{bmatrix} \begin{array}{c} 1 \quad\quad 2 \quad\quad 3 \\ \\ \\ \end{array} \frac{\text{W} \cdot \text{s}}{°\text{C}} \tag{16.8.25}$$

Using Eq. (16.8.16) and Eqs. (16.8.21) and (16.8.25), we obtain

$$\left(\frac{1}{\Delta t}\underline{M} + \beta\underline{K}\right) = \begin{bmatrix} 1.7374 & 0.36603 & 0 \\ 0.36603 & 3.4747 & 0.36603 \\ 0 & 0.36603 & 1.7374 \end{bmatrix} \frac{\text{W}}{°\text{C}} \tag{16.8.26}$$

and

$$\left[\frac{1}{\Delta t}\underline{M} - (1-\beta)\underline{K}\right] = \begin{bmatrix} 1.2280 & 0.8655 & 0 \\ 0.8655 & 2.457 & 0.8655 \\ 0 & 0.8655 & 1.2280 \end{bmatrix} \frac{\text{W}}{°\text{C}} \tag{16.8.27}$$

where $\beta = \frac{2}{3}$ and $\Delta t = 0.1$ s have been used to obtain Eqs. (16.8.26) and (16.8.27). For the first time step, $t = 0.1$ s, we then use Eqs. (16.8.23), (16.8.27), and (16.8.26) in Eq. (16.8.16) to obtain

$$\begin{bmatrix} 1.7374 & 0.36603 & 0 \\ 0.36603 & 3.4747 & 0.36603 \\ 0 & 0.36603 & 1.7374 \end{bmatrix} \begin{Bmatrix} 85\,°\text{C} \\ t_2 \\ t_3 \end{Bmatrix}$$

$$= \begin{bmatrix} 1.2280 & 0.8655 & 0 \\ 0.8655 & 2.457 & 0.8655 \\ 0 & 0.8655 & 1.2280 \end{bmatrix} \begin{Bmatrix} 25\,°\text{C} \\ 25\,°\text{C} \\ 25\,°\text{C} \end{Bmatrix} + \begin{Bmatrix} 0.23561 \\ 0.47122 \\ 0.23561 \end{Bmatrix} \tag{16.8.28}$$

In Eq. (16.8.28), we should note that because $\underline{F}_i = \underline{F}_{i+1}$ for all time, the sum of the terms is $(1-\beta)\underline{F}_i + \beta\underline{F}_{i+1} = \underline{F}_i$ for all time. This is the column matrix on the right side of Eq. (16.8.28). We now solve Eq. (16.8.28) in the usual manner by partitioning the second and third equations of Eq. (16.8.28) from the first equation and solving the second and third equations simultaneously for t_2 and t_3. The results are

$$t_2 = 18.534\,°\text{C} \qquad t_3 = 26.371\,°\text{C}$$

Table 16–4 Nodal temperatures at various times for Example 16.7

	Temperature of Node Numbers (°C)		
Time (s)	1	2	3
0.1	85	18.534	26.371
0.2	85	29.732	21.752
0.3	85	36.404	22.662
0.4	85	41.032	25.655
0.5	85	44.665	29.312
0.6	85	47.749	33.059
0.7	85	50.482	36.669
0.8	85	52.956	40.062
0.9	85	55.218	43.218
1.0	85	57.296	46.139
1.1	85	59.208	48.837
1.2	85	60.969	51.327
1.3	85	62.593	53.623
1.4	85	64.089	55.741
1.5	85	65.469	57.693
1.6	85	66.742	59.493
1.7	85	67.915	61.152
1.8	85	68.996	62.683
1.9	85	69.993	64.094
2.0	85	70.912	65.395
2.1	85	71.760	66.594
2.2	85	72.542	67.700
2.3	85	73.262	68.720
2.4	85	73.926	69.660
2.5	85	74.539	70.527
2.6	85	75.104	71.326
2.7	85	75.624	72.063
2.8	85	76.104	72.742
2.9	85	76.547	73.368
3.0	85	76.955	73.946

At time $t = 0.2$ s, Eq. (16.8.28) becomes

$$
\begin{bmatrix} 1.7374 & 0.36603 & 0 \\ 0.36603 & 3.4747 & 0.36603 \\ 0 & 0.36603 & 1.7374 \end{bmatrix} \begin{Bmatrix} 85\,°C \\ t_2 \\ t_3 \end{Bmatrix}
$$

$$
= \begin{bmatrix} 1.2280 & 0.8655 & 0 \\ 0.8655 & 2.457 & 0.8655 \\ 0 & 0.8655 & 1.2280 \end{bmatrix} \begin{Bmatrix} 85\,°C \\ 18.534\,°C \\ 26.371\,°C \end{Bmatrix} + \begin{Bmatrix} 0.23561 \\ 0.47122 \\ 0.23561 \end{Bmatrix} \qquad (16.8.29)
$$

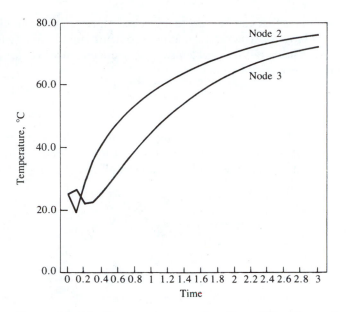

Figure 16–24 Temperature as a function of time for nodes 2 and 3 of Example 16.7

Solving Eq. (16.8.29) for t_2 and t_3, we obtain

$$t_2 = 29.732\,^\circ\mathrm{C} \qquad t_3 = 21.752\,^\circ\mathrm{C}$$

The results through a time of 3 s are tabulated in Table 16–4 and plotted in Figure 16–24. ■

▲ 16.9 Algor Example Solutions for Structural Dynamics and Transient Heat Transfer

In this section, we describe how to solve structural dynamics and transient heat-transfer problems using the Algor program. We first solve Example 16.3 to describe how Algor is used to determine natural frequencies of a bar. We then use Algor to solve Example 16.6 to determine natural frequencies in a cantilever beam. Next we explain how to find the natural frequencies of a fixed-fixed beam using the plane stress element in Algor and compare how many elements of this type are necessary to obtain correct results. We solve three structural dynamics problems, a bar, a beam, and a frame subjected to time-dependent loadings. Finally, we resolve the transient heat-transfer problem, Example 16.7.

These examples illustrate the use of the "Linear Transient Stress Analysis Using Direct Integration" feature of Algor. Using the direct integration feature within the "Global" screen and the "Load Curves Input" screen, we create a file of time-dependent loading, such as shown in Figures 16–30 and 16–33, for time history analysis. The load curves that can be created are constant, ramp, piecewise-linear, and sinusoidal [force loadings for structural dynamics problems and thermal (temperature) loadings for transient heat-transfer problems]. Also see Reference [16] for additional information on time-dependent loading.

Example 16.8

For the bar shown in Figure 16–25 with length 200 in., modulus of elasticity $E = 30 \times 10^6$ psi, weight density of 0.283 lb/in^3 (mass density of 0.00073 lb-s^2/in^4), and cross-sectional area 1 in^2, determine the first two natural frequencies.

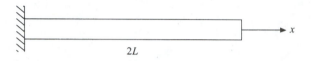

Figure 16–25 Bar used for determining natural frequencies

Step 1 Start Superdraw III

Select the "Start" Button of Windows NT/95/98 and then proceed to "Programs:Algor Software:Algor FEA".

Then click "Algor FEA" to start the program.

Or double-click on the "Algor FEA" icon on your desktop computer screen. The Superdraw III program appears. This is the Algor main graphics user interface. All your work will be done from this program. You will construct, analyze, and review your results from Superdraw III. Each menu option and a short example are available in the Superdraw III Reference Division (accessed through "Docutech", the on-line "Technical User Documentation".)

Step 2 Create the Bar Model

Select the "Add" menu from the top main menu bar.

Click "Line". A "Line" menu appears in the upper right part of the screen. Then enter the data points in succession.

`0<Tab>0<Tab>0`	Enter the first point of the model.
`<Enter>`	
`100<Tab>0<Tab>0`	Enter the second point of the model.
`<Enter>`	
`200<Tab>0<Tab>0`	Enter the last point of the model.
`<Enter>`	You are finished entering points.

Click "Done" on the menu in the upper right corner.

Select "View". This takes you to the draw options menu.

Click "Enclose". This encloses the model. Two elements appear as indicated by highlighted boxes.

Step 3 Eliminate Duplicate Lines

Click "Modify" on the main menu.

Select "Clean:Duplicate". This notation means to select "Clean" and then "Duplicate" in succession.

Click "Perform Cleaning" on the "Duplicate" menu. You see the following message in the lower part of the screen:

"2 Kept 0 Deleted. done."

Click "Done" on the "Duplicate" menu in the upper right corner of the screen.

Step 4 Add the Boundary Conditions

Click on the "FEA Add" menu on the main menu bar and highlight "Stress and Vibration Analysis".

Click "Boundary Conditions". The "Boundary Conditions" menu appears. The default boundary condition is "Use @ Symbol for Full". "Full" means full constraint or prevention of translations and rotations (when relevant for the element you choose). Move the cursor to node 1 and

Click on node 1. The "@" sign appears next to the node, indicating that node 1 is completely fixed.

Click "Change Values" on the "Boundary Condition" menu.

Left-click next to "Tx Constrained" to remove the x-translational boundary conditions.

Click "Done".

Right-click on nodes 2 and 3. It may be difficult to see the constraint notation at this time. Therefore

Click "Done" and then

Select "View", "Zoom", and "In" and create a box around node 2, for instance, to see the "TyzRxyz" boundary condition next to node 2. You can repeat this process of zooming in one or more times to see the "@" sign at node 1 or the other "TyzRxyz" at node 3.

Select "View" and "Enclose" to see the whole bar model again.

Step 5 Add the Material Properties Using the "Model Data Control" Window

After the mesh has been created and the nodal boundary conditions and forces have been applied, you can select: "Tools" and "Model Data Control" from Superdraw III to complete the data input for analysis. The "Model Data Control" window appears. Alternatively,

Click on the "Model Data" button in the lower part of the screen and highlighted in red.

From the "Analysis Type" line, scroll down to "Linear Mode Shapes and Natural Frequencies".

Click on the box below the "Element" column. A table of group 1 element types appears.

Click "Truss". You have now created a truss. All lines are truss elements.

Click "OK".

Click on the box below the "Data" column. It responds with "Please enter the model name first".

Click "OK" and then enter the model name in the "File_name" area that appears.

Ex168   The name of the file is *ex168*.

Click "Save". By default, the "Units Definition" menu appears.

Click "OK" to accept the default "English (in)" units system.

The "Element Definition" menu appears next.

Enter the cross-sectional area of the truss elements. All elements have the same area in this example. Hold the left click of the mouse over the default value "0" until a blue flashing box appears on the "0".

1.0   Enter a 1. This is the cross-sectional area of the truss elements. Leave the "Stress Free Reference Temperature" as "0".

Click "OK". This takes you back to the "Material Data Control" window.

Click on the box below the "Material" column.

Select "[Customer Defined]".

Select "Edit Properties" and enter the mass density and modules of elasticity.

0.00073   This is the mass density.

30e6   This is the modulus of elasticity.

Click "OK". This takes you to the "Element Material Selection" window.

Click "OK".

Step 6   Enter the Global Data

Select the "Global" button from the right-side menu of the "Model Data Control" window.

Click on the "Output" tab on the right side of the menu and then

Click on the boxes next to "Displacement/eigenvector data", "Element Data", and "Nodal Data". This creates nodal displacement output in the *.lm* file, and element data and nodal data written to the *.lm* file as well.

Click "OK". You are now out of the "Global" menu.

Step 7   Check the Model

Click "Check" under the "FEA Model" column on the right side of the "Model Data Control" window. This takes you to Superview's main menu, and you can check your model as explained in Section 4.1 and detailed by example in Appendix F. The truss model appears automatically showing the full constraint boundary condition at node 1 as a red triangle and the constraints at nodes 2 and 3 with translation in the x direction as red circles.

Click "Done". This takes you out of the Superview menu and back to the "Model Data Control" window.

Step 8   Run the Analysis

Click on the "Analysis" button under the "FEA Model" column in the "Model Data Control" window.

Click "Analyze" in the lower left corner of the "Linear Stress and Vibration-Mode Shapes and Natural Frequencies" menu. (See the green box next to "Analyze".) Wait a short time for the analysis to be performed.

Click "Done" on the lower right of the screen.

Step 9   Review the Results

Click on the "Results" button under the "FEA Model" column in the "Model Data Control" window.

Click "Displaced" on the main menu of Superview.

Click "Displ on" so that "*Displ on" appears. The asterisk means the displaced mode is active.

Click "With undi" so that "*With undi" appears.

Click "Nodes inq" on the "Displaced" menu.

Click "Get" to get a value at a node.

Click on node 3 (at the cantilevered end of the truss). Move the cursor to node 3 and click on it to see the x and y displacements appear at the bottom of the screen.

You see "DX=1.880e−001 DY=0.000e+00 DZ=1.129e−031"

<Esc>

<Esc>

Click "Stress-di"

Click "Post". The "Displaced Vector" legend comes up. You now see the model with colored boxes representing the different displacements in each element. For instance, red indicates modal displacement = 0.18798.

<Esc>

<Esc>

Click "Done". This takes you out of Superview.

Click "Close" to close the "Model Data Control" window.

You should be able to find the *ex168.lm* file (storing the echo check of your data and the nodal displacements) and the *ex168.frq* file (storing the element frequencies) and print a hard copy of each.

Table 16–5 First two natural frequencies for a fixed-free bar

Mode Number	Circular Frequency (rad/s)	Frequency (Hz)	Period (s)	Tolerance
1	1.5490E+03	2.4653E+02	4.0562E−03	7.7628E−16
2	3.7397E+03	5.9518E+02	1.6802E−03	2.9440E−07

The circular frequencies in Table 16–5 compare closely with those of Eq. (16.4.16) from the longhand solution. ∎

Example 16.9

Determine the first two natural frequencies of vibration of a cantilever beam with length 30 in., modulus of elasticity 30×10^6 psi, moment of inertia 0.0833 in^4, cross-sectional area 1 in^2, and mass density 0.00073 lb-s^2/in^4 (Figure 16–26).

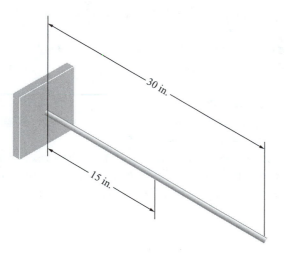

Figure 16–26 Fixed-free beam for natural frequency analysis

The Algor step-by-step solution follows.

Step 1 Start Superdraw III

Double-click on the "Algor FEA" icon on your desktop computer screen.

Step 2 Create the Beam Model

Select the "Add" menu from the top main menu bar.
Click "Line".
Enter the nodal data points.

```
Y
|
|Z   X
|_____
```

Figure 16–27 Two-element model of fixed-free beam

0<Tab>0   This is the left node of the beam.
<Enter>
15<Tab>0   This is the middle node of the beam.
<Enter>
30<Tab>0   This is the right end node of the beam.
<Enter>
Click "Done".
Select "View:Enclose". This encloses the model so that it is completely displayed on the screen. You now see two highlighted lines. The model will look like Figure 16–27.

Step 3   Apply the Boundary Conditions

Click on the "FEA Add" menu on the main menu bar and highlight "Stress and Vibration Analysis".
Click "Boundary Conditions". The default is "Use @ Symbol for Full".
Right-click on the left node. An "@" sign appears next to the node.
Click "Done".
Select "View:Zoom:In" and create a box around the left node to see the "@" sign better.
Click "Done".
Select "View:Enclose" again to see the whole beam.

Step 4   Enter the Material and Geometric Properties Using the "Model Data Control" Window

Click on the "Model Data" button in the lower part of the screen.
For "Analysis Type", select "Linear Mode Shapes and Natural Frequencies".
Click on the box below "Element". A table of "Group 1" element types appears.
Click "Beam". You now have created beam elements for both elements as they are both "Group 1" type elements.
Click "OK".
Click on the box below "Data". The "Save As" window appears. Give the model the filename *ex169* and select "Save". Next the "Units Definition" window appears. Select "OK" to keep the default units system. This takes you to "Beam Design Editor".
Click "Add/Mod" on the side menu.
Click "Sectional" to apply geometric properties to the beam.

Click "Value" to enter values for the geometric properties of the beam.

<Esc>	This clears the current cross-sectional area value.
1	This is the cross-sectional area of the beam.
<Tab>	over to "I2".
<Esc>	This clears the current value.
0.0833	
<Tab>	over to "I3"
<Esc>	This clears the current value.
0.0833	
<Enter>	
<Esc>	
<Esc>	

Click "Quit" and the "Save Current Model" window appears. Select "Yes" and this takes you back to the "Model Data Control" window.

Click on the box below "Material". This takes you to the "Element Material Selection" window.

Select "[Customer Defined]" and then click "View Properties". Then click "Unlock Properties".

Click "0" in the "Mass Density" box and change the mass density.

0.00073

Click "0" in the "Modulus of Elasticity" box and change the modulus.

30e6

Click "OK". This takes you back to the "Element Material Selection" window.

Click "OK". This takes you back to the "Model Data Control" window.

Step 5   Enter the Global Data

Click on the "Global" button on the right-side menu of the "Model Data Control" window.

Click "0" in the "Number of frequencies/modes to calculate" box.

6	This is the number of frequencies desired to be calculated.

Click on the "Output" button.

Click on the box next to "Displacement/eigenvector data".

Click "OK" to go back to the "Model Data Control" window.

Step 6   Check the Model

Click on the "Check" button on the right-side menu of the "Model Data Control" window. This verifies the geometry and finite element data of the model. Once verified, the Superview program opens, verifying the model.

Click "Done" to go back to the "Model Data Control" window.

Step 7 Run the Analysis

Click on the "Analysis" button on the right-side menu of the "Model Data Control" window.

Click "Analyze" in the lower left corner of the "Linear Stress and Vibration-Mode Shapes and Natural Frequencies" window. The analysis is now being performed. When it is finished, the "Algor Analysis" window appears.

Click "OK".

Click "Done".

Step 8 Use Superview to View the Results

Click on the "Results" button on the right-side menu of the "Model Data Control" window. This takes you to Superview again, and you can review the results.

Select "Displaced" from the main menu.

Click "Displ on" so that "*Displ on" appears. The asterisk means the displaced mode is active.

Click "With undi" so that "*With undi" appears. This allows the displaced and undisplaced models to be superimposed.

Click "Calc scal" if you want the program to enlarge or scale the displaced model. The model may not show any difference now, but this step will be necessary when viewing the frequency modes.

<Esc>   This takes you back to the main menu.

Select "Load Case". The first mode of vibration is now displayed on the screen, and the corresponding natural frequency appears below the graphics window. The window should look like Figure 16–28 for the first mode. Note that the natural frequency is 205.1862 rad/time.

Select "Next" to display the next mode and frequency. Remember that there are six modes.

Select "Previous" to display the previous mode.

<Esc>

Select "Quit" to exit the program.

The accompanying table shows the output for the two-element model. Modes one and two listed in the table are really the same mode as indicated by the identical frequencies. One mode represents the upward movement of the beam as shown in Figure 16–28, while the other mode represents the downward movement of the beam

Figure 16–28   First mode of vibration for fixed-free beam

as shown in Figure 16–18a. Similarly, modes three and four are really one mode as the frequencies indicate. This data can be obtained from the output file called *Example169.frq*.

FREQUENCIES =	6
MODE NUMBER	CIRCULAR FREQUENCY (RAD/SEC)
1	2.05186214303217D+02
2	2.05226851708000D+02
3	1.05693290598821D+03
4	1.05714223296699D+03
5	1.03437399737321D+04
6	2.49719973302448D+04

Table 16–3 on page 809 shows the output of the 2-element model, along with that of 6-, 10-, 30-, and 60-element models. The convergence toward the exact solution for the lower frequencies occurs by the 10-element solution. ■

Example 16.10

Use the plane stress element to model and then determine the first five natural frequencies for the fixed-fixed beam shown in Figure 16–29. Compare results found by using the plane stress element with those found by using the beam element. The beam is made of steel with modulus of elasticity 30e6 psi, Poisson's ratio 0.33, length 100 in., and cross-sectional area 1 $in.^2$.

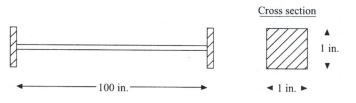

Figure 16–29 Fixed-fixed beam for natural frequency determination modeled using plane stress element

Step 1 Start Superdraw III

Click on the "Algor FEA" icon to start Superdraw III.

Step 2 Create the Beam Model

Click "Add".
Click "Rectangle".

0<Tab>0   This is the lower-left corner node of the rectangle.

<Enter>

100<Tab>1   This is the upper-right corner node of the rectangle.

<Enter>

Click "View".

Click "Enclose".

Click "FEA Mesh".

Click "Automatic Mesh".

Click "4 point...".

Click "Division Values...".

Click on the box below "AB Divisions".

1

<Enter>

Click on the box below "BC Divisions".

100

<Enter>

Click on the "Snap" button on the lower right side of the screen.

Click on the box next to "Point" on the pop-up menu.

Click on the lower left corner of the rectangle.

Click on the upper left corner of the rectangle.

Click on the upper right corner of the rectangle.

Click "Done".

The model then appears divided into 100 segments along the length of the rectangle.

Step 3   Eliminate Duplicate Lines

Click "Modify".

Click "Clean".

Click "Duplicate".

Click "Perform Cleaning".

This message appears at the bottom of the screen:

                     "301 Kept 0 Deleted. done."

Click "Done".

Step 4   Add the Boundary Conditions

Click "FEA Add:Stress and Vibration Analysis:Boundary Conditions".

Click "View:Zoom:In".

Create a box around the left side of the rectangle to zoom in.

Click "Done".

Click "Box Apply...".

Create a box around the left edge of the rectangle.

The "@" then appears in the top and bottom corners of the left side of the rectangle to signify that the nodes are fully constrained.

Click "Done".

Click "View:Enclose".

Click "View:Zoom:In".

Create a box around the right side of the rectangle to zoom in.

Click "Done".

Click "Box Apply".

Create a box around the right edge of the rectangle.

The "@" then appears in the top and bottom corners of the right side of the rectangle to signify that the nodes are fully constrained.

Click "View:Enclose".

Step 5 Add the Material Properties Using the "Model Data Control" Window

Click on the box next to "Analysis Type".

Select "Linear Mode Shapes and Natural Frequencies".

Click on the "Model Data" button.

Click on the box below "Element".

Click "2D".

Click "OK".

An "ATTENTION" window pops up saying that the two-dimensional element must be in the y-z plane.

Click "Yes" to transfer the model to the y-z plane.

Click on the box below "Data".

The Algor program then prompts you to enter a filename.

Enter the name of the file as

Ex16-10

Click "Save".

The "Units Definition" window then pops up.

Click "OK" to accept the default values (English units).

Then the "Element Definition" window pops up.

Click on the box to the right of "Thickness".

Enter 1 as the thickness.

Be sure that "Element Type" is set as "Plane stress", "Material Model" as "Isotropic", and "Compatibility" as "Not Enforced".

Click "OK".

Click on the box below "Material".

The "Element Material Selection" window then pops up.

Click "[Customer Defined]".

Click "Edit Properties".

The "Element Material Specification" window then pops up.

Click on the box to the right of "Modulus of Elasticity".

Enter 30e6 for the modulus of elasticity.

Click on the box to the right of "Poisson's Ratio".

Enter 0.33 for the Poisson's ratio.

Click on the box to the right of "Mass Density".

Enter 7.345e−4.

Click "OK".

Click "OK".

Step 6   Enter the Global Data

Click on the "Global" button on the right-side menu of the "Model Data Control" window.

Click on the "Number of frequencies/modes to calculate" box.

5   This is the number of frequencies desired to be calculated.

Click on the "Output" button.

Check the box next to "Displacement/eigenvector data".

Click "OK" to go back to the "Model Data Control" window.

Step 7   Check the Model

Click on the "Check" button on the right-side menu of the "Model Data Control" window. This verifies the geometry and finite element data of the model. Once verified, Superview opens, verifying the model.

Click "Done" to go back to the "Model Data Control" window.

Step 8   Run the Analysis

Click on the "Analysis" button on the right-side menu of the "Model Data Control" window.

Click "Analyze" in the lower left corner of the "Linear Stress and Vibration-Mode Shapes and Natural Frequencies" window. The analysis will now be performed. When it is finished, the "Algor Analysis" window appears.

Click "OK".

Click "Done".

Step 9   Use Superview to View the Results

Click on the "Results" button on the right-side menu of the "Model Data Control" window. This takes you to Superview again, and you can review the results.

Select "Displaced" on the main menu.

Click "Displ on" so that "*Displ on" appears. The asterisk means the displaced mode is active.

Click "With undi" so that "*With undi" appears. This allows the displaced and undisplaced models to be superimposed.

Table 16–6 Results for first five frequencies using 100 and 1000 elements and exact solution

ω (rad/s)	Analytical	100 Elements	Error (%)	1000 Elements	Error (%)
1	130.8	130.7	22	130.6	2
2	360.8	359.8	22	359.7	2
3	707.3	704.7	22	704.1	2
4	1169.2	1163.3	22	1161.6	2
5	1746.6	1734.5	21	1731.0	2

Click "Calc scal" if you want the program to enlarge or scale the displaced model. The model may not show any difference now, but this step will be necessary when viewing the frequency modes.

<Esc>   This takes you back to the main menu.

Select "Load Case". The first mode of vibration is now displayed on the screen, and the corresponding natural frequency appears below the graphics window. Note that the natural frequency is 130.65 rad/time.

Select "Next" to display the next mode and frequency. Remember that there are five modes.

Select "Previous" to display the previous mode.

<Esc>

Select "Quit" to exit the program.

Table 16–6 shows the first five natural frequencies for the fixed-fixed beam using 100 plane stress elements and then using 1000 elements. This data can be obtained from the output file called *Ex16-10.frq*. Comparisons to the analytical solutions from beam theory are shown. ▮

Example 16.11

For the bar shown in Figure 16–30, subjected to the time-dependent forcing function shown, determine the nodal displacements for five time steps using two finite elements.

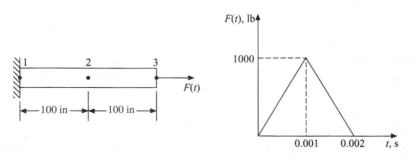

Figure 16–30 Bar subjected to forcing function shown

Let $E = 30 \times 10^6$ psi, mass density $= 0.00073$ lb-s^2/in^4, $A = 1$ in^2, and total length of bar $= 200$ in. Use two elements and a time step of integration of 0.00025 s. The time integration step is based on that recommended by Eq. (16.5.1) and determined in Example 16.4, as the bar has the same properties as that of Example 16.4.

The Algor step-by-step solution follows.

Step 1 Start Superdraw III

Double-click on the "Algor FEA" icon on your desktop computer screen.

Step 2 Create the Bar Model

Click on the "Add" menu on the top main menu bar.
Click "Line".
Enter the nodal data points.

0<Tab>0   This is the left node of the bar.

<Enter>

100<Tab>0   This is the middle node of the bar.

<Enter>

200<Tab>0   This is the right end node of the bar.

<Enter>

Click "Done".

Select "View:Enclose". This encloses the model so that it is completely displayed on the screen. You now see two highlighted lines.

Step 3 Apply Boundary Conditions

Click on the "FEA Add" menu on the main menu bar and highlight "Stress and Vibration Analysis".
Click "Boundary Conditions". The default is "Use @ Symbol for Full".
Right-click on the left node. An "@" sign appears next to the node.
Click "Change Values" and remove the check next to "Tx Constrained" to free up the x direction.
Click "Done" to return to the "Boundary Conditions" menu.
Right-click on nodes 2 and 3 to fix these nodes in the y and z directions.
Click "Done".
Select "View:Zoom:In" and create a box around the left node to see the "@" sign better.
Click "Done".
Select "View:Enclose" again to see the whole beam.
Click "Done".

Step 4 Enter the Material and Geometric Properties Using the "Model Data Control" Window

Click on the "Model Data" button located in the lower part of the screen.
For "Analysis Type" select "Linear Transient Stress (Direct Integration)".

Click on the box below "Element". A table of group 1 element types appears.

Click "Truss". You now have created truss elements for both elements as they are both group 1 element types.

Click "OK".

Click on the box below "Data". You are then asked for the model name.

Click "OK". The "Save As" window appears. Give the model the filename

ex1611

Click "Save". Next the "Units Definition" window appears.

Click "OK" to keep the default units system. Next the "Element Definition" window appears.

Click on the "0" next to "Cross-Sectional Area".

1   The cross-sectional area of the model is 1.

Click "OK" to return to the "Model Data Control" window.

Click on the box below "Material". This takes you to the "Element Material Selection" window.

Click "[Customer Defined]" and then

Click "View Properties". Then

Click "Edit Properties".

Click on the "0" in the "Mass Density" box.

0.00073   This is the mass density.

Click on the "0" in the "Modulus of Elasticity" box.

30e6   This is the modulus of elasticity.

Click "OK". This takes you back to the "Element Material Selection" window.

Click "OK". This takes you back to the "Model Data Control" window.

Step 5   Enter the Global Data

Click on the "Global" button on the right-side menu of the "Model Data Control" window. The "Linear Transient Stress Analysis Using Direct Integration" window appears.

Click on the "0" in the "Number of Time Steps" box and make it 40.

40   The number of time steps desired to be used is 40. This takes you through the total time of 0.001 s when the force reaches its maximum (Figure 16–30).

Click "0" in the "Time Step Size" box.

0.00025   The time step size desired to be used is 0.00025 s.

Leave the "Output interval" as "1" and neglect damping and leave alpha "0" and beta "0". See the discussion preceding Example 16.12 regarding damping.

Click "Load Curves" in the top right corner of the window.

This brings up the "Load Curve Input" window. This is where the forcing function ("Factor") in Figure 16–30 is created. Use the default "Piecewise Linear" for the "Load Curve Form".

Click on the box under "Time".

0   This is the x coordinate of the first data point (time starts at 0).

Click on the box under "Factor".

0   This is the y coordinate of the first data point (force starts at 0).

<Enter>   This takes you to the next line under "Time".

0.001   This is the x coordinate of the second data point (time is now 0.001).

Click on the new empty box under "Factor".

1000   This is the y coordinate of the second data point (the force is now 1000).

<Enter>   This takes you to the next line under "Time".

0.002   This is the x coordinate of the third (last) data point (time of forcing function ends at 0.002 s).

Click on the box under "Factor".

0   This is the y coordinate of the last data point (force is now back to 0).

You now see a graph of the forcing function matching that of Figures 16–30 and 16–31.

Click "OK" to go back to the "Global Data" window.

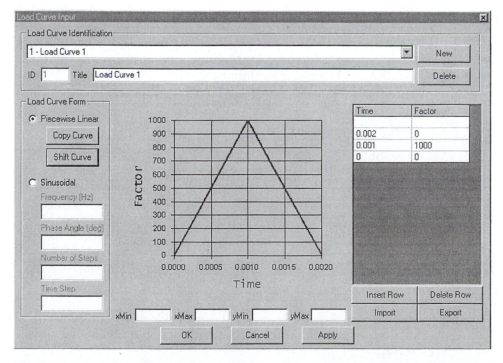

Figure 16–31 Forcing function created in "Load Curve Input" window for Example 16.11

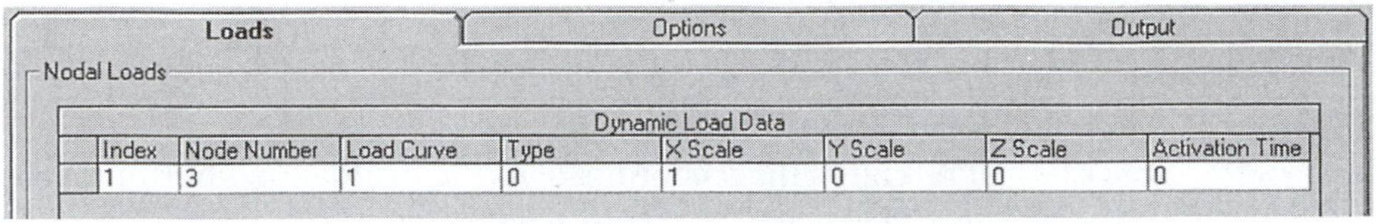

	Loads			Options			Output	

Nodal Loads

				Dynamic Load Data				
Index	Node Number	Load Curve	Type	X Scale	Y Scale	Z Scale	Activation Time	
1	3	1	0	1	0	0	0	

Figure 16–32 "Loads" tab screen used to identify the node number and direction of nodal force

Under the "Loads" tab, make the node number 3, as this is where the force is applied, and make the "X Scale" 1, as this is the scale factor to be used on the *x*-directed force. The default "0" for "Type" means the translational loading is applied (a "1" would mean a rotational loading is applied). The activation time is left as "0" to indicate that the load starts acting at 0 time. Figure 16–32 shows the entries in the table.

Click on the "Output" tab and check the box next to "Stress Data". Leave the default "Histories and Maxima" to create a table of time histories of displacement versus time.

Click "OK" to go back to the "Model Data Control" window.

Step 6 Check the Model

Click on the "Check" button on the right-side menu of the "Model Data Control" window. This verifies the geometry and finite element data of the model. Once verified, the Superview program opens, verifying the model. You see the model with a red triangle support at the left end to indicate complete fixity, and red circles at the middle and right end to indicate a constraint was freed.

Click "Done" to go back to the "Model Data Control" window.

Step 7 Run the Analysis

Click on the "Analysis" button on the right-side menu of the "Model Data Control" window.

Click "Analyze" in the lower left corner of the "Linear Stress and Vibration-Transient Stress…" window. The analysis will now be performed. When it is finished, the "Algor Analysis" window appears and indicates "Finished".

Click "OK".

Click "Done".

Step 8 Use Superview to View the Results

Click on the "Results" button on the right-side menu of the "Model Data Control" window. This takes you to the Superview program again, and you can review the results.

Select "Displaced" on the main menu.

Click "Displ on" so that "*Displ on" appears. The asterisk means the displaced mode is active.

Click "With undi" so that "*With undi" appears. This allows the displaced and undisplaced models to be superimposed.

Click "Calc scal" if you want the program to enlarge or scale the displaced model. The model may not show any difference now, but this step will be necessary when viewing the frequency modes.

<Esc>   This takes you back to the main menu.

Select "Load Case". The first position of the model is shown at time equal to zero; this is considered case 1 at the bottom of the screen.

Select "Next" to display the next position of the model as the input force increases with time. This can be done up to 40 times since this is the number of time steps.

Select "Previous" to display the previous case.

When in a particular case, you can escape back to the displacement menu under the main menu. Then under displacement, use "Node Inquiry" to obtain nodal displacement. For case 5 the nodal displacement was "DX = 4.88e−03" for this model. This can also be done for stress for the five cases. You can go back to "Displaced Menu" and "Nodes Inq:Find" and enter the node number (say, 3) for any load case to see the value of the displacement at that node.

<Esc>

<Esc>

Click "Done" to exit Superview.

Select "File" and then "Exit" to close the program.   ■

Table 16–7 lists the displacement/time history for nodes 2 and 3. The maximum displacement of node 3 is actually −0.007646 in. and occurs at time 0.00425 s.

A time step of 0.01 s is acceptable in Example 16.12 as it meets the recommended time step as suggested in Section 16.3. That is, $\Delta t < T_n/10$ to $T_n/20$ is recommended to provide accurate results for Wilson's direct integration scheme as used in the Algor program. From the frequency analysis (see the output in Table 16–8), the circular frequency $\omega_6 = 197.52$ or the natural frequency is $f_6 = \omega_6/(2\pi) = 31.44$ cycles/s or Hertz (Hz). Now we use $\Delta t = T_n/20 = 1/(20f_6) = 1/[20(31.43)] = 0.015$ s. Therefore, $\Delta t = 0.01$ s is acceptable. Using a time step greater than $T_n/10$ may result in loss of accuracy as some of the higher mode response contributions to the solution may be missed. Oftentimes a cut-off period or frequency is used to decide what largest natural frequency to use in the analysis. In many applications only a few lower modes contribute significantly to the response. The higher modes are then not necessary. The highest frequency used in the analysis is called the cut-off frequency. For machinery parts, the cut-off frequency is often taken as high as 250 Hz. In Example 16.12, we have selected a cut-off frequency of $f_6 = 31.44$ Hz in determining the time step of integration. This frequency is the highest flexural mode frequency computed for the four-element beam model.

Table 16–7   Displacement time history for nodes 2 and 3

	NODE NUMBER – (COMPONENT NUMBER)	
TIME	2-(2)	3-(2)
.00025	4.410E-06	6.156E-05
.00050	4.600E-05	4.668E-04
.00075	2.147E-04	1.425E-03
.00100	6.507E-04	2.967E-03
.00125	1.481E-03	4.873E-03
.00150	2.699E-03	6.439E-03
.00175	4.061E-03	7.143E-03
.00200	5.109E-03	6.860E-03
.00225	5.349E-03	5.793E-03
.00250	4.501E-03	4.385E-03
.00275	2.670E-03	2.862E-03
.00300	3.265E-04	1.141E-03
.00325	−1.907E-03	−9.441E-04
.00350	−3.538E-03	−3.354E-03
.00375	−4.376E-03	−5.694E-03
.00400	−4.530E-03	−7.319E-03
.00425	−4.232E-03	−7.646E-03
.00450	−3.645E-03	−6.463E-03
.00475	−2.772E-03	−4.057E-03
.00500	−1.514E-03	−1.083E-03
.00525	1.599E-04	1.740E-03
.00550	2.082E-03	3.921E-03
.00575	3.867E-03	5.313E-03
.00600	5.055E-03	6.021E-03
.00625	5.312E-03	6.185E-03
.00650	4.583E-03	5.814E-03
.00675	3.106E-03	4.776E-03
.00700	1.282E-03	2.947E-03
.00725	−5.031E-04	4.073E-04
.00750	−2.015E-03	−2.460E-03
.00775	−3.183E-03	−5.051E-03
.00800	−4.013E-03	−6.763E-03
.00825	−4.477E-03	−7.233E-03
.00850	−4.466E-03	−6.464E-03
.00875	−3.838E-03	−4.770E-03
.00900	−2.542E-03	−2.594E-03
.00925	−7.098E-04	−3.179E-04

MAXIMUM ABSOLUTE VALUES

MAXIMUM	5.349E-03	7.646E-03
TIME	2.250E-03	4.250E-03 ■

Damping will not be considered in any examples. However, Algor allows you to consider damping using Rayleigh damping in the direct integration method. For Rayleigh damping, the damping matrix is

$$[C] = \alpha[M] + \beta[K] \tag{16.9.1}$$

where the constants α and β are calculated from the system equations

$$\alpha + \beta\omega_i^2 = 2\omega_i\zeta_i \tag{16.9.2}$$

where ω_i are circular natural frequencies obtained through modal analysis, and ζ_i are damping ratios specified by the analyst. For instance, assume we assign damping ratios ζ_1 and ζ_2, from the above Eq. (16.9.2), we can show that α and β are

$$\alpha = \frac{2\omega_1\omega_2}{\omega_2^2 - \omega_1^2}(\omega_2\zeta_1 - \omega_1\zeta_2) \qquad \beta = \frac{2}{\omega_2^2 - \omega_1^2}(\omega_2\zeta_2 - \omega_1\zeta_1) \tag{16.9.3}$$

For $\beta = 0$, $[C] = \alpha[M]$ and the higher modes are only slightly damped, while for $\alpha = 0$, $[C] = \beta[K]$ and higher modes are heavily damped. To obtain α and β, we then necessarily run the modal analysis program first to obtain the frequencies. For instance, in Example 16.12 the first two different frequencies are $\omega_1 = 45.23$ rad/s and $\omega_3 = 120.16$ rad/s (ω_2 is the same as ω_3, so use ω_3). Now assume light damping ($\zeta \leqslant 0.05$). Therefore, let $\zeta_1 = \zeta_2 = 0.05$. Using these ω's and ζ's in Eqs. (16.9.3), we obtain $\alpha = 3.286$ and $\beta = 0.000605$. These values could be used in the Linear Transient Stress-Direct Integration Program in Algor for α and β if you want to include 5% damping ($\zeta = 0.05$).

Example 16.12

For the fixed-fixed beam subjected to the time-dependent forcing function shown in Figure 16–33, determine the displacement response for 0.2 s. Use a time step of integration of 0.01 s. Let $E = 6.58 \times 10^6$ psi, $I = 100$ in^4, mass density $= 0.1$ lb-s^2/in^4, and $L = 200$ in.

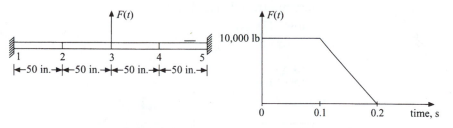

Figure 16–33 Fixed-fixed beam subjected to forcing function

Step 1   Start Superdraw III

Click on the "Algor FEA" icon on your desktop computer screen.

Step 2   Create the Beam Model

Click "Add".
Click "Line".

(0, 0, 0)   Enter the first data point.
<Enter>
(50, 0, 0)
<Enter>
(100, 0, 0)
<Enter>
(150, 0, 0)
<Enter>
(200, 0, 0)
<Enter>
Click "Done".
Click "View:Enclose" to see the model on the screen.
Click "Done".

Step 3   Add the Boundary Conditions

Click "FEA Add" and highlight "Stress and Vibration Analysis".

Click "Boundary Conditions". The default is "Use @ Symbol for Full".

Right-click on the left and right nodes. An "@" sign appears at each end of the beam.

Step 4   Enter the Material and Geometric Properties in the "Model Data Control" Window

Click on the "Model Data" button.

Select "Linear Transient Stress [Direct Integration]".

Click on the box below "Element".

Click on the box next to "Beam".

Click "OK".

Click on the box below "Data".

Ex1612   Enter the filename.

Click "Save".

Click "OK" to use the English units as default. This takes you to the main menu.

Click "Add/Mod".

Click "Sectional".

Click "Value" and enter values for the area and the moments of inertia.

1	The cross-sectional area is taken as 1.
100	Enter the moment of inertia in the "I2" location.
100	Enter the moment of inertia in the "I3" location. The default surface number 1 makes the bending occur about the local 3 axis.

<Esc>

<Esc>

Click "Quit".

Click "Yes" to save the current work and be sent back to the "Model Data Control" window.

Click "Material".

Click "[Customer Defined]".

Click "Edit Properties".

0.1   Enter the mass density.

6.58e6   Enter the modulus of elasticity.

Click "OK".

Click "OK".

Step 5   Enter the Global Data and Describe the Forcing Function

Click "Global" in the "Model Data Control" window. This takes you to "Linear Transient Stress Analysis:Using Direct Integration".

Enter the number of time steps (20) for the whole time you want the analysis to run, time step size (0.01) (the total time of the run will be $20 \times 0.01 = 0.2$ s), and output interval (1 for output at each time step) as follows:

20   Enter 20 next to "Number of time steps".

0.01   Enter 0.01 next to "Time step size".

1   Enter 1 next to "Output interval".

Click on the "Load Curves" button next. Use the default "Piecewise Linear".

Click under "Time" and enter 0. The start time is zero.

Click under "Factor" and enter 10000. The force acts on the beam at zero time and has a value of 10000.

<Enter>   This moves you to the next line. Enter 0.1 under "Time" and 10000 under "Factor". At time 0.1 s the force is 10000.

<Enter>   This moves you to the next line. Enter 0.2 under "Time" and enter 0 under "Factor". The $F(t)$ curve in Figure 16–33 should appear as a red line.

Click "OK". On the "Loads" menu and under "Node number" make the node 3. The force will be applied to the center node, which is node 3. And under "Y scale" type a 1. This is the scale factor for the load, and it is in the y direction.

Click on the "Output" tab and place a check next to "Stress data".

Click "OK". This takes you back to the "Model Data Control" window.

Step 6   Check the Model

Verify the model in the usual manner.

Step 7 Perform the Time History Direct Integration Analysis

Click "Analysis".
Click "Analyze".
Click "OK".
Click "Done". The analysis is complete.

Step 8 Review the Results

Click "Results". You see the beam appear. You can plot the deformed beam and the bending stress.
Click "Displaced" on the main menu.
Click "Disp on" so that "*Displ on" appears.
Click "With undi".
Click "Calc scale".
<Esc>   This takes you back to the main menu.
Click "Load Case". The first position of the model appears at time equal to zero. This is considered load case 1 at the bottom of the screen.
Click "Next" to display the next position of the model at time 0.01 s.
Click "Previous" if you want to go back to the previous case.
While in a particular case, you can escape back to the "Displacement" menu on the main menu. Then, under "Displacement" use "Node Inquiry" to obtain the nodal displacement. The nodal displacement at the load case corresponding to 0.08 s is 1.207 in.
Click "Quit" to exit Superview.
The maximum displacement appears in the *ex1612.L4* file. This maximum displacement is at node 3 at a time of 0.08 s with a value of 1.207 in. The minimum displacement is -0.2028 in. The static deflection for the beam with a concentrated load at midspan is 0.633 in. as obtained from the classical solution of $y = PL^3/192EI$. The time-dependent response oscillates about the static deflection.

   Table 16–8 lists the first six natural frequencies for the beam of Example 16.12 and the vertical displacement versus time for nodes 2 and 3 of the beam. The natural frequencies 1, 2, 3, 4, and 6 are flexural modes, while mode 5 is an axial mode. These modes are seen by looking at the modes in Superview from a frequency analysis of Example 16.12. The maximum displacement under the load (at node 3) compares with the solution in Reference [14].   ■

   Table 16–9 lists the element forces and moments for elements 1 and 2 up to time 0.08 s. This time corresponds to when the maximum displacement occurs and is also when the maximum moments occur. The largest element 1 bending moment is $M_3 = -509,800$ lb in. at the wall (node 1) at a time of 0.08 s [see the column "Case (Mode)," number 8]. The largest element 2 bending moment is $M_3 = -443,100$ lb in.

Table 16–8 Natural frequencies and displacement time history (nodes 2 and 3, Figure 16–33)

Frequencies =	6
mode number	circular frequency (rad/sec)
1	4.52276232074113D+01
2	4.52276232074113D+01
3	1.20159893475319D+02
4	1.20159893475319D+02
5	1.24168832797688D+02
6	1.97518763916263D+02

Y-DISPLACEMENT
NODE NUMBER – (COMPONENT NUMBER)

TIME	2-(2)	3-(2)
.01000	1.791E-02	4.050E-03
.02000	1.203E-01	3.458E-02
.03000	2.987E-01	1.197E-01
.04000	5.201E-01	2.542E-01
.05000	7.624E-01	3.978E-01
.06000	9.907E-01	5.143E-01
.07000	1.152E+00	5.916E-01
.08000	1.207E+00	6.246E-01
.09000	1.150E+00	6.024E-01
.10000	1.003E+00	5.217E-01
.11000	7.873E-01	3.989E-01
.12000	5.270E-01	2.601E-01
.13000	2.601E-01	1.241E-01
.14000	3.174E-02	4.247E-03
.15000	-1.267E-01	-8.361E-02
.16000	-2.028E-01	-1.244E-01
.17000	-1.962E-01	-1.153E-01

MAXIMUM ABSOLUTE VALUES

MAXIMUM	1.207E+00	6.246E-01
TIME	8.000E-02	8.000E-02 ■

Example 16.13

The plane frame shown in Figure 16–34(a) consists of six rigidly connected prismatic members. All members have the same values of E, ρ, A, and I. Dynamic forces $F(t)$ and $2F(t)$ are applied in the x direction at joints 6 and 4, and the time variation of $F(t)$ is shown in Figure 16–34(b). Assuming that the frame is steel, use the following numerical values for parameters: $E = 30 \times 10^6$ psi, mass density $\rho = 0.000735$ lb-s^2/in^4, $A = 30$ in^2, $I = 1000$ in^4, $L = 50$ in., and $F_1 = 10,000$ lb.

Table 16–9 Forces and moments versus time for elements 1 and 2 of Figure 16–33

1**** BEAM ELEMENT FORCES AND MOMENTS

ELEMENT NO.	CASE (MODE)	AXIAL FORCE R1	SHEAR FORCE R2	SHEAR FORCE R3	TORSION MOMENT M1	BENDING MOMENT M2	BENDING MOMENT M3
1	1	0.000E+00	1.685E+02	0.000E+00	0.000E+00	0.000E+00	6.764E+02
		0.000E+00	-1.685E+02	0.000E+00	0.000E+00	0.000E+00	7.748E+03
1	2	0.000E+00	6.662E+02	0.000E+00	0.000E+00	0.000E+00	-7.100E+03
		0.000E+00	-6.662E+02	0.000E+00	0.000E+00	0.000E+00	4.041E+04
1	3	0.000E+00	-4.880E+02	0.000E+00	0.000E+00	0.000E+00	-7.116E+04
		0.000E+00	4.880E+02	0.000E+00	0.000E+00	0.000E+00	4.676E+04
1	4	0.000E+00	-3.738E+03	0.000E+00	0.000E+00	0.000E+00	-1.961E+05
		0.000E+00	3.738E+03	0.000E+00	0.000E+00	0.000E+00	9.226E+03
1	5	0.000E+00	-7.069E+03	0.000E+00	0.000E+00	0.000E+00	-3.272E+05
		0.000E+00	7.069E+03	0.000E+00	0.000E+00	0.000E+00	-2.624E+04
1	6	0.000E+00	-9.022E+03	0.000E+00	0.000E+00	0.000E+00	-4.211E+05
		0.000E+00	9.022E+03	0.000E+00	0.000E+00	0.000E+00	-2.998E+04
1	7	0.000E+00	-1.008E+04	0.000E+00	0.000E+00	0.000E+00	-4.794E+05
		0.000E+00	1.008E+04	0.000E+00	0.000E+00	0.000E+00	-2.448E+04
1	8	0.000E+00	-1.086E+04	0.000E+00	0.000E+00	0.000E+00	-5.098E+05
		0.000E+00	1.086E+04	0.000E+00	0.000E+00	0.000E+00	-3.335E+04
2	1	0.000E+00	-4.514E+02	0.000E+00	0.000E+00	0.000E+00	-7.748E+03
		0.000E+00	4.514E+02	0.000E+00	0.000E+00	0.000E+00	-1.482E+04
2	2	0.000E+00	-2.566E+03	0.000E+00	0.000E+00	0.000E+00	-4.041E+04
		0.000E+00	2.566E+03	0.000E+00	0.000E+00	0.000E+00	-8.791E+04
2	3	0.000E+00	-4.229E+03	0.000E+00	0.000E+00	0.000E+00	-4.676E+04
		0.000E+00	4.229E+03	0.000E+00	0.000E+00	0.000E+00	-1.647E+05
2	4	0.000E+00	-4.476E+03	0.000E+00	0.000E+00	0.000E+00	-9.226E+03
		0.000E+00	4.476E+03	0.000E+00	0.000E+00	0.000E+00	-2.146E+05
2	5	0.000E+00	-4.970E+03	0.000E+00	0.000E+00	0.000E+00	2.624E+04
		0.000E+00	4.970E+03	0.000E+00	0.000E+00	0.000E+00	-2.747E+05
2	6	0.000E+00	-6.623E+03	0.000E+00	0.000E+00	0.000E+00	2.998E+04
		0.000E+00	6.623E+03	0.000E+00	0.000E+00	0.000E+00	-3.611E+05
2	7	0.000E+00	-8.118E+03	0.000E+00	0.000E+00	0.000E+00	2.448E+04
		0.000E+00	8.118E+03	0.000E+00	0.000E+00	0.000E+00	-4.304E+05
2	8	0.000E+00	-8.196E+03	0.000E+00	0.000E+00	0.000E+00	3.335E+04
		0.000E+00	8.196E+03	0.000E+00	0.000E+00	0.000E+00	-4.431E+05 ■

Determine the displacement in the horizontal direction at nodes 4, 5, and 6. Use a time step of integration of $\Delta t = 0.005$ s.

Step 1 Start Superdraw III

Click on the "Algor FEA" icon on your desktop computer screen.

Step 2 Create the Frame Model

Click "Add".
Click "Line". Begin typing in the coordinates of the frame.

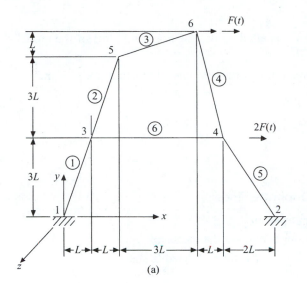

(a)

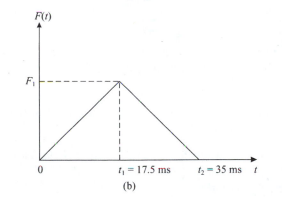

(b)

Figure 16–34   (a) Six-member plane frame; (b) dynamic load

(0, 0, 0)   This is the first coordinate of the frame.
<Enter>
(50, 150, 0)
<Enter>
(100, 300, 0)
<Enter>
(250, 350, 0)
<Enter>
(300, 150, 0)
<Enter>
(400, 0, 0)

```
<Enter>
```
Click "Done". You have created all but the horizontal line of the model.
Click "Add".
Click "Line". Now create the horizontal line connecting nodes 3 and 4.
```
(50, 150, 0)
<Enter>
(300, 150, 0)
<Enter>
```
Click "Done".

Step 3   Eliminate Duplicate Lines

Click "Modify:Clean:Duplicate".
Click "Perform Cleaning". You see the message:

"Kept 6 Deleted 0. done"

Click "Done".

Step 4   Add the Boundary Conditions

Click "FEA Add".
Click "Stress and Vibration Analysis:Boundary Conditions" and the on-screen menu appears. Use the default "@" symbol for full constraint.
Right-click on bottom nodes 1 and 2.

Step 5   Add the Material Properties in the "Model Data Control" Window

Click on the "Model Data" button. From the "Analysis Type" line scroll down to "Linear Transient Stress [Direct Integration]" and select it.
Click on the box below "Element".
Click "Beam".
Click "OK".
Click on the box below "Data". It responds with a message "Please enter a model name first".
Click "OK".

Ex1613   This is the name of the file.

Click "OK" to accept the default "English (in)" units system. This takes you to the "Beam Design Editor" window.
Click "Add/Mod".
Click "Sectional".
Click "Value".

30	Enter a value of 30 for the cross-sectional area.
1000	Move to the "I2" location and type 1000.
1000	Move to the "I3" location and type 1000.

```
<Enter>
<Esc>
<Esc>
```
Click "Quit". You see a message "Save Current Work?"

Click "Yes".

Click on the box below "Material". This takes you to the "Element Material Selection" window.

Click "[Customer Defined]".

Click "Edit Properties".

`0.000735`   Enter the mass density.

`30e6`   Enter the modulus of elasticity.

Click "OK".

Click "OK".

Step 6 Check the Model Node Numbers to Apply Correct Nodal Loading in the Global Screen

Because we do not know the way the nodes were numbered in Algor, return to the main menu and follow the steps below:

Click "Options:General:Node Num" and the node numbers appear on the model. These numbers are the ones in Figure 16–34. The forcing function will be applied to nodes 4 and 6 in the "Global" window.

Step 7 Enter the Global Data

Click "Global". A screen appears as "Linear Transient Stress Analysis Using Direct Integration".

Click on the "0" in the "Number of Time Steps" box and make it 10.

`10`   We will use 10 time steps in this run.

Click on the "0" in the "Time Step Size" box.

`0.005`   This is the time step.

Leave the "Output Interval" as "1", neglect damping, and leave alpha and beta as "0". See the discussion preceding Example 16.12 regarding damping.

Click "Load Curves" in the top right corner of the window.

This brings up the "Load Curve Input" window. This is where the forcing function in Figure 16–34(b) is created. Use the default "Piecewise Linear" for "Load Curve Form".

Click on the box under "Time".

`0`   Enter a 0 as the start time.

Click on the box under "Factor".

`0`   This is the starting force at time 0.

`<Enter>`   This takes you to the next line in the table.

`0.0175`   Enter the value of 0.0175 s for the next data point time value.

Click on the box under "Factor".

10000	This is the force at time 0.0175 s.
<Enter>	This takes you to the third line of the table.
0.035	Enter the third time data point. This is the time when the force is removed from the frame.

Click on the box under "Factor".

0	Enter a value of 0 for the force at time 0.035 s.
<Enter>	You see a load curve matching Figure 16–34(b).

Click "OK" to go back to the "Global Data" window. Under the "Loads" tab, make the node number 6 on the "Index 1" line and the "X Scale" 1 as this is the scale factor to be used on the load curve forcing function in the x direction. Also see Figure 16–32. Move down one line by clicking on the "Add Row" button and enter 4 for the node number on the "Index 2" line and "X Scale" 2 as the forcing function is now applied to node 4 as well with a scale factor of 2 as two times the load curve force is applied to node 4.

Click on the "Output" tab and place a check in the box next to "Stress Data".

Click "OK" to go back to the "Model Data Control" window.

Step 8   Run the Analysis

Click "Analysis".

Click "Analyze". The analysis will now be performed. When it is finished, the "Algor Analysis" window appears and indicates "Finished".

Click "OK".

Click "Done".

Step 9   Use Superview to View the Results

Click "Results". This takes you to Superview to view the results. Use the same steps as in Examples 16.11 and 16.12 to review the displaced frame. On the "Load Case" menu, click "Next" until "Case 7" appears. The largest displacement occurs for this time of 0.035 s. Figure 16–35 shows a plot of the displaced frame for case 7. The x displacement of node 6 for this time is 1.551e−01 in. as shown also in Table 16–10.

Table 16–10 lists the horizontal displacement versus time for nodes 4, 5, and 6 of the frame. The maximum displacement at node 6 compares closely with the solution in Reference [17]. ■

Example 16.14

The circular fin shown in Figure 16–36 is made of pure copper with a thermal conductivity of $K = 400$ W/(m · °C), $h = 150$ W/(m² · °C), mass density $\rho = 8900$ kg/m³, and specific heat $c = 375$ J/(kg · °C) (1 J = 1 W-s). The initial temperature of the fin is 25 °C. The fin length is 2 cm, and the diameter is 0.4 cm. The right tip of the fin is

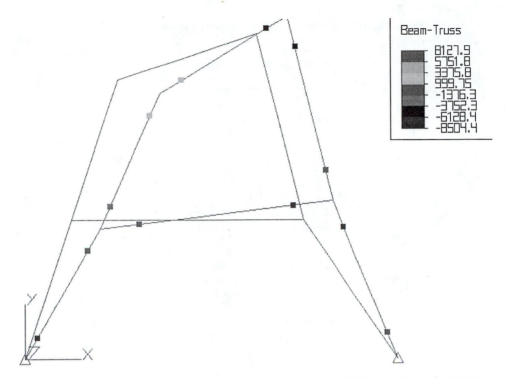

Figure 16–35 Displaced frame with worst stress at time 0.035 s for Example 16.13

insulated. The base of the fin is then suddenly increased to a temperature of 85 °C and maintained at this temperature. Use a time step of 0.1 s. Use two elements of equal length. Determine the temperature distribution up to 3 s.

Step 1 Start Superdraw III

Select the "Start" button of Windows NT/95/98 and then proceed to "Programs: Algor Software:Algor FEA".

Then click "Algor FEA" to start the program.

Or double-click on the "Algor FEA" icon on your desktop computer screen. The Superdraw III program appears. This is the Algor main graphics user interface. All your work will be done from this program. You will construct, analyze, and review your results from Superdraw III. Each menu option and a short example are available in the Superdraw III Reference Division (accessed through "Docutech", the on-line "Technical User Documentation").

Step 2 Create the Model

Click "Modify:Transform XY to YZ".

Select the "Add" menu from the top main menu bar.

Table 16–10 Displacement time history nodes 4, 5, and 6 Algor output

	NODE NUMBER – (COMPONENT NUMBER)		
TIME	4–(1)	5–(1)	6–(1)
.00500	1.416E–03	1.367E–03	1.265E–03
.01000	1.036E–02	1.097E–02	9.411E–03
.01500	3.074E–02	3.644E–02	2.868E–02
.02000	6.400E–02	8.040E–02	6.121E–02
.02500	1.016E–01	1.345E–01	1.005E–01
.03000	1.320E–01	1.826E–01	1.360E–01
.03500	1.434E–01	2.078E–01	1.551E–01
.04000	1.287E–01	1.972E–01	1.478E–01
.04500	9.148E–02	1.497E–01	1.135E–01
.05000	3.847E–02	7.298E–02	5.625E–02
.05500	–1.987E–02	–1.932E–02	–1.351E–02
.06000	–7.312E–02	–1.093E–01	–8.191E–02
.06500	–1.128E–01	–1.789E–01	–1.349E–01
.07000	–1.325E–01	–2.137E–01	–1.616E–01
.07500	–1.293E–01	–2.068E–01	–1.563E–01
.08000	–1.035E–01	–1.602E–01	–1.205E–01
.08500	–5.926E–02	–8.388E–02	–6.235E–02

MAXIMUM ABSOLUTE VALUES

MAXIMUM	1.434E–01	2.137E–01	1.616E–01
TIME	3.500E–02	7.000E–02	7.000E–02 ■

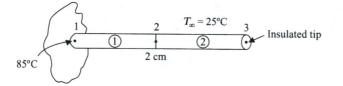

Figure 16–36 Rod subjected to time-dependent temperature

Click "Rectangle". The model will be made of two rectangles with cross-sectional areas defined later (by defining the thickness) to be equal to that of the rod with a diameter of 0.4 cm.

Note: Enter the data points for the first rectangle.

0<Tab>0<Tab>0

<Enter>

0<Tab>0.01<Tab>0.001

<Enter>

Select the "Add" menu from the top main menu bar.

Click "Rectangle".

Note: Enter the data points for the second rectangle.

0<Tab>0.01<Tab>0

<Enter>

0<Tab>0.02<Tab>0.001

<Enter>   All the data points have been entered.

Select "View". This takes you to the view options.

Click "Enclose". This encloses the model. You now see the model on the screen.

Select "Select". This takes you to the select options.

Click "None". This ensures that none of the previously drawn elements are selected.

Select "Select" again. This takes you back to the select options.

Click "Point".

Click "Toggle Mode" on the "Point Select" menu that appears. Right-click on each of the two top lines and each of the bottom two lines.

Click "Done".

Select "Modify".

Click "Update Object Parameters".

Click "Surface Number".

2   Enter a 2 to make the top and bottom boundary surfaces 2.

<Enter>

Step 3   Eliminate Duplicate Lines

Click "Modify" on the main menu.

Select "Clean:Duplicate". This notation means to select "Clean" and then click "Duplicate" in succession. This command brings up the on-screen "Duplicate" menu.

Click "Perform Cleaning" on the "Duplicate" menu. You see the following message on the lower part of the screen:

"7 Kept 1 Deleted. Done."

Click "Done" on the "Duplicate" menu located in the upper right corner of the screen.

Step 4   Apply the Temperature Boundary Conditions

Click "FEA Add".

Select "Heat Transfer Analysis".

Click "Applied Temperature".

Click "Temperature Value".

0   The left boundary has a temperature of $0\,°C$.

Click "Stiffness".

1e6   Make the stiffness 1e6 for good results.

Select "Box Apply" and create a box around the left vertical edge of the model. You see "TB;0;L1;1;1;1e+006."

Click "Done".

Step 5 Go to the "Model Data Control" Window to Enter the Material Properties

Click on the "Model Data" button at the bottom of the screen.

Select "Transient Heat Transfer" by scrolling down the "Analysis Type" menu.

Select the box below "Element".

Click "2D".

Click "OK".

Click on the box below "Data". A prompt appears: "Please enter model name first".

Click "OK".

Ex1614

Click "Save".

Now put in the units you want.

Select "Metric mks (SI)" from the top scroll menu.

Click "OK". This takes you to the "Element Definitions" window.

0.12566e-1   Enter 0.12566e−1 for the element thickness. This thickness makes the cross-sectional area equivalent to a circular one with a diameter of 0.4 cm.

Click "OK".

Click on the box below "Material".

Select "[Customer Defined]".

Click "Edit Properties".

8900   Enter the density of the material ρ in kg/m^3.

400   Enter the thermal conductivity of the material K in W/(m · °C).

375   Enter the specific heat of the material c_p in J/(kg · °C).

Click "OK".

Click "OK".

Click on the box below "Surface".

Click on the box next to "Surface 2".

150   Enter the convection coefficient h for surface 2 in W/(m^2 · °C).

25   Enter the ambient temperature T in °C.

Click "Apply".

Click "OK".

Click "Close" in the "Surface Definition" window.

Step 6 Enter the Global Data

Click on the "Global" box on the right side of the "Model Data Control" window.

Under "Multipliers" type a 1 next to "Boundary temperature multiplier" and next to "Convection Multiplier".

1

1

In the "Event" section enter a 30 for the number of time steps and a 0.1 for the time step size.

30

1

<Enter>

Click on the "Loads" tab.

In the "Index 1" row, set "Node #" equal to 1 and "Scale" equal to 1e6 to match the stiffness entered previously.

Click on the "Add Row" button. You are now in the "Index 2" row. Set "Node #" equal to 4 and "Scale" equal to 1e6 again.

<Enter>

Click on the "Options" tab.

In the space to the right of "Default Nodal Temperature" enter a 25.

25

Now select the "Load Curves" button in the "Load Curves Setup" section.

On the far right enter a 0 under "Time".

0

Then in the same row enter an 85 under "Factor".

85

Click "Apply".

Now in the second row, enter a 3 under "Time" and an 85 under "Factor".

3

85

Click "Apply".

Click "OK".

Click on the "Output" tab and click on the "Nodal temperature data" box.

Click "OK".

Step 7 Analyze the Model

Click "Analysis".

Click "Analyze".

Click "OK". You then see a message that you are finished.

Click "Done".

Step 8 Review the Results

Click "Results".

Click "Show T-ht".

Click "Temp-ht".

<Esc> You should be back in the main menu.

Click "Timestep".

Click "Next" to advance through each time step of the solution. The solution should match that of Figure 16–24 on page 821. ■

▲ **References**

[1] Thompson, W. T., *Theory of Vibrations with Applications*, 2nd ed., Prentice-Hall, Englewood Cliffs, NJ, 1981.

[2] Archer, J. S., "Consistent Matrix Formulations for Structural Analysis Using Finite Element Techniques," *Journal of the American Institute of Aeronautics and Astronautics*, Vol. 3, No. 10, pp. 1910–1918, 1965.

[3] James, M. L., Smith, G. M., and Wolford, J. C., *Applied Numerical Methods for Digital Computation*, 3rd ed., Harper & Row, New York, 1985.

[4] Biggs, J. M., *Introduction to Structural Dynamics*, McGraw-Hill, New York, 1964.

[5] Newmark, N. M., "A Method of Computation for Structural Dynamics," *Journal of the Engineering Mechanics Division*, American Society of Civil Engineers, Vol. 85, No. EM3, pp. 67–94, 1959.

[6] Clark, S. K., *Dynamics of Continuous Elements*, Prentice-Hall, Englewood Cliffs, NJ, 1972.

[7] Bathe, K. J., *Finite Element Procedures in Engineering Analysis*, Prentice-Hall, Englewood Cliffs, NJ, 1982.

[8] Bathe, K. J., and Wilson, E. L., *Numerical Methods in Finite Element Analysis*, Prentice-Hall, Englewood Cliffs, NJ, 1976.

[9] Fujii, H., "Finite Element Schemes: Stability and Convergence," *Advances in Computational Methods in Structural Mechanics and Design*, J. T. Oden, R. W. Clough, and Y. Yamamoto, Eds., University of Alabama Press, AL, pp. 201–218, 1972.

[10] Krieg, R. D., and Key, S. W., "Transient Shell Response by Numerical Time Integration," *International Journal of Numerical Methods in Engineering*, Vol. 17, pp. 273–286, 1973.

[11] Belytschko, T., "Transient Analysis," *Structural Mechanics Computer Programs, Surveys, Assessments, and Availability*, W. Pilkey, K. Saczalski, and H. Schaeffer, Eds., University of Virginia Press, Charlottesville, VA, pp. 255–276, 1974.

[12] Hughes, T. J. R., "Unconditionally Stable Algorithms for Nonlinear Heat Conduction," *Computational Methods in Applied Mechanical Engineering*, Vol. 10, No. 2, pp. 135–139, 1977.

[13] Hilber, H. M., Hughes, T. J. R., and Taylor, R. L., "Improved Numerical Dissipation for Time Integration Algorithms in Structural Dynamics," *Earthquake Engineering in Structural Dynamics*, Vol. 5, No. 3, pp. 283–292, 1977.

[14] Paz, M., *Structural Dynamics Theory and Computation*, 3rd ed., Van Nostrand Reinhold, New York, 1991.

[15] *Heat Transfer Reference Division*, Docutech On-line Documentation, Algor, Inc., Pittsburgh, PA, 1999.

[16] *Linear Stress and Dynamics Reference Division*, Docutech On-line Documentation, Algor, Inc., Pittsburgh, PA, 1999.

[17] Weaver, W., Jr., and Johnston, P. R., *Structural Dynamics by Finite Elements*, Prentice-Hall, Englewood Cliffs, NJ, 1987.

▲ Problems

16.1 Determine the consistent-mass matrix for the one-dimensional bar discretized into two elements as shown in Figure P16–1. Let the bar have modulus of elasticity E, mass density ρ, and cross-sectional area A.

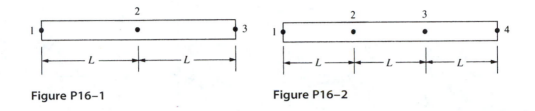

Figure P16–1   **Figure P16–2**

16.2 For the one-dimensional bar discretized into three elements as shown in Figure P16–2, determine the lumped- and consistent-mass matrices. Let the bar properties be E, ρ, and A throughout the bar.

16.3 For the one-dimensional bar shown in Figure P16–3, determine the natural frequencies of vibration, ω's, using two elements of equal length. Use the consistent-mass approach. Let the bar have modulus of elasticity E, mass density ρ, and cross-sectional area A. Compare your answers to those obtained using a lumped-mass matrix in Example 16.3.

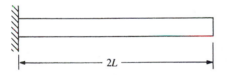

Figure P16–3

16.4 For the one-dimensional bar shown in Figure P16–4, determine the natural frequencies of longitudinal vibration using first two and then three elements of equal length. Let the bar have $E = 30 \times 10^6$ psi, $\rho = 0.00073$ lb-s²/in⁴, $A = 1$ in², and $L = 60$ in.

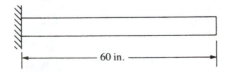

Figure P16–4

16.5 For the spring-mass system shown in Figure P16–5, determine the mass displacement, velocity, and acceleration for five time steps using the central difference method. Let $k = 2000$ lb/ft and $m = 2$ slugs. Use a time step of $\Delta t = 0.03$ s. You might want to write a computer program to solve this problem.

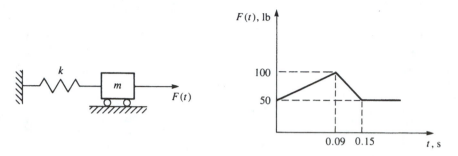

Figure P16–5

16.6 For the spring-mass system shown in Figure P16–6, determine the mass displacement, velocity, and acceleration for five time steps using (a) the central difference method, (b) Newmark's time integration method, and (c) Wilson's method. Let $k = 1200$ lb/ft and $m = 2$ slugs.

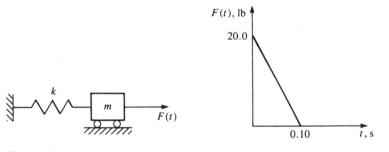

Figure P16–6

16.7 For the bar shown in Figure P16–7, determine the nodal displacements, velocities, and accelerations for five time steps using two finite elements. Let $E = 30 \times 10^6$ psi, $\rho = 0.00073$ lb-s^2/in^4, $A = 1$ in^2, and $L = 100$ in.

16.8 For the bar shown in Figure P16–8, determine the nodal displacements, velocities, and accelerations for five time steps using two finite elements. For simplicity of calculations, let $E = 1 \times 10^6$ psi, $\rho = 1$ lb-s^2/in^4, $A = 1$ in^2, and $L = 100$ in. Use Newmark's method and Wilson's method.

16.9– Rework Problems 16.7 and 16.8 using the Algor program.
16.10

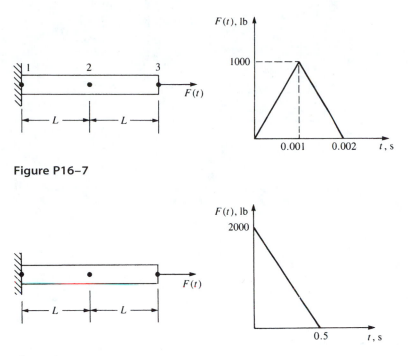

Figure P16-7

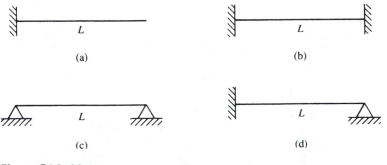

Figure P16-8

16.11 For the beams shown in Figure P16–11, determine the natural frequencies using first two and then three elements. Let E, ρ, and A be constant for the beams.

(a)

(b)

(c)

(d)

Figure P16-11

16.12 Rework Problem 16.11 using the Algor program with $E = 3 \times 10^7$ psi, $\rho = 0.00073$ lb-s^2/in^4, $A = 1$ in^2, and $I = 0.0833$ in^4.

16.13, 16.14 For the beams in Figures P16–13 and P16–14 subjected to the forcing functions shown, determine the maximum deflections, velocities, and accelerations. Use the Algor program.

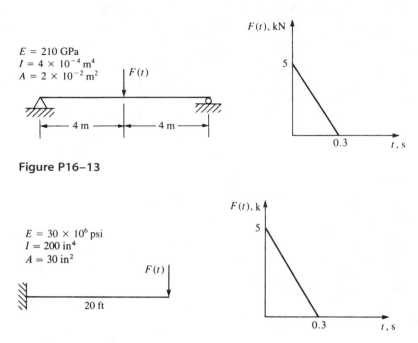

$E = 210$ GPa
$I = 4 \times 10^{-4}$ m⁴
$A = 2 \times 10^{-2}$ m²

Figure P16–13

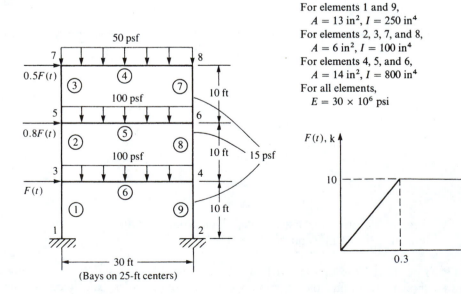

$E = 30 \times 10^6$ psi
$I = 200$ in⁴
$A = 30$ in²

Figure P16–14

16.15, For the rigid frames in Figures P16–15 and P16–16 subjected to the forcing functions
16.16 shown, determine the maximum displacements, velocities, and accelerations. Use the
Algor program.

For elements 1 and 9,
  $A = 13$ in², $I = 250$ in⁴
For elements 2, 3, 7, and 8,
  $A = 6$ in², $I = 100$ in⁴
For elements 4, 5, and 6,
  $A = 14$ in², $I = 800$ in⁴
For all elements,
  $E = 30 \times 10^6$ psi

Figure P16–15

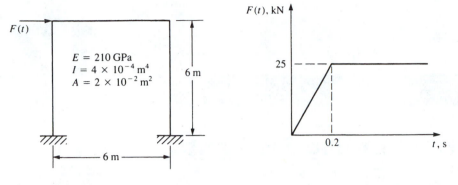

Figure P16–16

16.17 A marble slab with $k = 2$ W/(m · °C), $\rho = 2500$ kg/m^3, and $c = 800$ W · s/(kg · °C) is 2 cm thick and at an initial uniform temperature of $T_i = 200$ °C. The left surface is suddenly lowered to 0 °C and is maintained at that temperature while the other surface is kept insulated. Determine the temperature distribution in the slab for 40 s. Use $\beta = \frac{2}{3}$ and a time step of 8 s.

16.18 A circular fin is made of pure copper with a thermal conductivity of $k = 400$ W/ (m · °C), $h = 150$ W/(m^2 · °C), mass density $\rho = 8900$ kg/m^3, and specific heat $c = 375$ J/(kg · °C). The initial temperature of the fin is 25 °C. The fin length is 2 cm and the diameter is 0.4 cm. The right tip of the fin is insulated. See Figure P16–18. The base of the fin is then suddenly increased to a temperature of 85 °C and maintained at this temperature. Use the lumped form of the capacitance matrix, a time step of 0.1 s, and $\beta = \frac{2}{3}$. Use two elements of equal length. Determine the temperature distribution up to 3 s. Compare your results with Example 16.7, which used the consistent form of the capacitance matrix.

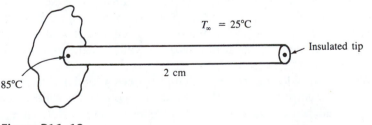

Figure P16–18

16.19– Rework Problems 16.17 and 16.18 using the Algor program.
16.20

17

Plate Bending Element

Introduction

In this chapter, we will begin by describing elementary concepts of plate bending behavior and theory. The plate element is one of the more important structural elements and is used to model and analyze such structures as pressure vessels, chimney stacks (Figure 1–5), and automobile parts. This description of plate bending is followed by a discussion of some commonly used plate finite elements. A large number of plate bending element formulations exist that would require a lengthy chapter to cover. Our purpose in this chapter is to present the derivation of the stiffness matrix for one of the most common plate bending finite elements and then to compare solutions to some classical problems from a variety of bending elements in the literature.

We finish the chapter with a solution to a plate bending problem using the Algor program. The step-by-step solutions provide the information we need to model and solve plate bending problems using a finite element program.

▲ 17.1 Basic Concepts of Plate Bending ▲

A plate can be considered the two-dimensional extension of a beam in simple bending. Both beams and plates support loads transverse or perpendicular to their plane and through bending action. A plate is flat (if it were curved, it would become a shell). A beam has a single bending moment resistance, while a plate resists bending about two axes and has a twisting moment.

We will consider the classical thin-plate theory or Kirchhoff plate theory [1]. Many of the assumptions of this theory are analogous to the classical beam theory or Euler–Bernoulli beam theory described in Chapter 5 and in Reference [2].

Basic Behavior of Geometry and Deformation

We begin the derivation of the basic thin-plate equations by considering the thin plate in the x-y plane and of thickness t measured in the z direction shown in Figure 17–1. The plate surfaces are at $z = \pm t/2$, and its midsurface is at $z = 0$. The assumed basic geometry of the plate is as follows: (1) The plate thickness is much smaller than its in-plane dimensions b and c (that is, $t \ll b$ or c). (If t is more than about one-tenth the

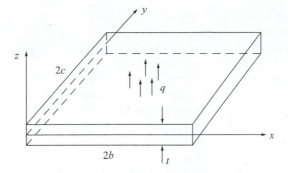

Figure 17–1 Basic thin plate showing transverse loading and dimensions

span of the plate, then transverse shear deformation must be accounted for and the plate is then said to be thick.) (2) The deflection w is much less than the thickness t (that is, $w/t \ll 1$).

Kirchhoff Assumptions

Consider a differential slice cut from the plate by planes perpendicular to the x axis as shown in Figure 17–2(a). Loading q causes the plate to deform laterally or upward in the z direction, and the deflection w of point P is assumed to be a function of x and y only; that is, $w = w(x, y)$ and the plate does not stretch in the z direction. A line a-b drawn perpendicular to the plate surfaces before loading remains perpendicular to the surfaces after loading [Figure 17–2(b)]. This is consistent with the Kirchhoff assumptions as follows:

1. Normals remain normal. This implies that transverse shear strains $\gamma_{yz} = 0$ and similarly $\gamma_{xz} = 0$. However, γ_{xy} does not equal 0; right angles in the plane of the plate may not remain right angles after loading. The plate may twist in the plane.

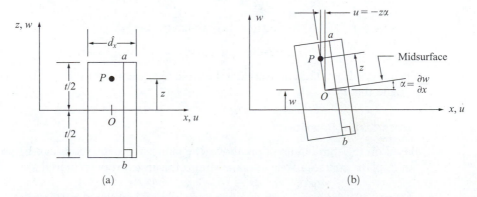

(a)   (b)

Figure 17–2 Differential slice of plate of thickness t (a) before loading and (b) displacements of point P after loading, based on Kirchhoff theory. Transverse shear deformation is neglected, and so right angles in the cross section remain right angles. Displacements in the y-z plane are similar

2. Thickness changes can be neglected and normals undergo no extension. This means normal strain, $\varepsilon_z = 0$.
3. Normal stress σ_z has no effect on in-plane strains ε_x and ε_y in the stress-strain equations and is considered negligible.
4. Membrane or in-plane forces are neglected here, and the plane stress resistance can be superimposed later (that is, the constant-strain triangle behavior of Chapter 7 can be superimposed with the basic plate bending element resistance). That is, the in-plane deformations in the x and y directions at the midsurface are assumed to be zero; $u(x, y, 0) = 0$ and $v(x, y, 0) = 0$.

Based on the Kirchhoff assumptions, any point P in Figure 17–2 has displacement in the x direction due to a small rotation α of

$$u = -z\alpha = -z\left(\frac{\partial w}{\partial x}\right) \tag{17.1.1}$$

and similarly the same point has displacement in the y direction of

$$v = -z\left(\frac{\partial w}{\partial y}\right) \tag{17.1.2}$$

The curvatures of the plate are then given as the rate of change of the angular displacements of the normals and are defined as

$$\kappa_x = -\frac{\partial^2 w}{\partial x^2} \qquad \kappa_y = -\frac{\partial^2 w}{\partial y^2} \qquad \kappa_{xy} = -\frac{2\partial^2 w}{\partial x \partial y} \tag{17.1.3}$$

The first of Eqs. (17.1.3) is used in beam theory [Eq. (5.1.1e)].

Using the definitions for the in-plane strains from Eq. (7.1.4), along with Eq. (17.1.3), the in-plane strain/displacement equations become

$$\varepsilon_x = -z\frac{\partial^2 w}{\partial x^2} \qquad \varepsilon_y = -z\frac{\partial^2 w}{\partial y^2} \qquad \gamma_{xy} = -2z\frac{\partial^2 w}{\partial x \partial y} \tag{17.1.4a}$$

or using Eq. (17.1.3) in Eq.(17.1.4a), we have

$$\varepsilon_x = -z\kappa_x \qquad \varepsilon_y = -z\kappa_y \qquad \gamma_{xy} = -z\kappa_{xy} \tag{17.1.4b}$$

The first of Eqs. (17.1.4a) is used in beam theory [see Eq. (5.1.10)]. The others are new to plate theory.

Stress/Strain Relations

Based on the third assumption above, the plane stress equations can be used to relate the in-plane stresses to the in-plane strains for an isotropic material as

$$\sigma_x = \frac{E}{1 - v^2}(\varepsilon_x + v\varepsilon_y)$$

$$\sigma_y = \frac{E}{1 - v^2}(\varepsilon_y + v\varepsilon_x) \tag{17.1.5}$$

$$\tau_{xy} = G\gamma_{xy}$$

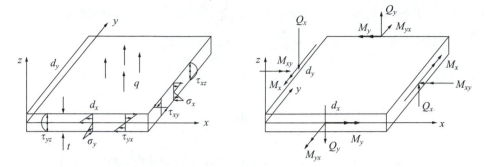

Figure 17–3 Differential element of a plate with (a) stresses shown on the edges of the plate and (b) differential moments and forces

The in-plane normal stresses and shear stress are shown acting on the edges of the plate in Figure 17–3(a). Similar to the stress variation in a beam, these stresses vary linearly in the z direction from the midsurface of the plate. The transverse shear stresses τ_{yz} and τ_{xz} are also present, even though transverse shear deformation is neglected. As in beam theory, these transverse stresses vary quadratically through the plate thickness. The stresses of Eq. (17.1.5) can be related to the bending moments M_x and M_y and to the twisting moment M_{xy} acting along the edges of the plate as shown in Figure 17–3(b).

The moments are actually functions of x and y and are computed per unit length in the plane of the plate. Therefore, the moments are

$$M_x = \int_{-t/2}^{t/2} z\sigma_x \, dz \qquad M_y = \int_{-t/2}^{t/2} z\sigma_y \, dz \qquad M_{xy} = \int_{-t/2}^{t/2} z\tau_{xy} \, dz \qquad (17.1.6)$$

The moments can be related to the curvatures by substituting Eqs. (17.1.4b) into Eqs. (17.1.5) and then using those stresses in Eq. (17.1.6) to obtain

$$M_x = D(\kappa_x + v\kappa_y) \qquad M_y = D(\kappa_y + v\kappa_x) \qquad M_{xy} = \frac{D(1-v)}{2}\kappa_{xy} \qquad (17.1.7)$$

where $D = Et^3/[12(1-v^2)]$ is called the bending rigidity of the plate.

The maximum magnitudes of the normal stresses on each edge of the plate are located at the top or bottom at $z = t/2$. For instance, it can be shown that

$$\sigma_x = \frac{6M_x}{t^2} \qquad (17.1.8)$$

This formula is similar to the flexure formula $\sigma_x = M_x c/I$ when applied to a unit width of plate and when $c = t/2$.

The governing equilibrium differential equation of plate bending is important in selecting the element displacement fields. The basis for this relationship is the equilibrium differential equations derived by the equilibrium of forces with respect to the z direction and by the equilibrium of moments about the x and y axes, respectively.

These equilibrium equations result in the following differential equations:

$$\frac{\partial Q_x}{\partial x} + \frac{\partial Q_y}{\partial y} + q = 0$$

$$\frac{\partial M_x}{\partial x} + \frac{\partial M_{xy}}{\partial y} - Q_x = 0 \qquad (17.1.9)$$

$$\frac{\partial M_y}{\partial y} + \frac{\partial M_{xy}}{\partial x} - Q_y = 0$$

where q is the transverse distributed loading and Q_x and Q_y are the transverse shear line loads shown in Figure 17–3(b).

Now substituting the moment/curvature relations from Eq. (17.1.7) into the second and third of Eqs. (17.1.9), then solving those equations for Q_x and Q_y, and finally substituting the resulting expressions into the first of Eqs. (17.1.9), we obtain the governing partial differential equation for an isotropic, thin-plate bending behavior as

$$D\left(\frac{\partial^4 w}{\partial x^4} + \frac{2\partial^4 w}{\partial x^2 \partial y^2} + \frac{\partial^4 w}{\partial y^4}\right) = q \qquad (17.1.10)$$

From Eq. (17.1.10), we observe that the solution of thin-plate bending using a displacement point of view depends on selection of the single-displacement component w, the transverse displacement.

If we neglect the differentiation with respect to the y coordinate, Eq. (17.1.10) simplifies to Eq. (5.1.1g) for a beam (where the flexural rigidity D of the plate reduces to EI of the beam when the Poisson effect is set to zero and the plate width becomes unity).

Potential Energy of a Plate

The total potential energy of a plate is given by

$$U = \frac{1}{2}\int (\sigma_x \varepsilon_x + \sigma_y \varepsilon_y + \tau_{xy}\gamma_{xy})\, dV \qquad (17.1.11)$$

The potential energy can be expressed in terms of the moments and curvatures by substituting Eqs. (17.1.4b) and (17.1.6) in Eq. (17.1.11) as

$$U = \frac{1}{2}\int (M_x \kappa_x + M_y \kappa_y + M_{xy}\kappa_{xy})\, dA \qquad (17.1.12)$$

▲ **17.2 Derivation of a Plate Bending Element Stiffness Matrix and Equations** ▲

Numerous finite elements for plate bending have been developed over the years, and Reference [3] cites 88 different elements. In this section we will introduce only one element formulation, the basic 12-degrees-of-freedom rectangular element shown in

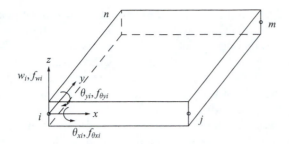

Figure 17–4 Basic rectangular plate element with nodal degrees of freedom

Figure 17–4. For more details of this formulation and of various other formulations including triangular elements, see References [4–18].

The formulation will be developed consistently with the stiffness matrix and equations for the bar, beam, plane stress/strain, axisymmetric, and solid elements of previous chapters.

Step 1 Select Element Type

We will consider the 12-degrees-of-freedom flat-plate bending element shown in Figure 17–4. Each node has 3 degrees of freedom—a transverse displacement w in the z direction, a rotation θ_x about the x axis, and a rotation θ_y about the y axis.

The nodal displacement matrix at node i is given by

$$\{d_i\} = \begin{Bmatrix} w_i \\ \theta_{xi} \\ \theta_{yi} \end{Bmatrix} \tag{17.2.1}$$

where the rotations are related to the transverse displacement by

$$\theta_x = +\frac{\partial w}{\partial y} \qquad \theta_y = -\frac{\partial w}{\partial x} \tag{17.2.2}$$

The negative sign on θ_y is due to the fact that a negative displacement w is required to produce a positive rotation about the y axis.

The total element displacement matrix is now given by

$$\{d\} = \{\underline{d_i} \quad \underline{d_j} \quad \underline{d_m} \quad \underline{d_n}\}^T \tag{17.2.3}$$

Step 2 Select the Displacement Function

Because there are 12 total degrees of freedom for the element, we select a 12-term polynomial in x and y as follows:

$$w = a_1 + a_2 x + a_3 y + a_4 x^2 + a_5 xy + a_6 y^2 + a_7 x^3 + a_8 x^2 y$$
$$+ a_9 xy^2 + a_{10} y^3 + a_{11} x^3 y + a_{12} xy^3 \tag{17.2.4}$$

Equation (17.2.4) is an incomplete quartic in the context of the Pascal triangle (Figure 9–2). The function is complete up to the third order (ten terms), and a choice of two more terms from the remaining five terms of a complete quartic must be made. The best choice is the x^3y and xy^3 terms as they ensure that we will have continuity in displacement among the interelement boundaries. (The x^4 and y^4 terms would yield discontinuities of displacement along interelement boundaries and so must be rejected. The x^2y^2 term is alone and cannot be paired with any other terms and so is also rejected.) The function [Eq. (17.2.4)] also satisfies the basic differential equation [Eq. (17.1.10)] over the unloaded part of the plate, although not a requirement in a minimum potential energy approximation.

Furthermore, the function allows for rigid-body motion and constant strain, as terms are present to account for these phenomena in a structure. However, interelement slope discontinuities along common boundaries of elements are not ensured.

To observe this discontinuity in slope, we evaluate the polynomial and its slopes along a side or edge (say, along side i-j, the x axis of Figure 17–4). We then obtain

$$w = a_1 + a_2 x + a_4 x^2 + a_7 x^3$$

$$\frac{\partial w}{\partial x} = a_2 + 2a_4 x + 3a_7 x^2 \qquad (17.2.5)$$

$$\frac{\partial w}{\partial y} = a_3 + a_5 x + a_8 x^2 + a_{12} x^3$$

The displacement w is a cubic as used for the beam element, while the slope $\partial w/\partial x$ is the same as in beam bending. Based on the beam element, we recall that the four constants $a_1, a_2, a_4,$ and a_7 can be defined by invoking the endpoint conditions of $(w_i, w_j, \theta_{yi}, \theta_{yj})$. Therefore, w and $\partial w/\partial x$ are completely defined along this edge. The normal slope $\partial w/\partial y$ is a cubic in x. However, only two degrees of freedom remain for definition of this slope, while four constants $(a_3, a_5, a_8,$ and $a_{12})$ exist. This slope is then not uniquely defined, and a slope discontinuity occurs. Thus, the function for w is said to be nonconforming. The solution obtained from the finite element analysis using this element will not be a minimum potential energy solution. However, this element has proven to give acceptable results, and proofs of its convergence have been shown [8].

The constants a_1 through a_{12} can be determined by expressing the 12 simultaneous equations linking the values of w and its slopes at the nodes when the coordinates take up their appropriate values. First, we write

$$\left\{ \begin{array}{c} w \\ +\dfrac{\partial w}{\partial y} \\ -\dfrac{\partial w}{\partial x} \end{array} \right\} = \begin{bmatrix} 1 & x & y & x^2 & xy & y^2 & x^3 & x^2y & xy^2 & y^3 & x^3y & xy^3 \\ 0 & 0 & +1 & 0 & +x & +2y & 0 & +x^2 & +2xy & +3y^2 & +x^3 & +3xy^2 \\ 0 & -1 & 0 & -2x & -y & 0 & -3x^2 & -2xy & -y^2 & 0 & -3x^2y & -y^3 \end{bmatrix} \times \left\{ \begin{array}{c} a_1 \\ a_2 \\ a_3 \\ \vdots \\ a_{12} \end{array} \right\} \qquad (17.2.6)$$

or in simple matrix form the degrees of freedom matrix is

$$\{\psi\} = [P]\{a\} \tag{17.2.7}$$

where $[P]$ is the 3×12 first matrix on the right side of Eq. (17.2.6).

Next, we evaluate Eq. (17.2.6) at each node point as follows

$$\{d\} = \begin{Bmatrix} w_i \\ \theta_{xi} \\ \theta_{yi} \\ w_j \\ \vdots \end{Bmatrix} \begin{bmatrix} 1 & x_i & y_i & x_i^2 & x_i y_i & y_i^2 & x_i^3 & x_i^2 y_i & x_i y_i^2 & y_i^3 & x_i^3 y_i & x_i y_i^3 \\ 0 & 0 & +1 & 0 & +x_i & +2y_i & 0 & +x_i^2 & +2x_i y_i & +3y_i^2 & +x_i^3 & +3x_i y_i^2 \\ \vdots & & & & & & & & & & & \\ \cdots & \cdots & \cdots & \cdots & \cdots & \cdots & \cdots & \cdots & \cdots & \cdots & \cdots & \cdots \end{bmatrix}$$

$$\times \begin{Bmatrix} a_1 \\ a_2 \\ \vdots \\ a_{12} \end{Bmatrix} \tag{17.2.8}$$

In compact matrix form, we express Eq. (17.2.8) as

$$\{d\} = [C]\{a\} \tag{17.2.9}$$

where $[C]$ is the 12×12 matrix on the right side of Eq. (17.2.8).

Therefore, the constants (a's) can be solved for by

$$\{a\} = [C]^{-1}\{d\} \tag{17.2.10}$$

Equation (17.2.7) can now be expressed as

$$\{\psi\} = [P][C]^{-1}\{d\} \tag{17.2.11}$$

or

$$\{\psi\} = [N]\{d\} \tag{17.2.12}$$

where $[N] = [P][C]^{-1}$ is the shape function matrix. A specific form of the shape functions $N_i, N_j, N_m,$ and N_n is given in Reference [9].

Step 3 Define the Strain (Curvature)/Displacement and Stress (Moment)/Curvature Relationships

The curvature matrix, based on the curvatures of Eq.(17.1.3), is

$$\{\kappa\} = \begin{Bmatrix} \kappa_x \\ \kappa_y \\ \kappa_{xy} \end{Bmatrix} = \begin{Bmatrix} -2a_4 - 6a_7 - 2a_8\,y - 6a_{11}xy \\ -2a_6 - 2a_9x - 6a_{10}y + 6a_{12}xy \\ -2a_5 - 4a_8 - 4a_9y - 6a_{11}x^2 - 6a_{12}y^2 \end{Bmatrix} \tag{17.2.13}$$

or expressing Eq. (17.2.13) in matrix form, we have

$$\{\kappa\} = [Q]\{a\} \tag{17.2.14}$$

where $[Q]$ is the coefficient matrix multiplied by the a's in Eq. (17.2.13). Using Eq. (17.2.10) for $\{a\}$, we express the curvature matrix as

$$\{\kappa\} = [B]\{d\} \tag{17.2.15}$$

where

$$[B] = [Q][C]^{-1} \tag{17.2.16}$$

is the gradient matrix.

The moment/curvature matrix for a plate is given by

$$\{M\} = \left\{ \begin{array}{c} M_x \\ M_y \\ M_{xy} \end{array} \right\} = [D] \left\{ \begin{array}{c} \kappa_x \\ \kappa_y \\ \kappa_{xy} \end{array} \right\} = [D][B]\{d\} \tag{17.2.17}$$

where the $[D]$ matrix is the constitutive matrix given for isotropic materials by

$$[D] = \frac{Et^3}{12(1 - v^2)} \begin{bmatrix} 1 & v & 0 \\ v & 1 & 0 \\ 0 & 0 & \dfrac{1-v}{2} \end{bmatrix} \tag{17.2.18}$$

and Eq. (17.2.15) has been used in the final expression for Eq. (17.2.17).

Step 4 Derive the Element Stiffness Matrix and Equations

The stiffness matrix is given by the usual form of the stiffness matrix as

$$[k] = \iint [B]^T [D][B] \, dx \, dy \tag{17.2.19}$$

where $[B]$ is defined by Eq. (17.2.16) and $[D]$ is defined by Eq. (17.2.18). The stiffness matrix for the four-noded rectangular element is of order 12×12. A specific expression for $[k]$ is given in References [4] and [5].

The surface force matrix due to distributed loading q acting per unit area in the z direction is obtained using the standard equation

$$\{F_s\} = \iint [N_s]^T q \, dx \, dy \tag{17.2.20}$$

For a uniform load q acting over the surface of an element of dimensions $2b \times 2c$, Eq. (17.2.20) yields the forces and moments at node i as

$$\left\{ \begin{array}{c} f_{wi} \\ f_{\theta xi} \\ f_{\theta yi} \end{array} \right\} = 4qcb \left\{ \begin{array}{c} 1/4 \\ -c/12 \\ b/12 \end{array} \right\} \tag{17.2.21}$$

with similar expressions at nodes j, m, and n. We should note that a uniform load yields applied couples at the nodes as part of the work-equivalent load replacement, just as was the case for the beam element (Section 5.4).

The element equations are given by

$$
\begin{Bmatrix} f_{wi} \\ f_{\theta xi} \\ f_{\theta yi} \\ \vdots \\ f_{\theta yn} \end{Bmatrix} = \begin{bmatrix} k_{11} & k_{12} & \cdots & k_{1,12} \\ k_{21} & k_{22} & \cdots & k_{2,12} \\ k_{31} & k_{32} & \cdots & k_{3,12} \\ \vdots & & \cdots & \cdots \\ k_{12,1} & & \cdots & k_{12,12} \end{bmatrix} \begin{Bmatrix} w_i \\ \theta_{xi} \\ \theta_{yi} \\ \vdots \\ \theta_{yn} \end{Bmatrix} \tag{17.2.22}
$$

The rest of the steps, including assembling the global equations, applying boundary conditions (now boundary conditions on w, θ_x, θ_y), and solving the equations for the nodal displacements and slopes (note three degrees of freedom per node), follow the standard procedures introduced in previous chapters.

17.3 Some Plate Element Numerical Comparisons

We now present some numerical comparisons of quadrilateral plate element formulations. Remember there are numerous plate element formulations in the literature. Figure 17–5 shows a number of plate element formulation results for a square plate simply supported all around and subjected to a concentrated vertical load applied at the center of the plate. The results are shown to illustrate the upper and lower bound solution behavior and demonstrate the convergence of solution for various plate element formulations. Included in these results is the 12-term polynomial described in Section 17.2. We note that the 12-term polynomial converges to the exact solution from above. It yields an upper bound solution. Because the interelement continuity of slopes is not ensured by the 12-term polynomial, the lower bound classical characteristic of a minimum potential energy formulation is not obtained. However, as more elements are used, the solution converges to the exact solution [1].

Figure 17–6 shows comparisons of triangular plate formulations for the same centrally loaded simply supported plate used to compare quadrilateral element formulations in Figure 17–5. We can observe from Figures 17–5 and 17–6 a number of different formulations with results that converge from above and below. Some of these elements produce better results than others.

The Algor program uses the Veubeke (after Baudoin Fraeijs de Veubeke) 16-degrees-of-freedom "subdomain" formulation [7] which converges from below, as it is based on a compatible displacement formulation. For more information on some of these formulations, consult the references at the end of the chapter.

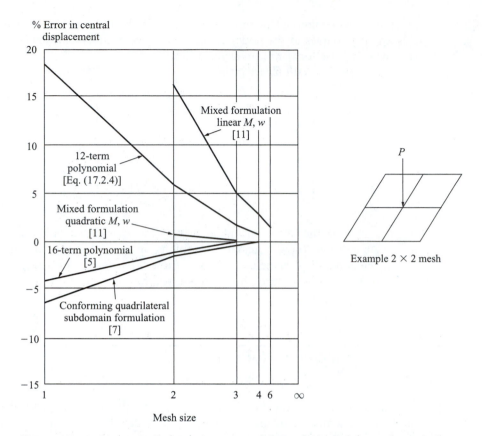

Figure 17–5 Numerical comparisons: quadrilateral plate element formulations. (Gallagher, R. H., *Finite Element Analysis Fundamentals*, 1975, p. 345. Reprinted by permission of Prentice-Hall, Inc., Upper Saddle River, NJ.)

Finally, Figure 17–7 shows results for some selected Mindlin plate theory elements. Mindlin plate elements account for bending deformation and for transverse shear deformation. For more on Mindlin plate theory, see Reference [6]. The "heterosis" element [10] is the best performing element in Figure 17–7.

▲ 17.4 Algor Example Solution for Plate Bending Problems ▲

To illustrate how to use the Algor computer program for plate bending problems [19], we now solve an example problem for a square plate fixed along all four edges and subjected to a concentrated load at its center.

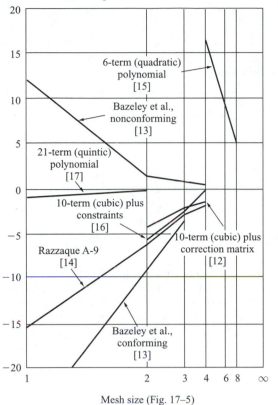

Mesh size (Fig. 17–5)

Figure 17–6 Numerical comparisons for a simply supported square plate subjected to center load triangular element formulations. (Gallagher, R. H., *Finite Element Analysis Fundamentals*, 1975, p. 350. Reprinted by permission of Prentice-Hall, Inc., Upper Saddle River, NJ.)

The model is drawn in Superdraw III. The *Group, Surface,* and *Layer* commands are the same as for the beam and frame element described in Chapters 5 and 6.

The plate element is a three- or four-noded element formulated in three-dimensional space. The element degrees of freedom allowed are all three translations ($u, v,$ and w) and in-plane rotations (θ_x and θ_y). The rotational degrees of freedom normal to the plate are undefined and must be constrained. The element formulated in the Algor program is the 16-term polynomial described in References [5] and [7]. This element is known as the Veubeke plate in the Algor program. The 16-node formulation converges from below for the displacement analysis, as it is based on a compatible displacement formulation. This is also shown in Figure 17–5 for a simply supported plate subjected to a concentrated center load.

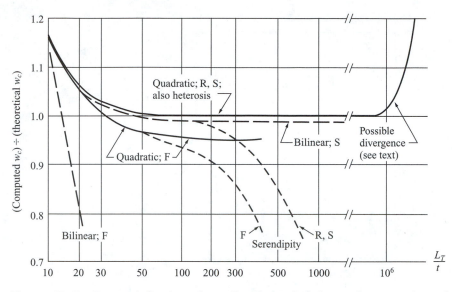

Figure 17–7 Center deflection of a uniformly loaded clamped square plate of side length L_T and thickness t. An 8×8 mesh is used in all cases. Thin plates correspond to large L_T/t. Transverse shear deformation becomes significant for small L_T/t. Integration rules are reduced (R), selective (S), and full (F) [18], based on Mindlin plate element formulations. (Cook, R., Malkus D., and Plesha, M. *Concepts and Applications of Finite Element Analysis*, 3rd ed., 1989, p. 326. Reprinted by permission of John Wiley & Sons, Inc., New York.)

Example 17.1

The 20-in. long, 20-in. wide plate with 0.1-in. thickness is fixed on all four edges and subjected to a concentrated load of 100 lb downward at its center (Figure 17–8). The plate is made of steel with $E = 30 \times 10^6$ psi and $v = 0.3$. Determine the deflection

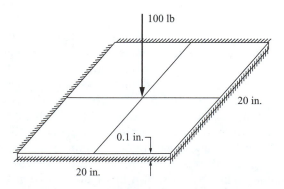

Figure 17–8 Square plate fixed along all four sides and subjected to a concentrated load at the center

under the load. Compare your results with the classical solution given in Reference [1]. Use four elements (2×2 mesh) in the model.

Step 1   Start Superdraw III

Double-click on the "Algor FEA" icon on your desktop computer screen to bring up the Superdraw III program.

Step 2   Create the Plate Model

We will create the 2×2 model (four elements).

Click "Add" on the main menu bar.

Click "Line" and enter the coordinates of the outer points of the plate.

(0, 0, 0)

<Enter>

(10, 0, 0)

<Enter>

(20, 0, 0)

<Enter>

(20, 10, 0)

<Enter>

(20, 20, 0)

<Enter>

(10, 20, 0)

<Enter>

(0, 20, 0)

<Enter>

(0, 10, 0)

<Enter>

(0, 0, 0)

<Enter>   The outer boundary is now complete.

Click "Done".

Click "Add".

Click "Line".

(10, 0, 0)

<Enter>

(10, 10, 0)

<Enter>

(20, 10, 0)

<Enter>   You now have drawn two of the interior lines.

Click "Done".

Click "Add".
Click "Line".
(0, 10, 0)
<Enter>
(10, 10, 0)
<Enter>
(10, 20, 0)
<Enter>   The model is now complete with four equal-sized square elements.
Click "Done".

Step 3   Eliminate Duplicate Lines

Eliminate duplicate lines in the usual manner.

Step 4   Add the Boundary Conditions and Applied Force

Click on the "FEA Add" menu on the main menu and highlight "Stress and Vibration Analysis".
Click "Boundary Conditions". The default boundary condition is full constraint or prevention of translation and rotation.
Click on the outer boundary nodes one at a time. This fixes the plate all around the outer edge.
Click "Done".
Click on the "FEA Add" menu again and move to "Stress and Vibration Analysis".
Click "Nodal Forces". You will now add the concentrated nodal force at the center of the plate.
Click "Vector". You will need to create the z direction for the force.
Click "Z Direction".
Click "Done".
Click "Magnitude". We will enter a force of -100 lb.
-100   This is the magnitude of the concentrated force.
<Enter>
Click on the center node of the plate. The force should appear.
Click "Done".

Step 5   Add the Material Properties Using the "Model Data Control" Window

Click on the "Model Data" button located in the lower part of screen.
For "Analysis Type", "Linear Static Stress" appears highlighted.
Click on the box below "Element". A table of group 1 element types appears.
Click "Plate". You now have created a plate bending structure.
Click "OK".
Click on the box below "Data". It responds with "Please enter the model name first".
Click "OK" and then enter the model name in the "File_name" area that appears.

Ex171

Click "Save".

Click "OK" to accept the default "English (in)" units system. The "Element Definition" menu appears next. Enter the thickness of the plate. Use the other defaults.

0.1 The thickness of the plate is 0.1 in.

Click "OK". This sends you back to the "Model Data Control" window.

Click on the box below "Material".

Click "[Customer Defined]".

Click "Edit Properties" and enter the modulus of elasticity and Poisson's ratio.

30e6 This is the modulus of elasticity.

0.3 This is Poisson's ratio.

Click "OK".

Click "OK". This takes you out of the "Material" menu and back to the "Model Data Control" window.

Step 6 Enter the Global Data

Click on the "Global" button on the right-side menu of the "Model Data Control" window.

Under "Load Case Multipliers" and under "Pressure" type a 1.

1 Typing a 1 under the "Pressure" column indicates that the load case is an applied force with a multiplier on the load of 1.

Click on the "Output" tab.

Click on the boxes next to "Displacement data" and "Stress data". This creates the nodal displacement output in the .*L* file and the element stresses in the .*S* file.

Click "OK" to exit the Global menu.

Step 7 Check the Model

Click "Check" under "FEA Model" on the right side of the "Model Data Control" window. This takes you to Superview where you can check the model.

Click "Done". This takes you out of Superview and back to the "Model Data Control" menu.

Step 8 Run the Analysis

Click on the "Analysis" button.

Click "Analyze" in the lower left corner of the "Linear Stress and Vibration-Static Stress" menu.

Click "Done" to go back to the "Model Data Control" window.

Step 9 Review the Results

Click on the "Results" button in the "Model Data Control" window to take you to Superview and review your results in the usual manner.

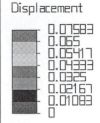

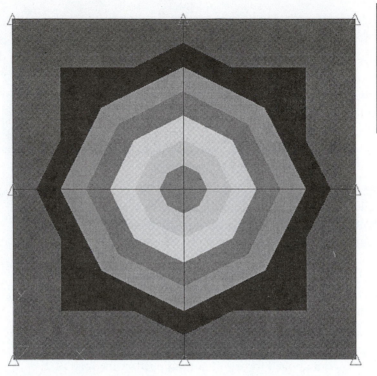

Figure 17–9 Displacement plot of the clamped plate of Example 17.1

Click "Displaced" on the main menu.

Click "Nodes inq".

Click "Get".

Click on the center node to see the displacement of -0.07583 in.

Select "Disp vec".

Select "Do Dither". The plot of the displacement of the plate appears as shown in Figure 17–9. ∎

The exact solution for the maximum displacement (which occurs under the concentrated load) is given in Reference [1] as $w = 0.0056 PL^2/D = 0.0056(-100 \text{ lb}) \cdot (20 \text{ in.})^2/(2.747 \times 10^3 \text{ lb-in.}) = -0.0815$ in., where $D = (30 \times 10^6 \text{ psi})(0.1 \text{ in})^3/[12(1-0.3^2)] = 2.747 \times 10^3$ lb-in.

▲ References

[1] Timoshenko, S. and Woinowsky-Krieger, S., *Theory of Plates and Shells*, 2nd ed., McGraw-Hill, New York, 1969.

[2] Gere, J. M., *Mechanics of Material*, 5th ed., Brooks/Cole Publishers, Pacific Grove, CA, 2001.

[3] Hrabok, M. M., and Hrudley, T. M., "A Review and Catalog of Plate Bending Finite Elements," *Computers & Structures*, Vol. 19, No. 3, 1984, pp. 479–495.

[4] Zienkiewicz, O. C., and Taylor R. L., *The Finite Element Method*, 4th ed., Vol. 2, McGraw-Hill, New York, 1991.

[5] Gallagher, R. H., *Finite Element Analysis Fundamentals*, Prentice-Hall, Englewood Cliffs, NJ, 1975.

[6] Cook, R. D., Malkus, D. S., and Plesha, M. E., *Concepts and Applications of Finite Element Analysis*, 3rd ed., Wiley, New York, 1989.

[7] Fraeijs De Veubeke, B., "A Conforming Finite Element for Plate Bending," *International Journal of Solids and Structures*, Vol. 4, No. 1, pp. 95–108, 1968.

[8] Walz, J. E., Fulton, R. E., and Cyrus N.J., "Accuracy and Convergence of Finite Element Approximations," Proceedings of the Second Conference on Matrix Method in Structural Mechanics, AFFDL TR 68-150, pp. 995–1027, Oct., 1968.

[9] Melosh, R. J., "Basis of Derivation of Matrices for the Direct Stiffness Method," *Journal of AIAA*, Vol. 1, pp. 1631–1637, 1963.

[10] Hughes, T. J. R., and Cohen, M., "The 'Heterosis' Finite Element for Plate Bending," *Computers & Structures*, Vol. 9, No. 5, 1978, pp. 445–450.

[11] Bron, J., and Dhatt, G., "Mixed Quadrilateral Elements for Bending," *Journal of AIAA*, Vol. 10, No. 10, pp. 1359–1361, Oct., 1972.

[12] Kikuchi, F., and Ando, Y., "Some Finite Element Solutions for Plate Bending Problems by Simplified Hybrid Displacement Method," *Nuclear Engineering Design*, Vol. 23, pp. 155–178, 1972.

[13] Bazeley, G., Cheung, Y., Irons, B., and Zienkiewicz, O., "Triangular Elements in Plate Bending—Conforming and Non-Conforming Solutions," Proceedings of the First Conference on Matrix Methods on Structural Mechanics, AFFDL TR 66-80, pp. 547–576, Oct., 1965.

[14] Razzaque, A. Q., "Program for Triangular Elements with Derivative Smoothing," *International Journal for Numerical Methods in Engineering*, Vol. 6, No. 3, pp. 333–344, 1973.

[15] Morley, L. S. D., "The Constant-Moment Plate Bending Element," *Journal of Strain Analysis*, Vol. 6, No. 1, pp. 20–24, 1971.

[16] Harvey, J. W., and Kelsey, S., "Triangular Plate Bending Elements with Enforced Compatibility", *AIAA Journal*, Vol. 9, pp. 1023–1026, 1971.

[17] Cowper, G. R., Kosko, E., Lindberg, G., and Olson M., "Static and Dynamic Applications of a High Precision Triangular Plate Bending Element", *AIAA Journal*, Vol. 7, No. 10, pp. 1957–1965, 1969.

[18] Hinton, E., and Huang, H. C., "A Family of Quadrilateral Mindlin Plate Elements with Substitute Shear Strain Fields," *Computers and Structures*, Vol. 23, No. 3, pp. 409–431, 1986.

[19] *Linear Stress and Dynamics Reference Division*, Docutech On-line Documentation, Algor, Inc., Pittsburgh, PA, 1999.

▲ Problems

Solve these problems using the plate element from the Algor computer program.

17.1 A square steel plate of dimensions 20 in. by 20 in. with thickness of 0.1 is clamped all around. The plate is subjected to a uniformly distributed loading of 1 lb/in^2. Using a 2

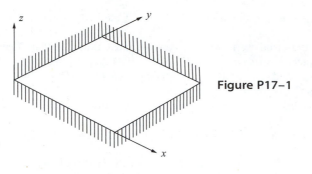

Figure P17–1

by 2 mesh and then a 4 by 4 mesh, determine the maximum deflection and maximum stress in the plate. Compare the finite element solution to the classical one in [1].

17.2 An L-shaped plate with thickness 0.1 in. is made of ASTM A-36 steel. Determine the deflection under the load and the maximum principal stress and its location using the plate element. Then model the plate as a grid with two beam elements with each beam having the stiffness of each L-portion of the plate and compare your answer.

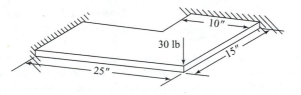

Figure P17–2

17.3 A square simply supported 20 in. by 20 in. steel plate with thickness 0.15 in. has a round hole of 4 in. diameter drilled through its center. The plate is uniformly loaded with a load of 2 lb/in². Determine the maximum principal stress in the plate.

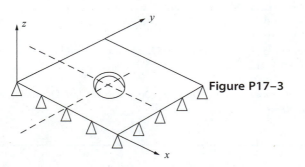

Figure P17–3

17.4 A C-channel section structural steel beam of 2 in. wide flanges, 3 in. depth and thickness of both flanges and web of 0.25 in. is loaded as shown with 100 lb acting in the y direction on the free end. Determine the free end deflection and angle of twist.

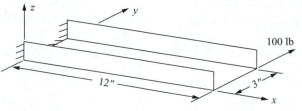

Figure P17–4

Now move the load in the z direction until the rotation (angle of twist) becomes zero. This distance is called the shear center (the location where the force can be placed so that the cross section will bend but not twist).

17.5 For the simply supported structural steel $W\ 14 \times 61$ wide flange beam shown, compare the plate element model results with the classical beam bending results for deflection and bending stress. The beam is subjected to a central vertical load of 22 kip. The cross-sectional area is 17.9 in.², depth is 13.89 in., flange width is 9.995 in., flange thickness of 0.645 in., web thickness of 0.375 in., and moment of inertia about the strong axis of 640 in.⁴

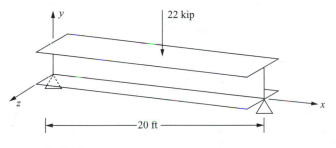

Figure P17–5

17.6 For the structural steel plate structure shown, determine the maximum principal stress and its location. If the stresses are unacceptably high, recommend any design changes.

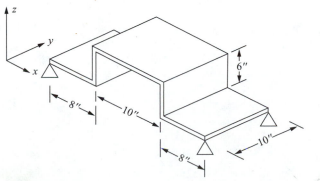

Figure P17–6

The initial thickness of each plate is 0.25 in. The left and right edges are simply supported. The load is a uniformly applied pressure of 10 lb/in.^{2} over the top plate.

17.7 Design a steel box structure 4 ft wide by 8 ft long made of plates to be used to protect construction workers while working in a trench. That is, determine a recommended thickness of each plate. The depth of the structure must be 8 ft. Assume the loading is from a side load acting along the long sides due to a wet soil (density of 62.4 lb/ft^3) and varies linearly with the depth. The allowable deflection of the plate type structure is 1 in. and the allowable stress is 20 ksi.

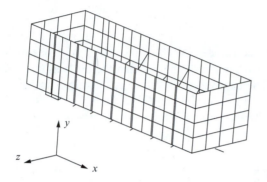

Figure P17–7

Matrix Algebra

Introduction

In this appendix, we provide an introduction to matrix algebra. We will consider the concepts relevant to the finite element method to provide an adequate background for the matrix algebra concepts used in this text.

A.1 Definition of a Matrix

*A **matrix** is an $m \times n$ array of numbers arranged in m rows and n columns.* The matrix is then described as being of order $m \times n$. Equation (A.1.1) illustrates a matrix with m rows and n columns.

$$[a] = \begin{bmatrix} a_{11} & a_{12} & a_{13} & a_{14} & \cdots & a_{1n} \\ a_{21} & a_{22} & a_{23} & a_{24} & \cdots & a_{2n} \\ a_{31} & a_{32} & a_{33} & a_{34} & \cdots & a_{3n} \\ \vdots & \vdots & \vdots & \vdots & & \vdots \\ a_{m1} & a_{m2} & a_{m3} & a_{m4} & \cdots & a_{mn} \end{bmatrix} \qquad (A.1.1)$$

If $m \neq n$ in matrix Eq. (A.1.1), the matrix is called **rectangular**. If $m = 1$ and $n > 1$, the elements of Eq. (A.1.1) form a single row called a **row matrix**. If $m > 1$ and $n = 1$, the elements form a single column called a **column matrix**. If $m = n$, the array is called a **square matrix**. Row matrices and rectangular matrices are denoted by using brackets [ ], and column matrices are denoted by using braces { }. For simplicity, matrices (row, column, or rectangular) are often denoted by using a line under a variable instead of surrounding it with brackets or braces. The order of the matrix should then be apparent from the context of its use. The force and displacement matrices used in structural analysis are column matrices, whereas the stiffness matrix is a square matrix.

To identify an element of matrix $\underline{a}$, we represent the element by a_{ij}, where the subscripts i and j indicate the row number and the column number, respectively, of $\underline{a}$. Hence, alternative notations for a matrix are given by

$$\underline{a} = [a] = [a_{ij}] \qquad (A.1.2)$$

Numerical examples of special types of matrices are given by Eqs. (A.1.3)–(A.1.6). A rectangular matrix $\underline{a}$ is given by

$$\underline{a} = \begin{bmatrix} 2 & 1 \\ 3 & 4 \\ 5 & 4 \end{bmatrix} \tag{A.1.3}$$

where $\underline{a}$ has three rows and two columns. In matrix $\underline{a}$ of Eq. (A.1.1), if $m = 1$, a row matrix results, such as

$$\underline{a} = \begin{bmatrix} 2 & 3 & 4 & -1 \end{bmatrix} \tag{A.1.4}$$

If $n = 1$ in Eq. (A.1.1), a column matrix results, such as

$$\underline{a} = \begin{Bmatrix} 2 \\ 3 \end{Bmatrix} \tag{A.1.5}$$

If $m = n$ in Eq. (A.1.1), a square matrix results, such as

$$\underline{a} = \begin{bmatrix} 2 & -1 \\ 3 & -2 \end{bmatrix} \tag{A.1.6}$$

Matrices and matrix notation are often used to express algebraic equations in compact form and are frequently used in the finite element formulation of equations. Matrix notation is also used to simplify the solution of a problem.

▲ A.2 Matrix Operations ▲

We will now present some common matrix operations that will be used in this text.

Multiplication of a Matrix by a Scalar

If we have a scalar k and a matrix $\underline{c}$, then the product $\underline{a} = k\underline{c}$ is given by

$$a_{ij} = kc_{ij} \tag{A.2.1}$$

—that is, every element of the matrix $\underline{c}$ is multiplied by the scalar k. As a numerical example, consider

$$\underline{c} = \begin{bmatrix} 1 & 2 \\ 3 & 1 \end{bmatrix} \qquad k = 4$$

The product $\underline{a} = k\underline{c}$ is

$$\underline{a} = 4 \begin{bmatrix} 1 & 2 \\ 3 & 1 \end{bmatrix} = \begin{bmatrix} 4 & 8 \\ 12 & 4 \end{bmatrix}$$

Note that if $\underline{c}$ is of order $m \times n$, then $\underline{a}$ is also of order $m \times n$.

Addition of Matrices

Matrices of the same order can be added together by summing corresponding elements of the matrices. Subtraction is performed in a similar manner. Matrices of unlike order cannot be added or subtracted. Matrices of the same order can be added (or subtracted) in any order (the commutative law for addition applies). That is,

$$\underline{c} = \underline{a} + \underline{b} = \underline{b} + \underline{a} \tag{A.2.2}$$

or, in subscript (index) notation, we have

$$[c_{ij}] = [a_{ij}] + [b_{ij}] = [b_{ij}] + [a_{ij}] \tag{A.2.3}$$

As a numerical example, let

$$\underline{a} = \begin{bmatrix} -1 & 2 \\ -3 & 2 \end{bmatrix} \qquad \underline{b} = \begin{bmatrix} 1 & 2 \\ 3 & 1 \end{bmatrix}$$

The sum $\underline{a} + \underline{b} = \underline{c}$ is given by

$$\underline{c} = \begin{bmatrix} -1 & 2 \\ -3 & 2 \end{bmatrix} + \begin{bmatrix} 1 & 2 \\ 3 & 1 \end{bmatrix} = \begin{bmatrix} 0 & 4 \\ 0 & 3 \end{bmatrix}$$

Again, remember that the matrices $\underline{a}$, $\underline{b}$, and $\underline{c}$ must all be of the same order. For instance, a 2×2 matrix cannot be added to a 3×3 matrix.

Multiplication of Matrices

For two matrices $\underline{a}$ and $\underline{b}$ to be multiplied in the order shown in Eq. (A.2.4), the number of columns in $\underline{a}$ must equal the number of rows in $\underline{b}$. For example, consider

$$\underline{c} = \underline{a}\underline{b} \tag{A.2.4}$$

If $\underline{a}$ is an $m \times n$ matrix, then $\underline{b}$ must have n rows. Using subscript notation, we can write the product of matrices $\underline{a}$ and $\underline{b}$ as

$$[c_{ij}] = \sum_{e=1}^{n} a_{ie}b_{ej} \tag{A.2.5}$$

where n is the total number of columns in $\underline{a}$ or of rows in $\underline{b}$. For matrix $\underline{a}$ of order 2×2 and matrix $\underline{b}$ of order 2×2, after multiplying the two matrices, we have

$$[c_{ij}] = \begin{bmatrix} a_{11}b_{11} + a_{12}b_{21} & a_{11}b_{12} + a_{12}b_{22} \\ a_{21}b_{11} + a_{22}b_{21} & a_{21}b_{12} + a_{22}b_{22} \end{bmatrix} \tag{A.2.6}$$

For example, let

$$\underline{a} = \begin{bmatrix} 2 & 1 \\ 3 & 2 \end{bmatrix} \qquad \underline{b} = \begin{bmatrix} 1 & -1 \\ 2 & 0 \end{bmatrix}$$

The product $\underline{a}\underline{b}$ is then

$$\underline{a}\underline{b} = \begin{bmatrix} 2(1) + 1(2) & 2(-1) + 1(0) \\ 3(1) + 2(2) & 3(-1) + 2(0) \end{bmatrix} = \begin{bmatrix} 4 & -2 \\ 7 & -3 \end{bmatrix}$$

In general, matrix multiplication is *not* commutative; that is,

$$\underline{ab} \neq \underline{ba} \tag{A.2.7}$$

The validity of the product of two matrices $\underline{a}$ and $\underline{b}$ is commonly illustrated by

$$\underset{(i \times e)}{\underline{a}} \quad \underset{(e \times j)}{\underline{b}} \quad = \quad \underset{(i \times j)}{\underline{c}} \tag{A.2.8}$$

where the product matrix $\underline{c}$ will be of order $i \times j$; that is, it will have the same number of rows as matrix $\underline{a}$ and the same number of columns as matrix $\underline{b}$.

Transpose of a Matrix

Any matrix, whether a row, column, or rectangular matrix, can be transposed. This operation is frequently used in finite element equation formulations. The transpose of a matrix $\underline{a}$ is commonly denoted by $\underline{a}^T$. The superscript T is used to denote the transpose of a matrix throughout this text. The transpose of a matrix is obtained by interchanging rows and columns; that is, the first row becomes the first column, the second row becomes the second column, and so on. For the transpose of matrix $\underline{a}$,

$$[a_{ij}] = [a_{ji}]^T \tag{A.2.9}$$

For example, if we let

$$\underline{a} = \begin{bmatrix} 2 & 1 \\ 3 & 2 \\ 4 & 5 \end{bmatrix}$$

then

$$\underline{a}^T = \begin{bmatrix} 2 & 3 & 4 \\ 1 & 2 & 5 \end{bmatrix}$$

where we have interchanged the rows and columns of $\underline{a}$ to obtain its transpose.

Another important relationship that involves the transpose is

$$(\underline{ab})^T = \underline{b}^T \underline{a}^T \tag{A.2.10}$$

That is, the transpose of the product of matrices $\underline{a}$ and $\underline{b}$ is equal to the transpose of the latter matrix $\underline{b}$ multiplied by the transpose of matrix $\underline{a}$ in that order, provided the order of the initial matrices continues to satisfy the rule for matrix multiplication, Eq. (A.2.8). In general, this property holds for any number of matrices; that is,

$$(\underline{abc} \ldots \underline{k})^T = \underline{k}^T \ldots \underline{c}^T \underline{b}^T \underline{a}^T \tag{A.2.11}$$

Note that the transpose of a column matrix is a row matrix.

As a numerical example of the use of Eq. (A.2.10), let

$$\underline{a} = \begin{bmatrix} 1 & 2 \\ 3 & 4 \end{bmatrix} \qquad \underline{b} = \begin{Bmatrix} 5 \\ 6 \end{Bmatrix}$$

First,

$$\underline{ab} = \begin{bmatrix} 1 & 2 \\ 3 & 4 \end{bmatrix} \begin{Bmatrix} 5 \\ 6 \end{Bmatrix} = \begin{Bmatrix} 17 \\ 39 \end{Bmatrix}$$

Then,

$$(\underline{ab})^T = [17 \quad 39] \tag{A.2.12}$$

Because $\underline{b}^T$ and $\underline{a}^T$ can be multiplied according to the rule for matrix multiplication, we have

$$\underline{b}^T \underline{a}^T = [5 \quad 6] \begin{bmatrix} 1 & 3 \\ 2 & 4 \end{bmatrix} = [17 \quad 39] \tag{A.2.13}$$

Hence, on comparing Eqs. (A.2.12) and (A.2.13), we have shown (for this case) the validity of Eq. (A.2.10). A simple proof of the general validity of Eq. (A.2.10) is left to your discretion.

Symmetric Matrices

If a square matrix is equal to its transpose, it is called a **symmetric matrix**; that is, if

$$\underline{a} = \underline{a}^T$$

then $\underline{a}$ is a symmetric matrix. As an example,

$$\underline{a} = \begin{bmatrix} 3 & 1 & 2 \\ 1 & 4 & 0 \\ 2 & 0 & 3 \end{bmatrix} \tag{A.2.14}$$

is a symmetric matrix because each element a_{ij} equals a_{ji} for $i \neq j$. In Eq. (A.2.14), note that the main diagonal running from the upper left corner to the lower right corner is the line of symmetry of the symmetric matrix $\underline{a}$. Remember that only a square matrix can be symmetric.

Unit Matrix

The **unit** (or **identity**) **matrix** $\underline{I}$ is such that

$$\underline{a}\underline{I} = \underline{I}\underline{a} = \underline{a} \tag{A.2.15}$$

The unit matrix acts in the same way that the number one acts in conventional multiplication. The unit matrix is always a square matrix of any possible order with each element of the main diagonal equal to one and all other elements equal to zero. For example, the 3×3 unit matrix is given by

$$\underline{I} = \begin{bmatrix} 1 & 0 & 0 \\ 0 & 1 & 0 \\ 0 & 0 & 1 \end{bmatrix}$$

Inverse of a Matrix

The **inverse of a matrix** is a matrix such that

$$\underline{a}^{-1}\underline{a} = \underline{a}\underline{a}^{-1} = \underline{I} \tag{A.2.16}$$

where the superscript, -1, denotes the inverse of $\underline{a}$ as $\underline{a}^{-1}$. Section A.3 provides more information regarding the properties of the inverse of a matrix and gives a method for determining it.

Orthogonal Matrix

A matrix $\underline{T}$ is an orthogonal matrix if

$$\underline{T}^T\underline{T} = \underline{T}\,\underline{T}^T = \underline{I} \tag{A.2.17}$$

Hence, for an orthogonal matrix, we have

$$\underline{T}^{-1} = \underline{T}^T \tag{A.2.18}$$

An orthogonal matrix frequently used is the *transformation* or *rotation* matrix $\underline{T}$. In two-dimensional space, the transformation matrix relates components of a vector in one coordinate system to components in another system. For instance, the displacement (and force as well) vector components of $\bar{\mathbf{d}}$ expressed in the x-y system are related to those in the $\hat{x}$-$\hat{y}$ system (Figure A–1 and Section 3.3) by

$$\underline{\hat{d}} = \underline{T}\underline{d} \tag{A.2.19}$$

or

$$\begin{Bmatrix} \hat{d}_x \\ \hat{d}_y \end{Bmatrix} = \begin{bmatrix} \cos\theta & \sin\theta \\ -\sin\theta & \cos\theta \end{bmatrix} \begin{Bmatrix} d_x \\ d_y \end{Bmatrix} \tag{A.2.20}$$

where $\underline{T}$ is the square matrix on the right side of Eq. (A.2.20).

Another use of an orthogonal matrix is to change from the local stiffness matrix to a global stiffness matrix for an element. That is, given a local stiffness matrix $\underline{\hat{k}}$ for an element, if the element is arbitrarily oriented in the x-y plane, then

$$\underline{k} = \underline{T}^T\underline{\hat{k}}\underline{T} = \underline{T}^{-1}\underline{\hat{k}}\underline{T} \tag{A.2.21}$$

Equation (A.2.21) is used throughout this text to express the stiffness matrix $\underline{k}$ in the x-y plane.

By further examination of $\underline{T}$, we see that the trigonometric terms in $\underline{T}$ can be interpreted as the direction cosines of lines $O\hat{x}$ and $O\hat{y}$ with respect to the x-y axes. Thus for $O\hat{x}$ or $\hat{d}_x$, we have from Eq. (A.2.20)

$$\langle t_{11} \quad t_{12} \rangle = \langle \cos\theta \quad \sin\theta \rangle \tag{A.2.22}$$

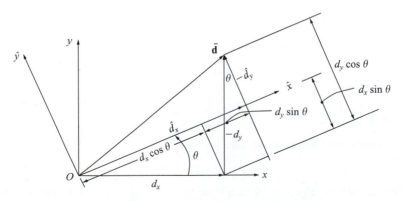

Figure A–1 Components of a vector in x-y and $\hat{x}$-$\hat{y}$ coordinates

and for $O\hat{y}$ or $\hat{d}_y$, we have

$$\langle t_{21} \quad t_{22} \rangle = \langle -\sin\theta \quad \cos\theta \rangle \tag{A.2.23}$$

or unit vectors $\bar{\mathbf{i}}$ and $\bar{\mathbf{j}}$ can be represented in terms of unit vectors $\hat{\mathbf{i}}$ and $\hat{\mathbf{j}}$ [also see Section 3.3 for proof of Eq. (A.2.24)] as

$$\hat{\mathbf{i}} = \mathbf{i}\cos\theta + \mathbf{j}\sin\theta$$

$$\hat{\mathbf{j}} = -\mathbf{i}\sin\theta + \mathbf{j}\cos\theta \tag{A.2.24}$$

and hence

$$t_{11}^2 + t_{12}^2 = 1 \qquad t_{21}^2 + t_{22}^2 = 1 \tag{A.2.25}$$

and since these vectors are orthogonal, by the dot product, we have

$$\langle t_{11} \quad t_{12} \rangle \cdot \langle t_{21} \quad t_{22} \rangle$$

or

$$t_{11}t_{21} + t_{12}t_{22} = 0 \tag{A.2.26}$$

or we say $\underline{T}$ is orthogonal and therefore $\underline{T}^T\underline{T} = \underline{T}\underline{T}^T = \underline{I}$ and that the transpose is its inverse. That is,

$$\underline{T}^T = \underline{T}^{-1} \tag{A.2.27}$$

Differentiating a Matrix

A matrix is differentiated by differentiating every element in the matrix in the conventional manner. For example, if

$$\underline{a} = \begin{bmatrix} x^3 & 2x^2 & 3x \\ 2x^2 & x^4 & x \\ 3x & x & x^5 \end{bmatrix} \tag{A.2.28}$$

the derivative $d\underline{a}/dx$ is given by

$$\frac{d\underline{a}}{dx} = \begin{bmatrix} 3x^2 & 4x & 3 \\ 4x & 4x^3 & 1 \\ 3 & 1 & 5x^4 \end{bmatrix} \tag{A.2.29}$$

Similarly, the partial derivative of a matrix is illustrated as follows:

$$\frac{\partial \underline{a}}{\partial x} = \frac{\partial}{\partial x}\begin{bmatrix} x^2 & xy & xz \\ xy & y^2 & yz \\ xz & yz & z^2 \end{bmatrix} = \begin{bmatrix} 2x & y & z \\ y & 0 & 0 \\ z & 0 & 0 \end{bmatrix} \tag{A.2.30}$$

In structural analysis theory, we sometimes differentiate an expression of the form

$$U = \tfrac{1}{2}[x \quad y]\begin{bmatrix} a_{11} & a_{12} \\ a_{12} & a_{22} \end{bmatrix}\begin{Bmatrix} x \\ y \end{Bmatrix} \tag{A.2.31}$$

where U might represent the strain energy in a bar. Expression (A.2.31) is known as a quadratic form. By matrix multiplication of Eq. (A.2.31), we obtain

$$U = \tfrac{1}{2}(a_{11}x^2 + 2a_{12}xy + a_{22}y^2) \tag{A.2.32}$$

Differentiating U now yields

$$\frac{\partial U}{\partial x} = a_{11}x + a_{12}y \tag{A.2.33}$$

$$\frac{\partial U}{\partial y} = a_{12}x + a_{22}y$$

Equation (A.2.33) in matrix form becomes

$$\left\{ \begin{array}{c} \dfrac{\partial U}{\partial x} \\[2mm] \dfrac{\partial U}{\partial y} \end{array} \right\} = \begin{bmatrix} a_{11} & a_{12} \\ a_{12} & a_{22} \end{bmatrix} \left\{ \begin{array}{c} x \\ y \end{array} \right\} \tag{A.2.34}$$

A general form of Eq. (A.2.31) is

$$U = \tfrac{1}{2}\{X\}^T[a]\{X\} \tag{A.2.35}$$

Then, by comparing Eq. (A.2.31) and (A.2.34), we obtain

$$\frac{\partial U}{\partial x_i} = [a]\{X\} \tag{A.2.36}$$

where x_i denotes x and y. Here Eq. (A.2.36) depends on matrix $\underline{a}$ in Eq. (A.2.35) being symmetric.

Integrating a Matrix

Just as in matrix differentiation, to integrate a matrix, we must integrate every element in the matrix in the conventional manner. For example, if

$$\underline{a} = \begin{bmatrix} 3x^2 & 4x & 3 \\ 4x & 4x^3 & 1 \\ 3 & 1 & 5x^4 \end{bmatrix}$$

we obtain the integration of $\underline{a}$ as

$$\int \underline{a}\, dx = \begin{bmatrix} x^3 & 2x^2 & 3x \\ 2x^2 & x^4 & x \\ 3x & x & x^5 \end{bmatrix}$$

In our finite element formulation of equations, we often integrate an expression of the form

$$\iint [X]^T[A][X]\, dx\, dy \tag{A.2.37}$$

The triple product in Eq. (A.2.37) will be symmetric if $\underline{A}$ is symmetric. The form $[X]^T[A][X]$ is also called a *quadratic form*. For example, letting

$$[A] = \begin{bmatrix} 9 & 2 & 3 \\ 2 & 8 & 0 \\ 3 & 0 & 5 \end{bmatrix} \qquad [X] = \begin{Bmatrix} x_1 \\ x_2 \\ x_3 \end{Bmatrix}$$

we obtain

$$\{X\}^T[A]\{X\} = \begin{bmatrix} x_1 & x_2 & x_3 \end{bmatrix} \begin{bmatrix} 9 & 2 & 3 \\ 2 & 8 & 0 \\ 3 & 0 & 5 \end{bmatrix} \begin{Bmatrix} x_1 \\ x_2 \\ x_3 \end{Bmatrix}$$

$$= 9x_1^2 + 4x_1 x_2 + 6x_1 x_3 + 8x_2^2 + 5x_3^2$$

which is in quadratic form.

▲ A.3 Cofactor or Adjoint Method to Determine the Inverse of a Matrix

We will now introduce a method for finding the inverse of a matrix. This method is useful for longhand determination of the inverse of smaller-order square matrices (preferably of order 4×4 or less). A matrix $\underline{a}$ must be square for us to determine its inverse.

We must first define the determinant of a matrix. This concept is necessary in determining the inverse of a matrix by the cofactor method. *A **determinant** is a square array of elements expressed by*

$$|\underline{a}| = |a_{ij}| \tag{A.3.1}$$

where the straight vertical bars, | |, on each side of the array denote the determinant. The resulting determinant of an array will be a single numerical value when the array is evaluated.

To evaluate the determinant of $\underline{a}$, we must first determine the cofactors of $[a_{ij}]$. The cofactors of $[a_{ij}]$ are given by

$$C_{ij} = (-1)^{i+j}|\underline{d}| \tag{A.3.2}$$

where the matrix $\underline{d}$, called the *first minor of* $[a_{ij}]$, is matrix $\underline{a}$ with row i and column j deleted. The inverse of matrix $\underline{a}$ is then given by

$$\underline{a}^{-1} = \frac{C^T}{|\underline{a}|} \tag{A.3.3}$$

where $\underline{C}$ is the **cofactor matrix** and $|\underline{a}|$ is the determinant of $\underline{a}$. To illustrate the method of cofactors, we will determine the inverse of a matrix $\underline{a}$ given by

$$\underline{a} = \begin{bmatrix} -1 & 3 & -2 \\ 2 & -4 & 2 \\ 0 & 4 & 1 \end{bmatrix} \tag{A.3.4}$$

Using Eq. (A.3.2), we find that the cofactors of matrix $\underline{a}$ are

$$C_{11} = (-1)^{1+1} \begin{vmatrix} -4 & 2 \\ 4 & 1 \end{vmatrix} = -12$$

$$C_{12} = (-1)^{1+2} \begin{vmatrix} 2 & 2 \\ 0 & 1 \end{vmatrix} = -2$$

$$C_{13} = (-1)^{1+3} \begin{vmatrix} 2 & -4 \\ 0 & 4 \end{vmatrix} = 8$$

$$C_{21} = (-1)^{2+1} \begin{vmatrix} 3 & -2 \\ 4 & 1 \end{vmatrix} = -11 \tag{A.3.5}$$

$$C_{22} = (-1)^{2+2} \begin{vmatrix} -1 & -2 \\ 0 & 1 \end{vmatrix} = -1$$

$$C_{23} = (-1)^{2+3} \begin{vmatrix} -1 & 3 \\ 0 & 4 \end{vmatrix} = 4$$

Similarly, $\qquad C_{31} = -2 \qquad C_{32} = -2 \qquad C_{33} = -2 \tag{A.3.6}$

Therefore, from Eqs. (A.3.5) and (A.3.6), we have

$$\underline{C} = \begin{bmatrix} -12 & -2 & 8 \\ -11 & -1 & 4 \\ -2 & -2 & -2 \end{bmatrix} \tag{A.3.7}$$

The determinant of $\underline{a}$ is then

$$|\underline{a}| = \sum_{j=1}^{n} a_{ij} C_{ij} \qquad \text{with } i \text{ any row number } (1 \leqslant i \leqslant n) \tag{A.3.8}$$

or $\qquad \displaystyle |\underline{a}| = \sum_{j=1}^{n} a_{ji} C_{ji} \qquad \text{with } i \text{ any column number } (1 \leqslant i \leqslant n) \tag{A.3.9}$

For instance, if we choose the first rows of $\underline{a}$ and $\underline{C}$, then $i = 1$ in Eq. (A.3.8), and j is summed from 1 to 3 such that

$$|\underline{a}| = a_{11} C_{11} + a_{12} C_{12} + a_{13} C_{13}$$

$$= (-1)(-12) + (3)(-2) + (-2)(8) = -10 \tag{A.3.10}$$

Using the definition of the inverse given by Eq. (A.3.3), we have

$$\underline{a}^{-1} = \frac{\underline{C}^T}{|\underline{a}|} = \frac{1}{-10} \begin{bmatrix} -12 & -11 & -2 \\ -2 & -1 & -2 \\ 8 & 4 & -2 \end{bmatrix} \tag{A.3.11}$$

We can then check that

$$\underline{a}\underline{a}^{-1} = \begin{bmatrix} 1 & 0 & 0 \\ 0 & 1 & 0 \\ 0 & 0 & 1 \end{bmatrix}$$

The transpose of the cofactor matrix is often defined as the **adjoint matrix**; that is,

$$\operatorname{adj} \underline{a} = \underline{C}^T$$

Therefore, an alternative equation for the inverse of $\underline{a}$ is

$$\underline{a}^{-1} = \frac{\operatorname{adj} \underline{a}}{|\underline{a}|} \qquad (A.3.12)$$

An important property associated with the determinant of a matrix is that if the determinant of a matrix is zero—that is, $|\underline{a}| = 0$—then the matrix is said to be **singular**. A singular matrix does not have an inverse. The stiffness matrices used in the finite element method are singular until sufficient boundary conditions (support conditions) are applied. This characteristic of the stiffness matrix is further discussed in the text.

▲ A.4 Inverse of a Matrix by Row Reduction

The inverse of a nonsingular square matrix $\underline{a}$ can be found by the method of row reduction (sometimes called the *Gauss–Jordan method*) by performing identical simultaneous operations on the matrix $\underline{a}$ and the identity matrix $\underline{I}$ (of the same order as $\underline{a}$) such that the matrix $\underline{a}$ becomes an identity matrix and the original identity matrix becomes the inverse of $\underline{a}$.

A numerical example will best illustrate the procedure. We begin by converting matrix $\underline{a}$ to an upper triangular form by setting all elements below the main diagonal equal to zero, starting with the first column and continuing with succeeding columns. We then proceed from the last column to the first, setting all elements above the main diagonal equal to zero.

We will invert the following matrix by row reduction.

$$\underline{a} = \begin{bmatrix} 2 & 2 & 1 \\ 2 & 1 & 0 \\ 1 & 1 & 1 \end{bmatrix} \qquad (A.4.1)$$

To find $\underline{a}^{-1}$, we need to find $\underline{x}$ such that $\underline{a}\underline{x} = \underline{I}$, where

$$\underline{x} = \begin{bmatrix} x_{11} & x_{12} & x_{13} \\ x_{21} & x_{22} & x_{23} \\ x_{31} & x_{32} & x_{33} \end{bmatrix}$$

That is, solve

$$\begin{bmatrix} 2 & 2 & 1 \\ 2 & 1 & 0 \\ 1 & 1 & 1 \end{bmatrix} \underline{x} = \begin{bmatrix} 1 & 0 & 0 \\ 0 & 1 & 0 \\ 0 & 0 & 1 \end{bmatrix}$$

We begin by writing $\underline{a}$ and $\underline{I}$ side by side as

$$
\begin{bmatrix}
2 & 2 & 1 & \vdots & 1 & 0 & 0 \\
2 & 1 & 0 & \vdots & 0 & 1 & 0 \\
1 & 1 & 1 & \vdots & 0 & 0 & 1
\end{bmatrix}
\tag{A.4.2}
$$

where the vertical dashed line separates $\underline{a}$ and $\underline{I}$.

1. Divide the first row of Eq. (A.4.2) by 2.

$$
\begin{bmatrix}
1 & 1 & \frac{1}{2} & \vdots & \frac{1}{2} & 0 & 0 \\
2 & 1 & 0 & \vdots & 0 & 1 & 0 \\
1 & 1 & 1 & \vdots & 0 & 0 & 1
\end{bmatrix}
\tag{A.4.3}
$$

2. Multiply the first row of Eq. (A.4.3) by -2 and add the result to the second row.

$$
\begin{bmatrix}
1 & 1 & \frac{1}{2} & \vdots & \frac{1}{2} & 0 & 0 \\
0 & -1 & -1 & \vdots & -1 & 1 & 0 \\
1 & 1 & 1 & \vdots & 0 & 0 & 1
\end{bmatrix}
\tag{A.4.4}
$$

3. Subtract the first row of Eq. (A.4.4) from the third row.

$$
\begin{bmatrix}
1 & 1 & \frac{1}{2} & \vdots & \frac{1}{2} & 0 & 0 \\
0 & -1 & -1 & \vdots & -1 & 1 & 0 \\
0 & 0 & \frac{1}{2} & \vdots & -\frac{1}{2} & 0 & 1
\end{bmatrix}
\tag{A.4.5}
$$

4. Multiply the second row of Eq. (A.4.5) by -1 and the third row by 2.

$$
\begin{bmatrix}
1 & 1 & \frac{1}{2} & \vdots & \frac{1}{2} & 0 & 0 \\
0 & 1 & 1 & \vdots & 1 & -1 & 0 \\
0 & 0 & 1 & \vdots & -1 & 0 & 2
\end{bmatrix}
\tag{A.4.6}
$$

5. Subtract the third row of Eq. (A.4.6) from the second row.

$$
\begin{bmatrix}
1 & 1 & \frac{1}{2} & \vdots & \frac{1}{2} & 0 & 0 \\
0 & 1 & 0 & \vdots & 2 & -1 & -2 \\
0 & 0 & 1 & \vdots & -1 & 0 & 2
\end{bmatrix}
\tag{A.4.7}
$$

6. Multiply the third row of Eq. (A.4.7) by $-\frac{1}{2}$ and add the result to the first row.

$$
\begin{bmatrix}
1 & 1 & 0 & \vdots & 1 & 0 & -1 \\
0 & 1 & 0 & \vdots & 2 & -1 & -2 \\
0 & 0 & 1 & \vdots & -1 & 0 & 2
\end{bmatrix}
\tag{A.4.8}
$$

7. Subtract the second row of Eq. (A.4.8) from the first row.

$$\left[\begin{array}{ccc|ccc} 1 & 0 & 0 & -1 & 1 & 1 \\ 0 & 1 & 0 & 2 & -1 & -2 \\ 0 & 0 & 1 & -1 & 0 & 2 \end{array}\right] \tag{A.4.9}$$

The replacement of $\underline{a}$ by the inverse matrix is now complete. The inverse of $\underline{a}$ is then the right side of Eq. (A.4.9); that is,

$$\underline{a}^{-1} = \left[\begin{array}{ccc} -1 & 1 & 1 \\ 2 & -1 & -2 \\ -1 & 0 & 2 \end{array}\right] \tag{A.4.10}$$

For additional information regarding matrix algebra, consult References [1] and [2].

References

[1] Gere, J. M., and Weaver, W., Jr., *Matrix Algebra for Engineers*, Van Nostrand, Princeton, NJ, 1966.

[2] Jennings, A., *Matrix Computation for Engineers and Scientists*, Wiley, New York, 1977.

Problems

Solve Problems A.1–A.6 using matrices $\underline{A}$, $\underline{B}$, $\underline{C}$, $\underline{D}$, and $\underline{E}$ given by

$$\underline{A} = \begin{bmatrix} 1 & 0 \\ -1 & 4 \end{bmatrix} \qquad \underline{B} = \begin{bmatrix} 2 & 0 \\ -2 & 8 \end{bmatrix} \qquad \underline{C} = \begin{bmatrix} 3 & 1 & 0 \\ -1 & 0 & 3 \end{bmatrix}$$

$$\underline{D} = \begin{bmatrix} 3 & 1 & 2 \\ 1 & 4 & 0 \\ 2 & 0 & 3 \end{bmatrix} \qquad \underline{E} = \begin{Bmatrix} 1 \\ 2 \\ 3 \end{Bmatrix}$$

(Write "nonsense" if the operation cannot be performed.)

A.1 (a) $\underline{A} + \underline{B}$ (b) $\underline{A} + \underline{C}$
        (c) $\underline{A}\underline{C}^T$ (d) $\underline{D}\underline{E}$
        (e) $\underline{D}\underline{C}$ (f) $\underline{C}\underline{D}$

A.2 Determine $\underline{A}^{-1}$ by the cofactor method.

A.3 Determine $\underline{D}^{-1}$ by the cofactor method.

A.4 Determine $\underline{C}^{-1}$.

A.5 Determine $\underline{B}^{-1}$ by row reduction.

A.6 Determine $\underline{D}^{-1}$ by row reduction.

A.7 Show that $(\underline{A}\underline{B})^T = \underline{B}^T\underline{A}^T$ by using

$$\underline{A} = \begin{bmatrix} a_{11} & a_{12} \\ a_{21} & a_{22} \end{bmatrix} \qquad \underline{B} = \begin{bmatrix} b_{11} & b_{12} & b_{13} \\ b_{21} & b_{22} & b_{23} \end{bmatrix}$$

A.8 Find $\underline{T}^{-1}$ given that

$$\underline{T} = \begin{bmatrix} \cos\theta & \sin\theta \\ -\sin\theta & \cos\theta \end{bmatrix}$$

and show that $\underline{T}^{-1} = \underline{T}^T$ and hence that $\underline{T}$ is an orthogonal matrix.

A.9 Given the matrices

$$\underline{X} = \begin{bmatrix} x & y \\ 1 & x \end{bmatrix} \qquad \underline{A} = \begin{bmatrix} a & b \\ b & c \end{bmatrix}$$

show that the triple matrix product $\underline{X}^T\underline{A}\underline{X}$ is symmetric.

A.10 Evaluate the following integral in explicit form:

$$\underline{k} = \int_0^L \underline{B}^T E \underline{B}\, dx$$

where

$$\underline{B} = \begin{bmatrix} -\dfrac{1}{L} & \dfrac{1}{L} \end{bmatrix}$$

[*Note:* This is the step needed to obtain Eq. (11.1.16) from Eq. (11.1.15).]

A.11 The following integral represents the strain energy in a bar:

$$U = \frac{A}{2} \int_0^L \underline{d}^T \underline{B}^T \underline{D} \underline{B} \underline{d}\, dx$$

where

$$\underline{d} = \left\{ \begin{matrix} d_1 \\ d_2 \end{matrix} \right\} \qquad \underline{B} = \begin{bmatrix} -\dfrac{1}{L} & \dfrac{1}{L} \end{bmatrix} \qquad \underline{D} = E.$$

Show that $dU/d\{d\}$ yields $\underline{k}\underline{d}$, where $\underline{k}$ is the bar stiffness matrix given by

$$\underline{k} = \frac{AE}{L} \begin{bmatrix} 1 & -1 \\ -1 & 1 \end{bmatrix}$$

Methods for Solution of Simultaneous Linear Equations

Introduction

Many problems in engineering and mathematical physics require the solution of a system of simultaneous linear algebraic equations. Stress analysis, heat transfer, and vibration analysis are engineering problems for which the finite element formulation for solution typically involves the solving of simultaneous linear equations. This appendix introduces methods applicable to both longhand and computer solutions of simultaneous linear equations. Many methods are available for the solution of equations; for brevity's sake, we will discuss only some of the more common methods.

B.1 General Form of the Equations

In general, the set of equations will have the form

$$a_{11}x_1 + a_{12}x_2 + \cdots + a_{1n}x_n = c_1$$

$$a_{21}x_1 + a_{22}x_2 + \cdots + a_{2n}x_n = c_2$$

$$\vdots \qquad \vdots \qquad \qquad \vdots \qquad \vdots \qquad\qquad\qquad \text{(B.1.1)}$$

$$a_{n1}x_1 + a_{n2}x_2 + \cdots + a_{nn}x_n = c_n$$

where the a_{ij}'s are the coefficients of the unknown x_j's, and the c_i's are the known right-side terms. In the structural analysis problem, the a_{ij}'s are the stiffness coefficients k_{ij}'s, the x_j's are the unknown nodal displacements d_i's, and the c_i's are the known nodal forces F_i's.

If the c's are not all zero, the set of equations is *nonhomogeneous*, and all equations must be independent to yield a unique solution. Stress analysis problems typically involve solving sets of nonhomogeneous equations.

If the c's are all zero, the set of equations in *homogeneous*, and nontrivial solutions exist only if all equations are not independent. Buckling and vibration problems typically involve homogeneous sets of equations.

▲ B.2 Uniqueness, Nonuniqueness, and Nonexistence of Solution

To solve a system of simultaneous linear equations means to determine a unique set of values (if they exist) for the unknowns that satisfy every equation of the set simultaneously. A unique solution exists if and only if the determinant of the square coefficient matrix is not equal to zero. (All of the engineering problems considered in this text result in square coefficient matrices.) The problems in this text usually result in a system of equations that has a unique solution. Here we will briefly illustrate the concepts of uniqueness, nonuniqueness, and nonexistence of solution for systems of equations.

Uniqueness of Solution

$$2x_1 + 1x_2 = 6$$
$$1x_1 + 4x_2 = 17$$

(B.2.1)

For Eqs. (B.2.1), the determinant of the coefficient matrix is not zero, and a unique solution exists, as shown by the single common point of intersection of the two Eqs. (B.2.1) in Figure B–1.

Nonuniqueness of Solution

$$2x_1 + 1x_2 = 6$$
$$4x_1 + 2x_2 = 12$$

(B.2.2)

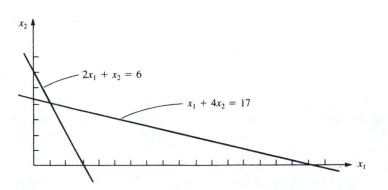

Figure B–1 Uniqueness of solution

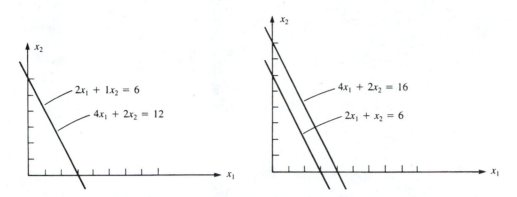

Figure B–2 Nonuniqueness of solution **Figure B–3** Nonexistence of solution

For Eqs. (B.2.2), the determinant of the coefficient matrix is zero; that is,

$$\begin{vmatrix} 2 & 1 \\ 4 & 2 \end{vmatrix} = 0$$

Hence the equations are called *singular*, and either the solution is not unique or it does not exist. In this case, the solution is not unique, as shown in Figure B–2.

Nonexistence of Solution

$$2x_1 + x_2 = 6$$
$$4x_1 + 2x_2 = 16$$
(B.2.3)

Again, the determinant of the coefficient matrix is zero. In this case, no solution exists because we have parallel lines (no common point of intersection), as shown in Figure B–3.

B.3 Methods for Solving Linear Algebraic Equations

We will now present some common methods for solving systems of linear algebraic equations that have unique solutions. Some of these methods work best for small sets of equations solved longhand, whereas others are well suited for computer application.

Cramer's Rule

We begin by introducing a method known as *Cramer's rule*, which is useful for the longhand solution of small numbers of simultaneous equations. Consider the set of equations

$$\underline{a}x = \underline{c}$$
(B.3.1)

or, in index notation,

$$\sum_{i=1}^{n} a_{ij}x_j = c_i \tag{B.3.2}$$

We first let $\underline{d}^{(i)}$ be the matrix $\underline{a}$ with column i replaced by the column matrix $\underline{c}$. Then the unknown x_i's are determined by

$$x_i = \frac{|\underline{d}^{(i)}|}{|\underline{a}|} \tag{B.3.3}$$

As an example of Cramer's rule, consider the following equations:

$$-x_1 + 3x_2 - 2x_3 = 2$$

$$2x_1 - 4x_2 + 2x_3 = 1 \tag{B.3.4}$$

$$4x_2 + x_3 = 3$$

In matrix form, Eqs. (B.3.4) become

$$\begin{bmatrix} -1 & 3 & -2 \\ 2 & -4 & 2 \\ 0 & 4 & 1 \end{bmatrix} \begin{Bmatrix} x_1 \\ x_2 \\ x_3 \end{Bmatrix} = \begin{Bmatrix} 2 \\ 1 \\ 3 \end{Bmatrix} \tag{B.3.5}$$

By Eq. (B.3.3), we can solve for the unknown x_i's as

$$x_1 = \frac{|\underline{d}^{(1)}|}{|\underline{a}|} = \frac{\begin{vmatrix} 2 & 3 & -2 \\ 1 & -4 & 2 \\ 3 & 4 & 1 \end{vmatrix}}{\begin{vmatrix} -1 & 3 & -2 \\ 2 & -4 & 2 \\ 0 & 4 & 1 \end{vmatrix}} = \frac{-41}{-10} = 4.1$$

$$x_2 = \frac{|\underline{d}^{(2)}|}{|\underline{a}|} = \frac{\begin{vmatrix} -1 & 2 & -2 \\ 2 & 1 & 2 \\ 0 & 3 & 1 \end{vmatrix}}{-10} = 1.1 \tag{B.3.6}$$

$$x_3 = \frac{|\underline{d}^{(3)}|}{|\underline{a}|} = \frac{\begin{vmatrix} -1 & 3 & 2 \\ 2 & -4 & 1 \\ 0 & 4 & 3 \end{vmatrix}}{-10} = -1.4$$

In general, to find the determinant of an $n \times n$ matrix, we must evaluate the determinants of n matrices of order $(n - 1) \times (n - 1)$. It has been shown that the solution of n simultaneous equations by Cramer's rule, evaluating determinants by expansion by minors, requires $(n - 1)(n + 1)!$ multiplications. Hence, this method takes large amounts of computer time and therefore is not used in solving large systems of simultaneous equations either longhand or by computer.

Inversion of the Coefficient Matrix

The set of equations $\underline{a}\underline{x} = \underline{c}$ can be solved for $\underline{x}$ by inverting the coefficient matrix $\underline{a}$ and premultiplying both sides of the original set of equations by $\underline{a}^{-1}$, such that

$$\underline{a}^{-1}\underline{a}\underline{x} = \underline{a}^{-1}\underline{c}$$

$$\underline{I}\underline{x} = \underline{a}^{-1}\underline{c} \tag{B.3.7}$$

$$\underline{x} = \underline{a}^{-1}\underline{c}$$

Two methods for determining the inverse of a matrix (the cofactor method and row reduction) were discussed in Appendix A.

The inverse method is much more time-consuming (because much time is required to determine the inverse of $\underline{a}$) than either the elimination method or the iteration method, which are discussed subsequently. Therefore, inversion is practical only for small systems of equations.

However, the concept of inversion is often used during the formulation of the finite element equations, even though elimination or iteration is used in achieving the final solution for the unknowns (such as nodal displacements).

Besides the tedious calculations necessary to obtain the inverse, the method usually involves determining the inverse of sparse, banded matrices (stiffness matrices in structural analysis usually contain many zeros with the nonzero coefficients located in a band around the main diagonal). This sparsity and banded nature can be used to advantage in terms of storage requirements and solution algorithms on the computer. The inverse results in a dense, full matrix with loss of the advantages resulting from the sparse, banded nature of the original coefficient matrix.

To illustrate the solution of a system of equations by the inverse method, consider the same equations that we solved previously by Cramer's rule. For convenience's sake, we repeat the equations here.

$$\begin{bmatrix} -1 & 3 & -2 \\ 2 & -4 & 2 \\ 0 & 4 & 1 \end{bmatrix} \begin{Bmatrix} x_1 \\ x_2 \\ x_3 \end{Bmatrix} = \begin{Bmatrix} 2 \\ 1 \\ 3 \end{Bmatrix} \tag{B.3.8}$$

The inverse of this coefficient matrix was found in Eq. (A.3.11) of Appendix A. The unknowns are then determined as

$$\begin{Bmatrix} x_1 \\ x_2 \\ x_3 \end{Bmatrix} = -\frac{1}{10} \begin{bmatrix} -12 & -11 & -2 \\ -2 & -1 & -2 \\ 8 & 4 & -2 \end{bmatrix} \begin{Bmatrix} 2 \\ 1 \\ 3 \end{Bmatrix} = \begin{Bmatrix} 4.1 \\ 1.1 \\ -1.4 \end{Bmatrix} \tag{B.3.9}$$

Gaussian Elimination

We will now consider a commonly used method called *Gaussian elimination* that is easily adapted to the computer for solving systems of simultaneous equations. It is based on triangularization of the coefficient matrix and evaluation of the unknowns by back-substitution starting from the last equation.

The general system of n equations with n unknowns given by

$$
\begin{bmatrix}
a_{11} & a_{12} & \cdots & a_{1n} \\
a_{21} & a_{22} & \cdots & a_{2n} \\
\vdots & \vdots & & \vdots \\
a_{n1} & a_{n2} & \cdots & a_{nn}
\end{bmatrix}
\begin{Bmatrix}
x_1 \\
x_2 \\
\vdots \\
x_n
\end{Bmatrix}
=
\begin{Bmatrix}
c_1 \\
c_2 \\
\vdots \\
c_n
\end{Bmatrix}
\tag{B.3.10}
$$

will be used to explain the Gaussian elimination method.

1. Eliminate the coefficient of x_1 in every equation except the first one.
   To do this, select a_{11} as the pivot, and
   a. Add the multiple $-a_{21}/a_{11}$ of the first row to the second row.
   b. Add the multiple $-a_{31}/a_{11}$ of the first row to the third row.
   c. Continue this procedure through the nth row.
   The system of equations will then be reduced to the following form:

$$
\begin{bmatrix}
a_{11} & a_{12} & \cdots & a_{1n} \\
0 & a'_{22} & \cdots & a'_{2n} \\
\vdots & & & \vdots \\
0 & a'_{n2} & \cdots & a'_{nn}
\end{bmatrix}
\begin{Bmatrix}
x_1 \\
x_2 \\
\vdots \\
x_n
\end{Bmatrix}
=
\begin{Bmatrix}
c_1 \\
c'_2 \\
\vdots \\
c'_n
\end{Bmatrix}
\tag{B.3.11}
$$

2. Eliminate the coefficient of x_2 in every equation below the second
   equation. To do this, select a'_{22} as the pivot, and
   a. Add the multiple $-a'_{32}/a'_{22}$ of the second row to the third row.
   b. Add the multiple $-a'_{42}/a'_{22}$ of the second row to the fourth row.
   c. Continue this procedure through the nth row.
   The system of equations will then be reduced to the following form:

$$
\begin{bmatrix}
a_{11} & a_{12} & a_{13} & \cdots & a_{1n} \\
0 & a'_{22} & a'_{23} & \cdots & a'_{2n} \\
0 & 0 & a''_{33} & \cdots & a''_{3n} \\
\vdots & & & & \vdots \\
0 & 0 & a''_{n3} & \cdots & a''_{nn}
\end{bmatrix}
\begin{Bmatrix}
x_1 \\
x_2 \\
x_3 \\
\vdots \\
x_n
\end{Bmatrix}
=
\begin{Bmatrix}
c_1 \\
c'_2 \\
c''_3 \\
\vdots \\
c''_n
\end{Bmatrix}
\tag{B.3.12}
$$

We repeat this process for the remaining rows until we have the
system of equations (called *triangularized*) as

$$
\begin{bmatrix}
a_{11} & a_{12} & a_{13} & a_{14} & \cdots & a_{1n} \\
0 & a'_{22} & a'_{23} & a'_{24} & \cdots & a'_{2n} \\
0 & 0 & a''_{33} & a''_{34} & \cdots & a''_{3n} \\
0 & 0 & 0 & a'''_{44} & \cdots & a'''_{4n} \\
\vdots & \vdots & \vdots & \vdots & & \vdots \\
0 & 0 & 0 & 0 & \cdots & a^{n-1}_{nn}
\end{bmatrix}
\begin{Bmatrix}
x_1 \\
x_2 \\
x_3 \\
x_4 \\
\vdots \\
x_n
\end{Bmatrix}
=
\begin{Bmatrix}
c_1 \\
c'_2 \\
c''_3 \\
c'''_4 \\
\vdots \\
c^{n-1}_n
\end{Bmatrix}
\tag{B.3.13}
$$

3. Determine x_n from the last equation as

$$
x_n = \frac{c^{n-1}_n}{a^{n-1}_{nn}}
\tag{B.3.14}
$$

and determine the other unknowns by back-substitution. These steps are summarized in general form by

$$k = 1, 2, \ldots, n - 1$$

$$a_{ij} = a_{ij} - a_{kj}\frac{a_{ik}}{a_{kk}} \qquad i = k+1, \ldots, n$$

$$j = k, \ldots, n+1 \tag{B.3.15}$$

$$x_i = \frac{1}{a_{ii}}\left(a_{i,n+1} - \sum_{r=i+1}^{n} a_{ir}x_r\right)$$

where $a_{i,n+1}$ represent the latest right side c's given by Eq. (B.3.13).

We will solve the following example to illustrate the Gaussian elimination method.

Example B.1

$$2x_1 + 2x_2 + 1x_3 = 9$$

$$2x_1 + 1x_2 \qquad\quad = 4 \tag{B.3.16}$$

$$1x_1 + 1x_2 + 1x_3 = 6$$

Step 1

Eliminate the coefficient of x_1 in every equation except the first one. Select $a_{11} = 2$ as the pivot, and

a. Add the multiple $-a_{21}/a_{11} = -2/2$ of the first row to the second row
b. Add the multiple $-a_{31}/a_{11} = -1/2$ of the first row to the third row.
We then obtain

$$2x_1 + 2x_2 + 1x_3 = 9$$

$$0x_1 - 1x_2 - 1x_3 = 4 - 9 = -5 \tag{B.3.17}$$

$$0x_1 + 0x_2 + \tfrac{1}{2}x_3 = 6 - \tfrac{9}{2} = \tfrac{3}{2}$$

Step 2

Eliminate the coefficient of x_2 in every equation below the second equation. In this case, we accomplished this in step 1.

Step 3

Solve for x_3 in the third of Eqs. (B.3.17) as

$$x_3 = \frac{\left(\frac{3}{2}\right)}{\left(\frac{1}{2}\right)} = 3$$

Solve for x_2 in the second of Eqs. (B.3.17) as

$$x_2 = \frac{-5 + 3}{-1} = 2$$

Solve for x_1 in the first of Eqs. (B.3.17) as

$$x_1 = \frac{9 - 2(2) - 3}{2} = 1$$

To illustrate the use of the index Eqs. (B.3.15), we re-solve the same example as follows. The ranges of the indexes in Eqs. (B.3.15) are $k = 1, 2$; $i = 2, 3$; and $j = 1, 2, 3, 4$.

Step 1

For $k = 1$, $i = 2$, and j indexing from 1 to 4,

$$a_{21} = a_{21} - a_{11} \frac{a_{21}}{a_{11}} = 2 - 2\left(\frac{2}{2}\right) = 0$$

$$a_{22} = a_{22} - a_{12} \frac{a_{21}}{a_{11}} = 1 - 2\left(\frac{2}{2}\right) = -1$$

(B.3.18)

$$a_{23} = a_{23} - a_{13} \frac{a_{21}}{a_{11}} = 0 - 1\left(\frac{2}{2}\right) = -1$$

$$a_{24} = a_{24} - a_{14} \frac{a_{21}}{a_{11}} = 4 - 9\left(\frac{2}{2}\right) = -5$$

Note that these new coefficients correspond to those of the second of Eqs. (B.3.17), where the right-side a's of Eqs. (B.3.18) are those from the previous step [here from Eqs. (B.3.16)], the right-side a_{24} is really $c_2 = 4$, and the left-side a_{24} is the new $c_2 = -5$.

For $k = 1$, $i = 3$, and j indexing from 1 to 4,

$$a_{31} = a_{31} - a_{11} \frac{a_{31}}{a_{11}} = 1 - 2\left(\frac{1}{2}\right) = 0$$

$$a_{32} = a_{32} - a_{12} \frac{a_{31}}{a_{11}} = 1 - 2\left(\frac{1}{2}\right) = 0$$

(B.3.19)

$$a_{33} = a_{33} - a_{13} \frac{a_{31}}{a_{11}} = 1 - 1\left(\frac{1}{2}\right) = \frac{1}{2}$$

$$a_{34} = a_{34} - a_{14} \frac{a_{31}}{a_{11}} = 6 - 9\left(\frac{1}{2}\right) = \frac{3}{2}$$

where these new coefficients correspond to those of the third of Eqs. (B.3.17) as previously explained.

Step 2

For $k = 2$, $i = 3$, and $j \,(= k)$ indexing from 2 to 4,

$$a_{32} = a_{32} - a_{22}\left(\frac{a_{32}}{a_{22}}\right) = 0 - (-1)\left(\frac{0}{-1}\right) = 0$$

$$a_{33} = a_{33} - a_{23}\left(\frac{a_{32}}{a_{22}}\right) = \frac{1}{2} - (-1)\left(\frac{0}{-1}\right) = \frac{1}{2} \qquad \text{(B.3.20)}$$

$$a_{34} = a_{34} - a_{24}\left(\frac{a_{32}}{a_{22}}\right) = \frac{3}{2} - (-5)\left(\frac{0}{-1}\right) = \frac{3}{2}$$

where the new coefficients again correspond to those of the third of Eqs. (B.3.17), because step 1 already eliminated the coefficients of x_2 as observed in the third of Eqs. (B.3.17), and the a's on the right side of Eqs. (B.3.20) are taken from Eqs. (B.3.18) and (B.3.19).

Step 3

By Eqs. (B.3.15), for x_3, we have

$$x_3 = \frac{1}{a_{33}}(a_{34} - 0)$$

or, using a_{33} and a_{34} from Eqs. (B.3.20),

$$x_3 = \frac{1}{\left(\frac{1}{2}\right)}\left(\frac{3}{2}\right) = 3$$

where the summation is interpreted as zero in the second of Eqs. (B.3.15) when $r > n$ (for x_3, $r = 4$, and $n = 3$). For x_2, we have

$$x_2 = \frac{1}{a_{22}}(a_{24} - a_{23}x_3)$$

or, using the appropriate a's from Eqs. (B.3.18),

$$x_2 = \frac{1}{-1}[-5 - (-1)(3)] = 2$$

and for x_1, we have

$$x_1 = \frac{1}{a_{11}}(a_{14} - a_{12}x_2 - a_{13}x_3)$$

or, using the a's from the first of Eqs. (B.3.16),

$$x_1 = \tfrac{1}{2}[9 - 2(2) - 1(3)] = 1$$

In summary, the latest a's from the previous steps have been used in Eqs. (B.3.15) to obtain the x's. ∎

Note that the pivot element was the diagonal element in each step. However, the diagonal element must be nonzero because we divide by it in each step. An original matrix with all nonzero diagonal elements does not ensure that the pivots in each step will remain nonzero, because we are adding numbers to equations below the pivot in each following step. Therefore, a test is necessary to determine whether the pivot a_{kk} at each step is zero. If it is zero, the current row (equation) must be interchanged with one of the following rows—usually with the next row unless that row has a zero at the position that would next become the pivot. Remember that the right-side corresponding element in c must also be interchanged. After making this test and, if necessary, interchanging the equations, continue the procedure in the usual manner.

An example will now illustrate the method for treating the occurrence of a zero pivot element.

Example B.2

Solve the following set of simultaneous equations.

$$2x_1 + 2x_2 + 1x_3 = 9$$
$$1x_1 + 1x_2 + 1x_3 = 6 \qquad \text{(B.3.21)}$$
$$2x_1 + 1x_2 \qquad = 4$$

It will often be convenient to set up the solution procedure by considering the coefficient matrix a plus the right-side matrix c in one matrix without writing down the unknown matrix x. This new matrix is called the *augmented matrix*. For the set of Eqs. (B.3.21), we have the augmented matrix written as

$$\begin{bmatrix} 2 & 2 & 1 & \vdots & 9 \\ 1 & 1 & 1 & \vdots & 6 \\ 2 & 1 & 0 & \vdots & 4 \end{bmatrix} \qquad \text{(B.3.22)}$$

We use the steps previously outlined as follows:

Step 1

We select $a_{11} = 2$ as the pivot and

a. Add the multiple $-a_{21}/a_{11} = -1/2$ of the first row to the second row of Eq. (B.3.22).

b. Add the multiple $-a_{31}/a_{11} = -2/2$ of the first row to the third row of Eq. (B.3.22) to obtain

$$\begin{bmatrix} 2 & 2 & 1 & \vdots & 9 \\ 0 & 0 & \frac{1}{2} & \vdots & \frac{3}{2} \\ 0 & -1 & -1 & \vdots & -5 \end{bmatrix} \qquad \text{(B.3.23)}$$

At the end of step 1, we would normally choose a_{22} as the next pivot. However, a_{22} is now equal to zero. If we interchange the second and third rows of Eq. (B.3.23), the new a_{22} will be nonzero and can be used as a pivot. Interchanging rows 2 and 3 results in

$$
\begin{bmatrix}
2 & 2 & 1 & | & 9 \\
0 & -1 & -1 & | & -5 \\
0 & 0 & \frac{1}{2} & | & \frac{3}{2}
\end{bmatrix}
\tag{B.3.24}
$$

For this special set of only three equations, the interchange has resulted in an upper-triangular coefficient matrix and concludes the elimination procedure. The back-substitution process of step 3 now yields

$$
x_3 = 3 \qquad x_2 = 2 \qquad x_1 = 1
$$ ■

A second problem when selecting the pivots in sequential manner without testing for the best possible pivot is that loss of accuracy due to rounding in the results can occur. In general, the pivots should be selected as the largest (in absolute value) of the elements in any column. For example, consider the set of equations given by

$$
0.002x_1 + 2.00x_2 = 2.00 \\
3.00x_1 + 1.50x_2 = 4.50
\tag{B.3.25}
$$

whose actual solution is given by

$$
x_1 = 1.0005 \qquad x_2 = 0.999
\tag{B.3.26}
$$

The solution by Gaussian elimination without testing for the largest absolute value of the element in any column is

$$
0.002x_1 + 2.00x_2 = 2.00 \\
-2998.5x_2 = -995.5 \\
x_2 = 0.3320 \\
x_1 = 668
\tag{B.3.27}
$$

This solution does not satisfy the second of Eqs. (B.3.25). The solution by interchanging equations is

$$
3.00x_1 + 1.50x_2 = 4.50 \\
0.002x_1 + 2.00x_2 = 2.00
$$

or

$$
3.00x_1 + 1.50x_2 = 4.50 \\
1.999x_2 = 1.997 \\
x_2 = 0.999 \\
x_1 = 1.0005
\tag{B.3.28}
$$

Equations (B.3.28) agree with the actual solution [Eqs. (B.3.26)].

Hence, in general, the pivots should be selected as the largest (in absolute value) of the elements in any column. This process is called *partial pivoting*. Even better results can be obtained by choosing the pivot as the largest element in the whole matrix of the remaining equations and performing appropriate interchanging of rows. This is called *complete pivoting*. Complete pivoting requires a large amount of testing, so it is not recommended in general.

The finite element equations generally involve coefficients with different orders of magnitude, so Gaussian elimination with partial pivoting is a useful method for solving the equations.

Finally, it has been shown that for n simultaneous equations, the number of arithmetic operations required in Gaussian elimination is n divisions, $\frac{1}{3}n^3 + n^2$ multiplications, and $\frac{1}{3}n^3 + n$ additions. If partial pivoting is included, the number of comparisons needed to select pivots is $n(n+1)/2$.

Other elimination methods, including the Gauss–Jordan and Cholesky methods, have some advantages over Gaussian elimination and are sometimes used to solve large systems of equations. For descriptions of other methods, see References [1–3].

Gauss–Seidel Iteration

Another general class of methods (other than the elimination methods) used to solve systems of linear algebraic equations is the *iterative methods*. Iterative methods work well when the system of equations is large and sparse (many zero coefficients). The Gauss–Seidel method starts with the original set of equations $\underline{a}x = \underline{c}$ written in the form

$$x_1 = \frac{1}{a_{11}}(c_1 - a_{12}x_2 - a_{13}x_3 - \cdots - a_{1n}x_n)$$

$$x_2 = \frac{1}{a_{22}}(c_2 - a_{21}x_1 - a_{23}x_3 - \cdots - a_{2n}x_n)$$

$$\vdots \tag{B.3.29}$$

$$x_n = \frac{1}{a_{nn}}(c_n - a_{n1}x_1 - a_{n2}x_2 - \cdots - a_{n,n-1}x_{n-1})$$

The following steps are then applied.

1. Assume a set of initial values for the unknowns $x_1, x_2, \ldots, x_n$, and substitute them into the right side of the first of Eqs. (B.3.29) to solve for the new x_1.
2. Use the latest value for x_1 obtained from step 1 and the initial values for $x_3, x_4, \ldots, x_n$ in the right side of the second of Eqs. (B.3.29) to solve for the new x_2.
3. Continue using the latest values of the x's obtained in the left side of Eqs. (B.3.29) as the next trial values in the right side for each succeeding step.
4. Iterate until convergence is satisfactory.

A good initial set of values (guesses) is often $x_i = c_i/a_{ii}$. An example will serve to illustrate the method.

Example B.3

Consider the set of linear simultaneous equations given by

$$
\begin{aligned}
4x_1 - x_2 &= 2 \\
-x_1 + 4x_2 - x_3 &= 5 \\
-x_2 + 4x_3 - x_4 &= 6 \\
-x_3 + 2x_4 &= -2
\end{aligned}
\tag{B.3.30}
$$

Using the initial guesses given by $x_i = c_i/a_{ii}$, we have

$$x_1 = \tfrac{2}{4} = \tfrac{1}{2} \qquad x_2 = \tfrac{5}{4} \approx 1 \qquad x_3 = \tfrac{6}{4} \approx 1 \qquad x_4 = -1$$

Solving the first of Eqs. (B.3.30) for x_1 yields

$$x_1 = \tfrac{1}{4}(2 + x_2) = \tfrac{1}{4}(2 + 1) = \tfrac{3}{4}$$

Solving the second of Eqs. (B.3.30) for x_2, we have

$$x_2 = \tfrac{1}{4}(5 + x_1 + x_3) = \tfrac{1}{4}(5 + \tfrac{3}{4} + 1) = 1.68$$

Solving the third of Eqs. (B.3.30) for x_3, we have

$$x_3 = \tfrac{1}{4}(6 + x_2 + x_4) = \tfrac{1}{4}[6 + 1.68 + (-1)] = 1.672$$

Solving the fourth of Eqs. (B.3.30) for x_4, we obtain

$$x_4 = \tfrac{1}{2}(-2 + x_3) = \tfrac{1}{2}(-2 + 1.67) = -0.16$$

The first iteration has now been completed. The second iteration yields

$$x_1 = \tfrac{1}{4}(2 + 1.68) = 0.922$$

$$x_2 = \tfrac{1}{4}(5 + 0.922 + 1.672) = 1.899$$

$$x_3 = \tfrac{1}{4}[6 + 1.899 + (-0.16)] = 1.944$$

$$x_4 = \tfrac{1}{2}(-2 + 1.944) = -0.028$$

Table B–1 lists the results of four iterations of the Gauss–Seidel method and the exact solution. From Table B–1, we observe that convergence to the exact solution has proceeded rapidly by the fourth iteration, and the accuracy of the solution is dependent on the number of iterations. ■

In general, iteration methods are self-correcting, such that an error made in calculations at one iteration will be corrected by later iterations. However, there are certain systems of equations for which iterative methods are not convergent. When the equations can be arranged such that the diagonal terms are greater than the off-diagonal terms, the possibility of convergence is usually enhanced.

Table B–1 Results of four iterations of the Gauss–Seidel method for Eqs. (B.3.30)

Iteration	x_1	x_2	x_3	x_4
0	0.5	1.0	1.0	−1.0
1	0.75	1.68	1.672	−0.16
2	0.922	1.899	1.944	−0.028
3	0.975	1.979	1.988	−0.006
4	0.9985	1.9945	1.9983	−0.0008
Exact	1.0	2.0	2.00	0

Finally, it has been shown that for n simultaneous equations, the number of arithmetic operations required by Gauss–Seidel iteration is n divisions, n^2 multiplications, and $n^2 - n$ additions for each iteration.

▲ B.4 Banded-Symmetric Matrices, Bandwidth, Skyline, and Wavefront Methods

The coefficient matrix (stiffness matrix) for the linear equations that occur in structural analysis is always symmetric and banded. Because a meaningful analysis generally requires the use of a large number of variables, the implementation of compressed storage of the stiffness matrix is desirable both from the standpoint of fitting into memory (immediate access portion of the computer) and for computational efficiency. We will discuss the banded-symmetric format, which is not necessarily the most efficient format but is relatively simple to implement on the computer.

Another method, based on the concept of the skyline of the stiffness matrix, is often used to improve the efficiency in solving the equations. *The **skyline** is an envelope that begins with the first nonzero coefficient in each column of the stiffness matrix* (Figure B–5). In skylining, only the coefficients between the main diagonal and the skyline are stored (normally by successive columns) in a one-dimensional array. In general, this procedure takes even less storage space in the computer and is more efficient in terms of equation solving than the conventional banded format. The Algor program used in this text is based on the skyline format for solving simultaneous equations. However, under the "Solution" tab in the "Model Data Control" window, you can choose to use the banded symmetric format for solving the equations if you desire. The banded symmetric concept, being the easiest to understand, will be discussed in more detail in this section. (For more information on skylining, consult References [10–12].)

A matrix is **banded** if the nonzero terms of the matrix are gathered about the main diagonal. To illustrate this concept, consider the plane truss of Figure B–4.

From Figure B–4, we see that element 2 connects nodes 1 and 4. Therefore, the 2×2 submatrices at positions 1–1, 1–4, 4–1, and 4–4 of Figure B–5 have nonzero coefficients. Figure B–5 represents the total stiffness matrix of the plane truss. The X's denote nonzero coefficients. From Figure B–5, we observe that the nonzero terms are

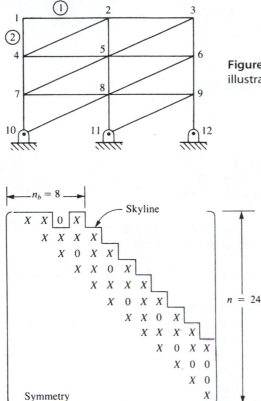

Figure B–4 Plane truss for bandwidth illustration

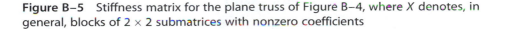

Figure B–5 Stiffness matrix for the plane truss of Figure B–4, where X denotes, in general, blocks of 2×2 submatrices with nonzero coefficients

within the band shown. When we use a banded storage format, only the main diagonal and the nonzero upper codiagonals need be stored as shown in Figure B–6. Note that any codiagonal with a nonzero term requires storage of the whole codiagonal *and* any codiagonals between it and the main diagonal. The use of banded storage is efficient for computational purposes. The Scientific Subroutine Package gives a more detailed explanation of banded compressed storage [4].

We now define the semibandwidth n_b as $n_b = n_d(m + 1)$, where n_d is the number of degrees of freedom per node and m is the maximum difference in node numbers determined by calculating the difference in node numbers for each element of a finite element model. In the example for the plane truss of Figure B–4, $m = 4 - 1 = 3$ and $n_d = 2$, so $n_b = 2(3 + 1) = 8$.

Execution time (primarily equation-solving time) is a function of the number of equations to be solved. It has been shown [5] that when banded storage of global stiffness matrix $\underline{K}$ is not used, execution time is proportional to $(1/3)n^3$, where n is the number of equations to be solved, or, equivalently, the size of $\underline{K}$. When banded storage of $\underline{K}$ is used, the execution time is proportional to $(n)n_b^2$. The ratio of time of execu-

$$\overleftarrow{\hspace{0.5em}n_b = 8\hspace{0.5em}}\overrightarrow{\hspace{0.5em}}$$

$$\begin{pmatrix} X & X & 0 & X \\ X & X & X & X \\ X & 0 & X & X \\ X & X & 0 & X \\ X & X & X & X \\ X & 0 & X & X \\ X & X & 0 & X \\ X & X & X & X \\ X & 0 & X & X \\ X & 0 & 0 & 0 \\ X & 0 & 0 & 0 \\ X & 0 & 0 & 0 \end{pmatrix}$$

Figure B–6 Banded storage format of the stiffness matrix of Figure B–5

tion without banded storage to that with banded storage is then $(1/3)(n/n_b)^2$. For the plane truss example, this ratio is $(1/3)(24/8)^2 = 3$. Therefore, it takes about three times as long to execute the solution of the example truss if banded storage is not used.

Hence, to reduce bandwidth we should number systematically and try to have a minimum difference between adjacent nodes. A small bandwidth is usually achieved by consecutive node numbering across the shorter dimension, as shown in Figure B–4. The computer program Algor uses the banded-symmetric format for storing the global stiffness matrix, $\underline{K}$. (We present a flowchart of the Algor finite element program for truss analysis in Section 4.1 and illustrate the use of the program in Section 4.2).

Several automatic node-renumbering schemes have been computerized [6]. This option is available in most general-purpose computer programs, including Algor. Alternatively, the wavefront or frontal method is becoming popular for optimizing equation solution time. In the **wavefront method**, elements, instead of nodes, are automatically renumbered.

In the wavefront method, the assembly of the equations alternates with their solution by Gauss elimination. The sequence in which the equations are processed is determined by element numbering rather than by node numbering. The first equations eliminated are those associated with element 1 only. Next, the contributions of stiffness coefficients of the adjacent element, element 2, are added to the system of equations. If any additional degrees of freedom are contributed by elements 1 and 2 only—that is, if no other elements contribute stiffness coefficients to specific degrees of freedom—these equations are eliminated (condensed) from the system of equations. As one or more additional elements make their contributions to the system of equations and additional degrees of freedom are contributed only by these elements, those degrees of freedom are eliminated from the solution. This repetitive alternation between assembly and solution was initially seen as a wavefront that sweeps over the structure in a pattern determined by the element numbering. For greater efficiency of this method, consecutive element numbering should be done across the structure in a direction that spans the smallest number of nodes.

The wavefront method, though somewhat more difficult to understand and to program than the banded-symmetric method, is computationally more efficient. A banded solver stores and processes any blocks of zeros created in assembling the stiffness matrix. In the wavefront method, these blocks of zero coefficients are not stored or processed. Many large-scale computer programs are now using the wavefront method to solve the system of equations. (For additional details of this method, see References [7–9].) Example B.4 illustrates the wavefront method for solution of a truss problem.

Example B.4

For the plane truss shown in Figure B–7, illustrate the wavefront solution procedure.

We will solve this problem in symbolic form. Merging k's for elements 1, 2, and 3 and enforcing boundary conditions at node 1, we have

$$
\begin{matrix}
\quad\quad d_{2x} & \quad\quad\quad d_{2y} & d_{3x} & d_{3y} & d_{4x} & d_{4y}
\end{matrix}
$$

$$
\left[\begin{array}{cc|cccc}
k_{33}^{(1)}+k_{11}^{(2)}+k_{11}^{(3)} & k_{34}^{(1)}+k_{12}^{(2)}+k_{12}^{(3)} & k_{13}^{(3)} & k_{14}^{(3)} & k_{13}^{(2)} & k_{14}^{(2)} \\
k_{43}^{(1)}+k_{21}^{(2)}+k_{21}^{(3)} & k_{44}^{(1)}+k_{22}^{(2)}+k_{22}^{(3)} & k_{23}^{(3)} & k_{24}^{(3)} & k_{23}^{(2)} & k_{24}^{(2)} \\
\hline
k_{31}^{(2)} & k_{32}^{(3)} & k_{33}^{(3)} & k_{34}^{(3)} & k_{33}^{(2)} & k_{34}^{(2)} \\
k_{41}^{(3)} & k_{42}^{(3)} & k_{43}^{(3)} & k_{44}^{(3)} & k_{43}^{(2)} & k_{44}^{(2)} \\
k_{31}^{(2)} & k_{32}^{(2)} & 0 & 0 & 0 & 0 \\
k_{41}^{(2)} & k_{42}^{(2)} & 0 & 0 & 0 & 0
\end{array}\right]
\left\{\begin{array}{c}
d_{2x} \\ d_{2y} \\ \hline d_{3x}' \\ d_{3y}' \\ d_{4x}' \\ d_{4y}'
\end{array}\right\}
=
\left\{\begin{array}{c}
0 \\ 0 \\ \hline 0 \\ -P \\ 0 \\ 0
\end{array}\right\}
$$

$$\text{(B.4.1)}$$

Eliminating d_{2x} and d_{2y} (all stiffness contributions from node 2 degrees of freedom have been included from these elements; these contributions are from elements 1–3) by static condensation or Gauss elimination yields

$$
[k_c']\left\{\begin{array}{c}
d_{3x}' \\ d_{3y}' \\ d_{4x}' \\ d_{4y}'
\end{array}\right\} = \{F_c'\}
$$

$$\text{(B.4.2)}$$

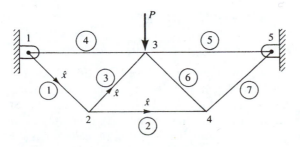

Figure B–7 Truss for wavefront solution

where the condensed stiffness and force matrices are (also see Section 8.5)

$$[k_c'] = [K_{22}'] - [K_{21}'][K_{11}']^{-1}[K_{12}'] \tag{B.4.3}$$

$$\{F_c'\} = \{F_2'\} - [K_{21}'][K_{11}']^{-1}\{F_1'\} \tag{B.4.4}$$

where primes on the degrees of freedom, such as d_{3x}' in Eq. (B.4.1), indicate that all stiffness coefficients associated with that degree of freedom have not yet been included. Now include elements 4–6 for degrees of freedom at node 3. The resulting equations are

$$
\begin{array}{cccc}
d_{3x} & d_{3y} & d_{4x} & d_{4y}
\end{array}
$$

$$
\left[
\begin{array}{cc:cc}
k_{c11}' + k_{33}^{(4)} + k_{11}^{(5)} + k_{11}^{(6)} & k_{34}^{(4)} + k_{12}^{(5)} + k_{12}^{(6)} + k_{c12}' & k_{13}^{(6)} + k_{c13}' & k_{14}^{(6)} + k_{c14}' \\
k_{c21}' + k_{34}^{(4)} + k_{21}^{(5)} + k_{21}^{(6)} & k_{44}^{(4)} + k_{22}^{(5)} + k_{22}^{(6)} + k_{c22}' & k_{23}^{(6)} + k_{c23}' & k_{24}^{(6)} + k_{c24}' \\ \hdashline
k_{c31}' + k_{31}^{(6)} & k_{c32}' + k_{32}^{(6)} & k_{c33}' + k_{33}^{(6)} & k_{c34}' + k_{34}^{(6)} \\
k_{c41}' + k_{41}^{(6)} & k_{c42}' + k_{42}^{(6)} & k_{c43}' + k_{43}^{(6)} & k_{c44}' + k_{44}^{(6)}
\end{array}
\right]
$$

$$
\times \left\{
\begin{array}{c}
d_{3x} \\
d_{3y} \\ \hdashline
d_{4x}' \\
d_{4y}'
\end{array}
\right\} = \left\{
\begin{array}{c}
0 \\
-P \\ \hdashline
0 \\
0
\end{array}
\right\} \tag{B.4.5}
$$

Using static condensation, we eliminate d_{3x} and d_{3y} (all contributions from node 3 degrees of freedom have been included from each element) to obtain

$$[k_c'']\left\{
\begin{array}{c}
d_{4x}' \\
d_{4y}'
\end{array}
\right\} = \{F_c''\} \tag{B.4.6}$$

where

$$[k_c''] = [K_{22}''] - [K_{21}''][K_{11}'']^{-1}[K_{12}''] \tag{B.4.7}$$

$$\{F_c''\} = \{F_2''\} - [K_{21}''][K_{11}'']^{-1}\{F_1''\} \tag{B.4.8}$$

Next we include element 7 contributions to the stiffness matrix. The condensed set of equations yield

$$[k_c''']\left\{
\begin{array}{c}
d_{4x} \\
d_{4y}
\end{array}
\right\} = \{F_c'''\} \tag{B.4.9}$$

$$[k_c'''] = [K_{22}'''] - [K_{21}'''][K_{11}''']^{-1}[K_{12}'''] \tag{B.4.10}$$

where

$$\{F_c'''\} = \{F_2'''\} - [K_{21}'''][K_{11}''']^{-1}\{F_1'''\} \tag{B.4.11}$$

The elimination procedure is now complete, and we solve Eq. (B.4.9) for d_{4x} and d_{4y}. Then we back-substitute d_{4x} and d_{4y} into Eq. (B.4.5) to obtain d_{3x} and d_{3y}. Finally, we back-substitute d_{3x} through d_{4y} into Eq. (B.4.1) to obtain d_{2x} and d_{2y}. Static condensation and Gauss elimination with back-substitution have been used to solve the set of equations for all the degrees of freedom. The solution procedure has then proceeded as though it were a wave sweeping over the structure, starting at node 2, engulfing node 2 and elements with degrees of freedom at node 2, and then sweeping through node 3 and finally node 4. ■

We now describe a practical computer scheme similar to the one used in Algor for the solution of the resulting system of algebraic equations. The significance of this scheme is that it takes advantage of the fact that the stiffness method produces a banded $\underline{K}$ matrix in which the nonzero elements occur about the main diagonal in $\underline{K}$. While the equations are solved, this banded format is maintained.

Example B.5

We will now use a simple example to illustrate this computer scheme. Consider the three-spring assemblage shown in Figure B–8. The assemblage is subjected to forces at node 2 of 100 lb in the x direction and 200 lb in the y direction. Node 1 is completely constrained from displacement in both the x and y directions, whereas node 3 is completely constrained in the y direction but is displaced a known amount δ in the x direction.

Our purpose here is not to obtain the actual $\underline{K}$ for the assemblage but rather to illustrate the scheme used for solution. The general solution can be shown to be given by

$$\begin{bmatrix} k_{11} & k_{12} & k_{13} & k_{14} & k_{15} & k_{16} \\ & k_{22} & k_{23} & k_{24} & k_{25} & k_{26} \\ & & k_{33} & k_{34} & k_{35} & k_{36} \\ & & & k_{44} & k_{45} & k_{46} \\ & & & & k_{55} & k_{56} \\ \text{Symmetry} & & & & & k_{66} \end{bmatrix} \begin{Bmatrix} d_{1x} \\ d_{1y} \\ d_{2x} \\ d_{2y} \\ d_{3x} \\ d_{3y} \end{Bmatrix} = \begin{Bmatrix} F_{1x} \\ F_{1y} \\ F_{2x} = 100 \\ F_{2y} = 200 \\ F_{3x} \\ F_{3y} \end{Bmatrix} \tag{B.4.12}$$

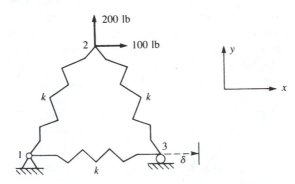

Figure B–8 Three-spring assemblage

where $\underline{K}$ has been left in general form. Upon our imposing the boundary conditions, the computer program transforms Eq. (B.4.12) to:

$$
\begin{bmatrix}
1 & 0 & 0 & 0 & 0 & 0 \\
0 & 1 & 0 & 0 & 0 & 0 \\
0 & 0 & k_{33} & k_{34} & 0 & 0 \\
0 & 0 & k_{43} & k_{44} & 0 & 0 \\
0 & 0 & 0 & 0 & 1 & 0 \\
0 & 0 & 0 & 0 & 0 & 1
\end{bmatrix}
\begin{Bmatrix}
d_{1x} \\
d_{1y} \\
d_{2x} \\
d_{2y} \\
d_{3x} \\
d_{3y}
\end{Bmatrix}
=
\begin{Bmatrix}
0 \\
0 \\
100 - k_{35}\delta \\
200 - k_{45}\delta \\
\delta \\
0
\end{Bmatrix}
\tag{B.4.13}
$$

From Eq. (B.4.13), we can see that $d_{1x} = 0$, $d_{1y} = 0$, $d_{3y} = 0$, and $d_{3x} = \delta$. These displacements are consistent with the imposed boundary conditions. The unknown displacements, d_{2x} and d_{2y}, can be determined routinely by solving Eq. (B.4.13).

We will now explain the computer scheme that is generally applicable to transform Eq. (B.4.12) to Eq. (B.4.13). First, the terms associated with the known displacement boundary condition(s) within each equation were transformed to the right side of those equations. In the third and fourth equations of Eq. (B.4.12), $k_{35}\delta$ and $k_{45}\delta$ were transformed to the right side, as shown in Eq. (B.4.13). Then the right-side force term corresponding to the known displacement row was equated to the known displacement. In the fifth equation of Eq. (B.4.12), where $d_{3x} = \delta$, the right-side, fifth-row force term F_{3x} was equated to the known displacement δ, as shown in Eq. (B.4.13). For the homogeneous boundary conditions, the affected rows of $\underline{F}$, corresponding to the zero-displacement rows, were replaced with zeros. Again, this is done in the computer scheme only to obtain the nodal displacements and does not imply that these nodal forces are zero. We obtain the unknown nodal forces by determining the nodal displacements and back-substituting these results into the original Eq. (B.4.12). Because $d_{1x} = 0$, $d_{1y} = 0$, and $d_{3y} = 0$ in Eq. (B.4.12), the first, second, and sixth rows of the force matrix of Eq. (B.4.13) were set to zero. Finally, for both nonhomogeneous and homogeneous boundary conditions, the rows and columns of $\underline{K}$ corresponding to these prescribed boundary conditions were set to zero except the main diagonal, which was made unity. That is, the first, second, fifth, and sixth rows and columns of $\underline{K}$ in Eq. (B.4.12) were set to zero, except for the main diagonal terms, which were made unity. Although doing so is not necessary, setting the main diagonal terms equal to 1 facilitates the simultaneous solution of the six equations in Eq. (B.4.13) by an elimination method used in the computer program. This modification is shown in the $\underline{K}$ matrix of Eq. (B.4.13). ■

References

[1] Southworth, R. W., and DeLeeuw, S. L., *Digital Computation and Numerical Methods*, McGraw-Hill, New York, 1965.
[2] James, M. L., Smith, G. M., and Wolford, J. C., *Applied Numerical Methods for Digital Computation*, 3rd ed., Harper & Row, New York, 1985.

[3] Bathe, K. J., and Wilson, E. L., *Numerical Methods in Finite Element Analysis*, Prentice-Hall, Englewood Cliffs, NJ, 1976.

[4] SYSTEM/360, Scientific Subroutine Package, IBM.

[5] Kardestuncer, H., *Elementary Matrix Analysis of Structures*, McGraw-Hill, New York, 1974.

[6] Collins, R. J., "Bandwidth Reduction by Automatic Renumbering," *International Journal For Numerical Methods in Engineering*, Vol. 6, pp. 345–356, 1973.

[7] Melosh, R. J., and Bamford, R. M., "Efficient Solution of Load-Deflection Equations," *Journal of the Structural Division*, American Society of Civil Engineers, No. ST4, pp. 661–676, April 1969.

[8] Irons, B. M., "A Frontal Solution Program for Finite Element Analysis," *International Journal for Numerical Methods in Engineering*, Vol. 2, No. 1, pp. 5–32, 1970.

[9] Meyer, C., "Solution of Linear Equations-State-of-the-Art," *Journal of the Structural Division*, American Society of Civil Engineers, Vol. 99, No. ST7, pp. 1507–1526, 1973.

[10] Jennings, A., *Matrix Computation for Engineers and Scientists*, Wiley, London, 1977.

[11] Cook, R. D., Malkus, D. S., and Plesha, M. E., *Concepts and Applications of Finite Element Analysis*, 3rd ed., Wiley, New York, 1989.

[12] Bathe, K. J., and Wilson, E. L., *Numerical Methods in Finite Element Analysis*, Prentice-Hall, Englewood Cliffs, NJ, 1976.

▲ Problems

B.1 Determine the solution of the following simultaneous equations by Cramer's rule.

$$1x_1 + 3x_2 = 5$$
$$4x_1 - 1x_2 = 12$$

B.2 Determine the solution of the following simultaneous equations by the inverse method.

$$1x_1 + 3x_2 = 5$$
$$4x_1 - 1x_2 = 12$$

B.3 Solve the following system of simultaneous equations by Gaussian elimination.

$$x_1 - 4x_2 - 5x_3 = 4$$
$$3x_2 + 4x_3 = -1$$
$$-2x_1 - 1x_2 + 2x_3 = -3$$

B.4 Solve the following system of simultaneous equations by Gaussian elimination.

$$2x_1 + 1x_2 - 3x_3 = 11$$
$$4x_1 - 2x_2 + 3x_3 = 8$$
$$-2x_1 + 2x_2 - 1x_3 = -6$$

B.5 Given that

$$x_1 = 2y_1 - y_2 \qquad z_1 = -x_1 - x_2$$
$$x_2 = y_1 - y_2 \qquad z_2 = 2x_1 + x_2$$

a. Write these relationships in matrix form.
b. Express z in terms of y.
c. Express y in terms of z.

B.6 Starting with the initial guess $\underline{X}^T = [1 \quad 1 \quad 1 \quad 1 \quad 1]$, perform five iterations of the Gauss–Seidel method on the following system of equations. On the basis of the results of these five iterations, what is the exact solution?

$$
\begin{aligned}
2x_1 - 1x_2 &= -1 \\
-1x_1 + 6x_2 - 1x_3 &= 4 \\
-2x_2 + 4x_3 - 1x_4 &= 4 \\
-1x_3 + 4x_4 - 1x_5 &= 6 \\
-1x_4 + 2x_5 &= -2
\end{aligned}
$$

B.7 Solve Problem B.1 by Gauss–Seidel iteration.

B.8 Classify the solutions to the following systems of equations according to Section B.2 as unique, nonunique, or nonexistent.

a. $\quad 2x_1 - 4x_2 = 2$
$\quad -9x_1 + 12x_2 = -6$

b. $10x_1 + 1x_2 = 0$
$\quad 5x_1 + \frac{1}{2}x_2 = 3$

c. $2x_1 + 1x_2 + 1x_3 = 6$
$\quad 3x_1 + 1x_2 - 1x_3 = 4$
$\quad 5x_1 + 2x_2 + 2x_3 = 8$

d. $1x_1 + 1x_2 + 1x_3 = 1$
$\quad 2x_1 + 2x_2 + 2x_3 = 2$
$\quad 3x_1 + 3x_2 + 3x_3 = 3$

B.9 Determine the bandwidths of the plane trusses shown in Figure PB–9. What conclusions can you draw regarding labeling of nodes?

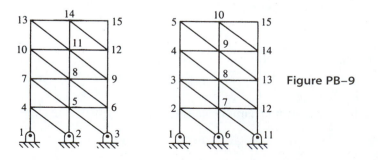

Figure PB–9

C

Equations from Elasticity Theory

Introduction

In this appendix, we will develop the basic equations of the theory of elasticity. These equations should be referred to frequently throughout the structural mechanics portions of this text.

There are three basic sets of equations included in theory of elasticity. These equations must be satisfied if an exact solution to a structural mechanics problem is to be obtained. These sets of equations are (1) the differential equations of equilibrium formulated here in terms of the stresses acting on a body, (2) the strain/displacement and compatibility differential equations, and (3) the stress/strain or material constitutive laws.

▲ C.1 Differential Equations of Equilibrium ▲

For simplicity, we initially consider the equilibrium of a plane element subjected to normal stresses σ_x and σ_y, in-plane shear stress τ_{xy} (in units of force per unit area), and body forces X_b and Y_b (in units of force per unit volume), as shown in Figure C–1. The stresses are assumed to be constant as they act on the width of each face. However, the stresses are assumed to vary from one face to the opposite. For example, we have σ_x acting on the left vertical face, whereas $\sigma_x + (\partial\sigma_x/\partial x)\,dx$ acts on the right vertical face. The element is assumed to have unit thickness.

Summing forces in the x direction, we have

$$\sum F_x = 0 = \left(\sigma_x + \frac{\partial\sigma_x}{\partial x}dx\right)dy(1) - \sigma_x\,dy(1) + X_b\,dx\,dy(1)$$

$$+ \left(\tau_{yx} + \frac{\partial\tau_{yx}}{\partial y}dy\right)dx(1) - \tau_{yx}\,dx(1) = 0 \tag{C.1.1}$$

After simplifying and canceling terms in Eq. (C.1.1), we obtain

$$\frac{\partial\sigma_x}{\partial x} + \frac{\partial\tau_{yx}}{\partial y} + X_b = 0 \tag{C.1.2}$$

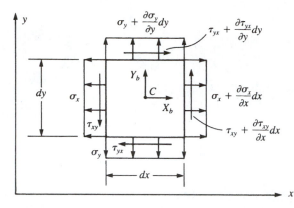

Figure C–1 Plane differential element subjected to stresses

Similarly, summing forces in the y direction, we obtain

$$\frac{\partial \sigma_y}{\partial y} + \frac{\partial \tau_{xy}}{\partial x} + Y_b = 0 \tag{C.1.3}$$

Because we are considering only the planar element, three equilibrium equations must be satisfied. The third equation is equilibrium of moments about an axis normal to the x-y plane; that is, taking moments about point C in Figure C–1, we have

$$\sum M_z = 0 = \tau_{xy}\,dy(1)\frac{dx}{2} + \left(\tau_{xy} + \frac{\partial \tau_{xy}}{\partial x}dx\right)\frac{dx}{2}$$

$$- \tau_{yx}\,dx(1)\frac{dy}{2} - \left(\tau_{yx} + \frac{\partial \tau_{yx}}{\partial y}dy\right)\frac{dy}{2} = 0 \tag{C.1.4}$$

Simplifying Eq. (C.1.4) and neglecting higher-order terms yields

$$\tau_{xy} = \tau_{yx} \tag{C.1.5}$$

We now consider the three-dimensional state of stress shown in Figure C–2, which shows the additional stresses σ_z, τ_{xz}, and τ_{yz}. For clarity, we show only the stresses on three mutually perpendicular planes. With a straightforward procedure, we can extend the two-dimensional equations (C.1.2), (C.1.3), and (C.1.5) to three dimensions. The resulting total set of equilibrium equations is

$$\frac{\partial \sigma_x}{\partial x} + \frac{\partial \tau_{xy}}{\partial y} + \frac{\partial \tau_{xz}}{\partial z} + X_b = 0$$

$$\frac{\partial \tau_{xy}}{\partial x} + \frac{\partial \sigma_y}{\partial y} + \frac{\partial \tau_{yz}}{\partial z} + Y_b = 0 \tag{C.1.6}$$

$$\frac{\partial \tau_{xz}}{\partial x} + \frac{\partial \tau_{yz}}{\partial y} + \frac{\partial \sigma_z}{\partial z} + Z_b = 0$$

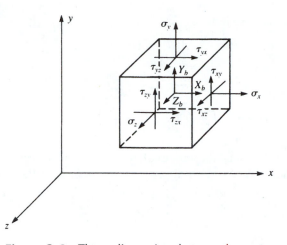

Figure C–2 Three-dimensional stress element

and $$\tau_{xy} = \tau_{yx} \qquad \tau_{xz} = \tau_{zx} \qquad \tau_{yz} = \tau_{zy} \tag{C.1.7}$$

▲ C.2 Strain/Displacement and Compatibility Equations ▲

We first obtain the strain/displacement or kinematic differential relationships for the two-dimensional case. We begin by considering the differential element shown in Figure C–3, where the undeformed state is represented by the dashed lines and the deformed shape (after straining takes place) is represented by the solid lines.

Considering line element AB in the x direction, we can see that it becomes $A'B'$ after deformation, where u and v represent the displacements in the x and y directions. By the definition of engineering normal strain (that is, the change in length divided by

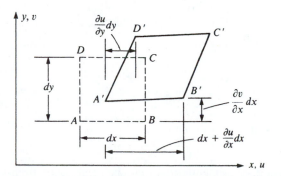

Figure C–3 Differential element before and after deformation

the original length of a line), we have

$$\varepsilon_x = \frac{A'B' - AB}{AB} \tag{C.2.1}$$

Now

$$AB = dx \tag{C.2.2}$$

and

$$(A'B')^2 = \left(dx + \frac{\partial u}{\partial x}dx\right)^2 + \left(\frac{\partial v}{\partial x}dx\right)^2 \tag{C.2.3}$$

Therefore, evaluating $A'B'$ using the binomial theorem and neglecting the higher-order terms $(\partial u/\partial x)^2$ and $(\partial v/\partial x)^2$ (an approach consistent with the assumption of small strains), we have

$$A'B' = dx + \frac{\partial u}{\partial x}dx \tag{C.2.4}$$

Using Eqs. (C.2.2) and (C.2.4) in Eq. (C.2.1), we obtain

$$\varepsilon_x = \frac{\partial u}{\partial x} \tag{C.2.5}$$

Similarly, considering line element AD in the y direction, we have

$$\varepsilon_y = \frac{\partial v}{\partial y} \tag{C.2.6}$$

The shear strain γ_{xy} is defined to be the change in the angle between two lines, such as AB and AD, that originally formed a right angle. Hence, from Figure C–3, we can see that γ_{xy} is the sum of two angles and is given by

$$\gamma_{xy} = \frac{\partial u}{\partial y} + \frac{\partial v}{\partial x} \tag{C.2.7}$$

Equations (C.2.5)–(C.2.7) represent the strain/displacement relationships for in-plane behavior.

For three-dimensional situations, we have a displacement w in the z direction. It then becomes straightforward to extend the two-dimensional derivations to the three-dimensional case to obtain the additional strain/displacement equations as

$$\varepsilon_z = \frac{\partial w}{\partial z} \tag{C.2.8}$$

$$\gamma_{xz} = \frac{\partial u}{\partial z} + \frac{\partial w}{\partial x} \tag{C.2.9}$$

$$\gamma_{yz} = \frac{\partial v}{\partial z} + \frac{\partial w}{\partial y} \tag{C.2.10}$$

Along with the strain/displacement equations, we need compatibility equations to ensure that the displacement components $u, v,$ and w are single-valued continuous functions so that tearing or overlap of elements does not occur. For the planar-elastic case, we obtain the compatibility equation by differentiating γ_{xy} with respect to both x and y and then using the definitions for ε_x and ε_y given by Eqs. (C.2.5) and (C.2.6). Hence,

$$\frac{\partial^2 \gamma_{xy}}{\partial x \partial y} = \frac{\partial^2}{\partial x \partial y}\frac{\partial u}{\partial y} + \frac{\partial^2}{\partial x \partial y}\frac{\partial v}{\partial x} = \frac{\partial^2 \varepsilon_x}{\partial y^2} + \frac{\partial^2 \varepsilon_y}{\partial x^2} \qquad (C.2.11)$$

where the second equation in terms of the strains on the right side is obtained by noting that single-valued continuity of displacements requires that the partial differentiations with respect to x and y be interchangeable in order. Therefore, we have $\partial^2/\partial x \partial y = \partial^2/\partial y \partial x$. Equation (C.2.11) is called the *condition of compatibility*, and it must be satisfied by the strain components in order for us to obtain unique expressions for u and v. Equations (C.2.5), (C.2.6), (C.2.7), and (C.2.11) together are then sufficient to obtain unique single-valued functions for u and v.

In three dimensions, we obtain five additional compatibility equations by differentiating γ_{xz} and γ_{yz} in a manner similar to that described above for γ_{xy}. We need not list these equations here; details of their derivation can be found in Reference [1].

In addition to the compatibility conditions that ensure single-valued continuous functions within the body, we must also satisfy displacement or kinematic boundary conditions. This simply means that the displacement functions must also satisfy prescribed or given displacements on the surface of the body. These conditions often occur as support conditions from rollers and/or pins. In general, we might have

$$u = u_0 \qquad v = v_0 \qquad w = w_0 \qquad (C.2.12)$$

at specified surface locations on the body. We may also have conditions other than displacements prescribed (for example, prescribed rotations).

▲ C.3 Stress/Strain Relationships

We will now develop the three-dimensional stress/strain relationships for an isotropic body only. This is done by considering the response of a body to imposed stresses. We subject the body to the stresses σ_x, σ_y, and σ_z independently as shown in Figure C–4.

We first consider the change in length of the element in the x direction due to the independent stresses σ_x, σ_y, and σ_z. We assume the principle of superposition to hold; that is, we assume that the resultant strain in a system due to several forces is the algebraic sum of their individual effects.

Considering Figure C–4(b), the stress in the x direction produces a positive strain

$$\varepsilon_x' = \frac{\sigma_x}{E} \qquad (C.3.1)$$

where Hooke's law, $\sigma = E\varepsilon$, has been used in writing Eq. (C.3.1), and E is defined as the *modulus of elasticity*. Considering Figure C–4(c), the positive stress in the y direc-

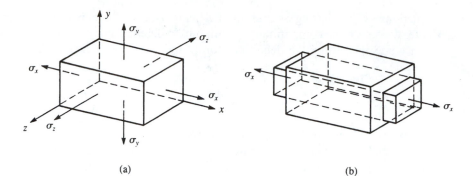

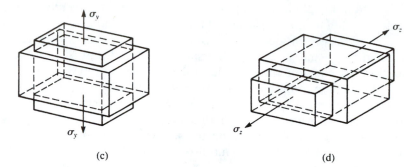

(a)

(b)

(c)

(d)

Figure C–4 Element subjected to normal stress acting in three mutually perpendicular directions

tion produces a negative strain in the x direction as a result of Poisson's effect given by

$$\varepsilon_x'' = -\frac{v\sigma_y}{E}$$ (C.3.2)

where v is Poisson's ratio. Similarly, considering Figure C–4(d), the stress in the z direction produces a negative strain in the x direction given by

$$\varepsilon_x''' = -\frac{v\sigma_z}{E}$$ (C.3.3)

Using superposition of Eqs. (C.3.1)–(C.3.3), we obtain

$$\varepsilon_x = \frac{\sigma_x}{E} - v\frac{\sigma_y}{E} - v\frac{\sigma_z}{E}$$ (C.3.4)

The strains in the y and z directions can be determined in a manner similar to that used to obtain Eq. (C.3.4) for the x direction. They are

$$\varepsilon_y = -v\frac{\sigma_x}{E} + \frac{\sigma_y}{E} - v\frac{\sigma_z}{E}$$

$$\varepsilon_z = -v\frac{\sigma_x}{E} - v\frac{\sigma_y}{E} + \frac{\sigma_z}{E}$$ (C.3.5)

Solving Eqs. (C.3.4) and (C.3.5) for the normal stresses, we obtain

$$\sigma_x = \frac{E}{(1+v)(1-2v)}[\varepsilon_x(1-v) + v\varepsilon_y + v\varepsilon_z]$$

$$\sigma_y = \frac{E}{(1+v)(1-2v)}[v\varepsilon_x + (1-v)\varepsilon_y + v\varepsilon_z] \tag{C.3.6}$$

$$\sigma_z = \frac{E}{(1+v)(1-2v)}[v\varepsilon_x + v\varepsilon_y + (1-v)\varepsilon_z]$$

The Hooke's law relationship, $\sigma = E\varepsilon$, used for normal stress also applies for shear stress and strain; that is,

$$\tau = G\gamma \tag{C.3.7}$$

where G is the *shear modulus*. Hence, the expressions for the three different sets of shear strains are

$$\gamma_{xy} = \frac{\tau_{xy}}{G} \qquad \gamma_{yz} = \frac{\tau_{yz}}{G} \qquad \gamma_{zx} = \frac{\tau_{zx}}{G} \tag{C.3.8}$$

Solving Eqs. (C.3.8) for the stresses, we have

$$\tau_{xy} = G\gamma_{xy} \qquad \tau_{yz} = G\gamma_{yz} \qquad \tau_{zx} = G\gamma_{zx} \tag{C.3.9}$$

In matrix form, we can express the stresses in Eqs. (C.3.6) and (C.3.9) as

$$
\begin{Bmatrix} \sigma_x \\ \sigma_y \\ \sigma_z \\ \tau_{xy} \\ \tau_{yz} \\ \tau_{zx} \end{Bmatrix} = \frac{E}{(1+v)(1-2v)}
$$

$$
\times \begin{bmatrix} 1-v & v & v & 0 & 0 & 0 \\ & 1-v & v & 0 & 0 & 0 \\ & & 1-v & 0 & 0 & 0 \\ & & & \frac{1-2v}{2} & 0 & 0 \\ & & & & \frac{1-2v}{2} & 0 \\ \text{Symmetry} & & & & & \frac{1-2v}{2} \end{bmatrix} \begin{Bmatrix} \varepsilon_x \\ \varepsilon_y \\ \varepsilon_z \\ \gamma_{xy} \\ \gamma_{yz} \\ \gamma_{zx} \end{Bmatrix} \tag{C.3.10}
$$

where we note that the relationship

$$G = \frac{E}{2(1+v)}$$

has been used in Eq. (C.3.10). The square matrix on the right side of Eq. (C.3.10) is called the *stress/strain* or *constitutive matrix* and is defined by $\underline{D}$, where $\underline{D}$ is

$$[D] = \frac{E}{(1+v)(1-2v)} \begin{bmatrix} 1-v & v & v & 0 & 0 & 0 \\ & 1-v & v & 0 & 0 & 0 \\ & & 1-v & 0 & 0 & 0 \\ & & & \dfrac{1-2v}{2} & 0 & 0 \\ & & & & \dfrac{1-2v}{2} & 0 \\ \text{Symmetry} & & & & & \dfrac{1-2v}{2} \end{bmatrix} \tag{C.3.11}$$

▲ Reference

[1] Timoshenko, S., and Goodier, J., *Theory of Elasticity*, 3rd ed., McGraw-Hill, New York, 1970.

D

Equivalent Nodal Forces

The equivalent nodal (or joint) forces for different types of loads on beam elements are shown in Table D–1.

Problems

D.1 Determine the equivalent joint or nodal forces for the beam elements shown in Figure PD–1.

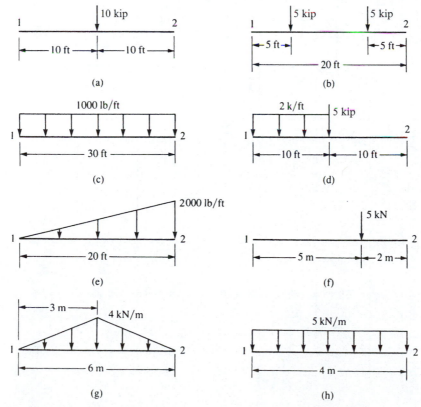

Figure PD–1

Table D–1   Equivalent joint forces f_0 for different types of loads

	f_{1y}	m_1	Loading case	f_{2y}	m_2
1.	$\dfrac{-P}{2}$	$\dfrac{-PL}{8}$		$\dfrac{-P}{2}$	$\dfrac{PL}{8}$
2.	$\dfrac{-Pb^2(L+2a)}{L^3}$	$\dfrac{-Pab^2}{L^2}$		$\dfrac{-Pa^2(L+2b)}{L^3}$	$\dfrac{Pa^2b}{L^2}$
3.	$-P$	$-\alpha(1-\alpha)PL$		$-P$	$\alpha(1-\alpha)PL$
4.	$\dfrac{-wL}{2}$	$\dfrac{-wL^2}{12}$		$\dfrac{-wL}{2}$	$\dfrac{wL^2}{12}$
5.	$\dfrac{-7wL}{20}$	$\dfrac{-wL^2}{20}$		$\dfrac{-3wL}{20}$	$\dfrac{wL^2}{30}$
6.	$\dfrac{-wL}{4}$	$\dfrac{-5wL^2}{96}$		$\dfrac{-wL}{4}$	$\dfrac{5wL^2}{96}$
7.	$\dfrac{-13wL}{32}$	$\dfrac{-11wL^2}{192}$		$\dfrac{-3wL}{32}$	$\dfrac{5wL^2}{192}$

Table D–1 (Continued)

	f_{1y}	m_1	Loading case	f_{2y}	m_2
8.	$\dfrac{-wL}{3}$	$\dfrac{-wL^2}{15}$	w (parabolic loading) L	$\dfrac{-wL}{3}$	$\dfrac{wL^2}{15}$
9.	$\dfrac{-M(a^2 + b^2 - 4ab + L^2)}{L^3}$	$\dfrac{Mb(2a - b)}{L^2}$	$a > L/3$ $\quad b > L/3$ M a b L	$\dfrac{M(a^2 + b^2 - 4ab + L^2)}{L^3}$	$\dfrac{Ma(2b - a)}{L^2}$

Positive nodal force conventions

E

Principle of Virtual Work

In this appendix, we will use the principle of virtual work to derive the general finite element equations for a dynamic system.

Strictly speaking, the principle of virtual work applies to a static system, but through the introduction of D'Alembert's principle, we will be able to use the principle of virtual work to derive the finite element equations applicable for a dynamic system.

The principle of virtual work is stated as follows:

> If a deformable body in equilibrium is subjected to arbitrary virtual (imaginary) displacements associated with a compatible deformation of the body, the virtual work of external forces on the body is equal to the virtual strain energy of the internal stresses.

In the principle, *compatible displacements* are those that satisfy the boundary conditions and ensure that no discontinuities, such as voids or overlaps, occur within the body. Figure E–1 shows the hypothetical actual displacement, a compatible (admissible) displacement, and an incompatible (inadmissible) displacement for a simply supported beam. Here δv represents the variation in the transverse displacement function v. In the finite element formulation, δv would be replaced by nodal degrees of freedom δd_i. The inadmissible displacements shown in Figure E–1(b) are the result when the support condition at the right end of the beam and the continuity of displacement and slope within the beam are not satisfied. For more details of this principle, consult structural mechanics references such as Reference [1]. Also, for additional descriptions of strain energy and work done by external forces (as applied to a bar), see Section 3.10.

Applying the principle to a finite element, we have

$$\delta U^{(e)} = \delta W^{(e)} \tag{E.1}$$

where $\delta U^{(e)}$ is the virtual strain energy due to internal stresses and $\delta W^{(e)}$ is the virtual work of external forces on the element. We can express the internal virtual strain energy using matrix notation as

$$\delta U^{(e)} = \iiint\limits_{V} \delta \underline{\varepsilon}^T \underline{\sigma} \, dV \tag{E.2}$$

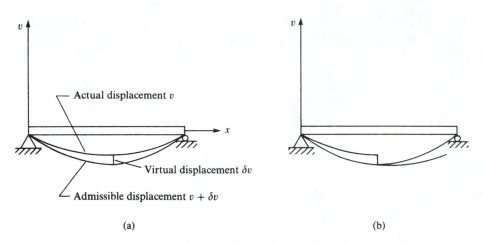

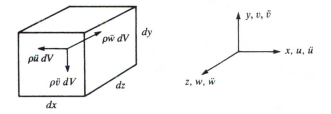

Figure E–1 (a) Admissible and (b) inadmissible virtual displacement functions

Figure E–2 Effective forces acting on an element

From Eq. (E.2), we can observe that internal strain energy is due to internal stresses moving through virtual strains $\delta\varepsilon$. The external virtual work is due to nodal, surface, and body forces. In addition, application of D'Alembert's principle yields effective or inertial forces $\rho\ddot{u}\,dV$, $\rho\ddot{v}\,dV$, and $\rho\ddot{w}\,dV$, where the double dots indicate second derivatives of the translations u, v, and w in the x, y, and z directions, respectively, with respect to time. These forces are shown in Figure E–2. According to D'Alembert's principle, these effective forces act in directions that are opposite to the assumed positive sense of the accelerations. We can now express the external virtual work as

$$\delta W^{(e)} = \delta\underline{d}^T\underline{P} + \iint_S \delta\underline{\psi}_s^T\underline{T}\,dS + \iiint_V \delta\underline{\psi}^T(\underline{X} - \rho\ddot{\underline{\psi}})\,dV \qquad (E.3)$$

where $\delta\underline{d}$ is the vector of virtual nodal displacements, $\delta\underline{\psi}$ is the vector of virtual displacement functions $\delta u, \delta v$, and δw, $\delta\underline{\psi}_s$ is the vector of virtual displacement functions acting over the surface where surface tractions occur, $\underline{P}$ is the nodal load matrix, $\underline{T}$ is the surface force per unit area matrix, and $\underline{X}$ is the body force per unit volume matrix.

Substituting Eqs. (E.2) and (E.3) into Eq. (E.1), we obtain

$$\iiint_V \delta\underline{\varepsilon}^T\underline{\sigma}\,dV = \delta\underline{d}^T\underline{P} + \iint_S \delta\underline{\psi}_s^T\underline{T}\,dS + \iiint_V \delta\ddot{\underline{\psi}}^T(\underline{X} - \rho\underline{\psi})\,dV \qquad (E.4)$$

As shown throughout this text, shape functions are used to relate displacement functions to nodal displacements as

$$\underline{\psi} = \underline{N}\underline{d} \qquad \underline{\psi}_s = \underline{N}_s\underline{d} \tag{E.5}$$

$\underline{N}_s$ is the shape function matrix evaluated on the surface where traction $\underline{T}$ occurs.

Strains are related to nodal displacements as

$$\underline{\varepsilon} = \underline{B}\underline{d} \tag{E.6}$$

and stresses are related to strains by

$$\underline{\sigma} = \underline{D}\underline{\varepsilon} \tag{E.7}$$

Hence, substituting Eqs. (E.5), (E.6), and (E.7) for $\underline{\psi}, \underline{\varepsilon},$ and $\underline{\sigma}$ into Eq. (E.4), we obtain

$$\iiint_V \delta\underline{d}^T \underline{B}^T \underline{D}\underline{B}\underline{d}\, dV = \delta\underline{d}^T \underline{P} + \iint_S \delta\underline{d}^T \underline{N}_s^T \underline{T}\, dS + \iiint_V \delta\underline{d}^T \underline{N}^T (\underline{X} - \rho\underline{N}\ddot{\underline{d}})\, dV \tag{E.8}$$

Note that the shape functions are independent of time. Because $\underline{d}$ (or $\underline{d}^T$) is the matrix of nodal displacements, which is independent of spatial integration, we can simplify Eq. (E.8) by taking the $\underline{d}^T$ terms from the integrals to obtain

$$\delta\underline{d}^T \iiint_V \underline{B}^T \underline{D}\underline{B}\, dV\underline{d} = \delta\underline{d}^T \underline{P} + \delta\underline{d}^T \iint_S \underline{N}_s^T \underline{T}\, dS + \delta\underline{d}^T \iiint_V \underline{N}^T (\underline{X} - \rho\underline{N}\ddot{\underline{d}})\, dV \tag{E.9}$$

Because $\delta\underline{d}^T$ is an arbitrary virtual nodal displacement vector common to each term in Eq. (E.9), the following relationship must be true.

$$\iiint_V \underline{B}^T \underline{D}\underline{B}\, dV\underline{d} = \underline{P} + \iint_S \underline{N}_s^T \underline{T}\, dS + \iiint_V \underline{N}^T \underline{X}\, dV - \iiint_V \rho\underline{N}^T \underline{N}\, dV\ddot{\underline{d}} \tag{E.10}$$

We now define

$$\underline{m} = \iiint_V \rho\underline{N}^T \underline{N}\, dV \tag{E.11}$$

$$\underline{k} = \iiint_V \underline{B}^T \underline{D}\underline{B}\, dV \tag{E.12}$$

$$\underline{f}_s = \iint_S \underline{N}_s^T \underline{T}\, dS \tag{E.13}$$

$$\underline{f}_b = \iiint_V \underline{N}^T \underline{X}\, dV \tag{E.14}$$

Using Eqs. (E.11)–(E.14) in Eq. (E.10) and moving the last term of Eq. (E.10) to the left side, we obtain

$$\underline{m}\ddot{\underline{d}} + \underline{k}\underline{d} = \underline{P} + \underline{f}_s + \underline{f}_b \tag{E.15}$$

The matrix $\underline{m}$ in Eq. (E.11) is the element consistent-mass matrix [2], $\underline{k}$ in Eq. (E.12) is the element stiffness matrix, f_s in Eq. (E.13) is the matrix of element equivalent nodal loads due to surface forces, and f_b in Eq. (E.14) is the matrix of element equivalent nodal loads due to body forces.

Specific applications of Eq. (E.15) are given in Chapter 16 for bars and beams subjected to dynamic (time-dependent) forces. For static problems, we set $\underline{\ddot{d}}$ equal to zero in Eq. (E.15) to obtain

$$\underline{k}\underline{d} = \underline{P} + \underline{f}_s + \underline{f}_b \tag{E.16}$$

Chapters 3–10 and 12 illustrate the use of Eq. (E.16) applied to bars, trusses, beams, frames, and plane stress, axisymmetric stress, and three-dimensional stress problems.

▲ References

[1] Oden, J. T., and Ripperger, E. A., *Mechanics of Elastic Structures*, 2nd ed., McGraw-Hill, New York, 1981.

[2] Archer, J. S., "Consistent Matrix Formulations for Structural Analysis Using Finite Element Techniques," *Journal of the American Institute of Aeronautics and Astronautics*, Vol. 3, No. 10, pp. 1910–1918, 1965.

F

Basics of Algor

Introduction

In this appendix, we will introduce some basic concepts necessary to use the Algor finite element analysis system. For more complete product and purchase information, please contact your representative at Algor, Inc. (412) 967–2700.

F.1 Hardware Requirements for Windows Installation

You will need the following hardware as a minimum to run Algor:

1. An Intel Pentium PC-compatible computer running Windows 95, 98, or NT is required.
2. A hard disk drive with 120 MB or more of free disk space. You may need more hard drive disk space, depending on the size of the models that you run. Very large models require 1 GB or more of available hard disk space.
3. 32 megabytes of total RAM memory. Additional RAM memory will allow the software to run faster and will enable you to run larger models.
4. A VGA or higher-resolution color graphics card that can achieve a minimum resolution of 800×600 in Windows with a minimum of 64K colors.
5. A mouse with an installed mouse driver is required.

F.2 Conventions

Certain conventions are employed to make it easier to follow the example solutions included throughout this book. The proper keystroke or command is listed at the left side of the page, followed by an explanation of the keystroke and/or results of the action you have taken.

When you see one of the following options listed on the left side of the page, perform the indicated action described next to the option:

⟨**Alt**⟩-**X**   Press the ⟨Alt⟩ key and continue holding it down while you press the key with "X" on it (or another letter indicated).

⟨**CR**⟩   Press the ⟨Enter⟩ key or the ⟨Return⟩ key ("CR" stands for "carriage return").

⟨**Esc**⟩   Press the ⟨Escape⟩ key (or choose "Esc" from the current menu).

⟨**F#**⟩   Press the corresponding function key, where # indicates the number of the function key. (Function keys are described in Section F.4.)

⟨**Mouse**⟩   Use the mouse to click at specified locations. Algor software is designed for a two-button mouse. "Click" or "left-click" means to press the left mouse button. The left mouse button is used for entering new points and making menu selections. The right mouse button is used to snap-to the nearest point in Algor graphics programs. In the menu system, the right button acts like the ⟨Esc⟩ key.

"nAme"   Choose this option from the menu by moving the mouse cursor over the option and clicking on it or by typing the first letter that is capitalized in the option name ("A" in this example).

⟨**Tab**⟩   Press the ⟨Tab⟩ key.

▲ F.3 Getting Around the Menu System   ▲

The Algor menu system contains a list of options. You can move to submenus, return to previous menus, and execute programs within menus.

To highlight an option, go to it by moving the mouse cursor over the option or use the arrow keys. You can switch between arrow keys and mouse selection whenever you like. To activate a highlighted option, press ⟨CR⟩ or left-click the mouse.

To return to a previous menu, press ⟨Esc⟩ or right-click the mouse.

To get help, type H.

To quit from any menu, type a Q or click "Quit".

▲ F.4 Function Keys   ▲

Function keys are defined for quick access to popular tasks within Algor. Currently supported function keys are described as follows:

F1—"1Help"   Press the ⟨F1⟩ key to get on-line help. Then select any command or menu item to receive a short description of that menu or command in the dialog region. After reading the help message, press ⟨CR⟩ to return to command selection mode. If more information is available than can be shown in the dialog region, press ⟨CR⟩ to see the remainder of the help information on a full-screen text display. When you finish reading the message and press ⟨CR⟩ again, the graphics screen is redrawn and you return to command mode.

F2—"2Undo" Press the ⟨CR⟩ key to undo the effects of the previous operation. If you press F2 a second time, you can undo or restore the effects of the previous operation. Remember that F2 or undo will work only if you have not changed the selected objects. The undo status is shown in the first item on the status line. If "F2 = U," then the last action or operation can be undone; if "F2 = *", then the last action cannot be undone.

F3—"3Inp" Press ⟨F3⟩ to access the input menu. This menu allows you to specify the current input mode, such as keyboard if you want to input coordinates using the keyboard, and coordinate system, such as Cartesian, polar, or spherical.

F4—"4Snap" Press ⟨F4⟩ to access the "Snap" menu. This is used with the grid display options set with the "Settings," then "Grid" commands to show coordinates so you can more accurately position the cursor. With snap-to grid set, the cursor will jump between grid points and allow you to select the coordinates you want. Without snap-to grid set, the points you enter when defining your geometry can be literally anywhere on your screen (for instance, x-coordinate 1.011 when you meant to select exactly 1.000). The following other options are available with F4:

None—This is the default and turns off the snap-to function.

Tgrid—Toggles snap-to the secondary grid points. The Tgrid is used to break down the increments of the grid, so you can select smaller defined increments between grid points.

Grid —Toggles snap-to the current grid setting.

Point—Toggles snap-to existing points. When this option is active, if an existing point is within two diameters of the cursor, the cursor will snap-to that point. If you do not have the equivalent of a right mouse button, this allows you automatically to connect new items to points on existing items in your model.

F5—"5Cur" Press ⟨F5⟩ to access the "Current" menu. In Superdraw III, this menu allows you to set the values for parameters that affect the way your model is currently displayed. These parameters include the current color (for modeling), the current height and style settings for text that is being added, the current layer and group shown, and options for modifying the display of the miniaxis. These settings are also found on the "Settings", then "Current" menu.

In Superview, ⟨F5⟩ accesses the "O General" menu. This menu allows you to toggle grid display, modify the miniaxis display or position, apply labels, and utilize other functions that change the model display. These settings are also found under the "Options," then "General" menu.

F6—"6Swtc" Press ⟨F6⟩ to access the "Switches" menu. This menu allows you to set the value of different display switches in Superdraw III and Superview. A switch is either active (on) or inactive (off). An asterisk (*) is shown before active switches. The "F6Swtc" menus in Superdraw III and Superview are different. See Docutech On-line Documentation for more information about these two versions.

F7—"7Big" Press ⟨F7⟩ to toggle the size of the cursor in the display window. The default cursor is shaped like a small plus (+) sign. Press ⟨F7⟩ once to change the small graphics cursor to a large cross-hair cursor. Press ⟨F7⟩ again to restore the cursor to its original size.

F8—"8Menu" Press ⟨F8⟩ to suspend dialog entry and select one menu item from the menu by its index character. This can be useful in keyboard entry if you want to select a menu item instead of entering coordinates.

When you are entering a filename, such as when loading a file, the ⟨F8⟩ key displays a list of files in the current directory with the appropriate extension (in Superdraw this is .ESD; in Superview it is .SST). With different commands, the file extension changes. For example, with the "File," then "Import," then "Autocad" command, files with a .DXF extension are displayed.

F9—"9Top" Press ⟨F9⟩ to return to the main menu in the graphics program. This "top" menu is the one that is first displayed when you execute the program, and it is the one from which you access most commands and submenus.

F10—"0Draw" Press ⟨F10⟩ to access the "Draw" menu. Options on this menu allow you to modify or set the parameters associated with the display window. These options include changing views, zooming in or out, setting the display window size, redrawing and enclosing the contents of the window, and Jetview.

▲ F.5 Algor Processor Names

After a model is created (normally by using Superdraw III), it may be run through the analysis processor to compute the output, such as displacements and stresses or nodal temperatures and heat fluxes. Depending on the element type and loading involved, there are approximately 20 processors in the Algor linear stress, vibration, and heat-transfer system. Table F–1 lists the processor filename along with a brief description of each of these processors. These processors are invoked from the "Model Data Control" window.

Note that the stresses are computed directly in the static stress processor. The heat-transfer processors compute only temperatures and associated fluxes. Stresses due to temperature distribution are computed by imposing the temperatures on the stress model and then running the static stress processor.

In general, the deflections are computed as the end result of the processor analysis for a stress analysis problem. After the deflections are computed, they are saved to a file, and an additional program (usually MKNS0) is called that computes the stresses on the basis of the model geometry and the computed deflections. This linking process is automatically performed by the processor.

Table F–1 Algor analysis processor names

Filename	Type of Processor
SSAP0	Static stress
SSAP0C	Static stress including composite elements
SSAP0G	Nonlinear cable/gap stress
SSAP1	Modal analysis
SSAP1C	Modal analysis including composite elements
SSAP2	Time-history-modal superposition
SSAP3	Response spectrum-modal superposition
SSAP4	Direct integration
SSAP5	Harmonic response-modal superposition
SSAP6	Buckling-beams only
SSAP6S	Buckling-beams and plates
SSAP7	Random response-modal superposition
SSAP8	Modal analysis with load stiffening—beams only
SSAP8S	Modal analysis with load stiffening—beams and plates
SSAP9	Weight, center of gravity, and moment of inertia
SSAP10	Steady-state heat transfer
SSAP11	Transient heat transfer

▲ ## F.6 File Extensions Generated by the Algor System ▲

During creation and analysis of the model, a number of different files are created by the Algor system. Each of these files has the model name assigned by the operator and an extension that is used to distinguish between the files. Table F–2 lists files created by the Algor system during model creation and analysis.

▲ ## F.7 Checking Model for Defects by Using Superview ▲

After your model has been successfully decoded, you are ready to run it through the processor. However, it is recommended that you check the model for defects before processing it. Defects typically occur in the following areas: geometry, topology or connectivity, applied boundary conditions, and applied loadings.

Some features of Superview will be demonstrated in this section. Superview is used only for displaying or viewing work that has already been done; no changes to the data are possible in Superview. Example 8.4, involving a plane stress plate with a hole and pressure loading along the right edge, will be used to demonstrate some of Superview's features. We assume you will enter Superview from the "Model Data Control" window after starting in Superdraw III and bringing up the model.

Table F–2   File extensions generated by the Algor system

File	Type	Notes
Filename.	ASCII	Processor input file created by decoder and/or other preprocessors.
Filename.esd	Binary	Input/graphics file created by Superdraw or other graphics programs; can be used by decoder, Superview, BEdit, and others.
Filename.bak	Binary	Backup file created by Superdraw.
Filename.mat	ASCII	Material input/properties file for decoder created through any editor.
Filename.mac	ASCII	Composite element material input/properties file created through any editor; used by composite decoder.
Filename.sst	Binary	Superview input/graphics file created by decoder or Mksv.
Filename.do	Binary	FEA node point displacement (or temperature) file created by processors.
Filename.nso	Binary	FEA node point stress file created by processors or MKNSO program.
Filename.svd	Binary	Parameter/graphics file which you can save for Superview and Superview-Electrostat programs, associated with a specific model.
Filename.stt	Binary	Thermal Superview input/graphics file created by decoder or Mksv.
Filename.std	Binary	Parameter/graphics file which you can save for Superview-Thermal program, associated with a specific model.
Filename.set	Binary	Superview-Electrostat input/graphics file created by decoder.
Filename.bed	Binary	BEdit input/graphics file.
Program.cfg	Binary	Configuration file which you can save for any graphics program. The algor.cfg file, modified using Setgraph and Setmode, contains graphics configuration data used by all Algor graphics programs.
Filename.1	ASCII	Input echo and results output file created by processors (.l10 for temperature).
Filename.s	ASCII	Stress output file created by processors.
Filename.su4	Binary	Input/graphics file for Supersurf surface modeling program.
ALGnn.tmp	Binary	Temporary file for graphics programs, automatically deleted.

Step 1   Start Superview

Click on the "Algor FEA" icon. You are now in the Superdraw III program.

Click on a file (*ex84.esd*) from the file list on your computer. Here we assume the file *ex84.esd* has been created previously. The model appears drawn on the screen in a few seconds. Red triangles appear along the left edge at each left edge node to indicate complete fixity of these nodes. Yellow arrows would appear to indicate nodal forces (in this example there are no nodal forces). Orange pressure arrows extending from the right edge indicate which elements have pressures applied and in which direction. Remember that arrows away from the edge indicate tension and were input as negative numbers on the "Surface" menu within the "Model Data Control" window. Refer to Example 8.4 for more on this subject. You should also see the message: "Data loaded from file: . . . *ex84.esd*".

Click on the "Model Data" button on the lower right side of the screen.

Click "Check" to enter Superview.

Step 2 Look at Model with the Hidden-Line View

For many models, particularly three-dimensional ones, you may want to look at hidden-line drawings. Here's how to do so.

Click "Stress-di" to enter the dithering section of the program.

Click "Hidden l" to bring up the hidden-line menu.

Click "*Mesh" to remove the asterisk.

Click "Display", and a hidden-line drawing appears. Geometric errors in modeling, if there are any, may be visible here.

Step 3 Change the Viewpoint

Click "View" or ⟨F10⟩. Go to the "Draw" options menu. Remember that this menu is available from almost anywhere in the Algor program. Use it to change the view or redraw the picture. There are *eight* preset views available under the "View" menu, and other views can be user-defined. The current settings are displayed on the status line, the bottom line on the screen. At the bottom right you see "Vu = 5" and some coordinates. This indicates that the user-defined view is number 5, with the coordinates shown. These parameters were loaded from the file *ex84.svd* when you started. You can store default views and settings from the "File" menu for a specific model or for all models.

Click "Jetview". Select "Jetview" to bring up the "JetView" menu. Move the cursor into the viewing window. The outline of a jet should appear. Do not click the mouse button yet. The jet represents the new viewing position based on the current view. When the mouse cursor is in the center of the screen, the jet is pointing toward the center of the screen from the front. This indicates that the new view will be from the same direction as the current view. As the mouse moves to the right or left, the jet moves horizontally around the center of the screen to indicate possible new view locations. As the mouse moves up and down, the jet moves vertically to indicate new viewpoints. To emphasize the difference between the top and bottom of the jet, the tail of the jet disappears when the bottom is being viewed. Take a moment to move the jet around.

The jet coordinates appear on the bottom of the screen just above the mouse or the view information. As you move the jets, the coordinates change.

Click "Big step mode". An asterisk appears in front of "Big step mode". If the "Big step mode" option is active, the jet's coordinates change by 5° steps. This makes it easy to specify common angles (such as 30°, 45°, and 90°) by using the mouse. The 90° angle is useful because it enables you to select a new side view based on the current view.

Click the mouse. Clicking the left mouse button shows the jet's viewing position, and you will see various views as you click and move the mouse around the model.

Click "lasT vu" to get back to the original view. Or if that does not get you there,

Click "View:Predefined:YZ right" to get back to the original *y-z* plane view.

Step 4 Check the Model Parameters

We will now examine the plate in more detail. Make sure you have restored the original drawing before you proceed. You may have to reload the *ex84.svd* Superview data file if you have lost the original drawing. We will assume you have the original drawing on the screen.

Click "Inquire" on the main menu to enter the "Inquire" menu section.

Click "Grp Stats". Look at a numerical summary of the model information in a dialogue area below the drawing window. This shows how many nodes there are and the numbers and types of element groups. In this example, there is only one element group. This group is "G1 ty = 4: 2262" in the statistics. There are 2334 nodes, 2262 elements, 1 "Ele" group, and one load case.

Click "Ele inq" to access the "Ele inq" menu.

Click "Elast" to activate the elastic element inquiry feature.

Click "Get" and note that an asterisk appears, making the line read "*Get". The following message appears in the dialogue region: "Click on center of element".

Click on the center of an element. Notice that the dialogue region now displays the group number ("GR#"), the element number ("ELAST#"), the material ID ("Mat=1"), the thickness ("THICK=1"), and the pressure ("pij*factor=−100" appears if you click on one of the right-side elements where a pressure of −100 psi was applied), as well as the node numbers that define the element. You can also use the "Find" command and type in the element number if you are trying to locate a specific element. You can click on other elements to obtain information about each.

Click "mat id" to activate the material ID display section. The material ID range prompt below the window defaults to the range "1 to 1".

<Enter>   All elements with a material ID of 1 will be highlighted with a small, white square at the center of the element. All elements are highlighted in the example as they all have material type 1.

Click "Next rng" to select the next range, and all elements with material ID "2 to 2" will be highlighted. Because there are no elements with material type 2, all highlights disappear.

<Esc>   Clear the current value. You can inspect the element thickness in the same manner.

Step 5 Check the Model Element Loads

Click "Press". To check the element pressures, try a different approach for the element range dialogue.

<Esc>

0   Enter a range beginning with 0.

<Tab>

<Esc>   This positions the cursor at the second place in the dialogue section and clears it.

-100 Enter the end range value and press ⟨Enter⟩ to highlight all elements with a pressure in the range 0 to −100. In this example, all elements should be highlighted.

Click "Next rng". This gives the next range from −100 to −200. The right-side elements appear highlighted. The other elements are not highlighted as they do not have the new range of pressure acting on them. You can check the reference temperature and temperature gradient in the same manner.

Step 6   Check the Nodal Forces

You can now look at the nodal forces and moments applied to the model.

<Esc>

<Esc> This takes you back to the "Inquire" menu.

Click "Force" to access the nodal force inquiry menu. Note that the top line of the dialogue region below the window indicates there are no forces.

Click "Hilite" to highlight nodes with nodal forces. In this example, no nodal forces exist. If a nodal force is present, do the following:

Click "Get" to activate the "Get" command, and the asterisk appears along with a message in the dialogue region saying "Click on node". Move the cursor to one of the nodes that has an applied force acting on it and click on it.

<Esc> Escape back to the "Inquire" menu.

Step 7   Check the Boundary Conditions

Click "Node BC" to enter the "Inq Node" menu.

Click "Get" to activate the individual node inquiry feature.

Click on a node that has a triangle around it. A "rubber band" shows which node has been chosen, and the dialogue region gives the node number, coordinate position, and degree-of-freedom setting for the node.

Click "Hilite" to activate the highlight capability.

Click "Use got". Then use a node selected at the left edge of the model by the "Get" feature as the example to highlight all nodes that share the same boundary conditions. All triangle nodes are highlighted, showing all translations on the left side of the model with "Tyz" as the boundary condition. This completes the model check.

<Esc>

<Esc>

Click "Done" to exit Superview.

Answers to Selected Problems

Chapter 2

2.1 **a.** $\underline{K} = \begin{bmatrix} k_1 & 0 & -k_1 & 0 \\ 0 & k_3 & 0 & -k_3 \\ -k_1 & 0 & k_1 + k_2 & -k_2 \\ 0 & -k_3 & -k_2 & k_2 + k_3 \end{bmatrix}$

b. $d_{3x} = \dfrac{k_2 P}{k_1 k_2 + k_1 k_3 + k_2 k_3}$, $\quad d_{4x} = \dfrac{(k_1 + k_2)P}{k_1 k_2 + k_1 k_3 + k_2 k_3}$

c. $F_{1x} = \dfrac{-k_1 k_2 P}{k_1 k_2 + k_1 k_3 + k_2 k_3}$, $\quad F_{2x} = \dfrac{-k_3(k_1 + k_2)P}{k_1 k_2 + k_1 k_3 + k_2 k_3}$

2.2 $d_{2x} = 0.5$ in., $\quad F_{3x} = 500$ lb, $\quad \hat{f}_{1x}^{(1)} = -\hat{f}_{2x}^{(1)} = -500$ lb, $\quad \hat{f}_{2x}^{(2)} = -\hat{f}_{3x}^{(2)} = -500$ lb

2.3 **a.** $\underline{K} = \begin{bmatrix} k & -k & 0 & 0 & 0 \\ -k & 2k & -k & 0 & 0 \\ 0 & -k & 2k & -k & 0 \\ 0 & 0 & -k & 2k & -k \\ 0 & 0 & 0 & -k & k \end{bmatrix}$

b. $d_{2x} = \dfrac{P}{2k}$, $\quad d_{3x} = \dfrac{P}{k}$, $\quad d_{4x} = \dfrac{P}{2k}$ $\quad$ **c.** $F_{1x} = -\dfrac{P}{2}$, $\quad F_{5x} = -\dfrac{P}{2}$

2.4 **a.** $\underline{K}$ same as 2.3a. $\quad$ **b.** $d_{2x} = \dfrac{\delta}{4}$, $\quad d_{3x} = \dfrac{\delta}{2}$, $\quad d_{4x} = \dfrac{3\delta}{4}$

c. $F_{1x} = \dfrac{-k\delta}{4}$, $\quad F_{5x} = \dfrac{k\delta}{4}$

2.5 $d_{2x} = 2$ in., $\quad d_{3x} = 4$ in.
$\hat{f}_{1x}^{(1)} = -\hat{f}_{2x}^{(1)} = -1000$ lb, $\quad \hat{f}_{2x}^{(2)} = -\hat{f}_{3x}^{(2)} = -1000$ lb, $\quad F_{1x} = -1000$ lb

2.6 $d_{1x} = 0$, $\quad d_{2x} = 3$ in., $\quad d_{3x} = 7$ in., $\quad d_{4x} = 11$ in.
$\hat{f}_{1x}^{(1)} = -\hat{f}_{2x}^{(1)} = -3000$ lb, $\quad \hat{f}_{2x}^{(2)} = -\hat{f}_{3x}^{(2)} = -4000$ lb
$\hat{f}_{3x}^{(3)} = -\hat{f}_{4x}^{(3)} = -4000$ lb, $\quad F_{1x} = -3000$ lb

2.7 $d_{2x} = -2$ in.
$\hat{f}_{1x}^{(1)} = -\hat{f}_{2x}^{(1)} = 2000$ lb, $\quad \hat{f}_{2x}^{(2)} = -\hat{f}_{3x}^{(2)} = -1000$ lb

$\hat{f}_{2x}^{(3)} = -\hat{f}_{4x}^{(3)} = -1000$ lb, $F_{1x} = 2000$ lb, $F_{3x} = F_{4x} = 1000$ lb

2.8 $d_{2x} = 0.005$ m, $\hat{f}_{1x}^{(1)} = -\hat{f}_{2x}^{(1)} = -10$ N
$\hat{f}_{2x}^{(2)} = -\hat{f}_{3x}^{(2)} = -10$ N, $F_{1x} = -10$ N

2.9 $d_{2x} = 0.027$ m, $d_{3x} = 0.018$ m
$\hat{f}_{1x}^{(1)} = -\hat{f}_{2x}^{(1)} = -270$ N, $\hat{f}_{2x}^{(2)} = -\hat{f}_{3x}^{(2)} = 180$ N
$\hat{f}_{3x}^{(3)} = -\hat{f}_{4x}^{(3)} = 180$ N, $F_{1x} = -270$ N, $F_{4x} = -180$ N

2.10 $d_{2x} = 0.25$ m, $d_{3x} = 0.5$ m, $d_{4x} = 0.25$ m
$\hat{f}_{1x}^{(1)} = -\hat{f}_{2x}^{(1)} = -5$ kN, $\hat{f}_{2x}^{(2)} = -\hat{f}_{3x}^{(2)} = -5$ kN
$\hat{f}_{3x}^{(3)} = -\hat{f}_{4x}^{(3)} = 5$ kN, $\hat{f}_{4x}^{(4)} = -\hat{f}_{5x}^{(4)} = 5$ kN
$F_{1x} = -5$ kN, $F_{5x} = -5$ kN

2.11 $d_{2x} = -0.25$ m, $d_{3x} = -0.75$ m
$\hat{f}_{1x}^{(1)} = -\hat{f}_{2x}^{(1)} = 100$ N, $\hat{f}_{2x}^{(2)} = -\hat{f}_{3x}^{(2)} = 200$ N
$F_{1x} = 100$ N

2.12 $d_{3x} = 0.001$ m, $\hat{f}_{1x}^{(1)} = -\hat{f}_{3x}^{(1)} = -0.5$ kN
$\hat{f}_{2x}^{(2)} = -\hat{f}_{3x}^{(2)} = -0.5$ kN, $\hat{f}_{3x}^{(3)} = -\hat{f}_{4x}^{(3)} = 1$ kN
$F_{1x} = -0.5$ kN, $F_{2x} = -0.5$ kN, $F_{4x} = -1$ kN

2.13 **a.** $x = 0.5$ in. ↓, $\pi_{p_{min}} = -125$ lb-in.
b. $x = 2.0$ in. ←, $\pi_{p_{min}} = -1000$ lb-in.
c. $x = 1.962$ mm ↓, $\pi_{p_{min}} = -3849$ N · mm
d. $x = 2.4525$ mm →, $\pi_{p_{min}} = -1203$ N · mm

2.14 $x = 2.0$ in. ↑

2.15 $x = 0.707$ in. ←, $\pi_{p_{min}} = -235.7$ in.-lb

2.16 Same as 2.7

2.17 Same as 2.12

Chapter 3

3.1 **a.** $\underline{K} = \begin{bmatrix} \dfrac{A_1 E_1}{L_1} & \dfrac{-A_1 E_1}{L_1} & 0 & 0 \\[2ex] \dfrac{-A_1 E_1}{L_1} & \dfrac{A_1 E_1}{L_1} + \dfrac{A_2 E_2}{L_2} & \dfrac{-A_2 E_2}{L_2} & 0 \\[2ex] 0 & \dfrac{-A_2 E_2}{L_2} & \dfrac{A_2 E_2}{L^2} + \dfrac{A_3 E_3}{L_3} & \dfrac{-A_3 E_3}{L_3} \\[2ex] 0 & 0 & \dfrac{-A_3 E_3}{L_3} & \dfrac{A_3 E_3}{L_3} \end{bmatrix}$

b. $d_{2x} = \dfrac{PL}{3AE}$, $d_{3x} = \dfrac{2PL}{3AE}$

c. **i.** $d_{2x} = 3.33 \times 10^{-4}$ in., $d_{3x} = 6.67 \times 10^{-4}$ in.
ii. $F_{1x} = -333$ lb, $F_{4x} = -667$ lb
iii. $\sigma^{(1)} = 333$ psi (T), $\sigma^{(2)} = 333$ psi (T), $\sigma^{(3)} = -667$ psi (C)

3.2 $d_{2x} = -1.19 \times 10^{-4}$ m, $d_{3x} = -2.38 \times 10^{-4}$ m, $F_{1x} = 10$ kN
$\hat{f}_{1x}^{(1)} = -\hat{f}_{2x}^{(1)} = 10$ kN, $\hat{f}_{2x}^{(2)} = -\hat{f}_{3x}^{(2)} = 10$ kN

3.3 $d_{2x} = 1.91 \times 10^{-3}$ in., $F_{1x} = -5715$ lb, $F_{3x} = -2286$ lb
$\hat{f}_{1x}^{(1)} = -\hat{f}_{2x}^{(1)} = -5715$ lb, $\hat{f}_{2x}^{(2)} = -\hat{f}_{3x}^{(2)} = 2286$ lb

3.4 $d_{2x} = -1.66 \times 10^{-4}$ in., $d_{3x} = -1.33 \times 10^{-3}$ in.
$F_{1x} = 667$ lb, $F_{4x} = 5333$ lb
$\hat{f}_{1x}^{(1)} = -\hat{f}_{2x}^{(1)} = 667$ lb, $\hat{f}_{2x}^{(2)} = \hat{f}_{3x}^{(2)} = 4667$ lb
$\hat{f}_{3x}^{(3)} = -\hat{f}_{4x}^{(3)} = -5333$ lb

3.5 $d_{2x} = 0.0015$ in., $d_{3x} = 0.0045$ in., $F_{1x} = -7500$ lb
$\hat{f}_{1x}^{(1)} = -\hat{f}_{2x}^{(1)} = \hat{f}_{2x}^{(2)} = -\hat{f}_{3x}^{(2)} = -7500$ lb

3.6 $d_{2x} = 3.16 \times 10^{-3}$ in., $F_{1x} = -3790$ lb, $F_{3x} = F_{4x} = -2105$ lb
$\hat{f}_{1x}^{(1)} = -\hat{f}_{2x}^{(1)} = -3790$ lb, $\hat{f}_{2x}^{(2)} = -\hat{f}_{3x}^{(2)} = \hat{f}_{2x}^{(3)} = -\hat{f}_{4x}^{(3)} = 2105$ lb

3.7 $d_{2x} = 2.21 \times 10^{-5}$ in., $d_{3x} = 6.65 \times 10^{-3}$ in.
$F_{1x} = -33.15$ lb, $F_{4x} = -9975$ lb
$\hat{f}_{1x}^{(1)} = -\hat{f}_{2x}^{(1)} = \hat{f}_{2x}^{(2)} = -\hat{f}_{3x}^{(2)} = -33.15$ lb, $\hat{f}_{3x}^{(3)} = -\hat{f}_{4x}^{(3)} = 9975$ lb

3.8 $d_{2x} = -0.125$ mm, $d_{3x} = -0.839$ mm

3.9 $d_{2x} = 0.01225$ m, $F_{1x} = -257.25$ kN, $F_{3x} = 267.75$ kN
$\hat{f}_{1x}^{(1)} = -\hat{f}_{2x}^{(1)} = -257.25$ kN, $\hat{f}_{2x}^{(2)} = -\hat{f}_{3x}^{(2)} = -267.75$ kN

3.10 $d_{2x} = 0.935 \times 10^{-3}$ m, $d_{3x} = 0.727 \times 10^{-3}$ m
$F_{1x} = -6.546$ kN, $F_{4x} = -1.455$ kN
$\hat{f}_{1x}^{(1)} = -\hat{f}_{2x}^{(1)} = -6.546$ kN, $\hat{f}_{2x}^{(2)} = -\hat{f}_{3x}^{(2)} = 1.455$ kN,
$\hat{f}_{3x}^{(3)} = -\hat{f}_{4x}^{(3)} = 1.455$ kN

3.11 $d_{2x} = 1.786 \times 10^{-4}$ m, $F_{1x} = -3.75$ kN, $F_{3x} = F_{4x} = F_{5x} = -3.75$ kN
$\hat{f}_{1x}^{(1)} = -\hat{f}_{2x}^{(1)} = -3.75$ kN,
$\hat{f}_{2x}^{(2)} = -\hat{f}_{3x}^{(2)} = \hat{f}_{2x}^{(3)} = -\hat{f}_{4x}^{(3)} = \hat{f}_{2x}^{(4)} = -\hat{f}_{5x}^{(4)} = 3.75$ kN

3.12 two-element solution, $d_{1x} = -0.686 \times 10^{-3}$ in.
one-element solution, $d_{1x} = -0.667 \times 10^{-3}$ in.

3.13 $B = \left[-\dfrac{1}{L} + \dfrac{4x}{L^2} \quad \dfrac{-8x}{L^2} \quad \dfrac{1}{L} + \dfrac{4x}{L^2} \right], \quad k = A \displaystyle\int_{-L/2}^{L/2} B^T E B \, dx$

3.15 a. $k = 2.25 \times 10^6 \begin{bmatrix} 1 & 1 & -1 & -1 \\ 1 & 1 & -1 & -1 \\ -1 & -1 & 1 & 1 \\ -1 & -1 & 1 & 1 \end{bmatrix}$ lb/in.

b. $k = \dfrac{10^6}{4} \begin{bmatrix} 1 & -\sqrt{3} & -1 & \sqrt{3} \\ -\sqrt{3} & 3 & \sqrt{3} & -3 \\ -1 & \sqrt{3} & 1 & -\sqrt{3} \\ \sqrt{3} & -3 & -\sqrt{3} & 3 \end{bmatrix}$ lb/in.

c. $k = 7000 \begin{bmatrix} 3 & -\sqrt{3} & -3 & \sqrt{3} \\ -\sqrt{3} & 1 & \sqrt{3} & -1 \\ -3 & \sqrt{3} & 3 & -\sqrt{3} \\ \sqrt{3} & -1 & -\sqrt{3} & 1 \end{bmatrix}$ kN/m

$$\textbf{d. } \underline{k} = 1.4 \times 10^4 \begin{bmatrix} 0.883 & 0.321 & -0.883 & -0.321 \\ 0.321 & 0.117 & -0.321 & -0.117 \\ -0.883 & -0.321 & 0.883 & 0.321 \\ -0.321 & -0.117 & 0.321 & 0.117 \end{bmatrix} \text{kN/m}$$

3.16 **a.** $\hat{d}_{1x} = 0.433$ in., $\hat{d}_{2x} = 0.592$ in.
b. $\hat{d}_{1x} = 0.433$ in., $\hat{d}_{2x} = -0.1585$ in.

3.17 **a.** $\hat{d}_{1x} = 2.165$ mm, $\hat{d}_{1y} = -1.25$ mm,
$\hat{d}_{2x} = 0.098$ mm, $\hat{d}_{2y} = -5.83$ mm
b. $\hat{d}_{1x} = -1.25$ mm, $\hat{d}_{1y} = 2.165$ mm,
$\hat{d}_{2x} = 3.03$ mm, $\hat{d}_{2y} = 5.098$ mm

3.18 **a.** $\sigma = 10{,}600$ psi, **b.** 45.47 MPa

3.19 **a.** $\underline{K} = k \begin{bmatrix} 2 & 0 & -\frac{1}{2} & \frac{1}{2} & -1 & 0 & -\frac{1}{2} & -\frac{1}{2} \\ 0 & 1 & \frac{1}{2} & -\frac{1}{2} & 0 & 0 & -\frac{1}{2} & -\frac{1}{2} \\ -\frac{1}{2} & \frac{1}{2} & \frac{1}{2} & -\frac{1}{2} & 0 & 0 & 0 & 0 \\ \frac{1}{2} & -\frac{1}{2} & -\frac{1}{2} & \frac{1}{2} & 0 & 0 & 0 & 0 \\ -1 & 0 & 0 & 0 & 1 & 0 & 0 & 0 \\ 0 & 0 & 0 & 0 & 0 & 0 & 0 & 0 \\ -\frac{1}{2} & -\frac{1}{2} & 0 & 0 & 0 & 0 & \frac{1}{2} & \frac{1}{2} \\ -\frac{1}{2} & -\frac{1}{2} & 0 & 0 & 0 & 0 & \frac{1}{2} & \frac{1}{2} \end{bmatrix}$

b. $d_{1x} = 0$, $d_{1y} = \dfrac{-10}{k}$

3.20 $d_{2x} = 0$, $d_{2y} = 0.285$ in., $\sigma^{(1)} = \sigma^{(2)} = 1414$ psi (T)

3.21 $d_{1x} = \dfrac{231L}{AE}$, $d_{1y} = \dfrac{43.5L}{AE}$

3.22 $d_{1x} = \dfrac{422L}{AE}$, $d_{1y} = \dfrac{1570L}{AE}$

$\sigma^{(1)} = \dfrac{574}{A}$ (C), $\sigma^{(2)} = \dfrac{422}{A}$ (T), $\sigma^{(3)} = \dfrac{996}{A}$ (T)

3.23 $d_{1x} = 0.12$ in., $d_{1y} = 0$, $\sigma^{(1)} = 6000$ psi

3.24 $d_{2x} = \dfrac{26{,}675}{AE}$, $d_{2y} = \dfrac{105{,}021}{AE}$, $d_{3x} = \dfrac{-26{,}675}{AE}$, $d_{3y} = \dfrac{105{,}021}{AE}$
$\hat{f}_{1x}^{(1)} = -\hat{f}_{2x}^{(1)} = -1333$ lb, $\hat{f}_{1x}^{(2)} = -\hat{f}_{3x}^{(2)} = -1667$ lb
$\hat{f}_{2x}^{(3)} = -\hat{f}_{4x}^{(3)} = 1667$ lb, $\hat{f}_{2x}^{(4)} = -\hat{f}_{3x}^{(4)} = 0$
$\hat{f}_{3x}^{(5)} = -\hat{f}_{4x}^{(5)} = 1333$ lb, $\hat{f}_{1x}^{(6)} = -\hat{f}_{4x}^{(6)} = 0$

3.25 $d_{2x} = 0$, $d_{2y} = \dfrac{225{,}000}{AE}$, $d_{3x} = \dfrac{-53{,}340}{AE}$, $d_{3y} = \dfrac{210{,}000}{AE}$
$\hat{f}_{1x}^{(1)} = -\hat{f}_{2x}^{(1)} = 0$, $\hat{f}_{1x}^{(2)} = -\hat{f}_{3x}^{(2)} = -3333$ lb
$\hat{f}_{2x}^{(4)} = -\hat{f}_{3x}^{(4)} = 1000$ lb, $\hat{f}_{3x}^{(5)} = -\hat{f}_{4x}^{(5)} = 2667$ lb
$\hat{f}_{1x}^{(6)} = -\hat{f}_{4x}^{(6)} = 0$

3.26 No, the truss is unstable, $|\underline{K}| = 0$.

3.27 $d_{3x} = 0.0463$ in., $d_{3y} = -0.0176$ in.

$\hat{f}_{1x}^{(1)} = -\hat{f}_{3x}^{(1)} = -2.055$ kip, $\hat{f}_{2x}^{(2)} = -\hat{f}_{3x}^{(2)} = 6.279$ kip

$\hat{f}_{3x}^{(3)} = -\hat{f}_{4x}^{(3)} = -6.6$ kip

3.28 $\underline{T}^T = \begin{bmatrix} C & -S & 0 & 0 \\ S & C & 0 & 0 \\ 0 & 0 & C & -S \\ 0 & 0 & S & C \end{bmatrix}$ and $\underline{T}\underline{T}^T = \begin{bmatrix} 1 & 0 & 0 & 0 \\ 0 & 1 & 0 & 0 \\ 0 & 0 & 1 & 0 \\ 0 & 0 & 0 & 1 \end{bmatrix}$

$\therefore \underline{T}^T = \underline{T}^{-1}$

3.29 $d_{1x} = -0.893 \times 10^{-4}$ m, $d_{1y} = -4.46 \times 10^{-4}$ m

$\sigma^{(1)} = 31.2$ MPa (T), $\sigma^{(2)} = 26.5$ MPa (T), $\sigma^{(3)} = 6.25$ MPa (T)

3.30 $d_{1x} = 1.71 \times 10^{-4}$ m, $d_{1y} = -7.55 \times 10^{-4}$ m

$\sigma^{(1)} = 79.28$ MPa (T), $\sigma^{(2)} = 11.97$ MPa (T), $\sigma^{(3)} = -23.87$ MPa (C)

3.31 $d_{1x} = 8.25 \times 10^{-4}$ m, $d_{1y} = -3.65 \times 10^{-3}$ m

$\sigma^{(2)} = 57.74$ MPa (T), $\sigma^{(3)} = -115.5$ MPa (C)

3.32 $d_{2x} = 0.135 \times 10^{-2}$ m, $d_{2y} = -0.850 \times 10^{-2}$ m,

$d_{3y} = -0.137 \times 10^{-1}$ m, $d_{4y} = -0.164 \times 10^{-1}$ m,

$\sigma^{(1)} = -198$ MPa (C), $\sigma^{(2)} = 0$, $\sigma^{(3)} = 44.6$ MPa (T)

$\sigma^{(4)} = -31.6$ MPa (C), $\sigma^{(5)} = -191$ MPa (C),

$\sigma^{(6)} = -63.1$ MPa (C)

3.33 $d_{1x} = -1.724 \times 10^{-3}$ m, $d_{1y} = -3.448 \times 10^{-3}$ m

$\sigma^{(1)} = 51.2$ MPa (T), $\sigma^{(2)} = -36.2$ MPa (C)

3.34 $d_{4x} = 9.93 \times 10^{-3}$ in., $d_{4y} = -2.46 \times 10^{-3}$ in.

$\sigma^{(1)} = 31.25$ ksi (T), $\sigma^{(2)} = 3.459$ ksi (T), $\sigma^{(3)} = -1.538$ ksi (C)

$\sigma^{(4)} = -3.103$ ksi (C), $\sigma^{(5)} = 0$

3.35 $d_{1y} = -0.5 \times 10^{-3}$ in., $\sigma^{(1)} = 250$ psi (T)

3.36 $\hat{d}_{1x} = 0.212$ in.

3.37 $\hat{d}_{1x} = 0.0397$ in.

3.38 $\hat{d}_{2x} = 16.98$ mm

3.39 $\hat{d}_{2x} = 1.71$ mm

3.40 $d_{1x} = -3.018 \times 10^{-5}$ m, $d_{1y} = -1.517 \times 10^{-5}$ m,

$d_{1z} = 2.684 \times 10^{-5}$ m, $\sigma^{(1)} = -338$ kN/m^2 (C),

$\sigma^{(2)} = -1690$ kN/m^2 (C), $\sigma^{(3)} = -7965$ kN/m^2 (C)

$\sigma^{(4)} = -2726$ kN/m^2 (C)

3.41 $d_{1x} = 1.383 \times 10^{-3}$ m, $d_{1y} = -5.119 \times 10^{-5}$ m

$d_{1x} = 6.015 \times 10^{-5}$ m, $\sigma^{(1)} = 20.51$ MPa (T),

$\sigma^{(2)} = 4.21$ MPa (T), $\sigma^{(3)} = -5.29$ MPa (C)

3.42 $d_{5x} = 0.0014$ in., $d_{5y} = 0$, $d_{5z} = -0.00042$ in.

$\sigma^{(1)} = \sigma^{(4)} = 180$ psi (T), $\sigma^{(2)} = \sigma^{(3)} = 140$ psi (C)

3.43 $d_{4x} = 0.00863$ in., $d_{4y} = 0$, $d_{4z} = -0.00683$ in.

$\sigma^{(1)} = -916$ psi (C)

3.46 $d_{2y} = -0.0192$ in., $\quad d_{3y} = -0.0168$ in.
$\quad\sigma^{(1)} = -1688$ psi (C), $\quad\sigma^{(2)} = 1332$ psi (T), $\quad\sigma^{(3)} = 1000$ psi (T)

3.47 $d_{1x} = \dfrac{-110P}{AE}$ in., $\quad d_{1y} = 0, \quad d_{2x} = 0, \quad d_{2y} = \dfrac{-405P}{AE}$ in.,

$\quad d_{3x} = 0, \quad d_{3y} = \dfrac{-433P}{AE}$ in., $\quad d_{4x} = \dfrac{50P}{AE}$ in., $\quad d_{4y} = \dfrac{-208P}{AE}$ in.

$\quad\sigma^{(1)} = -0.156\dfrac{P}{A}, \quad \sigma^{(2)} = -0.208\dfrac{P}{A}, \quad \sigma^{(3)} = -1.16\dfrac{P}{A}$

$\quad\sigma^{(4)} = 0.260\dfrac{P}{A}, \quad \sigma^{(5)} = -0.573\dfrac{P}{A}, \quad \sigma^{(6)} = 0.458\dfrac{P}{A}$

3.48 $d_{2y} = -0.955 \times 10^{-2}$ m, $\quad d_{4y} = -1.03 \times 10^{-2}$ m,
$\quad\sigma^{(1)} = 67.1$ MPa (C), $\quad\sigma^{(2)} = 60.0$ MPa (T), $\quad\sigma^{(3)} = 22.4$ MPa (C)
$\quad\sigma^{(4)} = 44.7$ MPa (C), $\quad\sigma^{(5)} = 20.0$ MPa (T)

3.49 $d'_{1x} = 0, \quad d_{2y} = -0.00283$ in., $\quad F_{2x} = 2000$ lb
$\quad\sigma^{(1)} = 0, \quad \sigma^{(2)} = 1414$ psi (T), $\quad \sigma^{(3)} = 0$

3.50 $d_{2y} = -0.00283$ in.

3.51 $d'_{2x} = -0.002$ in., $\quad f'_{1x} = -2800$ lb., $\quad f'_{2x} = -2000$ lb
$\quad F'_{2y} = -2828$ lb

3.52 **a.** $d_{1x} = 0.010$ in. $\downarrow, \quad \pi_{p_{min}} = -100$ lb-in.
   b. $d_{1x} = 0.00833$ in. $\rightarrow, \quad \pi_{p_{min}} = -41.67$ lb-in.

3.53 $\underline{k} = \dfrac{3A_0E}{2L}\begin{bmatrix} 1 & -1 \\ -1 & 1 \end{bmatrix}$

3.54 two-element solution: $d_{2x} = 0.00825$ in., $\quad d_{3x} = 0.012$ in., $\quad \sigma^{(1)} = 8250$ psi (T),
$\quad\sigma^{(2)} = 3750$ psi (T),

3.55 two-element solution: $d_{2x} = 6.75 \times 10^{-3}$ in., $\quad d_{3x} = 0.009$ in.
$\quad\sigma^{(1)} = 6750$ psi (T), $\quad \sigma^{(2)} = 2250$ psi (T)

3.56 $d_{2x} = 0.75 \times 10^{-3}$ in., $\quad \sigma^{(1)} = 750$ psi (T)

3.57 $d_{1x} = \gamma L^2/(2E), \quad d_{2x} = 3\gamma L^2/(8E), \quad \sigma^{(1)} = \gamma L/8, \quad \sigma^{(2)} = 3\gamma L/8$

3.58 **a.** $f_{1x} = 583.3$ lb, $\quad f_{2x} = 666.7$ lb
   b. $f_{1x} = 26.7$ kN, $\quad f_{2x} = 80$ kN

Chapter 4

4.1 $d_{2y} = -0.370 \times 10^{-3}$ m, $\quad d_{3y} = -0.307 \times 10^{-3}$ m,
$\quad d_{4y} = -0.386 \times 10^{-3}$ m, $\quad \sigma^{(1)} = 4.71$ MPa (C),
$\quad\sigma^{(2)} = \sigma^{(3)} = \sigma^{(6)} = 3.33$ MPa (T), $\quad \sigma^{(4)} = 3.33$ MPa (C)

4.2 $d_{2x} = 0.0223$ in., $\quad d_{2y} = 0.00528$ in., $\quad \sigma^{(1)} = 880$ psi (T),
$\quad\sigma^{(2)} = 1033$ psi (T), $\quad \sigma^{(3)} = 826$ psi (C), $\quad \sigma^{(4)} = 1467$ psi (C)

4.3 $A = 0.444$ in^2

4.4 $d_{1x} = 0.8342$ in., $\quad \sigma^{(1)} = \sigma^{(3)} = -12.37$ ksi (C), $\quad \sigma^{(2)} = \sigma^{(4)} = 9.74$ ksi (T)

4.5 $d_{1x} = -0.711 \times 10^{-1}$ in., $d_{1y} = 0$, $d_{1x} = -0.266$ in.
$\sigma^{(3)} = 2868$ psi (C), $\sigma^{(1)} = \sigma^{(4)} = 948$ psi (C)
$\sigma^{(2)} = \sigma^{(5)} = 1445$ psi (T)

4.6 $d_{1x} = 0.00391$ m, $d_{1y} = -0.0150$ m, $d_{1z} = -0.00458$ m
$\sigma^{(1)} = 41.1$ MPa (C), $\sigma^{(2)} = 43.0$ MPa (C), $\sigma^{(3)} = 19.65$ MPa (C),
$\sigma^{(4)} = 27.8$ MPa (T)

4.7 $d_{7x} = 0.102$ m, $d_{7y} = 0.0189$ m, $\sigma^{(1)} = 440$ MPa (T),
$\sigma^{(2)} = 227$ MPa (T), $\sigma^{(3)} = -245$ MPa (C)

4.8 $d_{6x} = 1.14$ mm, $d_{6y} = -0.571$ mm, $d_{6z} = 0.286$ mm
$\sigma^{(1)} = 0$, $\sigma^{(3)} = 20$ MPa (C), $\sigma^{(7)} = 40.0$ MPa (C)

Chapter 5

5.3 $d_{2y} = \dfrac{-7PL^3}{768EI}$, $\phi_1 = \dfrac{-PL^2}{32EI}$, $\phi_2 = \dfrac{PL^2}{128EI}$

$F_{1y} = \dfrac{5P}{16}$, $M_1 = 0$, $F_{3y} = \dfrac{11P}{16}$, $M_3 = \dfrac{-3PL}{16}$

5.4 $d_{1y} = \dfrac{-PL^3}{3EI}$, $\phi_1 = \dfrac{PL^3}{2EI}$, $F_{2y} = P$, $M_2 = -PL$

5.5 $d_{1y} = -2.688$ in., $\phi_1 = 0.0144$ rad, $\phi_2 = 0.0048$ rad
$F_{2y} = 2.5$ kip, $F_{3y} = -1.5$ kip, $M_3 = 10.0$ k-ft

5.6 $d_{3y} = -3.94$ in.

5.7 $d_{2y} = -0.105$ in., $\phi_2 = -0.003$ rad, $d_{3y} = -0.345$ in., $\phi_3 = -0.0045$ rad

5.8 $d_{2y} = -1.34 \times 10^{-4}$ m, $\phi_2 = 8.93 \times 10^{-5}$ rad
$F_{1y} = 10$ kN, $M_1 = 12.5$ kN $\cdot$ m, $F_{3y} = 1.87$ N, $M_3 = -2.5$ kN $\cdot$ m

5.9 $d_{3y} = -7.619 \times 10^{-4}$ m, $\phi_2 = -3.809 \times 10^{-4}$ rad, $\phi_1 = 1.904 \times 10^{-4}$ rad
$F_{1y} = -0.889$ kN, $F_{2y} = 4.889$ kN

5.10 $d_{2y} = -0.886$ in., $\phi_2 = -0.00554$ rad
$F_{1y} = 1115$ lb, $M_1 = -267$ k-in.

5.11 $d_{2y} = -7.934 \times 10^{-3}$ m, $\phi_1 = -2.975 \times 10^{-3}$ rad
$F_{1y} = 5.208$ kN, $F_{3y} = 5.208$ kN
$F_{\text{spring}} = 1.587$ kN

5.12 $d_{2y} = d_{4y} = \dfrac{-1wL^4}{607.5EI}$, $d_{3y} = \dfrac{-wL^4}{507EI}$

$\phi_2 = \dfrac{-1wL^3}{270EI}$, $\phi_4 = -\phi_2$

$F_{1y} = \dfrac{wL}{2}$, $M_1 = \dfrac{wL^2}{12}$

5.13 $d_{2y} = \dfrac{-wL^4}{384EI}$, $F_{1y} = \dfrac{wL}{2}$, $M_1 = \dfrac{wL^2}{12}$

5.14 $d_{2y} = \dfrac{-5wL^4}{384EI}, \quad \phi_1 = -\phi_3 = \dfrac{-wL^3}{24EI}, \quad F_{1y} = \dfrac{wL}{2}$

5.15 $d_{3y} = \dfrac{-wL^4}{4EI}, \quad \phi_2 = \dfrac{-wL^3}{8EI}, \quad \phi_3 = \dfrac{-7wL^3}{24EI}$

$\qquad F_{1y} = \dfrac{-3wL}{4}, \quad M_1 = \dfrac{-wL^2}{4}, \quad F_{2y} = \dfrac{7wL}{4}$

5.16 $\hat{f}_{1y} = \dfrac{-3wL}{20}, \quad \hat{m}_1 = \dfrac{-wL^2}{30}, \quad \hat{f}_{2y} = \dfrac{-7wL}{20}, \quad \hat{m}_2 = \dfrac{wL^2}{20}$

5.17 $F_{1y} = \dfrac{3wL}{20}, \quad M_1 = \dfrac{wL^2}{30}, \quad F_{3y} = \dfrac{7wL}{20}, \quad M_3 = \dfrac{-wL^2}{20}$

5.18 $\phi_2 = \dfrac{wL^3}{80EI}, \quad F_{1y} = \dfrac{9wL}{40}, \quad M_1 = \dfrac{7wL^2}{120}, \quad F_{2y} = \dfrac{11wL}{40}$

5.19 $d_{3y} = -0.0244$ m, $\quad \phi_3 = -0.0071$ rad, $\quad \phi_2 = -0.00305$ rad

$\qquad F_{1y} = -24$ kN, $\quad M_1 = -32$ kN $\cdot$ m, $\quad F_{2y} = 56$ kN

$\qquad \hat{f}_{1y}^{(1)} = -\hat{f}_{2y}^{(1)} = -24$ kN, $\quad \hat{m}_1^{(1)} = -32$ kN $\cdot$ m, $\quad \hat{m}_2^{(1)} = -64$ kN $\cdot$ m

$\qquad \hat{f}_{2y}^{(2)} = 32$ kN, $\quad \hat{m}_2^{(2)} = 64$ kN $\cdot$ m, $\quad \hat{f}_{3y}^{(2)} = 0, \quad \hat{m}_3^{(2)} = 0$

5.20 $\phi_1 = -0.0032$ rad, $\quad d_{2y} = -0.0115$ m, $\quad \phi_3 = 0.0032$ rad

$\qquad F_{1y} = 29.94$ kN, $\quad F_{2y} = 0.1152$ kN, $\quad F_{3y} = 29.94$ kN

$\qquad \hat{f}_{1y}^{(1)} = 29.94$ kN, $\quad \hat{m}_1^{(1)} = 0, \quad \hat{f}_{2y}^{(1)} = 0.058$ kN, $\quad \hat{m}_2^{(1)} = 59.65$ kN $\cdot$ m

5.21 $d_{2y} = -2.514$ in., $\quad \phi_2 = -0.00698$ rad, $\quad \phi_3 = 0.0279$ rad

$\qquad F_{1y} = 37.5$ kip, $\quad M_1 = 225$ k-ft, $\quad F_{3y} = 22.5$ kip

5.22 $d_{3y} = -3.277$ in., $\quad \phi_3 = -0.0323$ rad, $\quad \phi_2 = -0.0130$ rad

$\qquad F_{1y} = -20.5$ kip, $\quad M_1 = -71.67$ k-ft, $\quad F_{2y} = 60.5$ kip

5.23 $d_{2y} = -2.34$ in., $\quad F_{1y} = 5325$ lb $= F_{3y}, \quad M_1 = 19{,}900$ lb-ft $= -M_3$

5.24 $\phi_1 = -3.596 \times 10^{-4}$ rad, $\quad \phi_2 = 9.92 \times 10^{-5}$ rad, $\quad \phi_3 = 1.091 \times 10^{-4}$ rad

$\qquad F_{1y} = 9875$ N, $\quad F_{2y} = 28{,}406$ N, $\quad F_{3y} = 6719$ N

5.25 $d_{2y} = \dfrac{-PL^3}{192EI} - \dfrac{wL^4}{384EI}, \quad F_{1y} = \dfrac{P + wL}{2}, \quad M_1 = \dfrac{PL}{8} + \dfrac{wL^2}{12}$

5.26 $d_{2y} = \dfrac{-5PL^3}{648EI}$

5.27 $d_{2y} = \dfrac{-(25P + 22wL)L^3}{240EI}, \quad \phi_2 = \dfrac{-(PL^2 + wL^3)}{8EI}$

$\qquad F_{1y} = P + \dfrac{wL}{2}, \quad M_1 = \dfrac{PL}{2} + \dfrac{wL^2}{3}$

5.28 $d_{2y} = -1.57 \times 10^{-4}$ m, $\quad \phi_2 = 1.19 \times 10^{-4}$ rad

5.29 $d_{2y} = -3.18 \times 10^{-4}$ m, $\quad \phi_2 = 1.58 \times 10^{-4}$ rad, $\quad \phi_3 = 1.58 \times 10^{-4}$ rad

5.30 $d_{3y} = -4.05 \times 10^{-5}$ m, $\quad \phi_2 = -1.35 \times 10^{-5}$ rad, $\quad \phi_3 = 1.85 \times 10^{-5}$ rad

5.34 $\hat{\underline{k}} = EI \displaystyle\int_0^L [B]^T [B]\, d\hat{x} + k_f \int_0^L [N]^T [N]\, d\hat{x}$

Chapter 6

6.1 $d_{2x} = 0.0278$ in., $d_{2y} = 0$, $\phi_2 = -0.555 \times 10^{-4}$ rad
$\hat{f}_{1x}^{(1)} = -\hat{f}_{2x}^{(1)} = -8300$ lb, $\hat{f}_{1y}^{(1)} = -\hat{f}_{2y}^{(1)} = 4.6$ lb
$\hat{m}_1^{(1)} = 2775$ lb-in., $\hat{m}_2^{(1)} = 0$

6.2 $d_{2x} = d_{3x} = 0.688$ in., $d_{2y} = -d_{3y} = 0.00171$ in.
$\phi_2 = -\phi_3 = -0.00173$ rad
$\hat{f}_{1x}^{(1)} = -\hat{f}_{2x}^{(1)} = -2140$ lb, $\hat{f}_{1y}^{(1)} = -\hat{f}_{2y}^{(1)} = -2503$ lb
$\hat{m}_1^{(1)} = 343,600$ lb-in., $\hat{m}_2^{(1)} = 257,000$ lb-in.
$\hat{f}_{2x}^{(2)} = -\hat{f}_{3x}^{(2)} = 2497$ lb, $\hat{f}_{2y}^{(2)} = -\hat{f}_{3y}^{(2)} = -2140$ lb
$\hat{m}_2^{(2)} = -257,000$ lb-in., $\hat{m}_3^{(2)} = -256,600$ lb-in.
$\hat{f}_{3x}^{(3)} = -\hat{f}_{4x}^{(3)} = 2140$ lb, $\hat{f}_{3y}^{(3)} = -\hat{f}_{4y}^{(3)} = 2497$ lb
$\hat{m}_3^{(3)} = 256,600$ lb-in., $\hat{m}_4^{(3)} = 342,700$ lb-in.
$F_{1x} = F_{4x} = -2503$ lb, $F_{1y} = -F_{4y} = -2140$ lb
$M_1 = 343,600$ lb-in., $M_4 = 342,700$ lb-in.

6.3 Channel section 6×8.2 based on $M_{max} = 106,900$ lb-in.

6.4 $d_{4x} = 0.00445$ in., $d_{4y} = -0.0123$ in., $\phi_4 = -0.00290$ rad
$\hat{f}_{1x}^{(1)} = -\hat{f}_{4x}^{(1)} = 4.04$ kip, $\hat{f}_{1y}^{(1)} = -\hat{f}_{4y}^{(1)} = -1.43$ kip
$\hat{m}_1^{(1)} = -254$ k-in., $\hat{m}_4^{(1)} = -513$ k-in.
$\hat{f}_{2x}^{(2)} = -\hat{f}_{4x}^{(2)} = 5.82$ kip, $\hat{f}_{2y}^{(2)} = -\hat{f}_{4y}^{(2)} = -1.45$ kip
$\hat{m}_2^{(2)} = -260$ k-in., $\hat{m}_4^{(2)} = -519$ k-in.
$F_{1x} = 3.1$ kip, $F_{1y} = 2.96$ kip, $M_1 = -254$ k-in.
$F_{2x} = -1.31$ kip, $F_{2y} = 5.86$ kip, $M_2 = -260$ k-in.
$F_{3x} = -1.78$ kip, $F_{3y} = 11.17$ kip, $M_3 = -1736$ k-in.

6.5 $d_{2x} = 0.05618$ in., $d_{2y} = -0.1792$ in., $\phi_2 = -0.00965$ rad
$\hat{f}_{1x}^{(1)} = 90.07$ kip, $\hat{f}_{1y}^{(1)} = 3.83$ kip, $\hat{m}_1^{(1)} = 361$ k-in.
$\hat{f}_{2x}^{(1)} = -73.43$ kip, $\hat{f}_{2y}^{(1)} = 7.27$ kip, $\hat{m}_2^{(1)} = -1106$ k-in.
$\hat{f}_{2x}^{(2)} = -\hat{f}_{3x}^{(2)} = 46.8$ kip, $\hat{f}_{2y}^{(2)} = 17.05$ kip, $\hat{m}_2^{(2)} = 1107$ k-in.
$\hat{f}_{3y}^{(2)} = 22.95$ kip, $\hat{m}_3^{(2)} = -2171$ k-in.
$F_{1x} = F_{3x} = 46.8$ kip, $F_{1y} = 77.1$ kip, $M_1 = 361$ k-in.
$F_{3y} = 22.95$ kip, $M_3 = 2171$ k-in.

6.6 $d_{2x} = -0.000269$ in., $d_{2y} = -0.0363$ in., $\phi_2 = -0.00347$ rad
$\hat{f}_{1x}^{(1)} = 46.6$ kip, $\hat{f}_{1y}^{(1)} = 6.07$ kip, $\hat{m}_1^{(1)} = 491.3$ k-in.
$\hat{f}_{2x}^{(1)} = -32.4$ kip, $\hat{f}_{2y}^{(1)} = 8.07$ kip, $\hat{m}_2^{(1)} = -831.3$ k-in.
$\hat{f}_{2x}^{(2)} = -\hat{f}_{3x}^{(2)} = -0.28$ kip, $\hat{f}_{2y}^{(2)} = 58.31$ kip, $\hat{m}_2^{(2)} = 1123.9$ k-in.
$\hat{f}_{3y}^{(2)} = 21.69$ kip, $\hat{m}_3^{(2)} = -1611.8$ k-in.
$\hat{f}_{4x}^{(3)} = -\hat{f}_{2x}^{(3)} = 50.2$ kip, $\hat{f}_{4y}^{(3)} = -\hat{f}_{2y}^{(3)} = -1.49$ kip, $\hat{m}_4^{(3)} = -154.2$ k-in.
$\hat{m}_2^{(3)} = -293.2$ k-in.
$F_{1x} = 28.65$ kip, $F_{1y} = 37.24$ kip, $M_1 = 491.3$ k-in.
$F_{3x} = 0.28$ kip, $F_{3y} = 21.69$ kip, $M_3 = -1611.8$ k-in.
$F_{4x} = -28.93$ kip, $F_{4y} = 41.05$ kip, $M_4 = -154.2$ k-in.

6.7 $d_{2x} = 0.4308 \times 10^{-4}$ m, $d_{2y} = -0.9067 \times 10^{-4}$ m,
$\phi_2 = -0.1403 \times 10^{-2}$ rad
$\hat{f}_{1x}^{(1)} = -\hat{f}_{2x}^{(1)} = 23.8$ kN, $\hat{f}_{1y}^{(1)} = 17.26$ kN, $\hat{m}_1^{(1)} = 32.77$ kN · m
$\hat{f}_{2y}^{(1)} = 22.74$ kN, $\hat{m}_2^{(1)} = -54.64$ kN · m
$\hat{f}_{2x}^{(2)} = -\hat{f}_{3x}^{(2)} = 11.31$ kN, $\hat{f}_{2y}^{(2)} = 37.19$ kN, $\hat{m}_2^{(2)} = 65.09$ kN · m
$\hat{f}_{3y}^{(2)} = 42.81$ kN, $\hat{m}_3^{(2)} = -87.54$ kN · m
$\hat{f}_{2x}^{(3)} = -\hat{f}_{4x}^{(3)} = 17.55$ kN, $\hat{f}_{2y}^{(3)} = -\hat{f}_{4y}^{(3)} = 1.40$ kN
$\hat{m}_2^{(3)} = -10.51$ kN · m, $\hat{m}_4^{(3)} = -5.30$ kN · m
$F_{1x} = -17.26$ kN, $F_{1y} = 23.80$ kN, $M_1 = 32.77$ kN · m
$F_{3x} = -11.31$ kN, $F_{3y} = 42.81$ kN, $M_3 = -87.54$ kN · m
$F_{4x} = -11.42$ kN, $F_{4y} = 13.40$ kN, $M_4 = -5.30$ kN · m

6.9 $d_{2x} = -4.95 \times 10^{-5}$ m, $d_{2y} = -2.56 \times 10^{-5}$ m, $\phi_2 = 2.66 \times 10^{-3}$ rad
$\hat{f}_{1x}^{(1)} = -\hat{f}_{2x}^{(1)} = 26.9$ kN, $\hat{f}_{1y}^{(1)} = -\hat{f}_{2y}^{(1)} = -42.0$ kN
$\hat{m}_1^{(1)} = 55.9$ kN · m, $\hat{m}_2^{(1)} = 111.7$ kN · m
$\hat{f}_{2x}^{(2)} = -\hat{f}_{3x}^{(2)} = -42.0$ kN, $\hat{f}_{2y}^{(2)} = -\hat{f}_{3y}^{(2)} = 26.9$ kN
$M_1 = 55.9$ kN · m, $M_3 = 44.7$ kN · m

6.10 $d_{2y} = -0.1423 \times 10^{-2}$ m, $\phi_2 = -0.5917 \times 10^{-3}$ rad
$\hat{f}_{1x}^{(1)} = 0$, $\hat{f}_{1y}^{(1)} = 10$ kN, $\hat{m}_1^{(1)} = 23.3$ kN · m, $\hat{f}_{2x}^{(1)} = 0$,
$\hat{f}_{2y}^{(1)} = -10$ kN, $\hat{m}_2^{(1)} = 6.7$ kN · m

6.11 $d_{2y} = -0.928 \times 10^{-5}$ m, $F_{1x} = 1860$ N, $F_{1y} = 2500$ N, $M_1 = 28$ N · m

6.12 $d_{1x} = -0.2143$ m, $d_{1y} = -0.250$ m, $\phi_1 = 0.0893$ rad, $d_{2x} = -0.2143$ m,
$d_{2y} = -0.357 \times 10^{-4}$ m, $\phi_2 = 0.0714$ m

6.13 $d_{2x} = 0.0559$ in., $d_{2y} = 0.00382$ in., $\phi_2 = -0.000150$ rad
$d_{3x} = 0.0558$ in., $d_{3y} = -0.000133$ in., $\phi_3 = 0.000149$ rad
$F_{1x} = -198$ lb, $F_{1y} = -4770$ lb, $M_1 = 27460$ lb · in.
$F_{4x} = -4802$ lb, $F_{4y} = 4770$ lb, $M_4 = 27430$ lb · in.

6.14 $d_{2x} = 0.0174$ in., $d_{2y} = -0.0481$ in., $\phi_2 = -0.00165$ rad
$\hat{f}_{1x}^{(1)} = 19160$ lb, $\hat{f}_{1y}^{(1)} = -1385$ lb, $\hat{m}_1^{(1)} = -59050$ lb · in.
$\hat{f}_{2x}^{(1)} = -19160$ lb, $\hat{f}_{2y}^{(1)} = 1385$ lb, $\hat{m}_2^{(1)} = -176,000$ lb · in.

6.15 $d_{2x} = -1.76 \times 10^{-2}$ m, $d_{2y} = -1.87 \times 10^{-5}$ m, $\phi_2 = 5.00 \times 10^{-3}$ rad
$d_{3x} = -1.76 \times 10^{-2}$ m, $\phi_3 = -2.49 \times 10^{-3}$ rad
$F_{1x} = 20.0$ kN, $F_{1y} = 13.1$ kN, $M_1 = -57.4$ kN · m, $F_{3y} = -13.1$ kN

6.16 $d_{3y} = -2.83 \times 10^{-5}$ m, $d_{4x} = 1.0 \times 10^{-5}$ m, $d_{4y} = -2.03 \times 10^{-5}$ m

6.17 $d_{3y} = -0.397$ in., $\phi_3 = 0$

6.18 $d_{2x} = d_{2y} = -5 \times 10^{-6}$ m, $\phi_2 = 8.83 \times 10^{-5}$ rad

6.19 $d_{1x} = 0.702$ in., $d_{1y} = 0.00797$ in., $\phi_1 = -0.00446$ rad
$\hat{f}_{3x}^{(1)} = -\hat{f}_{1x}^{(1)} = -19.93$ kip, $\hat{f}_{3y}^{(1)} = -\hat{f}_{1y}^{(1)} = 18.1$ kip, $\hat{m}_3^{(1)} = 1309$ k · in.
$\hat{m}_1^{(1)} = 863$ k · in.

6.20 $d_{3x} = 1.24$ in., $d_{3y} = 0.00203$ in., $\phi_3 = -0.000556$ rad
$\hat{f}_{1x}^{(1)} = -2.76$ kip, $\hat{f}_{1y}^{(1)} = 1.79$ kip, $\hat{m}_1^{(1)} = 0$, $\hat{f}_{2x}^{(1)} = 2.76$ kip, $\hat{f}_{2y}^{(1)} = -1.79$ kip,
$\hat{m}_2^{(1)} = 322$ k · in.

6.21 Use a W16 × 31 for all sections

6.23 $d_{5x} = 0.0204$ in., $\quad d_{5y} = 0.00122$ in., $\quad \phi_5 = 0.000207$ rad

6.24 $d_{5x} = 2.82$ in., $\quad d_{5y} = 0.00266$ in., $\quad \phi_5 = -0.00139$ rad

6.25 **a.** $d_{2y} = -2.12 \times 10^{-3}$ in. $\qquad$ **b.** $d_{3y} = -6.07 \times 10^{-2}$ in.

6.26 $d_{2x} = 0.596 \times 10^{-5}$ in., $\quad d_{2y} = -0.332 \times 10^{-2}$ in., $\quad \phi_2 = -0.100 \times 10^{-3}$ rad
$F_{1x} = 130$ lb, $\quad F_{1y} = 10360$ lb, $\quad F_{4x} = -130$ lb, $\quad F_{4y} = 10360$ lb

6.27 $d_{3y} = -0.0128$ in., $\quad f_{1x}^{(1)} = 25$ kN, $\quad f_{1y}^{(1)} = -5.55$ kN, $\quad m_1^{(1)} = 0$

6.28 $d_{2x} = 5.70$ mm, $\quad d_{2y} = -0.0244$ mm, $\quad \phi_2 = 0.00523$ rad

6.29 $d_{3y} = -1.83$ in., $\quad d_{4y} = -1.22$ in.

6.30 $d_{3y} = 6.67$ in., $\quad d_{4y} = -6.67$ in., $\quad \phi_3 = -\phi_4 = -3.20$ rad
$F_{1x} = 11.69$ kN, $\quad F_{1y} = 30$ kN, $\quad M_1 = -1810$ kN · m
$F_{6x} = -11.69$ kN, $\quad F_{6y} = 30$ kN, $\quad M_6 = 1810$ kN · m

6.31 $d_{2y} = -1.58 \times 10^{-2}$ in.

6.32 $d_{2x} = 4.30$ mm, $\quad \phi_2 = -0.241 \times 10^{-3}$ rad
$F_{1x} = -8339$ N, $\quad F_{1y} = -4995$ N, $\quad M_1 = 26,700$ N · m,
$F_{4x} = -6661$ N, $\quad F_{4y} = 4995$ N, $\quad M_4 = 23,330$ N · m

6.33 $d_{7x} = 0.0264$ m, $\quad d_{7y} = 0.463 \times 10^{-4}$ m, $\quad \phi_7 = 0.171 \times 10^{-2}$ rad
$\hat{f}_{1x}^{(1)} = -21.1$ N, $\quad \hat{f}_{1y}^{(1)} = 30.4$ N, $\quad \hat{m}_1^{(1)} = 74.95$ N · m
$\hat{f}_{3x}^{(1)} = 21.1$ N, $\quad \hat{f}_{3y}^{(1)} = -30.4$ N, $\quad \hat{m}_3^{(1)} = 46.65$ N · m

6.35 $d_{9x} = 0.0174$ m, $\quad \hat{f}_{1x}^{(1)} = -22.6$ kN, $\quad \hat{f}_{1y}^{(1)} = 16.0$ kN, $\quad \hat{m}_1^{(1)} = 53.6$ kN · m
$\hat{f}_{3x}^{(1)} = 22.6$ kN, $\quad \hat{f}_{3y}^{(1)} = -16.0$ kN, $\quad \hat{m}_3^{(1)} = 42.4$ kN · m

6.36 $d_{6y} = -2.80 \times 10^{-7}$ m, $\quad d_{7y} = -4.87 \times 10^{-7}$ m

6.37 $d_{4x} = -0.495 \times 10^{-4}$ m, $\quad d_{4y} = -0.290 \times 10^{-3}$ m, $\quad \phi_4 = 0.1547 \times 10^{-2}$ rad

6.38 $d_{2x} = 1.396 \times 10^{-1}$ m

6.39 Truss: $d_{7x} = 0.0260$ m, $\quad d_{7y} = 0.00566$ m,
Frame: $d_{7x} = 0.0180$ m, $\quad d_{7y} = 0.00424$ m
Truss, element 1: $\hat{f}_{1x} = -49,730$ N, $\quad \hat{f}_{1y} = 0$
Frame, element 1: $\hat{f}_{1x} = -43,060$ N, $\quad \hat{f}_{1y} = 22670$ N

6.40 Tapered beam $n = 3$
one element: $\quad d_{1y} = -0.222 \times 10^{-1}$ in.
two elements: $\quad d_{1y} = -0.189 \times 10^{-1}$ in.
four elements: $\quad d_{1y} = -0.181 \times 10^{-1}$ in.
eight elements: $d_{1y} = -0.179 \times 10^{-1}$ in.

6.41 $\underline{K} = 15\dfrac{GJ_0}{L}\begin{bmatrix} 1 & -1 \\ -1 & 1 \end{bmatrix}$

6.43 $d_{2y} = -0.214$ in.

6.44 $d_{2y} = -0.729$ in.

6.46 $d_{1y} = -0.690 \times 10^{-2}$ m

6.47 $d_{5y} = -1.028 \times 10^{-4}$ in.

6.48 $d_{4y} = -1.026$ in.

6.50 $d_{3y} = -2.54 \times 10^{-3}$ m

6.52 $d_{5y} = -2.22 \times 10^{-2}$ m

6.53 $d_{2y} = 1.487$ in., $d_{3z} = -9.896 \times 1$

6.54 $d_{7z} = -8.80 \times 10^{-1}$ in.

Chapter 7

7.1 Use Eq. (7.2.10) in Eq. (7.2.18) to

7.3 **a.** $\underline{k} = 4.0 \times 10^6$

$$
\begin{bmatrix}
2.5 & 1.25 & -2 \\
 & 4.375 & -1. \\
 & & 4. \\
 & & \\
 & & \\
 & & \\
\text{Symmetry} & &
\end{bmatrix}
$$

b. $\underline{k} = 13.33 \times 10^6$

$$
\begin{bmatrix}
1.54 & 0.75 & \\
 & 1.815 & \\
 & & \\
 & & \\
\text{Symmetry} & &
\end{bmatrix}
$$

7.4 **a.** $\sigma_x = 19.2$ ksi, $\sigma_y = 4.8$ ksi, τ_x

$\sigma_1 = 28.6$ ksi, $\sigma_2 = -4.64$ ksi,

b. $\sigma_x = 32.0$ ksi, $\sigma_y = 8.0$ ksi, τ_x

$\sigma_1 = 47.7$ ksi, $\sigma_2 = -7.73$ ksi,

7.5 **a.** $\underline{k} = 2.074 \times 10^5$

$$
\begin{bmatrix}
8437.5 & 1687.5 & -7762.5 & -337.5 & -675 & -1350 \\
1687.5 & 3937.5 & 337.5 & -2137.5 & -2025 & -1800 \\
-7762.5 & 337.5 & 8437.5 & -1687.5 & -675 & 1350 \\
-337.5 & -2137.5 & -1687.5 & 3937.5 & 2025 & -1800 \\
-675 & -2025 & -675 & 2025 & 1350 & 0 \\
-1350 & -1800 & 1350 & -1800 & 0 & 3600
\end{bmatrix} \text{N/m}
$$

b. $\underline{k} = 4.48 \times 10^7$

$$
\begin{bmatrix}
25.0 & 0 & -12.5 & 6.25 & -12.5 & -6.25 \\
 & 9.375 & 9.375 & -4.6875 & -9.375 & -4.6875 \\
 & & 15.625 & -7.8125 & -3.125 & -1.5625 \\
 & & & 27.343 & 1.5625 & -3.125 \\
 & & & & 15.625 & 7.8125 \\
\text{Symmetry} & & & & & 27.343
\end{bmatrix} \text{N/m}
$$

7.6 **a.** $\sigma_x = -5.289$ GPa, $\sigma_y = -0.156$ GPa, $\tau_{xy} = 0.233$ GPa

$\sigma_1 = -0.1459$ GPa, $\sigma_2 = -5.30$ GPa, $\theta_p = -2.59°$

b. $\sigma_x = 0$, $\sigma_y = 42.0$ MPa, $\tau_{xy} = 33.6$ MPa

$\sigma_1 = 60.6$ MPa, $\sigma_2 = -18.6$ MPa, $\theta_p = -29°$

7.7 **a.** $\sigma_x = -15.0$ ksi, $\sigma_y = -45.0$ ksi, $\tau_{xy} = -18.0$ ksi
$\sigma_1 = -6.57$ ksi, $\sigma_2 = -53.4$ ksi, $\theta_p = -25.1°$
b. $\sigma_x = -15.0$ ksi, $\sigma_y = -45$ ksi, $\tau_{xy} = -21.0$ ksi
$\sigma_1 = -4.19$ ksi, $\sigma_2 = -55.8$ ksi, $\theta_p = -27.2°$

7.8 **a.** $\sigma_x = -52.5$ MPa, $\sigma_y = -32.8$ MPa, $\tau_{xy} = -5.38$ MPa
$\sigma_1 = -31.4$ MPa, $\sigma_2 = -53.9$ MPa, $\theta_p = -14.3°$
b. $\sigma_x = -31.4$ MPa, $\sigma_y = -13.5$ MPa, $\tau_{xy} = 5.38$ MPa
$\sigma_1 = -12.0$ MPa, $\sigma_2 = -32.9$ MPa, $\theta_p = -15.5°$
c. $\sigma_x = -27.6$ MPa, $\sigma_y = -19.5$ MPa, $\tau_{xy} = 4.04$ MPa
$\sigma_1 = 17.9$ MPa, $\sigma_2 = -29.3$ MPa, $\theta_p = -22.5°$
d. $\sigma_x = -31.6$ MPa, $\sigma_y = -28.9$ MPa, $\tau_{xy} = -6.73$ MPa
$\sigma_1 = -23.0$ MPa, $\sigma_2 = -38.0$ MPa, $\theta_p = 39°$

7.9 **a.** $f_{s1x} = 0$, $f_{s1y} = 0$, $f_{s2x} = p_0 Lt/6$, $f_{s2y} = 0$
$f_{s3x} = p_0 Lt/3$, $f_{s3y} = 0$
b. $f_{s1x} = 0$, $f_{s2x} = p_0 Lt/12$, $f_{s3x} = p_0 Lt/4$

7.10 $d_{3x} = 0.5 \times 10^{-3}$ in., $d_{3y} = -0.275 \times 10^{-2}$ in.
$d_{4x} = -0.609 \times 10^{-3}$ in., $d_{4y} = -0.293 \times 10^{-2}$ in.
$\sigma_x^{(1)} = 824$ psi, $\sigma_y^{(1)} = 247$ psi, $\tau_{xy}^{(1)} = -1587$ psi
$\sigma_1^{(1)} = 2149$ psi, $\sigma_2^{(1)} = -1077$ psi, $\theta_p^{(1)} = -40°$
$\sigma_x^{(2)} = -826$ psi, $\sigma_y^{(2)} = 292$ psi, $\tau_{xy}^{(2)} = -411$ psi
$\sigma_1^{(2)} = 426$ psi, $\sigma_2^{(2)} = -960$ psi, $\theta_p^{(2)} = 18.15°$

7.11 **a.** $d_{2x} = 0.281 \times 10^{-4}$ m, $d_{2y} = -0.330 \times 10^{-4}$ m
$d_{5x} = 0.115 \times 10^{-4}$ m, $d_{5y} = -0.103 \times 10^{-4}$ m
$\sigma_x^{(2)} = 16.4$ MPa, $\sigma_y^{(2)} = 15.2$ MPa
$\tau_{xy}^{(2)} = -6.99$ MPa, $\sigma_1^{(2)} = 22.8$ MPa
$\sigma_2^{(2)} = 8.80$ MPa, $\theta_p^{(2)} = -42.7°$
$\sigma_x^{(1)} = 10.6$ MPa, $\sigma_y^{(1)} = 3.18$ MPa
$\tau_{xy}^{(1)} = -3.34$ MPa, $\sigma_1^{(1)} = 11.9$ MPa
$\sigma_2^{(1)} = 1.90$ MPa, $\theta_p^{(1)} = -21.0°$
b. $d_{1x} = -d_{2x} = -0.165 \times 10^{-5}$ m, $d_{1y} = d_{2y} = -0.125 \times 10^{-4}$ m
$d_{5x} = 0.274 \times 10^{-12}$ m, $d_{5y} = -0.163 \times 10^{-4}$ m
$\sigma_x^{(1)} = 5.99 \times 10^5$ N/m², $\sigma_y^{(1)} = -3.78 \times 10^6$ N/m²
$\tau_{xy}^{(1)} = 4.05 \times 10^{-1}$ N/m², $\sigma_1^{(1)} = 5.99 \times 10^5$ N/m²
$\sigma_2^{(1)} = -3.78 \times 10^6$ N/m², $\theta_p^{(1)} = 0°$, $\sigma_x^{(3)} = 5.64 \times 10^6$ N/m²
$\sigma_y^{(3)} = 1.88 \times 10^7$ N/m², $\tau_{xy}^{(3)} = -1.11 \times 10^{-1}$ N/m²
$\sigma_1^{(3)} = 1.88 \times 10^7$ N/m², $\sigma_2^{(3)} = 5.64 \times 10^6$ N/m², $\theta_p^{(3)} = -90°$

7.12 All f_{bx}'s are equal to 0.
$f_{b1y} = f_{b2y} = f_{b3y} = f_{b4y} = 10.28$ N, $f_{b5y} = 20.56$ N

7.13 **a.** $n_b = 8$, **b.** $n_b = 12$

Chapter 8

8.6 $d_{2x} = d_{3x} = 0.647 \times 10^{-3}$ in., $d_{2y} = 0.666 \times 10^{-4}$ in.
$d_{3y} = -0.666 \times 10^{-4}$ in., skew effect

8.7 Stress approaches 2.5 psi near edge of whole for model of 70 nodes, 54 elements.

8.8 At depth 4 in. equal to width, stress approaches uniform $\sigma_y = -1000$ psi.

8.9 $\sigma_1 = 8836$ psi at top and bottom of hole

8.10 $\sigma_1 = 372$ psi at fillet

8.11 For 106-element model at re-entrant corner, $\sigma_1 = 3500$ psi, $\sigma_2 = 670$ psi.

8.12 $\sigma_{VM} = 93.7$ psi at load

8.14 For the model with 12 in. $\times \frac{1}{2}$ in. size elements, finite element solution yields free-end deflection of -0.499 in.; exact solution is -1.15 in. (See Table 8–1 in text for other results.)

8.16 $\sigma_1 = 3002$ psi

8.18 $\sigma_{VM} = 8.1$ MPa

8.19 a. $\sigma_1 = 58700$ psi

8.20 $\sigma_1 = 6.2$ MPa

8.22 Largest principal stress 30–40 MPa in support region (72-element, 91-node model)

8.24 Largest principal stress $\sigma_1 = 1005$ MPa at narrowest width of member (70-element, 94-node model)

Chapter 9

9.2 $\varepsilon_x = \dfrac{1}{3b}(-u_1 + u_2 + 4u_4 - 4u_5), \quad \varepsilon_y = \dfrac{1}{3h}(-v_1 + v_3 + 4v_4 - 4v_6)$

$\gamma_{xy} = \dfrac{1}{3h}(-u_1 + u_3 + 4u_4 - 4u_6) + \dfrac{1}{3b}(-v_1 + v_3 + 4v_4 - 4v_6)$

$\sigma_x = \dfrac{E}{1-v^2}(\varepsilon_x + v\varepsilon_y), \quad \sigma_y = \dfrac{E}{1-v^2}(\varepsilon_y + v\varepsilon_x), \quad \tau_{xy} = G\gamma_{xy}$

9.3 $f_{s1x} = f_{s3x} = \dfrac{-pth}{6}, \quad f_{s5x} = \dfrac{-2pth}{3}$

9.4 $f_{s1x} = 0, \quad f_{s3x} = \dfrac{-p_0 th}{6}, \quad f_{s5x} = \dfrac{-p_0 th}{3}$

9.5 a. $\varepsilon_x = -5 \times 10^{-5}y + 2.5 \times 10^{-4}, \quad \varepsilon_y = -1.67 \times 10^{-4}x + 3.33 \times 10^{-5},$
  $\gamma_{xy} = -5 \times 10^{-5}x - 1.11 \times 10^{-4}y + 4.17 \times 10^{-4}$
  $\sigma_x = 3290$ psi, $\sigma_y = -4850$ psi, $\tau_{xy} = 1540$ psi
b. $\varepsilon_x = -5 \times 10^{-5}y + 1.67 \times 10^{-4}, \quad \varepsilon_y = -1.67 \times 10^{-4}x + 5 \times 10^{-5}$
  $\gamma_{xy} = -5 \times 10^{-5}x - 4.17 \times 10^{-5}y + 2.08 \times 10^{-4}$
  $\sigma_x = 928$ psi, $\sigma_y = -8290$ psi, $\tau_{xy} = 632$ psi

9.6 $\varepsilon_x = 2.54 \times 10^{-3}$
  $\varepsilon_y = -7.62 \times 10^{-3}$
  $\gamma_{xy} = -7.04 \times 10^{-3}$

9.7 $N_1 = 1 - \dfrac{x}{20} + \dfrac{x^2}{1800}, \quad N_2 = \dfrac{-x+y}{60} + \dfrac{x^2+y^2}{1800} - \dfrac{xy}{900}$

  $N_3 = \dfrac{-y}{60} + \dfrac{y^2}{1800}, \quad N_4 = \dfrac{xy}{900} - \dfrac{y^2}{900}, \quad N_5 = \dfrac{y}{15} - \dfrac{xy}{900},$ etc.

Chapter 10

10.1   **a.** $K = 25.132 \times 10^6$
$$\begin{bmatrix} 5 & 1 & 0 & -1 & 1 & 0 \\ 1 & 4 & -2 & -1 & -2 & -3 \\ 0 & -2 & 8 & 0 & 4 & 2 \\ -1 & -1 & 0 & 1 & 1 & 0 \\ 1 & -2 & 4 & 1 & 4 & 1 \\ 0 & -3 & 2 & 0 & 1 & 3 \end{bmatrix} \text{lb/in.}$$

b. $K = 50.265 \times 10^6$
$$\begin{bmatrix} 2.75 & 0 & -2.25 & 0.5 & 0.25 & -0.5 \\ 0 & 1 & 1 & -1 & -1 & 0 \\ -2.25 & 1 & 5.75 & -2.5 & 0.25 & 1.5 \\ 0.5 & -1 & -2.5 & 4 & 0.5 & -3 \\ 0.25 & -1 & 0.25 & 0.5 & 1.75 & 0.5 \\ -0.5 & 0 & 1.5 & -3 & 0.5 & 3 \end{bmatrix} \text{lb/in.}$$

10.2 $f_{s2x} = \dfrac{2\pi bp_0 h}{6}, \quad f_{s3x} = \dfrac{2\pi bp_0 h}{3}$

10.3 $f_{b1r} = f_{b2r} = f_{b3r} = 0.382 \text{ lb}$
$f_{b1z} = f_{b2z} = f_{b3z} = -6.32 \text{ lb}$

10.4   **a.** $\sigma_r = 8000 \text{ psi}, \quad \sigma_z = 0, \quad \sigma_\theta = 8000 \text{ psi}, \quad \tau_{rz} = 1200 \text{ psi}$
b. $\sigma_r = 5830 \text{ psi}, \quad \sigma_z = -3770 \text{ psi}, \quad \sigma_\theta = 3090 \text{ psi}, \quad \tau_{rz} = 400 \text{ psi}$

10.6   **a.** $k = 7.037$
$$\begin{bmatrix} 3125 & 625 & 0 & -625 & 625 & 0 \\ & 2500 & -1250 & -625 & -1250 & -1875 \\ & & 5000 & 0 & 2500 & 1250 \\ & & & 625 & 625 & 0 \\ & & & & 2500 & 625 \\ \text{Symmetry} & & & & & 1875 \end{bmatrix} \text{kN/mm}$$

b. $k = 11.73$
$$\begin{bmatrix} 2475 & 0 & -2025 & 450 & 225 & -450 \\ & 900 & 900 & -900 & -900 & 0 \\ & & 5175 & -2250 & 225 & 1350 \\ & & & 3600 & 450 & -2700 \\ & & & & 1575 & 450 \\ \text{Symmetry} & & & & & 2700 \end{bmatrix} \text{kN/mm}$$

10.7   **a.** $\sigma_r = -84 \text{ MPa}, \quad \sigma_z = -84 \text{ MPa}, \quad \sigma_\theta = 252 \text{ MPa}, \quad \tau_{rz} = -101 \text{ MPa}$
b. $\sigma_r = -103 \text{ MPa}, \quad \sigma_z = -103 \text{ MPa}, \quad \sigma_\theta = 112 \text{ MPa}, \quad \tau_{rz} = -73 \text{ MPa}$

Chapter 11

11.2 $N_1 = \dfrac{s(s-1)}{2}, \quad N_2 = \dfrac{s(s+1)}{2}, \quad N_3 = 1 - s^2$

$[B] = \begin{bmatrix} \dfrac{2s-1}{L} & \dfrac{2s+1}{L} & \dfrac{-4s}{L} \end{bmatrix}$

11.3 $K = \dfrac{AE}{2L} \begin{bmatrix} 4.67 & 0.667 & -5.33 \\ 0.667 & 4.67 & -5.33 \\ -5.33 & -5.33 & 10.67 \end{bmatrix}$

11.4 $N_1 = \dfrac{-2s^3}{3} + \dfrac{2s^2}{3} + \dfrac{s}{6} - \dfrac{1}{6}$, $\quad N_2 = \dfrac{4s^3}{3} - \dfrac{2s^2}{3} - \dfrac{4s}{3} + \dfrac{2}{3}$

$N_3 = \dfrac{-4s^3}{3} - \dfrac{2s^2}{3} + \dfrac{4s}{3} + \dfrac{2}{3}$, $\quad N_4 = \dfrac{2s^3}{3} + \dfrac{2s^2}{3} - \dfrac{s}{6} - \dfrac{1}{6}$

$$[B] = \left[\dfrac{-12s^2 + 8s + 1}{3L} \quad \dfrac{12s^2 - 4s - 4}{1.5L} \quad \dfrac{-12s^2 - 4s + 4}{1.5L} \quad \dfrac{12s^2 + 8s - 1}{3L}\right]$$

11.6 $\varepsilon_x = 0.0009375$ in./in., $\quad \varepsilon_y = -0.00125$ in./in., $\quad \gamma_{xy} = -0.000625$ rad
$\sigma_x = 18.5$ ksi, $\quad \sigma_y = -31.9$ ksi, $\quad \tau_{xy} = -7.21$ ksi

11.11 **a.** $f_{s2s} = 500$ lb, $\quad f_{s3s} = 500$ lb, $\quad$ **b.** $f_{s1t} = 83.33$ lb, $\quad f_{s4t} = 41.67$ lb

11.12 **a.** 1.917, $\quad$ **b.** 0.667, $\quad$ **c.** 0.400, $\quad$ **d.** 2.87, $\quad$ **f.** 0

Chapter 12

12.1 $\underline{B} = \dfrac{1}{8}\begin{bmatrix} 0 & 0 & 0 & 0 & 0 & 0 & 4 & 0 & 0 & -4 & 0 & 0 \\ 0 & 0 & 0 & 0 & 4 & 0 & 0 & 0 & 0 & 0 & 4 & 0 \\ 0 & 0 & 4 & 0 & 0 & 0 & 0 & 0 & 0 & 0 & 0 & -4 \\ 0 & 0 & 0 & 4 & 0 & 0 & 0 & 4 & 0 & -4 & -4 & 0 \\ 0 & 4 & 0 & 0 & 0 & 4 & 0 & 0 & 0 & 0 & -4 & -4 \\ 4 & 0 & 0 & 0 & 0 & 0 & 0 & 0 & 4 & -4 & 0 & -4 \end{bmatrix}$

12.2 $\sigma_x = 77.9$ ksi, $\quad \sigma_y = 8.65$ ksi, $\quad \sigma_z = -49.0$ ksi
$\tau_{xy} = 11.5$ ksi, $\quad \tau_{yz} = -23.1$ ksi, $\quad \tau_{zx} = 5.77$ ksi

12.4 $\underline{B} = \dfrac{1}{18750}$

$$\times \begin{bmatrix} -625 & 0 & 0 & 0 & 0 & 0 & 0 & 0 & 0 & 625 & 0 & 0 \\ 0 & -375 & 0 & 0 & -375 & 0 & 0 & 0 & 0 & 0 & -375 & 0 \\ 0 & 0 & -375 & 0 & 0 & 0 & 0 & 0 & 750 & 0 & 0 & -375 \\ -375 & -625 & 0 & 750 & 0 & 0 & 0 & 0 & 0 & -375 & 625 & 0 \\ 0 & -375 & -375 & 0 & 0 & 750 & 0 & 750 & 0 & 0 & -375 & -375 \\ -375 & 0 & -625 & 0 & 0 & 0 & 750 & 0 & 0 & -375 & 0 & 625 \end{bmatrix}$$

12.5 $\sigma_x = 72.7$ MPa, $\quad \sigma_y = 169.6$ MPa, $\quad \sigma_z = 72.7$ MPa
$\tau_{xy} = 59.2$ MPa, $\quad \tau_{yz} = 32.3$ MPa, $\quad \tau_{zx} = 91.5$ MPa

12.6 $N_2 = \dfrac{(1-s)(1-t)(1-z')}{8}$, $\quad N_3 = \dfrac{(1-s)(1+t)(1-z')}{8}$,

$N_4 = \dfrac{(1-s)(1+t)(1+z')}{8}$,

$N_5 = \dfrac{(1+s)(1-t)(1+z')}{8}$, $\quad N_6 = \dfrac{(1+s)(1-t)(1-z')}{8}$,

$N_7 = \dfrac{(1+s)(1+t)(1-z')}{8}$, $\quad N_8 = \dfrac{(1+s)(1+t)(1+z')}{8}$

12.7 $N_1 = \dfrac{(1-s)(1-t)(1+z')(-s-t+z'-2)}{8}$,

$N_2 = \dfrac{(1-s)(1-t)(1-z')(-s-t-z'-2)}{8}$

12.9 $d_{\max} = -0.662$ in. under the load

Chapter 13

13.1 $t_2 = 166.7\,°C, \quad t_3 = 233.3\,°C$

13.2 $t_2 = 150\,°F, \quad t_3 = 100\,°F, \quad t_4 = 50\,°F$

13.3 $t_2 = 875\,°F, \quad t_3 = 1250\,°F$

13.4 $t_1 = 151\,°F, \quad t_2 = 148\,°F, \quad t_3 = 140\,°F, \quad t_4 = 125\,°F$

13.5 $t_2 = 183\,°F, \quad t_3 = 267\,°F, \quad t_4 = 350\,°F, \quad t_5 = 433\,°F$

13.6 $t_2 = 421\,°C, \quad t_3 = 121\,°C, \quad q^{(3)} = 3975 \text{ W/m}^2$

13.7 $t_2 = 418.2\,°C, \quad t_3 = 527.3\,°C$

13.9 $\underline{k} = \begin{bmatrix} 39.57 & 7.076 & -5.417 \\ & 35.82 & -1.667 \\ & & 7.083 \end{bmatrix}, \quad \underline{f} = \left\{ \begin{array}{c} 2936 \\ 2936 \\ 50 \end{array} \right\} \text{ Btu/h}$

13.10 $\underline{f} = \left\{ \begin{array}{c} 1291 \\ 27.3 \\ 1254 \end{array} \right\} \text{ W}$

13.13 $t_4 = 75\,°F, \quad t_5 = 25\,°F$

Chapter 14

14.1 $p_2 = 4.545 \text{ m}, \quad p_3 = 1.818 \text{ m}, \quad v_x^{(1)} = 10.91 \text{ m/s}, \quad Q_f^{(1)} = 21.82 \text{ m}^3/\text{s}$

14.2 $p_2 = -15 \text{ m}, \quad p_3 = -40 \text{ m}, \quad p_4 = -65 \text{ m}, \quad v_x^{(1)} = 25 \text{ m/s}, \quad Q_1 = 50 \text{ m}^3/\text{s}$

14.3 $p_2 = 8.182 \text{ in.}, \quad p_3 = 5.455 \text{ in.}, \quad v_x^{(1)} = 0.182 \text{ in./s}, \quad v_x^{(2)} = 0.273 \text{ in./s}, \\ v_x^{(3)} = 0.545 \text{ in./s}, \quad Q_f^{(1)} = 1.091 \text{ in}^3/\text{s}$

14.4 $p_2 = -3 \text{ cm}, \quad p_3 = -8 \text{ cm}, \quad v_x^{(1)} = 1.2 \text{ cm/s}, \quad v_x^{(2)} = 2 \text{ cm/s}, \\ Q_1 = Q_2 = 6 \text{ cm}^3/\text{s}$

14.6 $v^{(1)} = 2.0 \text{ in./s}, \quad v^{(2)} = 4.0 \text{ in./s}, \quad Q^{(1)} = Q^{(2)} = 4 \text{ in}^3/\text{s}$

14.7 $\underline{f_Q} = \left\{ \begin{array}{c} 54.76 \\ 28.57 \\ 16.67 \end{array} \right\} \text{ m}^3/\text{s}$

14.8 $f_1 = f_3 = 5 \text{ in}^3/\text{s}, \quad f_2 = 0$

14.9 $p_2 = p_3 = 12 \text{ m}, \quad p_5 = 11 \text{ m}$

Chapter 15

15.1 $d_{2x} = 0.021 \text{ in.}, \quad d_{3x} = 0.042 \text{ in.}, \quad \sigma_x = 0$

15.2 $d_{2x} = 0, \quad \sigma_x = 50.4 \text{ MPa}$

15.3 $d_{1x} = d_{1y} = -0.0175 \text{ in.}, \quad \sigma^{(1)} = 4350 \text{ psi (T)} \\ \sigma^{(2)} = -6150 \text{ psi (C)}, \quad \sigma^{(3)} = 4350 \text{ psi (T)}$

15.4 $d_{1x} = -0.0291 \text{ in.}, \quad d_{1y} = -0.0095 \text{ in.} \\ \sigma^{(1)} = -1370 \text{ psi (C)}, \quad \sigma^{(2)} = 2375 \text{ psi (T)}, \quad \sigma^{(3)} = -1370 \text{ psi (C)}$

15.5 $d_{2x} = 1.44 \times 10^{-4} \text{ m}, \quad \sigma^{(1)} = -20.2 \text{ MPa (C)}, \quad \sigma^{(2)} = \sigma^{(3)} = -10.1 \text{ MPa (C)}$

15.6 $d_{1x} = 0$, $d_{1y} = 6.0 \times 10^{-4}$ m, $\sigma^{(1)} = \sigma^{(3)} = -10.5$ MPa (C)
$\sigma^{(2)} = 18.2$ MPa (T)

15.7 $d_{1x} = 0$, $d_{1y} = -3.6 \times 10^{-4}$ m, $\sigma^{(1)} = \sigma^{(2)} = 0$

15.8 $d_{2x} = 0.0173$ in., $\sigma_{st} = 840$ psi (T), $\sigma_{br} = 1680$ psi (C)

15.9 $f_{T1x} = -4464$ lb, $f_{T1y} = -8929$ lb, $f_{T2x} = 4464$ lb
$f_{T2y} = -8929$ lb, $f_{T3x} = 0$, $f_{T3y} = 17,857$ lb

15.10 $f_{T1x} = -43.125$ kN, $f_{T1y} = 0$, $f_{T2x} = 43.125$ kN, $f_{T2y} = -86.250$ kN
$f_{T3x} = 0$, $f_{T3y} = 86.250$ kN

15.11 $f_{T1x} = -60.0$ kip, $f_{T1y} = -90$ kip, $f_{T2x} = 60$ kip, $f_{T2y} = 0$,
$f_{T3x} = 0$, $f_{T3y} = 90$ kip

15.12 $f_{T1x} = 134$ kN, $f_{T1y} = 134$ kN, $f_{T2x} = -134$ kN, $f_{T2y} = 0$
$f_{T3x} = 0$, $f_{T3y} = -134$ kN

15.13 $\sigma_x = \sigma_y = -8929$ psi (C), $\tau_{xy} = 0$

15.14 $\sigma_x = 67.2$ MPa, $\sigma_y = 67.2$ MPa, $\tau_{xy} = 0$

15.15 $\{f_T\} = \dfrac{AE\alpha_0}{6} \left\{ \begin{array}{c} -4t_1 - 5t_2 \\ 4t_1 + 5t_2 \end{array} \right\}$

15.16 $\dfrac{AE\alpha}{2} \left\{ \begin{array}{c} -t_1 - t_2 \\ t_1 + t_2 \end{array} \right\}$

15.17 $\{f_T\} = \dfrac{2\pi \bar{r} A E \alpha (\Delta T)[\bar{B}]^T}{1 - 2v} \left\{ \begin{array}{c} 1 \\ 1 \\ 1 \\ 0 \end{array} \right\}$

15.18 $d_{2x} = 0.8 \times 10^{-3}$ in., $d_{3x} = 0$, $d_{3y} = 0.8 \times 10^{-3}$ in.
$d_{4x} = d_{4y} = 0.8 \times 10^{-3}$ in.; stresses are zero

15.19 $d_{2x} = 0.989 \times 10^{-3}$ in., $d_{3x} = -0.756 \times 10^{-3}$ in.,
$d_{3y} = 0.989 \times 10^{-3}$ in., $d_{4x} = 0.132 \times 10^{-2}$ in.,
$d_{4y} = 0.2045 \times 10^{-2}$ in., $\sigma_1^{(1)} = 17$ ksi, $\sigma_2^{(2)} = -17$ ksi

Chapter 16

16.1 $[M] = \dfrac{\rho A L}{6} \begin{bmatrix} 2 & 1 & 0 \\ 1 & 4 & 1 \\ 0 & 1 & 2 \end{bmatrix}$

16.2 **a.** $[M] = \dfrac{\rho A L}{2} \begin{bmatrix} 1 & 0 & 0 & 0 \\ 0 & 2 & 0 & 0 \\ 0 & 0 & 2 & 0 \\ 0 & 0 & 0 & 1 \end{bmatrix}$

b. $[M] = \dfrac{\rho A L}{6} \begin{bmatrix} 2 & 1 & 0 & 0 \\ 1 & 4 & 1 & 0 \\ 0 & 1 & 4 & 1 \\ 0 & 0 & 1 & 2 \end{bmatrix}$

16.3 $\omega_1 = 0.806\sqrt{u}, \quad \omega_2 = 2.81\sqrt{\mu}$

16.4 $\omega_1 = 5.368 \times 10^3 \text{ rad/s}, \quad \omega_2 = 17.556 \times 10^3 \text{ rad/s}$

16.5 a.

t (s)	d_i (ft)	$\dot{d}_i$ (ft/s)	$\ddot{d}_i$ (ft/s^2)
0	0	0	25
0.03	0.01125	0.71	22.09
0.06	0.04238	1.03	−0.715
0.09	0.07287	0.67	−22.87
0.12	0.08278	−0.35	−45.28
0.15	0.05194	−1.43	−26.94

16.6 a.

t (s)	d_i (ft)	$\dot{d}_i$ (ft/s)	$\ddot{d}_i$ (ft/s^2)
0	0	0	10.00
0.02	0.0020	0.168	6.80
0.04	0.00672	0.256	1.968
0.06	0.01223	0.242	−3.338
0.08	0.01640	0.130	−7.84
0.10	0.01743	−0.053	−10.46

b.

t (s)	d_i (ft)	$\dot{d}_i$ (ft/s)	$\ddot{d}_i$ (ft/s^2)	$F(t)$ (lb)
0.00	0.00000	0.000	10.000	20.0
0.02	0.00179	0.169	6.923	16.0
0.04	0.00625	0.263	2.248	12.0
0.06	0.0115	0.254	−2.945	8.0
0.08	0.0157	0.150	−7.458	4.0
0.10	0.0169	−0.0147	−10.251	0.0

16.7

Node	t (s)	d_i (in.)	$\dot{d}_i$ (in./s)	$\ddot{d}_i$ (in./s^2)
2	0	0	0	0
	0.00025	2.6E-8	0	249.6
	0.00050	3.4E-5	0.284	1768.9
	0.00075	1.9E-4	1.085	4641.9
	0.0010	6.36E-4	2.605	7519.3
3	0	0	0	0
	0.00025	6.59E-5	0.791	6328.8
	0.00050	4.99E-4	2.817	9881.2
	0.00075	1.51E-3	5.265	9701.7
	0.0010	3.10E-3	7.369	7128.3

16.8 Using Newmark's method with $\gamma = \frac{1}{2}$, $\beta = \frac{1}{6}$

Node	t (s)	d_i (in.)	$\dot{d}_i$ (in./s)	$\ddot{d}_i$ (in./s^2)	$F(t)$ (lb)
2	0	0	0	0	0
	0.05	0.00172	0.103	4.131	0
	0.10	0.01544	0.513	12.27	0
3	0	0	0	40.0	2000
	0.05	0.0448	1.685	27.39	1800
	0.10	0.1536	2.479	4.37	1600

16.9 **a.** $\omega_1 = \dfrac{3.15}{L^2}\left(\dfrac{EI}{\rho A}\right)^{1/2}$, $\omega_2 = \dfrac{16.24}{L^2}\left(\dfrac{EI}{\rho A}\right)^{1/2}$, **c.** $\omega_1 = \dfrac{9.8}{L^2}\left(\dfrac{EI}{\rho A}\right)^{1/2}$

d. $\omega = \dfrac{14.8}{L^2}\left(\dfrac{EI}{\rho A}\right)^{1/2}$

16.14

		Node: 1	2	3	4	5	6
i	t (s)			Temperature (°C)			
0	0	200	200	200	200	200	200
1	8	0	159.0095	191.4441	198.2110	199.6110	199.8444
2	16	0	135.5852	178.1491	193.6620	198.2112	199.1445
3	24	0	120.2309	165.7003	187.3485	195.5379	197.5152
4	32	0	109.1993	154.9587	180.4038	191.7446	194.8115
5	40	0	100.7600	145.7784	173.4129	187.1268	191.1242
6	48	0	94.00311	137.8529	166.6182	181.9599	186.6590
7	56	0	88.39929	130.9034	160.1012	176.4598	181.6395
8	64	0	83.61745	124.7101	153.8759	170.7856	176.2620
9	72	0	79.43935	119.1075	147.9316	165.0508	170.6822
10	80	0	75.71603	113.9733	142.2502	159.3352	165.0171

16.15

Time (s)	Node 1	2	3	(using consistent capacitance matrix)
		Temperature (°C)		
0	25	25	25	
0.1	85	18.53611	26.36189	
0.2	85	29.61303	21.63526	
0.3	85	36.18435	22.42717	
0.4	85	40.72491	25.30428	
0.5	85	44.27834	28.85201	
0.6	85	47.29072	32.49614	
0.7	85	49.95809	36.01157	
0.8	85	52.37152	39.31761	
0.9	85	54.57756	42.39278	

16.15

Time (s)	1	2	3	
		Node		(using consistent capacitance matrix)
		Temperature (°C)		
1	85	56.60353	45.23933	
1.1	85	58.46814	47.86852	
1.2	85	60.1859	50.29457	
1.3	85	61.76908	52.53218	
1.4	85	63.22852	54.59557	
1.5	85	64.574	56.49814	
1.6	85	65.81448	58.25235	
1.7	85	66.95818	59.86974	

16.15

Time (s)	1	2	3
		Node	
		Temperature (°C)	
0	25	25	25
1.8	85	68.01265	61.36096
1.9	85	68.98485	62.73586
2	85	69.88121	64.0035
2.1	85	70.70765	65.17226
2.2	85	71.46961	66.24984
2.3	85	72.17214	67.24336
2.4	85	72.81986	68.15938
2.5	85	73.41705	69.00393
2.6	85	73.96766	69.78261
2.7	85	74.47531	70.50053
2.8	85	74.94336	71.16246
2.9	85	75.3749	71.77274
3	85	75.77277	72.33542

Appendix A

A1. **a.** $\begin{bmatrix} 3 & 0 \\ -3 & 12 \end{bmatrix}$ **b.** Nonsense **c.** Nonsense

d. $\begin{Bmatrix} 11 \\ 9 \\ 11 \end{Bmatrix}$ **e.** Nonsense **f.** $\begin{bmatrix} 10 & 7 & 6 \\ 3 & -1 & 7 \end{bmatrix}$

A2. $\begin{bmatrix} 1 & 0 \\ \frac{1}{4} & \frac{1}{4} \end{bmatrix}$

A3. $\dfrac{1}{17} \begin{bmatrix} 12 & -3 & -8 \\ -3 & 5 & 2 \\ -8 & 2 & 11 \end{bmatrix}$

A4. Nonsense

A5. $\begin{bmatrix} \frac{1}{2} & 0 \\ \frac{1}{8} & \frac{1}{8} \end{bmatrix}$

A6. Same as A3

A8. $\begin{bmatrix} \cos\theta & -\sin\theta \\ \sin\theta & \cos\theta \end{bmatrix}$

Appendix B

B1. $x_1 = 3.15, \quad x_2 = 0.62$

B2. $x_1 = 3.15, \quad x_2 = 0.62$

B3. $x_1 = 2.5, \quad x_2 = -1, \quad x_3 = 0.5$

B4. $x_1 = 3, \quad x_2 = -1, \quad x_3 = -2$

B5. **a.** $\begin{Bmatrix} x_1 \\ x_2 \end{Bmatrix} = \begin{bmatrix} 2 & -1 \\ 1 & -1 \end{bmatrix} \begin{Bmatrix} y_1 \\ y_2 \end{Bmatrix}$ **b.** $\begin{Bmatrix} z_1 \\ z_2 \end{Bmatrix} = \begin{bmatrix} -3 & 2 \\ 5 & -3 \end{bmatrix} \begin{Bmatrix} y_1 \\ y_2 \end{Bmatrix}$

B6. $x_1 = 0, \quad x_2 = 1, \quad x_3 = 2, \quad x_4 = 2, \quad x_5 = 0$

B7. $x_1 = 3.15, \quad x_2 = 0.62$

B8. **a.** Unique **b.** Nonexistent **c.** Unique **d.** Nonunique

Appendix D

D1. **a.** $f_{1y} = f_{2y} = -5$ kip, $m_1 = -m_2 = -100$ k-ft
b. $f_{1y} = f_{2y} = -5$ kip, $m_1 = -m_2 = -18.75$ k-ft
c. $f_{1y} = f_{2y} = -15$ kip, $m_1 = -m_2 = -75$ k-ft
d. $f_{1y} = -18.75$ kip, $f_{2y} = -6.25$ kip, $m_1 = -58.3$ k-ft, $m_2 = 33.3$ k-ft
e. $f_{1y} = -6$ kip, $f_{2y} = -14$ kip, $m_1 = -26.67$ k-ft, $m_2 = 40$ k-ft
f. $f_{1y} = -0.99$ kN, $f_{2y} = -4.0$ kN, $m_1 = -2.04$ kN · m, $m_2 = 5.10$ kN · m
g. $f_{1y} = f_{2y} = -6$ kN, $m_1 = -m_2 = -7.5$ kN · m
h. $f_{1y} = f_{2y} = -10$ kN, $m_1 = -m_2 = -6.67$ kN · m

Index